T0181276

Lecture Notes in Artificial Intelligence   10192

Subseries of Lecture Notes in Computer Science

More information about this series at http://www.springer.com/series/1244

Ngoc Thanh Nguyen · Satoshi Tojo
Le Minh Nguyen · Bogdan Trawiński (Eds.)

Intelligent Information and Database Systems

9th Asian Conference, ACIIDS 2017
Kanazawa, Japan, April 3–5, 2017
Proceedings, Part II

 Springer

Editors

Ngoc Thanh Nguyen
Wrocław University of Science
   and Technology
Wrocław
Poland

Satoshi Tojo
Japan Advanced Institute of Science
   and Technology
Nomi
Japan

Le Minh Nguyen
Japan Advanced Institute of Science
   and Technology
Nomi
Japan

Bogdan Trawiński
Wrocław University of Science
   and Technology
Wrocław
Poland

ISSN 0302-9743   ISSN 1611-3349   (electronic)
Lecture Notes in Artificial Intelligence
ISBN 978-3-319-54429-8   ISBN 978-3-319-54430-4   (eBook)
DOI 10.1007/978-3-319-54430-4

Library of Congress Control Number: 2017932640

LNCS Sublibrary: SL7 – Artificial Intelligence

Printed on acid-free paper

This Springer imprint is published by Springer Nature
The registered company is Springer International Publishing AG
The registered company address is: Gewerbestrasse 11, 6330 Cham, Switzerland

Preface

ACIIDS 2017 was the ninth event in a series of international scientific conferences on research and applications in the field of intelligent information and database systems. The aim of ACIIDS 2017 was to provide an international forum for scientific research in the technologies and applications of intelligent information and database systems. ACIIDS 2017 was co-organized by Japan Advanced Institute of Science and Technology (Japan) and Wrocław University of Science and Technology (Poland) in co-operation with IEEE SMC Technical Committee on Computational Collective Intelligence, Quang Binh University, Vietnam, Yeungnam University, South Korea, Bina Nusantara University, Indonesia, Universiti Teknologi Malaysia, and the University of Newcastle, Australia. It took place in Kanazawa (Japan) during April 3–5, 2017.

The conference series ACIIDS is well established. The first two events, ACIIDS 2009 and ACIIDS 2010, took place in Dong Hoi City and Hue City in Vietnam, respectively. The third event, ACIIDS 2011, took place in Daegu (Korea), followed by the fourth event, ACIIDS 2012, which took place in Kaohsiung (Taiwan). The fifth event, ACIIDS 2013, was held in Kuala Lumpur in Malaysia while the sixth event, ACIIDS 2014, was held in Bangkok in Thailand. The seventh event, ACIIDS 2015, took place in Bali (Indonesia). The last event, ACIIDS 2016 was held in Da Nang (Vietnam).

We received more than 400 papers from 42 countries all over the world. Each paper was peer reviewed by at least two members of the international Program Committee and international reviewer board. Only 154 papers with the highest quality were selected for oral presentation and publication in the two volumes of ACIIDS 2017 proceedings.

Papers included in the proceedings cover the following topics: knowledge engineering and Semantic Web, social networks and recommender systems, text processing and information retrieval, intelligent database systems, intelligent information systems, decision support and control systems, machine learning and data mining, computer vision techniques, advanced data mining techniques and applications, intelligent and context systems, multiple model approach to machine learning, applications of data science, artificial intelligence applications for e-services, automated reasoning and proving techniques with applications in intelligent systems, collective intelligence for service innovation, technology opportunity, e-learning and fuzzy intelligent systems, intelligent computer vision systems and applications, intelligent data analysis, applications and technologies for the Internet of Things, intelligent algorithms and brain functions, intelligent systems and algorithms in information sciences, IT in biomedicine, intelligent technologies in smart cities in the twenty-first century, analysis of image, video and motion data in life sciences, modern applications of machine learning for actionable knowledge extraction, mathematics of decision sciences and information science, scalable data analysis in bioinformatics and biomedical informatics, and technological perspectives of agile transformation in IT organizations.

The accepted and presented papers highlight new trends and challenges of intelligent information and database systems. The presenters showed how new research could lead to novel and innovative applications. We hope you will find these results useful and inspiring for your future research.

We would like to extend our heartfelt thanks to Jarosław Gowin, Deputy Prime Minister of the Republic of Poland and Minister of Science and Higher Education, for his support and honorary patronage of the conference.

We would like to express our sincere thanks to the honorary chairs, Prof. Testsuo Asano (President of JAIST, Japan) and Prof. Cezary Madryas (Rector of Wrocław University of Science and Technology, Poland), for their support.

Our special thanks go to the program chairs, special session chairs, organizing chairs, publicity chairs, liaison chairs, and local Organizing Committee for their work for the conference. We sincerely thank all the members of the international Program Committee for their valuable efforts in the review process, which helped us to guarantee the highest quality of the selected papers for the conference. We cordially thank the organizers and chairs of special sessions who contributed to the success of the conference.

We also would like to express our thanks to the keynote speakers (Prof. Tu-Bao Ho, Prof. Bernhard Pfahringer, Prof. Edward Szczerbicki, Prof. Hideyuki Takagi) for their interesting and informative talks of world-class standard.

We cordially thank our main sponsors, Japan Advanced Institute of Science and Technology (Japan), Wrocław University of Science and Technology (Poland), IEEE SMC Technical Committee on Computational Collective Intelligence, Quang Binh University (Vietnam), Yeungnam University (South Korea), Bina Nusantara University (Indonesia), Universiti Teknologi Malaysia (Malaysia), and the University of Newcastle (Australia). Our special thanks are due also to Springer for publishing the proceedings, and to all the other sponsors for their kind support.

We wish to thank the members of the Organizing Committee for their very significant work and the members of the local Organizing Committee for their excellent work.

We cordially thank all the authors, for their valuable contributions, and the other participants of this conference. The conference would not have been possible without their support.

Thanks are also due to many experts who contributed to making the event a success.

April 2017   Ngoc Thanh Nguyen
                                                                Satoshi Tojo
                                                            Le Minh Nguyen
                                                           Bogdan Trawiński

Organization

Honorary Chairs

Testsuo Asano   President of Japan Advanced Institute of Science
and Technology, Japan

Cezary Madryas   Rector of Wrocław University of Science
and Technology, Poland

General Chairs

Satoshi Tojo   Japan Advanced Institute of Science and Technology,
Japan

Ngoc Thanh Nguyen   Wrocław University of Science and Technology, Poland

Program Chairs

Tzung-Pei Hong   National University of Kaohsiung, Taiwan

Le Minh Nguyen   Japan Advanced Institute of Science and Technology,
Japan

Bogdan Trawiński   Wrocław University of Science and Technology, Poland

Steering Committee

Ngoc Thanh Nguyen (Chair)   Wrocław University of Science and Technology, Poland

Longbing Cao   University of Science and Technology Sydney, Australia

Suphamit Chittayasothorn   King Mongkut's Institute of Technology Ladkrabang,
Thailand

Ford Lumban Gaol   Bina Nusantara University, Indonesia

Tu Bao Ho   Japan Advanced Institute of Science and Technology,
Japan

Tzung-Pei Hong   National University of Kaohsiung, Taiwan

Dosam Hwang   Yeungnam University, Korea

Lakhmi C. Jain   University of South Australia, Australia

Geun-Sik Jo   Inha University, Korea

Hoai An Le-Thi   University Paul Verlaine, Metz, France

Toyoaki Nishida   Kyoto University, Japan

Leszek Rutkowski   Technical University of Czestochowa, Poland

Ali Selamat   Universiti Teknologi Malaysia, Malaysia

Special Session Chairs

Dariusz Król	Wrocław University of Science and Technology, Poland
Kiyoaki Shirai	Japan Advanced Institute of Science and Technology, Japan

Liaison Chairs

Ford Lumban Gaol	Bina Nusantara University, Indonesia
Bao Hung Hoang	Viethanit, Vietnam
Mong-Fong Horng	National Kaohsiung University of Applied Sciences, Taiwan
Dosam Hwang	Yeungnam University, Korea
Ali Selamat	Universiti Teknologi Malaysia, Malaysia

Organizing Chairs

Atsuo Yoshitaka	Japan Advanced Institute of Science and Technology, Japan
Adrianna Kozierkiewicz-Hetmańska	Wrocław University of Science and Technology, Poland

Publicity Chairs

Danilo S. Carvalho	Japan Advanced Institute of Science and Technology, Japan
Maciej Huk	Wrocław University of Science and Technology, Poland
Bernadetta Maleszka	Wrocław University of Science and Technology, Poland

Publication Chair

Marcin Maleszka	Wrocław University of Science and Technology, Poland

Webmaster

Marek Kopel	Wrocław University of Science and Technology, Poland

Keynote Speakers

Tu-Bao Ho	Japan Advanced Institute of Science and Technology, Japan
Bernhard Pfahringer	University of Waikato, New Zealand
Edward Szczerbicki	The University of Newcastle, Australia
Hideyuki Takagi	Kyushu University, Japan

Special Sessions Organizers

1. Special Session on Advanced Data Mining Techniques and Applications (ADMTA 2017)

Tzung-Pei Hong	National University of Kaohsiung, Taiwan
Bac Le	University of Science, VNU-HCM, Vietnam
Tran Minh Quang	Ho Chi Minh City University of Technology, Vietnam
Bay Vo	Ho Chi Minh City University of Technology, Vietnam

2. Special Session on Intelligent and Contextual Systems (ICxS 2017)

Maciej Huk	Wrocław University of Science and Technology, Poland
Goutam Chakraborty	Iwate Prefectural University, Japan
Basabi Chakraborty	Iwate Prefectural University, Japan
Qiangfu Zhao	University of Aizu, Japan

3. Multiple Model Approach to Machine Learning (MMAML 2017)

Tomasz Kajdanowicz	Wrocław University of Science and Technology, Poland
Edwin Lughofer	Johannes Kepler University Linz, Austria
Bogdan Trawiński	Wrocław University of Science and Technology, Poland

4. Special Session on Applications of Data Science (ADS 2017)

Fulufhelo Nelwamondo	Council for Scientific and Industrial Research (CSIR), South Africa
Vukosi Marivate	Council for Scientific and Industrial Research (CSIR), South Africa

5. Special Session on Artificial Intelligence Applications for E-services (AIAE 2017)

Chen-Shu Wang	National Taipei University of Technology, Taiwan
Deng-Yiv Chiu	Chung Hua University, Taiwan

6. Special Session on Automated Reasoning and Proving Techniques with Applications in Intelligent Systems (ARPTA 2017)

Jingde Cheng	Saitama University, Japan

7. Special Session on Collective Intelligence for Service Innovation, Technology Opportunity, E-Learning and Fuzzy Intelligent Systems (CISTEF 2017)

Chao-Fu Hong	Aletheia University, Taiwan
Kuo-Sui Lin	Aletheia University, Taiwan

8. Special Session on Intelligent Computer Vision Systems and Applications (ICVSA 2017)

Dariusz Frejlichowski	West Pomeranian University of Technology, Szczecin, Poland
Leszek J. Chmielewski	Warsaw University of Life Sciences, Poland
Piotr Czapiewski	West Pomeranian University of Technology, Szczecin, Poland

9. Special Session on Intelligent Data Analysis, Applications and Technologies for Internet of Things (IDAIoT 2017)

Shunzhi Zhu	University of Technology Xiamen, PR China
Rung Ching Chen	Chaoyang University of Technology, Taiwan
Yung-Fa Huang	Chaoyang University of Technology, Taiwan

10. Special Session on Intelligent Algorithms and Brain Functions (InBRAIN 2017)

Andrzej Przybyszewski	Polish-Japanese Academy of Information Technology, Poland
Tomasz Rutkowski	University of Tokyo, Japan

11. Special Session on Intelligent Systems and Algorithms in Information Sciences (ISAIS 2017)

Martin Kotyrba	University of Ostrava, Czech Republic
Eva Volna	University of Ostrava, Czech Republic
Ivan Zelinka	VŠB, Technical University of Ostrava, Czech Republic

12. Special Session on IT in Biomedicine (ITiB 2017)

Ondrej Krejcar	University of Hradec Kralove, Czech Republic
Ali Selamat	Universiti Teknologi Malaysia, Malaysia
Kamil Kuca	University of Hradec Kralove, Czech Republic
Dawit Assefa Haile	Addis Ababa University, Ethiopia
Tanos C.C. Franca	Military Institute of Engineering, Brazil

13. Special Session on Intelligent Technologies in Smart Cities in the 21st Century (ITSC 2017)

Cezary Orłowski	WSB University Gdańsk, Poland
Artur Ziółkowski	WSB University Gdańsk, Poland
Aleksander Orłowski	Gdansk University of Technology, Poland
Katarzyna Ossowska	Gdansk University of Technology, Poland
Arkadiusz Sarzyński	Gdansk University of Technology, Poland

14. Special Session on Analysis of Image, Video and Motion Data in Life Sciences (IVMLS 2017)

Kondrad Wojciechowski	Polish-Japanese Academy of Information Technology, Poland
Marek Kulbacki	Polish-Japanese Academy of Information Technology, Poland
Jakub Segen	Polish-Japanese Academy of Information Technology, Poland
Andrzej Polański	Silesian University of Technology, Poland

15. Special Session on Modern Applications of Machine Learning for Actionable Knowledge Extraction (MAMLAKE 2017)

Waseem Ahmad	Waiariki Institute of Technology, New Zealand
Paul Leong	Auckland University of Technology, New Zealand
Muhammad Usman	Shaheed Zulfiqar Ali Bhutto Institute of Science and Technology, Pakistan

16. Special Session on Mathematics of Decision Sciences and Information Science (MDSIS 2017)

Takashi Matsuhisa	Ibaraki Christian University, Japan
Vladimir Mazalov	Karelia Research Centre Russian Academy of Sciences, Russia
Pu-Yan Nie	Guangdong University of Finance and Economics, PR China

17. Special Session on Scalable Data Analysis in Bioinformatics and Biomedical Informatics (SDABBI 2017)

Dariusz Mrozek	Silesian University of Technology, Poland
Stanisław Kozielski	Silesian University of Technology, Poland
Bożena Małysiak-Mrozek	Silesian University of Technology, Poland

18. Special Session on Technological Perspective of Agile Transformation in IT organizations (TPATIT 2017)

Cezary Orłowski	WSB University Gdańsk, Poland
Artur Ziółkowski	WSB University Gdańsk, Poland
Miłosz Kurzawski	Blue Media Corporation, Poland
Tomasz Deręgowski	ACXIOM Corporation, Poland
Włodzimierz Wysocki	University of Technology Koszalin, Poland

Program Committee

Salim Abdulazeez	College of Engineering, Trivandrum, India
Ajith Abraham	Machine Intelligence Research Labs, USA

Kazimierz Choroś	Wrocław University of Science and Technology, Poland
Kun-Ta Chuang	National Cheng Kung University, Taiwan
Piotr Chynał	Wrocław University of Science and Technology, Poland
Robert Cierniak	Czestochowa University of Technology, Poland
Dorian Cojocaru	University of Craiova, Romania
Phan Cong-Vinh	Nguyen Tat Thanh University, Vietnam
Jose Alfredo Ferreira Costa	UFRN, Universidade Federal do Rio Grande do Norte, Brazil
Keeley Crockett	Manchester Metropolitan University, UK
Bogusław Cyganek	AGH University of Science and Technology, Poland
Piotr Czapiewski	West Pomeranian University of Technology, Szczecin, Poland
Ireneusz Czarnowski	Gdynia Maritime University, Poland
Piotr Czekalski	Silesian University of Technology, Poland
Paul Davidsson	Malmö University, Sweden
Mauricio C. de Souza	Universidade Federal de Minas Gerais, Brazil
Roberto De Virgilio	Università degli Studi Roma Tre, Italy
Tien V. Do	Budapest University of Technology and Economics, Hungary
Grzegorz Dobrowolski	AGH University of Science and Technology, Poland
Habiba Drias	University of Science and Technology Houari Boumediene, Algeria
Maciej Drwal	Wrocław University of Science and Technology, Poland
Ewa Dudek-Dyduch	AGH University of Science and Technology, Poland
El-Sayed M. El-Alfy	King Fahd University of Petroleum and Minerals, Saudi Arabia
Nadia Essoussi	University of Carthage, Tunisia
Rim Faiz	University of Carthage, Tunisia
Victor Felea	Alexandru Ioan Cuza University of Iasi, Romania
Thomas Fober	University of Marburg, Germany
Simon Fong	University of Macau, SAR China
Tanos C.C. Franca	Military Institute of Engineering, Brazil
Dariusz Frejlichowski	West Pomeranian University of Technology, Szczecin, Poland
Hamido Fujita	Iwate Prefectural University, Japan
Mohamed Gaber	Robert Gordon University, UK
Ford Lumban Gaol	Bina Nusantara University, Indonesia
Dariusz Gasior	Wrocław University of Science and Technology, Poland
Janusz Getta	University of Wollongong, Australia
Daniela Gifu	Romanian Academy, Iasi Branch, Romania
Dejan Gjorgjevikj	Ss. Cyril and Methodius University in Skopje, Macedonia
Daniela Godoy	ISISTAN Research Institute, Argentina
Gergő Gombos	Eötvös Loránd University, Hungary
Adam Gonczarek	Wrocław University of Science and Technology, Poland
Antonio Gonzalez-Pardo	Universidad Autonoma de Madrid, Spain

Manuel Graña	San Sebastián University, Spain
Janis Grundspenkis	Riga Technical University, Latvia
Quang-Thuy Ha	Vietnam National University, Hanoi (VNU), Vietnam
Sung Ho Ha	Kyungpook National University, Korea
Dawit Assefa Haile	Addis Ababa University, Ethiopia
Pei-Yi Hao	National Kaohsiung University of Applied Sciences, Taiwan
Ctibor Hatar	Constantine the Philosopher University, Slovakia
Marcin Hernes	Wrocław University of Economics, Poland
Bogumila Hnatkowska	Wrocław University of Science and Technology, Poland
Huu Hanh Hoang	Hue University, Vietnam
Quang Hoang	Hue University, Vietnam
Jaakko Hollmén	Aalto University School of Science, Finland
Chao-Fu Hong	Aletheia University, Taiwan
Tzung-Pei Hong	National University of Kaohsiung, Taiwan
Mong-Fong Horng	National Kaohsiung University of Applied Sciences, Taiwan
Jen-Wei Huang	National Cheng Kung University, Taiwan
Yung-Fa Huang	Chaoyang University of Technology, Taiwan
Maciej Huk	Wrocław University of Science and Technology, Poland
Zbigniew Huzar	Wrocław University of Science and Technology, Poland
Dosam Hwang	Yeungnam University, Korea
Roliana Ibrahim	Universiti Teknologi Malaysia, Malaysia
Dmitry Ignatov	National Research University Higher School of Economics, Russia
Lazaros Iliadis	Democritus University of Thrace, Greece
Hazra Imran	Athabasca University, Canada
Agnieszka Indyka-Piasecka	Wrocław University of Science and Technology, Poland
Mirjana Ivanovic	University of Novi Sad, Serbia
Sanjay Jain	National University of Singapore, Singapore
Jarosław Jankowski	West Pomeranian University of Technology, Szczecin, Poland
Chuleerat Jaruskulchai	Kasetsart University, Thailand
Khalid Jebari	LCS Rabat, Morocco
Joanna Jedrzejowicz	University of Gdansk, Poland
Piotr Jedrzejowicz	Gdynia Maritime University, Poland
Janusz Jezewski	Institute of Medical Technology and Equipment ITAM, Poland
Geun Sik Jo	Inha University, Korea
Kang-Hyun Jo	University of Ulsan, Korea
Jason J. Jung	Chung-Ang University, Korea
Janusz Kacprzyk	Systems Research Institute, Polish Academy of Sciences, Poland
Tomasz Kajdanowicz	Wrocław University of Science and Technology, Poland
Nadjet Kamel	Ferhat Abbas University of Setif, Algeria

Mehmet Karaata	Kuwait University, Kuwait
Ioannis Karydis	Ionian University, Greece
Nikola Kasabov	Auckland University of Technology, New Zealand
Arkadiusz Kawa	Poznan University of Economics, Poland
Rafal Kern	Wrocław University of Science and Technology, Poland
Chonggun Kim	Yeungnam University, Korea
Pan-Koo Kim	Chosun University, Korea
Attila Kiss	Eötvös Loránd University, Hungary
Jerzy Klamka	Silesian University of Technology, Poland
Goran Klepac	Raiffeisen Bank, Croatia
Blanka Klimova	University of Hradec Kralove, Czech Republic
Shinya Kobayashi	Ehime University, Japan
Joanna Kolodziej	Cracow University of Technology, Poland
Marek Kopel	Wrocław University of Science and Technology, Poland
Jozef Korbicz	University of Zielona Gora, Poland
Jacek Koronacki	Institute of Computer Science, Polish Academy of Sciences, Poland
Raymondus Kosala	Bina Nusantara University, Indonesia
Leszek Koszalka	Wrocław University of Science and Technology, Poland
Malgorzata Kotulska	Wrocław University of Science and Technology, Poland
Martin Kotyrba	University of Ostrava, Czech Republic
Zdzisław Kowalczuk	Gdańsk University of Technology, Poland
Jan Kozak	University of Silesia, Poland
Stanisław Kozielski	Silesian University of Technology, Poland
Adrianna Kozierkiewicz-Hetmańska	Wrocław University of Science and Technology, Poland
Bartosz Krawczyk	Wrocław University of Science and Technology, Poland
Ondrej Krejcar	University of Hradec Kralove, Czech Republic
Dalia Kriksciuniene	Vilnius University, Lithuania
Dariusz Krol	Wrocław University of Science and Technology, Poland
Marzena Kryszkiewicz	Warsaw University of Technology, Poland
Adam Krzyzak	Concordia University, Canada
Tetsuji Kuboyama	Gakushuin University, Japan
Kamil Kuca	University of Hradec Kralove, Czech Republic
Elżbieta Kukla	Wrocław University of Science and Technology, Poland
Marek Kulbacki	Polish-Japanese Academy of Information Technology, Poland
Kazuhiro Kuwabara	Ritsumeikan University, Japan
Halina Kwasnicka	Wrocław University of Science and Technology, Poland
Mark Last	Ben-Gurion University of the Negev, Israel
Annabel Latham	Manchester Metropolitan University, UK
Bac Le	University of Science, VNU-HCM, Vietnam
Hoai An Le Thi	Université de Lorraine, France
Kun Chang Lee	Sungkyunkwan University, Korea
Yue-Shi Lee	Ming Chuan University, Taiwan
Paul Leong	Auckland University of Technology, New Zealand

Chunshien Li	National Central University, Taiwan
Horst Lichter	RWTH Aachen University, Germany
Kuo-Sui Lin	Aletheia University, Taiwan
Sebastian Link	University of Auckland, New Zealand
Igor Litvinchev	Nuevo Leon State University, Mexico
Lian Liu	University of Kentucky, USA
Rey-Long Liu	Tzu Chi University, Taiwan
Edwin Lughofer	Johannes Kepler University Linz, Austria
Lech Madeyski	Wrocław University of Science and Technology, Poland
Bernadetta Maleszka	Wrocław University of Science and Technology, Poland
Marcin Maleszka	Wrocław University of Science and Technology, Poland
Bożena Małysiak-Mrozek	Silesian University of Technology, Poland
Neel Mani	Dublin City University, Ireland
Yannis Manolopoulos	Aristotle University of Thessaloniki, Greece
Vukosi Marivate	Council for Scientific and Industrial Research, South Africa
Karolina Marzantowicz	IBM Poland, Poland
Francesco Masulli	University of Genoa, Italy
Mustafa Mat Deris	Universiti Tun Hussein Onn Malaysia, Malaysia
Takashi Matsuhisa	Karelia Research Centre, Russian Academy of Science, Russia
Tamás Matuszka	Eötvös Loránd University, Hungary
Vladimir Mazalov	Karelia Research Centre, Russian Academy of Sciences, Russia
Joao Mendes-Moreira	University of Porto, Portugal
Héctor Menéndez	Autonomous University of Madrid, Spain
Jacek Mercik	Wrocław School of Banking, Poland
Radosław Michalski	Wrocław University of Science and Technology, Poland
Peter Mikulecky	University of Hradec Kralove, Czech Republic
Marek Milosz	Lublin University of Technology, Poland
Jolanta Mizera-Pietraszko	Opole University, Poland
Yang-Sae Moon	Kangwon National University, Korea
Dariusz Mrozek	Silesian University of Technology, Poland
Leo Mrsic	IN2data Ltd- Data Science Company, Croatia
Pawel Myszkowski	Wrocław University of Science and Technology, Poland
Grzegorz J. Nalepa	AGH University of Science and Technology, Poland
Mahyuddin K.M. Nasution	Universitas Sumatera Utara, Indonesia
Richi Nayak	Queensland University of Technology, Australia
Fulufhelo Nelwamondo	Council for Scientific and Industrial Research, South Africa
Huu-Tuan Nguyen	Vietnam Maritime University, Vietnam
Loan T.T. Nguyen	Nguyen Tat Thanh University, Vietnam
Ngoc Thanh Nguyen	Wrocław University of Science and Technology, Poland
Thai-Nghe Nguyen	Can Tho University, Vietnam
Vinh Nguyen	University of Melbourne, Australia

Pu-Yan Nie	Guangdong University of Finance and Economics, China
Yusuke Nojima	Osaka Prefecture University, Japan
Mariusz Nowostawski	University of Otago, New Zealand
Alberto Núñez	Universidad Complutense de Madrid, Spain
Manuel Núñez	Universidad Complutense de Madrid, Spain
Mariusz Ochla	IBM Center for Advances Studies, Poland
Richard Jayadi Oentaryo	Singapore Management University, Singapore
Kouzou Ohara	Aoyama Gakuin University, Japan
Tomasz Orczyk	University of Silesia, Poland
Cezary Orłowski	WSB University Gdańsk, Poland
Shingo Otsuka	Kanagawa Institute of Technology, Japan
Marcin Paprzycki	Systems Research Institute, Polish Academy of Sciences, Poland
Jakub Peksinski	West Pomeranian University of Technology, Szczecin, Poland
Danilo Pelusi	University of Teramo, Italy
Xuan Hau Pham	Quang Binh University, Vietnam
Tao Pham Dinh	National Institute for Applied Sciences, France
Maciej Piasecki	Wrocław University of Science and Technology, Poland
Bartłomiej Pierański	Poznan University of Economics and Business, Poland
Dariusz Pierzchala	Military University of Technology, Poland
Marcin Pietranik	Wrocław University of Science and Technology, Poland
Piotr Pietrzak	IBM Poland, Poland
Elias Pimenidis	University of the West of England, UK
Andrzej Polanski	Silesian University of Technology, Poland
Elvira Popescu	University of Craiova, Romania
Piotr Porwik	University of Silesia, Poland
Petra Poulova	University of Hradec Kralove, Czech Republic
Bhanu Prasad	Florida A&M University, USA
Andrzej Przybyszewski	University of Massachusetts Medical School, USA
Tran Minh Quang	Ho Chi Minh City University of Technology, Vietnam
Paulo Quaresma	Universidade de Evora, Portugal
Ngoc Quoc Ly	University of Science Ho Chi Minh City, Vietnam
Mohammad Rashedur Rahman	North South University, Bangladesh
Ewa Ratajczak-Ropel	Gdynia Maritime University, Poland
Patricia Riddle	University of Auckland, New Zealand
Manuel Roveri	Politecnico di Milano, Italy
Przemysław Różewski	West Pomeranian University of Technology, Szczecin, Poland
Leszek Rutkowski	Czestochowa University of Technology, Poland
Tomasz Rutkowski	University of Tokyo, Japan
Tiia Ruutmann	Tallinn University of Technology, Estonia
Alexander Ryjov	Lomonosov Moscow State University, Russia
Virgilijus Sakalauskas	Vilnius University, Lithuania

Daniel Sanchez	University of Granada, Spain
Cesar Sanin	University of Newcastle, Australia
Minoru Sasaki	Gifu University, Japan
Moamar Sayed-Mouchaweh	Ecole des Mines de Douai, France
Juergen Schmidhuber	Swiss AI Lab IDSIA, Switzerland
Björn Schuller	University of Passau, Germany
Jakub Segen	Gest3D, USA
Ali Selamat	Universiti Teknologi Malaysia, Malaysia
S.M.N. Arosha Senanayake	Universiti Brunei Darussalam, Brunei Darussalam
Natalya Shakhovska	Lviv Polytechnic National University, Ukraine
Andrzej Siemiński	Wrocław University of Science and Technology, Poland
Dragan Simic	University of Novi Sad, Serbia
Ivana Simonova	University of Hradec Kralove, Czech Republic
Bharat Singh	Universiti Teknology PETRONAS, Malaysia
Andrzej Skowron	Warsaw University, Poland
Leszek Sliwko	University of Westminster, UK
Adam Slowik	Koszalin University of Technology, Poland
Vladimir Sobeslav	University of Hradec Kralove, Czech Republic
Kulwadee Somboonviwat	King Mongkut's Institute of Technology Ladkrabang, Thailand
Zenon A. Sosnowski	Bialystok University of Technology, Poland
Jerzy Stefanowski	Poznan University of Technology, Poland
Serge Stinckwich	University of Caen-Lower Normandy, Vietnam
Ja-Hwung Su	Kainan University, Taiwan
Andrzej Swierniak	Silesian University of Technology, Poland
Edward Szczerbicki	University of Newcastle, Australia
Julian Szymanski	Gdansk University of Technology, Poland
Yasufumi Takama	Tokyo Metropolitan University, Japan
Zbigniew Telec	Wrocław University of Science and Technology, Poland
Krzysztof Tokarz	Silesian University of Technology, Poland
Jakub Tomczak	Wrocław University of Science and Technology, Poland
Diana Trandabat	Alexandru Ioan Cuza University of Iasi, Romania
Bogdan Trawinski	Wrocław University of Science and Technology, Poland
Hong-Linh Truong	Vienna University of Technology, Austria
Ualsher Tukeyev	al-Farabi Kazakh National University, Kazakhstan
Olgierd Unold	Wrocław University of Science and Technology, Poland
Muhammad Usman	Shaheed Zulfiqar Ali Bhutto Institute of Science and Technology, Pakistan
Pandian Vasant	Universiti Teknologi PETRONAS, Malaysia
Jorgen Villadsen	Technical University of Denmark, Denmark
Bay Vo	Ho Chi Minh City University of Technology, Vietnam
Ngoc Chau Vo Thi	Ho Chi Minh University of Technology, Vietnam
Eva Volna	University of Ostrava, Czech Republic
Gottfried Vossen	ERCIS Münster, Germany

Chen-Shu Wang	National Taipei University of Technology, Taiwan
Lipo Wang	Nanyang Technological University, Singapore
Xiaodong Wang	Fujian University of Technology, China
Yongkun Wang	University of Tokyo, Japan
Junzo Watada	Waseda University, Japan
Izabela Wierzbowska	Gdynia Maritime University, Poland
Konrad Wojciechowski	Silesian University of Technology, Poland
Michal Wozniak	Wrocław University of Science and Technology, Poland
Krzysztof Wrobel	University of Silesia, Poland
Tsu-Yang Wu	Harbin Institute of Technology Shenzhen Graduate School, China
Marian Wysocki	Rzeszow University of Technology, Poland
Farouk Yalaoui	University of Technology of Troyes, France
Xin-She Yang	Middlesex University, UK
Lina Yao	University of Adelaide, Australia
Slawomir Zadrozny	Systems Research Institute, Polish Academy of Sciences, Poland
Drago Žagar	University of Osijek, Croatia
Danuta Zakrzewska	Lodz University of Technology, Poland
Constantin-Bala Zamfirescu	Lucian Blaga University of Sibiu, Romania
Katerina Zdravkova	St. Cyril and Methodius University, Macedonia
Ivan Zelinka	VŠB, Technical University of Ostrava, Czech Republic
Vesna Zeljkovic	Lincoln University, USA
Aleksander Zgrzywa	Wrocław University of Science and Technology, Poland
Qiang Zhang	Dalian University, China
Zhongwei Zhang	University of Southern Queensland, Australia
Qiangfu Zhao	University of Aizu, Japan
Dongsheng Zhou	Dalian University, China
Zhi-Hua Zhou	Nanjing University, China
Shunzhi Zhu	University of Technology Xiamen, China
Maciej Zieba	Wrocław University of Science and Technology, Poland
Artur Ziółkowski	WSB University Gdańsk, Poland
Marta Zorrilla	University of Cantabria, Spain

Program Committees of Special Sessions

Advanced Data Mining Techniques and Applications (ADMTA 2017)

Tzung-Pei Hong	National University of Kaohsiung, Taiwan
Tran Minh Quang	Ho Chi Minh City University of Technology, Vietnam
Bac Le	University of Science, VNU-HCM, Vietnam
Bay Vo	Ho Chi Minh City University of Technology, Vietnam
Chun-Hao Chen	Tamkang University, Taiwan
Chun-Wei Lin	Harbin Institute of Technology Shenzhen Graduate School, China

Wen-Yang Lin	National University of Kaohsiung, Taiwan
Yeong-Chyi Lee	Cheng Shiu University, Taiwan
Le Hoang Son	University of Science, Ha Noi, Vietnam
Vo Thi Ngoc Chau	Ho Chi Minh City University of Technology, Ho Chi Minh City, Vietnam
Van Vo	Ho Chi Minh University of Industry, Ho Chi Minh City, Vietnam
Ja-Hwung Su	Cheng Shiu University, Taiwan
Ming-Tai Wu	University of Nevada, Las Vegas, USA
Kawuu W. Lin	National Kaohsiung University of Applied Sciences, Taiwan
Tho Le	Ho Chi Minh City University of Technology, Vietnam
Dang Nguyen	Deakin University, Geelong, Australia
Hau Le	Thuyloi University, Hanoi, Vietnam
Thien-Hoang Van	Ho Chi Minh City University of Technology, Vietnam
Tho Quan	Hochiminh City University of Technology, Vietnam
Ham Nguyen	University of People's Security Hochiminh City, Vietnam
Thiet Pham	Ho Chi Minh University of Industry, Vietnam

Intelligent and Contextual Systems (ICxS 2017)

Basabi Chakraborty	Iwate Prefectural University, Japan
Goutam Chakraborty	Iwate Prefectural University, Japan
Hideyuki Takahashi	RIEC, Tohoku University, Japan
Jerzy Świątek	Wrocław University of Science and Technology, Poland
Józef Korbicz	University of Zielona Gora, Poland
Keun Ho Ryu	Chungbuk National University, South Korea
Maciej Huk	Wrocław University of Science and Technology, Poland
Masafumi Matsuhara	Iwate Prefectural University, Japan
Michael Spratling	University of London, UK
Qiangfu Zhao	University of Aizu, Japan
Tetsuji Kubojama	Gakushuin University, Japan
Tetsuo Kinoshita	RIEC, Tohoku University, Japan
Thai-Nghe Nguyen	Can Tho University, Vietnam
Zhenni Li	University of Aizu, Japan

Multiple Model Approach to Machine Learning (MMAML 2017)

Emili Balaguer-Ballester	Bournemouth University, UK
Urszula Boryczka	University of Silesia, Poland
Abdelhamid Bouchachia	Bournemouth University, UK
Robert Burduk	Wrocław University of Science and Technology, Poland
Oscar Castillo	Tijuana Institute of Technology, Mexico
Rung-Ching Chen	Chaoyang University of Technology, Taiwan
Suphamit Chittayasothorn	King Mongkut's Institute of Technology Ladkrabang, Thailand

José Alfredo F. Costa   Federal University (UFRN), Brazil
Bogusław Cyganek   AGH University of Science and Technology, Poland
Ireneusz Czarnowski   Gdynia Maritime University, Poland
Patrick Gallinari   Pierre et Marie Curie University, France
Fernando Gomide   State University of Campinas, Brazil
Francisco Herrera   University of Granada, Spain
Tzung-Pei Hong   National University of Kaohsiung, Taiwan
Agnieszka   Wrocław University of Science and Technology, Poland
  Indyka-Piasecka
Konrad Jackowski   Wrocław University of Science and Technology, Poland
Piotr Jędrzejowicz   Gdynia Maritime University, Poland
Tomasz Kajdanowicz   Wrocław University of Science and Technology, Poland
Yong Seog Kim   Utah State University, USA
Bartosz Krawczyk   Wrocław University of Science and Technology, Poland
Kun Chang Lee   Sungkyunkwan University, Korea
Edwin Lughofer   Johannes Kepler University Linz, Austria
Bernadetta Maleszka   Wrocław University of Science and Technology, Poland
Hector Quintian   University of Salamanca, Spain
Andrzej Sieminski   Wrocław University of Science and Technology, Poland
Dragan Simic   University of Novi Sad, Serbia
Adam Słowik   Koszalin University of Technology, Poland
Zbigniew Telec   Wrocław University of Science and Technology, Poland
Bogdan Trawiński   Wrocław University of Science and Technology, Poland
Olgierd Unold   Wrocław University of Science and Technology, Poland
Pandian Vasant   University Technology Petronas, Malaysia
Michał Woźniak   Wrocław University of Science and Technology, Poland
Zhongwei Zhang   University of Southern Queensland, Australia
Zhi-Hua Zhou   Nanjing University, China

Special Session on Applications of Data Science (ADS 2017)

Partha Talukdar   Indian Institute of Science, India
Jp de Villiers   Council for Scientific and Industrial Research (CSIR),
                              South Africa
George Anderson   University of Botswana, Botswana
Vukosi Marivate   Council for Scientific and Industrial Research (CSIR),
                              South Africa
Bo Xing   University of Johannesburg, South Africa
Benjamin Rosman   Council for Scientific and Industrial Research (CSIR),
                              South Africa
Fulufhelo Nelwamondo   Council for Scientific and Industrial Research (CSIR),
                              South Africa

Special Session on Artificial Intelligence Applications for E-services (AIAE 2017)

Chi-Chung Lee   Chung Hua University, Taiwan
Mei-Yu Wu   Chung Hua University, Taiwan

Yuan-Chu   Hwang, National United University, Taiwan
Ming-Hsiung   Ying, Chung Hua University, Taiwan
Wei-Lun Chang   Tamkang University, Taiwan
Hsien Ting   National University of Kaohsiung, Taiwan
Duen-Ren Liu   National Chiao Tung University, Taiwan
Chih-Kun Ke   National Taichung University of Science
                            and Technology, Taiwan

Special Session on Automated Reasoning and Proving Techniques with Applications in Intelligent Systems (ARPTA 2017)

Shoichi Morimoto   Senshu University, Japan
Yuichi Goto   Saitama University, Japan
Hongbiao Gao   Saitama University, Japan
Shinsuke Nara   Muraoka Design Laboratory, Japan
Kai Shi   Northeastern University, China
Kazunori Wagatsuma   CIJ solutions, Japan

Special Session on Collective Intelligence for Service Innovation, Technology Opportunity, E-Learning and Fuzzy Intelligent Systems (CISTEF 2017)

Albim Y. Cabatingan   University of the Visayas, Philippines
Teh-Yuan Chang   Aletheia University, Taiwan
Chi-Min Chen   Aletheia University, Taiwan
Chih-Chung Chiu   Aletheia University, Taiwan
Wen-Min Chou   Aletheia University, Taiwan
Chao-Fu Hong   Aletheia University, Taiwan
Chia-Lin Hsieh   Aletheia University, Taiwan
Chia-Ling Hsu   Tamkang University, Taiwan
Chi-Cheng Huang   Aletheia University, Taiwan
Rahat Iqbal   Coventry University, UK
Huan-Ting Lin   The University of Tokyo, Japan
Kuo-Sui Lin   Aletheia University, Taiwan
Min-Huei Lin   Aletheia University, Taiwan
Yuh-Chang Lin   Aletheia University, Taiwan
Shin-Li Lu   Aletheia University, Taiwan
Janet Argot Pontevedra University of San Carlos, Philippines
Shu-Chin Su   Aletheia University, Taiwan
Pen-Choug Sun   Aletheia University, Taiwan
Chen-Fang Tsai   Aletheia University, Taiwan
Ai-Ling Wang   Tamkang University, Taiwan
Chia-Chen Wang   Aletheia University, Taiwan
Leuo-Hong Wang   Aletheia University, Taiwan
Hung-Ming Wu   Aletheia University, Taiwan
Feng-Sueng Yang   Aletheia University, Taiwan
Hsiao-Fang Yang   National Chengchi University, Taiwan
Sadayuki Yoshitomi   Toshiba Corporation, Japan

Special Session on Intelligent Computer Vision Systems and Applications (ICVSA 2017)

Ferran Reverter Comes	University of Barcelona, Spain
Michael Cree	University of Waikato, New Zealand
Piotr Dziurzański	University of York, UK
Paweł Forczmański	West Pomeranian University of Technology, Szczecin, Poland
Marcin Iwanowski	Warsaw University of Technology, Poland
Heikki Kälviäinen	Lappeenranta University of Technology, Finland
Tomasz Marciniak	UTP University of Science and Technology, Poland
Adam Nowosielski	West Pomeranian University of Technology, Szczecin, Poland
Krzysztof Okarma	West Pomeranian University of Technology, Szczecin, Poland
Arkadiusz Orłowski	Warsaw University of Life Sciences, Poland
Edward Półrolniczak	West Pomeranian University of Technology, Szczecin, Poland
Pilar Rosado Rodrigo	University of Barcelona, Spain
Khalid Saeed	AGH University of Science and Technology Cracow, Poland
Rafael Saracchini	Technological Institute of Castilla y León (ITCL), Spain
Samuel Silva	University of Aveiro, Portugal
Gregory Slabaugh	City University London, UK
Egon L. van den Broek	Utrecht University, Utrecht, The Netherlands
Ventzeslav Valev	Bulgarian Academy of Sciences, Bulgaria

Special Session on Intelligent Data Analysis, Applications and Technologies for Internet of Things (IDAIoT 2017)

Goutam Chakraborty	Iwate Prefectural University, Japan
Bin Dai	University of Technology Xiamen, China
Qiangfu Zhao	University of Aizu, Japan
David C. Chou	Eastern Michigan University, USA
Chin-Feng Lee	Chaoyang University of Technology, Taiwan
Lijuan Liu	University of Technology Xiamen, China
Kien A. Hua	Central Florida University, USA
Long-Sheng Chen	Chaoyang University of Technology, Taiwan
Xin Zhu	University of Aizu, Japan
David Wei	Fordham University, USA
Qun Jin	Waseda University, Japan
Jacek M. Zurada	University of Louisville, USA
Tsung-Chih Hsiao	Huaoiao University, China
Hsien-Wen Tseng	Chaoyang University of Technology, Taiwan

Nitasha Hasteer   Amity University Uttar Pradesh, India
Chuan-Bi Lin   Chaoyang University of Technology, Taiwan
Cliff Zou   Central Florida University, USA

Special Session on Intelligent Algorithms and Brain Functions (InBRAIN 2017)

Zbigniew Struzik   RIKEN Brain Science Institute, Japan
Zbigniew Ras   University of North Carolina at Charlotte, USA
Konrad Ciecierski   Warsaw University of Technology, Poland
Piotr Habela   Polish-Japanese Academy of Information Technology,
                              Warsaw, Poland
Peter Novak   Brigham and Women's Hospital, Boston, USA
Wieslaw Nowinski   Cardinal Stefan Wyszynski University, Warsaw, Poland
Andrei Barborica   Research & Compliance and Engineering, FHC, Inc.,
                              Bowdoin, USA
Alicja Wieczorkowska   Polish-Japanese Academy of Information Technology,
                              Warsaw, Poland
Majaz Moonis   UMass Medical School, Worcester, USA
Krzysztof Marasek   Polish-Japanese Academy of Information Technology,
                              Warsaw, Poland
Mark Kon   Boston University, Boston, USA
Rafal Zdunek   Wrocław University of Science and Technology, Poland
Lech Polkowski   Polish-Japanese Academy of Information Technology,
                              Warsaw, Poland
Andrzej Skowron   Computer Science and Mechanics, Warsaw University,
                              Poland
Ryszard Gubrynowicz   Polish-Japanese Academy of Information Technology,
                              Warsaw, Poland
Takeshi Okada   The University of Tokyo, Japan
Dominik Slezak   Warsaw University, Poland
Radoslaw Nielek   Polish-Japanese Academy of Information Technology,
                              Warsaw, Poland

Special Session on Intelligent Systems and Algorithms in Information Sciences (ISAIS 2017)

Martin Kotyrba   University of Ostrava, Czech Republic
Eva Volna   University of Ostrava, Czech Republic
Ivan Zelinka   VŠB-Technical University of Ostrava, Czech Republic
Hashim Habiballa   Institute for Research and Applications of Fuzzy
                              Modeling, Czech Republic
Alexej Kolcun   Institute of Geonics, AS CR, Czech Republic
Roman Senkerik   Tomas Bata University in Zlin, Czech Republic
Zuzana Kominkova   Tomas Bata University in Zlin, Czech Republic
  Oplatkova
Katerina Kostolanyova   University of Ostrava, Czech Republic
Antonin Jancarik   Charles University in Prague, Czech Republic

Igor Kostal	The University of Economics in Bratislava, Slovakia
Eva Kurekova	Slovak University of Technology in Bratislava, Slovakia
Leszek Cedro	Kielce University of Technology, Poland
Dagmar Janacova	Tomas Bata University in Zlin, Czech Republic
Martin Halaj	Slovak University of Technology in Bratislava, Slovakia
Radomil Matousek	Brno University of Technology, Czech Republic
Roman Jasek	Tomas Bata University in Zlin, Czech Republic
Petr Dostal	Brno University of Technology, Czech Republic
Jiri Pospichal	The University of Ss. Cyril and Methodius (UCM), Slovakia
Vladimir Bradac	University of Ostrava, Czech Republic
Roman Jasek	Tomas Bata University in Zlin, Czech Republic
Vaclav Skala	University of West Bohemia, Czech Republic

Special Session on IT in Biomedicine (ITiB 2017)

Golnoush Abae	Universiti Teknologi Malaysia (UTM), UTM Johor Bahru, Malaysia
Orcan Alpar	University of Hradec Kralove, Czech Republic
Dawit Assafa Haile	Addis Ababa University, Ethiopia
Branko Babusiak	University of Zilina, Slovakia
Pavel Blazek	University of Defense, Hradec Kralove, Czech Republic
Peter Brida	University of Zilina, Slovakia
Petr Cermak	Silesian University Opava, Czech Republic
Martin Cerny	VSB, Technical University of Ostrava, Czech Republic
Richard Cimler	University of Hradec Kralove, Czech Republic
Rafael Dolezal	University of Hradec Kralove, Czech Republic
Ricardo J. Ferrari	Federal University of Sao Carlos, Brazil
Tanos C.C. Franca	Military Institute of Engineering, Praça, Brazil
Michal Gala	University of Zilina, Slovakia
Jan Honegr	University Hospital Hradec Kralove, Czech Republic
Radovan Hudak	Technical University of Kosice, Slovakia
Roliana Ibrahim	Universiti Teknologi Malaysia (UTM), UTM Johor Bahru, Malaysia
Marek Kukucka	Slovak University of Technology in Bratislava, Slovakia
David Korpas	Silesian University Opava, Czech Republic
Ondrej Krejcar	University of Hradec Kralove, Czech Republic
Kamil Kuca	University of Hradec Kralove, Czech Republic
Juraj Machaj	University of Zilina, Slovakia
Jaroslav Majerník	Pavol Josef Safarik University in Kosice, Slovakia
Petra Maresova	University of Hradec Kralove, Czech Republic
Reza Masinchi	Universiti Teknologi Malaysia (UTM), UTM Johor Bahru, Malaysia
Marek Penhaker	VSB Technical University of Ostrava, Czech Republic
Jan Plavka	Technical University of Kosice, Slovakia
Teodorico C. Ramalho	Federal University of Lavras (UFLA), Brazil

Martin Rozanek	Czech Technical University in Prague, Czech Republic
Saber Salehi	Universiti Teknologi Malaysia (UTM),
	UTM Johor Bahru, Malaysia
Ali Selamat	Universiti Teknologi Malaysia (UTM),
	UTM Johor Bahru, Malaysia

Special Session on Intelligent Technologies in Smart Cities in the 21st Century (ITSC 2017)

Cezary Orłowski	WSB University Gdansk, Poland
Piotr Oskar Czechowski	Gdynia Maritime University, Poland
Ewa Glińska	Bialystok University of Technology, Poland
Joanna Godlewska	Bialystok University of Technology, Poland
Jarosław Hryszko	Wrocław University of Technology, Poland
Dariusz Kralewski	University of Gdansk, Poland
Kostas Karatzas	Aristotle University of Thessaloniki, Greece
Lech Madeyski	Wrocław University of Technology, Poland
Maciej Nowak	Jagiellonian University in Kraków, Poland
Cezary Orłowski	WSB University Gdansk, Poland
Helena Szczerbicka	Leibniz University Hannover, Germany
Paweł Węgrzyn	Jagiellonian University in Kraków, Poland
Artur Ziółkowski	WSB University Gdansk, Poland

Special Session on Analysis of Image, Video and Motion Data in Life Sciences (IVMLS 2017)

Artur Bąk	Polish-Japanese Academy of Information Technology, Poland
Leszek Chmielewski	Warsaw University of Life Sciences, Poland
Aldona Barbara Drabik	Polish-Japanese Academy of Information Technology, Poland
Marcin Fojcik	Sogn og Fjordane University College, Norway
Adam Gudyś	Silesian University of Technology, Poland
Celina Imielińska	Vesalius Technologies LLC, USA
Henryk Josiński	Silesian University of Technology, Poland
Ryszard Klempous	Wrocław University of Science and Technology, Poland
Ryszard Kozera	The University of Life Sciences, SGGW, Poland
Julita Kulbacka	Wrocław Medical University, Poland
Marek Kulbacki	Polish-Japanese Academy of Information Technology, Poland
Aleksander Nawrat	Silesian University of Technology, Poland
Jerzy Paweł Nowacki	Polish-Japanese Academy of Information Technology, Poland
Eric Petajan	LiveClips LLC, USA
Andrzej Polański	Silesian University of Technology, Poland
Joanna Rossowska	Polish Academy of Sciences, Institute of Immunology and Experimental Therapy, Poland

Jakub Segen	Gest3D LLC, USA
Aleksander Sieroń	Medical University of Silesia, Poland
Michał Staniszewski	Polish-Japanese Academy of Information Technology, Poland
Adam Świtoński	Silesian University of Technology, Poland
Agnieszka Szczęsna	Silesian University of Technology, Poland
Kamil Wereszczyński	Polish-Japanese Academy of Information Technology, Poland
Konrad Wojciechowski	Polish-Japanese Academy of Information Technology, Poland
Sławomir Wojciechowski	Polish-Japanese Academy of Information Technology, Poland

Special Session on Modern Applications of Machine Learning for Actionable Knowledge Extraction (MAMLAKE 2017)

Ajit Narayanan	AUT University, New Zealand
Simon Fong	University of Macau, SAR China
Parma Nand	AUT University, New Zealand
Muhammad Asif Naeem	AUT University, New Zealand
Philip Bright	Waiariki Institute of Technology, New Zealand
Akhtar Zaman	Waiariki Institute of Technology, New Zealand

Special Session on Mathematics of Decision Sciences and Information Science (MDSIS 2017)

Hakim Bendjenna	University of Tebessa, Algeria
Masahiro Hachimori	University of Tsukuba, Japan
Ryuichiro Ishikawa	Waseda University, Japan
Masami Ito	Kyoto Sangyo University, Japan
Yoshihiro Hoshino	Kagawa University, Japan
Evgeny Ivashko	IAMR KarRC RAS, Russia
Diang-yu Jiang	Huaihai Institute of Technology, PR China
Yuji Kobayashi	Toho University, Japan
Hidetoshi Komiya	Keio University, Japan
Michiro Kondo	Tokyo Denki University, Japan
Ridda Laouar	University of Tebessa, Algeria
Tieju Ma	CEEEM, East China University of Science and Technology, PR China
Mikio Nakayama	Keio University, Japan
Hiroyuki Ozaki	Keio University, Japan
Rohit Parikh	CUNY, USA
Leon A. Petrosjan	St. Petersburg University, Russia
Vincenzo Scalzo	University of Naples Federico II, Italy
Kunitaka Shyoji	Shimane University, Japan
Krzysztof Szajowski	Wrocław University of Science and Technology, Poland
Wataru Takahashi	Tokyo Institute of Technology, Japan

Stefano Vannucci	University of Siena, Italy
Alexander Vasin	Moscow State University, Russia
Hongbin Yan	CEEEM, East China University of Science and Technology, PR China
Jun Zhang	University of Kentucky, USA
Xingzhou Zhang	Dalian University of Technology, PR China

Special Session on Scalable Data Analysis in Bioinformatics and Biomedical Informatics (SDABBI 2017)

Hesham H. Ali	University of Nebraska, Omaha, USA
José P. Cerón-Carrasco	Universidad Católica San Antonio de Murcia (UCAM), Spain
Po-Yuan Chen	China Medical University, Taichung, Taiwan
Rudolf Fleischer	German University of Technology, Oman
Che-Lun Hung	Providence University, Taichung, Taiwan
Sergio Lifschitz	Pontificia Universidade Catolica do Rio de Janeiro, Brazil
Stanisław Kozielski	Silesian University of Technology, Poland
Xun Lan	Stanford University, USA
Jung-Hsin Lin	Academia Sinica, Taipei, Taiwan
Pradipta Maji	Indian Statistical Institute, Kolkata, India
Bożena Małysiak-Mrozek	Silesian University of Technology, Poland
Dariusz Mrozek	Silesian University of Technology, Poland
Alessandro S. Nascimento	IFSC, University of Sao Paulo, Brazil
Karin Verspoor	University of Melbourne, Australia
Quan Zou	Tianjin University, PR China

Special Session on Technological Perspective of Agile Transformation in IT organizations (TPATIT 2017)

Jakub Chabik	EBIT Company, Poland
Ireneusz Czarnowski	Gdynia Maritime University, Poland
Bogdan Franczyk	University of Leipzig, Germany
Anna Kosieradzka	Warsaw University of Technology, Poland
Dariusz Kralewski	University of Gdansk, Poland
Leszek Maciaszek	Macquarie University Sydney, Australia
Cezary Orłowski	WSB University Gdansk, Poland
Edward Szczerbicki	University of Newcastle, Australia
Artur Ziółkowski	WSB University Gdansk, Poland

Contents – Part II

Automated Reasoning and Proving Techniques with Applications in Intelligent Systems

Collective Intelligence for Service Innovation, Technology Opportunity, E-Learning and Fuzzy Intelligent Systems

Intelligent Computer Vision Systems and Applications

**Intelligent Data Analysis, Applications and Technologies for Internet
of Things**

Intelligent Technologies in the Smart Cities in the 21st Century

Analysis of Image, Video and Motion Data in Life Sciences

Modern Applications of Machine Learning for Actionable Knowledge Extraction

Mathematics of Decision Sciences and Information Science

Scalable Data Analysis in Bioinformatics and Biomedical Informatics

Technological Perspective of Agile Transformation in IT organizations

Contents – Part I

Text Processing and Information Retrieval

Intelligent Database Systems

Intelligent Information Systems

Computer Vision Techniques

Advanced Data Mining Techniques and Applications

Intelligent and Context Systems

Multiple Model Approach to Machine Learning

Applications of Data Science

Exploring Spatial and Social Factors of Crime: A Case Study of Taipei City

Nathan Kuo[1](✉), Chun-Ming Chang[2](✉), and Kuan-Ta Chen[2]

[1] Taipei American School, Taipei 11152, Taiwan
nathank17110888@tas.tw
[2] Institute of Information Science, Academia Sinica, Taipei 11529, Taiwan
{cmchang,swc}@iis.sinica.edu.tw

Abstract. Recognizing the significance of transparency and accessibility of government information, the Taipei Government recently published city-wide crime data to encourage relevant research. In this project, we explore the underlying relationships between crimes and various geographic, demographic and socioeconomic factors. First we collect a total of 25 datasets from the City and other publicly available sources, and select statistically significant features via correlation tests and feature selection techniques. With the selected features, we use machine learning techniques to build a data-driven model that is capable of describing the relationship between high crime rate and the various factors. Our results demonstrate the effectiveness of the proposed methodology by providing insights into interactions between key geographic, demographic and socioeconomic factors and city crime rate. The study shows the top three factors affecting crime rate are educational attainment, marital status, and distance to schools. The result is presented to the Taipei City officials for future government policy decision making.

Keywords: Crime factor analysis · Geographic information system · Demographics · Socio-economics · Crime hotspots

1 Introduction

Study of crime, criminology, has long been a key area of research spanning across multiple disciplines from behavioral science, sociology, government and education policy planning, to the more recent interdisciplinary data science research. The extreme complexity, and the multifaceted nature of the problem has lead to a recent trend to focus heavily on empirical data analysis approach, with the help of increasing availability of data, and advancements in machine learning techniques in data science.

The study of crime often are in two main branches, one focuses on human side of criminal patterns, be it an individual repeated offender, or group/gangs of criminal organizations, with the goal of assisting police investigators in criminal investigations and crime prevention. The other branch focuses on the

© Springer International Publishing AG 2017
N.T. Nguyen et al. (Eds.): ACIIDS 2017, Part II, LNAI 10192, pp. 3–13, 2017.
DOI: 10.1007/978-3-319-54430-4_1

geographical, spatial, and demographics feature[1] analysis, with the goal of better understanding key factors of why certain areas have higher crime rate (termed *hotspots*). The immediate benefits of this branch of research is more effective law enforcement resource allocation, while the medium and long term goals are to better assist government policy making, which is the focus of our research.

2   Related Work

Chen et al. [12] presents a general overview and a framework of using many classic data mining techniques in the first branch, the criminal patterns of human offenders. Wang et al. [21] expands on this with heavy use of machine learning techniques to better the results. On the second branch, there exists many previous works mostly on using geographic information system (GIS) and statistical models and tools like cluster analysis to better understand the correlation between high crime rates with various features [12,15,17,18]. In particular, Ratcliffe et al. [17] works on more accurate prediction of hotspots via better clustering mechanism, with manual inputs of parameters/features from experienced law enforcement agents. Spicer et al. [18] presents a new method for analyzing temporal and spacial crime patterns along major roadways, where linear spaces are analyzed instead of 2D spaces. This approach allows visualization of temporal variances, crime type comparisons and historical crime trends. Some works focus on analyzing demographic features such as age, education level and divorce rate [6,14,19]. Gary [19] in particular, focus on age and explanation of crime, with very detailed yet singleton analysis of age effects on crime behaviors. Bogomolov's work [7] deserves a special mentioning in that it combines the traditional demographic features, such as migrant population, ethnicity, employment and so on, with anonymized and aggregated human behavioral data computed from mobile network activity (called *smartsteps*) to better the prediction accuracy from averaging 50% to 70%.

All of the previous related work exhibits a common theme typical of the modeling and prediction of a complex problem based on incomplete datasets: *the devil is in the detail*, in all aspects of the process, from data collection, cleaning, feature selection/grouping, model building, to repetitive run-through specific model learning and approximation mechanisms. The paper takes the approach of combining spatial and demographic datasets, and uses SVM machine learning techniques to build the most accurate model. This work is summarized as follows: Sect. 3 details the dataset; Sect. 4 describes our proposed methodology; Sect. 5 reports our experimental results and applications; finally, Sect. 6 concludes our paper.

3   Data Description

Data used in this study are from various government open data repositories and publicly available sources [1–3,5]. We aggregate these data sources into

---

[1] Factor is the term used in Criminology while feature is the technical term used in data science.

three datasets, one describing the criminal cases, the other describing demographic characteristics of villages, and the last containing the physical locations of infrastructure and services in Taipei City.

3.1   Criminal Cases Dataset

The criminal cases dataset made available by the City covers three types of crime: burglary, car theft and bike theft[2], covering the time span from January 2015 to April 2016, with a total of 746 household theft, 132 car theft, and 452 bike theft records. Each record contains the crime ID, time (within 3 hours), date, and street address.

3.2   Village Profiling Dataset

Villages, under districts, are the fourth administrative subdivision of Taiwan, and there are, in total, 456 villages in Taipei. This dataset describes demographic and socioeconomic characteristics of the village and includes the following data fields: boundary, area, population, sex ratio[3], average age, ageing index[4], average income, number of households, average electricity usage per household, average number of people per household, ratios of education levels[5] and ratios of marital status[6]. The datasets from various sources are matched with their village ID.

3.3   Point of Interest Dataset

This dataset includes the Point of Interest of infrastructure and services with potential impact on crime factors, including parks, banks, bus stops, Metro stations, street surveillance cameras, public restrooms, convenience stores, factories/warehouses, public parking lots, street-side parking spaces, police stations and city street lamps. Note that the location of schools include the location of kindergarten, elementary, junior and senior high school. Because these datasets come from a variety of sources with different coordinate systems, as a part of the data cleaning process, we use Google Maps Geocoding API for converting street addresses to WGS84-coordinates, and tools from the Information Science Institute from Academia Sinica [4] for converting TWD97-coordinates to WGS84-coordinates.

---

[2] Taipei is a relatively safe city with extremely low violent crime rate.

[3] Sex ratio is defined as the ratio of males to females in a population.

[4] Ageing Index is calculated as the number of persons 60 years old or over per hundred persons under age 15.

[5] Ratios of education levels include the six percentages: The percentage of population illiterate, with elementary school, junior high, senior high, undergraduate and graduate education.

[6] Ratios of marital status include the four percentages: The percentage of population single, married, divorced or widowed.

4   Methodology

In this section, we apply the grip thematic mapping technique [8,10] to combine the different layers of spatial data, including criminal cases dataset, village profiling dataset and point of interest dataset to predefined 500 m by 500 m cells. We extract various features from the above datasets and exploit feature selection and ranking methods to identify high impact features. We then build a data-driven model using the selected features to predict the probability of high crime rate of each individual cell. The remaining section details this experimental process.

4.1   Granularity Definition and Data Preprocessing

Each field in our datasets, such as location of police stations and marital status of villages, can be thought of as individual spatial layers. However, these data are not valuable or informative unless we define a suitable referencing system to combine and relate the various spatial layers. In order to address this, we apply the grid thematic technique and draw a grid over Taipei that creates 500 m by 500 m cells as visualized in Fig. 1. The reasons are two fold:

1. Manually defining the grid ensures that each of our boundaries enclose the same area. Because the variance in the area of villages is too large, then it is more difficult to perform proximity analysis that can be generalized to each village.
2. This approach allow us to control the fidelity of our combined dataset. When the cells are too small, the village datasets obtained via village-wide statistics are too coarse to describe the characteristic of individual cells; when the cells are too large, the level of detail of cells deteriorates.

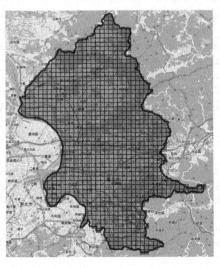

Fig. 1. Visualization of every 500 * 500 cell in the Taipei City

Furthermore, multiple studies have shown the effectiveness of this type of spatial analysis [7,10]. As the result, we defined 1081 cells in Taipei City, and, for each of the defined cells, we perform the following preprocessing steps.

Mapping Crime Locations to Cells. We georeference crime locations to cells by assigning it to its closest cell, and in order to avoid spatial autocorrelation [15], we calculated a crime *danger index* by normalizing the number of crimes by the number of buildings within each cell with crime occurrences. In order to build a binary classification model, we split the transformed crime dataset into two classes with respect to its median: low danger index(class "0") when the index is less than or equal to its median, and high danger index(class "1"), when the index is higher than its median. The empirical distribution function of the index is shown in Fig. 2 and the median is 0.019. In addition, because the dataset is split with respect to its median, the distribution of classes is balanced.

Mapping Village Profiles to Cells. We georeference the village dataset to each cell by calculating the percentage of the cell in nearby villages to create a weighted sum for the village dataset. In order to decrease computational cost, we created 100 points evenly spaced within the cell and sorted them into nearby villages to estimate the percentage of each cell in nearby villages. Furthermore, in order to avoid collinearity between education level ratios, we combine the six ratios into one metric: the percentage of population with undergraduate or graduate level of education. We also generate another set of features by transforming the non-ratio features into log-scale As the result, there are 19 features generated from the village profiling dataset.

Mapping Point of Interest Dataset to Cells. In order to relate the point of interest dataset to cells, we use proximity analysis where we associate spatial features through their physical proximity. For each datasets, there are two metrics: one is to count the number of objects within the boundary of the cell, and the other is to calculate the distance between the center of the cell and its nearest object. Furthermore, we generate another metrics by transforming the distance features into log-scale. Thus, there are 36 point of interest features created for a cell.

4.2   Feature Selection

After we combine our datasets, we normalize each feature by taking standard scores to increase the convergence speed of our algorithms [9], and ease our analysis later on. In order to perform out-of-sample validation with our model, we randomly separate one-fifth of our dataset from our sample. The result is a training set with 80% of our data and testing set with 20% of our data. We only use the training set in the following feature selection and model construction process.

In our preliminary data analysis, we perform a Pearson's correlation analysis to identify correlation between each feature, as well as with the response variable. Because of high correlation between several of our features and in order to avoid overfitting, we decide that feature reduction is necessary. As the result, we use a feature ranking generated from random forest, as well as Pearson's correlation test to perform an initial feature selection process. For each point of interest, we keep only one metric out of the three generated from Sect. 4.1 and for each characteristic of a village, we keep only one metric out of the two generated from Sect. 4.1 if a log-scaled feature is created. We also remove any features that has high correlation to other features and low correlation to our response variable. As the result, we reduce our feature space from 55 features to 29 features.

4.3 Model Construction

We then train a variety of binary classifiers with the training data using 5-fold cross validation, including logistic regression, naive bayesian, decision tree, random forest and support vector machine with different parameters and kernels if applicable. For each classifier, we determine its optimal set of features by using a SVM-based recursive feature elimination (RFE) algorithm [11]. In this algorithm, we rank the features by importance using the linear SVM feature ranking and recursively eliminated the feature with the lowest significance while calculating the performance of the model. In order to increase robustness, the linear SVM feature ranking is the average of our result from each iteration of our 5-fold cross validation process and the top 10 features are illustrated in Table 1. The RFE process allows us to graph the area under the curve (AUC) performance against the number of features used to identify the optimal set of features for each model. The SVM-based RFE graph for each binary classifier is illustrated in Fig. 3.

Table 1. Features ranked by SVM-RFE

Rank	Features	Min.	1st-quantile	Median	3rd-quantile	Max.
1	log.village.average.peopleperhousehold	0.694	0.925	0.986	1.028	1.131
2	log.village.average.population	7.050	8.466	8.684	8.911	9.254
3	log.nearestdistance.kindergarden	2.159	4.873	5.316	5.833	7.667
4	village.ratio.undergraduate	0.275	0.463	0.526	0.584	0.708
5	log.nearestdistance.anyschool	2.159	4.803	5.192	5.713	7.551
6	log.nearestdistance.highschool	3.454	6.033	6.526	6.888	8.274
7	village.ratio.married	0.373	0.452	0.469	0.484	0.643
8	log.nearestdistance.bank	1.902	4.920	5.600	6.272	8.407
9	log.nearestdistance.mrt	2.568	5.807	6.318	6.868	8.659
10	log.nearestdistance.policestation	3.685	5.708	6.235	6.573	7.763

Note that calculate (i) distances in unit of meters, and (ii) population in unit of thousands.

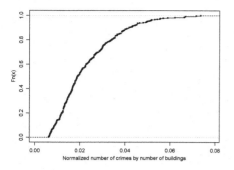

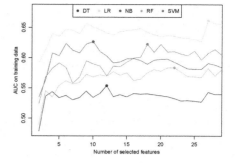

Fig. 2. The empirical cumulative distribution function of the danger index

Fig. 3. AUC curve of each model with respect to number of selected features

5   Experiment Results

In this section we compare our experimental results among the five models mentioned in Sect. 4.3. For each model, we use its accuracy, F1 score, and the area under the ROC curve (AUC) to evaluate the performance. We then analyze the implications of our model.

5.1   Model Comparison

Table 2 concludes the performance metrics of the five models we explore in our experiment. The SVM model outperforms the other models in all metrics, achieving 73.4% accuracy in predicting areas of high versus low danger index. Figure 5 depicts the spatial visualization of predicted results. Hence, we adopt this SVM model to analyze how the demographic, geographic and socioeconomic features influence the danger index of a cell.

In order to further analyze the interaction between danger index and the various features, we apply the Platt Scaling method, which trains the parameters of an additional sigmoid function that maps the output scores of our SVM classification model into posterior class probabilities [13, 16]. This transforms our

Table 2. Model performance comparison

Classification model	# Used features	Accuracy	F1 score	AUC
Decision tree (DT)	12	0.671	0.629	0.668
Logistic regression (LR)	27	0.646	0.650	0.660
Naive Bayesian (NB)	10	0.671	0.594	0.657
Random forest (RF)	21	0.696	0.684	0.705
Support vector machine (SVM)	18	**0.734**	**0.747**	**0.756**

binary classification model into a regression model that predicts the probability of a cell having high danger index, defined as *high danger probability*.

5.2   Quantification of Feature Impact

First of all, we conduct exploratory study on the correlation between each feature and *high danger probability* of cells and observe several critical implications. As Fig. 6 indicates, there is a strong negative correlation between education level and *high danger probability*. In Fig. 6 we also observe that decrease in average income would also accompany increase in *high danger probability*, and according to the study [20], this may be the result from the unaffordable financial cost of having burglary protection devices such as window grill bars. To further quantify the significance of each feature on *high danger probability*, we create a cell that contains the feature's median value, as listed in Table 1, for each feature. We then, for each feature, perform the following procedure:

1. Vary the feature's value from its median to 1st-quantile and to 3rd-quantile while constraining the rest of the features at their median.
2. Evaluate these two newly created cells in our model to calculate their respective *high danger probability*.

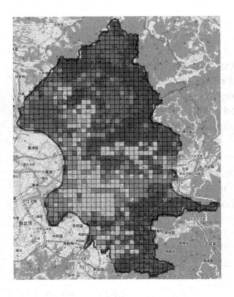

Fig. 4. Predicted by the SVM model   **Fig. 5.** Ground truth

The importance of a feature is therefore quantified as the difference between the *high danger probability* when varying it from its 1st-quantile to its 3rd-quantile. The calculated significance of each feature is illustrated in Fig. 7 and we make several observations: (i) increase in educational attainment best reduces *high danger probability* (ii) the further the distance away from more populous residential areas, marked by schools and parks, the higher the *high danger probability*, and (iii) decrease in stability of marital status, marked by being divorced or widowed, correlates to increased *high danger probability* (Fig. 4).

5.3   Applications

Our experimental results provide insight into how danger index of cells are influenced by their geographic, demographic and socioeconomic features. We can use this knowledge to reveal the underlying causes of crime occurrence and advise government officials accordingly. For instance, given the significant negative correlation between education and crime, this becomes something that the Taipei government can focus on. Furthermore, our predictive model can also be used

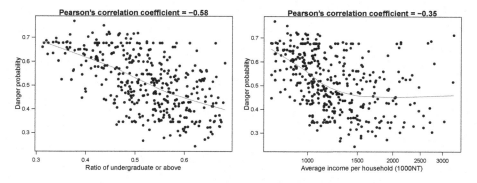

Fig. 6. Relationship between demographic and socioeconomic features and *high danger probability*

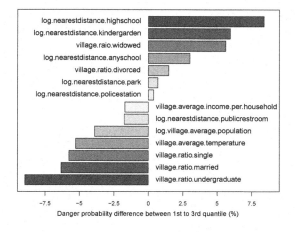

Fig. 7. The quantification of feature significance.

on those cells without crime occurrences yet to identify cells with similar characteristics as known places with high danger index.

6   Conclusion and Future Work

This paper documents our efforts in applying data science techniques to a specific social problem of analyzing factors of areas of high danger index. The results illustrates the unique effectiveness of combing spatial data and demographic, social and economic factors. Going forward, we would like to investigate further into the mechanism of the SVM machine learning internals to gain better insights as to the inter-relationships between the selected key factors. Applying the same methodology to other cities/metro areas in Taiwan, or Greater China areas will help solidifying the modeling process and correctness. In addition, collaboration with other parts of the world will help us better understand if culture difference plays a role in crime factor analysis.

References

1. Open data from directorate-general of budget, accounting and statistics. http://www.dgbas.gov.tw/mp.asp?mp=1
2. Open data ministry of interior. http://data.moi.gov.tw/MoiOD/default/Index.aspx
3. Open data Taipei. http://data.taipei/
4. Supporting toolset for GIS applications. http://gis.rchss.sinica.edu.tw/ISTIS/tools/
5. Taiwan government open data. http://data.gov.tw/
6. Anwer, M., Nasreen, S., Shahzadi, A.: Social and demographic determinants of crime in Pakistan: a panel data analysis of Province Punjab. Int. J. Econ. (2015)
7. Bogomolov, A., Lepri, B., Staiano, J., Oliver, N., Pianesi, F., Pentland, A.: Once upon a crime: towards crime prediction from demographics and mobile data. In: Proceedings of the 16th International Conference on Multimodal Interaction, pp. 427–434. ACM (2014)
8. Bowers, K., Newton, M., Nutter, R.: A GIS-linked database for monitoring repeat domestic burglary. In: Mapping and Analysing Crime Data-Lessons from Research and Practice, pp. 120–137 (2001)
9. Box, G.E., Cox, D.R.: An analysis of transformations. J. Roy. Stat. Soc. Ser. B (Methodol.) 211–252 (1964)
10. Chainey, S., Tompson, L., Uhlig, S.: The utility of hotspot mapping for predicting spatial patterns of crime. Secur. J. **21**(1–2), 4–28 (2008)
11. Chang, Y.-W., Lin, C.-J.: Feature ranking using linear SVM. In: WCCI Causation and Prediction, Challenge, pp. 53–64 (2008)
12. Chen, H., Chung, W., Xu, J.J., Wang, G., Qin, Y., Chau, M.: Crime data mining: a general framework and some examples. Computer **37**(4), 50–56 (2004)
13. Lin, H.-T., Lin, C.-J., Weng, R.C.: A note on Platt's probabilistic outputs for support vector machines. Mach. Learn. **68**(3), 267–276 (2007)
14. Lochner, L., Moretti, E.: The effect of education on crime: evidence from prison inmates, arrests, and self-reports. Am. Econ. Rev. **94**(1), 155–189 (2004)

15. Murray, A.T., McGuffog, I., Western, J.S., Mullins, P.: Exploratory spatial data analysis techniques for examining urban crime implications for evaluating treatment. Br. J. Criminol. **41**(2), 309–329 (2001)
16. Platt, J., et al.: Probabilistic outputs for support vector machines and comparisons to regularized likelihood methods. Adv. Larg Margin Classif. **10**(3), 61–74 (1999)
17. Ratcliffe, J.H., McCullagh, M.J.: Hotbeds of crime and the search for spatial accuracy. J. Geogr. Syst. **1**(4), 385–398 (1999)
18. Spicer, V., Song, J., Brantingham, P., Park, A., Andresen, M.A.: Street profile analysis: a new method for mapping crime on major roadways. Appl. Geogr. **69**, 65–74 (2016)
19. Sweeten, G., Piquero, A.R., Steinberg, L.: Age and the explanation of crime, revisited. J. Youth Adolesc. **42**(6), 921–938 (2013)
20. Tilley, N., Tseloni, A., Farrell, G.: Income disparities of burglary risk security availability during the crime drop. Br. J. Criminol. azr010 (2011)
21. Wang, T., Rudin, C., Wagner, D., Sevieri, R.: Learning to detect patterns of crime. In: Blockeel, H., Kersting, K., Nijssen, S., Železný, F. (eds.) ECML PKDD 2013. LNCS (LNAI), vol. 8190, pp. 515–530. Springer, Heidelberg (2013). doi:10.1007/978-3-642-40994-3_33

A Fuzzy Logic Based Network Intrusion Detection System for Predicting the TCP SYN Flooding Attack

Nenekazi Nokuthala Penelope Mkuzangwe[1,2](✉)
and Fulufhelo Vincent Nelwamondo[1,2]

[1] Modelling and Digital Science, Council for Scientific and Industrial Research,
CSIR Main Site, Meiring Naude Road, Pretoria, South Africa
[2] Department of Electrical and Electronic Engineering,
University of Johannesburg, Corner of Kingsway and University Road,
Auckland Park, Johannesburg, South Africa
mmkuza@gmail.com

Abstract. Fuzzy logic is one of the powerful tools for reasoning under uncertainty and since uncertainty is an intrinsic characteristic of intrusion analysis, Fuzzy logic is therefore an appropriate tool to use to analyze intrusions in a Network. This paper presents a fuzzy logic based network intrusion detection system to predict neptune which is a type of a Transmission Control Protocol Synchronized (TCP SYN) flooding attack. The performance of the proposed fuzzy logic based system is compared to that of a decision tree which is one of the well-known machine learning techniques. The results indicate that the performance difference, in terms of predicting the proportion of attacks in the data, of the proposed system with respect to the decision tree is negligible.

Keywords: Fuzzy logic · Intrusion detection · Network intrusion detection system · Neptune · Decision tree · TCP SYN flooding attack

1 Introduction

Intrusion detection systems (IDSes) are software applications or hardware systems that perform intrusion detection in information systems. They may be classified according to the source of those events that they monitor, for example, network events or host events. They may also be classified based on the method they use to perform detection. In general two detection methods exists, namely, misuse detection and anomaly detection. In misuse detection, all know attack patterns are defined and the IDS is trained to recognise them. The misuse detection method is good in detecting known attacks, however, it is unable to detect new attacks since it will not have defined patterns for the new attacks. In anomaly detection the normal behaviour of network traffic is modelled and any network traffic that deviates from the normal behaviour would be defined as anomalous which may mean the network is under attack. The anomaly based detection can detect new

N.T. Nguyen et al. (Eds.): ACIIDS 2017, Part II, LNAI 10192, pp. 14–22, 2017.
DOI: 10.1007/978-3-319-54430-4_2

attacks, however, it generates a lot of false alarms. The effectiveness of an intrusion detection system (IDS) is evaluated based on its ability to correctly classify events to be attacks or normal network behaviour [1].

Intrusion detection research community has proposed various intrusion detection systems. These systems differ from each other mainly based on the techniques they employ to analyse data in order to classify it as intrusive or normal. These techniques have their origins, to mention a few, in statistical methods [2], machine learning methods [3] and data mining techniques [4]. Researchers also compare the performance of their IDSes with IDSes that are based on different techniques [3,5].

In this study a fuzzy logic based network intrusion detection system for detecting neptune which is a type of a Transmission Control Protocol Synchronized (TCP SYN) flooding attack is presented. The TCP SYN flooding attack is of interest in this work since it is the most widespread denial of service (DOS) attack and a serious threat to organizations that provide online services. The performance of the proposed intrusion detection system is compared to that of a decision tree.

There rest of this paper is arranged as follows. In Sect. 2 the work done in detecting denial of service attack using fuzzy logic is presented. Section 3 gives a brief description of the decision tree and the fuzzy logic. Section 4 discusses the dataset that is used in this study. In Sect. 5 the proposed fuzzy logic based network intrusion detection system is presented. Sections 6, 7, 8 and 9 present the experimental work, results, discussion and conclusion.

2   Related Work

This section presents the fuzzy logic based techniques for detecting denial of service attacks. Tuncer and Tatar [5] propose a fuzzy logic based system for detecting SYN flooding attacks. They compared the performance of their proposed system with Cumulative Sum (CUSUM) algorithm. Their system yielded results that are comparable to existing IDS. Kanlayasiri and Sanguanpong [6] presented a BENEF (Behavior Statistic, Network Information Based and Fuzzy logic Decision) model. In this model the important features are extracted from the network packets. These features are used in the pre detector component to analyze events with the set of rules to determine if those events have intrusive patterns or not. The anomalous events are passed to the Decision Engine that uses fuzzy logic principle to calculate the possibility of intrusion. They used the model to detect the TCP SYN flooding attack. Their results were recorded as the percentage of intrusion possibility. It would have been interesting to know how far their prediction was from the ground truth. Gao and Zhou [7] proposed a method for detecting intrusion based on fuzzy rule-based technique. A Fuzzy Reasoning Petri Nets (FRPN) model is used to represent fuzzy rule base and derive the final detection decision. They implemented their model in detecting the TCP SYN flooding attack. Their results were recorded as the percentage of intrusion possibility but this does not tell much about the accuracy of their system i.e. how far was their prediction from the ground truth. Shanmugavadivu and Nagarajan [8] have developed an anomaly based intrusion detection system in detecting the intrusion behaviour within a network. A fuzzy decision-making module was designed using the fuzzy inference approach to detect the network attacks. They used automated strategy for the generation of fuzzy rules. They used KDD99 dataset to evaluate the performance of their system. The experimental results indicate that their system is able to effectively detect all four attacks types found in this dataset with accuracy of more than 90 percent. This supports the notion that fuzzy logic is appropriate for use in intrusion detection.

Our work implements a fuzzy logic based IDS to detect the TCP SYN flooding attack and it differs from the work done in [5–7] in terms of the dataset used. In this work the NSL KDD dataset is used whereas [5] observed the connection request coming to the Firat university web server for SYN flooding attack detection and TCP packets coming to port 80 were collected every specified seconds using ethereal application and [6,7] did not specify the dataset they have used. [8] implemented a fuzzy logic based IDS to detect all four categories of attacks that exist in the KDD99 dataset and their work differs from ours in that they detected a denial of service attack as a whole while we are proposing a system to detect a particular denial of service attack in the NSL KDD dataset.

3   A Brief Overview of the Decision Tree and Fuzzy Logic

In this work a fuzzy logic based system is presented and its performance is compared to that of a decision tree therefore this section gives a refresher on decision trees and fuzzy logic.

3.1   Decision Tree

Decision trees are among the well-known machine learning techniques. A decision tree consists of three basic elements. Namely a decision node specifying a test attribute, a branch corresponding to the one of the possible test attribute values and a leaf which contains the class to which the object belongs. In decision trees, two major phases should be ensured:

– Building the tree. A decision tree is built based on a given training set. It consists of selecting the appropriate test attribute for each decision node and also to define the class labelling each leaf.
– Classification. Classification of a new case starts from the root of the decision tree where the attribute specified by this node is tested. The result of this test allows to move down the tree branch relative to the attribute value of the given case. This process is repeated until a leaf is reached. The case is then classified in the class that is described by the reached leaf.

Numerous algorithms have been developed to construct decision trees. The ID3 and C4.5 algorithms developed by Quinlan [9] are probably the most popular ones. There is also the CART algorithm of Breiman et al. [10].

3.2   Fuzzy Logic

The Fuzzy logic concept was conceived by professor Lofti Zadeh as a way of processing data by allowing partial set membership rather than crisp set membership or non-membership. It provides a very valuable flexibility for reasoning that takes inaccuracies and uncertainties into account [11]. It provides a simple way of arriving to a definite conclusion based upon vague, ambiguous, noisy, imprecise or missing input information. The process of reaching this definite conclusion is described below.

- All input values are fuzzified into fuzzy membership functions.
- Fuzzy rules are generated. The rules are in the form of IF THEN statement.
- Given an instance, some of the fuzzy rules will be activated.
- The activated rules are combined in the rule base to compute the fuzzy output distribution.
- The fuzzy output distribution is defuzzified to obtain a crisp output value.

4   Dataset

The NSL KDD [12] is the dataset that was used in this study. The NSL KDD dataset was generated from the KDD99 dataset [13] by removing redundant and duplicate instances and reducing the size of the dataset. The KDD99 dataset is a revised version of the DARPA 98 MIT Linconln Lab dataset [14] that was summarized into network connections where each connection is a single row vector consisting of 41 features and is marked as either normal or an attack with exactly one particular type of attack. The network connections are referred to as cases in this work. The different attacks included in the dataset fall into four categories, namely, denial of service attack, remote to user attack, user to root attack and probes.

In this study the NSL KDD training and test data were filtered for normal and neptune (a type of a denial of service attack) connections that are referred to as normal and neptune cases respectively. The training data consisted of 108558 cases with 67343 normal cases and 41215 neptune cases. The test data consisted of 14368 cases with 9711 normal cases and 4657 neptune cases. Our interest is to predict the actual proportion of neptune cases in the test data where the actual proportion of neptune cases in the test data is the number of neptune cases in the test data divided by the number of cases in the test data.

The NSL KDD dataset has 41 attributes and we had to decide on the attributes to use to detect neptune. [15] recommended ten attributes as relevant in identifying neptune. In this studied we initially selected four of those attributes to train and test our system. These attributes were % of connections that have SYN errors, % of connections to the same service, % of connections to the different service and count of connections having the same destination host and using the same service. In this work they are denoted as SynErrorRate, SameSrvRate, DiffSrvRate and DStHostSrvCount respectively. However, we were not satisfied with the performance of our system. Furthermore, when we trained the decision tree using the four attributes the resulting decision tree did not include DstHostSrvCount attribute. This happens when the attribute does not improve the accuracy of the decision tree [16] which means this attribute is redundant and including a redundant attribute badly affects the accuracy of a classifier [17]. Based on the trained decision tree we decided to implement our proposed system using the attributes that were included in the trained decision tree and the performance of our system improved.

5   The Proposed Fuzzy Logic Based Network Intrusion Detection System

In this section the proposed fuzzy logic based network intrusion detection system used in detecting neptune is presented. In which the membership values of the three attributes and the output and their membership functions are defined, the rules are generated and the fuzzy inferencing and defuzzification methods are described. The implementation of this system was done in Matlab.

5.1   Fuzzification and Membership Functions

To derive the membership values for each of the attributes, we observed the range of values each attribute takes and calculated the average value for each attribute in the normal, attack and mixed (contains both normal and attack) data of the training data. Based on the minimum, average and maximum values of each attributes, three membership values L (low), M (medium) and H (high) were defined. The triangular membership function was used to define membership index associated with each membership value of the three attributes as shown in Fig. 1. The output is in the form of percentage of intrusion in the data (% Intrusion). It is also defined into three membership values, namely, L (low), M (medium) and H (high).

5.2   Fuzzy Rules Generation

The rules were created in the form of an IF THEN statement. We enumerate all possible permutations of the membership values of the three attributes. At first all twenty seven permutations were used to create the rules, however, for some permutations it was difficult to infer the consequent which led to a poor performance of the system. We decided to select only the permutations that led to an easy way to deduce the consequent of the rules and it resulted to only nine permutations. The permutations were used as the antecedent for each rule. From the training data we observed the average value for each attribute in the normal, attack and mixed data and noticed that for some attributes the average value decreased in the presence of attacks while it increased for some attributes. The consequent of each rule was then based on the behavior of the average value of each attribute in the presence or absence of an attack. An example of the rules is given below.

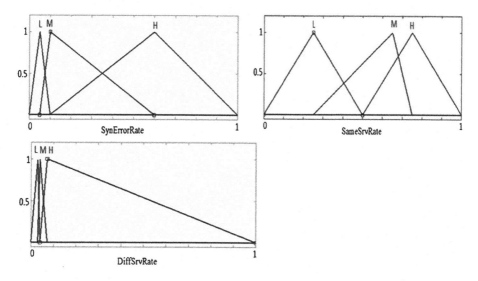

Fig. 1. The membership functions of the proposed system

IF Average of SynErrorRate = H AND Average SameSrvRate = L AND Average of DiffSrvRate = H THEN % Intrusion = H.

5.3 Fuzzy Inferencing and Defuzzification

The activated rules were aggregated using the Mamdani fuzzy inferencing that is explained in [18] and the crisp value of the output was obtained using the centroid approach that returns the centre of the area under the curve of a membership function.

6 Experimental Work

This section outlines how neptune was predicted using the decision tree and the proposed system and defines the performance metrics used in this study.

6.1 Decision Tree Construction

The decision tree was constructed in R Studio using the processed training data set as described in Sect. 4. The resulting decision tree is presented in Fig. 2. The constructed decision tree was used to predict neptune from the processed test data.

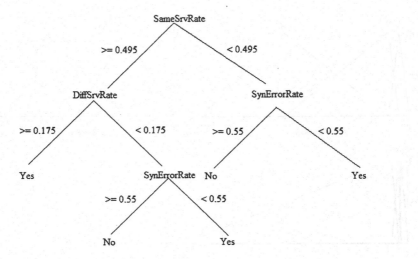

Fig. 2. The decision tree.

6.2   Prediction with the Proposed System

For each attribute an average was calculated from the test data. These averages were tested against the generated rules of the proposed system to predict the percentage of intrusion in the test data. Figure 3 illustrates the crisp value of the output variable (% intrusion) where the solid vertical black line corresponds to the crisp value of the output variable.

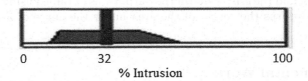

Fig. 3. The crisp value of the output variable, % intrusion.

6.3   Performance Metrics

The performance metrics that are used in this study are the proportion of attacks predicted by the two algorithms in the test data where each algorithm predicts the actual proportion of neptune cases in the test data and the accuracy of the two algorithms where accuracy is the sum of the true positives and true negatives divided by the sum of the true positives, false positives, false negatives and true negatives of the two algorithms.

7 Results

The actual proportion of attacks (neptune cases) in the test data is 0.3241. Table 1 presents the predicted proportion of attacks by the two algorithms and their accuracies.

Table 1. The results of the proposed system and the decision tree

Performance metrics	Proposed system	Decision tree
Predicted proportion of attacks	0.3200	0.3241
Accuracy	0.9324	0.9980

The results in Table 1 indicate that the performance difference, in terms of predicting the proportion of neptune cases in the test data, of the proposed system with respect to the decision tree is negligible and the accuracy of the decision tree is better than that of the proposed system.

8 Discussion

Our method generate the rules using a simple strategy where we enumerated permutation of the features of interest and decided on the output based on the behavior of each feature in the permutation. The results indicate that the performance difference, in terms of predicting the proportion of neptune cases in the test data, of the proposed system with respect to the decision tree is negligible and the accuracy of the decision tree is better than that of the proposed system. However, decision trees are known to be highly unstable with respect to minor changes in the training data whereas fuzzy logic based system may be more stable in the presence of minor changes in the training data due to the elasticity of the fuzzy sets [19]. This gives a fuzzy logic based system an advantage over a decision tree which means the proposed system has a higher chance of yielding the same accuracy if the training data can be slightly changed as compared to the decision tree.

9 Conclusion

We proposed a simple fuzzy logic network intrusion detection system to detect neptune which is a type of TCP SYN flooding attack. The NSL KDD dataset was used to train and evaluate our system. The performance of our system was compared to that of a decision tree. The results indicate that the performance difference, in terms of predicting the proportion of neptune cases in the test data, of the proposed system with respect to the decision tree is negligible.

Acknowledgments. I would like to thank my promoter Professor Fulufhelo Nelwamondo for his guidance and support and the CSIR: Modelling and Digital Science Unit for financially supporting my studies.

References

1. Wu, S.X., Banzhaf, W.: The use of computational intelligence in intrusion detection systems: a review. Appl. Soft Comput. J. **10**, 1–35 (2010)
2. Chin-Ling, C.: A new detection method for distributed denial of service attack traffic based on statistical test. J. Univ. Comput. Sci. **15**, 488–503 (2009)
3. Amor, N.B., Benferhat, S., Elouedi, Z.: Naive Bayes vs decision trees in intrusion detection systems. In: Proceedings of the 2004 ACM Symposium on Applied Computing, pp. 420–424. ACM (2004)
4. Kanwal, G., Rshma, C.: Detection of DDoS attacks using data mining. Int. J. Comput. Bus. Res. (IJCBR) **2**, 1–10 (2011)
5. Tuncer, T., Tatar, Y.: Detection SYN flooding attacks using fuzzy logic. In: Proceedings of the International Conference on Information Security and Assurance, ISA 2008, pp. 321–325. IEEE (2008)
6. Kanlayasiri, U., Sanguanpong, S.: Network-based intrusion detection model for detecting TCP SYN flooding. In: Proceedings of the 4th National Computer Science and Engineering Conference, Bangkok, Thailand, pp. 148–153 (2000)
7. Gao, M., Zhou, M.: Fuzzy intrusion detection based on fuzzy reasoning Petri nets. In: IEEE International Conference on Systems, Man and Cybernetics, vol. 2, pp. 1272–1277. IEEE (2003)
8. Shanmugavadivu, R., Nagarajan, N.: Network intrusion detection system using fuzzy logic. Indian J. Comput. Sci. Eng. (IJCSE) **2**(1), 101–111 (2011)
9. Quinlan, J.R.: C4.5: Programs for Machine Learning. Morgan Kaufmann, San Mateo (1993)
10. Breiman, L., Friedman, J.H., Olshen, R.A., Stone, C.J.: Classification and Regression Trees. Wadsworth and Brooks, Monterey (1984)
11. Introduction to fuzzy logic. http://www.francky.me/doc/course/fuzzy_logic.pdf
12. Tavallaee, M., Bagheri, E., Lu, W., Ghorbani, A.A.: A detailed analysis of the KDD CUP 99 data set. In: Proceedings of the Second IEEE Symposium on Computational Intelligence for Security and Defence Applications (2009)
13. KDD Cup 1999 Data. https://kdd.ics.uci.edu/databases/kddcup99/kddcup99.html
14. DARPA Intrusion Detection Data Sets. https://www.ll.mit.edu/ideval/data/
15. Kayacik, H.G., Zincir-Heywood, A.Z., Heywood, M.I.: Selecting features for intrusion detection: a feature relevance analysis on KDD 99 intrusion detection datasets. In: Proceedings of the Third Annual Conference on Privacy, Security and Trust (2005)
16. Decision Tree Models. https://www.ibm.com/support/knowledgecenter/SS3RA7_15.0.0/com.ibm.spss.modeler.help/nodes_treebuilding.htm
17. Choudhary, N.K., Shinde, Y., Kannan, R., Venkatraman, V.: Impact of attribute selection on the accuracy of Multilayer Perceptron. Int. J. IT Knowl. Manag. (IJITKM) **7**(2), 32–36 (2014)
18. Fuzzy inference systems. http://www.cs.princeton.edu/courses/archive/fall07/cos436/HIDDEN/Knapp/fuzzy004.htm
19. Engineering Decision Trees in Fuzzy Logic ABSTRACT - International. http://isindexing.com/isi/papers/1413180985.pdf

Analytical Ideas to Improve Daily Demand Forecasts: A Case Study

Sougata Deb[(⊠)]

Institute of Systems Science, National University of Singapore, Singapore, Singapore
deb.sougata@gmail.com

Abstract. With the growing popularity of app-based taxi aggregators, bike-sharing systems and supermarkets across the world, it is now important to forecast short-term (often daily) demand accurately. Imprecise forecasts generally result in daily losses due to over or under stocking. This paper proposes multiple analytical constructs for demand prediction using Capital Bikeshare's data as an example. The aim is to provide novel and business-justified ideas on feature engineering and subsequently using these features to create different analytical constructs for the actual prediction problem. A comparison of different modeling techniques in solving the same problem is also included. The findings demonstrate that a decomposed multi-stage prediction performs better than the *pure* forecasting or prediction approaches. Ensembling results show that a cross-construct ensemble may perform better than the traditional multiple-learner ensembles within the same construct.

1 Introduction

Short-term demand forecasting has always been an essential task for any business. Accurate prediction of demand plays a pivotal role in inventory management, resource allocation and maintenance planning across industries. Usage of predictive models is prevalent in call centres, manufacturing, and power companies [1] and is gaining popularity in industries like vehicle aggregators and bike-sharing systems [6,7].

A key challenge to precise prediction is the impact of event-based external factors [4] that keep affecting demand intermittently. Additionally, information on some of the important factors may not be available at the time of prediction, *e.g.* a heavy rain tomorrow will impact bicycle demand negatively but it is not known when the demand prediction is made today.

This paper discusses effect of such constraints on prediction and offers ideas on multiple aspects of modeling that can improve prediction performance in such scenarios. Capital Bikeshare (CaBi)'s data is used as an example to test the proposed ideas. This dataset was provided by Hadi Fanaee Tork using data from CaBi for a Kaggle contest in 2014-15 [3]. However, imposing the practical

S. Deb—Author is currently a postgraduate student at the Institute of Systems Science, National University of Singapore.

N.T. Nguyen et al. (Eds.): ACIIDS 2017, Part II, LNAI 10192, pp. 23–32, 2017.
DOI: 10.1007/978-3-319-54430-4_3

data availability constraints makes the business problem more challenging than the original task that was set for the contest [8]. Details of business problem formulation, along with a summary of literature review are presented in Sect. 2.

Section 3 describes the pre-processing and feature engineering steps. It includes the motivation and computation details of the new features and their impact on prediction performance. Section 4 provides ideas on target variable engineering to improve performance further using the same set of features.

Section 5 introduces three multi-stage analytical constructs using the key variables identified in Sects. 3 and 4. It also illustrates how an ensemble combining multiple constructs provides a better prediction than the traditional multi-learner ensembles. Section 6 concludes with a summary of insights derived from the empirical results. Furthermore, the future research scope is outlined briefly.

2 Business Problem

2.1 Problem Formulation

In order to evaluate the forecast models in a business context, certain constraints are placed on the problem as originally defined. Namely, CaBi (a fictitious company) is assumed to be an aggregator who sources its bicycles from multiple suppliers. On a daily basis, they communicate the expected demand for the next day with a certain lead time. Because of this, the total actual demand for current day remains unknown, hence cannot be used for prediction. Additionally, an asymmetric profit scenario is assumed where customers pay $3 per unit as rent to CaBi and CaBi pays $2 per unit to its suppliers for the daily inventory. Thus the daily profit function can be defined using **Act**ual and **Pred**icted demand as

$$Profit_t = min(Act_t, Pred_t)^*3 - Pred_t^*2 \qquad (t = 1, 2, \dots) \qquad (1)$$

This reflects typical real-life scenarios where demand is estimated with a lead time and profitability is asymmetric $w.r.t.$ over or under-estimation. This ensures that the ideas and constructs proposed in this paper are easily extendable to other domains. While it is expected that the key predictors will change depending on the domain, the analytical ideas will remain relevant and replicable for short-term demand forecasting purposes.

2.2 Literature Review and Specific Considerations

Raw daily dataset contains information on 16 variables for two years, 2011 and 2012. Out of 16, two (instant and dteday) variables are used for indexing the dataset. In this paper, *year* is used as partition where 2011 data is used for model development and 2012 for performance validation. The other 13 variables by types are listed in Table 1.

The broad domain of prediction and forecasting is a rather vast field. As a result, literature review has primarily been restricted to prior researches involving this specific dataset. Most of these studies involved predicting the total

Table 1. Raw variables and range of values (in bracket) for 2011

Day	Weather	Demand
Season (1/2/3/4)	Weathersit (1/2/3)	Casual (9–3065)
Month (1–12)	Temp (0.06–0.85)	Registered (416–4614)
Holiday (0/1)	Atemp (0.06–0.91)	cnt (431–6043)
Weekday (0–6)	Hum (0.18–0.97)	
Workingday (0/1)	Windspeed (0.02–0.51)	

demand (cnt) by using covariates from the same day [7,8] and with unrestricted usage of past demand up to the previous day. However, introduction of lead time allows only a subset of available information to be used in the models (Fig. 1). Furthermore, most of the published results are based on the hourly dataset while this paper aims to predict the daily aggregate demand. Hence, the studies can only be consulted for analytical ideas and insights on the underlying relationships, but not for benchmarking any actual result.

Published Results	✔	✔	✔	✔	✔	✔	✔	✔	
Information Type:	Day Variables			Weather Variables			Demand Variables		
Day Stamp:	Till Yesterday	Today	Tomorrow	Till Yesterday	Today	Tomorrow	Till Yesterday	Today	Tomorrow
In This Paper	✔	✔	✔	✔	✔		✔		

Fig. 1. Data availability and usage constraints

The studies have, in general, established the importance of Weather variables such as extreme temperature and precipitation [4]. Analytically, most studies found boosting techniques like Random Forest (RAF) [7,9] and Gradient Boosting Machines (GBM) [6] to be delivering the best results. This paper also uses both RAF and GBM for model building. Robust Linear Regression (RLR) using Huber's M-estimation with Bisquare weights is chosen as the third model. This helps contrast and establish a comparative view between tree-based and regression models. Furthermore, the overfitting bias inherent in RAF and GBM can be benchmarked against the robust predictions.

Variable selection for RLR is done using p-values (<0.01). The same variables are fed into RAF and GBM for appropriate comparison. In line with the published results, accuracy is measured using root mean square error (RMSE). Business performance is calculated as yearly sum of daily profits.

3 Pre-processing and Feature Engineering

3.1 Data Cleaning and Exploratory Analysis

Adhering to an accurate training approach, 2012 data is kept aside and only 2011 data is used in exploratory analysis. Since Day variables have fixed values for a day, no Statistical analysis is necessary. All other variables are evaluated using both line charts and box plots (Fig. 2). This ensures that both time series and distributional aspects of the variables are considered.

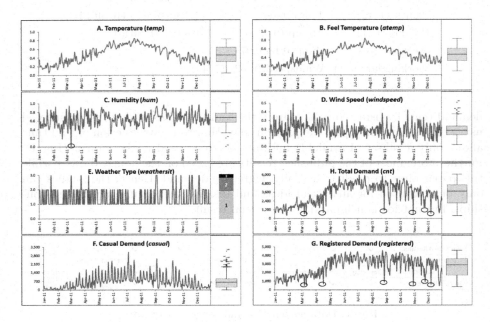

Fig. 2. Univariate analysis of weather and demand variables

No missing values are found. Except for one day showing 0 humidity, weather variables do not show any significant outliers or structural breaks. The 0 value is replaced by the average humidity for 2011.

Total and registered demands show several unusual drops. Publicly available information (newspaper articles and FEMA website) are consulted to identify any attributable causes. 9 of these days are found to be impacted by natural disasters or extreme weather (Table 2). As disaster occurrences are random in nature, demands for these days are imputed by a centered average of 7 neighboring days. This modified data is used in time series analysis to capture the *true* pattern. For other approaches and model performance analysis, however, the original demand data is used.

3.2 Feature Selection and Engineering

Day Variables. All Day variables are coverted to multiple indicator (0/1) variables using One Hot Encoding. Basic feature selection is done by analyzing the difference in mean demand across attribute levels. Between *season* and *month* (having a one-to-many relationship), *season* is retained because (1) *month* has

Table 2. Details of outlier days for 2011

Day	Reason	Day	Reason	Day	Reason
26–27 Jan	Blizzard	10 Mar	Tornado	16 Apr	Tornado
27 Aug	Hurricane Irene	7–8 Sep	Hurricane Lee	29 Oct	Winter storm
7 Dec	Rain storm				

too many levels and (2) *season* relates to demand more intuitively. As variations are observed for each variable, no other variable is discarded.

A new *Special Day* variable with 3 levels, +1 for Independence day (4 Jul), −1 for Thanksgiving (24 Nov) and Christmas time (24–26 Dec), and 0 for all other days is created. This captures the surge and dips driven by these special events which are recurring in nature. Day variables are used up to the forecast day (tomorrow) since these are known beforehand. Short term lags up to 2 days are created for *workingday* and *holiday*.

Weather Variables. Weather type (*weathersit*) is converted to indicators. Bivariate correlations (r) with total demand for temperature (*temp*), feel temperature (*atemp*) and *windspeed* come out to be statistically significant. Furthermore, *temp* and *atemp* are found to be highly correlated ($r = 0.99$). **Temp is dropped** to avoid multicollinearity as feel temperature relates to a bike-renting decision more intuitively.

Short-term lags are created for each weather variable up to 3 preceding days-today, yesterday and day before. Larger lags are discarded because past weather is not expected to impact demand for long in a busy city like Washington D.C. The average and range for these 3 days are also included. Once lags are created **the original variables are dropped** as weather condition for the prediction day cannot be used.

Additionally, ordinal variables (−1, 0, +1) are created to capture any sharp decrease (−1) or increase (+1) in Weather variables from yesterday to today. Finally, indicator variables are introduced to capture the extreme high (top 5%) and low (bottom 5%) values of each Weather variable.

Demand Variables. The three demand variables are collinear by design. Bivariate analysis suggests a stronger correlation (r: 0.92) between total and registered demand. Hence registered demand is not included in creation of lagged predictors. The maximum lag is chosen based on correlograms (Fig. 3C). Due to the strong correlation and a weekly repeating pattern, lagged variables for 7 preceding days up to yesterday are included (L02–L08). Similar to Weather variables, average, range and standard deviation of the lagged values (L02–L08) and ordinal variables for sharp changes between L02 and L03 are created. Once lags are created, **the original demand variables are dropped** from the predictors.

Fig. 3. Bivariate analysis of demand variables

A special lagged variable, latest actual demand for same day type (working or non-working) is created to capture the differentiated demand patterns observed by day type, more evidently for casual demand.

3.3   Time Series Analysis

The final stage of feature engineering is to build a time series model for total demand. Due to presence of a strong trend and a yearly seasonality (Fig. 2H), first step is to estimate this underlying pattern. A multiplicative decomposition method is chosen for this as CaBi is in its starting phase where organic growth is expected. To balance any excessive volatility in 2012, the trend is modeled as a power series instead of linear. This will translate to a slower rate of growth in 2012. Once the trend is identified, the seasonality is computed on the de-trended series as a quadratic function of days. This curve-fitting method is adopted as the data is limited to one year only.

Since decomposition models do not adjust to pattern shifts over time, an ARIMA model is built on the residual series ($\mathbf{R}_t$), where

$$R_t = \mathbf{Act}_t - \mathbf{Decomp}_t$$

Guided by the ACF/PACF plots (Fig. 4), the most appropriate model turns out to be ARIMA (1,0,0). It corroborates well with the findings from other studies [2]. Thus the final time series forecast for total demand becomes

$$\mathbf{AR}_t = \mathbf{Decomp}_t + 0.4427^* R_{t-1} \qquad (t = 1, 2, \dots) \qquad (2)$$

However, it should be noted that this **forecast cannot be used directly** as it uses immediate prior day's actual demand information (to calculate R_{t-1}). Hence, a *2-point ahead* forecast is used. Forecast values AR_t become an additional feature for model building.

Table 3 shows the improvement in accuracy and profit as new features are added to the models. Clear improvement in prediction performance establishes the effectiveness of feature engineering.

It should be noted that RAF and GBM performances are significantly worse than RLR since tree-based models fail to capture the temporal increasing trend between development and validation [7].

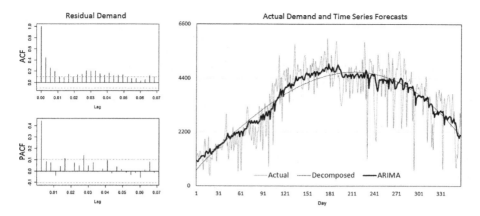

Fig. 4. ACF-PACF plots and time series forecasts

Table 3. Impact of feature engineering on prediction performance

Model	RAF		GBM		RLR	
Features	RMSE	Profit('000)	RMSE	Profit('000)	RMSE	Profit('000)
O	2140 ± 28	\$1317 $\pm$ 19	2068 ± 29	\$1332 $\pm$ 19	1738 ± 19	\$1454 $\pm$ 16
O, L	2139 ± 29	\$1319 $\pm$ 20	2016 ± 26	\$1351 $\pm$ 18	1574 ± 18	\$1500 $\pm$ 17
O, L, C	2218 ± 29	\$1290 $\pm$ 18	2056 ± 25	\$1339 $\pm$ 16	1171 ± 15	\$1562 $\pm$ 16
O, L, C, AR_t	2121 ± 28	\$1321 $\pm$ 20	1946 ± 18	\$1373 $\pm$ 16	1140 ± 15	\$1579 $\pm$ 15

O: Original i.e. latest available, L: Lagged, C: Calculated

4 Target Variable Engineering

As the features are finalized, this section deliberates ideas on creating alternate target variables to improve performance further. Given the strong trend stationarity observed in the demand series (Fig. 4C), a natural extension is to model the deviations from the time series forecast (AR_t) separately. This essentially assumes that the Day and Weather variables cause short-lived *shocks* to AR_t. Two variations of this *shock*, *viz.* the absolute (AG_t) and relative gap (RG_t) between Act_t and AR_t are modeled. Alternatives like logarithmic demand or change from the previous day can be tried depending on specific scenarios.

Table 4. Impact of target variable engineering on prediction performance

Model	RAF		GBM		RLR	
Target	RMSE	Profit('000)	RMSE	Profit('000)	RMSE	Profit('000)
AG_t	1127 ± 18	\$1606 $\pm$ 14	1147 ± 18	\$1596 $\pm$ 17	1138 ± 18	\$1603 $\pm$ 12
RG_t	1129 ± 18	\$1608 $\pm$ 14	1147 ± 16	\$1601 $\pm$ 18	1136 ± 18	\$1612 $\pm$ 13

The results (Table 4) suggest that a small improvement in performance can be obtained by using RG_t over AG_t. It is interesting to note the higher profit for RG_t even when the RMSEs are comparable. This can possibly be attributed to a more stable distribution for RG_t due to the implicit trend normalization.

5   Multi-stage Prediction

5.1   Analytical Constructs

Once the target and predictor variables get decided, this section explores ways of breaking the overall problem down into smaller sub-problems. It proposes three analytical constructs, each leading to a two-stage prediction as detailed below.

High-Low Day (HLD): Here the first stage is to predict if tomorrow will be a High/Regular/Low demand day. This day type (DT_t) is defined using $\pm 10\%$ cutoff on RG_t. Prediction of DT_t is done using a Multinomial Logistic Regression model. Subsequently 3 RG_t prediction models are built, 1 within each day type. Final prediction is derived as

$$HLD_t = AR_t \sum_{z \in \{L,R,H\}} P(DT_t = z)^*(1 + \hat{RG}_{t|DT=z}) \quad (t = 1, 2, \dots) \quad (3)$$

Day Clustering (DCL): This uses an unsupervised clustering of observations (days) in the first stage. Once the clusters are finalized, a supervised RG_t prediction model is built within each cluster. The aim here is to capture the localized relationships by restricting to *similar* days only.

K-means method is used for clustering. Goodness of fit for the clusters (C_i) is assessed by Davies-Bouldin index. Inverse Euclidean distances ($d_{C_i}^{-1}$) from respective cluster centroids are used as weights to combine predictions from the in-cluster RG_t models. Final prediction thus becomes

$$DCL_t = AR_t(\sum_{i=1}^{k}(1 + \hat{RG}_{t|C_i})^* d_{C_i}^{-1})/(\sum_{i=1}^{k} d_{C_i}^{-1}) \quad (t = 1, 2, \dots) \quad (4)$$

Sudden Drop Day (SDD): Here the motivation is to capture the low side outliers which often lead to high losses as a result of overstocking. Sudden drop (SD_t) days are defined as either a natural disaster day or a day with $> 20\%$ drop in RG_t both from its immediate prior and next days. Supervised models for RG_t built separately for both types of days. Logistic Regression is used for predicting SD_t. The final prediction is calculated as

$$SDD_t = AR_t \sum_{x \in \{Y,N\}} P(SD_t = x)^*(1 + \hat{RG}_{t|SD=x}) \quad (t = 1, 2, \dots) \quad (5)$$

The results (Table 5) demonstrate how each construct provides additional improvement over the predictions from a single model. It also illustrates that due to profit asymmetry, analytical accuracy alone may not always be sufficient to assess a model's applicability for business use.

Table 5. Impact of analytical constructs on prediction performance

Model	RAF		GBM		RLR	
Construct	RMSE	Profit('000)	RMSE	Profit('000)	RMSE	Profit('000)
HLD	1156 ± 18	$1613 $\pm$ 14	1165 ± 18	$1613 $\pm$ 14	1140 ± 16	$1619 $\pm$ 13
DCL	1133 ± 17	$1614 $\pm$ 14	1128 ± 14	$1614 $\pm$ 13	1106 ± 14	$1619 $\pm$ 14
SDD	1119 ± 16	$1624 $\pm$ 15	1113 ± 14	$1622 $\pm$ 13	1104 ± 14	$1626 $\pm$ 13

5.2 Ensemble Prediction Performance

Two types of ensembles are compared here. Traditional ensembles follow the prevalent approach where the predictions from different modeling techniques within the same analytical construct are combined. Separately, a cross-construct ensemble is proposed. This combines the predictions obtained by using the same modeling technique across different analytical constructs (Table 6).

Table 6. Impact of ensembling on prediction performance

Traditional			Proposed		
Ensemble	RMSE	Profit('000)	Ensemble	RMSE	Profit('000)
HLD	1146 ± 14	$1619 $\pm$ 12	RAF	1097 ± 12	$1632 $\pm$ 12
DCL	1115 ± 12	$1617 $\pm$ 12	GBM	1100 ± 12	$1630 $\pm$ 12
SDD	1097 ± 11	$1633 $\pm$ 13	RLR	1086 ± 12	$1636 $\pm$ 11

6   Conclusions and Next Steps

Most frequent predictors include high temperature, rainy day, special day and weather change. Rainy day and high temperature ($>35\,°C$) impact demand negatively as expected [5]. Independence day is found to push demand up by 60% while Christmas and Thanksgiving days lead to a 60% drop. A positive weather change drives an increase in demand by $\sim8\%$.

High-Low Day model shows how demand at different levels gets impacted by different factors. While low demand is impacted by high volatility in wind speed in past 3 days (possible indication of an upcoming storm or extreme weather), high and regular demand are generally determined by the recent demand trends as past actual demand comes out to be an important predictor.

Finally, it shows that a pure time series model, based on past demand itself can provide a powerful prediction. Demand movement is largely trend stationary where impact of other factors are limited and short-lived.

The results also suggest that the proposed cross-construct ensemble may lead to better performance. Disparate constructs will help control the correlation among the predictions from these constructs, which is essential for a powerful ensemble. Further work is, however, necessary in other domains and datasets to validate this.

Interestingly and contrary to previous studies, robust regression is found to outperform both RAF and GBM consistently. This can be attributed to the difference in volume of data ($1/24^{th}$ of hourly) used in this paper and the business problem formulation. Original task used both 2011 and 2012 days (first 20 of each month) for model development [8]. As a result, the models could learn the recent trends dynamically whereas models in this paper use only 2011 data and scale up to one year into future. This suggests that the choice of an appropriate modeling technique can often be context or usage driven.

Future studies should explore using adaptive machine learning techniques, *e.g.* neural networks or optimization approaches like MAP-EF [10] to solve for the profit asymmetry directly. Options to improve RAF and GBM performance in small datasets and separate predictions for registered and casual demand should also be evaluated. Finally, a disaster day prediction model using additional covariates can be tried to capture the extreme low days to avoid overstocking and reduce cost further.

Acknowledgement. This work was undertaken as part of the National University of Singapore, Master of Technology in Enterprise Business Analytics Program under the guidance of Dr. Barry Shepherd. Author would like to thank Dr. Shepherd and the two anonymous reviewers for their valuable suggestions.

References

1. Taylor, J.W., Buizza, R.: Using weather ensemble predictions in electricity demand forecasting. Int. J. Forecast. **19**(1), 57–70 (2003)
2. Kaltenbrunner, A., et al.: Urban cycles and mobility patterns: exploring and predicting trends in a bicycle-based public transport system. Pervasive Mob. Comput. **6**(4), 455–466 (2010). doi:10.1016/j.pmcj.2010.07.002
3. Fanaee-T, H., Gama, J.: Event labeling combining ensemble detectors and background knowledge. Prog. Artif. Intell. **2**, 113–127 (2014)
4. Gebhart, K., Noland, R.: The impact of weather conditions on bikeshare trips in Washington DC. Transportation **41**, 1205–1225 (2014)
5. Wadud, Z.: Cycling in a changed climate. J. Transp. Geogr. **35**, 12–20 (2014)
6. Li, Y., Zheng, Y., Zhang, H., Chen, L.: Traffic prediction in a bike-sharing system. In: Proceedings of the 23rd SIGSPATIAL International Conference on Advances in Geographic Information Systems, p. 33. ACM, November 2015
7. Cosp Arqué, O.: Demand forecast model for a bicycle sharing service (2015)
8. Lee, C., Wang, D., Wong, A.: Forecasting utilization in city bike-share program (2015)
9. Wang, W.: Forecasting bike rental demand using New York Citi Bike data (2016)
10. An, B., Chen, H., Park, N., Subrahmanian, V.S.: MAP: frequency-based maximization of airline profits based on an ensemble forecasting approach. In: Proceedings of the 22nd ACM SIGKDD International Conference on Knowledge Discovery and Data Mining, pp. 421–430. ACM, August 2016

Increasing the Detection of Minority Class Instances in Financial Statement Fraud

Stephen Obakeng Moepya[1,2]([⊠]), Fulufhelo V. Nelwamondo[1,2], and Bhekisipho Twala[2]

[1] CSIR Modelling and Digital Science, Pretoria, South Africa
{smoepya,fnelwamondo}@csir.co.za
[2] Faculty of Engineering and the Built Environment,
University of Johannesburg, Johannesburg, South Africa
btwala@uj.ac.za
http://www.csir.co.za

Abstract. Financial statement fraud has proven to be difficult to detect without the assistance of data analytical procedures. In the fraud detection domain, minority class instances cannot be readily found using standard machine learning algorithms. Moreover, incomplete instances or features tend to be removed from investigations, which could lead to greater class imbalance. In this study, a combination of imputation, feature selection and classification is shown to increase the identification of minority samples given severely imbalanced data.

Keywords: Financial statement fraud · Class imbalance · Feature selection · Imputation

1 Introduction

Fraudulent activities are prevalent in almost any scenario which involves money (either physical or electronic), whereby the perpetrator intends to/has obtained a monetary gain from an unsuspecting victim. Due to the relative ease in which money flows between institutions and people, the field of financial fraud has received much interest in the past two decades. As a natural consequence, financial fraud detection remains paramount to counter or capture this potentially criminal activity.

Fraud detection automates and helps reduce manual parts of a screening process [19]. It is often utilized as part of the overall fraud control. According to Richhariya et al. [21], fraud detection enables the development of targeted strategies by separating authentic and suspicious behavior. In general, companies which deliberately mislead investors and creditors using falsified financial statements are known to be committing financial statement fraud (FSF). Financial statements contain information about the financial position, performance and cash flows of a company [16]. In a practice, FSF may comprise many aspects including the misapplication of accounting policies and the non-disclosure of

© Springer International Publishing AG 2017
N.T. Nguyen et al. (Eds.): ACIIDS 2017, Part II, LNAI 10192, pp. 33–43, 2017.
DOI: 10.1007/978-3-319-54430-4_4

crucial transactions [24]. A common strategy of detecting fraudulent behavior, in vasts amounts of data, is through a process of knowledge discovery. This is commonly known as *data mining*.

Data Mining (DM) is the process of finding previously unknown patterns or trends in databases and using that information to build predictive models. This is important in a scenario where the target concept is difficult to understand, such as in the case of fraud. There are mainly two approaches to DM, *supervised* and *unsupervised*. Supervised machine learning takes a known set of inputs and makes predictions on unseen data. Examples of supervised techniques include Support Vector Machines (SVM), Artificial Neural Networks (ANN), Decision Trees (DT), Random Forests (RF), Logistic Regression (LR) and Naïve Bayes (NB). The unsupervised approach consists of machine learning methods that do not require labels to teach a concept. These methods cluster data objects by searching for common patterns and characteristics of sub groups.

The task of detecting financial statement fraud lies in the hands of auditors. Detecting FSF is non-trivial when one is using normal audit procedures. This is due to the lack of knowledge about the characteristics of fraudulent financial reporting. In some cases, the prevalence of FSF can be tiny compared to legitimate (non-fraudulent) cases. In addition, the complexity within the Generally Accepted Accounting Principles (GAAP) and International Financial Reporting Standards (IFRS) frameworks provide sufficient legitimate room for manipulation. This makes it extremely challenging to detect this type of fraud.

Previous research shows that machine learning has made major strides in detecting financial statement fraud [5,9,10]. However, there are two areas which have not been dealt with effectively: severe class-imbalance and missing values. Generally, studies have used balanced (or mildly imbalanced) data to fit models. This type of approach is not robust, since in reality a large number of listed companies receive unqualified reports more than those which commit fraud. Secondly, instances or features containing missing information tend to be removed from investigations altogether. The removal of information can lead to biased or incomplete conclusions. The aim of this paper is to use imputation, given data which contains a high class-imbalance, and feature selection to enhance the classification of minority instances. The remainder of this paper is organized as follows: Sect. 2 presents the literature review; Sect. 3 provides the methods that will be utilized; Sect. 4 describes the dataset and experimental results from the study and Sect. 5 gives concluding remarks.

2 Literature Review

Table 1 presents an overview of the studies that have been performed in the FSF detection domain. The explanation of the headings is as follows: 'Author' represents the authors of the study; 'Primary Aim' is the main objective of each paper; 'Dataset Imbalance' gives the ratio of non-fraudulent to fraudulent instances; and 'Handling Missing Data' indicates the manner in which incomplete information was dealt with. The 'Classifier' column represents the machine learning methods which were implemented to classify instances. These include: Probabilistic Neural Networks (PNN), Logistic Regression (LR), Support Vector Machines (SVM), Radial Basis Function (RBF), Group Method of Data Handling (GMDH), Linear Discriminant Analysis (LDA), Genetic Programming (GP), Multi-layer Perceptron (MLP), Naive Bayes (NB) and Neural Networks (NN). To summarize Table 1, most of the dataset which experiments were performed on have been fairly balanced. In addition, the authors either have not mentioned how missing data was treated or they removed instances or

Table 1. Summary of some literature in the field

Author	Classifier	Primary aim	Dataset imbalance	Handling missing data
Öğüt et al. [16] 2009	PNN, LR, LDA, SVM	Classifying the financial information manipulation with PNN and SVM	50:50	No Information Provided
Pai et al. [17] 2011	SVM, MLP, C4.5, LR RBF, LDA	Propose a support vector machine-based fraud warning (SVMFW) model	75:25	No Information Provided
Ravisankar et al. [20] 2011	MLP, SVM PNN, LR GP, GMDH	Use data mining techniques in order to flag fraudulent financial statements	50:50	No Information Provided
Persons [18] 2011	Step-wise LR	Develop parsimonious models to select attributes contributing to financial statement fraud	50:50	Removal of missing values
Li [23] 2010	SVM	Use support vector machines to detect financial statement fraud	50:50	No Information Provided
Gupta and Gill [6] 2012	GP, NB, DT	Proposing a data mining framework to prevent fraudulent reporting	75:25	No Information Provided
Deng [4] 2009	SVM	Building support vector machine models to detect financial statement fraud	50:50	No Information Provided
Amara et al. [1] 2013	LR	Testing the concept of the fraud triangle to detect financial statement fraud	50:50	No Information Provided
Roxas [22] 2011	Probit Benford's Law	Compare the effectiveness of two analytical procedures in detecting earnings management through revenue manipulation	67:33	Removal of missing values
Lou and Wang [12] 2011	LR	Develop and test a logistic regression model for classification to detect financial statement fraud	84:16	Removal of missing values
Lin et al. [11] 2015	LR, NN DT	Using public data to examine aspects of the fraud triangle using data mining to compare them with expert opinion	78:22	No Information Provided

features which contained incomplete information. In this study, missing information will be included in experiments and used to enhance the classification of minority instances.

3   Methodology

3.1   Feature Selection

Feature selection is the process of identifying and removing irrelevant and redundant information. In machine learning, feature selection can be generally broken up into two main categories: *wrapper* and *filter* methods. Wrapper methods add or remove features based on the classification performance of a chosen learner. The combination of features which achieve the least error rate is selected. Alternatively, filter methods do not utilize any prediction models to evaluate features. The relevance of features are determined by solely evaluating the features outside of any classifier. In this paper, both a wrapper (Recursive Feature Elimination) and filter (Correlation-based Feature Selection) method will be utilized to remove noisy features.

Correlation-based Feature Selection (CFS) [8] is a multivariate feature selection technique that ranks attribute subsets according to a correlation-based heuristic function defined by $Merit_s = \frac{k\bar{r}_{cf}}{\sqrt{k+k(k-1)\bar{r}_{ff}}}$, where s is a subset containing k features, $\bar{r}_{cf}$ is the mean feature-class correlation ($f \in S$), and $\bar{r}_{ff}$ is the average feature-feature inter-correlation. $Merit_s$ can be thought of as Pearson's correlation where all the variables have been standardised.

Recursive Feature Elimination (RFE) was proposed by Guyon et al. [7]. The aim of RFE is to recursively rank a subset of features by measuring their importance with respect to classification performance. At each step of RFE, variable importance is measured (using a classifier) and thereafter the least relevant feature is removed. The procedure continues, and the best (achieving the greatest accuracy for example) subset of features is used as the final variable ranking. In this paper, the Random Forest classifier will be used within the RFE.

3.2   Classification

The three base classifiers that will be used in this study are: Naive Bayes (NB), Decision Trees (DT) and Random Forest (RF). These methods were chosen based on the fact that they can be fit on data with missing values, hence benchmark classification results can be obtained.

The Naive Bayes (NB) [13] is a highly scalable classifier which is based on Bayes' Theorem. The aim of NB is to estimate the posterior probability $p(C_l|x)$ calculated using Bayes' Theorem, where x represents the dataset and C is the class of interest. The main advantages of the NB model are it's simplicity to implement and explain. A known caveat of NB is that the it makes a very strong assumption on the distribution shape (i.e., any two features are independent given the output class).

A single decision tree (DT) is composed of three basic elements [2]: a decision node specifying a test attribute; an edge or branch corresponding to one of the possible attribute values and a leaf which contains the class to which each object belongs. There are two steps that are required to classify instances using a DT. The first requires building the tree. This involves, given training data, selecting for each node the 'appropriate' test attribute and also defining the class label for each leaf. After the tree has been built, all test cases are split by the each attribute node from the root (in general the tree is graphically represented upside down) and successively split moving down the branches until reaching a termination leaf where the instances are assigned to a class (descend strategy).

The Random Forest (RF) classifier was originally proposed by Breiman [3]. The RF classifier is known as an ensemble, which generally tends to achieve superior classification performance when compared to one classification tree (weak learner). This is due to the fact that an ensemble of trees obtain a lower variance when compared to a single classifier. The RF selects strong, complex models which contribute a low proportion to the overall bias. In summary, the combination of independent models achieves improved classification rates.

3.3   Experimental Setup

The experimental setup is as follows. The data is separated into a training (data from years 2003 to 2010) and hold-out set (2011–2013). The training and test sets consisted of 2270 and 773 instances respectively. There are 48 numeric features in the dataset. In order to create benchmark results, the raw data is taken through a process whereby the above feature selection techniques are applied. Using the significant ratios found by each feature selection technique, the cost-sensitive DT, NB and RF (using a 'Cost' parameter of 20 guided by [14,15]) are fit on the training data using 10-fold cross-validation to reduce over-fitting. The DT's complexity parameter is varied between $cp \in (0, 1)$. The $n_{tree} \in \{50, 100, 150, 200\}$ and $m_{try} \in \{round(\sqrt{k})-1, \sqrt{k}, \sqrt{k}+1\}$ in the RF represent the number of trees and the number of features to randomly select respectively (i.e., m and k in the algorithm [3]). Thereafter, the best trained model is fit on the test. The above-mentioned procedure is repeated utilizing data which is imputed with zero, mean and k-Nearest Neighbor imputed data for comparison.

4   Results

4.1   Dataset Description

The data used in this experiment has been provided by INET BFA. The data comprises publicly listed companies trading on the Johannesburg Stock Exchange (JSE). The period of the listing ranges from January 2003 to July 2013. The JSE categorizes companies into many different sectors from Basic Materials to Utilities. The total number of instances in the dateset are 3043 and approximately 4.4% of them were given a qualified financial report by auditors. The attributes in the dataset contains missing information. Some features have more than 45% of their values missing. Since the pattern of missingness is traceable (predictable) from other variables in the data, a conclusion that the missing data mechanism is missing at random (MAR) is reached. For example, if a company did not issue any dividend for a given year, any ratio involving dividends is labeled 'NA'.

4.2   Empirical Results

Table 2 presents the features selected by Correlation-based Feature Selection (CFS) and Recursive Feature Elimination (RFE). The CFS and RFE selected three variables in common: Earnings Yield, Book Value/Share and Return on Capital Employed. The Score in Table 2 (RFE) represents the weight of importance of each feature. This implies that Return on Capital Employed was the most highly ranked feature by RFE algorithm.

Table 2. Features selected by correlation-based feature selection and recursive feature elimination on the entire dataset without imputation

Attribute	Attribute	Score
1 **Book Value/Share**	1 **Return on Capital Employed**	**0.10**
2 Debt/Assets	2 Return on Average Equity	0.09
3 Debt/Equity	3 Return On Equity	0.08
4 Dividend/Share	4 Inflation Adjusted	
5 **Earnings Yield**	Return on Average Equity	0.04
6 Operating Profit Margin	5 **Earnings Yield**	**0.01**
7 Quick Ratio	6 **Book Value/Share**	**0.01**
8 **Return on Capital Employed**	7 Return on Average Assets	0.01

(CFS)	(RFE)

The cross-validation (CV) training results using the features in Table 2 are presented in Table 3. The model accuracy is given by 'Acc', along with the sensitivity ('Sens') and specificity ('Spec'). The 'Sens' and 'Spec' are also known as the true positive (TP) and true negative (TN) rates respectively. Kappa is given by the formula $Kappa = \frac{O-E}{1-E}$, where O is the observed accuracy and E is the expected accuracy based on the marginal totals of the confusion matrix. From a previous study [14], given the high class imbalance, cost-sensitive (CS) classifiers were used instead of the standard learners. When encountered with high class imbalance, accuracy is not the only metric one should evaluate to rank performance. A good balance between 'Sens' and 'Spec' is generally acceptable (depending on the application domain). Table 3 (CFS) shows that NB and RF achieve TP and TN rates of above 75% when fit on CFS features. DT is the only classifier which achieves the equivalent when training on RFE features.

Table 3. Classification performance of features selected using correlation-based feature selection and recursive feature elimination on the training dataset without imputation

Method	Acc (%)	Sens (%)	Spec (%)	Kappa
DT	81.23	72.22	81.61	0.18
NB	**78.90**	**80.00**	**78.85**	**0.18**
RF	**82.69**	**76.67**	**82.94**	**0.21**

(CFS)

Method	Acc (%)	Sens (%)	Spec (%)	Kappa
DT	**79.21**	**76.67**	**79.31**	**0.17**
NB	77.09	74.44	77.20	0.15
RF	81.81	71.11	82.25	0.18

(RFE)

The results using the test data are presented in Fig. 1. The Area Under the Curve (AUC) values, in Fig. 1a for the DT, NB and RF are 0.8809, 0.9123 and 0.8542 respectively. Although the NB curve achieves the greatest AUC, it does not necessarily imply that it's the best classifier. If a target false positive rate is 20%, then the DT may be preferable. Figure 1b shows the poor performance of the RF classifier (AUC = 0.7096). This may be due to the fact that the underlying classifier in the RFE algorithm is a random forest. A possible reason is that there was over-fitting, even though CV was performed. The DT and NB were able to generalize better than RF, with AUC values of 0.8809 and 0.8914 respectively.

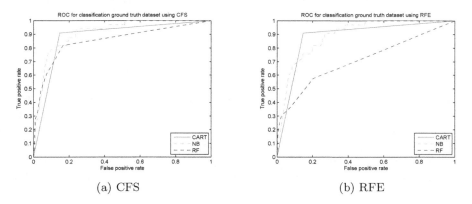

(a) CFS   (b) RFE

Fig. 1. Receiver Operating Characteristic (ROC) curves of the three classifiers using the test dataset and both feature subsets

Imputed datasets were previously shown (in [15]) not to substantially decrease classification performance. For this reason, zero, mean and kNN imputed datasets were used in conjunction with feature selection. Table 4 presents the classifier results fit on the holdout set. This table (compared to the CV training in Table 2) demonstrates an improvement in classification performance. Utilizing features from both CFS and RFE, DT obtains sensitivity and specificity rates of above 80%.

Table 4. Classification performance of using correlation-based feature selection and recursive feature elimination on the zero imputed test dataset

Method	Acc (%)	Sens (%)	Spec (%)	Kappa
DT	85.51	90.91	85.27	0.30
NB	79.04	90.91	78.51	0.21
RF	84.09	87.88	83.92	0.27

(CFS)

Method	Acc (%)	Sens (%)	Spec (%)	Kappa
DT	83.83	81.82	83.92	0.25
NB	76.84	90.91	76.22	0.19
RF	77.75	87.88	77.30	0.19

(RFE)

The results presented in Table 5 show only a slightly less optimistic performance when compared to Table 4. DT is again able to achieve an overall accuracy of 85.51% with both TP and TN rates being above 80%. The RF classifier fails to achieve a sensitivity of above 70% using features extracted by RFE.

Table 5. Classification performance of using correlation-based feature selection and recursive feature elimination on a mean imputed test dataset

Method	Acc (%)	Sens (%)	Spec (%)	Kappa
DT	85.51	90.91	85.27	0.30
NB	79.69	87.88	79.32	0.21
RF	83.70	78.79	83.92	0.24

(CFS)

Method	Acc (%)	Sens (%)	Spec (%)	Kappa
DT	85.38	72.73	85.95	0.25
NB	70.63	84.85	70.00	0.13
RF	80.47	69.70	80.95	0.17

(RFE)

The final imputation method that is used to estimate missing data is the kNN. Table 6 presents the most promising results using kNN imputed data varying the number of nearest neighbors. Both classifiers (DT and RF) in Table 6 show that kNN imputed dataset present some variance with respect to classification performance. The value $k = 9$ is preferred by both DT and RF, with the maximum of 85.12% (where the sensitivity is above 90%). In a previous study [15], it was suggested that the extension of the research involved varying the imputation parameters. Table 6 showed that careful consideration must be given to imputation methods that have parameters.

Another interesting finding is that filtering the data using CFS (at each level of k) did not make much difference to the classification performance. For almost all levels of k, irrespective of the classifier, the performance is very similar (if not the same). For example, Table 7 shows that at all the specified levels of k, RF fit on CFS features obtains very high levels of sensitivity and specificity rates of above 80%.

Table 6. Classification performance of using correlation-based feature selection and recursive feature elimination on kNN imputed test datasets

k	Acc (%)	Sens (%)	Spec (%)	Kappa
1	79.30	75.76	79.46	0.18
3	83.83	75.76	84.19	0.23
5	66.62	93.94	65.41	0.13
7	66.62	93.94	65.41	0.13
9	**85.12**	**96.97**	**84.59**	**0.31**
11	77.23	87.88	76.76	0.19

(DT + RFE)

k	Acc (%)	Sens (%)	Spec (%)	Kappa
1	78.01	75.76	78.11	0.17
3	**80.98**	**84.85**	**80.81**	**0.22**
5	79.69	81.82	79.59	0.20
7	73.61	90.91	72.84	0.16
9	**81.63**	**84.85**	**81.49**	**0.23**
11	77.62	84.85	77.30	0.18

(RF + RFE)

Table 7. Classification performance of using correlation-based feature selection and random forest fit on kNN imputed test datasets

k	Acc (%)	Sens (%)	Spec (%)	Kappa
1	83.83	93.94	83.38	0.28
3	83.70	93.94	83.24	0.28
5	**85.38**	**90.91**	**85.14**	**0.30**
7	83.70	93.94	83.24	0.28
9	82.79	93.94	82.30	0.27
11	84.48	90.91	84.19	0.28

5 Conclusion

In summary, the fact that data contains missing information has not been utilized effectively in detecting fraud statement fraud. This study showed that missing data imputation, cost-sensitive classification and multi-variate feature selection techniques can be used to increase the detection of minority class instances. Furthermore, imputation was shown to enhance classification accuracy with varying degrees of success. In order to extend this work, different datasets, imputation methods and classifiers need to be utilized. In addition, non-structured data needs to be included in experiments to detect fraud.

Acknowledgments. The current work is being supported by the Department of Science and Technology (DST) and Council for Scientific and Industrial Research (CSIR).

References

1. Amara, I., Amar, A.B., Jarboui, A.: Detection of fraud in financial statements: French companies as a case study. Int. J. Acad. Res. Account. Finance Manag. Sci. **3**(3), 40–51 (2013)

2. Amor, N.B., Benferhat, S., Elouedi, Z.: Naïve Bayes vs decision trees in intrusion detection systems. In: Proceedings of the 2004 ACM Symposium on Applied Computing, pp. 420–424. ACM (2004)
3. Breiman, L.: Random forests. Mach. Learn. **45**(1), 5–32 (2001)
4. Deng, Q.: Application of support vector machine in the detection of fraudulent financial statements. In: 4th International Conference on Computer Science & Education, ICCSE 2009, pp. 1056–1059. IEEE (2009)
5. Fanning, K.M., Cogger, K.O.: Neural network detection of management fraud using published financial data. Int. J. Intell. Syst. Acc. Finance Manag. **7**(1), 21–41 (1998)
6. Gupta, R., Gill, N.S.: Prevention and detection of financial statement fraud-an implementation of data mining framework. Editorial Preface **3**(8), 150–156 (2012)
7. Guyon, I., Weston, J., Barnhill, S., Vapnik, V.: Gene selection for cancer classification using support vector machines. Mach. Learn. **46**(1–3), 389–422 (2002)
8. Hall, M.A., Smith, L.A.: Feature selection for machine learning: comparing a correlation-based filter approach to the wrapper. In: Florida Artificial Intelligence Research Society Conference, pp. 235–239 (1999)
9. Kaminski, K.A., Sterling Wetzel, T., Guan, L.: Can financial ratios detect fraudulent financial reporting? Manag. Auditing J. **19**(1), 15–28 (2004)
10. Kirkos, E., Spathis, C., Manolopoulos, Y.: Data mining techniques for the detection of fraudulent financial statements. Expert Syst. Appl. **32**(4), 995–1003 (2007)
11. Lin, C.C., Chiu, A.A., Huang, S.Y., Yen, D.C.: Detecting the financial statement fraud: the analysis of the differences between data mining techniques and experts' judgments. Knowl.-Based Syst. **89**, 459–470 (2015)
12. Lou, Y.I., Wang, M.L.: Fraud risk factor of the fraud triangle assessing the likelihood of fraudulent financial reporting. J. Bus. Econ. Res. (JBER) **7**(2), 61–78 (2011)
13. Lowd, D., Domingos, P.: Naïve Bayes models for probability estimation. In: Proceedings of the 22nd International Conference on Machine Learning, pp. 529–536. ACM (2005)
14. Moepya, S.O., Akhoury, S.S., Nelwamondo, F.V.: Applying cost-sensitive classification for financial fraud detection under high class-imbalance. In: 2014 IEEE International Conference on Data Mining Workshop (ICDMW), pp. 183–192. IEEE (2014)
15. Moepya, S.O., Akhoury, S.S., Nelwamondo, F.V., Twala, B.: The role of imputation in detecting fraudulent financial reporting. Int. J. Innov. Comput. Inf. Control **12**(1), 333–356 (2016)
16. Öğüt, H., Aktaş, R., Alp, A., Doğanay, M.M.: Prediction of financial information manipulation by using support vector machine and probabilistic neural network. Expert Syst. Appl. **36**(3), 5419–5423 (2009)
17. Pai, P.F., Hsu, M.F., Wang, M.C.: A support vector machine-based model for detecting top management fraud. Knowl.-Based Syst. **24**(2), 314–321 (2011)
18. Persons, O.S.: Using financial statement data to identify factors associated with fraudulent financial reporting. J. Appl. Bus. Res. (JABR) **11**(3), 38–46 (2011)
19. Phua, C., Lee, V., Smith, K., Gayler, R.: A comprehensive survey of data mining-based fraud detection research. arXiv preprint arXiv:1009.6119 (2010)
20. Ravisankar, P., Ravi, V., Rao, G.R., Bose, I.: Detection of financial statement fraud and feature selection using data mining techniques. Decis. Support Syst. **50**(2), 491–500 (2011)
21. Richhariya, P., Singh, P.K.: A survey on financial fraud detection methodologies. Int. J. Comput. Appl. **45**(22), 15–22 (2012)

22. Roxas, M.L.: Financial statement fraud detection using ratio and digital analysis. J. Leadersh. Accountability Ethics **8**(4), 56–66 (2011)
23. Xiuzhi, L., Shuangshuang, Y.: Lib-SVMs detection model of regulating-profits financial statement fraud using data of Chinese listed companies. In: 2010 International Conference on E-Product E-Service and E-Entertainment (ICEEE), pp. 1–4. IEEE (2010)
24. Zhou, W., Kapoor, G.: Detecting evolutionary financial statement fraud. Dezhoucision Support Syst. **50**(3), 570–575 (2011)

Information Technology Services Evaluation Based ITIL V3 2011 and COBIT 5 in Center for Data and Information

Firman Hartawan[✉] and Jarot S. Suroso

Master in Information Systems Management, Bina Nusantara University,
Jl. Kebon Jeruk Raya No. 27, Jakarta Barat, Indonesia
firman@kemhan.go.id, jsembodo@binus.edu

Abstract. Increasing the role of IT is directly proportional to the increase in investment is accompanied also by an increase in expenses is great. Indicators of successful implementation of IT in the form of service excellence in Center for Data and Information Ministery of Defence Republic of Indonesia is reliable, available, fast, and accurate. By planning a mature IT governance implementation of IT services is expected to do well and the embodiment of good IT Governance. Success in providing information services can provide a positive impact for organizations and society in general. Thus, the investments made by the Ministry of Defence in the implementation of ICT should not contradict with the goals of the organization and to the expectations of stakeholders. This study aims to measure the capability of service information in the Ministry of Defence in order to improve stakeholder satisfaction. Measurement capability is used COBIT 5 with qualitative method and the case study method. Stages of this research is the analysis of the condition of the eighth COBIT 5 process, the target area of process improvement, gap analysis and the determination of strategies for achieving capability. The end result of this research activity in the form of recommendations adopted policies and procedures of ITILV3 2011, as well as Key Performance Indicator (KPI) recommendations for Center for Data and Information Ministery of Defence.

Keywords: IT governance · Capability level · COBIT 5 · ITIL V3 2011

1 Introduction

Information Communication Technology (ICT) is becoming the technology is widely adopted by almost all organizations and is expected to help meet the goals of an organization. Utilization of ICT has provided solutions and benefits through opportunities as a form of the strategic role of ICT in achieving the vision and mission of the organization. Opportunities are created from the optimization of ICT resources in the area of company resources including data, application systems, infrastructure and human resources.

There are two ways the use of a state defense force that defense to use military force and non military. To support the implementation of the main tasks of the Ministry of Defence in defending the sovereignty of the Homeland necessary data and information

© Springer International Publishing AG 2017
N.T. Nguyen et al. (Eds.): ACIIDS 2017, Part II, LNAI 10192, pp. 44–51, 2017.
DOI: 10.1007/978-3-319-54430-4_5

containing aspects of ideological, political, economic, social, cultural and security defense to put into force a comprehensive defense. In the era of globalization and information such as this, data and information management system that is easily accessible to become a necessity. Data and information containing the above aspect is data and information that is needed in order to formulate a national defense policy should be prepared in a comprehensive, systematic, fast and accurate. Thus the need to build and develop the national defense information system that can be used for decision making and the formulation of the national defense policy.

PUSDATIN (Center for Data and Information) as the supervisor of the IT function within the Ministry of Defence Republic of Indonesia has a vision that is: "Towards a society and state defense community who are knowledgeable with data and information.". In order to realize this vision, Pusdatin mission is: "Creating Media Centre as a center of excellence telematics field of national defense."

Governance or governance serves to ensure that the needs, conditions, and selection of stakeholders evaluated to fit the organization's goals, set priorities in decision making and monitor performance based on the purpose and direction. One of the ICT governance practices that can be done by the organization is to evaluate the Information and Communication Technology or information systems. Evaluation aims to assess, monitor, and ensure that the company can manage the information system data integrity is well and able to operate effectively in accordance with corporate goals and objectives of the ICT organization.

There are some fundamental issues that make IT governance is less influential on the performance of the IT organization. Human resource issues that have a formal education in the field of IT is very less when Pusdatin is a data management and informatics means at the Ministry of Defence. Furthermore, there is also a technical problem in the form of integrated data for each work unit have their respective data and managed independently. Then, the application utilization SI/IT at the Ministry of Defense is still low.

The purpose of this study is to evaluate in Pusdatin Kemhan IT services and provides recommendations to improve the activity of the IT services based on the COBIT 5 framework and ITIL V3 2011.

2 Methods

This research used Cobit 5, where Implementation of lifecycle is shown in Fig. 1 and also services lifecycle of ITIL V3 2011 as shown in Fig. 2.

Measurement capabilities of the organization today is done through a questionnaire process. The goal of the questionnaire is thirteen leaders or officials related to governance at the Center for Data and Information (Pusdatin), Ministry of Defence. Tabulation measurement capabilities of organizations in all relevant processes in accordance with organizational problems can be seen in Table 1.

Methods of targeting can be done in several ways, but in this study the target was determined by questionnaire technique is based on the expectations of the thirteen officers or chief information related to governance. It can be seen in Table 2. below.

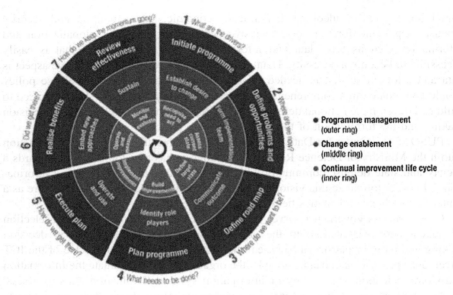

Fig. 1. Implementation lifecycle of Cobit 5

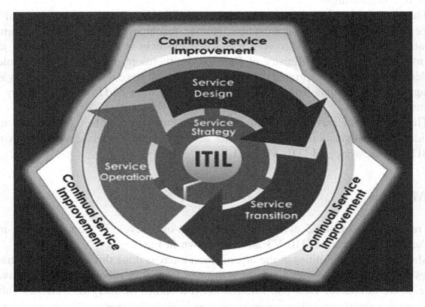

Fig. 2. Services lifecycle of ITIL V3 2011

Summary results of the target capability level ratings on the entire process of relevant COBIT 5 can be seen in Table 3.

Table 1. Capabilities level current condition.

Capabilities level current condition

Proses	1	2	3	4	5	6	7	8	9	10	11	12	13	Result
AP007 Manage human resource	3	2	2	2	2	2	2	2	2	2	2	2	2	2,1
AP009 Manage service agreements	2	1	1	1	1	1	1	2	1	2	1	2	2	1,4
AP013 Manage security	2	2	2	3	2	2	2	2	2	2	2	2	2	2,1
BAI02 Manage requirement definition	2	2	2	1	1	1	1	1	1	1	2	2	2	1,5
BAI04 Manage availability and capacity	3	3	2	2	2	3	2	2	2	3	2	1	1	2,2
BAI08 Manage knowledge	2	2	2	2	2	2	2	2	2	2	1	1	2	1,8
BAI10 Manage configuration	2	2	2	3	2	2	2	2	2	3	2	1	2	2,1
DSS01 Manage operation	1	2	2	1	1	1	1	1	1	1	1	2	2	1,3
DSS02 Manage service requests and incidents	2	1	1	1	1	1	1	1	1	1	1	2	1	1,2
DSS03 Manage problems	2	1	1	2	1	1	1	1	1	1	1	1	2	1,2
DSS04 Manage continuity	1	1	1	2	1	1	1	1	1	2	2	1	1	1,2

Table 2. Capabilities level target

Capability level target

Proses	1	2	3	4	5	6	7	8	9	10	11	12	13	Result
AP007 Manage human resource	4	4	4	4	4	4	4	4	4	4	4	4	4	4.0
AP009 Manage service agreements	4	3	3	4	3	3	3	3	3	4	4	3	4	3.4
AP013 Manage security	4	5	4	5	4	4	4	4	4	5	4	4	4	4.2
BAI02 Manage requirement definition	4	3	4	4	3	3	3	3	3	3	4	3	4	3.4
BAI04 Manage availability and capacity	5	5	4	4	4	4	4	4	4	4	4	4	5	4.2
BAI08 Manage knowledge	5	4	5	4	4	4	4	4	4	4	4	5	4	4.2
BAI10 Manage configuration	3	3	3	4	3	3	3	3	4	4	3	4	4	3.4
DSS01 Manage operation	4	4	4	4	3	3	3	3	3	3	3	3	4	3.4
DSS02 Manage service requests and incidents	4	4	4	4	3	3	3	3	3	4	4	4	4	3.6
DSS03 Manage problems	4	3	3	4	3	3	3	3	3	4	4	4	3	3.4
DSS04 Manage continuity	3	3	4	4	3	3	3	3	3	3	4	3	3	3.2

Gap Analysis was conducted to determine the extent of the gap occurs between the current conditions with the expected conditions and what efforts can be undertaken to minimize the gap. Gaps (Gap) is obtained through a comparison between the level of capability of governance in Media Centre, the Ministry of Defence on the current conditions with the level of capability targets to be achieved, With the help of spider chart or radar chart as Image, clearly visible gaps each process based on the categories of organizational objectives (Fig. 3).

Table 3. Capabilities level target

Process assessment result

Area process	Process capability Level						Level capability
	0	1	2	3	4	5	
APO07 Manage human resource					*		Predictable
APO09 Manage service agreements				*			Established
APO13 Manage security					*		Predictable
BAI02 Manage requirement definition				*			Established
BAI04 Manage availability and capacity					*		Predictable
BAI08 Manage knowledge					*		Predictable
BAI10 Manage configuration				*			Established
DSS01 Manage operation				*			Established
DSS02 Manage service requests and incidents				*			Established
DSS03 Manage problems				*			Established
DSS04 Manage continuity				*			Established

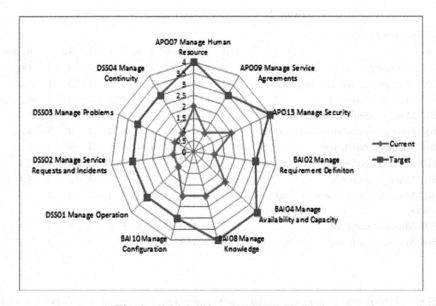

Fig. 3. Services lifecycle of ITIL V3 2011

3 Result and Discussion

After knowing the existence of gaps in any IT processes COBIT 5 according to the determination of strategies for achieving the target capabilities required for consideration in this study. Here is the strategy of concern in determining achievement of capabilities:

1. Improvements were made gradually in accordance with the priorities which the attribute with a value lower capability, received a higher priority to be repaired.
2. Referring to the first point above it, the process area with the current conditions capability level is 1, got the main priority corrective measures to achieve capability level 2 in advance. Some process areas referred to respectively APO09, BAI02, BAI10, DSS01, DSS02 DSS03 and DSS04.
3. The conditions under which the achievement of the equilibrium level of capability throughout the process is at a level 2 this time. Simultaneously (throughout the process) is done step towards improvement in the condition of the target level of capability, that capability level 3.
4. The entire area is currently in the process capability level 3, then some process areas targeted were on level 4 can be raised later, that area APO07 process, APO13, BAI04 and BAI08.
5. The mapping process carried out between 5 COBIT and ITIL V3 2011 came from guide COBIT 5: Enabling Process on the related guidance in every process of COBIT 5. In addition, the mapping process is also obtained from the mapping has been done by Glenfis AG - a consulting and training company in Switzerland specializing in Service Management and IT Governance. Glenfis AG is the first company to map the ITIL V3 2011 with COBIT 5 is licensed under the Cabinet Office.

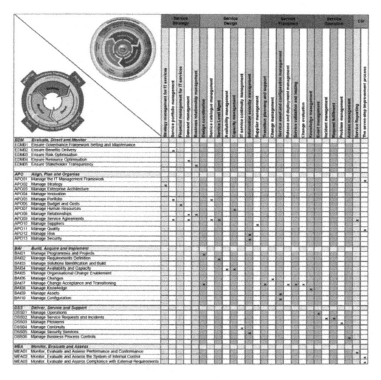

Fig. 4. Services lifecycle of ITIL V3 2011

The combination of these two processes framework, COBIT 5 can be done because only provides a way to measure and assess the level of process capability, while ITIL can provide best practice approaches along operationalin detail. All processes COBIT 5 with 5 domains including domain EDM (Evaluate, Direct and Monitor), DSS (Deliver, Service and Support), BAI (Build, Acquire and Implement), MEA (Monitor, Evaluate and Assess) and APO (Align, Plan and Organise) associated with ITIL V3 processes in 2011 will be given a cross (x) as in Figure below (Fig. 4).

The results of the above mapping (31 COBIT 5 process areas) are then been returned by eleven COBIT 5 process areas that have been adapted to organizations problems. Selection process relevant COBIT 5 shown in table below (Fig. 5).

Governance policy recommendations compiled in this study is intended as an input to the relevant stakeholders. After the measurement of the processes of COBIT 5, which have been based on issues and strategies for achieving specified determination capabilities. The next process is the development strategy into the draft recommendation activity, so it is important to determine the KPI as a next step. KPI (Key Performance Indicator) is a measuring instrument for the achievement or success of the processes.

Mapping of COBIT 5 to ITIL V3 2011		
No.	COBIT 5 Process	ITIL V3 2011 Process, Fungction (F) and Activity (a)
1.	APO07 Manage Human Resources	SD 4.5 Capacity management
2.	APO09 Manage Service Agreements	SS 4.2 Service portfolio management SS 4.4 Demand management SD 4.2 Service catalogue management SD 4.3 Service level management CSI 5.7 Service reporting (a)
3.	APO13 Manage Security	SD 4.7 Information security management
4.	BAI02 Manage Requirements Definition	SD 4.3 Service level management
5.	BAI04 Manage Availability and Capacity	SD 4.4 Availability management SD 4.5 Capacity management
6.	BAI08 Manage Knowledge	ST 4.7 Knowledge management
7.	BAI10 Manage Configuration	ST 4.3 Service asset and configuration management
8.	DSS01 Manage Operations	SO 4.1 Event management SO 6.2.1.3 IT Operations management (f)
9.	DSS02 Manage Service Requests and Incidents	SO 4.2 Incident management SO 4.3 Request fulfillment
10.	DSS03 Manage Problems	SO 4.4 Problem management
11.	DSS04 Manage Continuity	SD 4.6 IT Service vontinuity management (ITSCM)

Fig. 5. Services lifecycle of ITIL V3 2011

4 Conclusion

1. Processes area that measured the level of capability at this time the average was at level 1 (performed) in the area APO09, BAI02, BAI10, DSS01, DSS02, DSS03, DSS04 and level 2 (managed) in the area APO07, APO13, BAI04, BAI08.
2. Processes area expectation or target average is in level 3 (Established) in the area APO09, BAI02, BAI10, DSS01, DSS02, DSS03, DSS04 and level 4 (Predictable) in the area APO07, APO13, BAI04, BAI08.

3. Creation of strategies for achieving capability improvements done gradually by increasing the level area of the smallest first.
4. In this research process improvement efforts and goal achievement through a recommendation process given activity in the form of policies and procedures for data management and information.
5. KPI (Key Performance Indicator) as an indicator of the success obtained by the process of document COBIT 5, which then made the selection of KPIs that are relevant to the three strategic objectives of the organization (SS). The process of selecting the relevant KPI produces 53 of 66 KPI obtained from the document COBIT 5.

References

1. Brown, A.E., Grant, G.G.: Framing the frameworks: a review of IT governance research **15**, 696–712 (2005)
2. Cabinet Office: Government construction strategy. Construction **96**, 43 (2011). http://doi.org/Vol19
3. Cartlidge, A., Hanna, A., Rudd, C., Ivor, M., Stuart, R.: An introductory overview of ITIL V3. The UK Chapter of the itSMF (2007). http://doi.org/10.1080/13642818708208530
4. Isaca: COBIT: A Business Framework for the Governance and Management of Enterprise IT, pp. 1–94 (2013)
5. ITGI: Global status report on the Governance of Enterprise IT (GEIT)—2011. Gov. Int. J. Policy Adm. **70** (2011). http://www.isaca.org/Knowledge-Center/Research/Documents/Global-Status-Report-GEIT-10Jan2011-Research.pdf
6. Luftman, J., Luftman, J.: Assessing business-IT alignment maturity **4**, 1–51 (2000)
7. National Computing Centre: A best practice guide for decision makers in IT. The UK's Leading Provider of Expert Services for It Professionals (2005). https://www.isaca.org/Certification/CGEIT-Certified-in-the-Governance-of-Enterprise-IT/Prepare-for-the-Exam/Study-Materials/Documents/Developing-a-Successful-Governance-Strategy.pdf
8. General Platform Criterion Assessment Question: PinkVERIFY TM 2011 ITSM Tool Assessment Criteria Service Asset & Configuration Management (SACM), pp. 1–6 (2014)
9. Sambamurthy, V., Zmud, R.W.: Arrangements for information technology governance: a theory of multiple contingencies. Manag. Inf. Syst. Q. **23**(2), 261–290 (1999). http://doi.org/10.2307/249754

Artificial Intelligence Applications for E-services

To Solve the TDVRPTW via Hadoop MapReduce Parallel Computing

Bo-Yi Li and Chen-Shu Wang[✉]

Graduate Institute of Information and Finance Management,
National Taipei University of Technology, Taipei, Taiwan
a2595245512@gmail.com, wangcs@ntut.edu.tw

Abstract. The convenience of online shopping has made it common to everyone. With the increase of online transaction, optimization of VRP is an important issue in logistics and transportation. TDVRPTW is a crucial problem which considers a given time window in VRP. This paper targets solving TDVRPTW by using Hadoop MapReduce and compares the effectiveness of Hadoop with a single machine. We used an existing program to cluster the demand nodes and then calculated a route for every cluster by using random method and heuristic algorithm including nearest time window algorithm, nearest neighbor algorithm and 2-opt. After that, we executed parallel computing in Hadoop by implementing program on MapReduce. We used Solomon benchmarking problem as the base of experimental examples and made the experiments. This research proved that Hadoop MapReduce has better efficacy to calculate the best solution than a single machine.

Keywords: Hadoop · MapReduce · TDVRPTW · Heuristic algorithm

1 Introduction

On a daily basis, more and more people are shopping online because of the shopping process of the electronic commerce web site is getting common and easy to use [9]. According to Visa's customer survey of e-commerce behavior in Taiwan[1], 87% of customers in Taiwan have had shopping experience on the internet over the past 12 months. It shows that shopping online has become part of Taiwanese's daily life. It is convenient and multivariate way of payment plus, one of the main reason attracting customer to buy is the service of home delivery. We can expect that the demand of package delivery is increasing due to the trend of online shopping. Therefore, it is becoming important to optimize the route of logistics to it maximum efficiency.

A similar and well define problem is known as Vehicle Routing Problem (VRP). VRP is a classic routing problem that was first introduced by Dantzig and Ramse in 1959 [2]. According to the description of VRP, t fixed number of nodes with different demand is toned to be served by a known fleet. Similar to VRP, there is another kind of problem which is Time Dependent Vehicle Routing Problem with Time Window

---

[1] http://www.visa.com.tw/aboutvisa/mediacenter/NR_tw_122215.html, Visa's customer survey of e-commerce.

© Springer International Publishing AG 2017
N.T. Nguyen et al. (Eds.): ACIIDS 2017, Part II, LNAI 10192, pp. 55–64, 2017.
DOI: 10.1007/978-3-319-54430-4_6

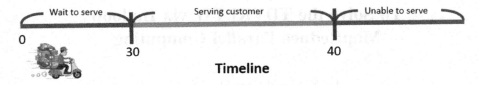

Fig. 1. TDVRPTW

(TDVRPTW). In TDVRPTW, some conditions, such as every node must consider the delivery time windows. It defines fleet can serve the node if the fleet arrives the node in the assigned time window range. For example, a node has a time window in thirty minutes to forty minutes. When fleet arrives before thirty minutes, the fleet has to wait until thirty minutes. When fleet arrives in the time window, fleet can immediately server the customers. When fleet arrives it after forty minutes, fleet can't server them. Figure 1 below shows the time window. Therefore, when the logistics order increases, the complexity of VRP is increases, in which increases data and processing time.

Because of the rapid development of information technology and network communication, there is more data produced now a day than in the past. Thereby, the transaction of online shopping has increased, and the data which is processed by logistics officer has increased in VRP. According to IDC[2], in 2013, there are 4.4 ZB data in the world. IDC thinks that there will be 44 ZB data in 2020. The amount of data will expand 10 times. When the amount of data substantial increases, how to increase effectiveness of computing by current models and tools would be important. Apache Hadoop is one of the solutions [10]. Hence, Apache Hadoop is also a solution in VRP. Apache Hadoop is an open source software framework for distribution store and big data processing. Its computing framework is called MapReduce. It can parallel process data by the concept of map and reduce. There are many well-known companies like Facebook, Yahoo, LinkedIn etc. which use Hadoop to store and process their big data[3].

Therefore, the aim of this paper is to solve TDVRPTW by using Hadoop MapReduce and self-writing program. I will compare the effectiveness of Hadoop with a single machine. The paper uses Solomon benchmarking problem [1] as base of experimental examples.

2   Literature Review

2.1   VRP

The vehicle routing problem (VRP) is a combinatorial optimization and integer programming problem which has a well-known special case called TSP (travelling salesman problem) and first presented in 1959 by Dantzig and Ramse. There are a set

[2] http://www.ithome.com.tw/article/87190, IDC, 2016.

[3] http://www.hadoopwizard.com/which-big-data-company-has-the-worlds-biggest-hadoop-cluster/, Hadoop, 2016.

of customers, a fleet of vehicles and one or more depots in VRP. We plan a set of routes including through all customers once for a fleet of vehicles which derived from depots. Every customer must be visited once. Because TSP is an NP-hard problem [3, 4, 11] so VRP is also an NP-hard problem.

2.2   TDVRPTW

Time Dependent Vehicle Routing Problem with Time Window (TDVRPTW) which is extension of VRP should serve customers within a given time window. Beside condition of given time window, there is a paper which investigates the condition of minimizing fuel consumption [5], and there is another paper which considers traffic condition of the road network. [6]

2.3   MapReduce

MapReduce is a cloud framework model which provided by Google. It is suitable for big data, because of its parallel and distributed architectures. MapReduce provides automated parallel processing, distributed computing, scheduling management and tolerance of fault of system [7] (Fig. 2).

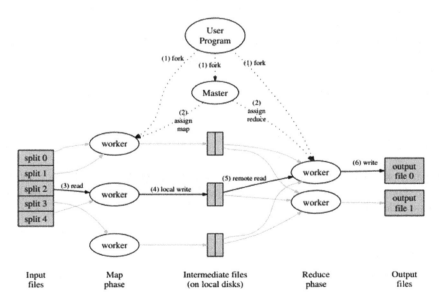

Fig. 2. MapReduce [7]

2.3.1 Map

Mapper reads data from HDFS (Hadoop Distributed File System). Mapper's main mission is generating key-value pairs by analyzing and splitting input data. After that, program store key-value pairs in buffer.

2.3.2 Reduce

Every Reducer reads key-value pairs data which had same key from buffer. According to reduce's program, it processes data. Every Reducer will output a file.

2.4 Heuristic Algorithm

2.4.1 Nearest Neighbor Algorithm

Nearest neighbor algorithm is a heuristic algorithm. Its operation steps of route construction procedure is as follow:

(1) Calculate the distance of every node and depot in every cluster.
(2) Sort distance by ascending arrangement for every cluster.
(3) According to sorting list, it generates route in every cluster.

2.4.2 2-Opt

2-opt is a optimizing local search algorithm and is first proposed by Croes [8] in 1958 for solving the traveling salesman problem. Its operation steps of route construction procedure is as follow:

For example: A => B => C => D => E => F => G => H

(1) I select two difficult number which is less than length of route (i.e.: 4, 7)
(2) I take the path (D => E => F => G)
(3) Generate a new route. (A => B => C => G => F => E => D => H)

3   Solve VRP by Hadoop

The paper will explain how to solve TDVRPTW by Hadoop MapReduce in this chapter. The chapter consist of design of experimental process and design of data set, and I provide pseudocode and mathematical function.

3.1 Definition of Constraints and Parameter

See Table 1.

Table 1. Definition and description of variables and the constraints

Variables	Description
P_id	Demand node's ID
X	x coordinate
y	y coordinate
bonus	Performance bonus. When you server a node, you can get performance bonus in this node
open_t	Start server time window
close_t	Close server time window
server_t	Service time
S	Shipments
l	The logistics officer of I
SorR	Receipt or delivery
cm	Customer's category
fs	Feasible solution of route
k	The demand nodes in k clustering
C_k	The number of demand nodes in k clustering
Constraints	
Fleet's capacity must more than shipment through every node in the route	
Logistics officer must arrive before close time window for every node	
When distance is shorter, the cost of route is lower	

3.2 Solving Architecture

The paper uses a solving strategies which is clustering first and routing later. Research process as show in Fig. 3 below:

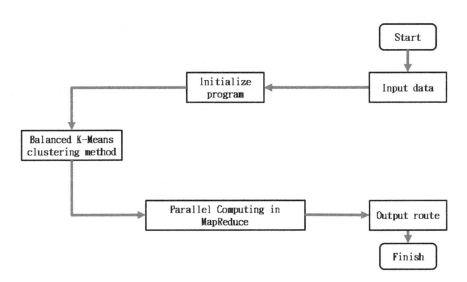

Fig. 3. Research process

I use existing program to cluster demand nodes. The program improves k-means algorithm which considers ability of logistics officer and balances logistics officer's performance bonus. After that, according to the number of logistics officer, demand nodes will be clustered.

After clustering, this paper will parallel computing by writing program. This paper show steps and architecture chart as below (Fig. 4).

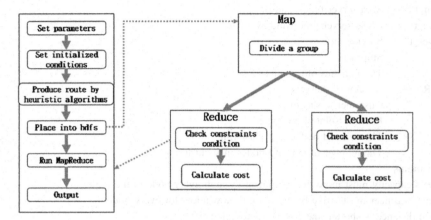

Fig. 4. MapReduce architecture

i. **Parameter Setting and Initialization**

In the stage, I will set parameter of MapReduce, parameter of constraints and initial Hadoop.

ii. **Initial Route by Heuristic Algorithms**

This study presents a nearest time window algorithm according to nearest neighbor algorithm. First, I get demand node's start time windows from a cluster. And then, I sort demand nodes by the start time windows. The sequence is a new route. Doing the above steps for every cluster produce new routes.

The study uses nearest time window algorithm and nearest neighbor algorithm and random function to produce three routes in every cluster. And then, using function as below calculate fs.

$$f_{S_k} = \begin{cases} 10000, C_k > 10 \\ 15 \times \sqrt{C_k!}, C_k \leqq 10 \end{cases}$$

Afterwards, according to fs, I use 2-opt to optimize routes. Finally, I get $f_{S_k} \times 3$. Initial solutions in k clustering.

iii. **Put Data into HDFS and Execute MapReduce**

Hadoop Distributed File System (HDFS) is dedicated file system of Hadoop. Because of that, all data must be placed HDFS before program runs.

MapReduce consist of map and reduce. The data show as key-value pairs in every stage.

Map's mission is generating key-value pairs by analyzing and splitting input data. And then, program store key-value pairs in buffer.

Every Reduce reads key-value pairs data which had same key from buffer. According to reduce's program, it examines constraints and calculate cost of route. After it compares new cost with old cost, it stores the smaller cost. Until no route, it finishes. There is some pseudocode in Fig. 5 below

```
1       public void reduce(Text key, Iterable<routes> values, Context context)
2              throws IOException, InterruptedException {
3           read_pointdata(context.getConfiguration());
4           read_dis(context.getConfiguration());
5           String better_route = "";
6           String[] better_sche = new String[5];
7           for (routes froutes : values) {
8               String sss = froutes.get();
9               String[] sp = sss.split("_");//check constraints conditions ' Calculate the cost
10              String[] sche = schedule(sp);
11              if (new_route is better than original route) {//store new route
12                  better_sche=sche;
13                  better_route=sp[1];
14              }
15          }
16          context.write(new Text("logistics officer"),new Text("P_id\tx\t\tty\t\t
17              cm\tarrival time\t\tdeparture time\t\tcumulative distance\t\tcumulative time"));
18          for(int i = 0;i<better_route.split("_").length;i++){//output route by sequence
19              F.set(point of route);
20              context.write(key, F);
21          }
22      }}
```

Fig. 5. Pseudocode

iv. **Output Result**

Every Reducer will output a file. File consist of sequence of route, arrival time, departure time, cumulative distance and cumulative time and so on.

3.3 Design of Data Set

Data set base on Solomon benchmarking problem, I add some parameter including shipment, customer's category and performance bonus, and I determine start server time window and close server time window by random. There are five data set in this experiment, detail is shown in Table 2 below.

Table 2. Data set

p_id	x	y	Bonus	Open_t	Close_t	Server_t	s	SorR	cm
0	35	35	0	0	0	0	0	−1	−1
1	41	49	87.5	0	204	10	5	0	0
2	35	17	66.5	0	202	10	5	1	0
3	55	45	59.5	0	197	10	5	1	1
4	55	20	80.5	149	159	10	5	1	1
5	15	30	56	0	199	10	5	0	1
6	25	30	87.5	0	208	10	5	1	0
7	20	50	66.5	0	198	10	5	1	0
8	10	43	77	95	105	10	5	0	0
9	55	60	56	97	107	10	5	0	1
10	30	60	59.5	0	194	10	5	0	0
11	20	65	84	67	77	10	5	1	1
12	50	35	59.5	0	205	10	5	0	0
13	30	25	52.5	159	169	10	5	1	1
14	15	10	84	0	187	10	5	0	0
15	30	5	73.5	61	71	10	5	1	1

3.4 Experimental Environment

This study build Hadoop distributed environment by oracle virtualbox. There are three virtual machines which consist of Quad-core processor and 4G memory. There are a master and two reducers in virtual machines. All environment is shown in Table 3 below.

Table 3. Experimental environment

CPU	Intel(R) Core(TM) i7-870 CPU
RAM	12G RAM
Hard disk drive	SATA 3.5" 500 GB
Operating system	Ubuntu-12.04
JVM	Java SE Runtime v.1.7
Hadoop	0.20.2
Programming language	Java
Number of machine	3

4 Experimental Results

This experiment investigates difference of execution time of single machine computing and hadoop MapReduce distribution computing. And then, it uses Solomon benchmarking problem as foundation to produce five data sets. There are one hundred demand nodes in every data set. Execution parameters is shown in Table 4 below.

Table 4. Execution parameters

Demand node	100
Logistics officer	4
Upper limit ability of logistics officer	26,26,26,26
Lower limit performance bonus	800
Capacity	1000

This study calculates the best solution for every data set by using optimization software Gurobi. I compare single machine with Hadoop by using best solution as answer, and the result is shown in Table 5 and Fig. 6 below.

Table 5. Experimental results (Units seconds)

	R100-2	R100-3	R100-4	R100-5	R100-6
Stand - alone	130.708	135.78	134.577	133.833	133.366
MapReduce	71	71	66	69	73

Fig. 6. Execution time

5   Conclusions and Future Research

The aim of this paper is to solve TDVRPTW by using Hadoop MapReduce and comparison between a single machine and Hadoop MapReduce. We can discover that execution time of Hadoop is almost half execution time of single machine by chart as above. Hence, this study proves that Hadoop MapReduce has better efficacy to calculate the best solution than a single machine, when it considers the condition of balance of performance bonus for logistics officer in TDVRPTW.

In future, beside using heuristic algorithms as above, I will use heuristic optimal algorithm like genetic algorithm (GA), Ant. Algorithm and so on. By using those heuristic optimal algorithms, I can get the best solution rapidly and reduce computing time. I hope to replace virtual machine with physical machine, I think it has better results.

References

1. Solomon, M.M.: Algorithms for the vehicle routing and scheduling problems with time window constraints. Oper. Res. **35**(2), 254–265 (1987)
2. Dantzig, G.B., Ramser, J.H.: The truck dispatching problem. Manag. Sci. **6**(1), 80–91 (1959)
3. Cook, S.A.: The complexity of theorem-proving procedures. In: Proceedings of the Third Annual ACM Symposium on Theory of Computing, pp. 151–158. ACM (1971)
4. Garey, M.R., Johnson, D.S.: Computers and Intractability, vol. 29. W. H. Freeman, New York (2002)
5. Yao, E., Lang, Z., Yang, Y., Zhang, Y.: Vehicle routing problem solution considering minimising fuel consumption. IET Intell. Transp. Syst. **9**(5), 523–529 (2015)
6. Kondekar, R., Gupta, A., Saluja, G., Maru, R., Rokde, A., Deshpande, P.: A MapReduce based hybrid genetic algorithm using island approach for solving time dependent vehicle routing problem. In: 2012 International Conference on Computer and Information Science (ICCIS), vol. 1, pp. 263–269. IEEE (2012)
7. Dean, J., Ghemawat, S.: MapReduce: simplified data processing on large clusters. Commun. ACM **51**(1), 107–113 (2008)
8. Croes, G.A.: A method for solving traveling-salesman problems. Oper. Res. **6**(6), 791–812 (1958)
9. Gunasekaran, A., Marri, H.B., McGaughey, R.E., Nebhwani, M.D.: E-commerce and its impact on operations management. Int. J. Prod. Econ. **75**(1), 185–197 (2002)
10. Patel, A.B., Birla, M., Nair, U.: Addressing big data problem using Hadoop and Map Reduce. In: 2012 Nirma University International Conference on Engineering (NUiCONE), pp. 1–5. IEEE (2012)
11. Karp, R.M.: Reducibility among combinatorial problems, pp. 85–103. Springer, New York (1972)

MapReduce-Based Frequent Pattern Mining Framework with Multiple Item Support

Chen-Shu Wang[1(✉)], Shiang-Lin Lin[2], and Jui-Yen Chang[1]

[1] Department of Information and Finance Management,
National Taipei University of Technology, Taipei, Taiwan
wangcs@ntut.edu.tw, ketrelo0225@gmail.com
[2] Department of Management Information System,
National Chengchi University, Taipei, Taiwan
shiang0623@gmail.com

Abstract. The analysis of big data mining for frequent patterns is become even more problematic. Many efficient itemset mining algorithms to set a multiple support values for each transaction which could seem feasible as real life applications. To solve problem of single support have been discovered in the past. Since, we know that parallel and distributed computing are valid approaches to deal with large datasets. In order to reduce the search space, we using MISFP-growth algorithm without the process of rebuilding and post pruning steps. Accordingly, in this paper we proposed a model to use of MapReduce framework for implement the parallelization under multi-sup values, thereby improving the overall performance of mining frequent patterns and rare items accurately and efficiently.

Keywords: Association rule · Hadoop MapReduce · Multiple supports

1 Introduction

For information system practitioners (from both academic and industry perspectives), it is a challenge era. Compared with past, the data amount grow exponential. According to Leavitt's research in 2013, the data amount of user generation increases 35% every year [1]. In addition, due to the user acceptance rate and usage rate of mobile devices (such as mobile phone and pad) is getting high, many users download so-called "SoLoMo"[1] (Social, Local, Mobile) Apps, thereafter, more log files are generated. For example, for apps of Facebook, users share their photo and files with friends and also mark where they go. These SoLoMo apps enable users share their news feed or feeling (such as "like" or emotional sticker) with their friends/fans and the digital behavior are recorded. Furthermore, the development of sensor technology is mature and stable, lots of applications of sensor technology become visible, such as Smartwatch or Sport Wristband. These device of sensors collect lots of passive monitor information, such as heartbeat and blood-pressure, and could be further analyzed. For both active and passive way, the data is generated 24/7 and the data amount is huge [2, 3].

---

[1] SoLoMo struck, http://blog.sina.com.cn/s/blog_56c35a550102dr4d.html, 2016.

© Springer International Publishing AG 2017
N.T. Nguyen et al. (Eds.): ACIIDS 2017, Part II, LNAI 10192, pp. 65–74, 2017.
DOI: 10.1007/978-3-319-54430-4_7

The trend of data tsunami is unavoidable [4]. Lots of data management issues become important, such as online data access, archiving data management, and big data mining [5]. For all these issues, compared with past, the core of these technology maybe similar, however the data amount is becoming very huge. Big data is going to change our daily life, we should figure out a way to apply these huge data more wise and more effective.

Big data mining got a lot of applications and attempt to promote people's health and daily life better and easier. For instance, we each have a different way of posture by function of leg length and the contours of back. Hence, we could create an index data which unique to you through a lot of sensors into of chairs to prevent the problem of car thefts. This idea is that the carnapper sits behind the wheel and tries to start the car, but the potential anti-theft device recognizes that a non-approved driver and engine stops automatically, unless user type in a password into the dashboard. Even we could identify signs to predict of car accident or fatigue of driver, to set an internal alarm alert: "pay more attention to the road." So we are going to need to analysis from big data, adjust it for our life carefully and must have to be the master of this technology, not its servant.

Association mining is the analyzing process of discovering interesting and useful association rules hidden from huge and complicated data in different databases. It focuses on finding out the frequent patterns (itemsets) that occurred together, and used to reduce costs and increase revenue. In real life applications of data mining, such as the recommendation of related products to customers by analyzing their purchase behavior and histories. Several researches have been devoted to mine frequent patterns in the past decades. According to the literature, Apriori algorithm [6, 7] is one of the most classical approaches to find association rules from datasets that have to be scanned many times to create all candidate items, it cause to consume a huge amount of operating hours. In order to improve this weakness, Han et al. in 2000 proposed the FP-growth algorithm that without candidate generation and could improve the insufficient of traditional Apriori algorithm [8]. Although Fp-growth algorithm can mine frequent patterns successful through lesser of scan time of database, however, use a single minimum item support value for all items are not sufficient since it could not reflect the characteristic of each item. When the minimum support value (MIS) is set too low, despite it would find rare items, similarly, it may generate a large number of meaningless patterns. On the other hand, if the minimum support value is set too high, we will lose useful rare patterns. Thus, how to set the threshold value of minimum support for each item to find out correlated patterns efficiently and accurately is essential.

Recently, some methods have been proven effective in mine association rules with multiple minimum supports to overcome the constraint of single min-sup. The MIS-Tree proposed by Hu et al. in 2006, it improved the FP-Tree that storing the crucial information and developing CFP-growth algorithm for mining all frequent itemsets of association rules with multiple minimum supports [9]. It focuses tuning minimum support on the maintenance of the MIS-Tree which to ensure correct status and unnecessary rescanning database through removing, moving and merging item phase. In addition, the MS-FP-Growth is implemented by Taktak et al. in 2014 [10], it is a new version of FP-Growth algorithm to convert the min-sup value by three ways of increasing and decreasing from user only, user assisted by a function and automatically to find association rules. On the other hand, similarly, according to FP-Growth algorithm, it making rare child node pruning step with multi-MS using maximum constraints to promote mining patterns effectively through Elgaml et al. in 2015 [11]. Nevertheless, in order to achieve the best performance possibly from mining frequent patterns with multiple supports, an efficient method of the multiple item support frequent pattern growth, MISFP-growth algorithm was proposed in 2016 [12]. It advocates that post pruning and reconstruction phases are not required since the architecture of the MISFP-Tree.

However efficient computing has been an active research issue of data mining in recent years. MapReduce [13] was first introduced by Google Jeffrey Dean and Sanjay Ghemawat in 2008, it could easier implement parallel algorithm to compute various kinds of derived data and reduce run-time. In the near future, based on MapReduce for mining association rules, it could enhance performance. And could improve FP-Growth algorithm when mining infrequent itemsets than before [14–16]. In view of the discussion above, how to valid and rapid to complete set of mining frequent patterns that generated for association rule, and how to using multiple support value, without losing rare items and creating interesting patterns are matter of great account. In view of these facts, this study motivates to propose a model of solution based on multi-support with MapReduce for frequent pattern mining.

The remainder of this paper is organized as follows. Section 2 reminds key methods and algorithm steps used in this paper. Our proposed conceptual model for frequent pattern mining using MapReduce under multiple support values at Sect. 3. And Sect. 4 we demonstrate of the proposed framework, Sect. 5 arrange the expected of experimental environment and result. Finally, we draw conclusions and the expected future works are stated in Sect. 6.

2 Literature Review

Answer to in this chapter, we arrange our current understanding of mining method using this paper called MISFP-growth algorithm, Multiple Item Support Frequent Pattern growth, as an example by Sadeq et al. in 2016 [12] to efficient with set multi-support values for frequent patterns mining completely.

In MISFP-growth algorithm method, frequent patterns can be mined through multiple item support frequent pattern tree, MISFP-Tree data structure which is used to store all key information besides rebuilding and post pruning phases are not required. Its approach is extended version of FP-growth algorithm [8] with main differences are used multiple item support thresholds to mine and create frequent patterns. Additionally, the items are sorted in descending order in terms of their support values in MISFP-growth. It can also reduce search space to enhance performance that based on the minimum of minimum item support threshold, MIN-MIS since any item to give up discouraging which support less than MIN-MIS and building up MISFP-Tree no more.

Now, we discuss how to mine frequent patterns with multiple supports using MISFP-growth algorithm of the following essential processes shown in Fig. 1. First, we find out the actual support of each item and the lowest minimum support threshold

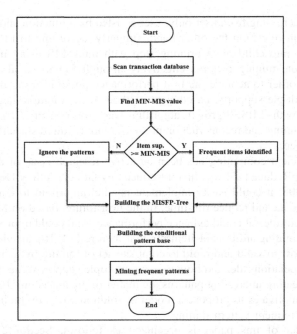

Fig. 1. Flow diagram for mining frequent pattern from MISFP-growth method

(MIN-MIS) among all items through scan transaction database once. After that, scan database again to collect and sort items in the descending order that satisfy MIN-MIS in each transaction. Next, the constructed of MISFP-Tree by insert these items according to their MIS values in non-increasing order which its count is increased by one if the appropriate node of an item exists, and the remaining items would be created as the new nodes. The step is similar to building the approach of FP-Tree. Further, in order to facilitate and simplify tree traversal, a MIN-MIS frequent header table of MISFP-Tree is built which each item points to its occurrences in the tree by the head of node-link. Finally, we build the conditional pattern base and the conditional MISFP-Tree of each suffix item until no conditional pattern base can be generated that the MISFP-Tree contains the complete information for frequent pattern mining with multiple items support whose support is equal or greater to its predefined MIS.

3 Research Design and Research Model

Most of the methods to enhance effective using multiple support values without set single support threshold for mining frequent pattern and rare itemsets, but no ways were based the structure of Hadoop MapReduce framework. Thus, our necessary to propose and make a conceptual model to find interesting infrequent patterns faster with multiple support values on distributed computing of MapReduce.

Problem Statement. Given an enormous transactional database like word, picture, video or social media data etc., the problem is to discovery significant rare Itemsets under Hadoop MapReduce framework.

Input phase. In the first input phase is divided and conquered into amount of each chunk is assigned to one node.

Mapping phase. The mapper function step at each node accepts more inputs datasets to find minimum of minimum item support, MIN-MIS value and sort items in the descending order that satisfy MIN-MIS in each transaction. If the support value count of an each itemsets are less than MIN-MIS would discarded and the other interest items satisfy threshold are assigned to reducer building up MISFP-Tree.

Reducing phase. After that, the reducer function procedure builds the conditional MISFP-Tree of each suffix item until no conditional pattern base can be generated and outputs the interest itemsets are greater than MIN-MIS that considered as a frequent pattern items.

Output phase. Finally, we conduct association analysis to obtain the result of association rules. Our conceptual flow model is shown in Fig. 2.

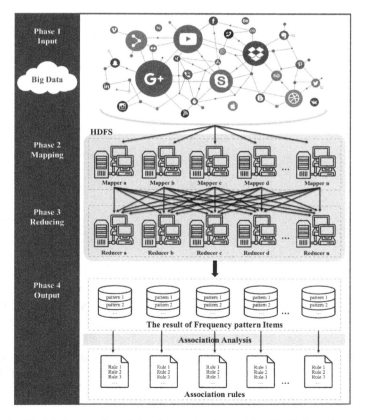

Fig. 2. Model diagram for mining frequent pattern under MapReduce framework

4 Demonstration of the Proposed Framework

Given for ten transaction record from database as shown in Table 1, which represent the glass cleaner, plastic gloves, rag, scraper, spray bottle, scouring pads, bucket, broom, baking soda and toothbrush, namely, item A, item B, item C, item D, item E, item F, item G, item H, item I and item J. Moreover, the multiple items supports of each items in Table 2.

Table 1. Transaction database

Transaction ID	Items	Items have support no less than MIN-MIS (ordered)
1	B, A, C	A, B, C
2	E, D, H, G, I	E, G, H, I
3	C, A, B, J	A, B, C, J
4	E, C, G	C, E, G
5	B, A, J, I	A, B, I, J
6	A, B	A, B
7	G, C, F, E, J	C, E, G, J
8	B, A, G	A, B, G
9	H, A, I	A, H, I
10	C, E, B	B, C, E

Table 2. MIS and actual support of each items

Items	MIS	Sup
A	4	6
B	4	6
C	4	5
D	5	1
E	3	4
F	4	1
G	3	4
H	2	2
I	2	3
J	2	3

In order to enhance the comprehensive of mining frequent patterns and without losing rare items accurately that using parallel mathematical operation of MapReduce framework with the multiple predefined minimum support values, the process works as follows.

Phase 1 input action, we scan transaction database once to find the support values of each items in Table 2. It also could discovery the least minimum support threshold among all minimum item support values of items, MIN-MIS = MIN{MIS(A), MIS(B), MIS(C), ..., MIS(J)} = 2. Next, our compare the actual support value of items with

MIN-MIS value that it satisfy MIN-MIS in each transaction and descending order. Hence, items {D, F} are discarded as shown at right column in Table 1. After that, we based on the identical of multiple support values to become a new transaction group and assign to different mappers.

Phase 2 mapping function, let us scan database once again for each new transaction group to construct MISFP-Tree. The process of inserting transactions into the tree works like FP-growth algorithm (FP-Tree) as follow depictions.

The first root of MISFP-Tree is created and labeled as "null" and inserting nodes for leading itemsets {E, G, H, I} of new transactions group which the multiple support value is 2 in Mapper A. Since it count of each node along the prefix is increased by one if shares the prefix nodes the same. By repeating steps until all transaction records are added to the tree afterwards. Furthermore, it need to build an item header table for facilitate tree traversal which nodes with exactly similar item-name are linked in sequence called node-links. After scanning all the transactions with associated node-links are shown Fig. 3.

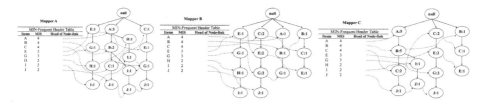

Fig. 3. MISFP-Tree with all new transactions group in parallel computing

Phase 3 reducing function, we start to mine frequent patterns that according to establish MISFP-Tree under different mappers. Following the node-link of conditional pattern base of item {J} is {A, B, I :1}, {A, B, C :1} and {C, E, G :1}. However, "mine (< A:2, B:2 > |J)" involves mining two items A and B in sequence that the MIS of item {J} is two. Thus, we could find the frequent patterns of item {J} are {B, J :2}, {A, B, J :2}, {A, J :2} and so forth to complete mine of all the patterns. It also conducts to remove repeat of items for output results exactly.

Phase 4 output results, according to this example of transaction purchase behavior, we find out the whole set of frequent patterns with multiple support threshold as shown in Table 3. However, it interesting that the results of frequent patterns except for common association rules like item A (glass cleaner), item B (plastic gloves) and item {B, J}, {A, B, J}, {A, J} that use single support value would be find, we discovery the rare pattern is item A (glass cleaner) and item I (baking soda). This might mean we use baking soda could help us make the glass or environment cleaner when we are cleaning house. It is the power and strength of the multiple support values.

Although this case analyzes transaction records of daily commodities with lower prices, it is possible to include high unit price items such as electric household appliances or 3C products, since items are often what decision makers are more concerned about. Therefore, this research proposes a model not only sets multiple supports

Table 3. The complete set of frequent patterns

Mapper	Suffix item	Min-sup	TID (New group)	Conditional pattern base	Conditional MISFP-Tree	Frequent patterns
A	J	2	2, 3, 5, 7, 9	{A, B, I :1}, {A, B, C :1}, {C, E, G :1}	{A, B:2}	{B, J :2}, {A, B, J :2}, {A, J :2}
	I	2		{E, G, H :1}, {A, H:1}, {A, B :1}	{A :2}	{A, I :2}
	H	2		{E, G :1}, {A :1}	-	-
	G	3		{E :1}, {C, E :1}	-	-
	E	3		{C :1}	-	-
	C	4		{A, B :1}	-	-
	B	4		{A :2}	-	-
	A	4		-	-	-
B	J	2	2, 4, 7, 8, 10	{C, E, G :1}	-	-
	I	2		{E, G, H :1}	-	-
	H	2		{E, G :1}	-	-
	G	3		{E :1}, {A, B :1}, {C, E :2}	-	-
	E	3		{C :2}, {B, C :1}	-	-
	C	4		{B :1}	-	-
	B	4		{A :1}	-	-
	A	4		-	-	-
C	J	2	1, 3, 4, 5, 6, 7, 8, 10	{A, B, C :1}, {A, B, I :1}, {C, E, G :1}	{A, B :2}	{B, J :2}, {A, B, J :2}, {A, J :2}
	I	2		{A, B :1}	-	-
	H	2		-	-	-
	G	3		{C, E :2}	-	-
	E	3		{C :2}, {B, C :1}	-	-
	C	4		{A, B :2}, {B :1}	-	-
	B	4		{A :5}	{A :5}	{A, B :5}
	A	4		-	-	-

for characteristic of each item but also uses parallel computing with Hadoop MapReduce framework to mine frequent patterns faster and without generating meaningless items.

5   The Hypothesized Experimental Result

In this section, we expect to present and measure experimental results that demonstrate the performance of MISFP-growth algorithm under MapReduce with parallel computing for solving frequent patterns mining of association rules on large datasets. All the experiments were conducted based on a 3-node Hadoop 2.2.0 cluster by Oracle VM VirtualBox environment where one of the node is set to be a master node and the remaining 2 nodes are set to serve as data node in which each has a 3.0 GHz quad core Intel processor with 8 GB memory and 1 TB of storage.

In addition, the data sets were often used in preceding studies or research of frequent patterns mining and were downloaded from website http://archive.ics.uci.edu/ml/. Furthermore, we would test our model to find the infrequent itemsets quickly and compare and analyze it with CFP-growth and MS-FP-Growth algorithm. The hypothesized experimental result would show that the algorithm with parallel computing mining of frequent itemsets takes less time even for huge datasets. Therefore, it would also clearly display the great advantage in execution time over parallel of MISFP-growth algorithm to generate frequent patterns using MapReduce than other methods, especially with a large data volume.

6   Conclusions and Future Research

In this era of information the general challenge is to find ways to utilize information from the huge amount of data accumulated every day. The main focus of data mining research work is to utilize mining frequent patterns understand how to extract useful knowledge from big data effectively. However, discover infrequent association rule is insufficient and unsatisfactory if a single min-sup is set too high or too low that it cannot reflect the real items in the database. It has become obvious that finding rare itemset is one of the most important data mining problems. Besides, high performance implement would require a lot of computation to find patterns from the large data. Thus, the architecture of MapReduce computing has been proposed to reduce the execution time fairly. We proposed to a concept model of solutions set multiple support value for each item and using MapReduce framework to find correlated patterns involving both of frequent and rare items accurately and efficiently. It would not require post pruning and rebuilding phases since each item are either promising more or equal to MIN-MIS.

For future research, we will conduct experiments with this method to construct the MISFP-Tree directly which combines the parallel computing mathematical to better produce increasing numbers of meaning items more quickly during its process. Furthermore, we also plan to compare the execution time of our new scheme with an existing method of using multi-sup threshold, such as CFP-growth and MS-FP-Growth algorithm to prove its effectiveness is significantly better than before.

References

1. Leavitt, N.: Storage challenge: where will all that big data go? Computer **46**(9), 22–25 (2013)
2. Fazio, M., Paone, M., Puliafito, A., Villari, M.: Huge amount of heterogeneous sensed data needs the cloud. In: 9th International Multi-Conference on IEEE Systems, Signals and Devices (SSD), pp. 1–6. IEEE (2012)
3. McAfee, A., Brynjolfsson, E., Davenport, T.H., Patil, D.J., Barton, D.: Big data. The Management Revolution. Harvard Bus. Rev. **90**(10), 61–67 (2012)
4. Aalst, W.V.D., Zhao, J.L., Wang, H.J.: Editorial: business process intelligence: connecting data and processes. ACM Trans. Manag. Inf. Syst. (TMIS) **5**(4), 18e.1–18e.7 (2015)
5. Wu, X., Zhu, X., Wu, G.Q., Ding, W.: Data mining with big data. IEEE Trans. Knowl. Data Eng. **26**(1), 97–107 (2014)
6. Agrawal, R., Srikant, R.: Fast algorithms for mining association rules. In: The 20th VLDB Conference, pp. 487–499 (1994)
7. Agrawal, R., Mannila, H., Srikant, R., Toivonen, H., Verkamo, A.I.: Fast discovery of association rules. Adv. Knowl. Discov. Data Min. **12**(1), 307–328 (1996)
8. Han, J., Pei, J., Yin, Y.: Mining frequent patterns without candidate generation. ACM Sigmod Rec. **29**(2), 1–12 (2000)
9. Hu, Y.H., Chen, Y.L.: Mining association rules with multiple minimum supports: a new mining algorithm and a support tuning mechanism. Decis. Support Syst. **42**(1), 1–24 (2006)
10. Taktak, W., Slimani, Y.: MS-FP-growth: a multi-support version of FP-growth agorithm. Int. J. Hybrid Inf. Technol. **7**(3), 155–166 (2014)
11. Elgaml, E.M., Ibrahim, D.M., Sallam, E.A.: Improved FP-growth Algorithm with Multiple Minimum Supports Using Maximum Constraints. World Academy of Science, Engineering and Technology. International Journal of Computer, Electrical, Automation, Control and Information. Engineering. **9**(5), 1087–1094 (2015)
12. Darrab, S., Ergenç, B.: Frequent pattern mining under multiple support thresholds. WSEAS Trans. Comput. Res. **4**, 1513–2415 (2016)
13. Dean, J., Ghemawat, S.: MapReduce: simplified data processing on large clusters. Commun. ACM **51**(1), 107–113 (2008)
14. Mitra, S., Bande, S., Kudale, S., Kulkarni, A., Deshpande, A.P.L.A.: Efficient FP growth using Hadoop-(improved parallel FP-growth). Int. J. Sci. Res. Publ. **4**(7), 283–285 (2014)
15. Ramakrishnudu, T., Subramanyam, R.B.V.: Mining interesting infrequent itemsets from very large data based on MapReduce framework. Int. J. Intell. Syst. Appl. **7**(7), 44 (2015)
16. Al-Hamodi, A.A., Lu, S., Al-Salhi, Y.E.: An enhanced frequent pattern growth based on MapReduce for mining association rules. Int. J. Data Min. Knowl. Manag. Process (IJDKP) **6**(2), 19–28 (2016)

Balanced k-Means

Chen-Ling Tai and Chen-Shu Wang[✉]

Department of Information and Finance Management,
National Taipei University of Technology, Taipei, Taiwan
{tl04ab8023,wangcs}@ntut.edu.tw

Abstract. K-Means is a very common method of unsupervised learning in data mining. It is introduced by Steinhaus in 1956. As time flies, many other enhanced methods of k-Means have been introduced and applied. One of the significant characteristic of k-Means is randomize. Thus, this paper proposes a balanced k-Means method, which means number of items distributed within clusters are more balanced, provide more equal-sized clusters. Cases those are suitable to apply this method are also discussed, such as Travelling Salesman Problem (TSP). In order to enhance the performance and usability, we are in the process of proposing a learning ability of this method in the future.

Keywords: k-Means · Clusters · Unsupervised learning · Balanced k-Means · Travelling Salesman Problem (TSP)

1 Introduction

Clustering problems are frequently discussed in decades, which they are in a variety of fields, such as pattern recognition, data mining and machine learning. These fields are also very popular in academic research. Basically, the clustering problem is defined as the problem of finding groups of data points or nodes in a given data set. These groups are called a cluster respectively and can be defined as a region in which the density of objects is locally higher than in other regions, stated by Jain et al. (1999).

As Likas et al. (2011) stated, the simplest form of clustering is partitioned clustering which aims at partitioning a given data set into clusters so that specific clustering criteria are optimized. Clustering error criterion is widely used for each point computes its squared distance from the corresponding cluster center and then takes the sum of these distances for all points in the data set.

K-Means algorithm is a common clustering method that minimizes the clustering error and was used in unsupervised learning in data mining. It is introduced by Steinhaus in 1956 and is rather a simple and efficiency algorithm. However, as we all know, the k-Means algorithm has a fatal weakness which is its performance heavily depends on the initial starting conditions and it is randomize indeed.

As time flies, many other enhanced methods of k-Means have been introduced and applied. This paper is to introduce a more balanced k-Means method. It is suitable to be applied into many cases, such as Travelling Salesman Problem (TSP) and Vehicle Routing Problem (VRP).

© Springer International Publishing AG 2017
N.T. Nguyen et al. (Eds.): ACIIDS 2017, Part II, LNAI 10192, pp. 75–82, 2017.
DOI: 10.1007/978-3-319-54430-4_8

The main function of Travelling Salesman Problem is to solve a question that, 'Given a list of cities and the distances between each pair of cities, what is the shortest possible route that visits each city exactly once and returns to the origin city?' It is an NP-hard problem in combinatorial optimization, important in operations research and theoretical computer science.[1] In another word, TSP is all about a salesman starts travel from a starting point, visiting all known demand points (places with customers or potential customers) by selling products or other promoting purpose, then return to the starting point. Mostly known methods are to solve TSP with a minimum cost, including minimizing time cost and with least effort.

On the other hand, salesmen's salary mostly is based on the number of customers are served and sales volume are achieved. The greater number of customers are served and sales volume are achieved as well as sales targets are hit, the better pay they get and vice versa. Imbalance volume of customers to serve is bringing to a problem of imbalance salary of each salesman as well.

As TSP is an NP-hard problem which is hard to be solved and requires a long period of time. Method that applied in this paper is kind of divide and conquer. We divide the problem into 2 parts, which is allocate orders based on the constrains into cluster first, then provide the routing based on time windows and other constraints. In this paper, we focus our research on the former but not the latter. Hence, we are discussing about applying a balanced k-Means method to a clustering analysis.

Therefore, TSP is a problem which needs a more balanced clusters allocation. While traditional k-Means results are produced through randomize method, thus TSP is suitable to apply a balanced k-Means method.

In the future, we are focusing on enhancing k-Means performance, which is by adding machine learning mechanism. By adding machine learning into k-Means, it will learn every time when the items in the clusters are allocated. Thus, the performance will improve time by time. Eventually, the algorithm will be enhanced and produce better performance.

2   Literature Review

2.1   Clustering

Jain (2010) stated that cluster analysis is the formal study of methods and algorithms for grouping or clustering, objects according to measured or perceived intrinsic characteristics or similarity. Cluster analysis does not use category labels that tag objects with prior identifiers, such as class labels. The absence of category information distinguishes data clustering (unsupervised learning) from classification or discriminant analysis (supervised learning). The aim of clustering is to find structure in data and is therefore exploratory in nature.

---

[1] Travelling salesman problem, https://en.wikipedia.org/wiki/Travelling_salesman_problem.

2.2   K-Means

Jain (2010) stated that K-Means has a rich and diverse history as it was independently discovered in different scientific fields by Steinhaus (1956), Lloyd (proposed in 1957, published in 1982), Ball and Hall (1965), and MacQueen (1967). Even though K-Means was first proposed over 50 years ago, it is still one of the most widely used algorithms for clustering. Ease of implementation, simplicity, efficiency, and empirical success are the main reasons for its popularity.

Minimizing this objective function is known to be an NP-hard problem (Drineas et al. 1999). Thus K-Means is a greedy algorithm that can only converge to a local minimum, although recent study has shown with a large probability K-Means could converge to the global optimum when clusters are well separated (Meila 2006). K-Means starts with an initial partition with K clusters and assign patterns to clusters so as to reduce the squared error. Since the squared error always decreases with the increasing number of clusters K, it can only be minimized for a fixed number of clusters.

The main steps of K-Means algorithm are as follows (Jain and Dubes 1988):

1. Select an initial partition with K clusters; repeat steps 2 and 3 until cluster membership stabilizes.
2. Generate a new partition by assigning each pattern to its closest cluster center.
3. Compute new cluster centers.

2.3   Balanced k-Means

He et al. (2009) stated that k-Means is used to partition the whole customers into several areas in the first stage and a border adjustment algorithm aims to adjust the unbalanced areas to be balanced in the second stage. The objective of partitioning areas is to design a group of geographically closed customers with balanced number of customers. The presented algorithm is specifically designed for large-scale problems.

3   Research Framework

In this study, we are proposing an enhanced k-Means method, which is a balanced k-Means method.

Basically, TSP is taking account into these conditions:

1. Salesmen are supposed to start the journey from the center.
2. Salesmen are supposed to visit all customers who are assigned to them.
3. The next customer to visit is supposed to be the nearest to the current customer.
4. Each customer is supposed to be visited once and by one salesman only.
5. The arrangement should be cost effective.
6. Salesmen are supposed to return to the center after visiting the final customers assigned to them.

In this study, in order to achieve our goal of proposing a balanced k-Means method, we included some constraints. The process starts by data input, which can be a hundred transactions including the following data:

Time Window

Customers are not likely to be at home all the time and not anytime is suitable to visit. Moreover, there are some specific period of time that customers are more likely to spend money, such as when they are free and in a good mood. Therefore, there must be a suitable time window to visit the customers.

Minimum Allowance

This mechanism is to ensure all salesmen are able to hit the sales target set by the company policies, so that they will at least earn the minimum allowance, for their sake.

Salesmen's Ability

1. For salesmen who are able to serve more customers and achieve better sales target to gain more profit, more customers will be assigned to them.
2. For salesmen who are still fresh and lack of experience, they will rather be assigned to serve fewer customers.

After considering all the constraints, we use Sum of Squared Error (SSE) to evaluate the performance of the nodes allocation into clusters. SSE with smaller value represents the distance among the nodes inside the clusters is closer, which will produce a better result while the greater SSE value meaning the result is worse.

In this experiment, we are seeking a clustering results with a smaller SSE value. We set the iteration to 10000 times to observe its performance. If the SSE value is greater, then the next step will be going back to the nodes clustering process, repeat the previous steps until a smaller SSE value is found, leading it to a convergence step. This process keeps repeating until it obtains a smaller SSE value, will then allocate these nodes to form a cluster, then lead it to the end of the process.

With the method we proposed, the research framework is shown as Fig. 1.

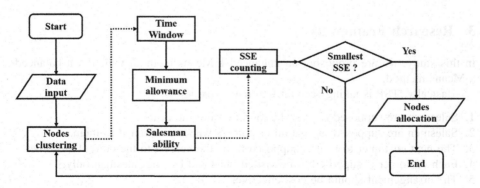

Fig. 1. Research framework

4   Results Analysis

With the proposed balanced k-Means method, we verify it through applying it to a test case, which is a Travelling Salesman Problem in this research.

Experimental schematic is shown in Fig. 2.

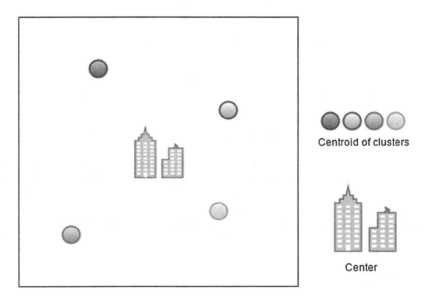

Fig. 2. Experimental schematic with centroid of clusters and center. (Color figure online)

The data points are allocated into 4 clusters, which are represented by red, black, blue and green circles. While applying to TSP, data points represent customers to be served, and clusters are travelling salesman. In other words, there are 4 salesmen (clusters) and every salesman must visit all the customers (data points) those assigned to them once. Basically, the next customer to visit is the one nearest to the current or previous customer. They will start the journey from the center (starting point which is the centroid of a cluster) and finish visiting all the customers then return to the center after completing the visit.

In this research, we apply Travelling Salesman Problem (TSP) to verify the algorithm. By applying balanced k-Means into TSP, number of items in every clusters are more equal if compare to clusters without applying balanced k-Means method. It means every travelling salesman will serve more equal number of customers than before. Problems of some salesmen are flooded with tones of appointments (customers to serve) while others with very little appointments will not happen by applying this balanced k-Means method.

Moreover, every travelling salesman is able to hit the sales target yet at least to gain the minimum sales allowance. On the other side, for well-experienced salesmen or with better sales performance, if they are able to finish visiting the assigned customers faster or within a shorter period of time, then return to the center (starting point), they might be assigned with another round of customer visiting appointment.

After completing balanced k-Means clusters allocation process, routing sequence will be arranged by a routing method. Travelling salesmen will follow the sequence arranged by the routing application to visit customers.

Parameter setting for experiment is shown in Table 1.

Table 1. Parameter setting for experiment.

Information	Parameters
Total customers to serve	100
Number of salesmen	4
Ability of each salesman to serve customers in a roundtrip	26, 27, 27, 28
Minimum allowance	1000

After taking account into constraints set to the algorithm, the process starts with finding the closest nodes with each other to form into a cluster. Besides nodes being allocated into clusters according to smaller SSE value, the clusters will stop the allocation process when it tends to be converge.

As a result, nodes are allocated into 4 clusters and the result is shown in Fig. 3.

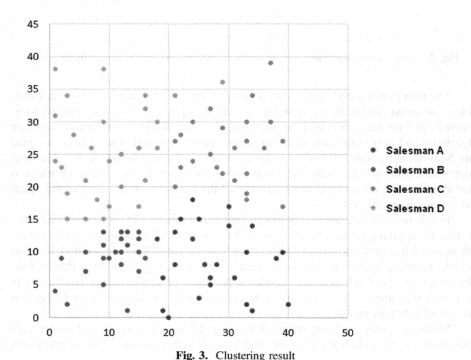

Fig. 3. Clustering result

The main object of the proposed balanced k-Means is to provide a more equal cluster and a better performance. The result is shown in Fig. 4 and it is moving towards to a more balanced result after thousands of iterations when it tends to converge.

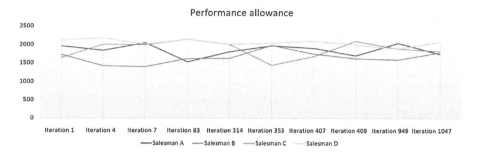

Fig. 4. Performance allowance of salesmen tends to be more equal

As a result, by applying a balanced k-Means algorithm to a Travelling Salesman Problem (TSP), salesmen will get more equal performance allowance among themselves.

5 Conclusions and Future Research

This research is to propose a balanced k-Means method that is suitable in applying to some cases. Case that we apply is Travelling Salesman Problem (TSP). By applying balanced k-Means into TSP, number of items in every clusters are more equal if compare with clusters without a balanced k-Means method. It means every travelling salesman will serve more equal number of customers than before. Moreover, every travelling salesman is able to hit the sales target yet at least to gain the minimal sales allowance.

In order to achieve the result, we include some constraint into this algorithm. Hence, the balanced k-Means function is not limited to find the closest nodes to form into clusters only.

After completing balanced k-Means clusters allocation, routing sequence will be arranged by a routing method. Travelling salesmen will follow the sequence arranged by the routing application to serve customers. Since this research is focus on the former, hence, thus, we are not discussing the latter in the meantime. The clustering result produces a solution that is with a minimal cost, effective and equal to every single travelling salesman.

In the future, we are planning to focus our research on applying Machine Learning to the balanced k-Means that we proposed. Meaning that we aim to provide a solution which the algorithm will perform self-learning mechanism every-time allocating items into clusters. After a period of learning process, or the more time the algorithm operates, it will become smarter and provide better solutions of clustering, in the terms of smaller SSE which leading to a better performance.

References

Likas, A., Vlassis, N., Verbeek, J.: The global k-Means clustering algorithm. Technical report 12 (2011)

Ball, G., Hall, D.: ISODATA, a novel method of data analysis and pattern classification. Technical report NTIS AD 699616. Stanford Research Institute, Stanford, CA (1965)

Drineas, P., Frieze, A., Kannan, R., Vempala, S., Vinay, V.: Clustering large graphs via the singular value decomposition. Mach. Learn. **56**(1–3), 9–33 (1999)

He, R., Xu, W., Sun, J., Zu, B.: Balanced k-Means algorithm for partitioning areas in large-scale vehicle routing problem. In: Third International Symposium on IEEE Intelligent Information Technology Application, IITA 2009, vol. 3, pp. 87–90 (2009)

Jain, A.K., Murty, M.N., Flynn, P.J.: Data clustering: a review. ACM Comput. Surv. (CSUR) **31** (3), 264–323 (1999)

Jain, A.K., Dubes, R.C.: Algorithms for Clustering Data. Prentice Hall, Upper Saddle River (1988)

Jain, A.K.: Data clustering: 50 years beyond k-Means. Pattern Recogn. Lett. **31**(8), 651–666 (2010)

Lloyd, S.: Least squares quantization in PCM. IEEE Trans. Inform. Theory **28**, 129–137 (1982)

MacQueen, J.: Some methods for classification and analysis of multivariate observations. In: Fifth Berkeley Symposium on Mathematics, Statistics and Probability, pp. 281–297. University of California Press (1967)

Meila, M.: The uniqueness of a good optimum for k-Means. In: Proceedings of 23rd International Conference Machine Learning, pp. 625–632 (2006)

Steinhaus, H.: Sur La Division des Corp Materiels En Parties. Bull. Acad. Polon. Sci. **IV**(C1.III), 801–804 (1956)

An Optimal Farmland Allocation E-Service Deployment

Wei-Feng Tung[✉] and Chun-Liang Pan

Department of Information Management,
Fu-Jen Catholic University, Taipei, Taiwan
076144@mail.fjuf.eud.tw, beggarandpoor@gmail.com

Abstract. This research develops an intelligent farmland allocation e-service. The objective of the research is to provide an effective production and cooperative model between peasants and social enterprises. Since the production and marketing manager needs to visit organic-certified farmlands and to offer advices, and help to decide what types of agricultural commodities will be planted. Social enterprises offer their peasants plant-to-order. Thus, the research creates an intelligent farmland allocation e-service for organic produce, and users also adopt an optimization model which is capable of searching for the best combination of agricultural commodities based on the predefined lowest planting costs. The research also can be viewed as an automatic recommendation for peasants and certain production and marketing managers of social enterprises.

In traditional, a company case of certain social enterprise uses the manual allocating and worksheet file management. The research introduces the innovative service for social enterprise and peasants to improve the pervious service process. The social enterprise can create a new way to collaborate and improve productivity. In order to embody the agricultural innovation, the peasants and social enterprise can achieve the value co-creation of business and social values. Even the proposed service can be used to support and land friendly and sustainable development of safe agricultural commodities.

Keywords: Farmland allocation · Genetic algorithm · Agricultural innovation · Social enterprise

1 Introduction

People think that agriculture is of labor intensive industry with minor machine automation. With the technology development, information technology is gradually applied to agriculture. Farmers learn to use smart phone to record growing progress or contact production managers to report production status. This indicates that IT has positive impact on traditional agriculture and brings convenience in framing work.

The case company provides job opportunity of organic agriculture to aborigines. Aborigines cultivate and sell organic crops for living. Peasants who contract with the case company have monthly meetings from history data as reference to allocate and arrange crop items artificially. This research intends to design an information system to improve the communication between farmers and production managers to save time

© Springer International Publishing AG 2017
N.T. Nguyen et al. (Eds.): ACIIDS 2017, Part II, LNAI 10192, pp. 83–90, 2017.
DOI: 10.1007/978-3-319-54430-4_9

and relevant costs. The main purpose is to break up traditional limitation, to innovate service process, and to increase co-creation value.

The case company is an organic agriculture social enterprise. Creating job opportunity for Taiwanese aborigines, promoting organic agriculture, improving tribute education and developing tribute agriculture are the funders' philosophy of the case company. The case company rents farmlands and arranges aborigines for organic agriculture in tributes. This action creates more job opportunities. Meanwhile, by offering training and consulting, the case company helps tribute aborigines grow crops with better planning and technology.

According to user's (peasant and production manager) requirement, the system which this research develops can calculate the related data and propose the recommending table of growing combination with minimum planting costs. This table is called growing allocation table. The table provides important information to production managers and peasants while making decision on what to grow in the coming season.

Originally, the communication between production manager and peasants in the case company is by routine oral meeting. However, with the implementation of information technology, an innovative collaboration model is built whenever production manager and peasant plan the new growing schedule. It improves work efficiency and reaches the purpose of service innovation.

2   Literature Review

2.1   Relative Information Platform of Organic Agriculture

Bio@gro. Karetsos et al. [1] proposed Bio@ago as a service platform primarily designed for organic agriculture. This platform contacts and links every stockholder in organic agriculture industry for communication through computer or cell phone network. This platform even provides the functions of e-commerce negotiation and digital learning, such as cultivate methods, skills, experience and person opinion sharing.

CerOrganic. Thanopoulos et al. [2] design CerOrganic, a network platform, for people who are engaged in organic agriculture industry to access relative skill and knowledge through internet any time and any place.

2.2   Management and Planning of Agriculture

Land Utilization for Agriculture. Population increase rapidly as new technology invented. Today, more people live in the same land area. So, maximum planning and utilization of land for least waste becomes important. The purpose of land utilization for agriculture is minimum cost and best economic benefit as well as thinking of environmental protection.

Carsjens and Van Der Knaap [3] use Geographical Information System (GIS) to collect geographic information to solve the problem of farmland utilization. Manos et al. [4] develop a framework to combine economic, environmental and even groundwater pollution factors as parameters to utilize land optimally. The framework provides not only optimal land utilization but also the measurement index of economic benefit.

Production Planning of Crop Species. Production planning problem is about, under limited farmland area, how many crop species and volumes to produce in very season for farmer's best benefit in every season from the annual viewpoint. People who work in agriculture industry would be very interested in this kind of seasonal allocation problem of crop production.

Biswas and Pal [5] utilize the multi-goal programming proposed on practical agriculture allocation. According to practical requirement, users can decide annual goals for their preference. The goals are not limited to financial benefit. Other forms of goal such as cost expense and land utilization are also applied. This research successfully provides a further solution for crop allocation problem.

3 Methodology

The case company provides training sessions of organic agriculture. Helping aborigines grow organic agriculture crops is another approach of employment. According the distributor's requirement, the case company allocates required crop species to every aborigine peasants to satisfy the requirement. Traditionally, production manager and peasant have meetings to communicate and discuss the new requirement according to historical data, then to allocate crops species and volume.

To the traditional allocation method, this research develops an intelligent crop allocation system for it. The system saves the history data of all popular crops and peasant capability in the case company as the reference data base. The genetic algorithm of artificial intelligence is applied to calculate the crop allocation table with minimum cost in every contracted farmland. After the table is generated, the system presents the result on the user interface as the main and important reference.

3.1 Research Framework

Figure 1 shows the research framework. The following is the description of main characters in the framework.

Farmer and production manager: main users of the system in this research.

User platform: when production manager conducts meeting with farmers, user platform can be applied to access the intelligent allocation system. The platform also displays the recommended allocation table which calculated by the system according to user's requirement. Later on, the recommended allocation table can be treated as basic information for both production manager and farmer to decide what species to grow in the coming season.

Intelligent crop allocation system: It is the system this research develops. According to user requirement, the system utilizes artificial intelligence algorithm, generic algorithm, with user data to calculate and generate recommended crop allocation table.

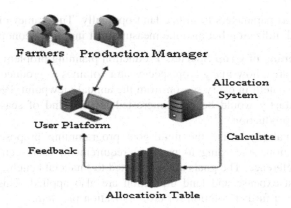

Fig. 1. Research framework.

3.2 Module Development

This system takes planting costs and harvest ratio as input unit, then apply generic algorithm to calculate the allocation table with minimum planting cost in the specific farmland. The design concept of the allocation table is that farmland area is the measurement unit instead of crop species.

Figure 2 shows the three representative steps of the system implementation. The following sections offer detail description.

Fig. 2. Three steps of the system implementation

3.2.1 Service Modeling

To illustrate the ratio of crop species, the system turns farmland area into several cube units and defines the unit as N (N = 1, 2… N). Meanwhile, this system assumes that there are K kinds of crop species (K = 1, 2… K). Every cube unit must contain one kind of species from K. The species cannot be repeated. The cube unit cannot be null. On the other hand, every cube unit of total N is allocated to one and only one species of K.

Nk is defined as the total number of cube units which are allocated with K species of crops. Under the allocation rule, the maximum of Nk equals to the total number of cube units.

Since every cube unit is allocated with one crop species, there is corresponding planting cost to every species. The cost type is defined as P (P = 1, 2). Two cost types are defined in the research. P = 1 represents seed-buying cost. P = 2 represents fertilizer cost. These two cost types have different influence values. The influence value is defined as αkp.

Under certain cost factor (p = 1 or p = 2), the ratio of one plating cost over highest planting cost is defined as the influence value (αkp) of different cost factor (P). Crop planting cost id defines as Ckp. Equation 1 represents the relationship.

$$\alpha kp = Ckp/\max \{Ckp\} \quad 0 < \alpha kp < 1 \tag{1}$$

With the influence value (αkp), the Eq. 2 shows the function.

$$fp(u) = \sum_{k=1}^{K} \alpha kp \ Nk \quad \text{for} \ p = 1, \ldots, P \tag{2}$$

According to the corresponding harvest ratio from every peasant's crop, we define Wk as non-harvest ratio (1 – harvest ratio) to measure the peasant's planting capability of every crop. We get another Wk Eq. 3 while Wk is put into Eq. 2.

$$gp \ (u) = \sum_{k=1}^{K} Wk \ \alpha kp \ Nk \quad \text{or} \ 0 < WK < 1 \tag{3}$$

When $0 < Wk < 1$, higher harvest ratio means lower Wk. The low Wk makes the result of cost equation lower. This means that planting cost will be reduced to the corresponding peasant. Wk is defined as 0.1 if the peasant's historical harvest ratio is equal or bigger than 0.9.

Different cost factor has different result of gp(u). The maximum value of gp(u) is defined as Y Eq. 4 represents the condition.

$$\max \{ gp(u) \} = Y \quad \text{for all} \ p = 1, \ldots, P \tag{4}$$

According the linear function proposed by Wierzbicki (1999), we can simplify the above equations into Eq. 5. This equation is applied to measure the planting cost of allocation Table

$$F(u) = \sum_{p=1}^{P} \left[\frac{gp(u)}{Y} \right]^4 \tag{5}$$

$\frac{\max \{ gp(u) \}}{Y}$ of F(u) equals to 1. $\frac{gp(u)}{Y}$ is smaller than 1. Hence, cost equation has the following relationship as shown in Eq. 6. Table 1 is the definition of parameter and equation.

$$F(u) \geq 1 \tag{6}$$

3.3   Optimization Algorithm

This research uses genetic algorithm. Figure 3 shows the process flow. The following is the detail description of each process.

Table 1. Definition of parameter and equation

N	Total land cube volumes
K	Corp species
N_k	Land cube volumes which are assigned crop species k
P	Cost factor
α_{kp}	Influence value
C_{kp}	Cost of crops
$f_p(u)$	Basic function
W_k	Non-harvest ratio
$g_p(u)$	Function which includes non-harvest ratio (W_k)
Y	Maximum number of cost function $g_p(u)$
$F(u)$	Cost function

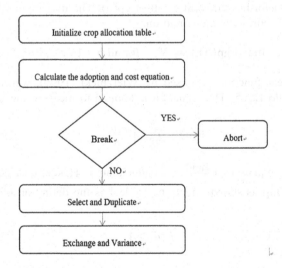

Fig. 3. The process flow of genetic algorithm

4 Develop and Test

Because the system is still under testing stage, peasant-related practical data such as farmland area, seed cost, and planting cost are simulated. The data has no practical unit yet. The algorithm uses formula ($\alpha_{kp} = C_{kp}/\max \{C_{kp}\}$) to transfer seed and fertilizer costs into ratio format. Therefore, simulation failure may not happen in the testing stage. The system can apply to practical world after cost calculation method is defended.

The system is under testing stage. The allocation table is set up with 50 cube units only. From data the case company provides, the crop species (K) that peasant can select is set up as 17. The following data are required to input by user: land area, crop species, crop volume, seed cost, fertilizer cost, and harvest ratio. This system uses random number to simulate the data from the case company for system testing. Table 2 show the input data of parameters.

This research inputs the above data which is randomly generated into testing system. The system will calculate the recommended result with minimum cost of allocation Table 2. Table 3 shows the recommended result of allocation table which is under 500 farmland cube units.

Table 2. Input data of variables

Item	Seed cost	Corp cost	Harvest ratio (%)
Green bean	73	61	80
Plum	82	59	90
Cucumber	77	63	85
Makino bamboo shoot	86	54	80
Beef tomato	71	68	83

Table 3. The allocation result under 500 farmland area units

Item	Ratio	Area
Green bean	18%	90
Plum	22%	110
Cucumber	22%	110
Makino bamboo shoot	26%	130
Beef tomato	20%	100

The crop allocation table recommended by the system will be offered to peasants and the production manager of the case company for the planning reference before planting.

5   Conclusion

This research, through optimal solution model of generic algorithm, develops an intelligent allocation service to calculate minimum planting costs in social enterprise environment such as the case company lives in. The system takes the crop harvest ratio as reference and combines the seed and fertilizer costs as key parameters to calculate the best allocation table with the minimum planting cost in farmland. The recommended result of allocation table will be provided to the production manager of the case company and peasants to make decision while planting.

The communication between peasant and production manager of the case company was traditionally manual and oral. Production manager has to visit farmland regularly and contact peasants personally. By implementing the service of mobile APP and network platform, this research intends to build up an innovative collaboration mode for the case company. By using the service innovation, work efficiency can be improved for both parties and to co-create business and social value.

This research intends to help peasants find out crops with high harvest ratio and low planting cost when planting organic agriculture through intelligent allocation system. It makes peasants sense that organic agriculture is not just only for living but also for better work achievement and self-recognition. Meanwhile, farmland can be utilized continuously with good sustainability. Peasants also make contribution to nature ecology.

References

1. Karetsos, S., Costopoulou, C., Sideridis, A., Patrikakis, C., Koukouli, M.: Bio@gro – an online multilingual organic agriculture e-services platform. Inf. Serv. Use 27, 123–132 (2007)
2. Thanopoulos, C., Protonotarios, V., Stoitsis, G.: Online web portal of competence-based training opportunities for organic agriculture. Agris Online Pap. Econ. Inform. 4(1), 49 (2012)
3. Carsjens, G.J., Van Der Knaap, W.: Strategic land-use allocation: dealing with spatial relationships and fragmentation of agriculture. Landscape Urban Plan. 58, 171–179 (2002)
4. Manos, B., Papathanasiou, J., Bournaris, T., Voudouris, K.: A multicriteria model for planning agricultural regions within a context of groundwater rational management. J. Environ. Manag. 91, 1593–1600 (2010)
5. Biswas, A., Pal, B.B.: Application of fuzzy goal programming technique to land use planning in agricultural system. Omega 33, 391–398 (2005)

Automated Reasoning and Proving Techniques with Applications in Intelligent Systems

Anticipatory Runway Incursion Prevention Based on Inaccurate Position Surveillance Information

Kai Shi[1], Hai Yu[1(✉)], Zhiliang Zhu[1], and Jingde Cheng[2]

[1] Software College, Northeastern University, Shenyang 110819, China
{shik,yuh,zhuzl}@swc.neu.edu.cn
[2] Department of Information and Computer Sciences,
Saitama University, Saitama 338-8570, Japan
cheng@ics.saitama-u.ac.jp

Abstract. To build a practical anticipatory runway incursion prevention system (ARIPS), it is necessary to predict runway incursions based on inaccurate position information of aircraft and vehicles. To this end, this paper proposes a series of improved methods to predict runway incursions based on inaccurate position surveillance information for ARIPSs. The evaluation shows that our system can handle different types of runway incursions based on inaccurate position information, deal with the momentary absence of surveillance data, and produce few false detection under non-runway incursion circumstances.

Keywords: Runway incursion prevention · Logic-based reasoning · Anticipatory computing · Anticipatory reasoning-reacting system

1 Introduction

An anticipatory runway incursion prevention system (ARIPS) cannot only predict and detect runway incursions (RI), but also give explicit instructions and/or suggestions to pilots/drivers to avoid the RI and collision beforehand [1]. As a new type of RI prevention systems, ARIPSs aim to remove human factors from the system operation loop as much as possible [1], because all causes leading to RIs belong to human factors [2]. Moreover, ARIPSs could give long-range prediction, and can be customized for different airports and different air traffic control policies by using flexible model [1].

In previous studies of ARIPSs, it was supposed that airport surveillance systems could provide accurate position information of aircraft/vehicles, moreover,

This work is supported by the National Natural Science Foundation of China (Grant Nos. 61374178, 61402092), the online education research fund of MOE research center for online education, China (Qtone education, Grant No. 2016ZD306), the Ph.D. Start-up Foundation of Liaoning Province, China (Grant Nos. 201601010, 201501141), and the Fundamental Research Funds for Central Universities (Grant No. N141703001).

N.T. Nguyen et al. (Eds.): ACIIDS 2017, Part II, LNAI 10192, pp. 93–104, 2017.
DOI: 10.1007/978-3-319-54430-4_10

the evaluation of original ARIPSs was based on this presupposition [1]. Because the previous work did not take inaccurate position data into account, the previous work did not show whether ARIPSs could work based on inaccurate position information, and did not show the performance of ARIPSs under the circumstances.

In practice, the position data derived from surveillance systems contain unavoidable amount of error, i.e., the position data at every second contains several meters of random error. For example, EUROCAE ED-117 specifies minimum operational performance of multilateration surveillance system, the accuracy of position, as 7.5 m on runways and taxiways [3]. Schönefeld and Möller reviewed the worst experimental accuracy is SMR(i), 15 m (90%) [2]. Miyazaki et al. showed the performance of MLAT was: in runway area, the accuracy was 6.5 m, and detection rate was 100%; in taxiway area near runway, the accuracy was 6.8 m, and detection rate was 100% [4]. In addition to position errors, current surveillance systems have other disadvantages for predicting/detecting RIs. First, surveillance data may momentarily be absent [5]. Second, different surveillance systems measure different position of the aircraft: SMR measures the position of the centroid of the aircraft; MLAT measures the position of the transponder antenna of the aircraft. Third, the position data contains unavoidable latency due to surveillance sensor intercommunication, processing, data fusion, etc. Fourth, the measured speed and track angle also contains avoidable errors [5].

This paper focuses on RI prediction based on inaccurate position information and shows that, when the runway's width is greater than or equal to 45 m, ARIPSs could work under the conditions of inaccurate position information. First, after we investigated the performance of original ARIPSs in different types of RIs, we found that, if the runway's width is greater than or equal to 45 m, then only for certain types of RIs, inaccurate position information could affect the prediction results. Second, for these affected types of RIs, we proposed improved filtering methods for ARIPSs to eliminate the influence of inaccurate position information on qualitative position information; we also consider the momentary absence of surveillance data and the difference between the aircraft's nose position and measured position, and the position errors exceeding the position accuracy. Third, to evaluate the correctness and performance of ARIPSs based on inaccurate position information, we evaluated our system by using both real historical scenarios and conventional fictional scenarios. The evaluation showed that our system could work based on inaccurate position information, i.e., to predict different types RIs stably, to give explicit instructions and/or suggestions for handling the incidents effectively, to deal with the momentary absence of surveillance data, and to produce few false detection under non-RI circumstances.

2   Analysis of Original ARIPSs Based on Inaccurate Position Surveillance Information

The ARIPS uses qualitative position information to predict and make decision. The qualitative position information is only about the relation between aircraft/vehicle and region, i.e., "which aircraft/vehicle" locates "which region".

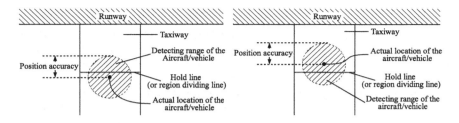

Fig. 1. Detecting aircraft/vehicles performance at taxiway near hold line

The key issues are whether the inaccurate numeric position information could affect the qualitative position information, and which types of qualitative position information could be affected.

When the runway's width is greater than or equal to 45 m, the inaccurate numeric position information could not affect the qualitative position information on runways. The common width of runway is 18 m, 23 m, 30 m, 45 m, or 60 m [6]. In most international airports, the width of runway is 45 m or 60 m; big aircraft such 747-400 and 747-8 require 45 m wide runway for safety [7]; in the airports equipped with ASDE-X or MLAT, the width of runway is 45 m or 60 m. Therefore, in this research, we only consider the airports with runways whose width is greater than or equal to 45 m. Even if the locating accuracy is the worst experimental accuracy, 15 m, if an aircraft is at a runway whose width is greater than or equal to 45 m, then the fact "the aircraft is at the runway" could not be affected by inaccuracy.

If we suppose that only airport position surveillance information is inaccurate, and the width of runway is greater than or equal to 45 m, then only the prediction about RI scenarios A/T (cr) and D/T (cr), could be affected by inaccurate position information in ARIPSs. For RI scenario A/D (tc), A/D (cr), D/D (tc), A/A (tc), D/D (cr), and A/A (cr), if we consider only the airport position information is inaccurate and the width of runway is greater than or equal to 45 m, then inaccurate position information cannot affect the qualitative position information, thus cannot affect the prediction of incursions in ARIPSs.

Inaccurate position information or random position errors of taxiing aircraft/vehicles could result in incorrect qualitative position information of these taxiing aircraft/vehicles. In ARIPSs, in order to predict RI for scenarios A/T (cr) and D/T (cr), we need to know the qualitative information about whether the aircraft/vehicle has passed the runway hold line or not-if there is no hold line at the runway's adjacent region, we consider the region dividing line of two adjacent regions of runway as the hold line, shown in Fig. 1. In Fig. 1, the left picture shows that the aircraft does not cross the hold line, i.e., the aircraft/vehicle is at the runway's non-adjacent region. However, due to the inaccurate detecting position or random position errors, the qualitative information of the aircraft/vehicle may be "the aircraft/vehicle is at the runway's adjacent region". Similarly, in Fig. 1, the right picture shows the aircraft/vehicle has passed the hold line, i.e., the aircraft is at the runway's adjacent region. Due to the inaccurate detecting position, the qualitative information of the aircraft/vehicle may be "the aircraft/vehicle is at the runway's non-adjacent region". Moreover, because the inaccurate detecting position or random position errors might be different for each detection, for both scenarios in Fig. 1, the detecting position could frequently shift between the runway's adjacent region and its non-adjacent region. Furthermore, for aircraft, the position measured by surveillance systems is not the position of the aircraft nose. SMR measures the position of the centroid of the aircraft; MLAT measures the position of the transponder antenna of the aircraft. Using the measured position in stead of the aircraft nose's position produces delay to determine whether the aircraft has passed the hold line to enter the runway or not. In summary, the original ARIPSs could not handle the detecting position of taxiing aircraft/vehicles, thus could not work under A/T (cr) and D/T (cr).

3   Improved Methods for Runway Incursions A/T (cr) and D/T (cr)

In order to predict RI A/T (cr) and D/T (cr) based on inaccurate position information in ARIPSs, we propose some improved filtering methods to eliminate the influence of the inaccurate position information, as well as to handle the difference between the aircraft's nose position and the measured position, the random position errors exceeding the position accuracy, and the momentary absence of surveillance data.

First, to handle the those position error which does not exceed the position accuracy, the filtering method must ensure the following requirements.

R1 If an aircraft/a vehicle does not cross the hold line, i.e., the aircraft/vehicle is at the runway's non-adjacent region, then the filter should not detect that the aircraft/vehicle has passed through the hold line, i.e., the aircraft/vehicle is at the runway's adjacent region.

R2 If an aircraft/a vehicle has passed through the hold line, i.e., the aircraft/vehicle is at the runway's adjacent region, then the filter should not detect the aircraft/vehicle is at the runway's non-adjacent region.

To ensure requirement **R1**, we decrease the size of adjacent region represented in ARIPSs. In Fig. 2, the left picture shows the actual scope of an adjacent region, which is a section of taxi way between runway and hold line (or region dividing line). In order to ensure that when an aircraft/a vehicle at the runway's non-adjacent region should not be detected at the runway's adjacent region, we reserve a blank area, which is next to the hold line, as well as between the runway and the hold line. The blank area's width is the runway's width, while its length is the position accuracy of surveillance system. Then we remove this blank area from the runway's adjacent region represented in ARIPSs. Therefore, for any aircraft/vehicle at runway's non-adjacent region, e.g., an aircraft waiting at hold line, could not be detected at runway's adjacent region in ARIPSs. However,

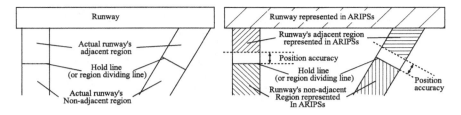

Fig. 2. Decreasing the size of adjacent region represented in ARIPSs

the drawback of this method is that when an aircraft/a vehicle has really crossed the hold line, we might not detect that immediately, thereby consuming more time to predict the RI.

To ensure requirement **R2**, we improve the position storage mechanism of the filter. For original ARIPSs, after an aircraft/a vehicle passes through the hold line, i.e., is at the blank area, the aircraft/vehicle might be detected at either the runway's non-adjacent region or adjacent region because of the inaccurate position information. In other words, it violates the requirement that an aircraft/a vehicle at the runway's adjacent region should not be detected at the runway's non-adjacent region. In original ARIPSs, if the detecting qualitative location of an aircraft/a vehicle changes, the *PastLocation* and *CurrentLocation* will update at once. By contrast, our improved method is to compare whether the current detecting location is as same as *PastLocation*'s locating region before updating. If they are same, it illustrates that the aircraft/vehicle has just come from that region, thus we ignore its current detecting location. Therefore, for any aircraft/vehicle has just passed through the hold line, i.e., it just came from runway's non-adjacent region, that aircraft/vehicle cannot be located at the runway's non-adjacent region in ARIPSs. Because the aircraft cannot draw back by itself, it is impossible for an aircraft, which has just passed through hold line, to draw back to runway's non-adjacent region. If a vehicle just passes through hold line then draws back to runway's non-adjacent region, then the ARIPS might not represent that the vehicle has drawn back immediately. Consequently, this might cause a prediction of RI, because the vehicle is potential to cause a collision before it draws back. However, after the vehicle goes to other regions, the ARIPS will represent its new current location normally.

Second, in order to eliminate the difference between the measured position of the aircraft and the nose of the aircraft, we propose a filtering method to get possible locations of the nose. Because the measured direction of aircraft also contains error [5], the proposed filtering method calculates the possible locations of the nose only based on the measured location and the offset instead of the direction of aircraft. The offset is the distance between the nose and the measured position. The measured position, i.e., the centroid or the transponder antenna of the aircraft, is determined by the type of the surveillance system; the offset between the nose and the centroid, or the offset between the nose and the transponder antenna, is determined by the aircraft model. The ARIPS

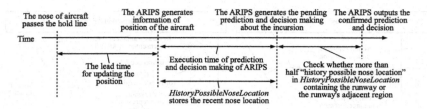

Fig. 3. Confirmation procedure to handle the position errors exceeding the accuracy

could obtain the aircraft model from the traffic information system [1]. Thus, the possible location of nose is located on a circle, whose center is the measure location, and radius is the offset. In the locating module, we add two data hash maps: $PossibleNoseLocation<aircraft\ identifier,\ Set<region>>$ stores the current possible locations of the nose; $HistoryPossibleNoseLocation<aircraft\ identifier,\ Queue<Set<region>>>$ stores several groups of possible locations of the nose in the last few seconds (5 s in default). Because the aircraft cannot turn round by itself, the possible locations of the nose exclude the region which the aircraft has just passed. The past location of the aircraft is stored in the hash map $PastLocation$ [1].

Third, we add the knowledge to predict RI A/T (cr) and D/T (cr) based on the locations of the nose, and propose a confirmation method to handle the position errors exceeding the accuracy. Because the position errors could exceed the accuracy, it is still possible to falsely detect an taxiing aircraft or a vehicle has passed the hold line, which is actually staying at the hold line. Therefore, we only make pending predictions and pending decisions about A/T (cr) and D/T (cr). Once the ARIPS generates the pending predictions and pending decisions about A/T (cr) and D/T (cr), the ARIPS check whether more than half "history possible nose location" in $HistoryPossibleNoseLocation$ containing the runway or the runway's adjacent region in calculating process, and only output these confirmed predictions and decisions, shown in Fig. 3. Here are some knowledge about pending predictions: $\forall o \forall r_1 \forall r_2 \forall r_3\ (C2(r_1, r_2, r_3) \wedge At(o, r_1) \wedge NosePossiblyAt\ (o, r_2)) \Rightarrow F(PossiblyAt(o, r_3)))$, $C2(r_1, r_2, r_3)$ means $Runway(r_3) \wedge C(r_1, r_2) \wedge C(r_2, r_3) \wedge Way(r_2); \forall o \forall r (F(PossiblyAt(o, r)) \wedge Active(r) \Rightarrow F(PendingRIby(o, r)))$, $Active(r)$ means runway r is occupied for taking off or landing.

Fourth, in order to handle the momentary absence of surveillance data, we propose a method to give alerts based on critical regions and critical events. Once the filter does not receive the update surveillance data in an update cycle, i.e., missing update, the filter generates a fact about missing update as $miss(AIRCRAFT\_ID)$. A hash map $MissSurveillance<aircraft\ identifier,\ Queue<boolean>>$ stores the facts of missing update in the last few seconds (5 s in default). Based on the empirical knowledge such as $\forall o_1 \forall o_2 \forall r_1 \forall r_2 (Miss(o_1) \wedge At(o_1, r_1) \wedge Intersecting(r_1, r_2) \wedge TakeOffFrom (o_2, r_2) \Rightarrow PedingMissWarningAt(o_1, r_1))$, the ARIPS could generate pending missing update warning about the certain aircraft. After that, the ARIPS check whether the certain aircraft has not updated in calculating process. The above procedure is similar with that in Fig. 3. In the ARIPS, only a small mount of missing updates in certain regions and certain events cause alerts; most of miss updates do not cause alerts.

4 Simulation Experiments

To show the performance and correctness of the improved ARIPS based on inaccurate position surveillance information, we did the simulation experiments using both real historical scenarios and conventional fictional scenarios. Table 1 describes these scenarios briefly. In both real historical scenarios and conventional fictional scenarios, the runway's width is greater than or equal to 45 m. The test scenarios are divided into four groups: A-G are used to compare the performance of the ARIPS with that of other RIPSs; H-M are used to show that the ARIPS could work under different types of RIs; N-Q are non-RI scenarios, i.e., the aircraft staying at the hold line for a few minutes, which is used to evaluate the probability of false detection of the ARIPS; R-W are used to show the ARIPS could handle missing updates, that aircraft's surveillance data is absent for both RI scenarios and non-RI scenarios.

We used the same simulation program and the similar experimental procedure of [1], except that (1) we rewrote the corresponding components and updated the empirical knowledge using our improved methods; (2) we built an inaccurate surveillant position simulator to generate the inaccurate sensory position of aircraft; (3) we set the probability of detection as 95%; detection means the error value does not exceed the accuracy value; (4) we set the offset between the measure position and the aircraft nose is 22 m (as same as [5]), which could be obtained from the traffic information system in practice; and (5) we upgraded the computer hardware. To generate the inaccurate position information of aircraft, we used the following algorithm.

1 Randomly generated a double-precision floating-point number $fitrand \in [0,1)$. If $fitrand \leq 0.95$, it means the error value does not exceed the accuracy value; if $fitrand > 0.95$, it means the error exceeding the accuracy value.
2 The *accuracy* is the accuracy of position surveillance system in meters. If $fitrand \leq 0.95$, randomly generated a double float $radius \in [0, accuracy]$; else if $fitrand > 0.95$, randomly generated a double float $radius \in (accuracy, 75)$.
3 Randomly generated a double float $radian \in [0, 2\pi)$.
4 The inaccurate horizontal and vertical ordinate of aircraft location are $x_{random} = x + radius \times cos(radian)$ and $y_{random} = y + radius \times sin(radian)$; x and y are the actual horizontal and vertical coordinate of the aircraft location.

All random numbers are generated based on current system millisecond time. All programs ran on a PC with Intel Core i7-2600 Processor (3.4 GHz, 4 cores, 8 threads), 8 Gbyte memory, and Scientific Linux release 6.2 for x86_64 (Linux

Table 1. Description of scenarios for simulation experiments

No.	RI type	Description
A	D/T (cr)	At BOS, JBL417 was taking off from 33L, while JBL1264 was taxiing at <u>M</u>, the pilot took a right instead of turning left. The ATC controller used 9 s to react, and 2 more s to give instructions. The RI prevention system did not work [8]
B	D/T (cr)	At CLT, JIA390 was taking off from 18L, while N409DR was taxiing at A to cross 18L. After the incident occurred 14 s, ASDE-X alert was given [9]
C	D/D (cr)	At LSZH, THA971 was taking off from 16, while BCI937 was taking off from 28. The crew of BAW713 found it was possible that two aircraft would take off simultaneously according to the radio, and informed the air traffic controller [10]
D	A/T (cr)	Test Scen. 1 at DFW. The vehicle was crossing the hold line on a taxiway near the runway threshold when the aircraft B757 was 2000 m from the threshold [5]
E	D/T (cr)	Test Scen. 2 at DFW. Once the aircraft began take off roll, the vehicle crossed the hold line. The vehicle was at least 3000 m from the aircraft's hold position [5]
F	D/T (cr)	Test Scen. 3 at DFW. The aircraft was positioned just behind a hold line of a taxiway that croses the runway. The vehicle accelerated to 70 mph from the departure end of the runway. The aircraft then began crosing the hold line [5]
G	A/D (tc)	Test Scen. 4 at DFW. The aircraft was 2000 m from the runway threshold, while the vehicle entered the runway and accelerated to 70 mph [5]
H	A/D (cr)	Based on Scen. C, an aircraft was taking off from 16, while an aircraft was 2000 m from the runway threshold of 28
I	D/D (tc)	Based on Scen. G, the first aircraft was taking off from 35C, while the s aircraft was also taking off from 35C, 3000 m from the first aircraft
J	A/A (tc)	Based on Scen. G, the first aircraft just landed and touched down on 35C, while the s aircraft was 2000 m from the runway threshold of 35C
K	D/T (ho)	Based on Scen.s G, the first aircraft was taking off from 35C, while the s aircraft was taxiing at 35C, 3000 m from the first aircraft
L	A/A (cr)	Based on Scen. C, an aircraft was 2000 m from the runway threshold of 16, while an aircraft was 2000 m from the runway threshold of 28
M	D/D (ho)	Based on Scen. G, two aircraft was taking off from 35C and 17C separately
N	n/a	Based on Scen. D, the arriving aircraft was 2000 m from the runway threshold, then landed at 35C, while the taxiing aircraft stayed at the hold line all the time

Table 1. (*continued*)

No.	RI type	Description
O	n/a	Based on Scen. E, the departure aircraft was taking off from 17C, while the taxiing aircraft stayed at the hold line. The taxiing aircraft was located at least 3000 m from the departure aircraft's take off hold position
P	n/a	Based on Scen. O. The departure aircraft was taking off from 17C, while the taxiing aircraft stayed at the hold line of EG. Then the taxiing aircraft entered the runway
Q	n/a	Based on Scen. C. THA971 was taking off from 16, while BCI937 held at 28
R	D/D (cr)	Based on Scen. C. THA971 was taking off, while the surveillance of BCI937 was absent for 10 s
S	D/T (cr)	Based on Scen. E. The aircraft stayed at the hold line, then its surveillance was absent. The aircraft took off, while the taxiing aircraft began to cross the hold line
T	n/a	Based on Scen. Q. THA971 was taking off from 16, while BCI937 held at 28. Then the surveillance data of THA 971 was absent for 10 s
U	n/a	Based on Scen. Q. THA971 was taking off from 16, while BCI937 held at 28. Then the surveillance data of BCI937 was absent for 10 s
V	n/a	Based on Scen. O. An aircraft was taking off from 17C, while the taxiing aircraft stayed at the hold line. The surveillance of taxiing aircraft was absent for 10 s
W	n/a	Based on Scen. O. An aircraft was taking off from 17C, while the taxiing aircraft stayed at the hold line. The surveillance of departure aircraft was absent for 10 s

kernel is 2.6.32-220). All programs are written in Java and running on OpenJDK Runtime Environment (IcedTea6 1.10.4), except the forward reasoning engine FreeEnCal [11] in C++.

For ARIPSs, the total time of prediction/detection is the sum of the lead time for updating surveillance information and the time for predicting the incursion; the total time of decision-making for instructions is the sum of the lead time for updating surveillance information, the time for prediction, and the time for generating the decision. In scenarios D-G, the average lead time is 3.5 s, while the lead time was not reported in scenarios A-C. Therefore, it is difficult to compare the prediction time of ARIPSs with that of other RIPSs directly. As an alternative, we did not simulate the process of generating surveillance information, but adopted the worst lead time 3.5 s [2], then added that lead time to the execution time of ARIPS additionally. Thus, the comparisons between ARIPSs and RIPSs are for reference only. For ARIPS, the total time is the worst time in the experiments, which we repeated 30 times for each scenarios.

Table 2. Experimental results

Scen. No.	Actor	Accuracy (m)	Total time including lead time (s)		Hit rate	False rate
			RI prediction	Decision		
A	ATC controllers	n/a	9	11	n/a	n/a
	Current RIPS	n/a	did not work	n/a	n/a	n/a
	ARIPS	0	6.0	6.0	100%	n/a
	ARIPS	8	7.8	7.8	100%	n/a
	ARIPS	15	9.8	9.8	100%	n/a
B	Current RIPS	n/a	14	n/a	n/a	n/a
	ARIPS	0	5.9	5.9	100%	n/a
	ARIPS	8	7.3	7.3	100%	n/a
	ARIPS	15	8.6	8.6	100%	n/a
C	ATC controller*	n/a	16	19	n/a	n/a
	Current RIPS	n/a	30	n/a	n/a	n/a
	ARIPS	0	7.7	7.7	100%	n/a
	ARIPS	8	7.8	7.8	100%	n/a
	ARIPS	15	8.2	8.2	100%	n/a
D	Current RIPS	n/a	7	n/a	n/a	n/a
	ARIPS	0	7.3	7.3	100%	n/a
	ARIPS	8	7.4	7.4	100%	n/a
	ARIPS	15	8.6	8.6	100%	n/a
E	Current RIPS	n/a	15	n/a	n/a	n/a
	ARIPS	0	6.8	6.8	100%	n/a
	ARIPS	8	8.2	8.2	100%	n/a
	ARIPS	15	9.5	9.5	100%	n/a
F	Current RIPS	n/a	10	n/a	n/a	n/a
	ARIPS	0	7.1	7.1	100%	n/a
	ARIPS	8	8.8	8.8	100%	n/a
	ARIPS	15	10.4	10.4	100%	n/a
G	Current RIPS	n/a	13	n/a	n/a	n/a
	ARIPS	0	5.8	8.1	100%	n/a
	ARIPS	8	5.5	7.4	100%	n/a
	ARIPS	15	5.9	7.9	100%	n/a
H	ARIPS	15	7.1	7.1	100%	n/a
I	ARIPS	15	6.1	8.0	100%	n/a
J	ARIPS	15	6.0	6.0	100%	n/a
K	ARIPS	15	6.7	6.7	100%	n/a
L	ARIPS	15	6.7	6.7	100%	n/a
M	ARIPS	15	6.0	8.0	100%	n/a
N	ARIPS	15	n/a	n/a	n/a	0%
O	ARIPS	15	n/a	n/a	n/a	0%
P	ARIPS	15	n/a	n/a	n/a	0%
Q	ARIPS	15	n/a	n/a	n/a	0%
R	ARIPS	15	9.0	n/a	100%	n/a
S	ARIPS	15	10.8	n/a	100%	n/a
T	ARIPS	15	n/a	n/a	n/a	0%
U	ARIPS	15	9.0	n/a	n/a	100%
V	ARIPS	15	10.6	n/a	n/a	100%
W	ARIPS	15	n/a	n/a	n/a	0%

*with the help of crew of aircraft.

Table 2 shows our experimental results. For ARIPSs based on inaccurate position information, the rate of correct prediction and decision for RIs was 100%; there was no runway collision if the aircraft/vehicle operated according to the instructions of ARIPSs; for RI types D/D (cr), A/D (tc), A/D (cr), D/D (tc), A/A (tc), A/A (cr), the performance of ARIPSs was not affected by inaccurate position information; for RI types D/T (cr) and A/T (cr), the performance decreased with the increase of accuracy value; for some RI scenarios such as D and F, the prediction time of ARIPSs based on inaccurate surveillance information was longer (1.6 s at most) than that of current RIPSs; for non-RI scenarios, the rate of false detection is 0%; ARIPSs could handle absence of surveillance data by generating alerts, however, for some certain scenarios such as U and V, ARIPSs gave false alerts.

5   Concluding Remarks

This paper proposed improved methods for ARIPSs to handle the inaccurate position information. The methods could handle the momentary absence of surveillance data, the difference between the aircraft's nose position and measured position, and the position error exceeding the position accuracy. The simulation experiments showed our system could handle different types of RIs based on inaccurate position information, deal with the momentary absence of surveillance data, and produce few false detection under non-RI circumstances.

We consider this work as a part of a system for automation of air traffic control procedures. To achieve that, we will consider using our system to prevent other runway safety problems, such as runway excursion and runway confusion, and automatic route planning and guiding based on logic-based reasoning.

References

1. Shi, K., Goto, Y., Zhu, Z., Cheng, J.: Anticipatory runway incursion prevention systems. IEICE Trans. Inf. Syst. **E96–D**(11), 2385–2396 (2013)
2. Schönefeld, J., Möller, D.: Runway incursion prevention systems: a review of runway incursion avoidance and alerting system approaches. Prog. Aerosp. Sci. **51**, 31–49 (2012)
3. EUROCAE: minimum operational performance specification for mode S multilateration systems for use in A-SMGCS, ED-117. Technical report, EUROCAE (2003)
4. Miyazaki, H., Koga, T., Ueda, E., Kakubari, Y., Nihei, S.: Evaluation results of airport surface multilateration. In: Proceedings of ENRI International Workshop on ATM/CNS, ENRI, pp. 41–46, Tokyo (2010)
5. Cassell, R., Evers, C., Esche, J., Sleep, B.: NASA runway incursion prevention system (RIPS) Dallas-Fort Worth demonstration performance analysis. Technical report, NASA (2002)
6. FAA: Aeronautical information manual. http://www.faa.gov/air_traffic/publications/atpubs/aim/index.htm. Accessed 14 Oct 2014

7. FAA: FAA eastern region modification of airport design standards. http://www.faa.gov/air_traffic/publications/atpubs/aim/index.htm. Accessed 16 Oct 2014
8. NATCA: Boston Logan controller's November 24 save. http://www.youtube.com/watch?v=V-dbEYk-ikU. Accessed 26 Feb 2013
9. FAA: May 29, 2009 OE/D operational error category A. http://www.faa.gov/airports/runway_safety/videos/media/simulation.html. Accessed 26 Feb 2013
10. AAIB: Final report No. 2113 by the Aircraft Accident Investigation Bureau. Technical report, AAIB (2010)
11. Cheng, J., Nara, S., Goto, Y.: FreeEnCal: a forward reasoning engine with general-purpose. In: Apolloni, B., Howlett, R.J., Jain, L. (eds.) KES 2007. LNCS (LNAI), vol. 4693, pp. 444–452. Springer, Heidelberg (2007). doi:10.1007/978-3-540-74827-4_56

The Relation Between Syntax Restriction of Temporal Logic and Properties of Reactive System Specification

Noriaki Yoshiura[✉]

Department of Information and Computer Sciences, Saitama University,
255, Shimo-ookubo, Sakura-ku, Saitama, Japan
yoshiura@fmx.ics.saitama-u.ac.jp

Abstract. Open reactive systems provide services to users by interacting the users and environments of the systems. There are several methods that describe formal specifications of reactive systems. Temporal logic is one of the methods. An open reactive system specification is defined to be realizable if and only if there is a program that satisfies the specification even if the environment and the users of the reactive system take any behaviors. There are several kinds of the methods of deciding realizability of open reactive system. These methods are based on automata theory and their complexities are at least double exponential times of the length of a specification. This paper shows the relation between a syntax and realizability properties of reactive system specifications. This relation can reduce the complexity of deciding the properties of reactive system specifications.

1 Introduction

Open reactive systems provide services to users by interacting with the users and environment of the systems. This interaction decides behaviors of Open reactive systems [3]. Description of specifications of open reactive systems require to distinguish between output events and input events. Output events are generated by reactive systems and input events are generated by the users and the environment of the reactive systems. Open reactive systems cannot control input events [4].

There are several methods of describing specifications of open reactive systems. Temporal logic is one of the methods. Temporal logic has an advantage of proving several properties of the specifications. Satisfiability of an open reactive system specification that is described by temporal logic only implies that we can synthesize a system that satisfies the specification for some behavior of users and environments. It is, however, necessary to synthesize a system that satisfies the specification for all environment behaviors. The realizability problem of open reactive system specification ϕ that is a temporal logic formula is to determine whether there exists a program that satisfies the specification ϕ for all behaviors of users and environments [2,6]. There are several kinds of

© Springer International Publishing AG 2017
N.T. Nguyen et al. (Eds.): ACIIDS 2017, Part II, LNAI 10192, pp. 105–114, 2017.
DOI: 10.1007/978-3-319-54430-4_11

the methods of deciding realizability of open reactive system. These methods are based on automata theory and have very high complexity such as at least double exponential times for the length of a specification. It is impractical to synthesize programs from reactive system specifications because synthesis of reactive system programs from specifications requires to determine the realizability of the specifications and it takes much time to determine the realizability [1,5].

This paper shows the relation between a syntax and realizability properties of reactive system specifications. One of the reasons why the complexity of determining the realizability of reactive system specifications is so high is expressive power of temporal logic. Restriction of syntax for description of reactive system specifications may reduce the complexity of determining realizability of specifications. This is the aim of this paper. This paper shows that some properties of realizability are the same under the restriction of syntax of temporal logic. From this result, some properties of realizability are characterized by syntax or operators of temporal logic.

2   Open Reactive System

This section provides a formal definition of an open reactive system based on references [7]. Let A be a finite set. A^+ and A^ω denote the set of *finite sequences* and the set of *infinite sequences* over A respectively. $A^\dagger$ denotes $A^+ \cup A^\omega$. Sequences in $A^\dagger$ are indicated by $\hat{a}, \hat{b}, \cdots$, sequences in A^+ by $\bar{a}, \bar{b}, \cdots$ and sequences in A^ω by $\tilde{a}, \tilde{b}, \cdots$. $|\hat{a}|$ denotes the *length* of $\hat{a}$ and $\hat{a}[i]$ denotes the *i-th element* of $\hat{a}$. If B is a set whose elements are also sets, '$\sqcup$' is defined over $B^\dagger \times B^\dagger$ by

$$\hat{a} \sqcup \hat{b} = \hat{a}[0] \cup \hat{b}[0], \; \hat{a}[1] \cup \hat{b}[1], \; \hat{a}[2] \cup \hat{b}[2], \cdots .$$

Finally, the *infinity set* of $\tilde{a} \in A^\omega$ is defined by

$$inf(\tilde{a}) = \{a \in A | \; a \text{ occurs infinitely often in } \tilde{a}\}.$$

Definition 1. *An open reactive system RS is a triple* $RS = \langle X, Y, r \rangle$ *where*

– X *is a finite set of input events produced by the environment.*
– Y $(X \cap Y = \emptyset)$ *is a finite set of output events produced by the system itself.*
– $r : (2^X)^+ \to 2^Y$ *is a reaction function.*

A subset of X is defined to be an *input set* and a sequence of input sets is defined to be an *input sequence*. Similarly, a subset of Y is defined to be an *output set* and a sequence of output sets is defined to be an *output sequence*. In this paper, a reaction function corresponds to an open reactive system program.

Definition 2. *Let* $RS = \langle X, Y, r \rangle$ *be an open reactive system and* $\hat{a} = a_0, a_1, a_2, \cdots \in (2^X)^\dagger$ *be an input sequence. The behavior of RS for* $\hat{a}$ *is the following sequence:*

$$behave_{RS}(\hat{a}) = \langle a_0, b_0 \rangle, \langle a_1, b_1 \rangle, \langle a_2, b_2 \rangle, \dots ,$$

where for any i $(0 \leq i < |\hat{a}|)$, $b_i = r(a_0, \dots, a_i) \in 2^Y$.

In the following, $behave_{RS}(\hat{a})$ is defined to be *behavior* of RS for $\hat{a}$.

3 Specification

This paper uses propositional linear time temporal logic with until operator $(f_1 \mathcal{U} f_2)$ to describe reactive system specifications. The semantics of the logic formula is defined on the behaviors. Let P be a set of events and $\mathcal{P}$ be a set of atomic propositions corresponding to each element of P. $\langle \sigma, i \rangle \models f$, denotes that a formula f over $\mathcal{P}$ holds at the i-th state of a behavior $\sigma \in (2^P)^\omega$. $\langle \sigma, i \rangle \models f$ is defined as follows.

- $\langle \sigma, i \rangle \models p$ **if and only if** $p' \in \sigma[i]$ ($p \in \mathcal{P}$ is an atomic proposition corresponding to $p' \in P$).
- $\langle \sigma, i \rangle \models \neg f$ **if and only if** $\langle \sigma, i \rangle \not\models f$.
- $\langle \sigma, i \rangle \models f_1 \wedge f_2$ **if and only if** $\langle \sigma, i \rangle \models f_1$ and $\langle \sigma, i \rangle \models f_2$.
- $\langle \sigma, i \rangle \models f_1 \mathcal{U} f_2$ **if and only if** $(\forall j > 0) \langle \sigma, i+j \rangle \models f_1$ or

$$(\exists j \geq 0) \left(\langle \sigma, i+j \rangle \models f_2 \text{ and } \forall k (0 < k < j) \langle \sigma, i+k \rangle \models f_1 \right).$$

σ is defined to be a model of f if and only if $\langle \sigma, 0 \rangle \models f$. $\sigma \models f$ is defined to be $\langle \sigma, 0 \rangle \models f$. A formula f is satisfiable if and only there is a model of f.

3.1 Specification

A specification for an open reactive system is defined to be a triple $Spec = \langle \mathcal{X}, \mathcal{Y}, \varphi \rangle$ where

- $\mathcal{X}$ is a set of *input propositions* that are atomic propositions corresponding to the input events of the intended open reactive system, i.e. the truth value of an input proposition represents the occurrence of the corresponding input event.
- $\mathcal{Y}$ is a set of *output propositions* that are atomic propositions corresponding to the output events of the intended open reactive system, i.e. the truth value of an output proposition represents the occurrence of the corresponding output event.
- φ is a formula that consists of $\mathcal{X} \cup \mathcal{Y}$.

For simplicity, we write $Spec = \langle \mathcal{X}, \mathcal{Y}, \varphi \rangle$ just as φ. Next, the realizability of open reactive system specifications is defined [7].

Definition 3. *An open reactive system RS is an* **implementation** *of a specification φ if and only behave$_{RS}(\tilde{a}) \models \varphi$ for every input sequence $\tilde{a}$. A specification is* **realizable** *if and only if it has an implementation.*

4 Properties of Reactive System Specification

This section describes realizability properties of specifications. The following defines *strong satisfiability*, *stepwise satisfiability* and *stepwise strong satisfiability* [8]. This section uses $\tilde{a}$, $\bar{a}$, $\tilde{b}$ and r to denote an infinite input sequence, a finite input sequence, an infinite output sequence and a reaction function.

Definition 4. φ *is strongly satisfiable if and only if* $\forall \tilde{a} \exists \tilde{b}, \ \tilde{a} \sqcup \tilde{b} \models \varphi$

In reactive systems, users and environment of the reactive systems generates input events. Realizable reactive system specifications have reactive systems that generates an infinite output event sequence for any infinite input event sequence. Therefore, the strong satisfiability is a necessary condition of realizability. On the other hand, strong satisfiability is not a sufficient condition; Suppose that φ is strong satisfiable. For each infinite input event sequence, there is an infinite output event sequence for each infinite input event sequence such that φ is satisfied. The reactive system, however, must produce an output event from the input event sequence until the current time but not from the infinite event sequence. Therefore strong satisfiability does not guarantee that reactive systems behave so that φ is satisfied.

Next, let me focus on stepwise satisfiability and stepwise strong satisfiability.

Definition 5. *A reactive system RS preserves satisfiability of* φ *if and only if* $\forall \tilde{a} \exists \tilde{a} \exists \tilde{b}, \ \tilde{a} \tilde{a} \sqcup behav_{RS}(\tilde{a}) \tilde{b} \models \varphi$. φ *is stepwise satisfiable if and only if a reactive system preserves the satisfiability of* φ.

Definition 6. *A reactive system RS preserves strong satisfiability of* φ *if and only if* $\forall \tilde{a} \forall \tilde{a} \exists \tilde{b}, \ \tilde{a} \tilde{a} \sqcup behav_{RS}(\tilde{a}) \tilde{b} \models \varphi$. φ *is stepwise strongly satisfiable if and if a reactive system preserves the strong satisfiability of* φ.

If φ is stepwise satisfiable, there is a reactive system $RS = \langle X, Y, r \rangle$ that preserves the satisfiability of φ. For any input event sequence, the reactive system RS can behave and there is always a possibility that φ is satisfied even though RS actually does not satisfy φ.

The meaning of stepwise strong satisfiability is the same as that of stepwise satisfiability. If φ is stepwise strong satisfiable, there is a reactive system $RS = \langle X, Y, r \rangle$ that preserves the strong satisfiability of φ. For any input event sequence, the reactive system RS can behave and there is always a possibility that φ is satisfied for any input event sequence even though RS actually does not satisfy φ for any input event sequence.

5   Restriction of Syntax

This section proposes the several restrictions on temporal logic and investigates the relation between properties of reactive system specifications and the restrictions of temporal logic syntax.

Using full syntax of temporal logic makes it possible to describe all reactive system specifications that belong to all five classes. It is very difficult to decide whether reactive system specifications are realizable, and however to decide the other properties of reactive system specifications is easier than to decide realizability. Restriction on syntax is one of the methods of reducing the complexity of deciding the realizability of reactive system specification. The following introduces restrictions of syntax.

Definition 7 (The zero type formula). *The zero type formula is of the form* $\square \wedge_i S_i$ *where* S_i *is a classical logic formula.*

Theorem 1. *If a reactive system specification is the zero type formula, stepwise satisfiability, strong satisfiability and realizability are equivalent.*

Proof. Because realizability implies stepwise satisfiability and strong satisfiability, we prove that each of stepwise satisfiability and strong satisfiability implies realizability. The zero type formula of reactive system specification can be defined to be $\square(F_1 \wedge F_2 \wedge \cdots \wedge F_n)$ where F_i $(1 \leq i \leq n)$ is a classical logic formula consisting of input and output propositions. Suppose that $\square(F_1 \wedge \cdots \wedge F_n)$ is strongly satisfiable. $F_1 \wedge \cdots \wedge F_n$ is also strongly satisfiable. At every time point, no matter what the truth values of input propositions are, there is an assignment of the truth values of output propositions in order to make $F_1 \wedge \cdots \wedge F_n$ satisfiable. This assignment can also be decided without the truth values of input and output propositions at future time points. It follows that $\square(F_1 \wedge \cdots \wedge F_n)$ is realizable.

We prove the stepwise satisfiability case similarly.   □

Definition 8 (The first type formula). *The first type formula is of the form* $\square \wedge_i S_i$ *where* S_i *is a disjunction formula consisting of the following formulas.*

1. *input proposition or negation of input proposition*
2. *output proposition or negation of output proposition*
3. $\square T$ *where* T *is a classical logic formula.*

The following theorem can be proved. This theorem shows the relation between the first type formulas and properties of reactive system specification.

Theorem 2. *For a reactive system specification is the first type formula, stepwise satisfiability, strong satisfiability and realizability are equivalent.*

Proof. The first type of reactive system specification can be defined to be $\square((X_1 \vee Y_1 \vee \square Z_1) \wedge (X_2 \vee Y_2 \vee \square Z_2) \wedge \cdots \wedge (X_n \vee Y_n \vee \square Z_n))$ where X_i $(1 \leq i \leq n)$ is a disjunction of input proposition and negation of input proposition, Y_i $(1 \leq i \leq n)$ is a disjunction of output proposition and negation of output proposition, and Z_i $(1 \leq i \leq n)$ is a classical logic formula.   □

Definition 9 (The second type formulas). *The first type formulas is of the form* $\square \wedge_i S_i$ *where* S_i *is a disjunction formulas consisting of the following formulas.*

1. *the classical logic formula consisting of only input propositions*
2. *the formula consisting of only output propositions and satisfying the following conditions*
   (a) Negation is attached to only atomic propositions.
   (b) $\square$ *operator does not appear.*
   (c) If $\lozenge T$ *exists in the formula,* T *is a classical logic formula.*

Theorem 3. *If a reactive system specification is the second type formula, strong satisfiability and realizability are equivalent. On other hand, stepwise satisfiability and realizability are not.*

Proof. The second type of reactive system specification can be defined to be $\Box((X_1 \rightarrow Y_1) \wedge (X_2 \rightarrow Y_2) \wedge \cdots \wedge (X_n \rightarrow Y_n))$ where X_i $(1 \leq i \leq n)$ consists of input propositions only with negation, disjunction and conjunction and Y_i $(1 \leq i \leq n)$ consists of input propositions and satisfies the conditions in Definition 9.

Because realizability implies strong satisfiability, we prove that strong satisfiability implies realizability. Suppose that $\Box((X_1 \rightarrow Y_1) \wedge (X_2 \rightarrow Y_2) \wedge \cdots \wedge (X_n \rightarrow Y_n))$ is strong satisfiable. Since we can decide the truth value of $Y_1 \cdots Y_n$ according to the current truth value of $X_1 \cdots Y_n$, this formula is also realizable. $\Box$

Definition 10 (The third type formulas). *The third type formulas is of the form $\Box \wedge_i S_i$ where S_i is a disjunction formulas consisting of the following formulas.*

1. *the classical logic formula consisting of only input propositions*
2. *the formula consisting of only output propositions and satisfying the following conditions*
   (a) Negation is attached to only atomic propositions.
   (b) Only $\Diamond\Box$ appears as modal operator.
   (c) If $\Diamond\Box T$ exists in the formula, T is a classical logic formula.

Theorem 4. *Suppose that a reactive system specification is the third type formula. Strong satisfiability, stepwise satisfiability, realizability of the specification are equivalent.*

Proof. We prove the theorem in the similar way of Theorem 3. $\Box$

Definition 11 (The fourth type formulas). *The fourth type formulas is of the form $\wedge_i \Box S_i$ where S_i is a disjunction formulas consisting of the following formulas.*

1. *input proposition or negation of input proposition*
2. *output proposition or negation of output proposition*
3. *formulas that are constructed of $\Diamond\Box Q$, or logical operators $\vee$ or $\wedge$ where Q is constructed of output propositions, negations of output propositions, or logical operators $\vee$ or $\wedge$.*

Theorem 5. *Suppose that a reactive system specification is the fourth type formula. Strong satisfiability, stepwise satisfiability, realizability of the specification are equivalent.*

Definition 12 (The fifth type formulas). *The fifth type formula is of the form $\wedge_i \Box S_i$ where S_i is a disjunction formula consisting of the following formulas.*

1. *formulas that are constructed of input propositions, negations of input propositions, or logical operators $\lor$ or $\land$*
2. *formulas that are constructed of output propositions, negations of output propositions, $\Box Q$, $\Diamond Q$, or logical operators $\lor$ or $\land$ where Q is constructed of output propositions, negations of output propositions, or logical operators $\lor$ or $\land$.*

Theorem 6. *There is a fifth type formula that is stepwise satisfiable but not strong satisfiable.*

Proof. $\Box(X_1 \rightarrow Y) \land \Box(X_2 \rightarrow \Diamond\neg Y)$ is a fourth type formula. Clearly this formula is stepwise satisfiable, but not strong satisfiable. □

Theorem 7. *Suppose that a reactive system specification is the fifth type formula. Strong satisfiable formulas are stepwise satisfiable.*

Definition 13 (The sixth type formulas). *The sixth type formula is of the form $\land_i \Box S_i$ where S_i is a disjunction formula consisting of the following formulas.*

1. *formulas that are constructed of input propositions, negations of input propositions, or logical operators $\lor$ or $\land$*
2. *formulas that are constructed of output propositions, negations of output propositions, $\Box Q$, $\Box \Diamond Q$, or logical operators $\lor$ or $\land$ where Q is constructed of output propositions, negations of output propositions, or logical operators $\lor$ or $\land$.*

Theorem 8. *There is a sixth type formula that is stepwise satisfiable but not strong satisfiable.*

Proof. $\Box(X_1 \rightarrow Y) \land \Box(X_2 \rightarrow \Box\Diamond\neg Y)$ is stepwise satisfiable, but not strong satisfiable. □

Theorem 9. *There is a sixth type formula that is strong satisfiable but not realizable.*

Proof. $\Box(X_1 \leftrightarrow Z) \land \Box(X_2 \rightarrow ((\neg Y \lor \Box\Diamond Z) \land (Y \lor \Box\neg Z)))$ is strong satisfiable, but not realizable. □

Theorem 10. *Suppose that a reactive system specification is the sixth type formula. Strong satisfiable formulas are stepwise satisfiable.*

Definition 14 (The sixth type formulas). *The seventh type formula is of the form $\land_i \Box S_i$ where S_i is a disjunction formula consisting of the following formulas.*

1. *formulas that are constructed of input propositions, negations of input propositions, or logical operators $\lor$ or $\land$*

2. *formulas that are constructed of output propositions, negations of output propositions, $\Box Q$, $\diamond \Box Q$, or logical operators $\lor$ or $\land$ where Q is constructed of output propositions, negations of output propositions, or logical operators $\lor$ or $\land$.*

Theorem 11. *There is a seventh type formula that is stepwise satisfiable but not strong satisfiable.*

Proof. $\Box(X_1 \rightarrow Y) \land \Box(X_2 \rightarrow \diamond \Box \neg Y)$ is stepwise satisfiable but not strong satisfiable. $\qquad \Box$

Theorem 12. *Suppose that a reactive system specification is the sixth type formula. Strong satisfiable formulas are stepwise satisfiable.*

Definition 15 (The eighth type formulas). *The eighth type formula is of the form $\land_i \Box S_i$ where S_i is a disjunction formula consisting of the following formulas.*

1. *formulas that are constructed of input propositions, negations of input propositions, or logical operators $\lor$ or $\land$*
2. *formulas that are constructed of output propositions, negations of output propositions, $\diamond Q$, $\Box \diamond Q$, or logical operators $\lor$ or $\land$ where Q is constructed of output propositions, negations of output propositions, or logical operators $\lor$ or $\land$.*

Theorem 13. *There is an eighth type formula that is stepwise satisfiable but not strong satisfiable.*

Proof. Trivially, $\Box(X_1 \rightarrow Y) \land \Box(X_2 \rightarrow \diamond \neg Y)$ is stepwise satisfiable, but not strong satisfiable. $\qquad \Box$

Theorem 14. *There is an eighth type formula that is strong satisfiable but not realizable.*

Proof. Trivially $\Box(X_1 \leftrightarrow Z) \land \Box(X_2 \rightarrow ((\neg Y \lor \diamond Z) \land (Y \lor \Box \diamond \neg Z)))$ is strong satisfiable but not realizable. $\qquad \Box$

Theorem 15. *Suppose that a reactive system specification is the eighth type formula. Strong satisfiable formulas are stepwise satisfiable.*

Proof. We prove the theorem in the similar way of Theorem 3. $\qquad \Box$

Definition 16 (The ninth type formulas). *The ninth type formula is of the form $\land_i \Box S_i$ where S_i is a disjunction formula consisting of the following formulas.*

1. *formulas that are constructed of input propositions, negations of input propositions, or logical operators $\lor$ or $\land$*

2. *formulas that are constructed of output propositions, negations of output propositions, $\Diamond Q$, $\Diamond \Box Q$, or logical operators $\vee$ or $\wedge$ where Q is constructed of output propositions, negations of output propositions, or logical operators $\vee$ or $\wedge$.*

Theorem 16. *There is a ninth type formula that is stepwise satisfiable but not strong satisfiable.*

Proof. Trivially, $\Box(X_1 \rightarrow Y) \wedge \Box(X_2 \rightarrow \Diamond \neg Y)$ is stepwise satisfiable, but not strong satisfiable.   □

Theorem 17. *There is a ninth type formula that is strong satisfiable, but not realizable.*

Proof. Trivially, $\Box(X_1 \leftrightarrow Z) \wedge \Box(X_2 \rightarrow ((\neg Y \vee \Diamond Z) \wedge (Y \vee \Diamond \Box \neg Z)))$ is strong satisfiable but not realizable.   □

Theorem 18. *Suppose that a reactive system specification is the ninth type formula. Strong satisfiable formulas are stepwise satisfiable.*

Proof. We prove the theorem in the similar way of Theorem 3.   □

Definition 17 (The tenth type formulas). *The tenth type formula is of the form $\wedge_i \Box S_i$ where S_i is a disjunction formula consisting of the following formulas.*

1. *formulas that are constructed of input propositions, negations of input propositions, or logical operators $\vee$ or $\wedge$*
2. *formulas that are constructed of output propositions, negations of output propositions, $\Diamond Q$, $\Diamond \Box Q$, or logical operators $\vee$ or $\wedge$ where Q is constructed of output propositions, negations of output propositions, or logical operators $\vee$ or $\wedge$.*

Theorem 19. *There is a tenth type formula that is stepwise satisfiable but not strong satisfiable.*

Proof. Trivially, $\Box(X_1 \rightarrow Y) \wedge \Box(X_2 \rightarrow \Diamond \Box \neg Y)$ is stepwise satisfiable but not strong satisfiable.   □

Theorem 20. *There is a tenth type formula that is strong satisfiable but not realizable.*

Proof. Trivially, $\Box(X_1 \leftrightarrow Z) \wedge \Box(X_2 \rightarrow ((\neg Y \vee \Box \Diamond Z) \wedge (Y \vee \Diamond \Box \neg Z)))$ is strong satisfiable but not realizable.   □

Theorem 21. *Suppose that a reactive system specification is the tenth type formula. Strong satisfiable formulas are stepwise satisfiable.*

Proof. We prove the theorem in the similar way of Theorem 3.   □

6   Conclusion

This paper shows the relation between syntax restriction and properties of realizability. This paper introduced ten kind types of formulas that consist of modal logic formula with □ and ◇. For these kind types of formulas, strong satisfiability, stepwise satisfiability and realizability are not always independent properties. Especially, stepwise satisfiability implies strong satisfiability for many types of formulas. The ten kind types of formulas do not use Until operator. Therefore, Until operator is important for separation of strong satisfiability and stepwise satisfiability.

References

1. Bouyer, P., Bozzelli, L., Chevalier, F.: Controller synthesis for MTL specifications. In: Baier, C., Hermanns, H. (eds.) CONCUR 2006. LNCS, vol. 4137, pp. 450–464. Springer, Heidelberg (2006). doi:10.1007/11817949_30
2. Abadi, M., Lamport, L., Wolper, P.: Realizable and unrealizable specifications of reactive systems. In: Ausiello, G., Dezani-Ciancaglini, M., Rocca, S.R. (eds.) ICALP 1989. LNCS, vol. 372, pp. 1–17. Springer, Heidelberg (1989). doi:10.1007/BFb0035748
3. Harel, D., Pnueli, A.: On the development of reactive systems. In: Apt, K.R. (ed.) Logics and Models of Concurrent Systems. NATO ASI Series, vol. 13, pp. 477–498. Springer, Heidelberg (1985)
4. Kupferman, O., Madhusudan, P., Thiagarajan, P.S., Vardi, M.Y.: Open systems in reactive environments: control and synthesis. In: Palamidessi, C. (ed.) CONCUR 2000. LNCS, vol. 1877, pp. 92–107. Springer, Heidelberg (2000). doi:10.1007/3-540-44618-4_9
5. Ouaknine, J., Worrell, J.: On the decidability of metric temporal logic. In: Proceedings of the 20th Annual IEEE Symposium on Logic in Computer Science, pp. 188–197 (2005)
6. Pnueli, A., Rosner, R.: On the synthesis of an asynchronous reactive module. In: Ausiello, G., Dezani-Ciancaglini, M., Rocca, S.R. (eds.) ICALP 1989. LNCS, vol. 372, pp. 652–671. Springer, Heidelberg (1989). doi:10.1007/BFb0035790
7. Pnueli, A., Rosner, R.: On the synthesis of a reactive module. In: Proceedings of 16th Annual ACM Symposium on the Principle of Programming Languages, pp. 179–190 (1989)
8. Mori, R., Yonezaki, N.: Several realizability concepts in reactive objects. In: Information Modeling and Knowledge Bases. IOS, Amsterdam (1993)

Measuring Interestingness of Theorems in Automated Theorem Finding by Forward Reasoning: A Case Study in Peano's Arithmetic

Hongbiao Gao and Jingde Cheng[✉]

Department of Information and Computer Sciences,
Saitama University, Saitama 338-8570, Japan
{gaohongbiao,cheng}@aise.ics.saitama-u.ac.jp

Abstract. Wos proposed 33 basic research problems for automated reasoning field, one of them is the problem of automated theorem finding. The problem has not been solved until now. The problem implicitly requires some metrics to be used for measuring interestingness of found theorems. We have proposed some metrics to measure interestingness of theorems found by using forward reasoning approach. We have measured interestingness of the theorems of NBG set theory by using those metrics. To confirm the generality of the proposed metrics, we have to apply them in other mathematical fields. This paper presents a case study in Peano's arithmetic to show the generality of proposed metrics. We evaluate the interestingness of theorems of Peano's arithmetic obtained by using forward reasoning approach, and confirm the effectiveness of the metrics.

Keywords: Metric · Automated theorem finding · Forward reasoning · NBG set theory · Peano's arithmetic

1 Introduction

In 1988, Wos proposed 33 basic research problems for automated reasoning field, one of them is the problem of automated theorem finding (ATF for short) [18,19]. The problem of ATF is "What properties can be identified to permit an automated reasoning program to find new and interesting theorems, as opposed to proving conjectured theorems?" [18,19].

Unlike well-developed automated theorem proving, the problem of ATF has not been solved until now [10]. The difficulty of the problem is that we have to provide the criteria for computer program to find those new and interesting theorems automatically. In fact, the interestingness of theorems are always measured by researchers of one certain field. However, the automated theorem proving has not that difficulty, because the proving theorems are supplied by the researchers of one certain field in advance. The significance of solving the problem of ATF is obvious, because it can provide great assistance for scientists in various fields [2–4].

© Springer International Publishing AG 2017
N.T. Nguyen et al. (Eds.): ACIIDS 2017, Part II, LNAI 10192, pp. 115–124, 2017.
DOI: 10.1007/978-3-319-54430-4_12

Some research called automated theorem discovery and automated theorem generation has been done [1,6–8,10,12–14,16,17]. However, those kinds of work do not need to provide the criteria for computer program to find new and interesting theorems, because they are based on the approach of automated theorem proving. Besides, those works focus on one certain mathematical field. To solve the ATF problem, we have to provide a systematic method such that we can find new and interesting theorems in various fields generally.

We have proposed a systematic methodology for ATF by using forward reasoning approach [2–4,9]. Following that, we have proposed some metrics to measure interestingness of theorems found by the methodology, and applied them to measure interestingness of the theorems in NBG set theory [11]. To confirm the generality of the proposed metrics, we have to apply them in other mathematical fields.

This paper presents a case study of ATF in Peano's arithmetic to show the generality of proposed metrics. In the case study, we use the metrics to measure interestingness of the theorems of Peano's arithmetic obtained by using forward reasoning approach, and confirm the generality of the metrics. The rest of the paper is organized as follows: Sect. 2 explains the terminology used in the case study. Section 3 presents the proposed metrics for measuring interestingness of theorems. Section 4 presents the case study in Peano's arithmetic. Finally, some concluding remarks are given in Sect. 5.

2   Basic Notions

A formal logic system L is an ordered pair $(F(L), \vdash_L)$ where $F(L)$ is the set of well formed formulas of L, and $\vdash_L$ is the consequence relation of L such that for a set P of formulas and a formula C, $P \vdash_L C$ means that within the framework of L taking P as premises we can obtain C as a valid conclusion. $Th(L)$ is the set of logical theorems of L such that $\phi \vdash_L T$ holds for any $T \in Th(L)$. According to the representation of the consequence relation of a logic, the logic can be represented as a Hilbert style system, a Gentzen sequent calculus system, a Gentzen natural deduction system, and so on [4].

Let $(F(L), \vdash_L)$ be a formal logic system and $P \subseteq F(L)$ be a non-empty set of sentences. A formal theory with premises P based on L, called a L-theory with premises P and denoted by $T_L(P)$, is defined as $T_L(P) =_{df} Th(L) \cup Th_L^e(P)$ where $Th_L^e(P) =_{df} \{A | P \vdash_L A \text{ and } A \notin Th(L)\}$, $Th(L)$ and $Th_L^e(P)$ are called the logical part and the empirical part of the formal theory, respectively, and any element of $Th_L^e(P)$ is called an empirical theorem of the formal theory [4].

Based on the definition above, the problem of ATF can be said as "for any given premises P, how to construct a meaningful formal theory $T_L(P)$ and then find new and interesting theorems in $Th_L^e(P)$ automatically?" [4].

3   Metrics for Measuring Interestingness of Theorems

We have proposed four factors which are involved in measuring the interestingness of empirical theorems obtained by the proposed methodology of ATF. The

Table 1. Frequent propositional schema in NBG set theory

Propositional schema	Appeared time	Appeared rate
A	186	43%
$A \Rightarrow B$	108	25%
$(A \wedge B) \Rightarrow C$	54	13%
$A \vee B$	26	6%
$A \Rightarrow (B \vee C)$	17	4%
$\neg A$	14	3%
$\neg(A \wedge B)$	10	2%
$(A \wedge B \wedge C) \Rightarrow D$	6	1%
$A \vee B \vee C$	5	1%
$(A \wedge B \wedge C \wedge D) \Rightarrow E$	2	<1%
$\neg(A \wedge B \wedge C)$	1	<1%

four factors are deduction distance of theorems, abstract level of theorems, propositional schema of theorems, and degree of logical connectives in theorems [11].

The first impact factor involved in measuring interestingness of theorem is deduction distance of theorems. It is not difficult to obtain those theorems which can be reasoned out by several steps. The interesting theorems are those theorems which are not easy to be obtained by our brain. We conjecture that the deduction distance of one theorem is longer, the possibility to be an interesting theorem is higher.

The abstract level of theorem is also a important factor involved in measuring interestingness of theorems. Predicates and functions are always defined from lower level to higher level by mathematicians. For example, the predicate "$\in$" is the most basic predicate in set theory. Then the mathematicians define the predicate "$\subseteq$" which is a higher level predicate than "$\in$", and abstracts from "$\in$" by the definition of "$\subseteq$": $\forall x \forall y (\forall u ((u \in x) \Rightarrow (u \in y)) \Leftrightarrow (x \subseteq y))$. Then the mathematicians define the predicate "$=$" which is a higher level predicate than "$\subseteq$", and abstracts from "$\subseteq$" by the axiom: $\forall x \forall y (((x \subseteq y) \wedge (y \subseteq x)) \Leftrightarrow (x = y))$. Based on the fact, we can consider that a theorem holds higher abstract level predicates and functions, the theorem is more interesting from the viewpoint of the meaning of the theorem.

The propositional schema of theorem is another factor to measure the interestingness of theorem. We have investigated 429 proved theorems of NBG set theory and 959 proved theorems of Peano's arithmetic [15], and we conjecture that the interesting theorems always hold several frequent propositional schemata. The most frequent schema of proved theorems is A type. The meaning of one theorem is always concise, if the theorem is an atomic formula. The second frequent schema is $A \Rightarrow B$. This type is very frequent to be used as an inference rule in various natural scientific filed. We have summarized the frequent propositional schemata of those theorems in NBG set theory and Peano's arithmetic, and showed our investigated results in Tables 1 and 2. Tables 1 and 2 show that existed theorems always hold

Table 2. Frequent propositional schema in Peano's arithmetic

Propositional schema	Appeared time	Appeared rate
A	321	33%
$A \Rightarrow B$	270	28%
$(A \wedge B) \Rightarrow C$	101	11%
$A \vee B$	72	8%
$A \Rightarrow (B \vee C)$	65	7%
$\neg A$	37	4%
$\neg(A \wedge B)$	35	4%
$(A \wedge B \wedge C) \Rightarrow D$	21	2%
$A \vee B \vee C$	15	2%
$(A \wedge B) \Rightarrow (C \vee D)$	8	<1%
$\neg(A \wedge B \wedge C)$	6	<1%
$(A \wedge B \wedge C \wedge D) \Rightarrow E$	3	<1%
$A \Rightarrow (B \vee C \vee D)$	2	<1%
$(A \wedge B \wedge C \wedge D) \Rightarrow (E \vee F \vee G)$	1	<1%
$(A \wedge B \wedge C \wedge D \wedge E) \Rightarrow F$	1	<1%
$(A \wedge B \wedge C) \Rightarrow (D \vee E)$	1	<1%

Table 3. Degree of logical connectives of theorems in NBG set theory

$\Rightarrow,0$	242	56%	$\wedge,0$	356	83%	$\vee,0$	381	89%	$\neg,0$	404	94%
$\Rightarrow,1$	187	44%	$\wedge,1$	64	15%	$\vee,1$	43	10%	$\neg,1$	25	6%
$\Rightarrow,2$	0	0%	$\wedge,2$	7	<2%	$\vee,2$	5	1%	$\neg,2$	0	0%
$\Rightarrow,3$	0	0%	$\wedge,3$	2	<1%	$\vee,3$	0	0%	$\neg,3$	0	0%
$\Rightarrow,4$	0	0%	$\wedge,4$	0	0%	$\vee,4$	0	0%	$\neg,4$	0	0%

Table 4. Degree of logical connectives of theorems in Peano's arithmetic

$\Rightarrow,0$	486	51%	$\wedge,0$	782	82%	$\vee,0$	795	83%	$\neg,0$	881	92%
$\Rightarrow,1$	473	49%	$\wedge,1$	144	15%	$\vee,1$	146	15%	$\neg,1$	78	8%
$\Rightarrow,2$	0	0%	$\wedge,2$	28	<3%	$\vee,2$	18	2%	$\neg,2$	0	0%
$\Rightarrow,3$	0	0%	$\wedge,3$	4	<1%	$\vee,3$	0	0%	$\neg,3$	0	0%
$\Rightarrow,4$	0	0%	$\wedge,4$	1	<1%	$\vee,4$	0	0%	$\neg,4$	0	0%

several frequent propositional schemata. Therefore, maybe the new and interesting theorems also hold those frequent propositional schemata.

The fourth factor involved in measuring interestingness of theorems is degree of logical connectives of theorems. We have investigated the degree of logical connectives in 429 proved theorems of NBG set theory and 959 proved theorems of Peano's arithmetic [15]. Tables 3 and 4 show that the degrees of the logical

Table 5. The value of interestingness of the degree of each logical connective

Degree	$Value_\Rightarrow$	Degree	$Value_\wedge$	Degree	$Value_\vee$	Degree	$Value_\neg$
$\Rightarrow,0$	1	$\wedge,0$	1	$\vee,0$	1	$\neg,0$	1
$\Rightarrow,1$	1	$\wedge,1$	1	$\vee,1$	1	$\neg,1$	1
$\Rightarrow,2$	1/2	$\wedge,2$	1/2	$\vee,2$	1/2	$\neg,2$	1/2
$\Rightarrow,3$	1/3	$\wedge,3$	1/3	$\vee,3$	1/3	$\neg,3$	1/3
$\Rightarrow,n$	1/n	$\wedge,n$	1/n	$\vee,n$	1/n	$\neg,n$	1/n

Table 6. The value of interestingness of propositional schema

Propositional schema	Value
A	3
$A \Rightarrow B$	3
$\neg A$	2
$\neg(A_1 \wedge ... \wedge A_n)$	2
$A_1 \vee ... \vee A_n$	2
$(A_1 \wedge ... \wedge A_n) \Rightarrow (B_1 \vee ... \vee B_n)$	2
Infrequent propositional schema	1
Propositional schema containing tautology	0

connectives in proved theorems of the two mathematical fields are almost lower than 2, and we can conjecture that maybe the degree of logical connectives of those new and interesting theorems is also low.

In order to measure the interestingness of theorems, we assign the value of interestingness of theorem according to deduction distance of theorems, abstract level of theorems, propositional schema of theorem, and degree of logical connectives of theorems. In detail, we use a four dimension array (Pd, Pp, Pa, Pe) to represent four dimensional values, in which Pd means the value of interestingness of the degree of logical connectives of theorems, Pp means the value of interestingness of the propositional schema, Pa means the value of interestingness of the abstract level of theorems, and Pe means the value of interestingness of deduction distance of theorems. The value of interestingness of the degree of logical connectives $Pd = Value_\Rightarrow * Value_\wedge * Value_\vee * Value_\neg$. Table 5 shows the value of interestingness of the degree of each logical connective. Table 6 shows the value of interestingness of propositional schema of formula. The value of interestingness of abstract level of theorems is counted by $Pa = k + m$, if the abstract level of one theorem is (k, m). Finally, the value of interestingness of deduction distance of one theorem Pe equals to the deduction distance of the theorem. Then, we defined the following 15 metrics to measure the interestingness of theorems found by reasoning approach. The value is larger, the theorem can be seen as more interesting. Pd, Pp, Pa, Pe, $Pd * Pp$, $Pd * Pa$, $Pd * Pe$, $Pp * Pa$, $Pp * Pe$, $Pa * Pe$, $Pd * Pp * Pa$, $Pd * Pp * Pe$, $Pp * Pa * Pe$, $Pd * Pa * Pe$, $Pd * Pp * Pa * Pe$.

4   Case Study in Peano's Arithmetic

To confirm the generality of our metrics, we performed a case study to measure the interestingness of empirical theorems of Peano's arithmetic obtained by reasoning approach. We collected all of definitions and axioms of Peano's arithmetic from Quaife's book [15]. We used all of them as empirical premises, used prepared logical fragments in the case study of NBG set theory [11] as logical premises, and performed automatically forward reasoning by FreeEnCal [5] which is a forward reasoning engine with general purpose. Then, we used proposed metrics to measure the interestingness of empirical theorems of Peano's arithmetic reasoned out by FreeEnCal. In detail, based on the value in Tables 5 and 6, we measured Pd and Pp of those empirical theorems. To measure Pa, we investigated the abstract levels of predicates of Peano's arithmetic in Quaife's book as shown in Table 7, and investigated the abstract levels of functions of Peano's arithmetic in Quaife's book as shown in Table 8 [9]. Finally, we counted

Table 7. Abstract level of predicates in Peano's arithmetic

Predicate	Abstract from	Level
=	None	1
<	=	2
DIV()	=	2
LD()	=	2
SET	=	2
SORTED	=	2
PERM()	=	2
On()	=	2
≡ ()	=	2
PR()	<, =	3
SUB	On	3
PP()	<, =	3
BSORTED()	<, =	3
INCONG()	<, =, ≡ ()	3
COMPLETE()	On(), =, ≡ ()	3
CRS()	PERM()	3
RP()	On(), =	3
RCOMPLETE()	On(), =, ≡ ()	3
RCRS()	PERM()	3
PRIMES()	=, PR()	4
PPOWERS()	PP, =	4

Table 8. Abstract level of functions in Peano's arithmetic

Function	Abstract from	Level
s()	none	1
+	none	1
*	none	1
gcd()	none	1
lf()	none	1
−	+	2
mod()	*	2
/	*	2
!	*, s()	2
x^y	*, s()	2
log	/, s()	3
min()	+, -	3
max()	+, -	3
ld1()	*, -	3
ld2()	*, -	3
lcm()	gcd(), *, /	3
gf()	/, lf()	3
[|]	+, *, /, s()	3
head()	[|]	4
tail()	[|]	4
app	[|]	4
len()	[|]	4
gcd1()	gcd(), ld1()	4
gcd2()	gcd(), ld2()	4
card()	s(), [|]	4
set()	[|]	4
del	[|]	4
$\sum$	[|], +	4
const	[|], s()	4
$\prod$	[|], *	4
pfact	lf(), gf(), [|]	4
lpp	lf(), log	4
times()	[|], *	4
red()	gcd(), [|]	4
ppfact()	[|], lpp(), /	5
init()	app(), [|], s()	5
rev	[|], app	5
at	s(), tail()	5
merge()	[|], head(), tail()	5
sort	[|], merge()	6
ht()	head(), at()	6
φ()	init(), red(), len()	6

Table 9. The value of interestingness of empirical theorems of Peano's arithmetic

	Range of value	Average value	Deviation
Pd	0.5–1	0.9	0.4
Pp	0–3	1.9	1.9
Pa	1–9	4.9	4.1
Pe	1–5	2.2	2.8
$Pd*Pp$	0–3	1.8	1.8
$Pd*Pa$	1–9	4.7	4.3
$Pd*Pe$	0.5–5	2.0	3.0
$Pp*Pa$	0–24	9.8	14.2
$Pp*Pe$	0–15	3.7	11.3
$Pa*Pe$	2–30	10.2	19.8
$Pd*Pp*Pa$	0–24	9.5	14.5
$Pd*Pp*Pe$	0–9	3.6	5.4
$Pp*Pa*Pe$	0–72	17.3	54.7
$Pd*Pa*Pe$	1.5–30	9.7	20.3
$Pd*Pp*Pa*Pe$	0–72	18.5	53.5

Pe according to the information provided by FreeEnCal, because FreeEnCal can show deduction distance for each reasoned out empirical theorem.

Table 9 shows the range of value of interestingness, average value and deviation by using those metrics in the case study. From the results, we can know that the metric $Pd*Pp*Pa*Pe$ is also the best choice as same as the case study in NBG set theory [11], because the range of value of interestingness is wide and deviation is higher. By using the combination, we can more easily find interesting empirical theorems from lots of reasoned out empirical theorems. Therefore, we also analyzed the number of empirical theorems on each value of interestingness for the combination. Figure 1 showed the analyzed result. According to the result, our metrics can generally filter uninteresting theorems from all of obtained empirical theorems in different mathematical fields (the case study of NBG set theory [11] and Peano's arithmetic). Besides, as same as the case study in NBG set theory, the measured values are near "Normal Distribution" in each deduction distance (note that Pe is from 1 to 5 in the case study), and the empirical theorems holding higher value are fewer in all of reasoned out theorems. Through the case study, we confirmed the generality of our metrics, especially the combination $Pd*Pp*Pa*Pe$ is hopeful to be used in ATF of different mathematical fields.

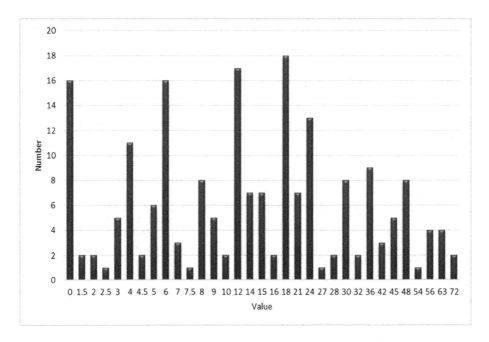

Fig. 1. The number of empirical theorems of Peano's arithmetic on each value of interestingness

5 Concluding Remarks

We have applied our metrics to measure the interestingness of empirical theorems of Peano's arithmetic reasoned out by forward reasoning approach, and confirmed the generality of those metrics to be used in ATF of different mathematical fields.

We also present some challenging research problems as future works. First, we want to apply our metrics to measure interestingness of proved theorems in mathematical books, and we will order those theorems from lower value to higher. Then, we will compare the order with the appearing order of those proved theorems in mathematical books. If they are almost same, we can confirm the effectiveness of our metrics firmly. Second, we will apply our metrics to measure the theorems in other mathematical fields, like geometry.

References

1. Bagai, R., Shanbhogue, V., Żytkow, J.M., Chou, S.C.: Automatic theorem generation in plane geometry. In: Komorowski, J., Raś, Z.W. (eds.) ISMIS 1993. LNCS, vol. 689, pp. 415–424. Springer, Heidelberg (1993). doi:10.1007/3-540-56804-2_39
2. Cheng, J.: A relevant logic approach to automated theorem finding. In: The Workshop on Automated Theorem Proving Attached to International Symposium on Fifth Generation Computer Systems, pp. 8–15 (1994)

3. Cheng, J.: Entailment calculus as the logical basis of automated theorem finding in scientific discovery. In: Systematic Methods of Scientific Discovery: Papers from the 1995 Spring Symposium, pp. 105–110. AAAI Press - American Association for Artificial Intelligence (1995)

4. Cheng, J.: A strong relevant logic model of epistemic processes in scientific discovery. In: Information Modelling and Knowledge Bases XI, Frontiers in Artificial Intelligence and Applications, vol. 61, pp. 136–159. IOS Press (2000)

5. Cheng, J., Nara, S., Goto, Y.: FreeEnCal: a forward reasoning engine with general-purpose. In: Apolloni, B., Howlett, R.J., Jain, L. (eds.) KES 2007. LNCS (LNAI), vol. 4693, pp. 444–452. Springer, Heidelberg (2007). doi:10.1007/978-3-540-74827-4_56

6. Colton, S.: Automated theorem discovery: a future direction for theorem provers. In: Proceeding of 1st Automated Reasoning: International Joint Conference, Workshop on Future Directions in Automated Reasoning, pp. 38–47 (2001)

7. Colton, S., Meier, A., Sorge, V., McCasland, R.: Automatic generation of classification theorems for finite algebras. In: Basin, D., Rusinowitch, M. (eds.) IJCAR 2004. LNCS (LNAI), vol. 3097, pp. 400–414. Springer, Heidelberg (2004). doi:10.1007/978-3-540-25984-8_30

8. Dalzotto, G., Recio, T.: On protocols for the automated discovery of theorems in elementary geometry. J. Autom. Reasoning 43(2), 203–236 (2009)

9. Gao, H., Goto, Y., Cheng, J.: A systematic methodology for automated theorem finding. Theoret. Comput. Sci. 554, 2–21 (2014). Elsevier

10. Gao, H., Goto, Y., Cheng, J.: Research on automated theorem finding: current state and future directions. In: Park, J.J.J.H., Pan, Y., Kim, C.-S., Yang, Y. (eds.) Future Information Technology. LNEE, vol. 309, pp. 105–110. Springer, Heidelberg (2014). doi:10.1007/978-3-642-55038-6_16

11. Gao, H., Goto, Y., Cheng, J.: A set of metrics for measuring interestingness of theorems in automated theorem finding by forward reasoning: a case study in NBG set theory. In: He, X., Gao, X., Zhang, Y., Zhou, Z.-H., Liu, Z.-Y., Fu, B., Hu, F., Zhang, Z. (eds.) IScIDE 2015. LNCS, vol. 9243, pp. 508–517. Springer, Heidelberg (2015). doi:10.1007/978-3-319-23862-3_50

12. McCasland, R., Bundy, A., Autexier, S.: Automated discovery of inductive theorems. J. Stud. Logic Grammar Rhetoric 10(23), 135–149 (2007)

13. Montes, A., Recio, T.: Automatic discovery of geometry theorems using minimal canonical comprehensive Gröbner systems. In: Botana, F., Recio, T. (eds.) ADG 2006. LNCS (LNAI), vol. 4869, pp. 113–138. Springer, Heidelberg (2007). doi:10.1007/978-3-540-77356-6_8

14. Puzis, Y., Gao, Y., Sutcliffe, G.: Automated generation of interesting theorems. In: Proceedings of 19th International Florida Artificial Intelligence Research Society Conference, pp. 49–54. AAAI press-The Association for the Advancement of Artificial Intelligence (2006)

15. Quaife, A.: Automated Development of Fundamental Mathematical Theories. Kluwer Academic, Norwell (1992)

16. Recio, T., Velez, M.Z.: Automatic discovery of theorems in elementary geometry. J. Autom. Reasoning 23(1), 63–82 (1999)

17. Tang, P., Lin, F.: Discovering theorems in game theory: two-person games with unique pure nash equilibrium payoffs. Artif. Intell. 175(14), 2010–2020 (2011)

18. Wos, L.: Automated Reasoning: 33 Basic Research Problem. Prentice-Hall, Upper Saddle River (1988)

19. Wos, L.: The problem of automated theorem finding. J. Autom. Reasoning 10(1), 137–138 (1993)

A Predicate Suggestion Algorithm for Automated Theorem Finding with Forward Reasoning

Yuichi Goto$^{(\boxtimes)}$, Hongbiao Gao, and Jingde Cheng

Department of Information and Computer Sciences,
Saitama University, Saitama 338-8570, Japan
{gotoh,gaohongbiao,cheng}@aise.ics.saitama-u.ac.jp

Abstract. The problem of automated theorem finding (ATF for short) is one of the 33 basic research problems in automated reasoning. To solve the ATF problem, an ATF method with forward reasoning based on strong relevant logics has been proposed and studied. In the method, predicate abstraction plays important role. However, in the current method, targets of predicate abstraction are predicates that an executor of ATF has already known. This paper presents a predicate suggestion algorithm to suggest previously unknown predicates and create abstraction rules for predicate abstraction in the ATF method with forward reasoning. The paper also shows that the proposed algorithm is effective through a case study.

Keywords: Predicate suggestion algorithm · Predicate abstraction · Automated theorem finding · Forward reasoning program

1 Introduction

The problem of automated theorem finding (ATF for short) is one of the 33 basic research problems in automated reasoning which was originally proposed by Wos in 1988 [9,10] and it is still an open problem until now. The problem of ATF is "What properties can be identified to permit an automated reasoning program to find new and interesting theorems, as opposed to proving conjectured theorems?" [9].

To solve the problem, an ATF method with forward reasoning based on strong relevant logics has been proposed and studied [1,5]. Cheng clarified (forward) reasoning, in contrast to proving (backward reasoning), is an indispensable process of ATF in scientific discovery [1], and also proposed strong relevant logics [2,3] as logic systems underlying the forward reasoning in ATF. Gao et al., then, proposed an ATF method with forward reasoning with strong relevant logics [5], "reasoning-based ATF method" for short. The reasoning-based ATF method consists of five phases: preparation of logical premises, preparation of empirical premises, forward reasoning, abstraction of deduced empirical theorems, and excavation of new and interesting empirical theorems.

© Springer International Publishing AG 2017
N.T. Nguyen et al. (Eds.): ACIIDS 2017, Part II, LNAI 10192, pp. 125–134, 2017.
DOI: 10.1007/978-3-319-54430-4_13

Predicate abstraction is important for the reasoning-based ATF method because, by predicate abstraction, an executor of the reasoning-based ATF can shorten the execution time of and save used memory space of automated forward reasoning in forward reasoning phase. Predicate abstraction is to replace subformulas with predicates in a logical formula according to abstraction rules. An abstraction rule is a pair of a logical formula and a predicate. In current reasoning-based ATF method, abstraction rules are given by the executor before doing reasoning-based ATF. In other word, only predicates that the executor has already known are targets of predicate abstraction. There is no study to automatically create previously unknown abstraction rules and suggest predicates that the executor has not known.

This paper presents a predicate suggestion algorithm to suggest previously unknown predicates and create abstraction rules for predicate abstraction in the reasoning-based ATF. The paper also shows a case study of creating abstraction rules from results of forward reasoning in von Neumann-Bernays-Godel (NBG) set theory and Peano's arithmetic, according to proposed algorithm.

The rest of the paper is organized as follows: Sect. 2 gives basic notions and notations. Section 3 explains overview of the reasoning-based ATF method, and relationship between predicate abstraction and the performance of forward reasoning programs. Section 4 presents a predicate suggestion algorithm, and shows a case study to confirm the effectiveness of the algorithm. Section 5 gives conclusion and future works.

2   Basic Notions and Notations

In general, a formal logic system L consists of a formal language, called the object language and denoted by $F(L)$, which is the set of all well-formed formulas of L, and a logical consequence relation, denoted by meta-linguistic symbol $\vdash_L$, such that $P \subseteq F(L)$ and $c \in F(L)$, $P \vdash_L c$ means that within the frame work of L, c is valid conclusion of premises P, i.e., c validly follows from P. For a formal logic system $(F(L), \vdash_L)$, a logical theorem t is a formula of L such that $\phi \vdash_L t$ where ϕ is empty set. Let $Th(L)$ denote the set of all logical theorems of L. $Th(L)$ is completely determined by the logical consequence relation $\vdash_L$. According to the representation of the logical consequence relation of a logic, the logic can be represented as a Hilbert style axiomatic system, a Gentzen natural deduction system, a Gentzen sequent calculus system, or other type of formal system.

A formal theory with premises P based on L, called a L-theory with premises P and denoted by $T_L(P)$, is defined as $T_L(P) := Th(L) \cup Th_L^e(P)$, and $Th_L^e(P) := \{et \mid P \vdash_L et \text{ and } et \notin Th(L)\}$ where $Th(L)$ and $Th_L^e(P)$ are called the logical part and the empirical part of the formal theory, respectively, and any element of $Th_L^e(P)$ is called an empirical theorem of the formal theory.

Both $Th(L)$ and $Th_L(P)$ are infinite sets in any logic system L and any formal theory based on L. To deal with finite subset of $Th(L)$ or $Th_L(P)$, the notion of degree of nested logical connectives and modal operators has been introduced [4]. Let θ be an arbitrary n-ary $(1 \leq n)$ connective or modal operator of logic L and A be a formula of L, the degree of θ in A, denoted by $D_\theta(A)$, is defined as follows: (1) $D_\theta(A) = 0$ if and only if there is no occurrence of θ in A, (2) if A is in the form $\theta(a_1, a_2, \ldots, a_n)$ where a_1, a_2, $\ldots$, a_n are formulas, then $D_\theta(A) = \max\{D_\theta(a_1), D_\theta(a_2), \ldots, D_\theta(a_n)\} + 1$, (3) if A is in the form $\sigma(a_1, a_2, \ldots, a_n)$ where σ is a connective or modal operator different from θ and a_1, a_2, $\ldots$, a_n are formulas, then $D_\theta(A) = \max\{D_\theta(a_1), D_\theta(a_2), \ldots, D_\theta(a_n)\}$, and (4) if A is in the form QB where B is a formula and Q is the quantifier prefix of B, then $D_\theta(A) = D_\theta(B)$.

By using notion of degree of nested logical connectives and modal operators, a formal logic system L can be divided into several fragments as follows: Let θ_1, θ_2, $\ldots$, θ_n be connectives or modal operators of logic L and $k_1, k_2, \ldots, k_n \in \mathbb{N}$, the fragment of L about θ_1, θ_2, $\ldots$, θ_n and their degrees k_1, k_2, $\ldots$, k_n, denoted by $Th^{(\theta_1, k_1, \theta_2, k_2, \ldots, \theta_n, k_n)}(L)$, is a set of logical theorems of L which is inductively defined as follows (in the terms of Hilbert style axiomatic system): (1) if A is an axiom of L and $D_{\theta_1}(A) \leq k_1$, $D_{\theta_2}(A) \leq k_2$, $\ldots$, $D_{\theta_n}(A) \leq k_n$, then $A \in Th^{(\theta_1, k_1, \theta_2, k_2, \ldots, \theta_n, k_n)}(L)$, (2) if A is the result of applying an inference rule of L to some members of $Th^{(\theta_1, k_1, \theta_2, k_2, \ldots, \theta_n, k_n)}(L)$ and $D_{\theta_1}(A) \leq k_1$, $D_{\theta_2}(A) \leq k_2$, $\leq$, $D_{\theta_n}(A) \leq k_n$, then $A \in Th^{(\theta_1, k_1, \theta_2, k_2, \ldots, \theta_n, k_n)}(L)$, and (3) Nothing else are members of $Th^{(\theta_1, k_1, \theta_2, k_2, \ldots, \theta_n, k_n)}(L)$. $Th^{(\theta_1, k_1, \theta_2, k_2, \ldots, \theta_n, k_n)}(L)$ is a subset of $Th(L)$.

Let $(F(L), \vdash_L)$ be a formal logic system, $P \subseteq F(L)$ ($P \neq \emptyset$), $\theta_1, \theta_2, \ldots$, $\theta_m, \eta_1, \eta_2, \ldots, \eta_n$ ($\eta_j = \theta_i$, $1 \leq i \leq m$, $1 \leq j \leq n$, $n \leq m$) be connectives or modal operators of logic L, and $k_1, k_2, \ldots, k_m, l_1, l_2, \ldots, l_n, p, q \in \mathbb{N}$. A formula A is said to be $(\eta_1, l_1, \eta_2, l_2, \ldots, \eta_n, l_n)$-degree-deducible from P based on $Th^{(\theta_1, k_1, \theta_2, k_2, \ldots, \theta_m, k_m)}(L)$ if and only if there is a finite sequence of formulas f_1, $\ldots$, f_p ($p \in \mathbb{N}$) such that $f_p = A$ and for all i ($i \leq p$), (1) $f_i \in Th^{(\theta_1, k_1, \theta_2, k_2, \ldots, \theta_m, k_m)}(L)$, or (2) $f_i \in P$, or (3) f_i is the result of applying an inference rule to some members f_{j_1}, $\ldots$, f_{j_m} ($j_1, \ldots, j_q < i$) of the sequence and $D_{\eta_1}(f_i) \leq l_1$, $D_{\eta_2}(f_i) \leq l_2$, $\ldots D_{\eta_n}(f_i) \leq l_n$. The set of all formulas which are $(\eta_1, l_1, \eta_2, l_2, \ldots, \eta_n, l_n)$-degree-deducible from P based on $Th^{(\theta_1, k_1, \theta_2, k_2, \ldots, \theta_m, k_m)}(L)$ is called the $(\eta_1, l_1, \eta_2, l_2, \ldots, \eta_n, l_n)$ degree fragment of the formal theory with premises P based on $Th^{(\theta_1, k_1, \theta_2, k_2, \ldots, \theta_m, k_m)}(L)$, denoted by $T^{(\eta_1, l_1, \eta_2, l_2, \ldots, \eta_n, l_n)}_{Th^{(\theta_1, k_1, \theta_2, k_2, \ldots, \theta_m, k_m)}(L)}(P)$. Note that in the above definitions, we do not require $l_i \leq k_j$ if $\eta_i = \theta_j$.

3 Automated Theorem Finding with Forward Reasoning

3.1 Overview

The reasoning-based ATF method consists of five phases: preparation of logical premises, preparation of empirical premises, forward reasoning, abstraction of deduced empirical theorems, and excavation of new and interesting theorems [5]. Figure 1 shows the flow of the reasoning-based ATF method. In the method, at first, executors choose logic premises, which is a set of logical theorems of strong relevant logics, as logic basis of the ATF, then prepare candidates of empirical

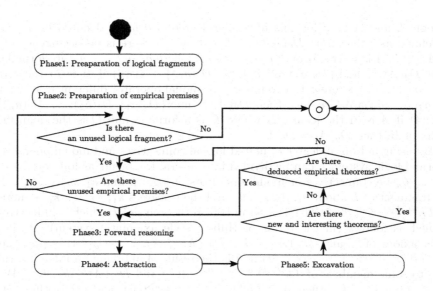

Fig. 1. Flow of reasoning-based ATF.

premises, which are logical formulas that denote definitions, axioms, and already known theorems in the target field. After that, the executors choose a certain set of candidates as empirical premises. They do forward reasoning, abstraction of deduced empirical theorems, and excavation of new and interesting theorems repeatedly until no new logical formulas are deduced at the forward reasoning phase. If new and interesting theorems are found, the ATF is succeeded. If not, then executors change logic premises and/or empirical premises and continue to try ATF.

The reasoning-based ATF method uses FreeEnCal [4] to do automated forward reasoning in forward reasoning phase. FreeEnCal is a forward reasoning engine for general purpose, that provides an easy way to customize reasoning task by providing different axioms, inference rules and facts. Users can set the degree of nested logical connectives and modal operators to make FreeEnCal to reason out in principle all logical theorem schemata of the fragment $Th^{(\theta_1,k_1,\theta_2,k_2,\ldots,\theta_m,k_m)}(L)$ where θ_1, θ_2, ..., θ_m are connectives or modal operators of logic L and $k_1, k_2, \ldots, k_m \in \mathbb{N}$. FreeEnCal can also reason out in principle all empirical theorems of $T_{Th^{(\theta_1,k_1,\ldots,\theta_m,k_m)}(L)}^{(\eta_1,j_1\ldots\eta_n,j_n)}(P)$ from $Th^{(\theta_1,k_1,\ldots,\theta_m,k_m)}(L)$ and P with inference rules of L where θ_1, θ_2, ..., θ_m, η_1, η_2, ..., η_n ($\eta_j = \theta_i$, $1 \leq i \leq m, 1 \leq j \leq n, n \leq m$) are connectives or modal operators of logic L, and $k_1, k_2, \ldots, k_m, l_1, l_2, \ldots, l_n \in \mathbb{N}$. In the reasoning-based ATF method, the above logic system L is a strong relevant logic, and the above P is a set of definitions, axioms, and already obtained theorems of the target field. An output file of FreeEnCal has three kinds of formulas: logical theorem schemata, intermediates, and empirical theorems. The logical theorem schemata are given as input data. An intermediate is a formula that is not a logical theorem schema and includes meta variables in the formula. A meta variable is a variable that can be replaced with any logical formula or logical schema. An empirical theorem is a logical formula that is an element of P or $T^{e\ (\eta_1, j_1 \dots \eta_n, j_n)}_{Th^{(\theta_1, k_1, \dots, \theta_m, k_m)}(L)}(P)$.

At abstraction phase, predicate abstraction is done according to following procedure [6]. An abstraction rule r is defined as $r = < p, w >$ where p is a predicate, w is a logical formula.

1. S is a set of all abstraction rules.
2. FOR $r = < p, w > \in S$ IF $S \neq \emptyset$.
   (a) IF any deduced empirical theorem and intermediate A includes w and does not include p, then create a new formula by replacing all w in A with p.
   (b) $S \leftarrow S - \{r\}$

Note that operation 2-(a) is different from the original operation in [6] because the original operation removes effective theorems that show the relationship between p and w by replacing w with p in those theorems.

In current reasoning-based ATF method, abstraction rules are created from definitions of predicates in a target field before doing ATF. For example, suppose p and q are different predicates, and r is defined as $\forall x_1 \dots \forall x_i ((p(x_1, \dots, x_j) \wedge q(x_{j+1}, \dots, x_i)) \Leftrightarrow r(x_1, \dots, x_i))$ where $i, j \in \mathbb{N}$, and $1 < j < i$. An abstraction rule is $< r(x_1, \dots, x_i), p(x_1, \dots, x_j) \wedge q(x_{j+1}, \dots, x_i) >$.

3.2 Predicate Abstraction and Performance of Forward Reasoning

Predicate abstraction is important for the reasoning-based ATF method. First, theorem finding processes in scientific discovery must include some concept/notion abstraction processes. For example, in any mathematical field, definitions and axioms mean simple concepts. Mathematicians continue to define more complex concepts by using previously given definitions and axioms, and already defined concepts. Then, mathematicians think, assume, prove propositions by using the defined complex concepts. After that, they obtain new theorems.

Second, predicate abstraction helps efficient forward reasoning in the reasoning-based ATF method. To do reasoning-based ATF method automatically, some forward reasoning engine is necessary for automated forward reasoning in forward reasoning phase. A forward reasoning engine is a computer program to automatically draw new conclusions by repeatedly applying inference rules to given premises and obtained conclusions until some previously specified conditions are satisfied. Thus, the execution time of any forward reasoning engine becomes longer explosively as the number of logical formulas that the engine deals with becomes larger. On the other hand, to find new and interesting theorems by reasoning-based ATF method, it is better to get the large number of logical formulas by forward reasoning phase. This is a dilemma.

By predicate abstraction, we can keep limitation of degree of nested logical connectives and modal operators low while we try to get complex logical formulas by automated forward reasoning. Current reasoning-based ATF method uses FreeEnCal as a forward reasoning engine. FreeEnCal controls the amount of deduced logical formulas by using limitation of degree of nested logical connectives and modal operators. As higher degree of nested logical connectives and modal operators is used as input of FreeEnCal, the execution time of FreeEnCal drastically becomes longer because the number of logical formulas in the fragment of a target logic system or a target formal theory under the higher degree drastically increases [7]. Thus, the low degree of nested logical connectives and modal operators is better for the execution of FreeEnCal rather than the high degree. On the other hand, the degree of nested logical connectives and modal operators of a logical formula can be regarded as an indicator of complexity of the logical formula. Although a complex theorem is not always interesting, an interesting theorem must be complex rather than simple in ATF. Predicate abstraction saves this dilemma. The degree of nested logical connectors of a logical formula becomes lower by replacing subformulas in deduced logical formulas with corresponding predicates. From viewpoint of meaning of logical formulas, we can try to get complex logical formulas with low degree of nested logical connectives by using predicate abstraction. Because it is possible to keep limitation of degree of nested logical connectives low, we can do efficient forward reasoning in the reasoning-based ATF method.

In current reasoning-based ATF method, abstraction rules are given by the executor before doing reasoning-based ATF. In other word, only predicates that the executor has already known are targets of predicate abstraction. There is no study to automatically suggest predicates that the executor has not known, and create previously unknown abstraction rules.

4   A Predicate Suggestion Algorithm

Key ideas of predicate suggestion are the patterns of subformulas and the frequency of occurrence of subformulas. Generally, the main logical connective of a logical formula that represent a definition of defined predicate is conjunction ($\land$) or entailment ($\Rightarrow$). A defined predicate is a predicate whose definition consists of other already defined predicates. The main logical connective of a logical formula is the first logical connective of a logical formula represented in the prefix (polish) notation. For example, the main connective of all logical formulas that represent defined predicate of von Neumann-Bernays-Godel (NBG) set theory denoted in Quaife's book [8] is conjunction or entailment. Thus, a subformula whose main connective is conjunction or entailment is a candidate of a new predicate. On the other hand, generally, a defined predicate of a certain mathematical field is often used in proofs in the field. If a concept that represented by several predicates is used in only a certain proof in a certain mathematical field, no one want to define a new predicate that represent the concept because it is waste. Thus, a subformula that frequently occurs in results of a forward reasoning is also a candidate of a new predicate.

According to above key ideas, we proposed a predicate suggestion algorithm for predicate abstraction in the reasoning-based ATF method. The input of the algorithm are a threshold $t \in \mathbb{N}$, a set of formulas ET that represent empirical theorems deduced by FreeEnCal, and a list of already given abstraction rules R_g. The output of the algorithm is a list of new abstraction rules R_n. An abstraction rule r is defined as $r =< p, s >$ where p is a predicate, and s is a subformula.

1. Count the number of occurrence of each subformula in ET
2. $SF \leftarrow$ all subformulas in ET.
3. FOR $s \in SF$ IF $SF \neq \emptyset$
   (a) IF ($t \leq$ the number of occurrence of s.)
       AND (s is not w_g of $r_g =< p_g, w_g >\in R_g$.)
       AND (the main logical connective of s is $\wedge$ or $\Rightarrow$.)
       i. Prepare new predicate symbol Q_n.
       ii. $i \leftarrow$ kinds of free variables in s.
       iii. Create new predicate $Q_n(x_1, \ldots, x_i)$.
       iv. Add new abstraction rule $< Q_n(x_1, \ldots, x_i), s >$ into R_n
   (b) $SF \leftarrow SF - \{s\}$

To confirm effectiveness of our proposed algorithm, we did a case study. The procedure of the case study is as follows: (1) Create abstraction rules from a fragment of formal theory deduced by FreeEnCal, according to our proposed algorithm. (2) Do predicate abstraction to the fragment of formal theory with the created abstraction rules. (3) Do forward reasoning with the fragment of formal theory that is done predicate abstraction.

We used two fragments as fragments of formal theories deduced by FreeEn-Cal. One is $T_{Th^{(\Rightarrow,2,\wedge,1,\neg,1)}(EcQ)}^{(\Rightarrow,2,\wedge,1,\neg,1)}(NBG)$, "NBG fragment" for short. Other is $T_{Th^{(\Rightarrow,2,\wedge,1,\neg,1)}(EcQ)}^{(\Rightarrow,2,\wedge,1,\neg,1)}(Peano)$, "PA fragment" for short. EcQ is a strong relevant logic [3]. "NBG" means a set of definitions and axioms of NBG set theory denoted in Quife's book [8]. "Peano" means a set of definitions and axioms of Peano's arithmetic denoted in Quife's book [8]. All $Th^{(\Rightarrow,2,\wedge,1,\neg,1)}(EcQ)$, NBG fragment, and PA fragment were prepared in [5]. The number of empirical premises, logical premises, intermediates, and deduced theorems in NBG fragment are 89, 311, 1,532, and 421, respectively. The number of empirical premises, logical premises, intermediates, and deduced theorems in PA fragment are 154, 311, 1,618, and 473, respectively.

At first, we counted all subformulas that satisfy all conditions at the proposed algorithm 2-(a), "suggestion condition" for short, in NBG fragment and PA fragment. Tables 1 and 2 are frequency tables of the number of occurrence and patterns of subformulas in the fragments. In those tables, "occur." means the number of occurrence of a subformula in the fragments. "subwff" means the number of subformulas that the number of occurrence is same. "($\Rightarrow$)" means the number of subformulas that have one $\Rightarrow$ and the number of occurrence is same. " ($\wedge$), ($\Rightarrow\Rightarrow$), ($\Rightarrow \wedge$), ($\wedge\wedge$), ($\Rightarrow \wedge\wedge$), ($\wedge\wedge\wedge$), and ($\Rightarrow \wedge\wedge\wedge$)" are similar meaning. Note that all $A \Rightarrow (B \wedge C)$, $A \wedge (B \Rightarrow C)$, $(A \Rightarrow B) \wedge C$, and $(A \wedge B) \Rightarrow C$ are regarded as elements of "($\Rightarrow \wedge$)." The tables show that there are dozen of subformula that satisfy the suggestion condition and the number of occurrence of them is more than 1. Those subformulas are candidates of new predicates.

Table 1. Frequency table of the number of occurrence and patterns of subformulas in a fragment of NBG set theory

Occur.	Subwff	(⇒)	(∧)	(⇒ ∧)	(∧∧)	(⇒ ∧∧)
1	77	46	7	20	2	2
3	4	0	4	0	0	0
5	3	1	2	0	0	0
6	7	0	7	0	0	0
8	1	0	1	0	0	0
9	2	0	2	0	0	0
10	4	0	4	0	0	0
12	1	0	1	0	0	0
13	1	1	0	0	0	0
18	1	0	1	0	0	0
19	0	0	0	0	0	0
21	1	0	1	0	0	0
22	1	0	1	0	0	0
Total	103	48	31	20	2	2

Table 2. Frequency table of the number of occurrence and patterns of subformulas in a fragment of Peano's arithmetic

Occur	Subwff	(⇒)	(∧)	(⇒⇒)	(⇒ ∧)	(∧∧)	(⇒ ∧∧)	(∧ ∧ ∧)	(⇒ ∧ ∧ ∧)
1	119	79	6	9	19	1	3	1	1
2	8	8	0	0	0	0	0	0	0
3	12	2	10	0	0	0	0	0	0
4	3	3	0	0	0	0	0	0	0
5	3	0	3	0	0	0	0	0	0
6	6	1	5	0	0	0	0	0	0
7	5	0	5	0	0	0	0	0	0
8	1	1	0	0	0	0	0	0	0
9	1	0	1	0	0	0	0	0	0
12	1	0	1	0	0	0	0	0	0
14	1	0	1	0	0	0	0	0	0
15	1	0	1	0	0	0	0	0	0
16	1	0	1	0	0	0	0	0	0
18	2	0	2	0	0	0	0	0	0
28	1	0	1	0	0	0	0	0	0
Total	165	94	37	9	19	1	3	1	1

Table 3. The number of abstracted and deduced formulas by using created abstraction rules

Kinds	NBG (3 ≤)	NBG (10 ≤)	PA (3 ≤)	PA (8 ≤)
Abstracted	185	141	231	114
Intermediate	110	88	124	76
Theorem	75	53	107	38
Deduced	1,220	1,005	1,379	301
Intermediate	1,168	962	1,310	275
Theorem	52	43	69	26

Second, we created abstraction rules according to our proposed algorithm. We used 3 and 10 as thresholds for NBG fragment, and 3 and 8 for PA fragment. 10 and 8 are about the median of the number of occurrence of subformulas in both fragments. From NBG fragment, 26 and 9 abstraction rules are created when thresholds are 3 and 10, respectively. From PA fragment, 38 and 9 abstraction rules are created when thresholds are 3 and 8, respectively.

Third, we did predicate abstraction to the fragment by using the created abstraction rules. The upper half of Table 3 shows the number of abstracted formulas. The first column of upper half of Table 3 means that we got 110 intermediates and 75 theorems (total 185) by predicate abstraction to NBG fragment according to created abstraction rules whose threshold is 3.

Finally, we add abstracted formulas into each fragment, then did forward reasoning under the same limitation of degree of nested logical connectives and modal operators, that is $(\Rightarrow, 2, \wedge, 1, \neg, 1)$. The lower half of Table 3 shows the number of deduced formulas. The first column of lower half of Table 3 means that output file of FreeEnCal includes 1,168 intermediates and 52 theorems (total 1,220) deduced from NBG fragment and abstracted formulas whose abstraction rules' threshold is 3.

For results of the case study, we can conclude that the proposed predicate suggestion algorithm is effective for the reasoning-based ATF method. First, there are actually dozen of subformula that satisfy suggestion condition and the number of occurrence of them is more than 1 in both NBG fragment and PA fragment. Second, after predicate abstraction by using the created abstraction rules according to the algorithm, FreeEnCal can deduced additional empirical theorems from the abstracted empirical theorems without changing the limitation of degree of nested logical connectives and modal operators. In this case study, using low threshold does not have so bad impact to additional forward reasoning.

5   Concluding Remarks

This paper has proposed the first predicate suggestion algorithm for predicate abstraction in automated theorem finding with forward reasoning. The proposed algorithm suggests predicates that are defined as logical formulas which consist of already known predicates, by finding subformulas that satisfy some conditions from deduced empirical theorem automatically. For predicate abstraction, the algorithm also provides abstraction rules to replace subformulas with suggested predicates in a logical formula.

The paper also has shown a case study. In the case study, the proposed algorithm can create abstraction rules from deduced empirical theorems in two different formal theories. After predicate abstraction by using the created abstraction rules, a forward reasoning engine can deduced additional empirical theorems from the abstracted empirical theorems without changing the limitation of degree of nested logical connectives and modal operators. For results of the case study, we can conclude that the proposed predicate suggestion algorithm is effective for automated theorem finding with forward reasoning.

On the other hand, from view point of meaning, we have not investigated whether the suggested predicates are meaningful or not in the target field. To propose how to investigate meaning of suggested predicates and to evaluate the predicates are future works.

References

1. Cheng, J.: Entailment calculus as the logical basis of automated theorem finding in scientific discovery. In: Raul, V. (ed.) Systematic Methods of Scientific Discovery: Papers from the 1995 Spring Symposium, pp. 105–110. AAI Press (1995)
2. Cheng, J.: A strong relevant logic model of epistemic processes in scientific discovery. In: Kawaguchi, E., Kangassalo, H., Jaakkola, H., Hamid, I.A. (eds.) Information Modeling and Knowledge Bases XI. Frontiers in Artificial Intelligence and Applications, vol. 61, pp. 136–159. IOS Press, Amsterdam (2000)
3. Cheng, J.: Strong relevant logic as the universal basis of various applied logics for knowledge representation and reasoning. In: Kiyoki, Y., Henno, J., Jaakkola, H., Kangassalo, H. (eds.) Information Modeling and Knowledge Bases XVII. Frontiers in Artificial Intelligence and Applications, vol. 136, pp. 310–320. IOS Press (2006)
4. Cheng, J., Nara, S., Goto, Y.: FreeEnCal: a forward reasoning engine with general-purpose. In: Apolloni, B., Howlett, R.J., Jain, L. (eds.) KES 2007. LNCS (LNAI), vol. 4693, pp. 444–452. Springer, Heidelberg (2007). doi:10.1007/978-3-540-74827-4_56
5. Gao, H., Goto, Y., Cheng, J.: A systematic methodology for automated theorem finding. Theoret. Comput. Sci. **554**, 2–21 (2014)
6. Gao, H., Goto, Y., Cheng, J.: Explicitly epistemic contraction by predicate abstraction in automated theorem finding: a case study in NBG set theory. In: Nguyen, N.T., Trawiński, B., Kosala, R. (eds.) ACIIDS 2015. LNCS (LNAI), vol. 9011, pp. 593–602. Springer, Heidelberg (2015). doi:10.1007/978-3-319-15702-3_57
7. Goto, Y., Cheng, J.: A quantitative analysis of implicational paradoxes in classical mathematical logic. Electron. Notes Theor. Comput. Sci. **169**, 87–97 (2007)
8. Quaife, A.: Automated Development of Fundamental Mathematical Theories. Kluwer Academic, Heidelberg (1992)
9. Wos, L.: Automated Reasoning: 33 Basic Research Problems. Prentice-Hall, Upper Saddle River (1988)
10. Wos, L.: The problem of automated theorem finding. J. Autom. Reason. **10**(1), 127–138 (1993)

Collective Intelligence for Service Innovation, Technology Opportunity, E-Learning and Fuzzy Intelligent Systems

Modeling a Multi-criteria Decision Support System for Capital Budgeting Project Selection

Kuo-Sui Lin[(⊠)] and Jui-Ching Pan

Department of Information Management,
Aletheia University, Taipei, Taiwan R.O.C.
{au4234,au1179}@mail.au.edu.tw

Abstract. Capital budgeting project selection is an important part of strategic decision-making in every enterprise because successful new investment projects essentially contribute to enterprise's financing growth, value proposition and strategic intent. Therefore, the main purpose of this study was to present a design philosophy and operation process for modeling a decision support system to handle capital budgeting project selection problems. In order to achieve this purpose, the goal of this study has been two-fold. The first objective was to propose a new fuzzy multi-criteria decision making method for project alternative comparison and selection. The second objective was to employ the new fuzzy multi-criteria decision-making method to model the computational architecture of the decision support system for capital budgeting project selection. Finally, an algorithm and a numerical example resumed the design philosophy and the operation process of the modeled multi-criteria DSS was illustrated and the results have indicated that the objectives of the study were achieved.

Keywords: Capital budgeting · Decision support system · Fuzzy multi-criteria decision-making · Project selection

1 Introduction

Enterprises need a software application or system to prioritize and select capital budgeting projects and to allocate limited resources for efficient investment decision-making. This need can be met by a decision support system (DSS). However, there has been little research done on issues pertaining to this topic and content. Therefore, the main purpose of this study was to present a design philosophy and operation process for modeling a DSS to handle capital budgeting project selection problems. The goal of this study has been two-fold. The first objective was to propose a new fuzzy multi-criteria decision-making (MCDM) method which has the ability to process quantitative and qualitative fuzzy data for project prioritization and selection. The second objective was to outline the computational architecture for the DSS system which employs the new fuzzy MCDM method for capital budgeting project selection.

© Springer International Publishing AG 2017
N.T. Nguyen et al. (Eds.): ACIIDS 2017, Part II, LNAI 10192, pp. 137–147, 2017.
DOI: 10.1007/978-3-319-54430-4_14

2 Related Works

2.1 Decision Support System

Accumulated knowledge, experience and expertise are precious. However, such knowledge, experience and expertise reside primarily in experts' minds and are seldom documented in a reusable form of case database (case base). A DSS is an interactive, computer-based system that helps the decision-makers in the use of data and models to solve various semi-structured and unstructured problems involving multiple attributes, objectives, or goals [1]. Like human decision-making process, DSSs with previously accumulated knowledge, experience and expertise have gained popularity for their efficacy to help individuals and organizations with their decision-making processes, typically resulting in ranking or choosing from among alternatives.

2.2 Capital Budgeting

Capital budgeting, also known as investment planning or capital budgeting planning, is the process of analyzing investment opportunities in long-term assets which are expected to produce benefits for more than one year [2]. Capital budgeting is often utilized as a means of directly or indirectly achieving objectives within an organization's long-term strategic plan. There are several types of capital budgeting project classification: (1) Cost reduction project; (2) Replacement project; (3) Expansion project; (4) Safety and environmental project. Booth et al. [3] defined capital budgeting decision making as a capital budgeting projects selection exercise by using quantitative criteria such as payback period, net present value, internal rate of return, profitability index and other qualitative criteria.

2.3 Multi-criteria Decision Making

A MCDM problem aims to find a desirable solution from a finite number of feasible alternatives, which are assessed on multiple criteria both quantitatively and qualitatively [4, 5]. Let A be a discrete set of m alternatives and let C be a set of n independent criteria, where $A = \{A_1, A_2, \ldots, A_m\}$, $C = \{C_1, C_2, \ldots, C_n\}$ respectively. A traditional MCDM problem can be expressed in a matrix format. As shown in Fig. 1, a decision matrix shows importance weightings and performance ratings of assessment of how

$$
R_{mn} = \begin{array}{c} \\ W \\ A_1 \\ A_2 \\ \vdots \\ A_i \\ \vdots \\ A_m \end{array}
\begin{array}{|cccccc|}
C_1 & C_2 & \ldots & C_j & \ldots & C_n \\
\omega_1 & \omega_2 & & \omega_j & & \omega_n \\
r_{11} & r_{12} & \ldots & r_{1j} & \ldots & r_{1n} \\
r_{21} & r_{22} & \ldots & r_{2j} & \ldots & r_{2n} \\
\vdots & \vdots & \ldots & \vdots & \ldots & \vdots \\
r_{i1} & r_{i2} & \ldots & r_{ij} & \ldots & r_{in} \\
\vdots & \vdots & \ldots & \ldots & \ldots & \vdots \\
r_{m1} & r_{m2} & \ldots & r_{mj} & \ldots & r_{mn}
\end{array}
$$

Fig. 1. Decision matrix showing importance weightings and performance ratings of assessment

each alternative satisfies each criterion. The case-criteria rating entries usually are described as an $m \times n$ decision matrix R_{mn}, where the row represents m cases and the column represents n criteria. The rating $r_{ij}(1 \leq i \leq m, 1 \leq j \leq n)$ describes the performance of alternative A_i against criterion C_j. Entries of the criteria rating matrix for alternative cases can be nominal, ordinal, interval or ratio scale. However, in real life situations preference ratings are usually subjective, vague and imprecise. The decision matrix must be transformed into fuzzy decision matrix to surpass these limitations.

3   The Proposed Multi-criteria Decision-Making Method

This study proposes a new fuzzy MCDM method to calculate the similarity measures between an ideal alternative and candidate alternatives and prioritize these alternatives. A stepwise description of the algorithm for the proposed fuzzy MCDM method can be summed up as a series of successive steps:

Step 1: Extracting Criteria
Project success criteria define key areas of performance that are essential for a project to accomplish its goal. Every alternative project can be characterized by a set of descriptive criteria C, which can be objective factual or subjective judgmental. $C = \{C_1, C_2, \ldots, C_j, \ldots, C_n\}, j = 1, 2, \ldots, n$. There are a number of appraisal criteria which may be recommended for evaluating capital budgeting projects. The criteria can be extracted and consensus can be reached during a focus group workshop.

Step2: Collecting Candidate Alternatives Cases and Ideal Alternative Case to the Case Base of the DSS
Possible alternative case solutions are brainstormed to satisfy the project case problem. The alternative case solutions can be expressed as $A_i = \{(C_1, r_{i1}), (C_2, r_{i2}), \ldots, (C_j, r_{ij}), \ldots (C_n, r_{in})\}, i = 1 \ldots m$. In multi-criteria decision-making environments, the concept of ideal point has been used to help the identification of the best alternative in the decision set. Although the ideal alternative does not exist in real world, it does provide a useful theoretical construct to evaluate alternatives. An fictitious ideal alternative to a MCDM problem can be described as the following criteria-rating pairs: $A_* = \{(C_1, r_1^*), (C_2, r_2^*), \ldots, (C_j, r_j^*), \ldots, (C_n, r_n^*)\}$. Each element of the alternatives can be evaluated by a traditional rating scale.

Step 3: Collecting Case-Criteria Weighting Vector for All Alternative Cases
Weightings reflect the relative importance attached by decision makers to various criteria. The respective weightings for the criteria can be described as the following criteria-weighting pairs: $W = \{(C_1, \omega_1), (C_2, \omega_2), \ldots, (C_j, \omega_j), \ldots, (C_n, \omega_n)\}$. Each element of the classical discrete set can be expressed by a traditional scale, which can be transformed into corresponding fuzzy number (fuzzy scale).

Step 4: Translating Traditional Scales into Fuzzy Measurement Scales
In practical application, ratings and weightings of criteria could be interval, ratio, ordinal, as well as nominal values. Besides, the larger the traditional scale value the higher the ranking is no more a common rule for decision making. Therefore,

multi-criteria decision-making requires commensurate scale so that traditional ratings and weightings of criteria can be transformed into a common, comparable same format. In order to treat uncertainty, by referring to several types of triangular fuzzy numbers in linguistic terms [6], the traditional measurement scales can be translated into commensurate fuzzy measurement scales. In the proposed MCDM method, the alternative case and ideal case which are rated and weighted by various traditional measurement scales will then be translated into their corresponding fuzzy linguistic terms (five-point linguistics cales) for representing its approximate measurement value. Each fuzzy linguistic term can be characterized and converted into their corresponding triangular fuzzy number (triangular fuzzy scale).

Step 5: Computing Similarity Measures
Two kinds of similarity measures used in this study are local similarity measure and global similarity measure. Local similarity is similarity on feature level; Global similarity is similarity on case level. For a criteria vector, a local similarity measure is computed by comparing each criteria value and a global similarity measure is obtained as a weighted calculation of the local similarity measures.

Substep 5.1: Computing Local Similarity Measure
The local similarity (criterion similarity) is the similarity measure on each criterion of the idea case vector A^* and the alternative case vector A_i, where i is an individual case from 1 to m. Let $sim(r_{*j}, r_{ij})$ denote the local similarity measure between the idea case vector A^* and the alternative case vector A_i with regard to criteria C_j, j is an individual criterion from 1 to n. r_{*j} and r_{ij} are the rating on C_j in the idea case A^* and the alternative case A_i, respectively. $sim(r_{*j}, r_{ij}) = \text{Dot}(r_{*j}, r_{ij}) / \|r_{*j}\| \|r_{ij}\| = r_{*j} \bullet r_{ij} / \|r_{*j}\| \|r_{ij}\|$ is a general equation to calculate the cosine similarity of the two vectors [7, 8].

(1) Cosine Similarity Measure for Fuzzy Nominal Scale
A nominal scale is a scale with no order in rank and an ordinal scale is a scale with an order in rank. Let $sim(r_{*j}, r_{ij})$ denote the feature similarity measure between the idea case A^* and the alternative case A_i with regard to the criteria C_j. If the criteria C_j is a nominal variable, such as Gender, the local similarity between two ratings r_{*j} and r_{ij} on the j-th criteria is given by binary similarity values: $sim(r_{*j}, r_{ij}) = 1$, if $r_{*j} = r_{ij}$; $sim(r_{*j}, r_{ij}) = 0$, if $r_{*j} \neq r_{ij}$.

(2) Cosine Similarity Measure for Fuzzy Ordinal, Interval and Ratio Scale
The ordinal, interval and ratio scales can be translated into their corresponding triangular fuzzy numbers. Thus, a cosine similarity measure for triangular fuzzy numbers is proposed in an analogous manner to the cosine similarity measure (angular coefficient) between fuzzy sets [9, 10]. Assume that there are two triangular fuzzy numbers: $r_{ij} = (\mu_{Ai}(x_1), \mu_{Ai}(x_2), \mu_{Ai}(x_3))$ and $r_{*j} = (\mu_{A*}(x_1), \mu_{A*}(x_2), \mu_{A*}(x_3))$, in the set of real numbers R, the three parameters in r_{*j} and r_{ij} can be considered as a vector representation with the three elements. Based on the extension of the cosine similarity measure for fuzzy sets, a cosine similarity measure between r_{*j} and r_{ij} is presented as:

Cosine similarity$(r_{*j},r_{ij}) = sim(r_{*j},r_{ij}) = \mathrm{Dot}(r_{*j},r_{ij})/\|r_{*j}\| \; \|r_{ij}\| = r_{*j}\cdot r_{ij}/\|r_{*j}\| \; \|r_{ij}\|$

$\sum_{k=1}^{3} (\mu_{*j}(x_k) \times \mu_{ij}(x_k))/(\mathrm{sqrt}(\sum_{k=1}^{3} \mu_{*j}(x_k)^{\wedge}2) \times \mathrm{sqrt}(\sum_{k=1}^{3} \mu_{ij}(x_k)^{\wedge}2)) = (\mu_{*j}(x_1) \times \mu_{ij}(x_1)$
$+\mu_{*j}(x_2) \times \mu_{ij}(x_2) + \mu_{*j}(x_3) \times \mu_{ij}(x_3))/(\mathrm{sqrt}(\mu_{*j}(x_1)^{\wedge}2 + \mu_{*j}(x_2)^{\wedge}2 + \mu_{*j}(x_3)^{\wedge}2) \times \mathrm{sqrt}$
$(\mu_{ij}(x_1)^{\wedge}2$
$+\mu_{ij}(x_2)^{\wedge}2 + \mu_{ij}(x_3)^{\wedge}2))$

Substep 5.2: Computing Global Similarity Measure

Let $Sim (A_*, A_i)$ denote the global similarity between the alternative case A_i and the idea case A_*, then a global similarity measure can be derived by the weighted summation of the local similarity matching measures: $Sim(A_*, A_i) = (\sum_{j=1}^{n} w_j \times sim(r_{*j}, r_{ij}))/\sum_{j=1}^{n} w_j$, where w_j is the local weights allocated to each feature (attribute) reflecting importance of the corresponding feature. The weightings for alternatives are represented as a weighting vector, $W = \{(C_1, \omega_1), (C_2, \omega_2),..., (C_j, \omega_j),..., (C_n, \omega_n)\}$. Comparing to an idea case $A_* = \{(C_1, r_{*1}), (C_2, r_{*2}),..., (C_j, r_{*j}),..., (C_n, r_{*n})\}$, if $Sim(A_*, A_i)$ is high enough, an alternative case $A_i = \{(C_1, r_{i1}), (C_2, r_{i2})...(C_j, r_{ij}),..., (C_n, r_{in})\}$ can be selected. This can be achieved from a weighted summation even if the local similarity matching measures $sim(r_{*j}, r_{ij})$ is low and w_j is high on criteria C_j.

Step 6: Ranking and Selecting the Alternative Cases

The larger the value of $sim(r_{*j}, r_{ij})$, the higher the priority of the alternative case A_i is.

4 Modeling a Multi-criteria Decision Support System: A Numerical Example

A multi-criteria DSS is designed to help structure and solve MCDM problems. This section demonstrates the practicability of the proposed multi-criteria decision making method for selecting capital budgeting projects. The following algorithm resumes the design philosophy and the operation process of the modeled DSS.

4.1 Architecture and Modules of the Multi-criteria Decision Support System

The proposed multi- criteria DSS consists of a database module, a model base module, a case base module, a man-machine interface module and a matching module. The modules of the modeled DSS can be completely automated, manual with partially automated support for information entry, retrieval and display. While different typologies of DSS architectures exist, the above modules are still typical in a variety of DSS architectures. The purpose of the model base module is to link the data files pre-stored in the database to the pre-built mathematical models, analytical tools and methods to perform various types of calculations and analyses. The case base module stores idea alternative case and a set of candidate alternative cases, representing lesson learned knowledge. The matching module calculates the similarity value of each alternative case to the ideal case and ranks the cases which are most similar to the ideal case. The interface module allows the user to enter input data (e.g., weightings, ratings), to obtain output results (e.g., numeric results, graphics) and to access to other modules. The data base module serves an infrastructure of the model base module, the case base module, the man-machine interface module and the matching module.

4.2   Algorithm of the Multi-criteria Decision Support System

Step 1: Building Modules and Reading Data to the Modules

Mathematical models required for performing calculations and measurements of the DSS, such as internal rate of return (IRR), Net Present Value (NPV) and Equivalent Annual Annuity (EAA), are pre-built in the model module. The criteria for measuring the success of a capital budgeting project are also solicited for case formulation in the case module and case matching. Project success criteria, which are beyond financial criteria and are both quantitative and qualitative, define key areas of performance that are essential for a project to accomplish its goal. According to PMI's PMBOK [11], a set of extracted criteria have been investigated for characterized the capital budget case, which are: $C = \{(C_1$, Payback Period (PP)), $(C_2$, Equivalent Annual Annuity (EAA)), $(C_3$, Profitability Index(PI)), $(C_4$, Internal Rate of Return (IRR)), $(C_5$, Market Demand), $(C_6$, Strategic Opportunity), $(C_7$, Social Need), $(C_8$, Environmental Consideration), $(C_9$, Customer Request), $(C_{10}$, Technological Advance), $(C_{11}$, Legal Requirement)}. The data and model parameters that the decision support system needs, such as initial investment, cash out-flow, cash in-flow, linguistic ratings etc., are read from the inside and the outside and have different measurement scales.

Step2: Reading Candidate Alternative Rating Vectors and Ideal Alternative Rating Vector to the Case Base Module

According to the criteria, Table 1 shows the ratio ratings of alternatives and Table 2 shows the ordinal ratings of alternatives. Ratings of the criteria rating matrix for alternative cases and ratings of the criteria rating vector for ideal case can be nominal, ordinal, interval or ratio scale. The idea case can be described as the following set of criteria-rating pairs: $A_* = \{(C_1, 5), (C_2, 15), (C_3, 1.6), (C_4, 15), (C_5, VS), (C_6, VH), (C_7, VS), (C_8, VS), (C_9, VS), (C_{10}, VH), (C_{11}, VH)\}$. The fictitious ideal alternative case will be compared with the performances of the real alternative cases.

Table 1. Collected candidate alternative cases and idea alternative case with ratio scale

Alternative ID	C_1, PB (Year)	C_2, EAA ($M)	C_3, PI (%,)	C_4, IRR (%)
	Ratio	Ratio	Ratio	Ratio
A_1	5	12	1.2	05
A_2	8	10	1.4	08
A_3	10	15	1.2	06
A_4	6	10	1.4	10
A_5	12	6	1.2	12
A_6	10	15	1.1	15
⋮	⋮	⋮	⋮	⋮
A_*	5	15	1.6	15

Step 3: Reading Case-Criteria Weighting Vector to the Case Base Module
Suppose the weightings are equally weighted and the weightings are represented as a weighting vector, $W = \{1, 1, \ldots, 1\}$.

Step 4: Translating Traditional Measurement Scales into Fuzzy Measurement Scales
By referring to several types of triangular fuzzy numbers in linguistic terms [6], the traditional measurement scales in Tables 1 and 2 can be quantified using commensurate triangular fuzzy numbers, ranged between 0 and 1 and donated as TFN(a_1, a_2, a_3) (Tables 3, 4 and 5 and Fig. 2).

Table 2. Collected candidate alternative cases and ideal alternative case with ordinal scale

Alternative ID	C_5 Market demand	C_6 Strategic opportunity	C_7 Social need	C_8 Environmental consideration	C_9 Customer request	C_{10} Technological advance	C_{11} Legal requirement
	Ordinal	Ordinal	Ordinal	Ordinal	Ordinal	Ordinal	Ordinal
A_1	Strong	Low	Weak	Strong	Strong	Low	Low
A_2	Strong	High	Weak	Weak	Strong	Low	Low
A_3	Strong	High	Strong	Weak	Strong	High	Low
A_4	Very strong	Very high	Strong	Weak	Very strong	Very high	Low
A_5	Medium	High	Medium	Weak	Medium	Low	Medium
A_6	Strong	Medium	Strong	Medium	Strong	Medium	High
⋮	⋮	⋮	⋮	⋮	⋮	⋮	⋮
A_*	Very strong	Very high	Very strong	Very strong	Very strong	Very high	Very high

Table 3. Linguistic terms and their corresponding TFNs for EAA, IRR and PI

Description ($M)	Description (%)	Description (%)	Linguistic term	Triangular fuzzy number
EAA $\leq$ 3	IRR $\leq$ 2	PI $\leq$ 80	Very low (V L)	TFN (0, 0, 0.25)
4 $\leq$ EAA $\leq$ 6	3 $\leq$ IRR $\leq$ 6	90 $\leq$ PI $\leq$ 100	Low (L)	TFN (0, 0.25, 0.5)
7 $\leq$ EAA $\leq$ 9	7 $\leq$ IRR $\leq$ 10	110 $\leq$ PI $\leq$ 120	Medium (M)	TFN (0.25, 0.5, 0.75)
10 $\leq$ EAA $\leq$ 12	11 $\leq$ IRR $\leq$ 14	130 $\leq$ PI $\leq$ 140	High (H)	TFN (0.5, 0.75, 1)
13 $\leq$ EAA	15 $\leq$ IRR	150 $\leq$ PI	Very High (VH)	TFN (0.75, 1, 1)

Table 4. Linguistic terms and their corresponding TFNs for PB period

Description (year)	Linguistic term	Triangular fuzzy number
13 $\leq$ PB	Very long (VL)	TFN (0, 0, 0.25)
10 $\leq$ PB $\leq$ 12	Long (L)	TFN (0, 0.25, 0.5)
7 $\leq$ PB $\leq$ 9	Medium (M)	TFN (0.25, 0.5, 0.75)
4 $\leq$ PB $\leq$ 6	Short (S)	TFN (0.5, 0.75, 1)
PB $\leq$ 3	Very short (VS)	TFN (0.75, 1, 1)

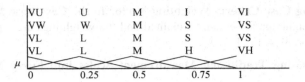

Fig. 2. Membership functions for different linguistic terms

Table 5. Linguistic terms and TFN for importance weightings

Weighting linguistic term	Ratings linguistic term	Triangular fuzzy number
Very unimportant (VU)	Very weak (VW)	TFN (0, 0, 0.25)
Unimportant (U)	Weak (W)	TFN (0, 0.25, 0.5)
Moderate (M)	Moderate (M)	TFN (0.25, 0.5, 0.75)
Important (I)	Strong (S)	TFN (0.5, 0.75, 1)
Very important (VI)	Very strong (VS)	TFN (0.75, 1, 1)

As shown in Tables 6 and 7, the linguistic scales and their corresponding fuzzy number scales can be obtained for evaluation (see Table 8).

Table 6. Transformed candidate cases with ordinal scale

Alternative ID	C_1:PB (Year)	C_2:EAA ($M)	C_3:PI (%)	C_4:IRR (%)
A_1	Short	High	Medium	Low
A_2	Medium	High	High	High
A_3	Long	Very high	Moderate	Low(L
A_4	Short	High	High	Moderate
A_5	Long	Low	Moderate	High
A_6	Long	Very high	Moderate	Very high
⋮	⋮	⋮	⋮	⋮
A_*	Very short	Very high	Very high	Very high

Table 7. Transformed cases with fuzzy numbers

Alternative ID	C_1: PB (Year)	C_2: EAA ($M)	C_3: PI (%)	C_4: IRR (%)
A_1	TFN (0.5, 0.75, 1)	TFN (0.5, 0.75, 1)	TFN (0.25, 0.5, 0.75)	TFN (0, 0.25, 0.5)
A_2	TFN (0.25, 0.5, 0.75)	TFN (0.5, 0.75, 1)	TFN (0.5, 0.75, 1)	TFN (0.5, 0.75, 1)
A_3	TFN (0, 0.25, 0.5)	TFN (0.75, 1, 1)	TFN (0.25, 0.5, 0.75)	TFN (0, 0.25, 0.5)
A_4	TFN (0.5, 0.75, 1)	TFN (0.5, 0.75, 1)	TFN (0.5, 0.75, 1)	TFN (0.25, 0.5, 0.75)
A_5	TFN (0, 0.25, 0.5)	TFN (0, 0.25, 0.5)	TFN (0.25, 0.5, 0.75)	TFN (0.5, 0.75, 1)
A_6	TFN (0, 0.25, 0.5)	TFN (0.75, 1, 1)	TFN (0.25, 0.5, 0.75)	TFN (0.75, 1, 1)
⋮	⋮	⋮	⋮	⋮
A_*	TFN (0.75, 1, 1)	TFN (0.75, 1, 1)	TFN (0.75, 1, 1)	TFN (0.75, 1, 1)

Step 5: Computing Similarity Measures in the Matching Module

Substep 5.1: Computing Local Similarity Measure

The local similarity (criteria similarities) calculates the similarity for each criterion C_j of the alternative case A_i and the ideal case A_*. By applying the local similarity measure $sim(r_{*j}, r_{ij})$, the similarity on criteria C_j between alternative case A_i and the idea case A^* can be calculated. Taking the idea case A_* and the alternative case A_i for example, calculation sheet for translated fuzzy numbers and calculated local similarity measures is shown in Table 8.

Table 8. Calculation sheet for translated fuzzy number and calculated local similarity measures

	C_1 PB	C_2 EAA	C_3 PI	C_4 IRR	C_5 MD	C_6 SO	C_7 SN	C_8 EC	C_9 CR	C_{10} TA	C_{11} LR
A_1	Short	High	Medium	Low	Strong	Low	Weak	Strong	Strong	Low	Low
A_2	Medium	High	High	High	Strong	High	Weak	Weak	High	Low	Low
A_3	Long	Very high	Medium	Low	Strong	High	Strong	Weak	Strong	High	Low
A_4	Short	High	High	Medium	VS	Very high	Strong	Weak	VS	Very high	Low
A_5	Long	Low	Medium	High	Medium	High	Medium	Weak	Medium	Low	Medium
A_6	Long	Very high	Medium	Very high	Strong	Medium	Strong	Medium	Strong	Medium	High
A^*	Very short	Very high	Very high	Very high	Very strong	Very high	Very strong	Very strong	Very strong	Very high	Very low
A_1	TFN (0.5, 0.75, 1)	TFN (0.5, 0.75, 1)	TFN (0.25, 0.5, 0.75)	TFN (0, 0.25, 0.5)	TFN (0.5, 0.75, 1)	TFN (0, 0.25, 0.5)	TFN (0, 0.25, 0.5)	TFN (0.5, 0.75, 1)	TFN (0.5, 0.75, 1)	TFN (0, 0.25, 0.5)	TFN (0, 0.25, 0.5)
A_2	TFN (0.25, 0.5, 0.75)	TFN (0.5, 0.75, 1)	TFN (0.5, 0.75, 1)	TFN (0.5, 0.75, 1)	TFN (0.5, 0.75, 1)	TFN (0.5, 0.75, 1))	TFN (0, 0.25, 0.5)	TFN (0, 0.25, 0.5)	TFN (0.5, 0.75, 1)	TFN (0, 0.25, 0.5)	TFN (0, 0.25, 0.5)
A_3	TFN (0, 0.25, 0.5)	TFN (0.75, 1, 1)	TFN (0.25, 0.5, 0.75)	TFN (0, 0.25, 0.5)	TFN (0,0.25, 0.5)	TFN (0.5, 0.75, 1)	TFN (0.5, 0.75, 1)	TFN (0.5, 0.75, 1)	TFN (0, 0.25, 0.5)	TFN (0.5, 0.75, 1)	TFN (0.5, 0.75, 1)
A_4	TFN (0.5, 0.75, 1)	TFN (0.5, 0.75, 1)	TFN (0.5, 0.75, 1)	TFN(0.25, 0.5, 0.75)	TFN (0.5,0.75, 1)	TFN (0.75, 1, 1)	TFN (0.75, 1, 1)	TFN (0.5, 0.75, 1)	TFN (0, 0.25, 0.5)	TFN (0.75, 1, 1)	TFN (0.75, 1, 1)
A_5	TFN (0,0.25, 0.5)	TFN (0,0.25, 0.5)	TFN (0.25, 0.5, 0.75)	TFN (0.5, 0.75, 1)	TFN (0,0.25, 0.5)	TFN (0.25, 0.5, 0.75)	TFN (0.5, 0.75, 1)	TFN (0.25, 0.5, 0.75)	TFN (0, 0.25, 0.5)	TFN (0.25, 0.5, 0.75)	TFN (0, 0.25, 0.5)
A_6	TFN (0,0.25, 0.5)	TFN (0.75, 1, 1)	TFN (0.25, 0.5, 0.75)	TFN (0.75, 1, 1)	TFN (0,0.25, 0.5)	TFN (0.5, 0.75, 1)	TFN (0.25, 0.5, 0.75)	TFN (0.5, 0.75, 1)	TFN (0.25, 0.5, 0.75)	TFN (0.5, 0.75, 1)	TFN (0.25, 0.5, 0.75)
A_*	TFN (0.75, 1, 1)	TFN (0.75, 1, 1)	TFN (0.75, 1, 1)	TFN (0.75, 1, 1)	TFN (0.75, 1, 1)	TFN (0.75, 1, 1)	TFN (0.75, 1, 1)	TFN (0.75, 1, 1)	TFN (0.75, 1, 1)	TFN (0.75, 1, 1)	TFN (0.75, 1, 1)
$sim(r_{*j}, r_{1j})$	0.984	0.984	0.956	0.84	0.984	0.84	0.84	0.984	0.984	0.84	0.84
$sim(r_{*j}, r_{2j})$	0.956	0.984	0.984	0.984	0.984	0.984	0.84	0.84	0.984	0.84	0.84
$sim(r_{*j}, r_{3j})$	0.84	1	0.956	0.84	0.984	0.984	0.984	0.84	0.984	0.984	0.84
$sim(r_{*j}, r_{4j})$	0.984	0.984	0.984	0.956	1	1	0.984	0.84	1	1	0.84
$sim(r_{*j}, r_{5j})$	0.84	0.84	0.956	0.984	0.956	0.984	0.956	0.84	0.956	0.84	0.956
$sim(r_{*j}, r_{6j})$	0.84	1	0.956	1	0.984	0.956	0.984	0.956	0.984	0.956	0.984

cosine similarity$(r_{*1}, r_{11}) = sim(r_{*1}, r_{11}) = Dot(r_{*1}, r_{11})/\|r_{*I}\|\|r_{11}\| = r_{*I}\bullet r_{11}/\|r_{*1}\|\|r_{11}\|$
$= \sum_{k=1}^{3} (\mu *_3 (x_k) \times \mu_{11}(xk))/(\text{sqrt}(\sum_{k=1}^{3} \mu_{*1}(xk)^\wedge 2) \times \text{sqrt}(\sum_{k=1}^{3} \mu_{11}(xk)^\wedge 2))$
$=(\mu_{*1}(x_I) \times \mu_{11}(x_I) + \mu_{*I}(x_2) \times \mu_{11}(x_2) + \mu_{*1}(x_3) \times \mu_{11}(x_3))/(\text{sqrt}(\mu_{*1}(x_I)$
$^\wedge 2 + \mu_{*I}(x_2)^\wedge 2 + \mu_{*I}(x_3)^\wedge 2) \times \text{sqrt}(\mu_{11}(x_I)^\wedge 2 + \mu_{11}(x_2)^\wedge 2 + \mu_{11}(x_3)^\wedge 2))$

cosine similarity$(r_{*1}, r_{12}) = sim(r_{*1}, r_{12}) = Dot(r_{*1}, r_{12})/\|r_{*I}\|\|r_{11}\| = r_{*I}\bullet r_{11}/\|r_{*1}\|\|r_{11}\|$
$= \sum_{k=1}^{3} (\mu_{*3}(x_k) \times \mu_{11}(x_k))/(\text{sqrt}(\sum_{k=1}^{3} \mu_{*1}(x_k)^\wedge 2) \times \text{sqrt}(\sum_{k=1}^{3} \mu_{11}(x_k)^\wedge 2))$
$= (0.75 \times 0.5 + 1\times 0.75 + 1\times 1)/(\text{sqrt}(0.75^\wedge 2 + 1^\wedge 2 + 1^\wedge 2) \times \text{sqrt}(0.5^\wedge 2 + 0.75^\wedge 2 + 1^\wedge 2)) = 0.984$

cosine similarity$(r_{*3}, r_{13}) = sim(r_{*3}, r_{13}) = Dot(r_{*3}, r_{13})/\|r_{*3}\|\|r_{13}\| = r_{*3}\bullet r_{13}/\|r_{*3}\|\|r_{13}\|$
$\sum_{k=1}^{3} (\mu_{*3}(x_k)\mu_{13}(x_k))/(\text{sqrt}(\sum_{k=1}^{3} \mu_{*3}(x_k)^\wedge 2) \times \text{sqrt}(\sum_{k=1}^{3} \mu_{13}(x_k)^\wedge 2))$
$=(0.75 \times 0.25 + 1\times 0.5 + 1\times 0.75)/(\text{sqrt}(0.75^\wedge 2 + 1^\wedge 2 + 1^\wedge 2) \times \text{sqrt}(0.25^\wedge 2 + 0.5^\wedge 2 + 0.75^\wedge 2)) = 0.956$

cosine similarity$(r_{*4}, r_{14}) = sim(r_{*4}, r_{14}) = Dot(r_{*4}, r_{14})/\|r_{*4}\|\|r_{14}\| = r_{*4}\bullet r_{14}/\|r_{*4}\|\|r_{14}\|$
$\sum_{k=1}^{3} (\mu_{*4}(x_k) \times \mu_{14}(x_k))/(\text{sqrt}(\sum_{k=1}^{3} \mu_{*4}(x_k)^\wedge 2) \times \text{sqrt}(\sum_{k=1}^{3} \mu_{14}(x_k)^\wedge 2))$
$=(0.75 \times 0+1 \times 0.25 + 1\times 0.5)/(\text{sqrt}(0.75^\wedge 2 + 1^\wedge 2 + 1^\wedge 2) \times \text{sqrt}(0^\wedge 2 + 0.25^\wedge 2 + 0.5^\wedge 2)) = 0.84$

cosine similarity$(r_{*1}, r_{31}) = sim(r_{*1}, r_{31}) = Dot(r_{*1}, r_{31})/\|r_{*I}\|\|r_{31}\| = r_{*I}\bullet r_{31}/\|r_{*1}\|\|r_{31}\|$
$\sum_{k=1}^{3} (\mu_{*3}(x_k) \times \mu_{31}(x_k))/(\text{sqrt}(\sum_{k=1}^{3} \mu_{*3}(x_k)^\wedge 2) \times \text{sqrt}(\sum_{k=1}^{3} \mu_{31}(x_k)^\wedge 2))$
$=(0.75 \times 0+1 \times 0.25 + 1\times 0.5)/(\text{sqrt}(0.75^\wedge 2 + 1^\wedge 2 + 1^\wedge 2) \times \text{sqrt}(0^\wedge 2 + 0.25^\wedge 2 + 0.5^\wedge 2)) = 0.84$

cosine similarity$(r_{*2}, r_{32}) = sim(r_{*2}, r_{32}) = Dot(r_{*2}, r_{32})/\|r_{*2}\|\|r_{32}\| = r_{*2}\bullet r_{32}/\|r_{*2}\|\|r_{32}\|$
$\sum_{k=1}^{3} (\mu_{*3}(x_k) \times \mu_{32}(x_k))/(\text{sqrt}(\sum_{k=1}^{3} \mu_{*3}(x_k)^\wedge 2) \times \text{sqrt}(\sum_{k=1}^{3} \mu_{32}(x_k)^\wedge 2))$
$= (0.75 \times 0.75 + 1 \times 1 + 1\times 1)/(\text{sqrt}(0.75^\wedge 2 + 1^\wedge 2 + 1^\wedge 2) \times \text{sqrt}(0.75^\wedge 2 + 1^\wedge 2 + 1^\wedge 2)) = 1$

Substep 5.2: Computing Global Similarity Measure

Suppose the weightings are equally weighted, and each of the global similarity measures can be calculated, representing the aggregated result of local similarity measures for each alternative.

$Sim(A_1, A^*) = (\Sigma_{j=1}^{n} w_j \times sim(r_{*j}, r_{1j}))/\Sigma_{j=1}^{n} w_j = 0.911$

$Sim(A_2, A^*) = (\Sigma_{j=1}^{n} w_j \times sim(r_{*j}, r_{2j}))/\Sigma_{j=1}^{n} w_j = 0.925$

$Sim(A_3, A^*) = (\Sigma_{j=1}^{n} w_j \times sim(r_{*j}, r_{3j}))/\Sigma_{j=1}^{n} w_j = 0.931$

$Sim(A_4, A^*) = (\Sigma_{j=1}^{n} w_j \times sim(r_{*j}, r_{4j}))/\Sigma_{j=1}^{n} w_j = 0.961$

$Sim(A_5, A^*) = (\Sigma_{j=1}^{n} w_j \times sim(r_{*j}, r_{5j}))/\Sigma_{j=1}^{n} w_j = 0.918$

$Sim(A_6, A^*) = (\Sigma_{j=1}^{n} w_j \times sim(r_{*j}, r_{6j}))/\Sigma_{j=1}^{n} w_j = 0.964$

Step 6: Ranking and Selecting the Alternative Project Cases

The alternative projects are ranked in decreasing order according to their similarity values, giving as follows: $A_6 > A_4 > A_3 > A_2 > A_5 > A_1$. The results demonstrated that the propose algorithm is potentially effective for implementing the proposed DSS. By using aforementioned decision making method, the modeled multi-criteria DSS offers a theoretically correct and appealing way for handling uncertainty capital budgeting project selection problems.

5 Conclusions and Future Works

We have proposed a new fuzzy MCDM method for project alternative comparison and selection. We also have outlined a design philosophy and operation process for a DSS system which employs the new fuzzy MCDM method to handle project selection problems. An algorithm and a numerical example were conducted to show that the MCDM method to the modeled DSS is efficient and realistic. The modeled multi-criteria DSS offers a theoretically correct and appealing way of handling uncertain capital budgeting project selection problems. The results are preliminary but have provided a fertile ground for future research. Further, more practical improvements and extensions will be made for further implementation of the multi-criteria DSS. In the future, with the increasing number of alternative projects collected, the concept of project portfolio management will be incorporated to further model the best subset selection of projects given the company's strategy goal, objectives and constraints.

References

1. Turban, E., Volonino, L.: Information Technology for Management: Transforming Organizations in the Digital Economy, 7th edn. Wiley, New York (2010)
2. Peterson, P.P., Fabozzi, F.J.: Capital Budgeting: Theory and Practice. Wiley, New York (2002)
3. Booth, L., Cleary, S., Drake, P.P.: Corporate Finance. Wiley, New York (2014)
4. Lin, K.S.: A novel vague set based score function for multi-criteria fuzzy decision making. WSEAS Trans. Math. **15**, 1–12 (2016)
5. Safari, H., Ebrahimi, E.: Using modified similarity multiple criteria decision making technique to rank countries in terms of human development index. J. Ind. Eng. Manag. **7**(1), 254–275 (2014)
6. Chen, T.Y., Ku, T.C.: Importance-assessing method with fuzzy number-valued fuzzy measures and discussions on TFNs and TrFNs. Int. J. Fuzzy Syst. **10**(2), 92–103 (2008)
7. Lin, K.S.: Fuzzy similarity matching method for interior design drawing recommendation. Rev. Socionetwork Strat. **10**(1), 17–32 (2016)
8. Lin, K.-S., Ke, M.-C.: A new cosine similarity matching model for interior design drawing case reasoning. In: Nguyen, N.T., Trawiński, B., Fujita, H., Hong, T.-P. (eds.) ACIIDS 2016. LNCS (LNAI), vol. 9622, pp. 307–318. Springer, Heidelberg (2016). doi:10.1007/978-3-662-49390-8_30
9. Salton, G., McGill, M.J.: Introduction to Modern Information Retrieval. McGraw-Hill, New York (1987)
10. Ye, J.: Multicriteria decision-making method based on a cosine similarity measure between trapezoidal fuzzy numbers. Int. J. Eng. Sci. Technol. **3**(1), 272–278 (2011)
11. PMI: A Guide to the Project Management Body of Knowledge (PMBOK® Guide), 5th Edition, Project Management Institute, Upper Darby, Pennsylvania (2013)

Innovative Diffusion Chance Discovery

Chao-Fu Hong[1]([⊠]) and Mu-Hua Lin[2]

[1] Department of Information Management, Aletheia University, Taipei, Taiwan
au4076@au.edu.tw
[2] Innovative DigiTech-Enabled Applications & Services Institute,
Institute for Information Industry, Taipei, Taiwan
jessicalin@iii.org.tw

Abstract. The purpose of this study is to explore the innovative use value created by early adopters existing on the Internet world. It is to affect the purchasing intention of the early majority when there is no social relationship network between early adopters and early majority so as to build a new innovative diffusion model. In order to achieve this, the innovative use value created in the Internet by the early adopters was designed as commercial fliers to directly stimulate the early majority. It is used to observe whether new products are accepted by the early majority, and to predict the new product's diffusion chance. The experimental results proved that this method surely can move the early majority toward groups with high intention and observe how many social networks of early majority will work to influence late majority.

Keywords: Innovation diffusion · Chance discovery · Innovative value

1 Introduction

The new product adoption model is always used to evaluate the consumers how to choose and buy the innovative product (Holak 1988). In technology acceptance model (TAM) the used the ease of use and useful are used to evaluate the innovative product's acceptance of consumers (Davis 1989). In addition, an interesting social consuming phenomenon is discovered, the consumers will get together on the same cognitions or consuming values, such as the consuming tribe. So, a consumer is belonged multi-tribes, and based on his/her consume need to different tribe for collecting consuming information.

Furthermore, according to Rogers, innovation diffusion refers to the diffusion of an innovative product all the way through innovators, early adopters, an early majority, and a late majority to laggards (Rogers 2003). Early adopters take on the traits of easily accepting innovative products and creating a use value out of these innovative products. The early majority is a group of practical people. And then, the early adopters could influence early majority to cross the chasm for accepting the innovative product.

The consuming tribe as previous discussion, for consumers buying the same product is a symbol to accept the tribal values. This social phenomenon is the same as the early adopters use innovative product to create use values, which use value could influence early majority to buy innovative product. Therefore, if the innovative product use values are created, these values could be broadcasted to majority, and to evaluate

N.T. Nguyen et al. (Eds.): ACIIDS 2017, Part II, LNAI 10192, pp. 148–156, 2017.
DOI: 10.1007/978-3-319-54430-4_15

how many early majority could accept the innovative product. Therefore, this method is very important to help us to predict how many early majority's social network could be used to influence late majority. Of course, this experimental result could be used to predict the selling of new product. So, our method is very important for solving new product diffusion problem.

2 Literature Reviews

Literature related to this study will be reviewed from two perspectives in this section. One perspective arises from aspects of how adopters perceive the uses of innovative products and how these uses take on power to influence the public, such as formatting consuming tribe. The other perspective centers are how to use text mining methods to extract the innovative use values on new product.

2.1 Innovative Diffusion Model and the Chasm to Format Consuming Tribe

Rogers (2003), proposed an Innovative Diffusion Model (IDM), and showed how innovation diffusion begins with innovators' creation of innovative products, and how this is diffused to early adopters, an early majority and a late majority. Additionally, Rogers' research studies showed that if early adopters accept the innovative product, the early majority will also easily accept the innovative product and consequently, the number of accepted customers will increase quickly. However, Rogers believed that the uses of innovative products for early adopters and the early majority have very different expectations, and argued that there is a chasm between early adopters and the early majority. Therefore, if the tribe is formatted only by early adopters, this tribe will be very small such as a neon-tribe. But if the innovative use cold attract early majority to buy a new product, so this tribal scale will be bigger than neon-tribe, which will arrive the scale of niche-market.

For this reason, we need to define "early adopters". Ram and Jung (1994) showed that early adopters had a significantly higher use of innovation with a product when compared to the early majority. Furthermore, early adopters may possess the specialized skills and abilities required to use innovative products in a wide variety of ways. In Morrison et al.'s experiment (2000), only about 26% of the consumers who have "leading edge status" and in-house technical capabilities freely share their innovations with others. Also, if a product is innovative and discontinuous, only some creative consumers may have a chance to discover creative ways of using the product. Therefore, it is hardly to discover early adopters, in this study we think that many people appear themselves on Internet, and that will be chance for us to find early adopters from Internet.

2.2 Data Mining and Knowledge Extraction

Data mining technology is used to analyze data collected from early adopters' postings concerning creative ways of using a product. For example, the algorithm of Apriori is used in Agrawal and Srikant (1994). It is an algorithm that utilizes a traditional consumption database to mine association rules. By satisfying the criteria of support and confidence we obtained rules for the uses of innovative products. These rules can be employed not only to realize consumption patterns, but also to infer meaningful consumption behaviors. However, low-frequency data that fail to pass the criteria tend to be ignored. Not all low-frequency data is noisy data. For example, in the innovation diffusion model only a few early adopters use innovative product data generated by the early adopters, but this innovative use of product data is an important opportunity for crossing the chasm. These kinds of data cannot be taken as noisy data. In order to prevent data with lower frequency from being sifted out, Liu et al. (1999) developed a rare association analysis, aimed at data with low support and high confidence. Even though the data is limited, rare association analysis can extract useful patterns. However, the data derived from early adopters' postings concerning the innovative uses of a product are isolated and limited, and the risks of using this method to analyze those data are very high.

2.3   Extracting and Visualizing the Consuming Values: Chance Discovery and Qualitative Chance Discovery

Ohsawa et al. (1998) put forward a KeyGraph algorithm as a way to calculate the frequency of products in a shopping cart and the frequency for co-occurrence between products, which lead to the emersion of clusters. After calculating the connections between the product and clusters, experts obtain all of the products' Key value, and then find a low frequency, but a high connection between products. Next, based on the linkages between those products and the clusters experts develop a reasonable scenario, which is called Chance Discovery (CD) (Ohsawa 2003). This CD method has been successful in discovering a product's innovative uses. However, when experts use the CD method to discover the uses of an innovative product, the result will lack the linkages between low-frequency products and clusters because there is not much data about innovation.

As previously discussed, the CD is not suited to solving the small data problem. Researchers repeatedly ask questions in studies based on grounded theory (GT) (Strauss and Corbin 1998), such as "What is going on?" and "What is the main problem of the participants and how are they trying to solve it?" The questions can help researchers carry out inductive and deductive analysis to discover something new to generate new theories. If GT, for example, could be used to analyze the data of how to find new uses for innovative products, then researchers should be able to find various uses for innovative products. Nevertheless, the validity in its traditional sense is consequently not an issue in GT, which instead judges by fit, relevance, workability, and modifiability. Therefore, how to develop workable information technology to reduce human power on GT analysis will become an important research issue.

Furthermore, Hong (2009) proposed a Qualitative Chance Discovery model (QCD), combining the theory of data mining and chance discovery with GT, and the theme of keywords, such as new ideas for using innovative products, are used to collect sentences that include keywords from all documentary data. Therefore, experts can focus on one theme to build one innovative use for a product, and repeat the steps mentioned above until all new uses of innovative products emerge. Finally, based on a reasonable scenario that integrates the relative uses of innovative products, the use of innovation emerges to complete the entire scenario. Although QCD is a kind of qualitative analysis approach to establish a scenario, it still needs to define research issues and use a quantitative approach to evaluate the successful possibilities.

3   Research Methods

Based on IDM, business managers explore novel ways of using a product by early adopters and recommend the product to the early majority. Managers want to enhance their use of product demand and achieve the purpose of expanding the market of innovative products. This study employs grounded theory and text-mining techniques to construct a Human-Centered Computing System (HCCS) by collecting and analyzing the articles posted on Internet by early adopters (Hong et al. 2015). This research tries to adopt the innovative uses of products from Internet by early adopters and

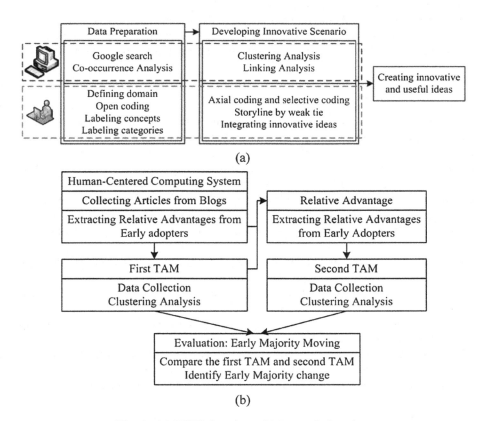

Fig. 1. (a) HCCS flowchart; (b) Research flowchart

broadcast these uses to real world early majority to investigate their acceptance, such as the new innovation diffusion model. Therefore, the questionnaire developed by Karahanna and Straub (1999) will be modified by the uses of innovative products to test the public's understanding of these products and to see their initial acceptance of innovative products first as TAM tests, as shown in Fig. 1. Next, the uses of innovative products by the Internet early adopters are used to create an advertising scenario, to broadcast to and influence the real world majority. Finally, a second TAM test is conducted to evaluate the acceptance of the real world majority again. The number of customers who expressed their willingness to accept the products in the two TAM tests may be compared to explore to what extent the early majority has crossed the chasm. The research flowchart is shown as follows.

4 A Case Study

In 2007, ASUSTeK Computer Incorporated (ASUS) and Intel Corporation (Intel) collaboratively introduced to the market a Netbook less than 10 inches. Their purpose was to provide more people with inexpensive computers. Contrarily, industries once again refer to netbooks with the characteristics of being light, thin, small, and as easy to carry as a consumers' netbook. Furthermore, this paper's main purpose is to investigate even there are no any social relationship between early adopters and early majority, and the early adopters' innovative use may influence early majority to accept netbook.

4.1 Data Resources

After observation of netbook circulation messages in the market, the researchers also found that there is a time difference in transmitting speed on the Internet between the time the new products appeared and the time it took consumers to utilize this information. This study was focused on the target innovative product: ASUS EeePC901. The researchers collected data posted on blogs relevant to netbook use. These data ranged from October 1, 2008 to October 31; six months after netbooks were introduced to the market. Using Google blogs (http://blogsearch.google.com/blogsearch) and the keywords, netbook + thought, to search for the data, the researchers obtained 63 related data from blog articles.

Excerpts from Rogers and Shoemaker (1971) define purchase innovation. Morrison et al. (2000) extend the concept of innovation to two categories - use innovation and vicarious innovation. The extension of these concepts was used to explain that relative uniqueness is a key factor affecting the majority's acceptance of innovative products. But they did not investigate these concepts could be influenced no social relationship majority? However, we carefully read the collected data and removed the articles that did not contain innovative uses, there were seven articles with new information about using the netbook. Then we executed a value extraction from these seven articles, and estimated no social relationship majority to understand their accepting haviors.

4.2   Experimental Results and Discussion

Based on HCCS, Fig. 2 illustrates some innovative uses of netbooks posted on Internet by early adopters. The data were collected and categorized by the researchers. Basically, the researchers noticed that early adopters could easily carry a netbook anywhere at anytime to have immediate access to the Internet because the netbook is light and small in size. These new uses of innovation may impact no social relationship consumers such that they agree that when they travel they can share immediate information with others in written form, photos, and audio-visual data. In the same vein, in the consumer's daily life, they may use netbooks to share movies with their good friends, discuss their travels, and also to discuss the products they have bought. The consumers also receive information shared by their friends. These scenarios help the researchers perceive how further innovative usages of the netbook's ease-of-use could improve no social relationship consumers' acceptance of the netbook to influence the majority's lifestyles.

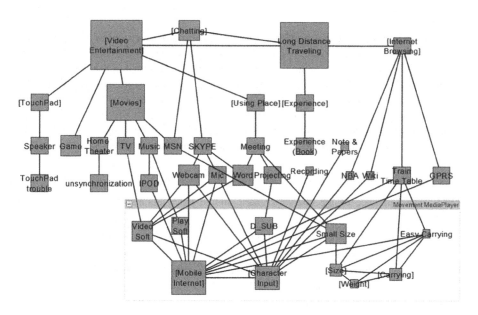

Fig. 2. Uses of the netbook by early adopters

4.3   Modification of the TAM Questionnaire

Although netbooks have been introduced to the market since 2008, experts believe they are still an innovative product for college freshmen. Moreover, the analyzing results shown in Fig. 2, which were derived from early adopters, can help researchers recognize characteristics of the netbook, such as its being light, small, and having a long-lasting battery. These characteristics are generally referred to as 'ease-of-use'. Additionally, from the data collected, the researchers also found that the mobile nature of netbooks allow people to carry a netbook anywhere because they can get information from the GPRS, train time tables, NBA news, Wiki, and blog articles. This information satisfies their diverse needs and represents innovative uses often created by the early adopters. This inference allowed researchers to use the newly discovered innovative usages of the netbook to modify its ease of use as well as to answer questions on Davis's TAM questionnaire to investigate the netbook's acceptance by freshmen.

4.4   Empirical Study

The two TAM tests was that the freshman had to take two investigative tests. Before the second TAM test the freshman had to read the advertising scenario provided by the system. The advertising scenario contained some innovative usages of netbooks to investigate the freshmen's acceptance of netbooks. A university located in the northern part of Taiwan and who the 119 freshmen are randomly sampled attending. After the first TAM test, only 99 freshmen are the valid questionnaires. The major difference between.

In this research, a Paired-Samples t-test was applied to examine the difference between these two tests, and the results of the study are shown in Table 1. The mean field is the difference of the average score in one dimension between the first TAM test and the second TAM test. All dimensional differences are negative. The negative score reveals that after reading the advertising scenario, the consumers' intention to buy netbooks might increase. Furthermore, all dimensions are significant ($p < 0.05$) except the physical dimension. Therefore, H1 was accepted and it became evident that the advertising scenario could positively influence the consumers' intention of buying the netbook.

Table 1. The Influence of the chance scenario in the twice-administered TAM test and a Paired-Samples t-test

Dimension		Mean	Std. deviation	t	Sig. (2-tailed)
Ease of use	Physical	−0.475	2.553	−1.850	0.067
	Operating	−0.455	2.072	−2.183	0.031*
Useful	Use environment	−1.131	2.958	−3.806	0.000***
	Learning & leisure	−1.707	2.844	−5.973	0.000***
	Use effect	−1.374	2.791	−4.898	0.000***
	Use efficiency	−1.030	2.663	−3.850	0.000***

Next, the researchers applied the two-step clustering algorithm to decide on the optimal numbers of the cluster; the K-means clustering algorithm was used to assign consumers to clusters. The result of the experiment is shown in Table 2. After the first TAM test, the customers' answers were used to cluster and two clusters could be classified; these two clusters are similar to the innovator and imitator proposed by Bass (1969). After the second TAM test, as customers were classified into three clusters: A2,

Table 2. The influence performance of early adopters, the early majority, and the late majority

	First time TAM		Second time TAM		
	High (A1)	Low (B1)	High (A2)	Middle (a + b)	Low (B2)
Mean	4.20	3.29	4.75	3.94	3.22
N	53	46	35	18 + 22 = 40	24
Percentage	54%	46%	35%	40%	24%

a + b, and B2, the researchers assigned A2 as innovators and early adopters, and a + b as the early majority. B2 represents the late majority and the laggards. After reading the advising scenario, innovators and early adopters showed an average score in all dimensions that was higher than what they received before by approximately 0.55. This result indicates that people are easily influenced by advertising scenarios designed to motivate people to buy because these advertisements present some innovative ways of using the products. This result is consistent with Rogers' IDM, i.e., suitable information is presented to the majority, and the early majority move closer to the early adopters. Furthermore, A2 cluster has 35% freshmen, which is larger than 15%, which is the sum of innovators and early adopters. This result also means that about half early majority accept innovative use and buy a new product, and because half early majority could use their social network to influence half late majority. So, this new product may be easily accepted by majority, at least half majority, to format a niche market.

From Table 2, the experimental result explores that even the early majority they have no any social relationship network with majority, about half of early majority are more closing the early adopters. This result evidence that the innovative use could affect the purchasing intention so as same as the innovative diffusion model. Furthermore, it also can help company to predict how easy the new product can be sold.

5 Conclusion

In Rogers' innovative diffusion model (Rogers 2003), if early adopters accepted new products and shared their innovative use value with others, and from our experimental results even there is no social relationship network between early adopters and early majority, the early adopters' creating innovative uses are the important value to influence early majority to accept the new products quickly and readily. This social phenomenon points out that new product could be successful diffusion, it depends on how many and how useful the innovative uses could be created by early adopters. Of course this study also intends to use information technology could extract innovative use information employed by the early adopters from Internet world. Finally, two TAM testes model is used to stimulate how useful the innovative use could influence majority. The experimental results evidence that over half early majority are influenced to buy the new product. Therefore, this result also pointes that our method can predict how successful the business will be.

References

Holak, S.L.: Determinants of innovative durables adoption an empirical study with implications for early product screening. J. Prod. Innov. Manag. **5**(1), 50–69 (1988). doi:10.1111/1540-5885.510050

Davis, F.D.: Perceived usefulness, perceived ease of use, and user acceptance of information technology. MIS Q. **13**(3), 319–340 (1989). doi:10.2307/249008

Rogers, E.M.: Diffusion of Innovations. Free Press, New York (2003)

Ram, S., Jung, H.S.: Innovativeness in product usage: a comparison of early adopters and early majority. Psychol. Mark. **11**(1), 57–67 (1994)

Morrison, P.D., Roberts, J.H., von Hippel, E.: Determinants of user innovation and innovation sharing in a local market. Manag. Sci. **46**(12), 1513–1527 (2000)

Agrawal, R., Srikant, R.: Fast algorithms for mining association rules in large database. In: Proceedings of the 20th International Conference on Very Large Data Bases, pp. 484–499 (1994)

Liu, B., Hsu, W., Ma, Y.: Mining association rules with multiple minimum supports. In: Proceedings of the Fifth ACM SIGKDD International Conference on Knowledge Discovery and Data Mining, pp. 337–341. ACM Press, New York (1999)

Ohsawa, Y., Benson, N.E., Yachida, M.: KeyGraph: automatic indexing by co-occurrence graph based on building construction metaphor. In: Proceedings of the Advances in Digital Libraries Conference, pp. 12–18 (1998). doi:10.1007/978-3-662-06230-2

Ohsawa, Y.: Chance discovery: the current states of art. In: Ohsawa, Y., Tsumoto, S. (eds.) Chance Discoveries in Real World Decision Making, vol. 30, pp. 3–20. Springer, New York (2003). doi:10.1007/978-3-662-06230-2

Strauss, A.C., Corbin, J.M.: Basics of Qualitative Research: Techniques and Procedures for Developing Grounded Theory, 2nd edn. Sage Publications Inc., London (1998)

Hong, C.F.: Qualitative chance discovery: Extracting competitive advantages. Inf. Sci. **179**(11), 1570–1583 (2009)

Hong, C.-F., Lin, M.-H., Yang, H.-F.: A novel framework of consumer co-creation for new service development. In: Nguyen, N.T., Trawiński, B., Kosala, R. (eds.) ACIIDS 2015. LNCS (LNAI), vol. 9012, pp. 149–158. Springer, Heidelberg (2015). doi:10.1007/978-3-319-15705-4_15

Karahanna, E., Straub, D.W.: The psychological origins of perceived usefulness and ease of use. Inf. Manag. **35**, 237–250 (1999)

Rogers, E.M., Shoemaker, F.F.: Communication of Innovations: A Cross-Culture Approach. Free Press, New York (1971)

Bass, F.M.: A new product growth for model consumer durables. Manag. Sci. **15**(5), 215–227 (1969)

Chance Discovery in a Group-Trading Model — Creating an Innovative Tour Package with Freshwater Fish Farms at Yilan

Pen-Choug Sun[1(✉)], Chao-Fu Hong[1], Tsu-Feng Kuo[2], and Rahat Iqbal[3]

[1] Department of Information Management, Aletheia University,
No. 32, Zhenli Street, Tamsui District, New Taipei City 25103, Taiwan R.O.C.
{aul159,au4076}@au.edu.tw
[2] Originate Science & Technology Co., Ltd, No. 183, Sec. 1, Tonghe E. Street,
Shilin District, Taipei City 11169, Taiwan R.O.C.
tfkuo@mail.tku.edu.tw
[3] Department of Computing, Coventry University,
Priory Street, Coventry CV1 5FB, UK
aa0535@coventry.ac.uk

Abstract. Information Technology in modern days is developing very fast, which has brought out new opportunities for Electrical Commercial Businesses in Tourism Industry. Crossing the chasm between early adopters and an early majority is an important issue for tourism innovation diffusion. In this paper, a group-trading model called the Core Broker Model is used to create an innovative tour product and to diffuse the innovation through E-markets on the Internet. A core broker organizes an innovation team, using the technique of text mining, to create a featured tour package and initiate a joint-selling project for different type of providers. The core broker then recruits market brokers to form a market team to promote the tourism innovation, and put it on the websites for consumers to group-purchase e-coupons for the trip. With the advantages of group-trading in the model, the chance for innovation diffusion to cross the chasm is high.

Keywords: Chance discovery · Crossing the chasm · Group-trading model · Text mining · User innovation

1 Introduction

The Tourism Industry has always been known as the Tertiary Industries. Its activities include restaurants, hotels, transport, tourism and many other industries. In the recent years, due to the rapid economic growth and the increase on national income, the population of people who does leisure activities are increasing, which had made the tourism industry grow substantially. With the popularity of Internet, a wide range of commercial activities are spreading on the Internet, and the Internet business models and services are updating day by day. From the new online business, there are also a lot of virtual stores created by the tourism and leisure industry rise up to meet the

© Springer International Publishing AG 2017
N.T. Nguyen et al. (Eds.): ACIIDS 2017, Part II, LNAI 10192, pp. 157–169, 2017.
DOI: 10.1007/978-3-319-54430-4_16

customers' needs. "If it can revive the service of tourism industry, it will help the local traditional industries grow [1]". Regarding the economical aspect, it not only stimulates the new consumption income and enhances the tourism value, but the large crowd that is attracted by it, can also bring additional business opportunities for the local industry. Therefore, the long-term contribution to the local economic environment cannot be underestimated.

In recent years, the government has put a lot of effort to promote the tourism industry. The population of foreign visitors in Taiwan is growing year by year, see Fig. 1, there is 36% in 2001, 43% in 2004 and 68% in 2014. More and more people came to Taiwan for sightseeing and it shows that the tourism strategy of Taiwanese government worked. According to the latest statistical report of the United Nations World Tourism Organization (UNWTO), in the first half of 2014, the number of the international tourists' annual growing rate has gone up to 26.7% in Taiwan overtaken Japan, at the top of the world. As for International tourist income wise, the increasing rate for the first half of 2014 is of 18.5%. It is in the third place, preceded by Japan, with the rate of 27.5%, and South Korea, 25.2% [2]. In Fig. 2, it shows that in the last decades, even when the domestic tourism expenditure made by R.O.C. citizens is going up and down, the foreign exchange income has always been increasing in the last 10 years [3]. Also from 2008 onwards, foreign exchange income has overtaken the domestic tourism income, and became the largest income source of tourism industry in Taiwan. It is for sure that developing the international tourism in Taiwan is a profitable business.

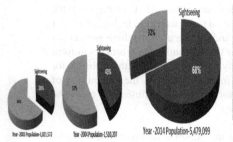

Fig. 1. Tourism industry in Taiwan   **Fig. 2.** Tourism income 2004–2015

There are too many benefits to be gained by running the tourism and leisure business on the Internet. It does not have the limitation of the traditional business, such as the leasing of physical stores, to find the best location, decoration costs, etc., as well as for the business hours, the online store can open 24 h a day throughout a year, no matter it is holiday time or not. Because of the nature of the network, there is no geographical limitation, the store can be expanded throughout the world, and the tourism is commercial service-oriented, so that the products do not require physical transportation. The Internet-based tourism and leisure industry is definitely a profitable business model.

Tourism and leisure site is full of potential, but the down-side is that it has too much false information on the large network, and this makes people hesitate. The tragedy of consumers not finding any product or service they want, after they have wasted too much time on the Internet, will cause them do not want to look for tourism products any more. Eventually, this will make most of tourist sites, facing the situation of loss or even shutting down, especially when they cannot lure consumers' eyes to their sites.

The main motivation of this paper is to use the advantages of the travel and tourism industry, with the help of modern computer technology, developing high-quality tourist sites to attract large number of consumers, in order to drive the local leisure activities, and help the local industry grow. This study focuses on the way to create innovative featured vacation packages which are attractive enough to both Taiwanese and international tourists on the Internet using technique of text mining.

According to Rogers, innovation diffusion refers to the diffusion of an innovative service or product all the way through innovators, early adopters, an early majority, and a late majority to laggards [4, 5]. Early adopters take on the traits of easily acceptance of innovations and create a new value through the use these new products or services. The early majority is a group of practical people, so they could be too stubborn to accept an innovation. Therefore, Moore discussed a chasm theory referring to the chasm existing between the early adopters and the early majority, which could disrupt and even stop diffusion of innovations [6].

The Core Broking Model is introduced here to help selling the innovations. It involves brokers to produce attractive selling methods in order to increase the supplier's profit. At the same time, through the group-buying of the packages, the brokers earn the commission on transactions, reduce consumers' purchase price, and thus form a win-win-win situation. And most important, this may help a diffusion of tourism innovations crossing the chasm between early adopters and the early majority.

2   Literature Review

Besides the Core Broking Model, there are some theories, which have been used in this study: the theory of the User Innovation on inviting enterprises, experts and consumers to work together to create new services or products. The theory of Chance Discovery and the technology of Text Mining for capturing consumers' preferences in order to discover the chance of innovations.

2.1   The Core

In economics, the core indicates the set of imputations under which no coalition has a value greater than the sum of its members' profits [7]. Therefore, every member of the core stays and has no incentive to leave, because he/she receives a larger profit. A core in a cooperative game $<N, v>$ with n players, denoted as $C(v)$ of a characteristic function $v : 2^N \rightarrow \Re$, which describes the collective profit a set of players can gain by forming a coalition, is derived from the following function [8],

$$C(v) := \left\{ x \in \Re^n | x(N) = v(N), x(S) \geq v(S) \textbf{ for all } S \in 2^N \right\},$$

here N is the finite set of participants with individual gains $\{x_1, x_2, \ldots x_n\}$. The $x(N) = \sum_{i \in N} x_i$ represents the total profit of each individual element in N by adding xi, which denotes the amount assigned to individual i, whereas the distribution of $v(N)$ can denote the joint profit of the grand coalition N. Suppose S, T are a set of pair wise disjoint non-empty subsets of coalition N. A cooperative game N is said to be convex, if

$$v(S) + v(T) \leq v(S \cup T) + v(S \cap T)$$

whenever S, T $\subseteq$ N and for N $\in 2^N$.

A convex core is always a stable set [6]. A game with $x(N) = v(N)$ is regarded as efficient [8]. An allocation is inefficient if there is at least one person who can do better, though none is worse off. The definition can be summarised as "The core of a cooperative game consists of all un-dominated allocations in the game" [10]. The profit of the allocations in a core should dominate other possible solutions, meaning that no subgroup or individual within the coalition can do better by deserting the coalition.

2.2 User Innovation

Von Hippel has brought a "lead user method", by selecting of a number of lead users, set up workshops to identify market needs and trends for the public, help to design the innovative goods or services that reaches the goal of user innovation [11]. Here the so-called "lead users", are the same thing as innovators and the early adopters in the Diffusion of Innovation Theory. The path to victory in business is to create values. The enterprises have always been created their own values, and then exchange it with consumers. This kind of traditional operating, which has deep-rooted in the society, have no more use, as the social environment evolves. In the present days, there are so many kind of sales model, the consumer has many choices, but satisfaction was getting lower and lower. The list of operating policy is very long, but the valuable outcome is far too little. The enterprises nowadays should "be able to understand the new trends, and create the experience with consumers, in order to stand out in this harsh market with the vast of products" [12]. The tradition commercial system of "treating the product as the center" has been seriously challenged. If an enterprise wants to survive in the modern environment, it would have transformed into new generation system of "treating the experience of consumer as the center", creating values with consumers.

2.3 Chance Discovery

Ohsawa *et al.* [13] put forward a Key Graph algorithm as a way to calculate the frequency of products in a shopping cart and the frequency for co-occurrence between products, which lead to the emersion of clusters. After calculating the connections between the product and clusters, experts obtain all of the products' Key value, and then find a low frequency, but a high connection between products. Next, based on the linkages between those products and the clusters experts develop a reasonable scenario, which is called Chance Discovery [14]. This method has been successful in discovering a product's innovative uses.

2.4   Text Mining

Sullivan defines Text Mining as a process editing, organizing and analyzing a large amount of unstructured documents, to meet the specific information needs for users [15]. It is also often being used to find the association between certain features. Through searching in documents, it can find the relationship between various concepts, such as the key information, people, things, times, places, objects, keywords, etc., and put them to different classes. It discovers unknown or implicit and useful information in unstructured or semi-structured documents, and the user can extract the hidden information, and to process, store, transform or reuse it.

3   The Core Broking Model

An innovative group-trading model, called the Core Broking Model (CBM) was proposed in 2012 [16]. The CBM inherits the core concept and a core in a coalition can also be derived from the same function as the core:

$$C(v) := \left\{ x \in \Re^n \,|\, x(N) = v(N), x(S) \geq v(S) \, for \, all \, S \in 2^N \right\},$$

however, the CBM creates a virtual market on the Internet by involving multi e-markets. In this virtual e-market, there can be many group-trading projects in the model. Many providers join in the projects and perform joint-selling and sufficient buyers gather in e-markets because they may get the high discounts available in the projects. In this study, the CBM is being used on the group transactions of innovative products or services. In Fig. 3, it shows a core broker recruits market brokers to form a "market team" to help innovations diffuse into the market as long as the innovations have been created.

The main task for the market team is to sell the innovations on the Internet. The members the team contain the core broker and several market brokers, whom are also an internet sales professionals. When the innovation team has produced an innovation, the team will then set up a group-trading project to diffuse this innovation into the market. When the commission rate and the group-purchase price has been decided, the market brokers can call out to the customers about the product in their websites, whether to put it on B2C or C2C auction sites, in order to increase the exposure of goods or the services rate. The Business to Consumer (B2C) e-commerce is where the enterprise facing the individual buyers, the famous "Amazon.com" and "Taobao" are the sites to serve this type of trading method. The Consumer to Consumer (C2C) e-commerce is where consumers sell products directly to the other consumers. The auction sites like "Open Auction", "eBay" or "Lynx", etc. are the most common examples. When the market team decides the starting date and closing date for batch trading, the batch trading will then be officially launched rounds by rounds. Each round usually lasts for a week, the diffusion mechanism and the tasks are shown in Fig. 4.

(1) **Commencing** – the core broker recruits market brokers to form a market team and begins sessions of group trading in a project.

(2) **Gathering** – the market brokers attract buyers to their websites and submit the coalitions of buyers they have formed to the core broker.

(3) **Combining** – the core broker combines the coalitions together, decides the final prices for the items, checks the stability of the coalitions and sends acceptance notices to the market brokers.

(4) **Closing** – the market brokers close the deal with the buyers. The final orders and invoices for each customer will be prepared by the market brokers. After the buyers have paid for their purchases, the profits of the providers and the commission for all the brokers will be calculated. At this point, the core broker may choose to have a new session of trading or stop the project for good.

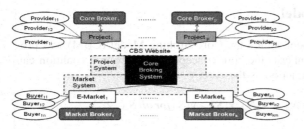

Fig. 3. The core broking model

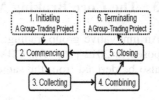

Fig. 4. Mechanism of user innovation

Each project may contain many rounds of batch trading session, and the trading results for each round are completely independent. In addition, the group exchanges required information, such as the number of different deals available discounts, commission proportion, will also be published, after the new services or products have been decided. The market brokers will then advertise in their e-shops, according to the information in the core brokers' website, to attract customers to do the online ordering. There are two factors that bring consumers to the brokers, one is the innovative products or services themselves. Another is the high discounts produced in group-buying, which should be quite attractive, because customers are unable to get such a low price in other e-shops.

Before the submission day of a round of a session, the market brokers submit the collected group-buy invoices, which they get from the customer, to the core broker. Also, when they have collected the payment made by the customers, they will pay for the bills to the core broker here. The supplier will also receive the detail of the e-voucher, when the core broker e-mailed it to the market broker, and wait for customers to shop with these e-vouchers.

4   Create an Innovative Vacation Package

Before the market team is set up, the Core Brokers, being a professional lead user, initiates projects of user innovation by setting up an "innovation team" first to focus on the innovation. The ultimate goal for the innovation team is to create an innovative products or services. Its members include the Core Broker, other product experts from enterprises and some "Lead Users." The aim here is to create an innovative tourism package, in which the fish farm plays a decisive role, because the natural ecosystem at Yilan has remained fairly successful. Yilan County was decided to for this study, mainly because it is a tourism-oriented area. The reason Yilan's tourism industry start to grow is because of the opening of the Hsuehshan Tunnel in 2006. Yilan County Government has also organized many large-scale cultural activities such as International Children's Folklore and Folkgame Festival, Green Expo, etc., which brings a lot of tourists into the area.

A report by the Tourism Bureau, sorted out the yearly statistic of number of visitors touring Yilan County in Fig. 5. From 2004 to 2014, the number of tourists is rising, and there is only three years where the number of the tourist falls, 2005, 2008 and 2012, respectively [18]. The number of tourists in 2014 is 2.8 times of 2005, the overall long-term trend is clearly a rising trend.

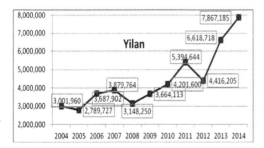

Fig. 5. Visitors to Yilan

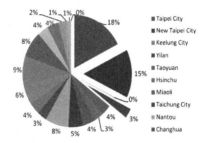

Fig. 6. Visitor to city [17]

Every January to March, October-December was classified as the off-season by Yilan hotel industry, and there is a large gap between the number of tourists during off-season and peak season [19]. How to increase the tourist in the off-season, has always been a major subject for Yilan tourism industry. Figure 6 shows that there were only 3% of the population who chose Yilan as their touring location. Its market share is very small, compared to the country's most popular tourist place, Taipei City, with 18%, and the second popular place, New Taipei City, with 15%. The tourism industry at Yilan is a new market in a high growing rate, and should be considered with an expanding method, and may return very high feedback in the future.

The water quality in Yilan is with less pollution, compared to other places in Taiwan. Therefore, there are many freshwater fish farms, and became one of Yilan's features. There is a need to explore the background of Taiwan's fish farm, in order to understand their current situation for the industry. In recent years, Taiwan seems to retired from the inland aquaculture industry, the total catches have a gradually downward trend. According to information released by the Fisheries Agency, early in 2004, there has already a diverted trend from farming fishermen. From the nine years between 2003 to 2011, retired households has more than 8,100 (see Fig. 7), it is almost equal to the sum of the decreasing amount of the other fisheries industry in Taiwan in five years. The freshwater fishing farm's catch is constantly decreasing, from the 20 million tonnes in 2003 to less than 15 million tonnes in 2012, a loss for more than 25% of the output value [20]. From the interpretation of the figures, it describes the fact of the serious deduction. If there is no improvement on the domestic aquaculture industry, and modifies the way enterprises runs. It would be difficult to survive from the threat of cheap cost by other strong foreign country, and facing the tragic fate of elimination [21].

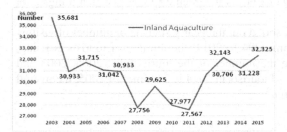

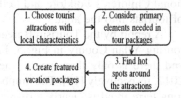

Fig. 8. Steps for creating innovations

Fig. 7. Statistics for population of fishermen

This paper takes advantages of information technology, to integrate an effective strategy for the traditional fish farming. Firstly, from the consumer's point of view, to find a quality fish farming industry, combine it with the local leisure, delicacy, education, industry, natural ecology, cultural practices, etc., make full use of fish farm resources, added with its close attractions and local tourism resources, and pack it into an integrated leisure tourism fish farm, intend to provide consumers with a complete leisure fish farm tourism experience. With the era of web 2.0 nowadays, consumers can easily pass on their own consuming experience to the Internet, resulting a lot of experiential articles are filled on the Internet. The study is to collect the data from this large network, and identify the preference of consumers'. In the present study, the innovation team performs the data analysis and integration, the four steps that generate the ideas are: discover the innovative highlights, establish innovative model, generate innovative combination and determine innovative products or services. An innovation team should follow the process shown in Fig. 8, to produce the tourism packages with special features. The details of the steps are described as follows:

4.1   Discover Innovative Highlights

The main task of this stage is to choose a fine tourist attraction with the characteristics of Yilan as the highlight of the innovative tour package. At this stage, the technique of text mining is used to create innovations. After articles about the innovations are collected, a pre-process on the text is conducted and then some word co-occurrence analyses are being taken. The detailed process is listed as follows: H labels indicate the steps done by human experts, and C labels indicate the steps to be executed by the computer system.

Stage 1: Data Pre-process

> Step 1-1 (H) Define relevant keywords about the domain that is being addressed.
> Step 1-2 (C) Select the texts correspond to keywords on the Internet.
> Step 1-3 (H) Segment texts into words. Based on experts' domain knowledge, irrelevant words are removed.

Stage 2: Word Co-occurrence Analysis

> Step 2-1 (C) Calculate the association values of the words with following equation,
> $$i = 1 \, to \, N - 1$$
> $$j = i + 1 \, to \, N$$
> $$assoc(w_i, w_j) = \sum_{S \in all \, T} min(|w_i|_S, |w_j|_S), \, T \text{ is text, } N \text{ is the number of } T.$$
> where S represents the co-occurrence of words in the sentence of T.
> Step 2-2 (C) Draw an associative diagram of the co-occurrence words based on above result.
> Step 2-3 (H) Put words into clusters, which are each marked with meaningful conceptual label.

Steps taken in this study is to search through web blogs in a time period through Google Search using specific keywords, select strong relevant articles, and then manually segment and collect the useful words in the texts, combine the texts into one article, unify the synonyms in it, and finally feed it into the value-focus system, which statistics the frequency of each word, and find the relation values between words. High-frequency words draw high concerns in the relation graph, it represents that many consumers have mentioned those words frequently in their articles. The relation value refers that the rate of two words appears in the same sentence at the same time, the number of time those two words appears, the stronger the relation between the two. Finally, the results of the data generated by the value-focus system, is visualized as a graph through the yEd software for further analyses.

Firstly, insert keyword "Yilan, fish farm," (in Chinese) section from a year after 5th March 2013, there were 91 consumers' article searched out from Google blog, filter the invalid article, and still 22 articles left. Surprisingly, only three fish farms revealed in the graph. By doing a conceptual analysis, the consumers' experiences about these farms were classified in four categories: leisure culture, attractions, restaurants and aquaculture. These three fish farms are no longer traditional and operating in multi-function. The concepts they used can be classified into six categories: restaurant/good food, leisure cultural, general fish farming, tropical fish farming, nearby attractions and internal attractions. By comparing the elements of these three fish farms, "Sheng-Yang Fish Farm" (SYFF) is better than the other two fish farms, and is elected as the winning highlight.

4.2   Establish an Framework of Innovations

At this stage, the essential elements in the innovative package are considered. Because Leisure Farms is the most successful tour product in Taiwan [22], the most popular Leisure Farms are studied to build up a framework for the innovation. Since the start of "Delicate Agricultural Health Excellence Action Plan" in May 2009, until the end of 2014, the delicate agricultural output value reached 1,365 billion, compared with 2008, when the plan has not started, growth about 46.1% [23]. Among all of the delicate agriculture, Leisure Farm stands out prominently. In 2014, Leisure Farm's tourism output value 102 billion, and the number of tourists are more than 23 million passengers [23] Compared with 5.7 billion in 2006 and 959 million passengers [24], it grows 81% and 140%. No wonder that the Government of Taiwan classifies the Leisure Agriculture as the highlight of the depth tourism.

After taking Quality Leisure Agriculture Experience Activities and the List of General Selected Qualified Farms as reference [25], the articles in the website of the top 3 leisure farms are gathered to analyze under text mining. Combining the concepts of fish farm and of leisure farms, accommodation, a new category is added. The model for innovations contains seven categories: restaurant (good food), leisure cultural, general fish farming, tropical fish farming, nearby attractions, internal attractions and accommodation.

4.3   Generate Innovative Combination

The main goal for this stage is to create virtual leisure farm by selecting suitable hot spots around the SYFF, the highlight. Using the technique of text mining, with the winning fish farm as principle, and the nearby hot spots and fine accommodations in most tourists mind are revealed to be supplement, the innovation team plans a cycling tour trip with Yilan's special feature. After considering many elements, such as how many snack-store nearby, how good is it stated in consumers' evaluation, etc., the tourism experts have selected eight hotspot of leisure attractions: King Car Whiskey Distillery, Bee Museum, Fifi Vanilla Aromatic Plant Museum, King Car Orchid Garden, Ba-Jia Leisure Fish Farm, Memorial Hall of Founding of Yilan Administration, Yilan Museum of Literature and Beneficial Microbes Museum, etc. With the help of tourism experts, 4 B&B near the bee farm are also listed.

The attractions for the entire two-day overnight trip, are mainly in Yilan County. Bicycles which are used for the whole trip will be rented and returned at Yilan train station. Most of the activities for the first day is taking place at the SYFF, and will be the major highlight of the trip. There will be some interacting activities and games for the guest to familiarize others. The customers can have opportunities to explore the beauty of Yilan by bicycle. As for foreign visitors, this two-day tour is only part of their vacation package, which combines different tour packages with hot spots around Taiwan.

4.4   Pack Innovative Products or Services

The main purpose of this phase is to pack products or services, so that the market team can put it on websites, and let the customer group-purchase freely. Before packing the vacation package, the innovation team need to send out questionnaire for the schedule of the featured tour for consumer to assess.

When all the tasks in the above 4 steps are completed, a joint-selling cooperation subsystem of the project system in the CBM is put to work. The list of the joint-selling individuals in this Tour Package includes four B & Bs, eight attractions, insurance company, bicycle rental shops. The core broker will need to negotiate with each of them, and signed the contract to joining them into the cooperation. The core broker will outline and publish of this innovative package as "Recommended Package" at the CBS website in the CBM. Its details will also be published in "Introducing Yilan", and becoming recommendation for tourist, the must visit attractions, cuisine and the quality B & Bs. In addition, the group exchanges required information, such as the number of different deals available discounts, commissions proportion, will also be published. The market broker will then advertise in their e-shops, according to the information in CBS website, to attract customers to do the online ordering.

5   Conclusion

In this paper, the core brokers in the CBM act as lead users to create a new-type of business model to diffuse their own tourist innovations in e-commerce. A group-trading e-commerce contains joint-selling and group-buying. The core broker is responsible for organizing the innovation team and the market team, where the member of the innovation team includes the core broker, product experts and leading users; and the member of the market team includes the core broker and the market broker. The job for innovation team is to pack a package of an innovative two-day family trip, using Yilan as an example, to create a Leisure Farm Tour, taking industry with special feature as the centre, and supported with the nearby attractions.

Through joining many vendors, the innovation team may easily create innovations full of attraction and potentials. After the core broker publishes the details of the innovative product or services, including the detail, pricing, and group-buy discount, and group-buying activity date on the CBS website in the CBM. The market brokers then sells this new tourism package in their e-shops, providing consumer discounts in group-buying, by selling the e-vouchers, and let the customers enjoy their family trip without spending too much effort in it.

As a result, the innovative products or services with joint-selling are the main attractions to consumers. The high discounts produced in group-buying is another motivation for them to visit the e-shops. With the advantages of group-trading in the

CBM for tourism innovations, there are so many opportunities to create innovative products or services and make good sales in e-markets, therefore the chance for the diffusion of the tourism innovation to cross the chasm between early adopters and the early majority is high.

References

1. Chu, P.-J.: An analysis on how Maokong gondola affects the economy and tourism in Maokong area. Unpublished Master Thesis, Graduate Institute of Land Economics, National Chengchi University, Taipei (2007)
2. Tseng, J.-K.: Crazy about Taiwan, My Growth in Tourism on the Top of the World. 2014/10/03 Economic Daily News (2014)
3. Tourism Bureau, Ministry of Transportation and Communications, Executive Yuan: The Graph for Visitor Statistics (2015). http://admin.taiwan.net.tw/public/public.aspx?no=315
4. Rogers, E.M., Shoemaker, F.F.: Communication of Innovations: A Cross-Cultural Approach, 2nd edn. The Free Press, New York (1971)
5. Rogers, E.M.: Diffusion of Innovations. Free Press, New York (2003)
6. Moore, G.A.: Crossing the Chasm - Marketing and Selling Technology Products to Mainstream Customers. Harper Collins, New York (1991)
7. Gillies, D.: Some theorems on n-person games. Unpublished Ph.D. thesis, Princeton University (1953)
8. Osborne, M., Rubenstein, A.: A Course in Game Theory. MIT Press, London (1994)
9. Shapley, L.: Cores of convex games. Int. J. Game Theory 1, 11–26 (1971)
10. McCain, R.A.: Cooperative Games and the Core, William King Server (2005). http://william-king.www.drexel.edu/top/eco/game/game-toc.html
11. Von Hippel, E.: Lead users: a source of novel product concepts. Manag. Sci. 32(7), 791–806 (1986)
12. Prahalad, C.K., Ramaswamy, V.: The Future of Competition: Co-Creating Unique Value with Customers. Common Wealth Magazine, Taipei (2004). (Shu-Hsin Ku, Trans.)
13. Ohsawa, Y., Nels, E.B., Yachida, M.: KeyGraph: automatic indexing by co-occurrence graph based on building construction metaphor. In: Proceedings of Advances in Digital Libraries, pp. 12–18 (1998)
14. Ohsawa, Y.: Chance discovery: the current states of art. In: Ohsawa, Y., Tsumoto, S. (eds.) Chance Discoveries in Real World Decision Making, vol. 30, pp. 3–20. Springer, Heidelberg (2003)
15. Sullivan, D.: Document Warehousing and Text Mining: Techniques for Improving Business Operations, Marketing, and Sales. Wiley, New York (2001)
16. Sun, P.: A core broking model for E-markets. Unpublished Ph.D. thesis, Coventry University (2011)
17. Chen, T.-S.: The analysis on tourist overview and development in the main tourist attractions in Taiwan. Taiwan Econ. Financ. Mon. Mag. 49(2), 1–15 (2013)
18. Tourism Bureau, Ministry of Transportation and Communications, Executive Yuan: The Year Report of the Statistics on the Number of Tourists at the main Tourist Attractions at Yilan County (2004–2014)
19. Budget, Accounting and Statistics Department, Yilan County: The Analysis on Tourism at Yilan County in 2008 (2014)

20. Fisheries Agency, Council of Agriculture, Executive Yuan: The Year Report on Statistics for Fishery Industry (2016). http://www.fa.gov.tw/cht/PublicationsFishYeares. Accessed 23 Oct 2016

21. Jian, M.-C.: The Quoted Price and Trade Practice for U.S.A. Water Products. Spreading Fisherie Ind. Mon. Mag. **196**, 42–50 (2003)

22. Chen, T.-S.: The analysis on market overview and trends in tourism entertainment industry. Taiwan Econ. Financ. Mon. Mag. **42**(8), 54–68 (2006)

23. Department of Economic, Energy and Agriculture, Executive Yuan: The Specific Accomplishment for Implementation of Quality Agriculture (2015)

24. Chen, B.-J.: The Future of Taiwanese Agriculture Under Regional Trading Agreements (2014)

25. Leisure Agriculture of Yilan County Site: Quality Leisure Agriculture Experience Activities and the List of General Selected Qualified Farms (2012)

Using Sentiment Analysis to Explore the Association Between News and Housing Prices

Hsiao-Fang Yang and Jia-Lang Seng[(✉)]

Department of Accounting, College of Commerce,
National Chengchi University, Taipei City 11605, Taiwan (R.O.C.)
hfyang.wang@gmail.com, seng@nccu.edu.tw

Abstract. In recent years, semi-structured and unstructured data have received substantial attention. Previous studies on sentiment analysis and opinion mining have indicated that media information features sentiment factors that can affect investor decisions. However, few studies have explored the correlation between news sentiment and housing prices; hence, the present study was conducted to investigate this correlation. A method was proposed to collect and filter news information and analyze the correlation between news sentiment and housing prices. The results indicate that news sentiment can serve as a reference for evaluating housing price trends.

Keywords: Sentiment analysis · Opinion mining · Housing price · News sentiment scoring model

1 Introduction

Currently, the average housing price in Taiwan is several times the annual income of residents [5]. Chinloy [6] indicated that real estate demand changes rapidly. Researchers [23] asserted that inefficient real estate market is mainly attributable to price information asymmetry. Lacking housing purchasing experience and associated knowledge, most people formulate their housing purchasing decisions according to the suggestions of housing agents. The negotiation processes can be classified into bidding and bargaining [7]. The difficulty of the bargaining process lies in learning the best value, which the opponent still would accept, and in obtaining this value. The prevalence of information transmission allows for easy acquisition of diverse data through the Internet. However, circulating a large amount of information online indirectly increases the difficulty of searching and processing relevant data. Therefore, managing and compiling information valuable to housing buyers and sellers has become a challenge [19]. The Internet is the most common tool for searching information [34]. Among the various information sources available online, the news is perceived as one of the most trustworthy sources because of its regular publication intervals and credibility. News information is often presented as semi-structured or unstructured data. Previous studies [24, 37, 39] have used news and financial reports as information sources to analyze the effects of financial news and corporate annual reports on the stock market [24, 37]. Researchers, including [24, 37], have confirmed that media

© Springer International Publishing AG 2017
N.T. Nguyen et al. (Eds.): ACIIDS 2017, Part II, LNAI 10192, pp. 170–179, 2017.
DOI: 10.1007/978-3-319-54430-4_17

information influences investor sentiment and behavior. Furthermore, researchers [17] also verified that negative news can decrease consumer confidence and influence stock prices. Yu et al. [43] indicated that information disclosed through Google blogs, Boardreader forums, and Twitter microblogs can substantially affect stock prices. Overall, these findings motivated the authors to investigate whether the sentiment orientation of housing price news can be used to estimate subsequent housing price trends, and to explore how news related to housing prices can be extracted and filtered from large amount of daily news information and how they can be analyzed and organized into information that facilitates housing price decisions.

Sentiment analysis can be used to evaluate the emotions expressed in the text [32] and determine the associated sentiment orientations. Accurate housing price information enables house buyers and sellers to accelerate the bargaining process. However, Yan et al. [41] investigated a housing price forecasting model, asserting that sentiment analysis can reveal the sentiment orientation of housing price news, but not the extent of housing price fluctuations, thus providing limited assistance to house buyers and sellers. Nevertheless, sentiment analysis can be used as a core technique in constructing a housing price news sentiment scoring model that incorporates a sentiment intensity value for assessing housing price fluctuations. Thus, valuable information can be extracted from this model. Therefore, the first topic of this study pertains to keyword (domain-specific lexicon) acquisition and news article filtering. The second and third research topics of this study pertain to creating a comprehensive dictionary and constructing a housing price news sentiment scoring model, respectively.

According to the aforementioned research topics, the remaining sections are organized as follows: Sect. 2 introduces information related to sentiment analysis, opinion mining, domain-specific and sentiment lexicons, the scoring model, and the sentiment analysis system. The associated methods are explained in Sect. 3. Section 4 pertains to experimental designs, research results, and the system prototype. The study discussion and conclusion are presented in the final section.

2 Literature Review

2.1 Sentiment Analysis and Opinion Mining

As studies [12, 18, 20, 27, 28, 33, 42, 44] mentioned, sentiment analysis is to determine positive or negative sentiment, opinions, attitudes, and emotions toward an entity (document or text). From the point of view of Medhat et al. [27], there are two approaches to sentiment analysis: machine learning and lexicon-based. Machine learning approach can be classified as supervised technique and unsupervised technique [27, 40]. The lexicon-based approach can be classified as dictionary-based and corpus-based. Zhang et al. [44] pointed that the sentiment analysis often conducted at one of three levels (article, sentence, or attribute). Di Caro [13] said that from a procedural point of view, sentiment analysis can be done at different levels, i.e., at the word level, sentences level, or document level. Serrano-Guerrero et al. [35] pointed that the sentiment analysis has three levels (document, sentence, entity) of analysis. In this paper, we use document level because of data characteristic.

Recent studies on sentiment analysis are discussed. Chen et al. [4] indicate that a system embedded with sentiment analysis technology can facilitate decision making. Das and Chen [11] explored the effect of stock chat board messages on stock prices. O'Leary [31] analyzed blog content by exploring the correlations of blogs and message boards according to financial accounting information. Eirinaki et al. [14] developed an opinion search engine that mines the opinions and extracts useful information related to the item's features and use it to rate them as positive, neutral, or negative. Lau et al. [22] proposed a prototype system of product recommendation to predict the polarities of aspect-level sentiments without requiring expensive manual labeling of training examples. Li et al. [26] analyzed the effect of financial news on the stock prices of Hong Kong firms. Kim et al. [21] examined the effect of financial news on Korean firms. Smailović et al. [36] investigated the effect of Twitter messages on firm stock prices. da Silva et al. [10] classified Twitter messages into various sentiment types. The aforementioned literature indicates that previous studies on sentiment analysis have mostly focused on the effect of news media on stock prices, and few have explored the correlation between news media and housing prices.

2.2   Sentiment Lexicon, Scoring Model and System

As sentiment analysis studies [27, 35, 40] said that the supervised techniques use the labeled data to train some classifiers to predict the unlabeled data. Although the drawback [35] of dictionary-based approach is the incapability to deal with domain and context specific orientations, domain-specific lexicons can be established beforehand to improve the performance of this sentiment analysis method. Regarding dictionary sources, previous studies on sentiment analysis have used hand-selected-based dictionaries [11], external dictionaries [26], established dictionaries such as Harvard IV-4[1] and Loughran–McDonald[2] [10], Website built-in sentiment tags [31], dictionaries constructed according to the calculated sentiment scores of a word and the probability that stock price will rise or fall [21], and term-frequency-based dictionaries [36]. Few studies have adopted dictionaries based on domain-specific lexicons because the accuracy of sentiment classification can be affected by the features of the target domain [33]. Therefore, the present study applied the hand-selected-based approach to establish domain-specific lexicons. During sentiment analysis, the presence of positive and negative words can affect the final sentiment orientation of a sentence, thus influencing the overall orientation of a document. Das [11] improved the performance of sentiment analysis by using grammar rules to capture the pulse of the financial market. Bagheri et al. [1] incorporated multi-words to enhance sentiment analysis performance. Therefore, multiword phrases were used to represent the lexicons in the dictionary of the present study, thus allowing for accurate acquisition of news content most related to housing prices.

---

[1] http://www.wjh.harvard.edu/~inquirer/homecat.htm.

[2] http://www3.nd.edu/~mcdonald/Word_Lists.html.

There are two techniques of natural language processing: statistical and linguistic [8, 9, 38], and dictionary-based [15]. The dictionary-based technique involves using a dictionary to identify the most relevant words. Generally, sentiment analysis is used to categorize text as having a positive or negative orientation [16]. Regarding the method for calculating the sentiment orientation of text, as Zhao et al. [45] stated that for each sentence to acquire polarity words through matching the words in the same sentiment lexicon. According to the textual features and stock price prediction method integrated by Li et al. [25], a news sentiment scoring model was constructed.

3 Research Method

3.1 Filtering, Pre-processing, Screening, and Dictionary

News articles were filtered to obtain information related to housing prices. A list of domain-specific lexicons was provided by a specialist to filter news articles related to housing prices. This filtering process mainly involved examining the title and content of each news article. If either the title or content contained any domain-specific lexicons, then the article was considered relevant to housing prices.

After the relevant news articles were acquired through the list of domain-specific lexicons provided by the specialist, the natural language processing technique was applied to organize these articles. First, the articles were converted to TXT files and UTF-8 encode format, and special characters, including Web site links, segmented words, and stop-words, were removed. When the articles were preprocessed, the processed articles were stored in a corpus database. Previous studies have indicated that using the bag-of-words model is an easy and effective method of presenting texts [15, 30]. Therefore, this method was also applied to present the processed news articles. After the aforementioned process, each news article was converted to a vector space of lexicon frequency, thus quantifying the qualitative data of the articles.

Representative news articles were screened for to increase the effectiveness of the dictionary used in this study. According to the frequency-based approach proposed by Berry and Kogan [2], two indicators were selected to sort the news article in the corpus database. Combining the two indicator values facilitates sorting the articles in descending order and screening for representative articles. According to Biber et al. [3], setting a high threshold value is advantageous for analyzing a large number of articles. Therefore, a top-30 query method was used to screen for the representative news articles.

According to Nation [29], who constructed a domain-specific dictionary, the present study also created a dictionary containing two types of lexicons (domain-specific lexicons and sentiment lexicons). The domain-specific lexicons were provided by the specialist. The domain-specific lexicons were mainly composed of nouns. According to the features of housing price-related terms, the words pertaining to people, events, time, locations, and objects were selected as the domain-specific lexicons. In contrast to domain-specific lexicons, sentiment lexicons are used to determine the sentiment orientation of a sentence. Because multiword phrases contain more information than single words do, these phrases are advantageous for determining the sentiment orientation of a sentence. Hence, multiword phrases were used to establish the sentiment lexicons in this study. After the domain-specific and sentiment lexicons were selected, they were categorized, trimmed, and revised. Categorizing lexicons can facilitate increasing the explicability of housing prices through news sentiment. Therefore, the collected domain-specific lexicons were sorted into multiple categories, whereas the sentiment lexicons were sorted into only two categories (positive and negative).

3.2   Scoring Criteria and Scoring Model

Four graduate students were invited to evaluate the screened Top-30 news articles between 2012 and 2014. Their scores were used to determine the dependent variable of the multiple regression models. News sentiment was scored on a scale of -10 to 10 points. A positive (negative) score indicated that the news content reflected a rise (fall) in housing prices. After numerous discussion sessions, the graduate students established initial scoring criteria. Criterion I: When the title or content of a news article involves factors including governmental policies and large-scale regional development as well as positively affects housing prices, the sentiment score is ≥ 8 points; otherwise, the score is ≤ -8 points. Criterion II: When the title or content of a news article involves factors including governmental policies and small-scale regional development and positively affects housing prices, the sentiment score ranges between 5 and 7 points; otherwise, the score is between -5 and -7 points. Criterion III: When the title or content of a news article involves uncertain statements made by governmental personnel and positively affects housing prices, the sentiment score ranges between 3 and 5 points; otherwise, the score is between -3 and -5 points. Criterion IV: When the title or content of a news article involves uncertain statements made by academic and business experts and positively affects housing prices, the sentiment score ranges between 1 and 3 points; otherwise, the score is between -1 and -3 points. Criterion V: When the title and content of a news article are unrelated to changes in housing prices, the sentiment score is 0 points. The scoring model is described in (1).

$$\hat{y}_i = b_0 + b_1 builder_i + b_2 city_i + b_3 neutral_i + b_4 parkArea_i + b_5 businessMan_i$$
$$+ b_6 gsPerson_i + b_7 realtor_i + b_8 shoppingDistrict_i + b_9 positive_i + b_{10} negative_i$$

$$(1)$$

In (1), $\hat{y}_i$ is the sentiment score of the i news article. b_0 is the intercept. $b_1, b_2 \ldots b_n$ are regression coefficients. *builder, city, neutral, parkArea, businessMan, gsPerson, realtor, shoppintDistrict, positive,* and *negative* represent the quantity of each news content feature. Because the sentiment lexicons were mostly composed of multiword phrases, incorporating a sentiment lexicon with another lexicon could have resulted in news content features being repeatedly calculated. Therefore, such an overlap must be subtracted from the calculation to prevent this complication.

3.3   Data

This study applied the news knowledge bank[3] that archives news articles. Specifically, news articles published between August 1, 2012 and June 30, 2014 were collected with titles or content containing any housing price-related lexicons. The collected articles have six categories: investment (46,692 articles), politics (32,488 articles), information technology (32,042 articles), finance (15,878 articles), industry (52,377 articles), and economics and trading (17,813 articles). From Top-30 articles (a total of 540) between 2012 and 2014, the number of lexicons in ten categories: builder (106), city (95), neutral (230), park area (5), businessman (171), government (or school) person (114), realtor (9), shopping district (33), positive (805), and negative (405).

4   Results

The estimated multiple regression models constructed from the 90 training data of 2012 was used to predict the sentiment scores of these data. Next, the 90 observed data (> 0 = positive orientation; < 0 = negative orientation) are compared with the predicted data. The data of 2013 and 2014 are processed through the same procedures as those for the data of 2012 (Table 1).

Table 1. Performance values of each year

Year	Accuracy	Precision	Recall	F-measure
2012	78.37%	84.375%	90%	87.09%
2013	89.36%	92.1%	94.59%	93.33%
2014	78%	79.31%	82.14%	80.7%
Average	81.91%	85.26%	88.91%	87.04%

Note: accuracy = (true positive + true negative)/total population, precision = true positive/(true positive + false positive), recall = true positive/(true positive + false negative), F-measure = 2 * ((precision * recall)/(precision + recall))

A line chart is adopted to display the variation trends between the observed and predicted values. Figure 1 depicts the variation trends of the observed and predicted values for 2012, 2013, and 2014, respectively.

The independent sample t-test was used to compare the observed and predicted values. Table 2 lists the fundamental descriptive statistics of the observed and predicted values for 2012, 2013, and 2014. Table 3 illustrates the t values and significance values of the data. Because the significance values exceeded 0.05, the null hypothesis was not rejected. In other words, the observed and predicted values differed non-significantly.

---

[3] http://kmw.chinatimes.com/.

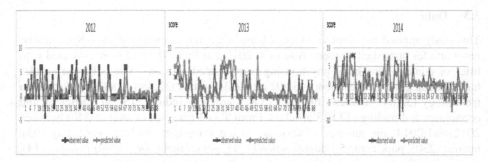

Fig. 1. The comparison between observed value and predicted value in 2012, 2013, and 2014

Table 2. Group statistics

Data	Group	Numbers	Mean	Standard deviation
2012	Observed value	90	1.2000	2.55809
	Predicted value	90	1.0339	1.41123
2013	Observed value	90	1.3000	2.67090
	Predicted value	90	1.6137	2.60328
2014	Observed value	90	0.5111	3.94412
	Predicted value	90	0.3409	2.63573

Table 3. Independent t-test

Data	Group	t-value	Significant
2012	Observed vs. predicted	0.539	0.591
2013	Observed vs. predicted	−0.798	0.426
2014	Observed vs. predicted	0.340	0.734

5   Conclusion and Discussion

In this study, we explored three research topics: (1) acquiring housing price-related lexicons to filter the associated news articles from a news database; (2) creating an applicable dictionary and applying sentiment analysis to investigate the sentiment orientations of housing price news; and (3) constructing a scoring model to assess the sentiment of housing price news. The results revealed that the accuracy, precision, recall, and F-measure values of the three yearly models are greater than 0.8, 0.85, 0.88, and 0.87, respectively, indicating high model performance. The variation trends between the observed and predicted values indicated that our models exhibited high accuracy. The results of which confirmed that the predicted values calculated through the models approximated the observed values.

Few studies have explored the effect of news sentiment on housing prices. Our proposed methodology can be used to evaluate the sentiment of news readers and how their reactions toward news content can affect housing price fluctuations. Additionally, we innovatively proposed two types of lexicons, used multiword phrases to establish sentiment lexicons, and applied a numerical method to represent news sentiment. The news sentiment scoring model proposed in this study can be used not only to present the current housing price trends but also as a reference for house buyers and sellers in bargaining decisions. If the model reveals a negative news sentiment score and a decreasing (increasing) housing price trend, then prospective house buyers (sellers) can consider delaying their purchase (selling) decisions or bargaining for lower (higher) prices. Overall, our methodology still requires further improvement. Additionally, we suggest that the correlation between news sentiment and housing prices should be further investigated. Subsequent researchers can conduct regional studies and integrate housing features, thus increasing the prevalence of news sentiment scoring models.

Acknowledgement. This research is supported by NSC 102-2627-E-004-001, MOST 103-2627-E-004-001, MOST 104-2627-E-004-001, MOST 105-2811-H-004-035.

References

1. Bagheri, A., Saraee, M., de Jong, F.: Care more about customers: unsupervised domain-independent aspect detection for sentiment analysis of customer reviews. Knowl.-Based Syst. **52**, 201–213 (2013)
2. Berry, M.W., Kogan, J.: Text Mining: Applications and Theory. Wiley, Hoboken (2010)
3. Biber, D., Conrad, S., Cortes, V.: If you look at…: lexical bundles in university teaching and textbooks. Appl. Linguist. **25**(3), 371–405 (2004)
4. Chen, L., Wang, F., Qi, L., Liang, F.: Experiment on sentiment embedded comparison interface. Knowl.-Based Syst. **64**, 44–58 (2014)
5. Chen, P.F., Chien, M.S., Lee, C.C.: Dynamic modeling of regional house price diffusion in Taiwan. J. Hous. Econ. **20**(4), 315–332 (2011)
6. Chinloy, P.: Real estate cycles: theory and empirical evidence. J. Hous. Res. **7**, 173–190 (1996)
7. Chiu, D.K.W., Cheung, S.C., Hung, P.C.K., Chiu, S.Y.Y., Chung, A.K.K.: Developing e-negotiation support with a meta-modeling approach in a web services environment. Decis. Support Syst. **40**(1), 51–69 (2005)
8. Chung, T.M., Nation, P.: Identifying technical vocabulary. System **32**, 251–263 (2004)
9. Conrado, M.S., Pardo, T.A.S., Rezende, S.O.: The main challenge of semi-automatic term extraction methods. In: 11th International Workshop on Natural Language Processing and Cognitive Science (NLPCS), Venice, Italy (2014)
10. da Silva, N.F.F., Hruschka, E.R., Hruschka Jr., E.R.: Tweet sentiment analysis with classifier ensembles. Decis. Support Syst. **66**, 170–179 (2014)
11. Das, S., Chen, M.: Yahoo! for Amazon: sentiment parsing from small talk on the web. In: Proceedings of the Eighth Asia Pacific Finance Association Annual Conference (2001)
12. Dave, K., Lawrence, S., Pennock, D.M.: Mining the peanut gallery: opinion extraction and semantic classification of product reviews. In: Proceedings of WWW, pp. 519–528 (2003)
13. Di Caro, L., Grella, M.: Sentiment analysis via dependency parsing. Comput. Stand. Interfaces **35**(5), 442–453 (2013)
14. Eirinaki, M., Pisal, S., Singh, J.: Feature-based opinion mining and ranking. J. Comput. Syst. Sci. **78**(4), 1175–1184 (2012)

15. Hagenau, M., Liebmann, M., Neumann, D.: Automated news reading: stock price prediction based on financial news using context-capturing features. Decis. Support Syst. **55**(3), 685–697 (2013)
16. Hogenboom, A., Frasincar, F., de Jong, F., Kaymak, U.: Polarity classification using structure-based vector representations of text. Decis. Support Syst. **74**, 46–56 (2015)
17. Hollanders, D., Vliegenthart, R.: The influence of negative newspaper coverage on consumer confidence: the dutch case. J. Econ. Psychol. **32**(3), 367–373 (2011)
18. Hu, N., Bose, I., Koh, N.S., Liu, L.: Manipulation of online reviews: an analysis of ratings, readability, and sentiments. Decis. Support Syst. **52**(3), 674–684 (2012)
19. Jin, X., Wah, B.W., Cheng, X., Wang, Y.: Significance and challenges of big data research. Big Data Res. **2**(2), 59–64 (2015)
20. Kennedy, H.: Perspectives on sentiment analysis. J. Broadcast. Electron. Media **56**(4), 435–450 (2012)
21. Kim, Y., Jeong, S.R., Ghani, I.: Text opinion mining to analyze news for stock market prediction. Int. J. Adv. Soft Comput. Appl. **6**(1), 1–13 (2014)
22. Lau, R.Y.K., Li, C., Liao, S.S.Y.: Social analytics: learning fuzzy product ontologies for aspect-oriented sentiment analysis. Decis. Support Syst. **65**, 80–94 (2014)
23. Lee, T.H., Liao, C.F.: The development and literature survey on property derivatives: the growing challenge for Taiwan housing index derivatives. J. Hous. Stud. **20**(1), 85–108 (2011)
24. Li, F.: Do stock market investors understand the risk sentiment of corporate annual reports? University of Michigan Working Paper (2006). http://www.cis.upenn.edu/~mkearns/finread/sentiment.pdf
25. Li, Q., Wang, T., Li, P., Liu, L., Gong, Q., Chen, Y.: The effect of news and public mood on stock movements. Inf. Sci. **278**, 826–840 (2014)
26. Li, X., Xie, H., Chen, L., Wang, J., Deng, X.: News impact on stock price return via sentiment analysis. Knowl.-Based Syst. **69**, 14–23 (2014)
27. Medhat, W., Hassan, A., Korashy, H.: Sentiment analysis algorithms and applications: a survey. Ain Shams Eng. J. **5**(4), 1093–1113 (2014)
28. Nasukawa, T., Yi, J.: Sentiment analysis: capturing favorability using natural language processing. In: Proceedings of the Conference on Knowledge Capture (K-CAP) (2003)
29. Nation, P.: Learning Vocabulary in Another Language. Cambridge University Press, Cambridge (2001)
30. Nuntiyagul, A., Naruedomkul, K., Cercone, N., Wongsawang, D.: Keyword extraction strategy for item banks text categorization. Comput. Intell. **23**(1), 28–44 (2007)
31. O'Leary, D.E.: Blog mining-review and extensions: "from each according to his opinion". Decis. Support Syst. **51**(4), 821–830 (2011)
32. Ortigosa-Hernández, J., Rodríguez, J.D., Alzate, L., Lucania, M., Inza, I., Lozano, J.A.: Approaching sentiment analysis by using semi-supervised learning of multi-dimensional classifiers. Neurocomputing **92**, 98–115 (2012)
33. Pang, B., Lee, L.: Opinion mining and sentiment analysis. Found. Trends Inf. Retrieval **2**(1–2), 1–135 (2008)
34. Savolainen, R.: The role of the internet in information seeking. Putting the networked services in context. Inf. Process. Manag. **35**(6), 765–782 (1999)
35. Serrano-Guerrero, J., Olivas, J.A., Romero, F.P., Herrera-Viedma, E.: Sentiment analysis: a review and comparative analysis of web services. Inf. Sci. **311**, 18–38 (2015)
36. Smailović, J., Grčar, M., Lavrač, N., Žnidaršič, M.: Stream-based active learning for sentiment analysis in the financial domain. Inf. Sci. **285**, 181–203 (2014)
37. Tetlock, P.C.: Giving content to investor sentiment: the role of media in the stock market. J. Finance **62**, 1139–1168 (2007)

38. Vivaldi, J., Màrquez, L., Rodríguez, H.: Improving term extraction by system combination using boosting. In: Raedt, L., Flach, P. (eds.) ECML 2001. LNCS (LNAI), vol. 2167, pp. 515–526. Springer, Heidelberg (2001). doi:10.1007/3-540-44795-4_44

39. Wei, C.P., Lee, Y.H.: Event detection from online news documents for supporting environmental scanning. Decis. Support Syst. 36(4), 385–401 (2004)

40. Xu, K., Liao, S.S., Li, J., Song, Y.: Mining comparative opinions from customer reviews for competitive intelligence. Decis. Support Syst. 50(4), 743–754 (2011)

41. Yan, Y., Xu, W., Bu, H., Song, Y., Zhang, W., Yuan, H., Wang, S.-Y.: Method for housing price forecasting based on TEI@I methodology. Syst. Eng. Theory Pract. 27(7), 1–9 (2007)

42. Yi, J., Nasukawa, T., Bunescu, R., Niblack, W.: Sentiment analyzer: extracting sentiments about a given topic using natural language processing techniques. In: Proceedings of the IEEE International Conference on Data Mining (ICDM) (2003)

43. Yu, Y., Duan, W., Cao, Q.: The impact of social and conventional media on firm equity value: a sentiment analysis approach. Decis. Support Syst. 55(4), 919–926 (2013)

44. Zhang, Y., Dang, Y., Chen, H.: Research note: examining gender emotional differences in web forum communication. Decis. Support Syst. 55(3), 851–860 (2013)

45. Zhao, Y.-Y., Qin, B., Liu, T.: Integrating intra-and inter-document evidences for improving sentence sentiment classification. Acta Automatica Sinica 36(10), 1417–1425 (2010)

Importance-Performance Analysis Based Evaluation Method for Security Incident Management Capability

Chih-Chung Chiu[✉] and Kuo-Sui Lin

Department of Information Management, Aletheia University, Taiwan, R.O.C.
{au4229, au4234}@mail.au.edu.tw

Abstract. SEI's Incident Management Capability Metrics provides an overview of how the metrics can be used to evaluate and improve organizations' information security incident management capability. However, there still exist several deficiencies when using SEI's Metrics to measure the function areas of security incident management capability. An importance-performance analysis based evaluation method for measuring and improving organizations' information security incident management capability was proposed in this paper. The evaluation method produces a four-quadrant IPA matrix that considers both importance and performance simultaneously for better identifying function areas needing improvement. A numerical example of the evaluation method showed that the proposed method is efficient for deploying continuous improvement program and better allocating limited resources.

Keywords: Fuzzy number · Graded mean integration representation method · Importance-performance analysis · Security incident management capability

1 Introduction

Prioritizing the function areas of security incident management capability is a critical decision making point in the incident management process. This can result in better allocating organizational resources and deploying measurement and improvement program of incident management capability. Focusing on just the importance function areas could lead to very different conclusions, or even opposite results. A function area is important does not mean that resources should be expended in that area because performance of the area may be adequate or good enough. Similarly, looking only at the performance measures and focusing only on function areas of low performance measures may be of little value, if these function areas are not important. That is to say, both importance and performance measures on function areas must be considered simultaneously. However, there is a lack of previously stated research for existing incident management capability metrics. Traditional evaluation method employs performance rating as the only criterion for improvement. Therefore the purpose of this study was to propose an importance-performance analysis based security incident management capability evaluation method that considers both importance and performance simultaneously and identifies function areas needing improvement. The IPA

© Springer International Publishing AG 2017
N.T. Nguyen et al. (Eds.): ACIIDS 2017, Part II, LNAI 10192, pp. 180–194, 2017.
DOI: 10.1007/978-3-319-54430-4_18

matrix intends to prioritize different improvement strategies for function areas of incident management capability. A numerical example of the evaluation method showed that the proposed method is efficient for deployment of remedy actions and better allocating limited resources.

2 Related Works

2.1 Importance-Performance Analysis

Importance-Performance analysis (IPA) is a simple but effective, easy-descriptive graphical tool developed by Martilla and James [1] to prioritize service attributes and to further develop effective marketing strategies, which simultaneously consider both the importance and performance dimensions. The IPA is characterized by a four quadrant grid where each quadrant has its' own strategy for future action (see Fig. 1). Typical steps taken for conducting Importance-Performance analysis are: (1) a set of key attributes or characteristics of the product or service being evaluated are identified through document content analysis, focus group interview, statistical analysis or expert judgment; (2) a survey of evaluators is made to ask two questions about each attribute: "How important is the attribute?" and "How well did the provided product or service perform on each attribute?" Importance weightings and performance ratings of each attribute are collected and calculated; (3) a two-dimensional plot called the action grid or importance-performance grid is developed with a vertical axis of importance values and a horizontal axis of performance values. The above weightings and ratings are then plotted into the grid, providing x and y coordinates of the grid.

As shown in Fig. 1, the four quadrant grid is formed by vertical axis/horizontal axis and vertical hairline/horizontal hairline. The crossed hairlines divide the importance-importance measurement area into four separate cells or quadrants. For positioning the cross hairlines of the grid, the overall means of the importance and performance score were used. The vertical hairline is placed at the overall mean score for all attributes/items on the performance scale and the horizontal hairline is placed at the overall mean score for all attributes/items on the importance scale. In addition to using the mean scores for crosshair positioning, middle points or arbitrary idea scores of the

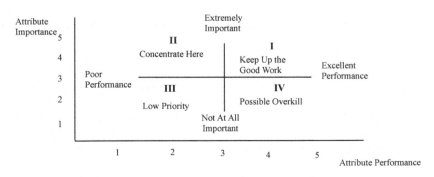

Fig. 1. The action grid (modified after Shia et al. [3])

x, y scale can also be used as dividing point to form cross hairlines. The grid hairlines are established that reflect thresholds or baselines of the management goals. Importance weightings and performance ratings attained from survey instrument (*e.g.* Likert 5 point rating scale, *etc.*) are plotted onto a two-dimensional matrix.

The action grid offers a visual display of analyzed results according to the different position in the grid of the value pairs. As indicated in Fig. 1, different quadrants in the grid suggest different marketing strategies:

(I)   High Importance/High Performance Quadrant: Attributes appearing in the upper right quadrant have been rated high on both importance and performance. This would suggest that a "keep up the good work" strategy;

(II) High Importance/Low Performance Quadrant: Attributes located in the upper left quadrant have been considered high in importance, but have been rated substandard in performance. This would suggest that a "Concentrate Here" strategy should be taken and major efforts should be focused on these attributes that require immediate corrective action and should be given top priority;

(III) Low Importance/Low Performance Quadrant: Attributes in the lower left quadrant have been rated low in terms of both performance and importance. Since evaluators do not perceive these attributes as important, they are considered "low priority" and do not require additional resources;

(IV) Low Importance/High Performance Quadrant: Attributes falling in the lower right quadrant have been rated high in performance, but are considered low in importance. This would suggest that a "Possible Overkill" strategy should be taken and some of the resources committed to these attributes could be more effectively allocated elsewhere [2, 3]:

2.2   Information Security Incident

ITIL [4] defines information security incidents as "those events that can cause damage to confidentiality, integrity or availability of information or information processing." An event is an observable occurrence in a system or network. Information confidentiality maintains the secrecy and privacy of services and information. An information security incident is defined as any real or suspected adverse that could lead to loss of, or disruption to, an organization's operations, services or functions in relation to the security of computer systems or computer networks. An information security incident is indicated by a single or a series of unwanted or unexpected information security events that have a significant probability of compromising business operations and threatening information security. Information security incidents may be deliberate or accidental (*e.g.* caused by error or acts of nature), and may be caused by technical or physical means. Their consequences include such events as information being disclosed or modified in an unauthorized manner, destroyed or otherwise being made unavailable or organizational assets being damaged or stolen. One or more unwanted or unexpected information security events could very likely compromise the security of organization's information and weaken or impair business operations. Five types of computer security incidents include Denial of Service, Malicious Code, Unauthorized

Access and Inappropriate Usage [5, 6]. An incident is an event that could lead to loss of, or disruption to, an organization's operations, services or functions. If not managed, an incident could escalate into an emergency, crisis or a disaster. Incident management is therefore the process of limiting the potential disruption caused by such an event, followed by a return to business as usual.

2.2.1 Information Security Incident Management Capability

The main purpose of the information security incident management is to restore normal service operation as quickly as possible with the intention of minimizing the consequential impact. An information security incident management capability (ISIMC) is the ability to provide management of computer security events and incidents. It implies end-to-end management for controlling or directing how security events and incidents should be handled. ISIMC involves: (1) defining a process to follow with supporting policies and procedures in place, (2) assigning roles and responsibilities, (3) having appropriate equipment, infrastructure, tools, and supporting materials ready, and (4) having qualified staff identified and trained to perform the work in a consistent, high-quality, and repeatable way [7–9]. The incident management roadmap covers ICT Readiness for Business Continuity Planning (BCP), Cyber Security, Network Security, Application Security, Trusted Third Party (TTP) Services Security, Forensic Investigation, *etc.* More detailed views of current advances of information security incident are referred to in [10–17].

ISIMC help the system owners, data owners, and operators to assure that their incident management services are being delivered with a high standard of quality and success, and within acceptable levels of risk. Therefore, it is critical to prioritize the function areas of incident management capability so as to make resource allocation recommendations more efficiently. ISIMC is instantiated in a set of services considered essential to protecting, defending, and sustaining an organization's computing environment, in addition to conducting appropriate response actions [10]. Such services can be provided internally by security or network operators, outsourced to Managed Security Service Providers, or they can also be provided and managed by a computer security incident response team (CSIRT) or Information Security Incident Response Team (ISIRT). Incident response team (IRT) is a team of appropriately skilled and trusted members of the organization, which will handle information security incidents during their lifecycle [18, 19].

2.2.2 SEI's Incident Management Capability Metrics

SEI's Incident Management Capability Metrics (IMCM) provides organizations with a baseline against which they can benchmark their current incident management processes or services [10]. The goal of this incident management capability evaluation is to help organizations assemble the right set of people, processes, and technology that enables them to protect and sustain their critical data, assets, and systems, and to conduct appropriate response and coordination actions for handling events and incidents when they occur. The IMCM can be used to evaluate an existing capability, identify areas for process improvement in an existing capability. The results obtained from the IMCM help an organization determine the maturity of its incident management capability.

The incident management functions, which are provided in a series of questions and indicators, are grouped into four basic functional categories: Protect, Detect, Respond and Sustain. The "Protect" process relates to actions taken to prevent attacks from happening, and mitigate the impact of those that do occur. In the "Detect" process, information about current events, potential incidents, vulnerabilities, or other computer security or incident management information is gathered both proactively and reactively. The "Respond" process includes the steps taken to analyze, resolve, or mitigate an event or incident. Finally, the "Sustain" process focuses on maintaining and improving the CSIRT or incident management capability, itself.

The incident management functions define the actual benchmark. This benchmark can be used by an organization to assess how its current security incident management capability is defined, managed, measured, and improved. The IMCM metrics are questions that can be used to benchmark or evaluate an incident management capability. Each function or service within the capability has a set of goals, tasks, and activities that must be completed to support the overall strategic mission of the organization. The questions explore different aspects of incident management activities for protecting, defending, and sustaining an organization's computing environment in addition to conducting appropriate response actions. There are five sections, representing the four main service categories of metrics as well as an additional category at the beginning for common metrics. These sections are: (1) Common: Section 0 of the metrics, (2) Protect: Section 1 of the metrics, (3) Detect: Section 2 of the metrics, (4) Respond: Section 3 of the metrics, (5) Sustain: Section 4 of the metrics.

An illustrative list of the incident management function questions with ratings is provided in the Appendix. A complete list of the incident management function questions is not provided in this study because of space limitations. Interested readers are encouraged to refer to this list provided in [10].

Table 1 provides an overview of the activities conducted in the Protect, Detect, Respond, and Sustain categories. Each category contains a range of subcategories with a set of one or more functions. Each function includes a question about the performance of that function and several indicators that essentially describe the activities leading to adequate benchmark performance of that function. Indicators include the prerequisites,

Table 1. Function categories

Protect	Detect	Respond	Sustain
-Risk assessment support	-Network security monitoring	-Incident reporting	-MOUs and contracts
-Malware protection support	-Indicators, warning, and situational awareness	-Incident response	-Project/program management
-CND operational exercises		-Incident analysis	-CND technology development, Evaluation, and implementation
-Constituent protection Support and training			-Personnel
-Information assurance/vulnerability management			-Security administration
			-CND information systems
			-Threat level implementation

controls, activities, supporting mechanisms, artifacts and quality indicators the evaluators can see or examine during the evaluation to help them determine whether the metric is being met.

3 Mathematics of the Research

Zadeh [20] used fuzzy sets theory and adopted the fuzzy logical concepts to treat uncertainty and ambiguous existence of traditional scientific methods. Mathematical backgrounds on the fuzzy set theory as well as its extensions used in the present study are briefly described in this section.

Definition 1: Fuzzy Linguistic Variables [21].

A linguistic variable represents natural language expressions whose values are not expressed in numerical values, but are expressed in a set of linguistic terms. Each linguistic term (linguistic value) can be modeled and quantified by different membership functions.

3.1 The Algebraic Operations on Fuzzy Numbers

By the extension principle [20], the fuzzy addition and subtraction of any two triangular fuzzy numbers are closed and are also triangular fuzzy numbers. However, the algebraic operation of fuzzy multiplication and division are not closed and are only approximate triangular fuzzy numbers.

Definition 2: The algebraic operations on fuzzy numbers [20].

Let $\tilde{A} = (a_1, b_1, c_1)$ and $\tilde{B} = (a_2, b_2, c_2)$ be any two triangular fuzzy numbers. According to the extension principle, the algebraic operations of any two fuzzy numbers $\tilde{A}$ and $\tilde{B}$ can be expressed as follows [20]:

Fuzzy addition of $\tilde{A}$ and $\tilde{B}$:
$\tilde{A} \oplus \tilde{B} = (a_1, a_2, a_3) \, (+) \, (b_1, b_2, b_3) = (a_1 + b_1, a_2 + b_2, a_3 + a_3)$, where a_1, a_2, a_3, b_1, b_2, b_3 are any real numbers.

Fuzzy subtraction of $\tilde{A}$ and $\tilde{B}$:
$\tilde{A} \ominus \tilde{B} = (a_1, a_2, a_3)(-)(b_1, b_2, b_3) = (a_1 - b_1, a_2 - b_2, a_3 - b_3)$,
Fuzzy multiplication of $\tilde{A}$ and $\tilde{B}$:
$\tilde{A} \otimes \tilde{B} \approx (a_1, a_2, a_3) \, (\times) \, (b_1, b_2, b_3) = (a_1 \times b_1, a_2 \times b_2, a_3 \times b_3)$, if $a_1 \geq 0$ and $b_1 \geq 0$
Fuzzy division of $\tilde{A}$ and $\tilde{B}$:
$\tilde{A} \oplus \tilde{B} \approx (a_1, a_2, a_3) \, (\div) \, (b_1, b_2, b_3) = (a_1 \div b_3, a_2 \div b_2, a_3 \div b_1)$,
Scalar Multiplication:
$k\tilde{A} = k \times (a_1, b_1, c_1) = (ka_1, ka_2, ka_3)$, if $k \geq 0$
$k\tilde{A} = k \times (a_1, b_1, c_1) = (ka_3, ka_2, ka_1)$, if $k < 0$

3.2 Graded Mean Integration Representation Method

Hsieh and Chen [22] proposed the Graded Mean Integration Representation (GMIR) method to defuzzify generalized fuzzy member and a similarity measure to rank fuzzy numbers. The degree of similarity $S(\tilde{A}, \tilde{B})$ between fuzzy numbers A, B can be calculated with following equation: $S(\tilde{A}, \tilde{B}) = 1/(1 + d|\tilde{A}, \tilde{B}|)$, where $d(\tilde{A}, \tilde{B}) = |P(\tilde{A}\tilde{A}) - P(\tilde{B})|$; $P(\tilde{A})$ and $P(\tilde{B})$ are the graded mean integration representations of $\tilde{A}$ and $\tilde{B}$, respectively.

Definition 3: GMIR approach for defuzzification [22].

Let $\tilde{A} = \text{TFN}(a_1, a_2, a_3)$, and $\tilde{B} = \text{TFN}(b_1, b_2, b_3)$ be two triangular fuzzy numbers. By the GMIR method, the arithmetic defuzzification for triangular fuzzy number $\tilde{A}$ and triangular fuzzy number and $\tilde{B}$ can be defined as $P(\tilde{A}) = (a_1 + 4a_2 + a_3)/6$ and $P(\tilde{B}) = (b_1 + 4b_2 + b_3)/6$, respectively.

The graded mean integration representation $P(\tilde{A})$ and $P(\tilde{B})$ can be regarded as ranking values of $\tilde{A}$ and $\tilde{B}$, and the ranking order of $\tilde{A}$ and $\tilde{B}$ can be determined as: $\text{TFN}\tilde{A} > \text{TFN}\tilde{B} \Leftrightarrow P(\tilde{A}) > P(\tilde{B})$; $\text{TFN}\tilde{A} = \text{TFN}\tilde{B} \Leftrightarrow P(\tilde{A}) = P(\tilde{B})$; $\text{TFN}\tilde{A} < \text{TFN}\tilde{B} \Leftrightarrow P(\tilde{A}) < P(\tilde{B})$.

3.3 Membership Function Construction

Membership Functions (MFs) are the mapping functions to capture the uncertainty and vagueness of the fuzzy data containing linguistic variables, linguistic terms, fuzzy numbers and linguistic values. During membership function construction, the traditional measurement ratings can be converted into labeled values and then translated into corresponding fuzzy values [23, 24]. The labeled values are also called as fuzzy linguistic terms.

Symmetric membership functions can reflect universal duality and natural relativity of linguistic terms [25, 26] and increase understandability and interpretability of fuzzy linguistic variables. Triangular fuzzy numbers (TFNs) have the advantage of simplicity, popularity and are easier to handle than the more complex trapezoidal or bell-shaped fuzzy numbers [27]. Considering simplicity, understandability and interpretability and referring to several types of TFNs in linguistic terms in [28], the TFNs associated with linguistic terms are used in this study. As shown in Fig. 2, the 50% overlap ratio between two neighboring membership functions is adopted in his study. Quantification of linguistic terms is done through membership functions which obtain values in the interval [0, 1].

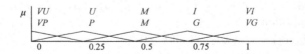

Fig. 2. Membership functions for different linguistic terms

With respect to the number of linguistic terms, the granularity of the universe of discourse of the fuzzy set that has to be taken into account, larger numbers implying more knowledge and higher accuracy of granularity. For example, the linguistic variables were defined by five triangular shaped fuzzy sets, evenly distributed in the domains, according to the equalized universe method [29]. The linguistic terms set is defined as S = {VP, P, M, G, VG}, where VP = Very Poor, P = Poor, M = Medium, G = Good, and VG = Very Good. Therefore, the spectrum of the above linguistic terms can be defined as the following TFNs: VP = TFN(0, 0, 0.25), P = TFN(0, 0.25, 0.5), M = TFN(0.25, 0.5, 0.75), G = TFN(0.5, 0.75, 1), VG = TFN(0.75, 1, 1).

4 Importance-Performance Analysis Based Incident Management Capability Evaluation Method

The purpose of this study was to propose an importance-performance analysis based incident management capability evaluation method as a feasible and efficient management tool to measure quality of incident management functions in organizations. The importance-performance-matrix is used as a special type of graphical presentation of traditional importance-performance analysis method by using SEI's 61 function areas as target of evaluation. Using the four-quadrant IPA matrix that identifies areas needing improvement as well as areas of effective performance, this IPA method can also be used to prioritize the organization's information security incident management capability functions and take actions for practical improvements. Working steps of the evaluation method is introduced and a numerical example is also incorporated for deploying efficient improvement program and better allocating limited resources.

Step 1. Establishing a Baseline Set of Thresholds for Function Areas
Once the evaluated groups and activities are identified, function areas from the Protect, Detect, Respond, and Sustain categories can be allocated to each group. For example, allocate "Protect" function areas to the groups performing "Malware Protection Support". The incident management functions define the actual benchmark. This benchmark can be started with some established set of thresholds across each function area. For example using the mean scores 0.6 as thresholds for each function areas.

Step 2. Evaluating How Each Group is Effectively Performing its Functions
Conduct interviews or group discussions and ask the assembled individuals about each function that is applicable to their group. Artifacts related to the functions can be requested and reviewed, and where necessary, activities can be observed.

Sub-step 2.1. Collecting Linguistic Ratings and Linguistic Weightings for All Function Areas
Each of incident management function areas can be regarded as a linguistic variable whose importance weightings and performance ratings can be defined by a set of importance linguistic terms and a set of performance linguistic terms, respectively. The evaluators used the fuzzified linguistic term set {Very Unimportant (VU), Unimportant (U), Moderate (M), Important (I), Very Important (VI)} for importance weightings.

Table 2. Linguistic term set and transformed fuzzy numbers

Linguistic variable	Linguistic term set for importance weighting	Linguistic term set for performance rating	Transformed fuzzy number
Incident management functions	Very Unimportant (VU)	Very Poor (VP)	(0, 0, 0.25)
	Unimportant (U)	Poor (P)	(0, 0.25, 0.5)
	Moderate (M)	Medium (M)	(0.25, 0.5, 0.75)
	Important (I)	Good (G)	(0.5, 0.75, 1)
	Very Important (VI)	Very Good (VG)	(0.75, 1, 1)

Likewise, the evaluators also used the fuzzified linguistic term set {Very Poor (VP), Poor (P), Medium (M), Good (G), Very Good (VG)} for performance ratings (see Table 2).

Sub-step 2.2. Membership Function Generation of the Linguistic Terms for the Function Areas

Membership functions (MFs) are often related to the perception by humans. Each of the above linguistic terms can be characterized, quantified and converted into their corresponding fuzzy number, ranged between 0 and 1 and denoted as TFN (a_1, a_2, a_3) (see Fig. 2).

Step 3. Prioritizing the Function Areas and Take Improvement Actions

The matrix with four quadrants helps establish and define action plans to minimize the differences between expectations and perceptions. The importance-performance (IPA) method is a specific type of graphical presentation of survey results in assessment and improvement of function areas. All function areas that need to be analyzed are arranged within the matrix according to their degrees of importance and their degrees of satisfaction to the assessment group.

A four step procedure is proposed in this paper, which are

(1) Listing every function areas and then developing a performance check list based on these listed function areas: Performance questions may read as "How did we perform on the following function aspects?"

(2) Giving an importance rating and a performance rating for each of those function areas: The Importance score is to reflect degree of importance priority to each function area. Ratings of importance priority are pre-assigned by SEI. By the graded mean integration representation, the crisp ranking values of triangular fuzzy numbers derived from previous step can be easily calculated and the ranking of the derived triangular fuzzy numbers can be effectively determined.

(3) Dividing the entire diagram into four quadrants and placing each of those function areas into one of the four quadrants: Gridlines divide the entire diagram into four quadrants or cells, using either mean scores on each axle or perceived importance and performance mean scores as the dividing point. For example, If mean scores on each axle are chosen for study, gridlines were placed at values of 0.5 to reflect mean scores of "important" and "performance"; If the importance and performance mean scores are chosen for study, the 4-cell grid was developed by plotting

the mean importance scores and performance scores of each function areas on the grid according to its assigned importance priority and perceived performance.

(4) Proposing different priority strategies for function items: Looking at the assessment results and judging from the resulted scores plotted in different quadrants, priority strategies can be taken by the organization to take improvement actions.

The outcome is a matrix that exhibits the performance and importance of the function areas, which can be evaluated by the performance rating (r_i) and the importance weighting (w_i) of each function area A_i, $i = 1...m$. r_i is the defuzzified performance score and w_i is the defuzzified importance score by using Graded Mean Integration Representation (GMIR) method. Let $d_i = r_i - r^*_i$, where d_i is the performance variance, r^*_i is the baseline performance rating and w^*_i is the baseline importance weighting:

(1) If $d_i < 0$, that means the performance expectation has not been achieved and therefore the evaluator dissatisfied with the performance of the implemented function area;
(2) If $d_i = 0$, that means the performance expectation is met;
(3) If $d_i > 0$, that means the performance expectation has exceeded the performance baseline and therefore the evaluator satisfied with the performance of the implemented function area.

As indicated in Fig. 3, different quadrants in the grid suggest different priority strategies:

Quadrant I: In "Keep Up the Good Work" Quadrant, the importance weightings and performance ratings both meet or exceed function threshold values; It means that evaluators care high importance and express a high level of satisfaction about the function areas in this cell.

Quadrant II: In Concentrate Here Quadrant, the importance weightings and performance rating both fall short of threshold values; It means that evaluators care high importance but they are not satisfied with the performance about the function areas in this cell. In other words, these function areas are the apparent weak points. Organizations need to pay more their attentions on areas in this region and identify the improvement actions as soon as possible.

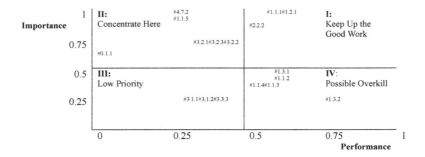

Fig. 3. The action grid for information security incident management capability

Quadrant III: In "Low Priority" Quadrant, the performance ratings do not meet the threshold values, and evaluators do not place a high level of importance on these function areas; It means that evaluators care less about the function areas in this cell, and they are not satisfied with these function performance either. It appears that function areas in this cell are not important issues, but there exists some potential to change evaluators' attitude if improvement solutions can be identified.

Quadrant IV: In "Possible Overkill" Quadrant, the performance rating scores meet or exceed threshold values, but a low level of importance is assigned to these function areas. It means that evaluators care less about the function items, but they satisfy more about the function areas in this cell. It means that the organization may over emphasize the role of those areas.

In the plot of the IPA matrix (Fig. 3), looking only at the importance scale, one would conclude that resources should be focused on those areas deemed important. For our example, these are function areas #0.1.1, #1.1.1, #1.1.5, #2.2.2, #3.2.2 and #4.7.2. Looking only at the performance scale, one would conclude that resources should be focused on those areas that are in need of improvement. For our example, these are function areas #0.1.1, #1.1.5, #3.1.1, #3.1.2, #3.3.3, #3.2.2 and #4.7.2. This study found that three functions are in the Concentrate Here Quadrant, so that the greatest concentration should be put in the#0.1.1, #1.1.5, #3.2.2 and #4.7.2 function areas.

The problem of how to allocate limited resources on the just and equitable ground has been the subject of concerted discussion and analysis. The above resultant importance-performance information of the Protect, Detect, and Respond function areas have been analyzed and identified to help prioritize actions for the incident management operational processes. Considering limited available resources and having prioritized order of these function areas, the organization can thus follow SEI's IMCM procedures and methods to perform the incident management functions areas of Protect, Detect, Respond and Sustain. The infrastructure can thus be established to provide appropriate software, hardware, equipment, and methodologies to support these incident management functions areas.

5   Conclusions and Future Works

SEI's Incident Management Capability Metrics provides an overview of how the metrics can be used to evaluate and improve an organization's information security incident management capability. However, there still exist several deficiencies when using SEI's IMCM to measure the quality of security incident management capability functions. This study proposed an importance-performance analysis based evaluation method to augment SEI's IMCM for assessing organization's status of incident management capability with respect to some critical function areas. The proposed method identifies function areas that should be emphasized or de-emphasized, guiding the prioritization and development of action plans to minimize mismatches between importance and performance. This can result in improved deployment of remedy actions and better allocating organizational resources. A numerical example has shown that the proposed evaluation method can be successfully used to define priorities and guide resource allocation decisions. As stated previously in SEI's report, metrics are not a precisely defined path for every organization to build the perfect incident management capability; metrics are a starting place served as a baseline for improving incident management capabilities. While results of this study could have a theoretical contribution to the research questions, the authors suggest that the practicality of the proposed method be demonstrated through more future empirical case studies. In the future, with the likely release of critical function areas of incident management capability metrics, a new study will be conducted to further evolve the new importance-performance analysis based evaluation method.

Appendix

The incident management function areas and evaluation ratings

	Graded mean of performance ratings	Graded mean of importance weightings
Interfaces		
0.1.1 Have well-defined, formal interfaces for conducting agency incident management activities been established and maintained?	0.47	0.78
Risk assessment support:		
1.1.1 Are Risk Assessments (RAs) performed on constituent systems?	0.52	0.78
1.1.2 Are the constituents assisted with correcting problems identified by Risk Assessment (RA) activities?	0.56	0.48
1.1.3 Is proactive vulnerability scanning (VS) performed on constituent networks and systems?	0.56	0.41
1.1.4 Is the constituent assisted with correcting problems identified by vulnerability scanning (VS) activities?	0.53	0.41
1.1.5 Is trend analysis supported and conducted?	0.32	0.78
Malware protection support:		
1.2.1 Is there an institutionalized Malware/Anti-Virus (AV) Program?	0.57	0.78
Computer network defense operational exercises:		

(continued)

(continued)

	Graded mean of performance ratings	Graded mean of importance weightings
1.3.1 Are operational exercises conducted to assess the security posture of the organization?	0.58	0.43
1.3.2 Are lessons learned from operational exercises incorporated into the constituents' network defenses?	0.74	0.33
Constituent protection support and training:		
⋮		
Respond		
Incident reporting:		
3.1.1 Are incidents reported to and coordinated with appropriate external organizations or groups in accordance with organizational guidelines?	0.32	0.47
3.1.2 Are incidents reported to appropriate organization management in accordance with organizational guidelines?	0.37	0. 47
Incident response:		
3.2.1 Is there an event/incident handling capability?	0.32	0.78
3.2.2 Is there an operations log or record of daily operational activity?	0.37	0.78
3.2.3 Is information on all events/incidents collected and retained in support of future analytical efforts and situational awareness?	0.36	0.78
⋮		
4.7.2 Is the constituency assisted with decisions regarding changes to local threat levels?	0.46	0.68

References

1. Martilla, J.A., James, J.C.: Importance-performance analysis. J. Mark. **41**(1), 77–79 (1997)
2. Crompton, J.L., Duray, N.A.: An investigation of the relative efficacy of four alternative approaches to importance-performance analysis. J. Acad. Mark. Sci. **13**(4), 69–80 (1985)
3. Shia, B.C., Chen, M., Ramdansyah, A.D.: Measuring customer satisfaction toward localization website by WebQual and importance performance analysis (case study on AliexPress site in Indonesia). Am. J. Ind. Bus. Manag. **6**(2), 117–128 (2016)
4. Office of Government Commerce (OGC): ITIL V3-Service Design Book, The Stationery Office, UK (2007)

5. Howard, J.D., Longstaff, T.A.: A Common Language for Computer Security Incidents, SANDIA REPORT, SAND98-8667, Sandia National Laboratories, CA, USA, October 1998
6. Sahibudin, S., Sharifi, M., Ayat, M.: Combining ITIL, COBIT and ISO/IEC 27002 in order to design a comprehensive IT framework in organizations. In: Proceedings of the 2nd Asia International Conference on Modelling & Simulation (2008)
7. Alberts, C., Dorofee, A., Killcrece, G., Ruefle, R., Zajicek, M.: Defining incident management processes for CSIRTs: a work in progress, Technical report CMU/SEI-2004-TR-015, ADA453378, Software Engineering Institute, Carnegie Mellon University, Pittsburgh, PA, USA (2004)
8. Alberts, C.J., Dorofee, A.J., Ruefle, R., Zajicek, M.: An introduction to the mission risk diagnostic for incident management capabilities (MRD-IMC), Technical report CMU/SEI-2014-TN-005, Software Engineering Institute, Carnegie Mellon University, Pittsburgh, PA, USA, May 2014
9. West-Brown, M.J., Stikvoort, D., Kossakowski, K.P., Killcrece, G., Ruefle, R., Zajicek, M.: Handbook for computer security incident response teams (CSIRTs), CMU/SEI-2003-HB-002, Software Engineering Institute, Carnegie Mellon University, Pittsburgh, PA, USA (2003)
10. Dorofee, A., Killcrece, G., Ruefle, R., Zajicek, M.: Incident management capability metrics, version 0.1, Technical report CMU/SEI-2007-TR-008, ESC-TR-2007–008, CERT Program, Software Engineering Institute, Carnegie Mellon University, Pittsburgh, PA, USA, April 2007
11. ISO/IEC 13335-1: Information technology—Security techniques - Management of information and communications technology security - Part 1: Concepts and models for information and communications technology security management (2004)
12. ISO/IEC TR 18044: Information technology—Security techniques—Information security incident management, International Organization for Standardization/International Electrotechnical Commission (2004)
13. ISO/IEC 27001:2013(E): Information technology–Security techniques–Information security management systems–Requirements, International Organization for Standardization/ International Electrotechnical Commission (2013)
14. ISO/IEC 27002: Information Technology—Security Techniques—Code of Practice for Information Security Management, International Organization for Standardization/International Electrotechnical Commission (2013)
15. ISO/IEC 27035: Information Technology—Security Techniques—Information Security Incident Management, International Organization for Standardization/International Electrotechnical Commission (2011)
16. NIST SP 800-61 Revision 2: Computer Security Incident Handling Guide, National Institute of Standards and Technology, U.S. Department of Commerce, August 2012
17. ISO/IEC JTC 1 SC27: IT Security techniques, International Organization for Standardization/International Electrotechnical Commission (2015)
18. Killcrece, G., Kossakowski, K.P., Ruefle, R., Zajicek, M.: Organizational models for computer security incident response teams (CSIRTs), Technical report CMU/SEI-2003-HB-001, ADA421684, Software Engineering Institute, Carnegie Mellon University, Pittsburgh, PA, USA (2003)
19. Killcrece, G., Kossakowski, K.P., Ruefle, R., Zajicek, M.: State of the practice of computer security incident response teams (CSIRTs), Technical report CMU/SEI-2003-TR-001, ESC-TR-2003-001, Software Engineering Institute, Carnegie Mellon University, Pittsburgh, PA, USA, October 2003
20. Zadeh, L.A.: Fuzzy sets. Inf. Control **8**, 338–356 (1965)

21. Zadeh, L.A.: The concept of a linguistic variable and its application to approximate reasoning. Part I. Inf. Sci. **8**, 199–249 (1975). Part II. **8**, 301–357, Part III. **9**, 4301–4308
22. Chen, S.H., Hsieh, C.H.: Graded mean integration representation of generalized fuzzy number. J. Chin. Fuzzy Syst. **5**(2), 1–7 (1999)
23. Lin, K.S.: Fuzzy similarity matching method for interior design drawing recommendation. Rev. Socionetwork Strat. **10**(1), 17–32 (2016)
24. Lin, K.S., Chiu, C.C.: A fuzzy similarity matching model for interior design drawing recommendation. In: Proceedings of the ASE BigData 2015 & SocialInformatics (2015)
25. Gacto, M.J., Alcalá, R., Herrera, F.: Interpretability of linguistic fuzzy rule-based systems: an overview of interpretability measures. Inf. Sci. **181**, 4340–4360 (2011)
26. Chang, C.H., Wu, Y.C.: The genetic algorithm based tuning method for symmetric membership functions of fuzzy logic control systems. In: Proceedings of the IEEE Conference on Industrial Automation and Control: Emerging Technologies, pp. 421–428, May 1995
27. Nachtmann, H., Needy, K.L.: Fuzzy activity based costing: a methodology for handling uncertainty in activity based costing systems. Eng. Econ. **46**(4), 245–273 (2001)
28. Chen, T.Y., Ku, T.C.: Importance-assessing method with fuzzy number-valued fuzzy measures and discussions on TFNs And TrFNs. Int. J. Fuzzy Syst. **10**(2), 92–103 (2008)
29. Chen, M.S., Wang, S.W.: Fuzzy clustering analysis for optimizing fuzzy membership functions. Fuzzy Sets Syst. **103**, 239–254 (1999)

Intelligent Computer Vision Systems and Applications

Moment Shape Descriptors Applied for Action Recognition in Video Sequences

Katarzyna Gościewska and Dariusz Frejlichowski[(✉)]

Faculty of Computer Science and Information Technology,
West Pomeranian University of Technology, Szczecin,
Żołnierska 52, 71-210 Szczecin, Poland
{kgosciewska,dfrejlichowski}@wi.zut.edu.pl

Abstract. Algorithms for recognition of human activities have found application in many computer vision systems, for example in visual content analysis approaches and in video surveillance systems, where they can be employed for the recognition of single gestures, simple actions, interactions and even behaviour. In this paper an approach for human action recognition based on shape analysis is presented. Set of binary silhouettes extracted from video sequences representing a person performing an action are used as input data. The developed approach is composed of several algorithms including those for shape representation and matching. It can deal with sequences of different number of frames and none of them has to be removed. The paper provides some initial experimental results on classification using proposed approach and moment shape description algorithms, namely the Zernike Moments, Moment Invariants and Contour Sequence Moments.

Keywords: Action recognition · Shape descriptors · Video sequences · Binary silhouettes

1 Introduction

The category of human activities includes gestures (elementary human body movements executed for a short time), actions (composed of multiple temporarily organized gestures performed by a single person, such as walking, running, bending or waving), interactions (human-human or human-object activities such as carrying/abandoning/stealing an object) and group activities (i.e. activities performed by groups consisting of multiple objects) [1]. If silhouettes are used for action classification then the human movement can be represented as a continuous pose change and silhouettes extracted from consecutive video frames can be applied to obtain action descriptors used in traditional classification approaches [2]. The approach presented in this paper addresses the problem of recognising an action of a single person and uses information contained in a sequence of binary silhouettes. Each silhouette is firstly processed individually using particular shape description algorithm. Shape representations are

© Springer International Publishing AG 2017
N.T. Nguyen et al. (Eds.): ACIIDS 2017, Part II, LNAI 10192, pp. 197–206, 2017.
DOI: 10.1007/978-3-319-54430-4_19

compared within a sequence to obtain sequence representation. Then sequence representations are processed and final representations are obtained. These are further subjected to the process of classification based on the template matching approach.

The variety of surveillance system applications affects the diversity of activity recognition approaches. In [1] methods for activity recognition were classified into hierarchical (statistical, syntactic and description-based) and non-hierarchical (spatio-temporal and sequential). Hierarchical methodologies enable the recognition of complicated and complex human activities, including interactions and group activities. In turn, non-hierarchical solutions aim to recognize short, primitive actions and repetitive activities—the recognition process is based on the analysis of unknown sequences using an algorithm that matches data to the predefined activity classes. One of the common spatio-temporal technique using accumulated silhouettes was proposed in [3] and is based on motion energy image (MEI, shows where the movement occurs) and motion history image (MHI, shows how the object is moving). MEI and MHI are used to create a static vector-image (temporal template), scale and shift invariant descriptor is used for representation and Mahalanobis distance is used for matching. Another space-time approach was introduced in [4]. It utilizes Poisson equation for feature extraction and human actions are represented as three-dimensional shapes (silhouettes accumulated in the space-time volume). In [5] action sequence is represented by a History Trace Template composed of the set of Trace Transforms extracted for each silhouette. In [6] an action is represented by a set of SAX (Symbolic Aggregate approXimation) vectors which are based on one-dimensional representations of each silhouette. Some other action recognition approaches utilizes only characteristic frames—action is recognized based on selected key poses, e.g. [7–9].

The rest of the paper is organized as follows: Sect. 2 describes the developed approach, Sect. 3 presents algorithms used for shape representation, Sect. 4 presents experimental results on action classification, and Sect. 5 concludes the paper.

2   Proposed Approach

The developed approach is composed of several steps, starting from representing shape information of each silhouette in the dataset and ending on preparing final representation of each sequence, which is then used for action classification. In our approach it is assumed that one video sequence represents one action and it corresponds to a set of binary images—one image contains one silhouette and can be understood as a foreground mask.

In step 1 each silhouette is represented by one shape descriptor using information about its contour or region. In many cases, one shape description algorithm can be employed to obtain representations of different size, e.g. by calculating various order of moments for Zernike Moments or taking various subparts of Fourier Transform-based descriptor. Thank to this we have an opportunity to investigate many shape representations and to select the smallest one which simultaneously carries the most information. If needed, the resultant shape representation is transformed into a vector.

Step 2, for a single sequence, includes calculation of dissimilarities between first frame and the rest of frames using Euclidean distance. The resulting vector containing distance values (normalized to interval $[0, 1]$) is a one-dimensional descriptor of a sequence (a distance vector). The number of its elements equals the number of silhouettes in the input sequence. This vector can be plotted and analysed visually in terms of similarities between actions and their characteristics.

Step 3 aims to convert distance vectors into the form and size that enables the calculation of similarity between them. Therefore, distance vectors were treated as signals and it turned out that the best way to transform such a signal was to use the magnitude of the fast Fourier Transform and a periodogram. Periodogram is a spectral density estimation of a signal and it can determine hidden periodicities in data [10]. Moreover, the periodogram helped to equalize the size of final representations.

Step 4 includes the process of action classification based on sequence descriptors using template matching approach and correlation coefficient. Here template matching is understood as a process that compares each test object with all templates and indicates the most similar one, which corresponds to the probable class of a particular test object. This is a traditional classification solution when only one template set is used. However, some initial tests showed that final results depend on templates. Therefore, we have decided to perform the experiment several times using k-fold cross-validation technique [11] and different set of templates in each iteration. The final recognition effectiveness is then the average of all iterations. For instance, in the first iteration objects with numbers from 1 to k are used as templates and objects with numbers from $k + 1$ to n as test objects, then in the second iteration objects with numbers from $k + 1$ to $2 * k$ are used as templates and remaining objects are used for testing, and so on. Then the results can be interpreted and analysed in three different ways (considering only correct classifications—'true positive'):

1. Recognition effectiveness for each shape descriptor, averaged for all classes and all iterations,
2. Recognition effectiveness for each iteration, each shape descriptor and averaged for all classes,
3. Classification accuracy for each class, each shape descriptor and averaged for all iterations (or for one selected iteration only).

3 Selected Shape Descriptors

3.1 Zernike Moments

The Zernike Moments are orthogonal moments which can be derived using Zernike orthogonal polynomials and the following formula [12]:

$$V_{nm}(x, y) = V_{nm}(r \cos \theta, \sin \theta) = R_{nm}(r) \exp(jm\theta), \tag{1}$$

where $R_{nm}(r)$ is the orthogonal radial polynomial [12]:

$$R_{nm}(r) = \sum_{s=0}^{(n-|m|)/2} (-1)^s \frac{(n-s)!}{s! \times \left(\frac{n-2s+|m|}{2}\right)! \left(\frac{n-2s-|m|}{2}\right)!} r^{n-2s}, \tag{2}$$

where $n = 0, 1, 2, \ldots$; $0 \leq |m| \leq n$; $n - |m|$ is even.

The Zernike polynomials are a complete set of functions orthogonal over the unit disk $x^2 + y^2 < 1$. The Zernike Moments are rotation invariant and resistant to noise and minor variations in shape. The Zernike Moments of order n and repetition m of a region shape $f(x, y)$ are derived using the following formula [12]:

$$Z_{nm} = \frac{n+1}{\pi} \sum_r \sum_\theta f(r \cos \theta, r \sin \theta) \cdot R_{nm}(r) \cdot \exp(jm\theta), \ r \leq 1. \tag{3}$$

3.2   Moment Invariants

Moment Invariants can be applied for grayscale images or objects (both region and contour) and are described below based on [13–15]. Firstly, general geometrical moments are calculated using a following formula (discrete version):

$$m_{pq} = \sum_x \sum_y x^p y^q f(x, y). \tag{4}$$

The value of function $f(x, y)$ equals 1 if a pixel belongs to the object (silhouette) and 0 for background. Then, to make representation invariant to translation, the centroid is calculated:

$$x_c = \frac{m_{10}}{m_{00}}, \quad y_c = \frac{m_{01}}{m_{00}}. \tag{5}$$

In the next step, Central Moments are calculated using the centroid:

$$\mu_{pq} = \sum_x \sum_y (x - x_c)^p (y - y_c)^q f(x, y). \tag{6}$$

Then, the invariance to scaling is obtained by central normalised moments:

$$\eta_{pq} = \frac{\mu_{pq}}{\mu_{00}^{\frac{p+q+2}{2}}}. \tag{7}$$

Ultimately, Moment Invariants are derived (usually seven first values are used in pattern recognition applications):

$$\phi_1 = \eta_{20} + \eta_{02}$$

$$\phi_2 = (\eta_{20} + \eta_{02})^2 + 4\eta_{11}^2$$

$$\phi_3 = (\eta_{30} - 3\eta_{12})^2 + (3\eta_{21} - \eta_{03})^2$$

$$\phi_4 = (\eta_{30} + \eta_{12})^2 + (\eta_{21} + \eta_{03})^2$$

$$\phi_5 = (\eta_{30} - 3\eta_{12})(\eta_{30} + \eta_{12})[(\eta_{30} + \eta_{12})^2 - 3(\eta_{03} + \eta_{21})^2]$$
$$+(3\eta_{21} - \eta_{03})(\eta_{03} + \eta_{21})[3(\eta_{30} + \eta_{12})^2 - (\eta_{03} + \eta_{21})^2]$$

$$\phi_6 = (\eta_{20} - \eta_{02})[(\eta_{30} + \eta_{12})^2 - (\eta_{21} + \eta_{03}^2]$$
$$+4\eta_{11}(\eta_{30} + \eta_{12})(\eta_{03} + \eta_{21})$$

$$\phi_7 = (3\eta_{21} - \eta_{03})(\eta_{30} + \eta_{12})[(\eta_{30} + \eta_{12})^2 - 3(\eta_{03} + \eta_{21})^2]$$
$$-(\eta_{30} - 3\eta_{12})(\eta_{03} + \eta_{21})[3(\eta_{30} + \eta_{12})^2 - (\eta_{03} + \eta_{21})^2]$$

$$(8)$$

3.3 Contour Sequence Moments

Another moment shape descriptor is Contour Sequence Moments (based on shape contour only). The method is described below based on [16]. In the first step, the contour is represented as ordered sequence $z(i)$ which elements are the Euclidean distances from the centroid to particular N points of the shape contour. The one-dimensional normalised contour sequence moments are calculated using following formulas:

$$m_r = \frac{1}{N} \sum_{i=1}^{N} [z(i)]^r, \tag{9}$$

$$\mu_r = \frac{1}{N} \sum_{i=1}^{N} [z(i) - m_1]^r. \tag{10}$$

The r-th normalised contour sequence moment and normalised central sequence moment are:

$$\bar{m}_r = \frac{m_r}{(\mu_2)^{r/2}}, \quad \bar{\mu}_r = \frac{\mu_r}{(\mu_2)^{r/2}}. \tag{11}$$

The final shape description consists four values:

$$F_1 = \frac{(\mu_2)^{1/2}}{m_1}, \quad F_2 = \frac{\mu_3}{(\mu_2)^{3/2}}, \quad F_3 = \frac{\mu_4}{(\mu_2)^2}, \quad F_4 = \bar{\mu}_5. \tag{12}$$

4 Experimental Conditions and Results

4.1 Data and Conditions

Several experiments have been carried out in order to verify the effectiveness and accuracy of the proposed approach using moment shape descriptors. Each experiment consisted of four steps described in Sect. 2, except that for each experiment different shape description algorithm was used: Moment Invariants, Contour Sequence Moments and Zernike Moments (orders from 1st to 15th). The experiments were performed using a part of the Weizmann dataset [17]. The original Weizmann dataset contains 90 low-resolution (180×144, 50 fps) video sequences of 9 actors performing 10 actions. The corresponding binary masks extracted using background subtraction are available and were used as input data. We have selected five types of actions for the experiments: run, walk, bend, jump and one-hand wave (see Fig. 1 for exemplary frames and Fig. 2 for exemplary silhouettes). Therefore, our experimental database consisted of 45 silhouette sequences of 9 actors performing 5 actions. The number of frames (silhouettes) in a sequence varied from 28 to 125. During the experiment each subgroup of 5 action sequences was iteratively used as a template set. Experimental results are presented and described in the next subsection.

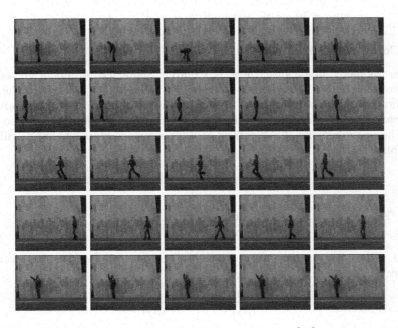

Fig. 1. Exemplary video frames from the Weizmann dataset [17]—images in rows correspond to (from the top) bending, jumping, running, walking and waving actions respectively.

4.2 Results

In this subsection some initial experimental results are provided. The goal of the first experiment was to identify the best approach for further work. Therefore, the results correspond to the average recognition effectiveness values for each shape descriptor, all classes and all iterations, and are as follows:

Fig. 2. Exemplary silhouettes extracted from video sequences presented in Fig. 1—silhouettes in rows correspond to (from the top) bending, jumping, running, walking and waving actions (images come from the Weizmann dataset [17]).

- 38.05% for Moment Invariants;
- 40.55% for Contour Sequence Moments;
- 32.5%, 37.22%, 29.44%, 33.61%, 30.55%, 34.16%, 32.22%, 40.55%, 43.05%, 45.83%, 47.5%, 46.66%, 50.27%, 47.77% and 48.88%, for Zernike Moments of orders from 1st to 15th respectively;

For Zernike Moments the use of 13th order gives the best results and only this order for feature representation will be further analysed. Table 1 contains the results of the second experiment—percentage recognition effectiveness values obtained in each iteration, for each shape descriptor and averaged for all classes. It can be seen that the classification accuracy values vary between iterations and that the best result is obtained in iteration no. 7. This can be interpreted in such a way that templates used in this iteration are represented by the most distinctive features enabling proper class indication.

In Table 2 the results of the third experiment are provided—the averaged classification values for all iterations. It can be clearly seen that 'bend' action is the most recognizable one, while the 'jump' action is the least distinctive. Based on Table 2, Zernike Moments can be selected as the best shape descriptor, except for the classification to 'wave' class which was more effective while Contour Sequence Moments descriptor was used. The same dependencies can be indicated for Table 3 where the classification accuracy values for iteration no. 7 are depicted.

Table 1. Recognition effectiveness for each iteration, each shape descriptor and averaged for all classes

Iteration no.	Moment Invariants	Contour Sequence Moments	Zernike Moments
1	27.5%	32.5%	27.5%
2	42.5%	35.0%	55.0%
3	25.0%	35.0%	47.5%
4	42.5%	50.0%	37.5%
5	35.0%	42.5%	37.5%
6	50.0%	40.0%	60.0%
7	50.0%	42.5%	67.5%
8	40.0%	50.0%	60.0%
9	30.0%	37.5%	60.0%

Table 2. Classification accuracy averaged for all iterations

Class	Moment Invariants	Contour Sequence Moments	Zernike Moments
'bend'	56.9%	48.6%	90.3%
'jump'	18.0%	22.2%	23.6%
'run'	26.3%	36.1%	40.3%
'walk'	43.0%	40.2%	48.6%
'wave'	45.8%	55.5%	48.6%

Table 3. Classification accuracy for iteration no. 7

Class	Moment Invariants	Contour Sequence Moments	Zernike Moments
'bend'	75.0%	37.5%	100%
'jump'	0.0%	12.5%	37.5%
'run'	37.5%	25.0%	37.5%
'walk'	37.5%	37.5%	75%
'wave'	100.0%	100.0%	87.5%

There is another interesting element of the approach—normalized distance vectors can be plotted and compared visually. Figure 3 contains five plots of distance vectors corresponding to five silhouette sequences used in Fig. 2 to illustrate exemplary frames. The differences between plots are clearly visible which relates to variability of silhouettes within a sequence and reveals periodicities in actions. Low peaks correspond to silhouettes that are most similar (due to the use of Euclidean distance) to the first silhouette in a sequence. Some distinctive features of the plots could be further employed to improve classification results. For instance, the faster the action, the more densely arranged low or high peaks are obtained.

Fig. 3. Exemplary plots of distance vectors for five actions performed by the same actor: bending, jumping, running, walking and waving respectively. The exemplary distance vectors were obtained using Zernike Moments of order 13th.

5   Summary and Conclusions

In the paper, an approach for action recognition based on silhouette sequences has been presented. It uses various shape description algorithms to represent silhouettes and Euclidean distance to estimate dissimilarity between first and the rest of frames. Normalized distance vectors are further processed using fast Fourier transform and periodogram in order to obtain final sequence representations. These representations are compared using template matching approach and correlation coefficient. The best experimental results in terms of classification accuracy were obtained using Zernike Moments of order 13th.

Generally, the initial results are promising, although the proposed approach requires improvements and should be examined using more data. To make the approach more effective, future works include experimental verification of other shape representation algorithms and matching measures. Moreover, the problems which cause lower classification accuracy should be identified and solved—the use of additional processing step, another shape feature or different classification process should be investigated. Also, not only shape descriptors may be used, but other features, such as centroid of an object which may help to distinguish e.g. jumping and running actions. Any modifications should be verified for their influence on final effectiveness of the approach and time consumption.

References

1. Vishwakarma, S., Agrawal, A.: A survey on activity recognition and behavior understanding in video surveillance. Vis. Comput. **29**, 983–1009 (2012)
2. Borges, P.V.K., Conci, N., Cavallaro, A.: Video-based human behavior understanding: a survey. IEEE Trans. Circuits Syst. Video Technol. **23**, 1993–2008 (2013)
3. Bobick, A.F., Davis, J.W.: The recognition of human movement using temporal templates. IEEE Trans. Pattern Anal. Mach. Intell. **23**, 257–267 (2001)
4. Gorelick, L., Blank, M., Shechtman, E., Irani, M., Basri, R.: Actions as space-time shapes. IEEE Trans. Pattern Anal. Mach. Intell. **29**, 2247–2253 (2007)

5. Goudelis, G., Karpouzis, K., Kollias, S.: Exploring trace transform for robust human action recognition. Pattern Recogn. **46**, 3238–3248 (2013)
6. Junejo, I.N., Junejo, K.N., Aghbari, Z.A.: Silhouette-based human action recognition using SAX-shapes. Vis. Comput. **30**, 259–269 (2014)
7. Baysal, S., Kurt, M.C., Duygulu, P.: Recognizing human actions using key poses. In: 20th International Conference on Pattern Recognition, pp. 1727–1730 (2010)
8. Liu, L., Shao, L., Zhen, X., Li, X.: Learning discriminative key poses for action recognition. IEEE Trans. Cybern. **43**, 1860–1870 (2013)
9. Chaaraoui, A.A., Climent-Pérez, P., Flórez-Revuelta, F.: Silhouette-based human action recognition using sequences of key poses. Pattern Recogn. Lett. **34**, 1799–1807 (2013)
10. Chitode, J.: Digital Signal Processing. Technical Publications, Berlin (2009)
11. Kohavi, R.: A study of cross-validation and bootstrap for accuracy estimation and model selection. In: Proceedings of 14th International Joint Conference on Artificial Intelligence, vol. 2, pp. 1137–1143 (1995)
12. Zhang, D., Lu, G.: Shape-based image retrieval using generic Fourier descriptor. Sig. Proc.: Image Commun. **17**, 825–848 (2002)
13. Rothe, I., Susse, H., Voss, K.: The method of normalization to determine invariants. IEEE T. Pattern Anal. **18**, 366–376 (1996)
14. Hupkens, T.M., de Clippeleir, J.: Noise and intensity invariant moments. Pattern Recogn. Lett. **16**, 371–376 (1995)
15. Liu, C.B., Ahuja, N.: Vision based fire detection. In: Proceedings of 17th International Conference on Pattern Recognition, vol. 4, pp. 134–137 (2004)
16. Sonka, M., Hlavac, V., Boyle, R.: Image Processing, Analysis, and Machine Vision. PWS—an Imprint of Brooks and Cole Publishing, Belmont (1998)
17. Blank, M., Gorelick, L., Shechtman, E., Irani, M., Basri, R.: Actions as space-time shapes. In: 10th IEEE International Conference on Computer Vision, pp. 1395–1402 (2005)

Automatic Detection of Singular Points in Fingerprint Images Using Convolution Neural Networks

Hong Hai Le[1,2(✉)], Ngoc Hoa Nguyen[1,2], and Tri-Thanh Nguyen[1,2]

[1] Vietnam National University, Hanoi, Vietnam
{hailh, hoa.nguyen, ntthanh}@vnu.edu.vn
[2] University of Engineering and Technology (UET), Hanoi, Vietnam

Abstract. Minutiae based matching, the most popular approach used in fingerprint matching algorithms, is to calculate the similarity by finding the maximum number of matched minutiae pairs in two given fingerprints. With no prior knowledge about anchor/clue to match, this becomes a combinatorial problem. Global features of the fingerprints (e.g., singular core and delta points) can be used as the anchor to speed up the matching process. Most approaches use the conventional Poincare Index method with additional techniques to improve the detection of the core and delta points. Our approach uses Convolution Neural Networks which gained state-of-the-art results in many computer vision tasks to automatically detect those points. With the experimental results on FVC2002 database, we achieved the accuracy and false alarm of (96%, 7.5%) and (90%, 6%) for detecting core, and delta points, correspondingly. These results are comparative to those of the detection algorithms with human knowledge.

Keywords: Fingerprint · Singular point · Minutiae · Matching · Convolution Neural Networks

1 Introduction

Approaches to fingerprint matching algorithms which compare two given fingerprints and return a degree of similarity are often classified into three types: correlation-based, minutiae-based, and ridge feature-based matching [1]. The Fingerprint Verification Competitions (FVC) [2] shows that the minutiae-based matching is the most popular approach. Figure 1 shows matches between two fingerprints based on minutiae.

With no prior knowledge about anchor points of the two fingerprints, the matching process is a combinatorial problem. Global features of the fingerprints such as the core and the delta points (called singular points (Fig. 2)) can be used as the clue to speed up the matching process. The core point is defined as the topmost point on the innermost upward ridge of a fingerprint and the delta point is defined as the point of bifurcation on a ridge splitting in two branches of a fingerprint. Besides being used to speed up the matching process, these points can be exploited to classify fingerprints (i.e., into six main classes: arch, tented arch, right loop, left loop, whorl and twin loop). Consequently, fingerprint classification helps filter out fingerprint candidates before matching process [3, 15].

© Springer International Publishing AG 2017
N.T. Nguyen et al. (Eds.): ACIIDS 2017, Part II, LNAI 10192, pp. 207–216, 2017.
DOI: 10.1007/978-3-319-54430-4_20

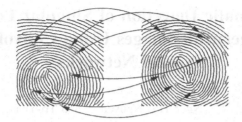

Fig. 1. Fingerprint matching based on minutiae

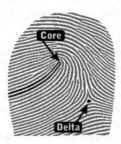

Fig. 2. Core and delta point of fingerprint image.

Common trend in singular point detection algorithms bases on Poincare index (PI) [8–10]. However, these algorithms require a lot of human knowledge to heuristically recognize the singular points. Recently, Convolution Neural Networks (CNNs) [6, 7] have greatly advanced the performance of state-of-the-art visual recognition systems. Sermanet and LeCun [6] obtained the best performance against various other approaches in a traffic sign classification challenge. One of the promises of CNNs is to replace handcrafted features with automatic hierarchical feature extraction. The CNNs do not need to have period knowledge about the objects (e.g. their features or hints) to recognize, or expertise to develop algorithm to detect the object.

Our proposal is to apply CNNs to detect singular points in fingerprint images. Our experimental results indicate that the performance of CNNs is comparative to the systems with human efforts. The rest of the paper is organized as follows: in Sect. 2, we review previous work on detecting singular points from fingerprint images; in Sect. 3, we present the CNN architectures and reasons why they work well; Sect. 4 describes our process to apply CNNs to detect singular points from fingerprint images and the experimental results on FVC 2002 DB database. Final section is the conclusion and future work.

2   Previous Work

By far, the most popular methods to detect singular points are based on the computation of the Poincare index (PI). For each point (i, j) of the orientation map, the Poincare index $P(i, j)$ is computed as the sum of the difference between the orientation of the point (i,j) and each of its neighbors (Fig. 3). Depending of the value of $P(i,j)$ the point may be identified as core or delta, i.e., if $P(i,j)$ is equal to 360° or 180° the point is core, but if the $P(i,j)$ is equal to $-180°$ then the point is a delta, otherwise it is not a singular point.

However, to improve the reliability of orientation field image is still a challenging task because of the noise or poor quality of fingerprint image. The wrong orientation of points can result in many spurious singular points. Heuristic post-processing steps are therefore usually necessary in algorithms. An example of singular point detection based on pointcare index is depicted in Fig. 3.

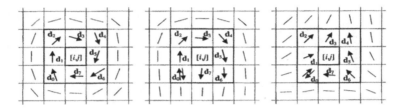

Fig. 3. Poincare based singular point detection.

After an image enhancement process, Chikkerur and Ratha [10] presented significant improvements to core and delta point detection algorithm based on complex filtering principles originally proposed by Nilsson and Bigun [14] as shown in Fig. 4.

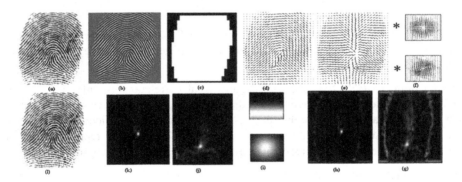

Fig. 4. Singular point detection by Chikkerur et al. [10].

Zhou et al. [9] introduced a novel so-called Differences of the ORIentation values along a Circle (DORIC) feature and proposed to use the DORIC feature for singular point verification, which can provide more discriminative information to remove spurious detections.

The above methods involve designing sophisticated filters and applying heuristics to detect singular points from fingerprint images. Section 3 will present Convolution Neural Networks (CNNs). One of the promises of CNNs is to replace handcrafted features with automatically hierarchical feature extraction, so it can reduce the complicated tasks in detecting singular point problem.

3   Convolution Neural Networks

Traditional multilayer perceptron (MLP) models use the full connectivity between nodes, thus they do not scale well on higher resolution images. This network architecture does not take into account the spatial structure of data, treating input pixels which are far apart and close to each other on exactly the same way.

Convolution neural networks (CNNs) are biologically inspired variants of multilayer perceptrons [5], designed to emulate the behavior of a visual cortex with simple cells which respond maximally to specific edge-like patterns within their receptive field, and complex cells which have larger receptive fields.

Figure 5 shows a CNN which has 2 stages of feature extraction and a two-layer non-linear classifier. The first convolution layer produces 4 features with 5 × 5 convolution filters while the second convolution layer outputs 6 features with 7 × 7 filters.

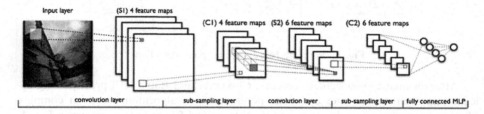

Fig. 5. A 2-stage ConvNet architecture.

CNNs overcome the challenges posed by the MLP architecture by exploiting the strong spatially local correlation present in natural images. CNNs have distinguishing features:

Local Connectivity: CNNs exploit spatially local correlation by enforcing a local connectivity pattern among neurons of adjacent layers. The architecture thus ensures that the learnt features produce the strongest response to a spatially local input pattern. Stacking many such layers allows the network to first create good representations of small parts of the input, then assemble representations of larger areas from them.

Shared Weights: In CNNs, each convolution is replicated across the entire visual field. These replicated units share the same parameterization and form a feature map. Replicating units in this way allows for features to be detected regardless of their position in the visual field. Weight sharing also helps dramatically reducing the number of free parameters being learnt which allows the training of larger, more powerful networks.

Compared to other image classification algorithms, CNNs use relatively little pre-processing. This means that the network is responsible for learning the filters that in traditional algorithms were hand-engineered. The lack of dependence on prior knowledge and human effort in designing features is a major advantage for CNNs. These properties allow CNNs to achieve better generalization on vision problems. CNNs are often used in image recognition systems. For example:

Ciresan et al. achieved an error rate of 0.23 percent on the MNIST database, which is the lowest achieved on the database [12].

In the ILSVRC 2014, a benchmark in object classification and detection, with millions of images and hundreds of object classes almost every highly ranked team used CNNs as their basic framework. The winner GoogLeNet network applied more than 30 layers [4], and its performance is now close to that of human systems.

Farfade et al. [13] demonstrated a many-layered CNN that can spot faces from a wide range of angles, including upside down, even when partially occluded with competitive performance.

From this analysis, we propose to apply CNNs to automatically detect singular core and delta points from fingerprint image. Section 4 demonstrates our detailed process.

4   Applying CNNs to Detect Singular Points

Figure 6 shows our steps to detect singular points (i.e. core and delta): the first main step is to train the CNN model for *detection module* to recognize singular points in new fingerprint images.

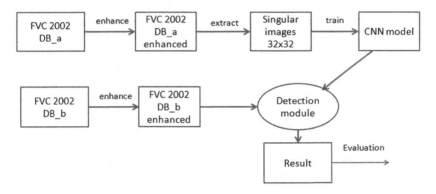

Fig. 6. Our model for singular point detection

To train a CNN model to detect singular points from fingerprint images, 2 main tasks need to be carried out: the first is to choose a CNN architecture to train the model, and the second one is to prepare data for training the CNN model.

For the CNN model, we used the multistage feature CNN provided by Sermanet et al. [7]. The network augmented the traditional CNN architecture by learning Multi-Stage (MS) features by using Lp pooling. Multistage features are obtained by branching out outputs of all stages into the final layer of the network (Fig. 7). They provide richer representations compared to Single-Stage features (SS) by adding complementary information such as local textures and fine details lost by higher levels. MS features have consistently improved the performance of the recognition tasks. The

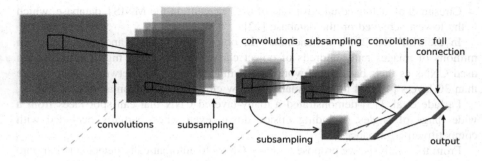

Fig. 7. A 2-stage ConvNet architecture with Multi-Stage features (MS) [7]

network establishes a new state-of-the-art of 94.85% accuracy on the SVHN[1] dataset (45.2% error improvement).

For the data preparation for training the model, we used FVC 2002 DB database, which consists of 4 subsets, namely db1, db2, db3, db4. The image quality of the 4 subset is different. Each subset is divided into db_a and db_b containing 800 and 80 fingerprints, correspondingly. We collected all db_a subsets to create the training sets, and all db_b subsets to form the testing set. Since CNNs are supervised learning algorithms, we need to annotate/mark the singular points in the training images. The training data set has 2738 core points and 731 delta points.

VeriFinger SDK tool [11] was used to enhance the quality of the fingerprint images. A square with the size of size of 32 × 32 pixels around core and delta points is manually extracted from fingerprint images to compatible with the input of the CNN. To increase the training data size, we rotated the training images 15° clock-wise and anti-clock wise.

Figures 8 and 9 show some core and delta images from training set used to train the CNN model.

The testing images do not need to be changed into the size of 32 × 32 pixels. The whole image was used as the input for detection module. Table 1 shows the experimental results.

Our results are comparative to the best reported results. Note that, the image quality of the 4 image subsets (DB1, DB2, DB3, and DB4) is different, and our testing set is all db_b sets in these subsets, which is not the same testing set as that of other authors. Tables 2 and 3 show some results from other authors that are tested on the whole FVC 2002 DB1, and FVC 2002 DB2. The authors' algorithm is heuristic, thus it does not need to have training set as ours so they evaluate on the whole DB1 and DB2 subsets.

Though our testing set includes lower quality from all subsets (i.e., the db_b parts), the false detection rate for core and delta points are 7.5% and 6.0%, correspondingly. The results are much better than those of Zhou and Chikkerur on FVC 2002 DB2. This indicates that CNNs work well on both high and lower quality images.

---

[1] http://ufldl.stanford.edu/housenumbers/.

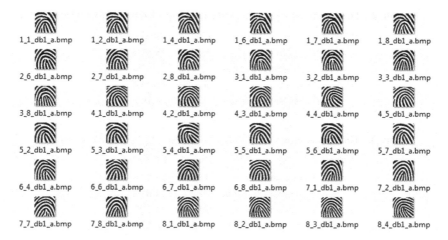

Fig. 8. Some core images used for the training.

Fig. 9. Some delta images used for the training.

Table 1. The results with FVC db_b 2002 database

FVC 2002 DB-b		Our method
Cores	Detection rate	95.94%
	False Detection rate	7.5%
Deltas	Detection rate	90.36%
	False Detection rate	6.0%

Another note, our method needs only the training image set with the mark of the singular points, and no other knowledge from human. We hope that with bigger training data, the trained model can be more robust.

Table 2. The results on FVC 2002 DB1 database of Zhou and Chikkerur

FVC 2002 DB1		Zhou's	Chikkerur's
Cores	Detection rate	95.78%	95.89%
	False Detection rate	2.27%	6.93%
Deltas	Detection rate	96.98%	92.75%
	False Detection rate	9.97%	8.16%

Table 3. The results on FVC 2002 DB2 database of Zhou and Chikkerur

FVC 2002 DB2		Zhou's	Chikkerur's
Cores	Detection rate	94.51%	93.23%
	False Detection rate	9.6%	13.87%
Deltas	Detection rate	90.88%	94.20%
	False Detection rate	12.54%	28.62%

Figures 10 and 12 show some correct core and delta detections from testing fingerprint images. Figures 11 and 13 show some false detections. Some false detections are due to much rotation or ambiguity of the ridges in the fingerprint images.

Fig. 10. Some true core detections.

Fig. 11. Some false core detections.

Fig. 12. Some true delta detections.

Fig. 13. Some false delta detections.

5 Conclusions

Unlike the traditional methods, which require a lot of heuristic processing steps to recognize the singular points in fingerprint images, this paper proposed to apply CNNs which gained state of the art result in many computer vision tasks to automatically detect core and delta points. CNNs do not need to have prior knowledge or guideline on how to detect the objects. The experiments on FVC 2002 DB, using a CNN architecture provided by Sermanet gave the accuracy of 96% in detecting core points and 90% in detecting delta points. The accuracy is comparative to the best result reported, while the false detection rate of 7.2% and 6.0% on core and delta points prevails other system on low quality image.

Some of our directions can be investigated in the future to increase the accuracy rate are to use more training data and try to use other CNNs with more layers.

References

1. Maltoni, D., Maio, D., Jain, A.K., Prabhakar, S.: Handbook of Fingerprint Recognition. Springer, London (2009)
2. Cappelli, R., Maio, D.: State-of-the-art in fingerprint classification. In: Ratha, N., Bolle, R. (eds.) Automatic Fingerprint Recognition Systems, pp. 183–205. Springer, New York (2004)
3. Hong, J.H., Min, J.K., Cho, U.K., Cho, S.B.: Fingerprint classification using one-vs-all support vector machines dynamically ordered with naive Bayes classifiers. Pattern Recogn. **41**(2), 662–671 (2008)
4. Szegedy, C., Liu, W., Jia, Y., Sermanet, P., Reed, S., Anguelov, D., Erhan, D., Vanhoucke, V., Rabinovich, A.: Going deeper with convolutions. Computing Research Repository (2014)
5. Fukushima, K.: Neocognitron: a self-organizing neural network model for a mechanism of pattern recognition. Biol. Cybern. **36**(4), 193–202 (1980)
6. Sermanet, P., LeCun, Y.: Traffic sign recognition with multi-scale convolutional networks. In: Proceedings of International Joint Conference on Neural Networks (2011)
7. Sermanet, P., Chintala, S., LeCun, Y.: Convolutional neural networks applied to house numbers digit classification. arXiv (2012)
8. Nilsson, K., Bigun, J.: Localization of corresponding points in fingerprints by complex filtering. Pattern Recogn. Lett. **24**, 2135–2144 (2003)
9. Zhou, J., Chen, F., Jinwei, G.: A novel algorithm for detecting singular points from fingerprint images. IEEE Trans. Pattern Anal. Mach. Intell. **31**(7), 1239–1250 (2009)

10. Chikkerur, S., Ratha, N.K.: Impact of singular point detection on fingerprint matching performance. In: Proceedings of 4th IEEE Workshop Automatic Identification Advanced Technologies, pp. 207–212 (2005)
11. VeriFinger SDK. www.neurotechnology.com/verifinger.html
12. Ciresan, D.C., Meier, U., Schmidhuber, J.: Multi-column deep neural networks for image classification. In: CVPR 2012, pp. 3642–3649 (2012)
13. Farfade, S.S., Saberian, M.J., Li, L.-J.: Multi-view face detection using deep convolutional neural networks. In: ICMR 2015, pp. 643–650 (2015)
14. Nilsson, K., Bigun, J.: Complex filters applied to fingerprint images detecting prominent symmetry points used for alignment. In: Tistarelli, M., Bigun, J., Jain, Anil, K. (eds.) BioAW 2002. LNCS, vol. 2359, pp. 39–47. Springer, Heidelberg (2002). doi:10.1007/3-540-47917-1_5
15. Zhang, Q., Yan, H.: Fingerprint classification based on extraction and analysis of singularities and pseudo ridges. Pattern Recogn. 37(11), 2233–2243 (2004)

Vehicle Detection in Hsuehshan Tunnel Using Background Subtraction and Deep Belief Network

Bo-Jhen Huang[1], Jun-Wei Hsieh[1], and Chun-Ming Tsai[2(✉)]

[1] Department of Computer Science and Engineering,
National Taiwan Ocean University, Keelung 202, Taiwan, ROC
`joe.huang74@gmail.com`, `shieh@ntou.edu.tw`
[2] Department of Computer Science, University of Taipei,
No. 1, Ai-Kuo W. Road, Taipei 100, Taiwan, ROC
`cmtsai2009@gmail.com`

Abstract. This paper proposes a method to detect vehicle in the Hsuehshan Tunnel. Vehicle detection in the Tunnel is a challenging problem due to use of heterogeneous cameras, varied camera setup locations, low resolution videos, poor tunnel illumination, and reflected lights on the tunnel wall. Furthermore, the vehicles to be detected vary greatly in shape, color, size, and appearance. The proposed method is based on background subtraction and Deep Belief Network (DBN) with three hidden layers architecture. Experimental results show that it can detect vehicles in he Tunnel effectively. The experimental accuracy rate is 96.59%.

Keywords: Vehicle detection · Long tunnel · Background subtraction · Deep belief network

1 Introduction

Hsuehshan Tunnel, opened in 2006, is the longest tunnel in Taiwan, located on the Taipei-Yilan Freeway (Taiwan National Highway No. 5). Its total length is 12.942 km, making the Hsuehshan Tunnel the second longest road tunnel in East Asia and the fifth longest in the world. The tunnel was designed to reduce severe traffic jams [1–3]. However, in this very long tunnel, road safety is a major concern. On May 7, 2012, a car burned in the tunnel, and two passengers were burned to death; 31 people were injured choking smoke, and smoke clouds from the burning vehicle caused hundreds of people to be trapped in the tunnel. This accident was the most serious and caused the most casualties, but there had been eight prior accidents since the tunnel opened [4].

Vehicle detection in Hsuehshan Tunnel is a challenging problem because of use of heterogeneous cameras, varied camera set up locations, low resolution videos, poor tunnel illumination, and reflection of lights on the tunnel wall. Furthermore, the vehicles to be detected vary greatly in shape, color, size, and appearance. To make this task even more difficult, the ego vehicles and target vehicles are generally in motion so that the size, location, and view of target vehicles mapped to the captured video are very diverse.

© Springer International Publishing AG 2017
N.T. Nguyen et al. (Eds.): ACIIDS 2017, Part II, LNAI 10192, pp. 217–226, 2017.
DOI: 10.1007/978-3-319-54430-4_21

Currently, there are many cameras in the Hsuehshan Tunnel to monitor and record the traffic flows for the traffic control center. When the traffic is congested, the traffic center notifies the vehicle drivers by broadcasting. However, the operators monitoring the traffic must keep their eyes on many cameras. Their role is therefore a passive one which makes constant high attention problematic and produces fatigue. To help the operators, an intelligent traffic monitoring system is necessary [1].

Vehicle detection is the first stage of such a system. Its accuracy will be affected by the quality of the cameras and the environment illumination in the tunnel both of which affect detecting informative features in the scene. Color, the most widely used feature in traffic applications, is not reliable in tunnels since the artificial lighting affects the obtained colors of vehicles. Texture information is also very limited, due not only to poor illumination conditions, but also to the low resolution of the surveillance cameras.

Only a few works have addressed the problem of detecting vehicles in tunnel scenes. Pflugfelder et al. [5] presented a study of camera properties and requirements for tunnel safety management. However, this study only focused on hardware technical details and did not evaluate software algorithms. Schwabach et al. [6] based their approach on digital video image analysis to recognize alarm situations, alert tunnel personnel, and store traffic incident video sequences. However, they did not propose an integrated method for tunnel surveillance [8]. Wu et al. [2] presented a method to detect vehicles in highways, roadways, or tunnels, using a block-based background extraction method combined with motion information and a maximum probability density function. Frías-Velázquez et al. [7] proposed an approach that used a local higher-order statistic filter to solve illumination changes for vehicle detection. Rios Cabrera et al. [8] proposed a tunnel security and surveillance application to detect vehicle by using multiple camera. However, many vehicles in the field of view of the camera could not be detected and the detected objects were often missed or wrongly classified.

The method used for background subtraction (BS) is a crucial step in vehicle detection. The BS method needs to detect moving objects within a video stream without any *a priori* knowledge about the objects [9–11]. Sobral and Vacavant [11] used background models to challenge a dataset that included 20 synthetic videos and 9 real videos. They compared the effectiveness of 29 methods [12], finding that PFinder [13] and GMM [14] are very good methods for BS. Algorithms like PBAS [15] and SOBS [16] have also shown an interesting robustness.

Recently, Xu et al. [17] analyzed advantages and disadvantages of various background modeling methods for video analysis, comparing their performance in quality and computational cost. Their evaluation of eight state-of-the-art background modeling methods showed AGMM [19], SOBS [16], Vibe [20], and PBAS [15] were the most promising. St-Charles et al. [18] proposed a novel non-parametric, pixel-level background modeling approach (PAWCS) based on world dictionaries to address the issue of inconsistent performance across different scenarios. Their method is superior to most others for meeting background subtraction challenges.

Deep belief network (DBN), first proposed by Hinton et al. [21], has demonstrated its success in MNIST image classification. DBN has been modified and used in a 3D object recognition [22]. Recently, Wu and Tsai [23] have applied DBN to classify pedestrians, bikes, motorcycles, and vehicles. Wang et al. [24] proposed a 2D-DBN to detect vehicles; their experimental results demonstrate their algorithm performs better in their testing data sets than other state-of-the-art vehicle detection algorithms.

In this paper, a vehicle detection method based on background subtraction and DBN is proposed to solve the above-mentioned problems in detecting vehicles in the Hsuehshan Tunnel.

2 Deep Belief Network

DBN is a probabilistic and generative graphical model composed of multiple layers of stochastic and hidden variables used to learn a set of examples in an unsupervised way and reconstruct the inputs. A DBN can be simply viewed as a simple learning module of Restricted Boltzmann Machine (RBM). The training procedure of DBN can be divided into two steps: first, generative learning to abstract information layer by layer with unlabeled samples and second, discriminative learning to fine-tune the whole deep network with labeled samples to the ultimate learning target [21].

In a DBN, every RBM contains n hidden units that can be viewed as a parametric model of the joint distribution between hidden variables h_i and visible variables v_i, as follows:

$$p(v, h) = \frac{1}{Z} e^{-E(v,h)} \tag{1}$$

where $E(v, h)$ is the energy function of RBM, defined as:

$$E(v, h) = -\sum_{i \in V} a_i v_i - \sum_{i \in H} b_j h_j - \sum_{i,j} v_i h_i w_{ij} \tag{2}$$

where v_i, h_j, a_i, b_j are the status and the bias of visible variables and hidden variables respectively. w_{ij} is the weight between the observed and hidden variables, and Z is the partition function.

According to RBM, there are symmetric connections between the visible layer and the hidden layer; but no connection for variables within the same layer. This characteristic makes it easy to compute the conditional probability distributions:

$$p(v_i = 1|h) = \sigma\left(a_i + \sum_j h_j w_{ij}\right) \tag{3}$$

$$p(h_j = 1|v) = \sigma\left(b_j + \sum_i v_i w_{ij}\right) \tag{4}$$

where $\sigma(x) = 1/(1 + \exp(-x))$ is the sigmoid function. The contrastive divergence (CD) algorithm [21] is used to learn the parameters w_{ij}, a_i, b_j. The CD procedure works as follows:

1. Initialize the symmetric weights, w, visible layer bias, a, and hidden layer bias, b, and define the learning rates.
2. Apply the forward propagation to the variables v_i^0 of the visible layer, and use Eq. (4) to compute the hidden layer output variables h_j^0.

3. Apply the back propagation to the variables h_j^0 of hidden layer using Eq. (3) to compute the visible layer variables v_i^0.
4. Similarly, apply the forward propagation to v_i^1 to get variables h_j^1.
5. Use the following equations to update the weights and the biases:

$$\Delta w_{ij} = \lambda_w \left(E\left[v_i^0 h_j^0\right] - E\left[v_i^1 h_j^1\right] \right) \tag{5}$$

$$\Delta a_i = \lambda_a \left(E\left[v_i^0\right] - E\left[v_i^1\right] \right) \tag{6}$$

$$\Delta b_j = \lambda_b \left(E\left[h_j^0\right] - E\left[h_j^1\right] \right) \tag{7}$$

where $E[.]$ was mathematical expectation.
6. Repeat step 1 to 5 until the parameters converge.

Figure 1 shows the architecture of the DBN classifier. From bottom to top are a visible input layer V^1, N hidden layers H^1 to H^N, and an output layer O^1. There are T units on the visible layer V^1, and 2 units representing the 2 categories on the output layer O^1.

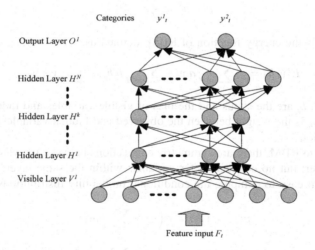

Fig. 1. The architecture of the DBN classifier.

The DBN classifier can pre-train one layer at a time effectively, the process being viewed as an unsupervised RBM. After the unsupervised pre-training, the back propagation is used to fine-tune the parameters to achieve a smaller classification error. More detail about DBN can be found at [21–24].

3   Vehicle Detection in Hsuehshan Tunnel

Figure 2 shows the flow diagram of the proposed system, which has both training and testing steps. In the training step, the algorithm includes input videos, background subtraction, connected component labeling, size and aspect ratio filters, selecting and labeling, image normalization, and DBN. In the testing step, the algorithm is similar to that of the training step except for the labeling procedure.

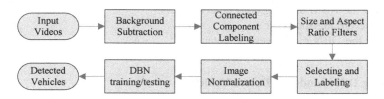

Fig. 2. The flow diagram of the proposed system

3.1   Collection the Positive Vehicle Samples

In order to collect the positive vehicle samples, the background subtraction method from [11, 17, 18] are used to compare and evaluate, the capability of AGMM, SOBS, Vibe, PBAS, and PAWCS to build their background models, respectively. Based on these background models, differential images are produced by using image subtraction. After applying the fixed threshold value, the foreground binary objects are extracted. Then connected component labeling is used to label the vehicle candidates within bounding-boxes. To remove the non-vehicle bounding-boxes, size and aspect ratio filters are used. Finally, the complete bounding-boxes of the vehicles are selected as the vehicle samples and collected for positive training samples.

3.2   Image Normalization

To reduce computational complexity, the color vehicle input images are transformed into gray level images. Next, histogram equalization is used to reduce the influence of environment illumination, and the Otsu thresholding method is applied to threshold the vehicle samples into binary images. However, the size of the collected vehicle samples still varies, so, in order to train the DBN model of vehicle detector, the sizes of the collected vehicle samples are normalized to 28 × 28 by using bi-linear interpolation as the training set.

3.3   DBN Model of Vehicle Detector

To train the DBN model, a DBN [25] model is modified for use on vehicle detection. To obtain a more accurate detection rate, two and three hidden layer DBN architectures are learned. The procedure for training DBN includes pre-training and fine-tuning. In pre-training, RBM and Sigmoid hidden layers are trained, while in fine-tuning, the logistic layer is trained. After training with the vehicle samples, a DBN model for vehicle detection is produced. This DBN model is applied to predict the independent test set to verify its generalization performance. The algorithm to train the DBN model is described as follows:

1. Set the architecture of the training DBN.
2. Initialize the parameters of the training DBN.
3. Input the vehicle and non-vehicle training samples.
4. Pre-train the RBM and Sigmoid hidden layers.
5. Fine-tune the logistic layer.
6. Produce the DBN model vehicle detector.

3.4   Using DBN Model to Detect Vehicles

The DBN model of vehicle detector having been produced, it is used to detect vehicles in the Hsuehshan Tunnel. The procedures for vehicle detection are:

1. Load the DBN model vehicle detector.
2. Input the testing vehicle samples.
3. Use background subtraction to extract the foreground objects.
4. Use connected component labeling to detect the vehicle candidates.
5. Use size and aspect ratio to remove noise candidates.
6. Transform the color vehicle candidates into grayscale.
7. Compensate the illumination by using histogram equalization.
8. Convert the vehicle candidates into binary images.
9. Resize the vehicle candidates to size 28×28.
10. Feed this vehicle candidate to the DBN model to identify whether it is a vehicle or not.

4   Experimental Results

The performance of proposed method was tested experimentally using Java 8 on a 64-bit Windows 8.1 platform of Intel® Pentium® CPU N3530 @ 2.16 GHz ASUS laptop.

To demonstrate the performance of the proposed method, vehicle and non-vehicle samples are collected as described in Subsect. 3.1 at which point the bounding-boxes include both vehicles and non-vehicles. The training set has 6,430 images – 2,087 positive and 4,473 negative samples. In the testing set, there are a total of 440 images – 130 positive and 310 negative samples. Figure 3 shows some representative positive and negative training and testing images.

Using the proposed method, two different DBN architectures are applied. Both include one visible layer and one output layer, but one includes two hidden layers, while the other uses three. In order to derive the best model, several model parameters are tried.

(a)   (b)

Fig. 3. Some positive and negative training and testing samples. (a) Positive samples. (b) Negative samples (Color figure online).

4.1 Two Hidden Layers DBN

In this DBN architecture, the two hidden layers have sizes 50 and 10, respectively. In the pre-training stage, the learning rate is set as 0.1. The parameter corresponding to contrastive divergence k is varied from 1 to 5 and increase 1. The training epochs are set as 3,000 and 10,000. In the fine-tuning stage, the learning rate is also set as 0.1, but the training epochs are set as 10,000 and 30,000. Table 1 shows the training results of the DBN architecture with two hidden layers [50, 10], where learning rate is 0.1, and k has values of 1, 2, 3, 4, and 5. As shown in Table 1, when the training epochs are 3,000 and 10,000 in pre-training and fine-tuning, respectively, k is 2 and testing accuracy is 0.8705. However, when the training epochs are 10,000 and 30,000 in the pre-training and fine-tuning, respectively, k is 2 and testing accuracy is 0.9114, the highest accuracy rate in Table 1. The false positive and the false negative results are only 9 and 30, respectively, and there are fewer false positives than false negatives. This reflects that in the collected data set the number of vehicles is fewer than the number of non-vehicles.

Table 1. Training results of the DBN architecture with two hidden layers.

Pre-training							
RBM layers	[50, 10]	[50, 10]	[50, 10]	[50, 10]	[50, 10]	[50, 10]	[50, 10]
Learning rate	0.1	0.1	0.1	0.1	0.1	0.1	0.1
k	1	2	3	4	5	1	2
Training epochs	3000	3000	3000	3000	3000	10000	10000
Fine-tune							
Learning rate	0.1	0.1	0.1	0.1	0.1	0.1	0.1
Training epochs	10000	10000	10000	10000	10000	30000	30000
Testing accuracy	0.8545	0.8705	0.8568	0.8432	0.8273	0.8909	**0.9114**
False Positive	11	14	22	17	29	15	**9**
False Negative	53	43	41	52	47	33	**30**

Table 2. Training results of the DBN architecture with three hidden layers.

Pre-training				
RBM layers	[392, 196, 98]	[392, 196, 98]	[392, 196, 98]	[392, 196, 98]
Learning rate	0.1	0.1	0.1	0.1
k	1	2	2	2
Training epochs	3000	3000	10000	20000
Fine-tune				
Learning rate	0.1	0.1	0.1	0.1
Training epochs	10000	10000	30000	30000
Testing accuracy	0.9386	0.9386	0.9500	**0.9659**
False Positive	3	7	2	**1**
False Negative	24	20	20	**14**

4.2   Three Hidden Layers DBN

In this DBN architecture, the three hidden layers have sizes 392, 196, and 98, respectively. In the pre-training stage, the learning rate is set as 0.1. The parameter corresponding to contrastive divergence k is set as 1 and 2. The training epochs are set as 3,000, 10,000, and 20,000. In the fine-tuning stage, the learning rate is also set as 0.1, but the training epochs are set as 10,000 and 30,000. Table 2 shows the training results of this DBN architecture with three hidden layers [392, 196, 98], learning rate 0.1, and k having values of 1 and 2. From this table, when the training epochs are 20,000 and 30,000 in pre-training and fine-tuning, respectively, k is 2 and testing accuracy is 0.9659 – the highest accuracy rate in Table 2. False positives and false negatives are only 1 and 14, respectively. Comparison to Table 1 shows that, the DBN with three hidden layers and training epochs 20,000 and 30,000 in the pre-training and fine-tuning, respectively, has the highest accuracy rate. Furthermore, the best k value is 2. However, the DBN with three hidden layers has more computational time.

5   Conclusions

In this research, we proposed a vehicle detection system for Taiwan's Hsuehshan tunnel based on background subtraction and used of a deep belief network. Experiment using the proposed method and a sample collected data set from monitoring cameras in the Hsuehshan Tunnel showed a classification accuracy of 96.59% and demonstrate that proposed method is feasible.

In the future, more training data will be collected and used to train the model to obtain a still higher classification accuracy rate, using a novel moving object detection method. Additionally, in further work, the proposed method will be enhanced to detect the vehicles in real-time.

Acknowledgements. The authors would like to express his gratitude to Walter Slocombe and Dr. Jeffrey Lee, who assisted editing the English language for this article. This work was supported by the Ministry of Science and Technology, R.O.C., under Grants MOST 104-2221-E-845-003-.

References

1. Tsai, C.M., Hsieh, J.W., Shih, F.Y.: Motion-based vehicle detection in Hsuehshan Tunnel. In: 8th International Conference on Advanced Computational Intelligence, Chiang Mai, Thailand, pp. 385–389 (2016)
2. Wu, B.F., Kao, C.C., Liu, C.C., Fan, C.J., Chen, C.J.: The vision-based vehicle detection and incident detection system in Hsuehshan Tunnel. In: 2008 IEEE International Symposium on Industrial Electronics, pp. 1394–1399 (2008)
3. Hsuehshan Tunnel. https://en.wikipedia.org/wiki/Hsuehshan_Tunnel
4. Hsuehshan Tunnel accidents. https://commons.wikimedia.org/wiki/Category:Hsuehshan_tunnel_accidents
5. Pflugfelder, R., Bischof, H., Fernandez Dominguez, G., Nolle, M., Schwabach, H.: Influence of camera properties on image analysis in visual tunnel surveillance. In: Proceedings of 8th International Conference on Intelligent Transportation Systems (ITSC), ITSS, pp. 868–873. IEEE Computer Society (2005)
6. Schwabach, H., Harrer, M., Holzmann, W., Bischof, H., Fernandez Dominguez, G., Nolle, M., Pflugfelder, R., Strobl, B., Tacke, A., Waltl, A.: Video based image analysis for tunnel safety – VITUS-1: a tunnel video surveillance and traffic control system. In: 12th World Congress on Intelligent Transport Systems, pp. 1–12 (2005)
7. Frías-Velázquez, A., NiñoCastañeda, J.O., Jelač, V., Pižurica, A., Philips, W.: A mathematical morphology based approach for vehicle detection in road tunnels. In: Proceedings of SPIE, vol. 8135, id. 81351V (2011)
8. Rios Cabrera, R., Tuytelaars, T., Van Gool, L.: Efficient multi-camera vehicle detection, tracking, and identification in a tunnel surveillance application. Comput. Vis. Image Underst. **116**(6), 742–753 (2012)
9. Toyama, K., Krumm, J., Brumitt, B., Meyers, B.: Wallflower: principles and practice of background maintenance. In: International Conference on Computer Vision, pp. 255–261 (1999)
10. Tsai, C.-M., Yeh, Z.-M.: Intelligent moving objects detection via adaptive frame differencing method. In: Selamat, A., Nguyen, N.T., Haron, H. (eds.) ACIIDS 2013. LNCS (LNAI), vol. 7802, pp. 1–11. Springer, Heidelberg (2013). doi:10.1007/978-3-642-36546-1_1
11. Sobral, A., Vacavant, A.: A comprehensive review of background subtraction algorithms evaluated with synthetic and real videos. Comput. Vis. Image Underst. **122**, 4–21 (2014)
12. Sobral, A.: BGSLibrary: an OpenCV C++ background subtraction library. In: IX Workshop de Visao Computacional, Rio de Janeiro, Brazil (2013). http://code.google.com/p/bgslibrary/
13. Wren, C., Azarbayejani, A., Darrell, T., Pentland, A.: Pfinder: real-time tracking of the human body. IEEE Trans. Pattern Anal. Mach. Intell. **19**(7), 780–785 (1997)
14. Kaewtrakulpong, P., Bowden, R.: An improved adaptive background mixture model for real-time tracking with shadow detection. In: European Workshop on Advanced Video Based Surveillance Systems (AVSS), pp. 1–5 (2001)
15. Hofmann, M., Tiefenbacher, P., Rigoll, G.: Background segmentation with feedback: the pixel-based adaptive segmenter. In: IEEE Computer Society Conference on Computer Vision and Pattern Recognition Workshops (CVPRW), pp. 38–43 (2012)
16. Maddalena, L., Petrosino, A.: A self-organizing approach to background subtraction for visual surveillance applications. IEEE Trans. Image Process. **17**(7), 1168–1177 (2008)

17. Xu, Y., Dong, J., Zhang, B., Xu, D.: Background modeling methods in video analysis: a review and comparative evaluation. CAAI Trans. Intell. Technol. **1**(1), 43–60 (2016)
18. St-Charles, P.L., Bilodeau, G.L., Bergevin, R.: Universal background subtraction using world consensus models. IEEE Trans. Image Process. **25**(10), 4768–4781 (2016)
19. Zivkovic, Z.: Improved adaptive Gaussian mixture model for background subtraction. In: Proceedings of 17th International Conference on Pattern Recognition, vol. 2, pp. 28–31 (2004)
20. Barnich, O., Droogenbroeck, M.V.: ViBe: a universal background subtraction algorithm for video sequences. IEEE Trans. Image Process. **20**(6), 1709–1724 (2011)
21. Hinton, G.E., Osindero, S., Teh, Y.W.: A fast learning algorithm for deep belief nets. Neural Comput. **18**(7), 1527–1554 (2006)
22. Nair V., Hinton, G.E.: 3D object recognition with deep belief nets. In: Proceedings of 23rd Annual Conference on Neural Information Processing Systems (NIPS 2009), pp. 1339–1347 (2009)
23. Wu, Y.Y., Tsai, C.M.: Pedestrian, bike, motorcycle, and vehicle classification via deep learning: deep belief network and small training set. In: 2016 International Conference on Applied System Innovation (ICASI 2016), Okinawa, Japan, pp. 1–4 (2016)
24. Wang, H., Cai, Y., Chen, L.: A vehicle detection algorithm based on deep belief network. Sci. World J. **2014**, 1–7 (2014)
25. Sugomori, Y.: Java Deep Learning Essentials. Packt Publishing, May 2016. https://www.packtpub.com/big-data-and-business-intelligence/java-deep-learning-essentials

Segment Counting Versus Prime Counting in the Ulam Square

Leszek J. Chmielewski$^{(\boxtimes)}$, Arkadiusz Orłowski, and Grzegorz Gawdzik

Faculty of Applied Informatics and Mathematics – WZIM,
Warsaw University of Life Sciences – SGGW,
ul. Nowoursynowska 159, 02-775 Warsaw, Poland
{leszek_chmielewski,arkadiusz_orlowski}@sggw.pl
http://www.wzim.sggw.pl

Abstract. Points that correspond to prime numbers in the Ulam square form straight line segments of various lengths. It is shown that the relation between the number of the segments and the number of primes present in a given square steadily grows together with the growing values of the prime numbers in the studied range up to 25 009 991 and is close to double log. These observations were tested also on random dot images to see if the findings were not a result of merely the abundance of data. In random images the densities of the longer segments and the lengths of the longest ones are considerably smaller than those in the Ulam square, while for the shorter segments it is the opposite. This could lead to a cautious presumption that the structure of the set of primes might contain long-range relations which do not depend on scale.

Keywords: Prime counting function · pi function · Ulam spiral · Ulam square · Line segments · Segment counting · Hough transform

1 Introduction

The Ulam spiral has been the object of interest as one of the tools of study on the fascinating set of prime numbers since its first presentation in [7]. It presents the set of the prime numbers as a two-dimensional pattern [9], so it has made it possible to investigate some of the properties of this set with the use of the image processing methods. The Ulam spiral will be considered as the part of the square in which it is embedded, which will be called the Ulam square. This square will be considered as an image having its pixels, sometimes called points.

In our previous papers [3,4,6] we have presented an image processing approach to the analysis of the regular line patterns present in the Ulam spiral, with use of the Hough transform. In [3] we have proposed a variant of the Hough transform capable of detecting the sequences of points which form regular segments of straight lines. As it is in the Ulam square, the slope of these lines is defined by a ratio of co-prime numbers. In [4] we have shown in more detail which segments

© Springer International Publishing AG 2017
N.T. Nguyen et al. (Eds.): ACIIDS 2017, Part II, LNAI 10192, pp. 227–236, 2017.
DOI: 10.1007/978-3-319-54430-4_22

are actually found and what is the distribution of their lengths. In [6] we have investigated the directional structure of the Ulam spiral. We have shown that the image of the spiral indeed has some strongly preferred directions and that the general directionality is far from symmetrical with respect to the vertical and horizontal axes. This structure differs much from that of a random dot image of similar dot density, which has a different relation of dominating directions and has the symmetries which the Ulam spiral does not have.

Long segments necessitate for a more ordered structure than the short ones. It is expected that the longer segments will be less numerous than the shorter ones. In [4] we have shown that in the squares considered the longest segment had 16 points. Interestingly, there were no segments of lengths 15 and 14 points. There is only one segment of length 16, one of length 13, two of 12, and the number of shorter segments grows rapidly with the decrease of their length. The question arises what is the relation of the number of segments of various lengths formed in the Ulam spiral containing numbers less than n to the function $\pi(n)$, that is, the number of primes less than or equal to n (cf. for example [1,8]). Other questions also arise, for example, will the just stated limits for segment length be exceeded, and if so, for how large numbers? In the present paper we concentrate on the relation of the numbers of segments to the number $\pi(n)$.

The next parts of this paper are organized as follows. In Sect. 2 the method of finding line segments, described in detail elsewhere, is briefly reminded. In Sect. 3 the results obtained for the Ulam square are presented. In Sect. 4 they are compared to those obtained for dot images created by randomizing the Ulam square, to see if these results are specific for the Ulam square or are typical for any unstructured dot images. The paper is concluded in the last Section.

2   Finding Line Segments

The method of finding line segments described in [3] is illustrated in Fig. 1 where the central part of the Ulam spiral of dimensions 21×21 is shown. This spiral contains 85 primes, of which the largest one is 439. Grey and color pixels correspond to primes written in black. Black pixels correspond to composite numbers written in grey. The origin of the coordinate system Opq is located in the center of the spiral corresponding to the unit 1. The image presents three segments of length 5 and one of length 6. There are no more segments of these and larger lengths in this image, but there are 21 segments of length 4, 110 segments of length 3 and 515 segments of length 2, not displayed in this image.

The slope of a line is defined by numbers Δp, Δq denoted here $\Delta p = i$, $\Delta q = j$, limited to $i \in [0 : N]$ and $j \in [-N, N] \setminus \{0\}$. In the current implementation, for practical reasons, it is $N = 10$. This defines the dimensions of the accumulator array D_{ij} in the Hough transform, called the *direction array* (see [3] for the detailed description of the method). During the evidence accumulation stage of the line detection algorithm, which can be viewed as the process of voting for straight lines, the elemental voting subset is a pair of pixels representing a pair of primes. If the slope represented by the pair corresponds to an element of D,

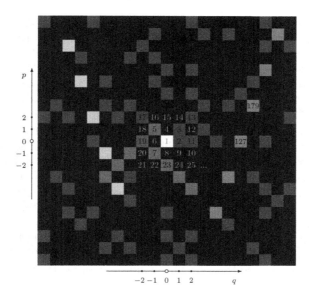

Fig. 1. Central part of the Ulam spiral of dimensions 21×21 and with segments of lengths 5 and 6 marked in colors. Notations: (p, q) – coordinates with origin in the first number of the spiral; primes: black on grey or color background; other numbers: grey on black or white. Blue: two segments of length 5, slopes $(\Delta p, \Delta q) = (1, 1)$ and $(1, -1)$, respectively, with one common point corresponding to prime number 19 marked with darker blue. Cyan: segment of length 5, slope $(\Delta p, \Delta q) = (3, -1)$. Magenta: segment of length 6, slope $(3, 1)$ (color images in the electronic edition).

the pair is stored in D_{ij}, where i/j is the reduced fraction $\Delta p/\Delta q$; otherwise it is neglected. To store the votes, in each D_{ij} a one-dimensional data structure is formed (a vector or a list, whichever is more convenient). For each vote the line offset, defined as the intercept with the axis Op for horizontal lines and with Oq for the remaining ones, is stored. Additionally, the locations of voting points are stored and the pairs and their primes are counted.

After the accumulation process the accumulator can be analyzed in various ways. In [4] we were interested in finding the numbers of contiguous segments of different lengths. In [6] we have analyzed the directional properties of the segments in the Ulam spiral. The accumulated results make it possible to look for other regularities in the set of segments present in the Ulam spiral.

Assume the square is $S \times S$, so it contains $P = \pi(S^2)$ primes. The complexity of the accumulation is $O(PN^2)$ as it needs the search in neighborhood proportional to $N \times N$ around each prime. Complexity of the analysis depends on a number proportional to the number of directions N^2, number of offsets S and needs sorting up to S points with quick sort, giving $O(N^2S^2log(S))$. Time requirements are not crucial, as the analysis is made only once. The results presented here required several minutes to complete on a general-purpose PC. Memory requirements are $O(N^2S^2)$ and can be considerable.

In the present analysis and also in the previous papers we have looked for contiguous segments, in the sense corresponding to the way the slope is represented in the direction array D. Let us consider an example of pairs of pixels $P_k(p, q)$: $P_1(-1, -1)$, $P_2(-2, 0)$ representing the primes $7, 23$, and $P_3(0, 6)$, $P_4(3, 7)$ representing the primes $127, 179$ (see Fig. 1). The pixels P_1, P_2 have $(\Delta p, \Delta q) = (i, j) = (1, 1)$ and are neighbors in the classical sense as well as in the sense of fitting to a direction conforming with D. The pixels P_3, P_4 have $(i, j) = (3, 1)$ and are neighbors in the sense of D, but not in the classical sense. However, they will be considered neighbors, as for this direction their distance is minimum. This explains in what sense the contiguity is considered here.

The fact that only the contiguous segments are considered here explains, among others, why in Fig. 1 there are only 515 segments of length 2, and not all the possible pairs of 85 primes, which would be $85 \times 84/2 = 3570$.

It is due to practical reasons that at present we have restricted our investigations to the central part of the Ulam squares of size up to 5001×5001 points, which contains $1\,566\,540$ primes, the largest of which is $25\,009\,991$. This is very little in comparison to the largest primes known at present, with the number of digits exceeding 20 millions [2]. However, it is only the question of time that this purely secondary limit will be overcome.

3   Results for the Ulam Square

Let us now come back to the question of our interest in this paper, that is, what is the relation between the number of segments of various lengths and the number of primes in the Ulam square? First, let us compare the number of segments to the prime counting function $\pi(n)$. This relation is shown in Fig. 2.

The graph of the relation of the number of segments to the largest prime number has a similar shape as the prime counting function $\pi(n)$. In the double log scale it is very close to linear, although some discrepancies from the linear shape can be seen with an unarmed eye for all segments having more than four points. For the longest segments, the number of which is small, the relation is less regular. It is striking that for the smallest square considered, that of size 51×51, the number of three-point straight segments is larger than the number of primes on which these segments are spanned. Actually, there are 378 primes and 411 three-point segments. The fact that the number of two-point segments is larger than the number of primes is not disappointing.

From now on we shall focus on the relations of the number of segments and the number of primes. This relation is shown in Fig. 3 for all the lengths of segments present in the images. It can be seen that this relation is always close to linear in the double log scale, provided that the number of segments involved is large. The longest segments are rare and for them the values differ not only quantitatively, but also qualitatively with the growth of the number of primes, so a steady relation can not be expected in the range of their large variability.

It is interesting to see that the relation is steady throughout the whole investigated range of the prime numbers, and its nature is the same for small as

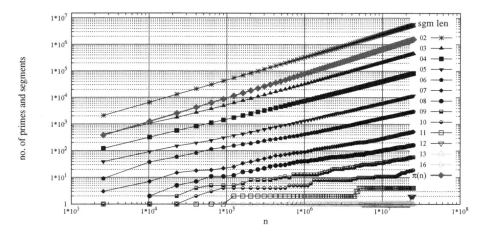

Fig. 2. Number of primes less than or equal to n, that is, the $\pi(n)$ function, and the numbers of segments of lengths encountered in the images considered, versus n. If the number of segments is zero the data point is absent. Color used to enhance discernibility (color images in the electronic edition).

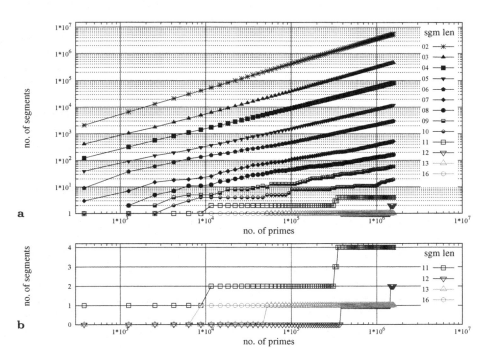

Fig. 3. Number of segments versus number of primes in the square. (**a**) Results for all the segments; (**b**) results for the longest segments.

well as for large values of the largest prime. The primes become rare together with the growth of their value n, in the sense that the distance between them increases with n. Such structures as regular linear patterns necessitate for not only the presence of the primes, but also the existence of some constant relations between them, which is a more restrictive or demanding condition than the presence itself. So, it might well be expected that at some point the number of regular structures could fall down. No such phenomenon is observed in the range of primes investigated until now. A modest believe can be formulated that the structure of the set of the prime numbers might contain the regularities or relation which do not depend on the scale and span the set of the prime numbers in its entirety.

4   Comparison with a Random Square

It is important to find out to what extent the regularities presented above are specific for the Ulam square itself, or are related to that in nearly every image, even that formed randomly, some regularities can be found, as a consequence of the abundance of data. We shall consider the dot images having the density of dots very close to that observed in the Ulam square, but displaced randomly.

The way of randomizing the Ulam square used in this study will be more sophisticated than that applied in [6], where in a square the dots were displaced with uniform probability density, while their number was kept equal to that in the Ulam square of equal size. However, actually the density of primes decreases with the distance from the center, as fixed by the random counting function $\pi(n)$. The randomization method should account for this. In other words, in the randomization process the prime numbers should be changed locally.

Here it has been done in the following way. Let us form a vector V_i of integers from 1 to the number M of pixels in the square. It represents the new set of numbers which will replace the set ot primes. The elements of V are initially empty. Let us take three subsequent primes, p_l, p_m and p_r. Now, let us draw a random number p'_m belonging to the closed interval $[p_l, p_r]$, with uniform probability density. This number is saved in V in the element with index p'_m, so $V_{p'_m} := p'_m$; but if $V_{p'_m}$ was not empty, another number is drawn. In this way, p'_m has replaced p_m in the set. This process is performed for all triplets of subsequent primes, from the smallest to the largest ones. The boundary elements of V require attention. In the first triplet of primes, p_l is replaced by 1, so $p_m = 2$ and $p_r = 3$. Therefore, p'_m can equally probably become 1, 2 or 3. In the last triplet, p_r is replaced by M. Finally, the set represented by the nonempty elements of V is used instead of prime numbers to fill the now randomized Ulam square. The proposed randomization method makes the primes move only in the direction according to the way the numbers are written down into the Ulam spiral. One realization of the random image is shown in Fig. 4 and two others, as a side effect of other visualization, in Fig. 6.

The calculations in this study are repeated for the squares of growing sizes. To keep the segments in their place during the calculations, when the size of

Fig. 4. The actual Ulam square of size 351×351 (left) and one of its randomized realizations (right). In each image there are 123 201 pixels, of which 11 578 represent primes. The largest prime is 123 176. The structure of the randomized square differs from that of the actual one even in visual inspection.

the square is increased, only the newly added pixels are drawn in this random process, and the middle ones remain.

The result for randomized data can be seen in Fig. 5a. To make it easier to compare the graphs to those of Fig. 3, they are mixed together in one plot.

The graphs received for the Ulam square and those for the corresponding random square are different. Among the most conspicuous differences, the following two can be pointed out.

1. In the Ulam square the longest segment contains 16 points, while in the considered realization of the random square it has 8 points. This suggests that in the Ulam square the regularity is stronger than that in a random square.
2. In the random square the numbers of the shorter segments are larger than those in the Ulam square – this holds for 2 and 3-point segments, while the numbers of the longer segments is smaller – this holds for the 4, 5, 6-point and longer segments. These relations do not depend on the size of the squares. This seems to suggest the existence of some factor which promotes the appearance of longer structures in the Ulam square.

It should be stressed that the graphs in Fig. 5a were found for a single realization of the randomization process of the square. To check how the results change from one realization to another, we have run the program for the same square of a specified size ten times each. The numbers of points were chosen to place the data in a possibly even way on the horizontal axis of the graph. These were the smallest sizes for which the number of primes exceeded 200, 1 000, 2 000, 10 000, 20 000, 100 000, 200 000, 1 000 000, which occurred for squares of size 51,

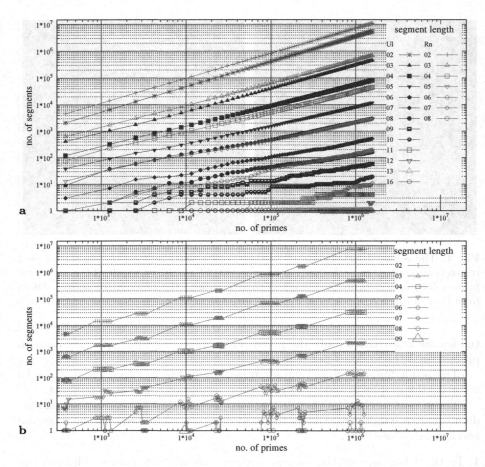

Fig. 5. (**a**) Number of segments vs. number of primes in the Ulam square (Ul: thick blue or green lines, full symbols for lengths present also in random squares) and in random square (Rn: thin red lines, empty symbols). (**b**) Fragments of graphs of Fig. **a** for random data: results for 10 realizations of the random process (dithered horizontally for better visibility). Note a single 9-point segment near no. of primes $= 10^4$ (color images in the electronic edition).

101, 151, 351, 501, 1 151, 1 701 and 3 951. The results are shown in Fig. 5b. It can be seen that the longer and hence the less numerous the segments, the more variable the results. This could be expected, due to that the variabilities of the detailed results for shorter segments are muted in the summing processes. For shorter segments the differences can be of the same order of magnitude as the values themselves (segments of lengths 6 and more), but the location of their graphs is much lower than of those of the actual Ulam square. Therefore, this variability does not change the relations between the graphs in a qualitative way.

It should be noted that the results are stable for shorter segments in randomized squares, besides that the locations of the segments in the realizations

Fig. 6. Segments of length 5 points in two randomized calculations for the square 351×351. Between 1000 and 1040 (approx.) such segments are present in an image. Colors were altered in a loop: {R, G, B, C, M, Y}. The realizations shown here are different from that in the right part of Fig. 4 (color images in the electronic edition).

are entirely different. This has been shown in Fig. 6 using the example of two random calculations for the square 351×351 which are small enough to keep the details visible. Segments of length 5 form different patterns in these realizations. For other lengths the results are similar.

In one of the realizations a single segment of length 9 points appeared in the square of size 1151 and its existence is shown in Fig. 5b with a large triangle. This indicates that the limit of 8 for the segment length which appeared in the calculations summarized in Fig. 5a is not general.

5   Summary and Prospects

The points which correspond to prime numbers in the Ulam square form numerous straight line segments of various lengths. The question of what is the relation between the number of such segments and the number of primes has been studied. It has been demonstrated that the relation between the number of segments of a specified length and the value of the largest prime involved in forming these segments is double *log*. This has been demonstrated within the range of numbers up to 25 010 001, which form the square of 5001×5001. The increase of the number of line segments with the growing values of the prime numbers has been observed, irrespective of that the density of primes in the set of integers goes down. It can be cautiously presumed that the structure of the set of the prime numbers might contain the regularities which do not depend on the scale and span the whole set of these numbers.

Very long segments which are rare in the Ulam square, like single segments of 16 and 13 points, two segments of 12 points, and a relatively small number of other longer segments do not follow the same tendency as the shorter segments, due to their small number.

It has been checked if the regularities found are actually specific for the Ulam square itself, or are typical for dot images. It is known that even in random dot images some regularities can be found due to the abundance of data. As a result, it has been found out that in the Ulam square the number of long segments is considerably larger than that in random dot images, and that the longest segments observed in the Ulam square are absent in them. However, the shorter segments, namely the ones of lengths 2 and 3, are more numerous in the random images than in the Ulam square. This suggests that the Ulam square is an object in which the relations between the points are more long-range that those in random dot images.

This research was carried out in hope that the application of the image processing techniques to the Ulam spiral will make it possible to broaden the set of tools and methods used in the analysis of the properties of the fascinating set of prime numbers.

The data used in this and our other papers related to the prime numbers and the detailed results in a downloadable form can be found in the web page [5].

References

1. Caldwell, C.K.: How many primes are there? In: The Prime Pages [2]. http://primes. utm.edu/howmany.html. Accessed 15 Oct 2016
2. Caldwell, C.K. (ed.): The Prime Pages (2016). http://primes.utm.edu. Accessed 15 Oct 2016
3. Chmielewski, L.J., Orłowski, A.: Hough transform for lines with slope defined by a pair of co-primes. Mach. Graph. Vis. **22**(1/4), 17–25 (2013). Open access: http://mgv.wzim.sggw.pl/MGV22.html
4. Chmielewski, L.J., Orłowski, A.: Finding line segments in the Ulam square with the Hough transform. In: Chmielewski, L.J., Datta, A., Kozera, R., Wojciechowski, K. (eds.) ICCVG 2016. LNCS, vol. 9972, pp. 617–626. Springer, Heidelberg (2016). doi:10.1007/978-3-319-46418-3_55
5. Chmielewski, L.J., Orłowski, A.: Prime numbers in the Ulam square (2016). http://www.lchmiel.pl/primes. Accessed 14 Oct 2016
6. Chmielewski, L.J., Orłowski, A., Janowicz, M.: A study on directionality in the Ulam square with the use of the Hough transform. In: Kobayashi, S., Piegat, A., Pejaś, J., El Fray, I., Kacprzyk, J. (eds.) ACS 2016. AISC, vol. 534, pp. 81–90. Springer, Heidelberg (2017). doi:10.1007/978-3-319-48429-7_8
7. Stein, M.L., Ulam, S.M., Wells, M.B.: A visual display of some properties of the distribution of primes. Am. Math. Mon. **71**(5), 516–520 (1964). doi:10.2307/2312588
8. Weisstein, E.W.: Prime counting function. From MathWorld–AWolfram Web Resource (2016). http://mathworld.wolfram.com/PrimeCountingFunction.html. Accessed 15 Oct 2016
9. Weisstein, E.W.: Prime spiral. From MathWorld–A Wolfram Web Resource (2016). http://mathworld.wolfram.com/PrimeSpiral.html. Accessed 15 Oct 2016

Improving Traffic Sign Recognition Using Low Dimensional Features

Laksono Kurnianggoro[1], Wahyono[1], and Kang-Hyun Jo[2]([✉])

[1] The Graduate School of Electrical Engineering, University of Ulsan,
Ulsan 680749, Korea
{laksono,wahyono}@islab.ulsan.ac.kr
[2] The School of Electrical Engineering, University of Ulsan, Ulsan 680749, Korea
acejo@ulsan.ac.kr

Abstract. In the recent decades, researches of the autonomous vehicle
are getting popular in the computer vision society, since such vehicle is
equipped with cameras for sensing the environment in helping navigation movement. Cameras give a lot of information and are low-cost device
sensor rather than the other sensors which can be mounted on the vehicle. One of the visual information which can be acquired by autonomous
vehicle for its navigation is traffic sign. Thus, this work addresses a traffic sign recognition framework as part of the autonomous vehicle. For
recognizing the traffic sign, it is assumed that the traffic sign regions
have been extracted using maximally extremal stable region (MSER).
Using a heuristic rule of geometry properties, the false detections will
be excluded. Furthermore, traffic sign images are classified using low
dimensional features which were encoded using Adversarial Auto-encoder
technique. Using this strategy, classification task can be performed using
2-dimensional features while improving the classification results over the
high dimensional grayscale features. Extensive experiments were carried
out over German traffic sign recognition database show that the proposed
method provides reliable results.

Keywords: Autonomous vehicle · Traffic sign recognition · MSER

1 Introduction

Vehicles can sense their surroundings and navigate autonomously by using several devices and techniques, such as radar, laser, global positioning system
(GPS), and computer vision. These vehicles analyze the environment to identify
its appropriate navigation path, as well as obstacles and relevant traffic sign [1].
Traffic signs carry a lot of significant information for driving, such as to drive
the vehicle in the correct lane and at the right speed; to avoid obstacles and
potential risks; and to notice roadway access, position, and trajectory of other
vehicles, pedestrians movement, road lanes, and statuses [2]. Thus, traffic sign
recognition (TSR) system performs an important part for autonomous vehicle
systems [3].

© Springer International Publishing AG 2017
N.T. Nguyen et al. (Eds.): ACIIDS 2017, Part II, LNAI 10192, pp. 237–244, 2017.
DOI: 10.1007/978-3-319-54430-4_23

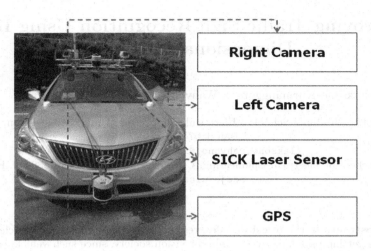

Fig. 1. Our autonomous vehicle equipped with two camera, GPS, and SICK laser sensors.

There have been many approaches for traffic sign recognition with significant results. Most of them divide the process into three main steps: (1) traffic sign region extraction using color information, (2) Traffic sign shape classification, and (3) traffic sign recognition for identifying the meaning of those sign. Specifically, the recognition step can be done by several classification techniques, such as k-nearest neighbor (KNN) and its variants [6,7,14], artificial neural network (ANN) [4,5], support vector machine (SVM) [8,9], and random forest (RF) [10,12]. Reference [13] reported that KNN is the best classifier among these classification techniques. However, it requires high dimensional features to obtain a good accuracy. To overcome the aforementioned problems, this paper proposed to incorporate the Adversarial Auto-encoder (AAE) [17] for encoding the features into a lower dimension. The AAE is very useful to encode features since the encoded representation can be guided to follow the distribution of the guide function.

In this work, a system to recognize traffic sign using low dimensional feature is presented, as part of our autonomous vehicle [11], which is shown in Fig. 1. It is equipped with several sensors such as two lasers, two cameras, and one GPS sensor. In this case, only one camera will be utilized to capture the image specifically for analyzing the traffic sign regions.

2   Candidate Region Segmentation

This paper focuses on the task of recognizing traffic sign. It means that we assume that the traffic sign regions have been obtained the previous step. However, in this section, one of the best methods for doing this task, Maximally Stable Extremal Region (MSER) [15], is presented. Even though, this strategy

Fig. 2. Traffic sign region segmentation sample in our previous work [14]. (Color figure online)

is not main focus in this paper, we briefly describe the MSER as in our previous work [14]. This method works on gray valued image. The RGB image should be converted into gray image before parsed into the MSER method. Since red and blue colors are the main color of the traffic sign, the gray image conversion is defined as (1):

$$I = \max\{\frac{R}{R+G+B}, \frac{B}{R+G+B}\} \tag{1}$$

The pixel values in the gray image are obtained by selecting the maximal value between two calculated values from the original image. The first value is the red component divided by the sum of red, green and blue value on a corresponding pixel in the original image. The second value is the blue component divided by the sum of the color component value on the corresponding pixel from the original image. MSER method can extract the stable regions from the picture. In this case, a stable region means the pixel values in the corresponding area are not changing too much or in other words, has distribution with small variance. Figure 2 shows the illustration of region segmentation that was implemented in our system.

After candidate regions are obtained, a preliminary filter will be carried out for rejecting false detection based on geometry properties of the traffic sign. Let define w and h be width and height of the region, respectively. The candidate regions will be filtered out if it does not meet the rules as shown in Table 1.

Table 1. Geometry properties rules of traffic sign

Properties	Formula	Rule
Aspect ratio	w/h	Between 0.5 and 1.5
Area	$w \times h$	Between 20 and 5000

3   Feature Encoding

This part presents the main contribution of the paper for reducing the feature dimension using encoding strategy. There are various methods of feature encoding, including principal component analysis (PCA), auto-encoder, and Siamese networks. PCA transforms the features into a new representation which are ordered from the most important to the least important based on their variance about the principal axis as represented by the eigenvalues. Meanwhile,

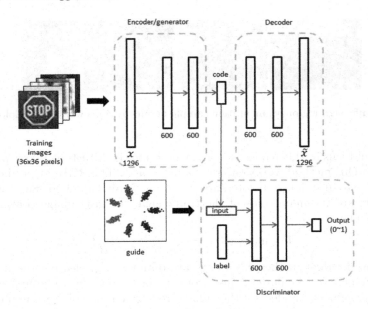

Fig. 3. Structure of the adversarial auto-encoder.

auto-encoder encodes the features using multi layers perceptron which is trained to minimize the reconstruction error, ensuring that the encoded features can be reconstructed again into its original form. In this work, Adversarial Auto-encoder (AAE), which is one of the auto-encoder variants is used to perform the feature encoding task due to its unique function, which is encoding the features into lower dimension in guided way where the distribution of the encoded representation follows the distribution of the guide function.

The structure of AAE network that used in this paper is shown in Fig. 3. This network is basically a combination of an auto-encoder [18] and generative adversarial network (GAN) [19]. The auto-encoder parts are including encoder and decoder networks, while the GAN parts consist of discriminator network and generator network. The generator network shares its properties with the encoder network in the auto-encoder part.

As described earlier, the auto-encoder aims to produce features that can be reconstructed into its original form. Hence, the training objective is to minimize the reconstruction error as described in (2) where M is the number of samples in the dataset. The reconstruction error is defined as the difference between input sample x to its reconstructed form $\tilde{x}$ which is the output of the auto-encoder network. To obtains the reconstructed from, an auto-encoder encodes the input sample into a code which is generated by $E(x)$ and then decoded it as shown in (3). Typically, $E(x)$ has lower dimension compared to the original feature size which later is being reconstructed by the decoder to replicate its original form.

$$\epsilon_{AE} = \sum_{i}^{M} \|x_i - \tilde{x}_i\| \tag{2}$$

$$\widetilde{x} = D\left(E\left(x\right)\right) \tag{3}$$

The GAN networks are useful to generate encoded features that follow the distribution of the guide function. This system consisted of 2 parts, generator, and discriminator. The discriminator is trained to recognize whether the input is from the guide function or not while the generator is trained to encode the input into a new representation with a distribution similar to the guide function. In this case, the generator tries to create data that will confuse the discriminator and fools the discriminator to think that the input produced by the generator is the distribution of the guide function. When the training is converged, the output of the generator, which is the encoded features, will have similar distribution with the guide function.

Since the generator should be trained to fools the discriminator, it's should produce an output that will be recognized as 1 by the discriminator. In this case, the loss value is shown as (4).

$$\epsilon_G = \sum_{i}^{M} log\left(1 - D\left(G\left(x_i\right)\right)\right) \tag{4}$$

Meanwhile, for the discriminator, it should be trained to discriminate whether the input belongs to the guide function or not. Hence, in this case, the loss function is described as (5)) where g_i is the sample taken from the guide function.

$$\epsilon_D = \sum_{i}^{M} \left[logD\left(g_i\right) + log\left(1 - D\left(G\left(x_i\right)\right)\right)\right] \tag{5}$$

4   Experiment

In evaluation, the German Traffic Sign Recognition (GTSR) database [16] was used. This database contains 43 classes, with 27,640 images for training, and 12,630 images for testing the system. However, since the purpose of our application is for the autonomous vehicle system, we limit the number of classes that used in video experiment. The training data contains seven categories (as presented in Fig. 4), where each type consists of more than 500 sample images, combining from GTSR database and our dataset. These seven classes are chosen since they are useful for car navigation.

In the experiment, the classification result by feature encoding using AAE was compared against the classification result by the original gray-values and features obtained by encoding using PCA. For the classifier itself, there are four

Fig. 4. Traffic sign classes used for navigating vehicle.

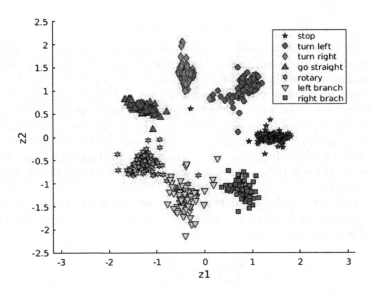

Fig. 5. Visualization of the encoded feature with 2 dimension.

classifiers were considered including support vector machine (SVM), non-linear SVM with Gaussian Kernel, decision tree with maximum 100 splits, and nearest neighbor classifier (NN).

The AAE networks were configured as shown in Fig. 3. All of the encoder/generator, decoder, and discriminator uses two hidden layers with 600 neurons in each layer. The encoded feature was set to 2 to make the features follow the distribution generated by guide function which is a mixture of Gaussian with $\sigma_1 = 0.2$ and $\sigma_2 = 0.08$ rotated every $51.4°$. The input image was converted to gray-scale and resized to 36×26 pixels, resulting the input and output of the auto-encoder part to be 1296 elements.

In training step, in a total of 3321 samples were used to train the network and 700 samples were used as test data where each class is tested with 100 samples. The visualization of the encoded feature for test data is shown in Fig. 5. Most of the encoded representation follow the distribution of the guide function with several features are located far away from where it should be. In this case, we are expecting to get good classification result since most of the data are located in a structured manner.

During the training of the classifiers, the training data were encoded using AAE and PCA. The original gray values also used to train the classifiers as the baseline. The classification result using the test data is shown in Table 2. As expected, the AAE method produces better results compared to the baseline. The AAE method achieve 29.71% of average improvement over the baseline with average classification performance 97.35% using four different classifiers. Moreover, the feature encoded by AAE also produce better results compared to the feature encoded using PCA. This is due to the fact that the encoded

Table 2. Performance evaluation of the classification result (%). The number in brackets show the performance change over the full feature.

	AAE	PCA	Baseline
Linear SVM	98.14 (+31.00)	14.28 (−52.86)	67.14
Non-linear SVM	95.28 (+27.14)	31.57 (−36.57)	68.14
Decision tree	97.86 (+32.15)	32.57 (−33.14)	65.71
kNN	98.14 (+28.57)	38.42 (−31.15)	69.57
Average	97.35 (+29.71)	29.21 (−38.43)	67.64

representation in AAE was guided to form distinctive clusters. In addition, since the AAE can work better using only two dimensional features, it is expected that the method may perform faster processing time for the real-time application.

5   Conclusion

A framework for recognizing traffic sign using low dimensional feature has been presented. The proposed method demonstrate that the classification of the traffic sign image can be done using the 2-dimensional feature which is obtained by encoding the input image using encoder network from adversarial auto-encoder. According to the experiments that were performed on a public dataset, the proposed strategy achieve significant improvement of 29.71% comparing to the baseline. This work shows promising results, make it suitable to be implemented in the real-time application as part of the autonomous vehicle system.

Acknowledgment. This work was supported by the National Research Foundation of Korea (NRF) Grant funded by the Korean Government (2016R1D1A1A02937579).

References

1. Anderson, J.M., et al.: Autonomous Vehicle Technology: A Guide for Policymakers. Rand Corporation, Santa Monica (2014). ISBN 9780833083982
2. Jesmin, F.K., Sharif, M.A.B., Reza, R.A.: Image segmentation and shape analysis for road-sign detection. IEEE Trans. Intell. Transp. Syst. **12**, 83–96 (2011)
3. Andrey, V., Jo, K.H.: Automatic detection and recognition of traffic signs using geometric structure analysis. In: Proceedings of International Joint Conference on SICE-ICASE, pp. 1451–1456 (2006)
4. Saha, S.K., Chakraborty, D., Bhuiyan, M.A.: Neural network based road sign recognition. Int. J. Comput. Appl. **50**(10), 35–41 (2012)
5. Rajesh, R., et al.: Coherence vector of oriented gradients for traffic sign recognition using neural networks. In: Proceedings of International Joint Conference on Neural Networks, pp. 907–910, July 2011
6. Zaklouta, F., Stanciulescu, B.: Warning traffic sign recognition using a HOG-based K-d tree. In: Proceedings of IEEE Intelligent Vehicles Symposium (IV), pp. 1019–1024, June 2011

7. Gu, M., Cai, Z.: Traffic sign recognition using dual tree-complex wavelet transform and 2D independent component analysis. In: Proceedings of World Congress on Intelligent Control and Automation, pp. 4623–4627, July 2012

8. Greenhalgh, J., Mirmehdi, M.: Real-time detection and recognition of road traffic signs. IEEE Trans. Intell. Transp. Syst. **13**(4), 1498–1506 (2012)

9. Hilario, G.-M., Saturnino, M.-B., Pedro, G.-J., Sergio, L.-A.: Goal evaluation of segmentation algorithms for traffic sign recognition. IEEE Trans. Intell. Transp. Syst. **11**(4), 917–930 (2010)

10. Kouzani, A.: Road-sign identification using ensemble learning. In: Proceedings of IEEE Intelligent Vehicles Symposium, pp. 438–443 (2007)

11. Wahyono, Kurnianggoro, L., Seo, D.-W., Jo, K.-H.: Visual perception of traffic sign for autonomous vehicle using k-nearest cluster neighbor classifier. In: Proceedings of URAI (2014)

12. Greenhalgh, J., Mirmehdi, M.: Traffic sign recognition using MSER and random forests. In: Proceedings of 20th European Signal Processing Conference (EUSIPCO), pp. 1935–1939 (2012)

13. Wahyono, Jo, K.-H.: A comparative study of classification methods for traffic sign recognition. In: Proceedings of IEEE International Conference on Industrial Technology (2014)

14. Wahyono, Kurnianggoro, L., Hariyono, J., Jo, K.-H.: Traffic sign recognition system for autonomous vehicle using cascade SVM classifier. In: Proceedings of IECON (2014)

15. Matas, J., Chum, O., Urban, M., Pajdla, T.: Robust wide baseline stereo from maximally stable extremal regions. In: Proceedings of British Machine Vision Conference, pp. 384–396 (2002)

16. Stallkamp, J., Schlipsing, M., Salmen, J., Igel, C.: Man vs. computer: benchmarking machine learning algorithms for traffic sign recognition. Neural Netw. **32**, 323–332 (2012)

17. Makhzani, A., et al.: Adversarial autoencoders. In: Proceedings of ICLR (2016)

18. Baldi, P.: Autoencoders, unsupervised learning, and deep architectures. In: Proceedings of ICML (2012)

19. Goodfellow, I., et al.: Generative adversarial nets. In: Proceedings of NIPS (2014)

Image Processing Approach to Diagnose Eye Diseases

M. Prashasthi, K.S. Shravya, Ankit Deepak, Manjunath Mulimani$^{(\boxtimes)}$, and Koolagudi G. Shashidhar

Department of Computer Science and Engineering,
National Institute of Technology Karnataka, Surathkal, Mangalore, India
prash.mavin@gmail.com, shravya.ks0@gmail.com, adadeepak8@gmail.com,
manjunath.gec@gmail.com, koolagudi@yahoo.com

Abstract. Image processing and machine learning techniques are used for automatic detection of abnormalities in eye. The proposed methodology requires a clear photograph of eye (not necessarily a fundoscopic image) from which the chromatic and spatial property of the sclera and iris is extracted. These features are used in the diagnosis of various diseases considered. The changes in the colour of iris is a symptom for corneal infections and cataract, the spatial distribution of different colours distinguishes diseases like subconjunctival haemorrhage and conjunctivitis, and the spatial arrangement of iris and sclera is an indicator of palsy. We used various classifiers of which adaboost classifier which was found to give a substantially high accuracy i.e., about 95% accuracy when compared to others (k-NN and naive-Bayes). To enumerate the accuracy of the method proposed, we used 150 samples in which 23% were used for testing and 77% were used for training.

Keywords: Image processing · Machine learning · Classifiers · Iris · Sclera

1 Introduction

Eye is a vital organ which is responsible for sense of vision in a human body. Symptoms of several disorders in the human body include abnormalities in the normal structure of eye which is manifested through change in the colour of eye or change in the orientation of iris, etc. Often many diseases have very similar effects on eyes and it becomes difficult to diagnose the exact cause of the abnormality.

Diagnosis of the cause of the abnormalities in the eye is usually done by the opthamologist, who is a specialist in the field of diagnosing disorders and diseases of the eye, with the help of sophisticated devices. Often, these devices are quite expensive and not affordable. Also, these facilities are inaccessible in rural areas. Thus, due to lack of the amenities and resources, people neglect the disorders or might use self diagnosis and take wrong medication and may worsen their health

© Springer International Publishing AG 2017
N.T. Nguyen et al. (Eds.): ACIIDS 2017, Part II, LNAI 10192, pp. 245–254, 2017.
DOI: 10.1007/978-3-319-54430-4_24

condition. Thus, an alternative should be provided which is cost-effective and easily accessible to people residing in remote regions lack of medical knowledge.

Various methods have been developed in order to provide better medical assistance. Some state-of-the-art methods use the concepts of deep learning, neural networks, etc. to diagnose the disease by collecting the symptoms from the patient as input. Few existing methods diagnose a particular abnormality in the eye using image processing on retinal images which have been captured using fundoscope, which is an expensive tool. The proposed method is different from the state-of-the-art methods as our tool can diagnose various eye abnormalities using images captured from a general purpose camera. Also, the proposed method uses colour based diagnosis of eye diseases.

The eye consists of three layers, i.e. sclera and cornea constitute the outer layer; choroid, ciliary body and iris constitute the middle layer and retina constitutes the innermost layer. For our diagnosis, we mainly concentrate on the outer and middle layers. The ideal colour of sclera of healthy eye is white and that of pupil is black but the colour of iris varies among individuals depending on their genes. However, the iris of a healthy eye is not clouded. Also, the alignment of iris of both the eyes are in the same direction. When there is variation from the characteristics as mentioned in Table 1, it signifies that the patient may have an abnormality in eye which may be due to any disorder in the human physiology.

The major contributions of this paper are:

- The proposed work is inexpensive and can be used by people in remote areas.
- Assistance from a specialist for using the proposed method is not required.
- The proposed method can accurately detect disorders in eye by operating on images captured by a normal and inexpensive camera. There is no necessity for any special instrument or special techniques which are expensive.
- The proposed algorithm can be adapted to detect and diagnose any eye disorder which affects the normal colour composition of eyes.

The rest of the paper is structured as follows. Section 2 presents literature survey highlighting the techniques used, merits and demerits. Section 3 gives the detailed description of the proposed algorithm. Section 4 presents the experimentation and results obtained. Section 5 presents the conclusion of our paper along with a brief proposal of future work.

2   Related Works

Machine learning and Image processing have been used in the diagnosis of eye diseases. Some of the methodologies that have proposed a system that uses these two techniques are mentioned as follows. In [1], the authors have proposed diagnosis of diseases by using the techniques of Naive Bayes and CBR. The advantage of the state of art is that the system shows 82% similarities with the results of the diagnosis by expert. However, the state of art primarily relies on the description of symptoms by patients which might lead to inaccuracies. Also, the testing phase involved just 11 cases. In [2], the authors have proposed

Table 1. Distinguishing features of diseases

Disease	Distinguishing feature
Jaundice from other diseases	Colour of sclera
Subconjunctival haemorrhage from conjuctivitis and corneal infections	Intensity and density of redness in sclera
Corneal infection from conjuctivitis	Redness of sclera
Leukocoria from cataract	Asymmetry in reflex
Palsy from others	Orientation of iris
Leukocoria from coat's disease	Colour of the reflex

Diagnosis of Age Related Macular Degeneration (ARMD) using threshold and geographical features. The advantages of the state of art is that the system gave 92% accuracy when tested on 50 fundus coloured images. However, the system requires automated and accurate fundoscopic images for diagnosis which makes the preprocessing step complex. In [3], the authors have used the technique of Convolutional-Recursive Neural Networks where they have tested the tool on a dataset of 5378 images and they have achieved 70.7% exact integral agreement ratio. However, the system focuses only on cataract and also, requires slit-lamp images. In [4], the authors have performed Comparison between Artificial Neural Network, Fuzzy classifier and Neuro-Fuzzy classifier to classify three diseases namely Cataract, Iridocyclitis and Corneal Haze. Big Ring Area, Small Ring Area, BWMorph and Homogeneity as features for image processing and extract these features and pass it to classifier. They have obtained an accuracy of 85% from the dataset of diseases associated with corneal discolouring which comprised of 135 images. However, the authors have restricted their diagnosis to just 3 diseases namely cataract, corneal haze and iridocyclitis. The above mentioned state of works concentrate on a limited number of diseases and they need very good quality images from the fundoscope. Capturing such images increases the cost of the approach. Besides, they rely on description of symptoms by parents which might cause inaccuracies. We have tried to develop a simple cost efficient model which can predict a wide range of diseases based on colour, structure and texture of eyes with better accuracy using the images captured by any simple camera.

3   Proposed Methodology

For the detailed working of the proposed methodology, the workflow diagram presented in Fig. 1 can be referred.

3.1   Preprocessing

The accuracy of the proposed method depends on the quality of the image of the eye given as an input to the system after preprocessing. The image given is

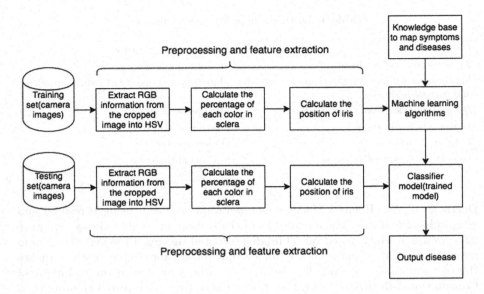

Fig. 1. Diagram of proposed methodology

a rectangular image containing the parts of surrounding skin if it is taken from any general purpose camera. The image should be cropped before it is used as an input sample. To remove the skin part, we have used a simple clipper. The clipper uses mouse click inputs by user for the boundary of the eye, the clipper then creates a matrix equal to the pixel size of the image. The matrix has binary value of $(255, 255, 255)$ for the polygon which encloses the eye and that of $(0, 0, 0)$ for the rest of the areas. A bitwise AND (&) operation is performed taking the matrix and the eye image, leaving the eye as the only unmasked area. We then move on to extract pixel wise information from the image. We convert RGB (Red, Green and Blue) colour format to HSV (Hue, Saturation and value) format because unlike RGB, HSV separates the image intensity called luma, from the

Table 2. HSV range of colour (range of values implies shades of the mentioned colour, - there is no one-to-one colour mapping of the value shown in the table)

Colour	H	S	V
Black	0	>80	<20
White	0–360	<20	>80
Red	[0–20] U [340–360]	[20–80]	[20–80]
Pink (it is a subset of red)	[340–360]	[20–80]	[20–80]
Yellow	[45–65]	[20–80]	[20–80]
Brown	[25–40]	[20–80]	[20–80]

colour information called chroma [5]. The HSV information is then interpreted in the way specified in Table 2. The values calculated for iris are subtracted from that of the sclera to improve the clarity of colours in sclera. In order to extract the position of the iris, we enumerate the amount of pixels from the origin of sclera to the iris in the middle of the sclera. Figure 2 shows the extraction of various colours from a normal eye.

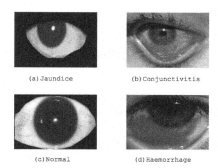

Fig. 2. (a) Eyes suffering from jaundice (yellow discolouration), (b) eyes suffering from conjunctivitis (pink discolouration), (c) normal eye (white), (d) eyes suffering from subconjunctival haemorrhage (deep red discolouration) (Color figure online)

Fig. 3. (a) Extracted white colour from a normal eye (b) extracted red colour from a normal eye (Color figure online)

3.2 Feature Extraction

The features needed for our proposed method are the different colours of the different layers of the eyes. In addition to these, the position of the iris is also considered for prognosis of a disease. Figure 2 shows images of normal eyes and few images of eyes with disorders. Figure 3 shows the extraction of white and red colors from normal eye shown in Fig. 2(c) whereas Fig. 4 shows the extraction of colors from eyes with abnormalities shown in Fig. 2(a), (b) and (d). The various disorders with their respective symptoms are mentioned below:

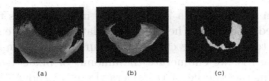

(a)   (b)   (c)

Fig. 4. (a) Extracted redness from eyes with subconjunctival haemorrhage, (b) extracted redness from eyes with conjunctivitis, (c) extracted yellowness from eyes with jaundice (Color figure online)

Detection of Eye-Disorders Using Color

- **Conjunctivitis:** Conjunctivitis is an eye disease which is characterised by the inflammation of the thin clear tissue called conjunctiva that lies over the sclera. The sclera is pink in color while conjunctiva is light red.
- **Jaundice:** Jaundice is a disorder which is characterised by the prominent yellow discoloration of the skin caused by the increased levels of bilirubin. The membrane covering the sclera shows a yellowish pigmentation.
- **Subconjunctival-Haemorrhage:** Subconjunctival Haemorrhage is characterised by the rupture of the numerous blood vessels embedded in the conjunctival membrane. A deep red dense region in the eye signifying the effects of conjunctival bleeding.
- **Corneal Infection:** Corneal Infection is characterised by the significant swelling of the cornea which is due to the infection caused by the fungus. This disease leads to inflammation of the cornea of the infected eye along with redness.
- **Cataract:** Cataract is commonly a late age disorder characterised by the cloudiness of the lens of the eye which may lead to partial or complete loss of vision. The pupil region which is concentric to the pupil starts developing cloudiness.

Detection of Eye-Disorders Using Orientation

- **Microvascular Cranial Nerve Palsy:** MCNP is a neurological disorder characterised by the irregular movement of eyes because of the blocking of flow of blood to nerves between the muscles of eye and brain.
- **Strabismus:** Strabismus commonly known as cross-eye is a eye disorder which is characterised by the inability of the individual to focus both the eyes on a point simultaneously or in other words it can also be said that the alignment of both eyes are not ideal while directing the vision to observe an object.

Detection of Eye-Disorders Using Red-Reflex Test

- **Leukocoria:** Leukocoria is an abnormal eye condition which is characterised by the white glow that is observed during the red reflex test. It is detected by Red reflex test which yields an image of an eye with a white shine.

– **Coat's Disease:** Coats' disease is characterised by partial or complete blindness caused due to irregularity observed in the blood vessels behind the retinal region. It is detected by the Red reflex test which yields an image of an eye with a yellow glow which occurs due to the reflection of the light rays from the flash, off the cholesterol deposits.

Various eye diseases have different implications on eyes. Hence, it is important to decide which feature differentiates one disease from the other.

3.3 Processing Features

The features extracted are in the form of 24 bits RGB colours, which are converted into HSV triplets for further processing. The HSV triplets are classified into colours as follows.

The presence of any colour is found by:

$$degree_i = \frac{\sum_{x,y}[h_{x,y} \, \epsilon \, h_i] * [s_{x,y} \, \epsilon \, s_i] * [v_{x,y} \, \epsilon \, v_i]}{\sum_{x,y} 1} \tag{1}$$

where i is the colour whose degree is found
x, y are the pixel indices
h, s and v are the respective colour values

3.4 Training

After the features are extracted, we use the classifier to learn the properties of each disease separately, i.e., model for every disease is independent. We use Naive Bayes [6], AdaBoost [7] and kNN [8] classifiers for the classification. The independent feature vector to the classifier is the set of all extracted features in the form {Chromatic Features of Sclera, Chromatic Features of Iris, Spatial Features}. The dependent variable is a binary value with respect to every disease, i.e., 1 implies the presence of disease and 0 implies the absence of disease.

3.5 Classifiers

Different classifiers used in the proposed methodology are discussed below in brief:

k-Nearest Neighbors. KNN is one of the simplest classifiers in machine learning and pattern recognition. As the name suggests, it ensures that the contribution by closest neighbours is given more priority than farther neighbours by assigning weights to the values contributed by the neighbors. Although, higher values of K yields results with reduced noise in classification, but it also reduces the distinctness in the boundaries. Hence, heuristic techniques should be applied to select the appropriate value of k. We have allowed maximum number of neighbors as 3 in our model. We have used this classifier under the assumption that two diseases of same class will have similar features, and hence will be the closest neighbour of the each other, and hence the classifier should be able to predict based on similarity of overall features of test data and our training data.

Naive Bayes' Classifier. It is one of the classifiers which belongs to set of simple probabilistic classifiers using Bayes' theorem with strong independence considering few assumptions. It is extremely scalable and requires a couple of parameters equal to the number of variables in the training problem. This classifier was used under assumption that each features will affect the disease independently, for instance, the pink discoloration of sclera will be more instrumental in the prediction of conjunctivitis while the yellow will be for jaundice, hence the classifier should predict based on all the prediction independently and give the final predictions.

AdaBoost Classifier. AdaBoost is classified under Ensemble Learning process where several combination of learning algorithms is used to come up with an efficient algorithm, where each classifier predicts based on disjoint set of features and hence it combines the advantage of both K-nearest neighbours and Naive Bayes by working on non-overlaping set of various features.

4   Results and Discussion

For the developing our model of eye disease detection, we used the images of eye handed over to us by Dr. Punith Kumar. We have collected 150 images of various eye diseases of which 23% are used for testing and 77% are used for training. There are various image samples of different diseases used for diagnosis. Out of 35 test samples, 4 suffered from Subconjunctival Haemorrhage, 5 suffered from jaundice, 7 from cataract, 3 from corneal infection, 7 from conjunctivitis and 4 from palsy, while 5 images were of healthy eye.

Our experimentation for the methodology was done using Python. For the preprocessing step, we used openCV library of python to crop the images. The features were extracted using the PIL library of python by extracting the pixel information. We then converted the extracted pixel information to HSV values. The classifiers we used were the ones implemented in scikit library of Python under the skLearn module.

The confusion matrix of different classifiers is shown in Table 3. Table 4 shows the accuracy value of all the classifiers for different diseases. It is observed that AdaBoost works better than all the other classifiers because it is able to boost the performance of several weak classifiers and is able to overcome settings and fittings. Adaboost is able to combine the advantages of both K-nearest neighbours and naive Bayes classifiers, while K-nearest neighbours is unable to take into account the independence of unrelated features, naive Bayes takes total independence of features and fails to takes into account closely associated features.

Table 3. Confusion matrix for the model[a]

	Naive Bayes							k-Nearest Neighbor							AdaBoost						
	H	J	C	C.I	Cn.	P	N	H	J	C	C.I	Cn.	P	N	H	J	C	C.I	Cn.	P	N
H	3	0	0	1	0	0	0	2	0	0	0	0	0	2	3	0	0	0	0	0	1
J	0	5	0	0	0	0	0	0	4	0	0	0	0	1	0	5	0	0	0	0	0
C	0	3	2	0	0	0	2	0	0	6	0	0	0	1	0	0	6	0	0	0	1
C.I	0	0	0	3	0	0	0	0	0	0	3	0	0	0	0	0	0	3	0	0	0
Cn.	1	0	0	1	5	0	0	0	0	0	0	4	3	0	2	0	0	0	7	0	0
P	0	0	0	0	0	4	0	0	0	0	0	0	4	0	0	0	0	0	0	4	0
N	0	0	0	0	0	0	5	0	0	0	0	0	0	5	0	0	0	0	0	0	5

[a]LEGEND: H: Haemorrhage; J: Jaundice; C: Cataract; C.I: Corneal Infection; Cn.: Conjunctivitis; P: Palsy; N: Normal;

Table 4. Percentage accuracy of the classifiers

Classifier	Diseases							
	Haem.	Jaun.	Cat.	Cor.Infn.	Conjn.	Palsy	Normal	Overall
Accuracy of Naive Bayes (%)	75	100	40	100	83.33	100	100	77.14
Accuracy of kNN (%)	50	50	85.71	100	33.33	100	100	77.14
Accuracy of AdaBoost (%)	75	100	100	85.71	100	100	100	94.28

Moreover, the results predicted by all the classifiers implemented were scrutinised by the doctor and the results predicted by the classifiers exactly matched with his analysis.

5 Conclusion

This paper discusses a technique for diagnosis of eye disorders and abnormalities. The proposed technique doesn't need high resolution images or images captured by sophisticated equipments like fundoscope. As the preprocessing phase holds an important role, we need to capture coloured images that have uniform illumination and sufficient resolution which can be precisely cropped to eliminate the unnecessary portions of the image and extract the features, taking these features as inputs, this classifier diagnoses each disease and classifies the image under one of the disease categories. The scope of the work is confined

1. To the diagnosis of a limited set of diseases and abnormalities.
2. To the image which have clear distinction based on colouring, sufficient illumination and quality.
3. To the use of only coloured images as an input to the proposed system.

Further above mentioned scope is to be expanded to remove the restrictions on data, number of diseases etc. However the obtained results of around 94% are very promising compared to other results from the state of the art works.

Acknowledgments. We would like to thank Dr. Punith Kumar, MBBS, MS (OPHTHOL), Varun Eye Clinic, Bangalore for his valuable suggestions and timely feedbacks given to us in building this model.

References

1. Kurniawan, R., Yant, N., Nazri, M.Z.: Expert systems for self-diagnosing of eye diseases using Naïve Bayes. In: International Conference of Advanced Informatics: Concept, Theory and Application (2014)
2. Kabari, G., Nwachukwu, O.: Chapter 3. Neural Networks and Decision Trees for Eye Diseases Diagnosis (2012)
3. Gao, X., Lin, S., Wong, T.: Automatic feature learning to grade nuclear cataracts based on deep learning. IEEE Trans. Biomed. Eng. **62**, 2693–2701 (2015)
4. Rayudu, M., Jain, V., Kunda, M.: Review of image processing techniques for automatic detection of eye diseases. In: 6th International Conference on Sensing Technology (2012)
5. Salunkhel, R., Pati, F.: Image processing for mango ripening stage detection: RGB and HSV method. In: 3rd International Conference on Image Information Processing (2015)
6. Klawonn, F., Angelov, P.: Evolving extended Naive Bayes classifiers. In: 6th IEEE International Conference on ICDM Workshops, pp. 643–647. IEEE, Hong Kong (2006)
7. An, T., Kim, M.: A new diverse AdaBoost classifier. In: International Conference on Artificial Intelligence and Computational Intelligence (2010)
8. Li, S., Harner, E., Adjeroh, D.: Random KNN. In: IEEE International Conference on Data Mining Workshop (2014)

Weakly-Labelled Semantic Segmentation of Fish Objects in Underwater Videos Using a Deep Residual Network

Alfonso B. Labao and Prospero C. Naval Jr.[✉]

Computer Vision and Machine Intelligence Group, Department of Computer Science,
College of Engineering, University of the Philippines Diliman,
Quezon City, Philippines
pcnaval@dcs.upd.edu.ph

Abstract. We propose the use of a 152-layer Fully Convolutional Residual Network (ResNet-FCN) for non motion-based semantic segmentation of fish objects in underwater videos that is robust to varying backgrounds and changes in illumination. For supervised training, we use weakly-labelled ground truth derived from motion-based adaptive Mixture of Gaussians Background Subtraction. Segmentation results of videos taken from six different sites at a benthic depth of around 10 m using ResNet-FCN provide a fish object average precision of 65.91%, and average recall of 83.99%. The network is able to correctly segment fish objects solely through color-based input features, without need for motion cues, and it could detect fish objects even in frames that have strong changes in illumination due to wave motion at the sea surface. It can segment fish objects that are located far from the camera despite varying benthic background appearance and differences in aquatic hues.

Keywords: Fish segmentation · Convolutional Neural Networks

1 Introduction

Semantic segmentation assigns class labels to each pixel in an image. In this paper we describe how semantic segmentation is performed in underwater videos, with the goal of labelling all fish object pixels as foreground, and all other underwater object pixels as background. It is a two-class pixel-wise classification process that serves as a preprocessing step for other underwater computer vision tasks such as fish counting or fish species classification. Semantic segmentation in underwater videos is a nontrivial problem, and has unique characteristics that can confound popular segmentation procedures which work well in a surface air environment. The underwater medium is characterized by light absorption and scattering caused by dissolved constituents and suspended particulates [13] that can produce "marine snow" artifacts resulting in false positives. Other problems involve illumination changes due to wave movement at the sea surface.

© Springer International Publishing AG 2017
N.T. Nguyen et al. (Eds.): ACIIDS 2017, Part II, LNAI 10192, pp. 255–265, 2017.
DOI: 10.1007/978-3-319-54430-4_25

Traditional segmentation procedures for underwater videos are based on background subtraction techniques that rely on motion [5]. Background subtraction has proven to be useful in fish segmentation since in an underwater environment most of the moving objects are comprised of fish. However, there are several limitations with these methods. Primarily, it has to operate under a stationary camera and non-moving background. If the camera is moved even slightly or if the background changes rapidly, background subtraction loses much of its precision. It is also not useful for videos that have camera movement such as those taken by underwater robots. Moreover, background subtraction cannot handle well several underwater imaging problems such as varying or sudden illumination caused by wave movement or changes in hue or color. To address these limitations of motion-based segmentation, one approach is to remove reliance on static background features, by making use of more motion-robust features such as edges or shapes. This way, changes in illumination or color do not form the primary basis for pixel segmentation. Recent segmentation techniques have been developed that are based on deep learning methods, such as Convolutional Neural Networks (CNNs) which first gained widespread popularity with Krizhevsky's state-of-the-art results [10] in the 2012 ImageNet challenge [16]. Some deep learning based segmentation techniques [1,11,12,14] perform pixel-wise classification without relying on static backgrounds since the features that these methods extract are not based on motion allowing the network to perform pixel-wise classification on moving or changing backgrounds. An additional advantage of CNN-based segmentation is the generality of the model. If sampling is done correctly, the resulting single model can be applied to varying environments and backgrounds without need for readjusting background information. Segmentation using CNNs however is not a straightforward task. Among the layers in most CNNs are max-pooling layers that shrink activation maps to very small sizes such as 7×7 pixels. As pixel classification has to be done in the original image size through interpolation, much information is lost in pooling - resulting in blocky shapes. One technique that addresses this problem adds the max-pooling features after every interpolation layer to regain information, leading to the fully-convolutional layer or FCN [12]. Segmentation in this paper uses ResNet-FCN proposed by Wu et al. [18], which uses a 152-layer Deep Residual Network followed by interpolation layers with max-pooling additions as suggested by Long et al. [12]. It is a fully-convolutional network or FCN with a final output of the same size as the original image containing label probabilities for segmentation. In training the ResNet-FCN, it is impractical to manually label ground-truth samples given the large number of frames and pixels. As a work-around, we settled with weakly-labeled ground truth derived from a motion-based segmentation technique [19] that implements a component-adaptive Mixture of Gaussians (MoG-BGS) to model per-pixel features. It has been found that CNNs can improve weakly-labeled ground truth similar to MRF's [11], resulting in lower false positives. In addition, in several frames that have strong illumination changes, ResNet-FCN is able to correctly segment fish objects despite strong variations in brightness intensities.

2 Methods

2.1 Data Sources

Data used in this paper is composed of several underwater videos taken in six different sites at the Verde Island Passage, Philippines. The frame rate for each video is 24 fps with a duration of 8 to 10 min. Each frame has size 1080×1920 pixels. All videos are taken at benthic depth which is the usual habitat for a wide variety of fish and coral species. A disadvantage of collecting videos at this depth is their susceptibility to sharp illumination changes as the benthic background remains relatively close to surface. In addition, the close similarity of fish objects with coral objects or sea-floor background poses serious segmentation challenges that could easily confuse less powerful classifiers. The number of fish objects in each video varies (Table 1). The video with the most number of fishes is in Site 6, where several of the fishes are located far from the camera. Almost all videos have coral objects in the background except for Site 4. Among the videos, Site 6 has the strongest changes in illumination.

Table 1. Maximum number of fish objects per frame

	Number of fish objects	Presence of corals	Environment specs
Site 1	11	Yes	B
Site 2	30	Yes	C
Site 3	23	Yes	B, C
Site 4	15	No	C, D
Site 5	56	Yes	B, C, D
Site 6	152	Yes	A, B

A - sudden illumination, B - small fish sizes, C - marine snow, D - poor visibility

2.2 Weakly-Labelled Ground Truth

Deep learning is a supervised machine learning approach which needs per frame ground truth. The best ground truth frames are manually-labeled by an expert. However the quantity of frames and fish pixels in our data makes it impractical to manually label fish objects on a per frame basis with a good accuracy. To compensate for the lack of manual annotations, this paper uses weakly-labeled ground truth derived from a motion-based background subtraction (BGS) technique [19] implemented in OpenCV 3.1 [8]. It models a Mixture of Gaussians (MoG) over the RGB color values of each pixel, and can automatically adapt the number of components in the MoG. For every new RGB color feature that appears far from statistically learned pixel means, a new component is added. We set the MoG-BGS to learn from a moving history of 1000 frames. This history gives leeway for the MoG to incorporate slight non-fish movements, such as those caused by noise or gradual changes in illumination, into the background class. After segmentation with the MoG-BGS, the binary mask image is passed through two iterations of image erosion using a 2×2 kernel to remove noisy foreground pixels. No other processing is done afterwards. The result is a weakly-labeled ground truth mask of the same size as the original image. To guard against cases where illumination or camera movement changes affect the weakly-labeled ground truth greatly, we manually remove frames with a large number of false positives. Most of the removed frames are from Site 6. Using weakly-labeled ground truth results in imperfect training since many of the motion-based ground truth frames are rife with false positives due to marine snow particles. We expect that CNN convolutions should be able to perform a smoothing process over the ground truth similar to MRFs [11], which would lower the number of false positives.

2.3   Residual Fully Convolutional Network (ResNet-FCN)

The ResNet-FCN variant implemented in this paper is based on Wu et al. [18], which achieved state-of-the art results in the 2015 PASCAL Segmentation Challenge [4]. ResNet-FCN is composed of a 152-layer Residual Network trunk proposed by He et al. [6], followed by four upsampling fully-convolutional layers to achieve per-pixel labeling over the original image. Upsampling layers are commonly termed as "deconvolution" layers, following Long et al. [12]. These layers use bilinear interpolation to restore reduced activation maps to the original image size. The 152-layer trunk has the following architecture (Table 2) [6], and achieved state-of-the-art results in the ILSVRC challenge [16]. It has served as the backbone for many object localization frameworks [15].

Table 2. ResNet-152 residual blocks

Layer name	Residual-block	No. of blocks
conv1	7×7 (stride 2)	$\times 1$
max-pool	3×3 (stride 2)	–
conv2	1×1: 3×3: 1×1	$\times 3$
conv3	1×1: 3×3: 1×1	$\times 8$
conv4	1×1: 3×3: 1×1	$\times 36$
conv5	1×1: 3×3: 1×1	$\times 3$
Average pool		

Each 3-layer residual block outputs an activation map that has four times the feature depth of the input-map in its first layer [6]. For the first layer of the residual block, we use zero-padding instead of projection to compensate for the discrepancy between its input and output maps, and the main trunk does not incorporate bias terms. Zero-padding leads to a slight degradation in precision compared to convolution-based projections, but the difference is not significant, and it also saves GPU memory since it forgoes additional projection weights. The last layer of each residual block is applied with global batch normalization [7], with the beta, and gamma parameters being set to trainable all throughout. Following most CNN architectures, Rectified Linear units (ReLU) [3] are added after the batch-normalized layers. Activation map sizes in the main trunk are gradually reduced with a stride of 2 through max-pooling, analogous to the down sampling layers in VGG-16 [17]. The resulting activation map after the main trunk has a size of $60 \times 34 \times 2048$, which is 1/32-th of the original input size. To prepare the last layer for upsampling, we perform average pooling across the 2048-depth dimension of conv5 for subsequent upsampling. The average pool layer thus has an activation map of $60 \times 34 \times 1$. Four upsampling layers follow average-pooling. Each upsampling layer restores the activation map size to a corresponding max-pooling size in the trunk using bilinear interpolation. Eventually, the last output activation map has a size similar to the original input image, but with 2 feature dimensions, corresponding to the softmax probabilities of each pixel for either the background or foreground class (Table 3).

Table 3. ResNet-152 activation map sizes

Layer name	Activation map size ($H \times W$)
Input image	1920×1080
conv1	960×540
3×3 max-pool	480×270
conv2	480×270
conv3	240×135
conv4	120×68
conv5	60×34
FC32	120×68
FC8	240×135
FC4	480×270
UpScore	1920×1080

For each up-sampling layer, their corresponding max-pooling terms are added to restore information. Following [18], upsampling up to FC4 restores much detail for increasing accuracy. However, for the upsampling layers, [18] shifts the max-pooling additions to approximate FC4 and save GPU memory. For this experiment, we implement an actual FC4 layer, but limit the batch-size to one image per iteration to compensate for added memory requirements. We train the ResNet-FCN from scratch, end-to-end, for 240 thousand iterations with a weight decay of 0.0001. Initialization is done using Gaussian random seeds with a standard deviation of 0.01. Starting learning rate is 0.001 for weights, with a decay of 0.04 every 25,000 steps. Optimization is performed using Stochastic Gradient Descent (SGD), with a Nesterov momentum of 0.9. The network was trained using TensorFlow running on an NVIDIA Titan X GPU.

3 Results and Discussion

3.1 Precision and Recall Performance

Test images were formed by randomly choosing 30 images from the thousands of images from each site. Table 4 presents the average precision and recall measures of ResNet-FCN segmentation, and the average number of fish detected per frame.

Table 4. Average precision, recall, fish detection results

	Precision	Recall	Number of fish detected
Site 1	29.58	100.00	11
Site 2	80.03	83.74	25
Site 3	77.67	81.56	18
Site 4	38.05	96.64	15
Site 5	74.94	70.16	40
Site 6	95.18	71.85	116
Average	65.91	83.99	

Precision and Recall measures are counted on a per fish object basis, i.e., a fish object that is marked by a positive blob in its segmentation mask is counted as only one true positive. This is opposed to counting accuracy measures on a per pixel basis. We follow the rule that fish objects whose blob masks comprise only a small portion of its total area are not to be counted as true positives. Blobs that mark non-fish objects are counted as false positives despite their small sizes. Blobs that are close together are counted as a single blob. In marking false positives, we have been quite strict, by marking even very small blobs on non-fish objects as false positives. Accuracy results indicate that recall measures, which measures the quantity of fish objects detected against missed fish objects, are higher in general than precision measures. There is also a difference in the accuracy measures for each site. This is just to be expected since each site presents a different aquatic environment, with different backgrounds and lighting conditions.

Among the sites, Sites 1 and 4 have the lower precision results. For Site 1, several coral edges were mistakenly identified as fish, while for Site 4, several "marine snow" particles came out as false positives. These errors can partly be blamed on the weak-labeling of ground truth using background subtraction. However, the resulting false positives for ResNet-FCN remain lower in general than segmentation from background subtraction. This can be attributed to the

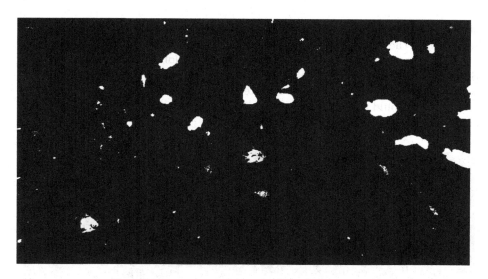

Fig. 1. Background subtraction segmentation results for frame without MRF or CRF post-processing.

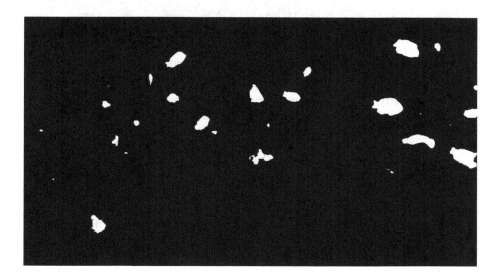

Fig. 2. Segmentation results for the same frame in Fig. 1 using ResNet-FCN where most of the false positive blobs or specks are removed.

automatic smoothing process performed by convolutions in a deep network. The smoothing process is similar to MRFs or dense-CRF post-processing [2, 9], which shares pixel potentials to its immediate neighbors. As shown in Fig. 1, segmentation with the ResNet-FCN removed several noisy particles from the original image in Fig. 2, and even created a more "whole" blob for the bottom-left fish. Some methods such as [2], take a further post-processing CRF-smoothing. The same method could also be applied to ResNet-FCN, but we choose to forego this step since it would take up a large amount of computational time given the large sizes of our image inputs (Figs. 3 and 4).

Fig. 3. Background subtraction result for frame with sudden illumination (Site 6).

For all six sites, recall measures are relatively high at above 70%. This is observed even in Site 6, where several of the identified fishes are already located far from the background and could be observed with difficulty by the human eye as shown in Fig. 5. The same could be said regarding Sites 2 and 3. The trained network performs well in identifying potential fish at different backgrounds and environments, even though its precision is somewhat lower than its recall. Among the six sites, the video that suffered most from irregular or sudden illumination is Site 6 since it was taken at a time when wave movements at surface were significant. Despite this, a full segmentation of the video shows that ResNet-FCN is able to robustly identify fish objects despite strong changes in illumination. For many frames, segmentation from a motion-based background subtractor created large flashes of white pixels in the segmentation mask which is the common result of sudden illumination changes. However, the ResNet-FCN segmentation of the same frames do not show this problem. This shows that the trained network is able to create motion-independent features, which are not as affected by brightness levels compared to background subtraction.

Fig. 4. Segmentation result for the same frame using ResNet-FCN where mistakes caused by sudden illumination are removed.

Fig. 5. Segmentation overlayed (Site 6): TP (Green), FP (Blue) and FN (Red). (Color figure online)

4 Conclusion

The ResNet-FCN network in this project performs semantic segmentation of fish objects using non-motion based features with good precision and recall after training using weakly-labelled ground truth. It is able to work around the weaknesses of motion-based segmentation procedures, and could robustly segment fish objects despite sudden changes in illumination which are usually present benthic-depth underwater videos. In addition, the model can detect fish objects despite changes in sea-floor/coral background.

References

1. Badrinarayanan, V., Kendall, A., Cipolla, R.: SegNet: a deep convolutional encoder-decoder architecture for image segmentation (2015). arXiv preprint arXiv: 1511.00561
2. Chen, L.C., Papandreou, G., Kokkinos, I., Murphy, K., Yuille, A.L.: Semantic image segmentation with deep convolutional nets and fully connected CRFs (2014). arXiv preprint arXiv:1412.7062
3. Dahl, G.E., Sainath, T.N., Hinton, G.E.: Improving deep neural networks for LVCSR using rectified linear units and dropout. In: 2013 IEEE International Conference on Acoustics, Speech and Signal Processing, pp. 8609–8613. IEEE (2013)
4. Everingham, M., Gool, L., Williams, C.K., Winn, J., Zisserman, A.: The Pascal visual object classes (VOC) challenge. Int. J. Comput. Vis. **88**(2), 303–338 (2010)
5. Garcia, R., Nicosevici, T., Cufí, X.: On the way to solve lighting problems in underwater imaging. In: OCEANS 2002 MTS/IEEE, vol. 2, pp. 1018–1024. IEEE (2002)
6. He, K., Zhang, X., Ren, S., Sun, J.: Deep residual learning for image recognition (2015). arXiv preprint arXiv:1512.03385
7. Ioffe, S., Szegedy, C.: Batch normalization: accelerating deep network training by reducing internal covariate shift (2015). arXiv preprint arXiv:1502.03167
8. Itseez: Open source computer vision library (2015). https://github.com/itseez/opencv
9. Krahenbuhl, P., Koltun, V.: Efficient inference in fully connected CRFs with Gaussian edge potentials, vol. 117. Curran Associates (2011)
10. Krizhevsky, A., Sutskever, I., Hinton, G.E.: ImageNet classification with deep convolutional neural networks. In: Advances in Neural Information Processing Systems, pp. 1097–1105 (2012)
11. Liu, Z., Li, X., Luo, P., Loy, C.C., Tang, X.: Semantic image segmentation via deep parsing network. In: Proceedings of IEEE International Conference on Computer Vision, pp. 1377–1385 (2015)
12. Long, J., Shelhamer, E., Darrell, T.: Fully convolutional networks for semantic segmentation. In: Proceedings of IEEE Conference on Computer Vision and Pattern Recognition, pp. 3431–3440 (2015)
13. Negahdaripour, S., Yu, C.H.: On shape and range recovery from image shading for underwater applications. Underw. Robot. Veh.: Des. Control 221–250 (1995)
14. Noh, H., Hong, S., Han, B.: Learning deconvolution network for semantic segmentation. In: Proceedings of IEEE International Conference on Computer Vision, pp. 1520–1528 (2015)

15. Ren, S., He, K., Girshick, R., Sun, J.: Faster R-CNN: towards real-time object detection with region proposal networks. In: Advances in Neural Information Processing Systems, pp. 91–99 (2015)

16. Russakovsky, O., Deng, J., Su, H., Krause, J., Satheesh, S., Ma, S., Huang, Z., Karpathy, A., Khosla, A., Bernstein, M., et al.: Imagenet large scale visual recognition challenge. Int. J. Comput. Vis. **115**(3), 211–252 (2015)

17. Simonyan, K., Zisserman, A.: Very deep convolutional networks for large-scale image recognition (2014). arXiv preprint arXiv:1409.1556

18. Wu, Z., Shen, C., Van den Hengel, A.: High-performance semantic segmentation using very deep fully convolutional networks (2016). arXiv preprint arXiv:1604.04339

19. Zivkovic, Z.: Improved adaptive Gaussian mixture model for background subtraction. In: Proceedings of 17th International Conference on Pattern Recognition, ICPR 2004, vol. 2, pp. 28–31. IEEE (2004)

A New Feature Extraction in Dorsal Hand Recognition by Chromatic Imaging

Orcan Alpar and Ondrej Krejcar[✉]

Faculty of Informatics and Management, Center for Basic and Applied Research,
University of Hradec Kralove,
Rokitanskeho 62, 500 03 Hradec Kralove, Czech Republic
orcanalpar@hotmail.com, ondrej@krejcar.org

Abstract. Biometric authentication is a trending topic in biometrics that increases the security of personal data. The clients effortlessly authenticate any kind of system, since the controllers, which could be one-time or continuous, stealthily validate the biometric characteristics. The enhancements could be used as the main authentication protocol or implemented as a second security layer. We therefore propose a dorsal hand recognition and validation procedure to increase the security of personal computers. The camera mounted on top a laptop monitor captures the hands on the keyboard and identifies the frames by adaptive chromatic method. Afterwards, new geometrical features are extracted as the key biometric traits to be analyzed in Levenberg-Marquardt based neural controller for validation.

Keywords: Authentication · Biometrics · Dorsal hand identification · Chromatic imaging · Levenberg-Marquardt based neural networks

1 Introduction

Biometrics is a discipline that deals with extraction of human characteristics which are unique and thus hard to mimic. These characteristics could be physical like face, iris, fingerprint, hand geometry and finger veins; or biological like DNA; or habitual like keystrokes and signatures. Given the uniqueness they could also be used as primary or secondary trait for authentication systems since all of them are very hard to imitate. Biometric authentication is a sub-branch of biometrics which enables the users authenticate the systems without any effort. Some of the traits are so useful for controlling, such as; face, iris or hand shape. The key features of these physical characteristics are extracted in training session for further classification. The main difference between the one-time and continuous approaches is the controlling methodology so that one successful login is enough in one-time authentication. On the contrary, continuous authentication systems continuously and stealthily validates the features within a desired time interval which increase the security.

Although the majority of emerging technologies are proposed for smart phones like innovative ear shape recognition [1] or for touchscreens like novel behavioral touchalytics [2], these devices indeed have a conceptual safety and the expected benefit of continuous authentication is limited. Moreover, there are several login interfaces which

N.T. Nguyen et al. (Eds.): ACIIDS 2017, Part II, LNAI 10192, pp. 266–275, 2017.
DOI: 10.1007/978-3-319-54430-4_26

all are easy to use and implement such as pattern passwords. However, personal computers and laptops have the security deficit that it is easier to use an already authenticated device for stealing the personal information.

Given these facts, we introduce a novel feature extraction methodology which could shed light on new authentication systems using chromatic images and Levenberg-Marquardt (LM-ANN) based neural controller. Firstly, we captured frames using a web-cam mounted on a laptop monitor to recognize the hands on the keyboard. Furthermore, the hands were identified and segmented using a version of the adaptive chromatic approach that we have already presented in [3]. Afterwards the geometrical features of the main user's hands are extracted from the binary dorsal hand images for training session. As the classifier, neural classifier trained by LM optimization is used for training and validating the future attempts.

Initially, our methodology starts with dorsal hand recognition while the user's hands are on the keyboard. Although the conventional hand recognition deals with gesture analysis for human-computer interaction and palm print identification, dorsal hand recognition is a trending topic in biometrics such as Zhang et al. proposed three articles regarding dorsal hand recognition with multispectral wavelength bands [4–6]. The primal features extracted in recent researches could be skin color [7] or geometrical characteristics [8]. In this research, we combined the hand recognition by skin color and classified the inputs by geometrical similarities. Regardless of luminance layer Y, we extracted the Cb (blue difference chroma) and Cr (red difference chroma) layers of YCbCr video frames. Since the YCbCr images are the most efficient and rather simple way to identify skin color, numerous researches could be found in the literature such as [9–12]. Kaur and Kranthi [10] use YCbCr imaging with CIE-Lab color for skin color segmentation in face and hand recognition while Chitra and Balakrishnan [11] proposed a similar technique with HSCbCr and YCbCr color spaces. Shen and Wu [12] dealt with segmentation of mouth and lips by YCbCr imaging as well. In particular, we presented the adaptive approach, which we used in this research, to identify dorsal hands on keyboards with YCbCr color space [3].

Even though the majority of geometric analysis on hand recognition is focused on ratios and lengths on palm including fingers like [8], this approach is not feasible for our concern. The main objective to discriminate the hands is still the hand geometry however the users are using the fingers while typing, therefore the lengths of phalange region (fingers) are always changing on the frames. Nonetheless, the metacarpal region (palm) stays the same while using the keyboard even the carpal angles are changing. Therefore, we focused on the intersection points of phalanges and metacarpals to identify the unique lengths and angles between. The intersection points, however, cannot be extracted in analytical plane due to change in lateral wrist angle.

Subsequent to feature extraction from the main user's hand, the angles and lengths are used in training session. We used LM-ANN classifier for differentiating the hands of the main user vs others, as we had promising results in our previous researches [13–16]. The graphical workflow, summarizing the image recognition and geometrical feature extraction, is presented in Fig. 1. The following sections start with the infrastructure of dorsal hand recognition using adaptive chromatic approach. Afterwards, the geometrical feature extraction is presented as well as the training and classification with LM-ANN. Finally this paper concludes with experiments, results and discussion.

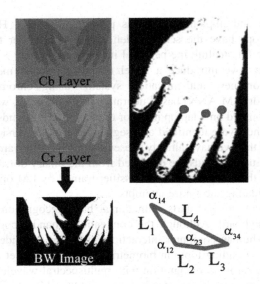

Fig. 1. Graphical workflow (Color figure online)

2 Dorsal Hand Recognition

Dorsal hand recognition algorithm starts with capturing the keyboard image F^1 with 640×480 resolution in RGB format by A4tech PK-333E LED Lighting Web Cam, mounted on Dell Inspiron 15R laptop. Afterwards the images F^2 with the hands on the keyboard are captured for hand segmentation. Each pixel on these frames are mathematically stated as follows: $p^1_{i,j,k} \in F^1_{i,j,k}(i = [1:w], j = [1:h], k = [1:3])$ and $p^2_{i,j,k} \in F^2_{i,j,k}(i = [1:w], j = [1:h], k = [1:3])$, where w is the width, h is the height of the image; i is the row j is the column number and k is the layer of color channel. Skin color is identified by turning these images into grayscale, namely: $g_{i,j} \in G_{i,j}(i = [1:w], j = [1:h]), g_{i,j} \in [0,1]$ and finding the ROI by thresholding

$$g_{i,j} \in G_{i,j} \begin{cases} 1, & \left| p^1_{i,j,k} - p^2_{i,j,k} \right| > T \\ 0, & otherwise \end{cases} \tag{1}$$

where T = 50 for this research since we found it optimal for grayscale images. The ROI is eroded to omit the noise and reach the skin tissue coordinates by $\dot{g}_{i,j} \in E_{i,j} = min(I(i-1,j), I(i,j), I(i+1,j))$ and $\dot{g}_{i,j} \in E_{i,j} = min(I(i,j-1), I(i,j), I(i,j+1))$. Moreover $F^2_{i,j,k}$ is turned to YCbCr Image by: $\tilde{p}^2_{i,j,k} \in F^2_{i,j,k}(i = [1:w], j = [1:h], k = [Y, Cb, Cr])$. Afterwards, the ROI of grayscale is applied to Cb and Cr layers to find the average CbCr values of skin tissue by

$$H_{Cb} = \sum_{i=1}^{640} \sum_{j=1}^{480} \tilde{p}_{i,j,Cb}^2 \cdot \dot{g}_{i,j} / \sum_{i=1}^{640} \sum_{j=1}^{480} \dot{b}_{i,j} \tag{2}$$

$$H_{Cr} = \sum_{i=1}^{640} \sum_{j=1}^{480} \tilde{p}_{i,j,Cr}^2 \cdot \dot{g}_{i,j} / \sum_{i=1}^{640} \sum_{j=1}^{480} \dot{b}_{i,j} \tag{3}$$

The averages H_{Cb} and H_{Cr} are used to define the confidence interval, yet in this research, narrow intervals are not feasible, therefore the intervals are decided as $H_{Cb} \pm 20 - H_{Cr} \pm 20$ for better identification. Each pixel $g_{i,j}$ on the image $G_{i,j}$ are turned to zero is outside of this interval while all pixels inside are turned to one and new image $\hat{B}_{i,j}$ is retrieved by:

$$\hat{b}_{i,j} \in \hat{B}_{i,j} \begin{cases} 1, & H_{Cb} - 20 > \tilde{p}_{i,j,Cb}^2 > H_{Cb} + 20 \\ 0, & otherwise \end{cases} \tag{4}$$

$$\hat{b}_{i,j} \in \hat{B}_{i,j} \begin{cases} 1, & H_{Cr} - 20 > \tilde{p}_{i,j,Cr}^2 > H_{Cr} + 20 \\ 0, & otherwise \end{cases} \tag{5}$$

Consequently the binary/black&white image is achieved to analyze. The details of mathematical infrastructure and the examples of this algorithm could be found in [3].

3   Geometric Feature Extraction

The feature extraction from the hand shape, disregarding the fingers, needs some stable points for identification. While using the keyboard, the wrist is always on the move as well as the fingers, therefore the most stable region seems to be the palm. Given this fact, the points we focused on are the intersection points that are easy to find from the segmented images. However, the segmented images are still noisy to find and extract these point even though we used a very wide interval like in Fig. 2 below. Applying the low-pass filter processed for both edges, with the dilation algorithm below also gives the left hand for feature extraction, namely: $\hat{b}_{i,j} \in \hat{B}_{i,j} = max(I(i-1,j), I(i,j), I(i+1,j))$ $\hat{b}_{i,j} \in \hat{B}_{i,j} = max(I(i,j-1), I(i,j), I(i,j+1))$. Low pass filter works like moving average algorithm for smoothing and thresholding to separate the hands. Firstly the horizontal and vertical pixel values are accumulated to find the contrasts with 250 thresholding since the image is binary and the contrast of black and white pixels is 255.

$$H_i = \left(\sum_{n=2}^{480} |I(i,n) - I(i,n-1)|; if |I(i,n) - I(i,n-1)| > 250 \right) \Big|_{i=1}^{640} \tag{6}$$

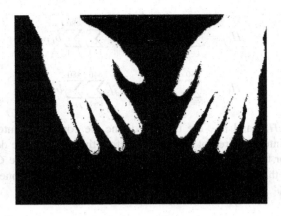

Fig. 2. Recognition process by adaptive chromatic method

$$V_j = \left(\sum_{m=2}^{640} |I(m,j) - I(m-1,j)|; \mathbf{\textit{if}} |I(m,j) - I(m-1,j)| > 250 \right) \Bigg|_{j=1}^{480} \quad (7)$$

Afterwards, the moving averages are calculated for 40 adjacent pixels for both axes, namely:

$$NH_i = \frac{\left(\sum_{i-20}^{i+20} H_i \right) \Big|_{i=21}^{620}}{41}, NV_j = \frac{\left(\sum_{j-20}^{j+20} V_j \right) \Big|_{j=21}^{460}}{41} \quad (8)$$

Although we used the whole frame in segmentation to enable various further analysis, we only concentrated on left hand. Averages are computed and below this

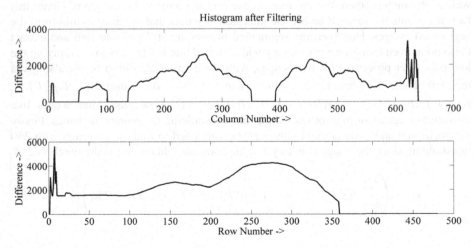

Fig. 3. Low-pass filter with dynamic threshold for horizontal axis on the top; for vertical axis on the right

average values all pixels are turned to zero. The details could be found in [17, 18] The results of this process is shown below in Fig. 3.

After segmentation, we start searching the unique features on the image, which are the intersection points of phalanges and metacarpals. The characteristics of these intersection points on an image are the black pixels surrounded by the white ones, however there still should be black pixels on the bottom. On the contrary, this characteristic cannot be perfectly found on an image therefore we focused on a searching the regions. We wrote a search algorithm to find the characteristics of four intersection points which is a specific square matrix. The unique feature in these 10×10 matrices is the matrix consisting of mathematical inequalities, namely:

$$
S = \begin{bmatrix} 1 & 1 & \ldots & 1 & 1 \\ 1 & & & & 1 \\ 1 & & & & 1 \\ 1 & & & & 1 \\ \ldots & & & & \ldots \\ 1 & & & & 1 \end{bmatrix} \begin{matrix} = 10 \\ < 10 \\ < 10 \\ < 10 \\ < 10 \\ < 10 \end{matrix}
$$
$$
\underset{10}{=} \qquad \underset{10}{=}
$$
(9)

The north-west and east edges should be white therefore the sum of these should be 10, yet there should be black pixels inside therefore the sum of these vectors should be below. Although the algorithm works perfectly, it however brings more than 4 matrices, therefore we extracted the most northwest black pixel. Each matrix brings one black pixel yet the number of extracted black pixel coordinates are as many as the matrices. Finally, we filtered the unique results for the coordinates and obtained the final 4 pixels. The example matrix is shown below in Fig. 4.

Fig. 4. Segmented left hand (on the left) and an example of intersection point extraction (on the right)

4 LM-ANN Classifier

The coordinates extracted in the previous section are the primal inputs in training and testing sessions of LM-ANN classifier yet these coordinates cannot be used as they are. While using the keyboard, the coordinates are changing while fingers and wrists are moving. In addition, the distances cannot be used either due to perspective effect. On the other hand, the angles of the lines between the points should stay same. Considering the four points, they form an irregular quadrilateral like in Fig. 5, yet the area and the edge length could vary. The angles between the edges are calculated with the analytical trigonometry as the main inputs of LM-ANN classifier.

Subsequent to calculation of the angles, these inputs are inserted to LM-ANN system for curve fitting. The LM-ANN actually is a neural network however it is a bit different than traditional weight tuning approach since the weights are adjusted without back-propagation namely; $\Delta w = -[\nabla^2 V(w)]^{-1} \nabla V(w)$ where $\nabla^2 V(w)$ is the Hessian matrix and $\nabla V(w)$ is the gradient. The main purpose is minimizing the errors in

$$V(w) = \sum_{i=1}^{N} e_i^2(w) \tag{10}$$

Therefore, if we substitute the Jacobian matrix as:

$$\nabla V(w) = J^T(w)e(w) \tag{11}$$

$$\nabla^2 V(w) = J^T(w)J(w) + S(w) \tag{12}$$

where:

$$S(w) = \sum_{i=1}^{N} e_i(w)\nabla^2 e_i(w) \tag{13}$$

and $J(w)$ is the Jacobian matrix therefore we obtain: $\Delta w = [J^T(w)J(w) + \lambda I]^{-1} J^T e(w)$ where I is the identity matrix and λ is a multiplier which is adaptive for forcing Hessian

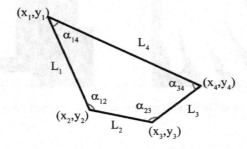

Fig. 5. An example of an irregular quadrilateral formed by four extracted intersection points.

matrix to be positive-definite. We chose LM-ANN since it perfectly fits to array and matrix inputs, which have inter-dependency.

5 Experimental Results

We made the experiments with one real user and a fraud set of four subject, two males and two females. The network is trained by LM for $\lambda = 0.1$ to reach the desired output $y = 1$. The results are presented in following Table 1, with the zoomed images for better presentation, angles and y values of each extracted feature set.

Where all y values represent the output of the angles for each set after running the network for these values. The angles of the training set seem inter-consistent while the

Table 1. Experimental results

Train				
α14	26.29582	29.84463	33.13161	34.47098
α12	153.7876	146.0572	150.181	142.1759
α23	140.8494	146.8458	139.5235	141.7397
α34	39.06715	37.25232	37.16393	41.61339
y	0.9170	0.9694	0.9854	1.0534
Real				
α14	33.04743	25.87313	29.07467	33.66934
α12	138.9097	133.8505	136.4505	142.9663
α23	142.8598	161.2788	151.3374	146.24025
α34	45.18305	38.99764	43.13748	37.12412
y	0.9755	0.8897	0.9315	1.0158
Fraud				
α14	21.64373	37.01606	17.80621	24.96837
α12	167.6783	129.7557	161.8889	139.8536
α23	135.6139	142.107	130.6013	167.0163
α34	35.06408	51.12132	49.70361	28.16173
y	0.6215	1.2344	0.6215	1.0331

values are not identical. Although the classification is not included in this research it is seen that $0.9170 \le y \le 1.0534$ interval could be determined as trusted region. As expected the y values of real set are inside or close, while the outputs of fraud set are discriminated.

6   Conclusions and Discussions

The brief summary of this research is searching for the consistent patterns on hands while a user is typing and extraction of geometrical features for biometric authentication. The first results seem promising and there could be a future of this type of feature extraction when fingers are invisible or images are unreliable. Although we obtained the results that we had expected, there are many assumptions and drawbacks. Above all, the points should form a concave polygon or there should be controller to identify prior to angle estimation. Eroding the images is mandatory for noise reduction however it also causes the key points disappear by enlarging the hands. Moreover, the 10×1 matrix is optimal for this research yet it could be altered depending on the camera angle of view or focus. In addition, all subjects are white, keyboards are not illuminated which actually are some limitations of this research as well. Finally using low-res frames unfortunately causes high sensitivity in angles that alters the angles of real and training set.

As future research, it is necessary to concentrate on less-noisy extraction to omit the possible eroding errors. The angle method has a potential while dealing with the images when the fingers are not fully visible and it is useful when perspective effect will be an issue, therefore it could be used in other types of authentication protocols.

Acknowledgment. This work and the contribution were supported by project "SP/2017 - Smart Solutions for Ubiquitous Computing Environments" from University of Hradec Kralove. We are also grateful for the support of Ph.D. students of our team (Richard Cimler and Jan Trejbal) in consultations regarding application aspects.

References

1. Fahmi, P.A., Kodirov, E., Choi, D.J., Lee, G.S., Azli, A.M.F., Sayeed, S.: Implicit authentication based on ear shape biometrics using smartphone camera during a call. In: 2012 IEEE International Conference on Systems, Man, and Cybernetics (SMC), pp. 2272–2276. IEEE (2012)
2. Frank, M., Biedert, R., Ma, E., Martinovic, I., Song, D.: Touchalytics: on the applicability of touchscreen input as a behavioral biometric for continuous authentication. IEEE Trans. Inf. Forensics Secur. **8**(1), 136–148 (2013)
3. Alpar, O., Krejcar, O.: Dorsal hand recognition through adaptive YCbCr imaging technique. In: Nguyen, N.-T., Manolopoulos, Y., Iliadis, L., Trawiński, B. (eds.) ICCCI 2016. LNCS (LNAI), vol. 9876, pp. 262–270. Springer, Heidelberg (2016). doi:10.1007/978-3-319-45246-3_25

4. Zhang, D., Guo, Z., Gong, Y.: Dorsal hand recognition. In: Zhang, D., Guo, Z., Gong, Y. (eds.) Multispectral Biometrics, pp. 165–186. Springer, Heidelberg (2016)
5. Zhang, D., Guo, Z., Gong, Y.: Comparison of palm and dorsal hand recognition. In: Zhang, D., Guo, Z., Gong, Y. (eds.) Multispectral Biometrics, pp. 207–220. Springer, Heidelberg (2016)
6. Zhang, D., Guo, Z., Gong, Y.: Multiple band selection of multispectral dorsal hand. In: Zhang, D., Guo, Z., Gong, Y. (eds.) Multispectral Biometrics, pp. 187–206. Springer, Heidelberg (2016)
7. Jeong, J., Jang, Y.: Max–min hand cropping method for robust hand region extraction in the image-based hand gesture recognition. Soft. Comput. **19**(4), 815–818 (2015)
8. Ahmad, I., Jan, Z., Shah, I.A., Ahmad, J.: Hand recognition using palm and hand geometry features. Sci. Int. **27**(2), 1177–1181 (2015)
9. Qiu-yu, Z., Jun-chi, L., Mo-yi, Z., Hong-xiang, D., Lu, L.: Hand gesture segmentation method based on YCbCr color space and K-means clustering. Int. J. Sig. Process. Image Process. Pattern Recogn. **8**(5), 105–116 (2015)
10. Kaur, A., Kranthi, B.V.: Comparison between YCbCr color space and CIELab color space for skin color segmentation. IJAIS **3**(4), 30–33 (2012)
11. Chitra, S., Balakrishnan, G.: Comparative study for two color spaces HSCbCr and YCbCr in skin color detection. Appl. Math. Sci. **6**(85), 4229–4238 (2012)
12. Shen, X.G., Wu, W.: An algorithm of lips secondary positioning and feature extraction based on YCbCr color space. In: International Conference on Advances in Mechanical Engineering and Industrial Informatics, pp. 1472–1478. Atlantis Press (2015)
13. Alpar, O.: Intelligent biometric pattern password authentication systems for touchscreens. Exp. Syst. Appl. **42**(17), 6286–6294 (2015)
14. Alpar, O.: Keystroke recognition in user authentication using ANN based RGB histogram technique. Eng. Appl. Artif. Intell. **32**, 213–217 (2014)
15. Alpar, O., Krejcar, O.: Biometric swiping on touchscreens. In: Saeed, K., Homenda, W. (eds.) CISIM 2015. LNCS, vol. 9339, pp. 193–203. Springer, Heidelberg (2015). doi:10. 1007/978-3-319-24369-6_16
16. Alpar, O., Krejcar, O.: Pattern password authentication based on touching location. In: Jackowski, K., Burduk, R., Walkowiak, K., Woźniak, M., Yin, H. (eds.) IDEAL 2015. LNCS, vol. 9375, pp. 395–403. Springer, Heidelberg (2015). doi:10.1007/978-3-319-24834-9_46
17. Alpar, O.: Corona segmentation for nighttime brake light detection. IET Intell. Transp. Syst. **10**(2), 97–105 (2015)
18. Alpar, O., Stojic, R.: Intelligent collision warning using license plate segmentation. J. Intell. Transp. Syst. **20**(6), 487–499 (2015)

Testing the Limits of Detection of the 'Orange Skin' Defect in Furniture Elements with the HOG Features

Leszek J. Chmielewski[1]([✉]), Arkadiusz Orłowski[1], Grzegorz Wieczorek[1], Katarzyna Śmietańska[2], and Jarosław Górski[2]

[1] Faculty of Applied Informatics and Mathematics – WZIM, Warsaw University of Life Sciences – SGGW, ul. Nowoursynowska 159, 02-775 Warsaw, Poland
{leszek_chmielewski,grzegorz_wieczorek}@sggw.pl
http://www.wzim.sggw.pl
[2] Faculty of Wood Technology – WTD, Warsaw University of Life Sciences – SGGW, ul. Nowoursynowska 159, 02-775 Warsaw, Poland
katarzyna_laszewicz@sggw.pl
http://www.wtd.sggw.pl

Abstract. In principle, the *orange skin* surface defect can be successfully detected with the use of a set of relatively simple image processing techniques. To assess the technical possibilities of classifying relatively small surfaces the Histogram of Oriented Gradients (HOG) and the Support Vector Machine were used for two sets of about 400 surface patches in each. Color, grey and binarized images were used in tests. For grey images the worst classification accuracy was 91% and for binarized images it was 99%. For color image the results were generally worse. The experiments have shown that the cell size in the HOG feature extractor should be not more than 4 by 4 pixels which corresponds to 0.12 by 0.12 mm on the object surface.

Keywords: Orange skin · Orange peel · Surface defect · Detection · Quality inspection · Furniture · Histogram of Oriented Gradients (HOG)

1 Introduction

Vision systems make it possible to effectively inspect the machining quality of the selected furniture elements. They provide a possibility of quick and sufficiently precise testing of the correctness of the manufacturing of the product in terms of its relation to the prescribed tolerances [10,11]. One of the most effective ways of improving the quality of furniture elements is the system of correcting the dimensions of the element under treatment [12]. In the production of furniture, the dimensional and shape accuracy are equally important as the visual appearance of the elements, which is the aesthetic aspect of the product.

In our previous papers we assessed the viability of the image processing methods in primary measurements [4,5]. The 3D images from the structured light scanner, suitable for shape measurements, were used. It appeared that a common surface defect called *orange skin* which can emerge in the painted

© Springer International Publishing AG 2017
N.T. Nguyen et al. (Eds.): ACIIDS 2017, Part II, LNAI 10192, pp. 276–286, 2017.
DOI: 10.1007/978-3-319-54430-4_27

surfaces is beyond the range of applicability of such images. In [5] we have shown that this defect should be easy to detect with simple methods.

As we have mentioned in the previous papers, in the domain of furniture elements quality control the image processing methods are very rarely if not at all mentioned in the literature. This is in contrast to the status in the timber industry, where image-based analysis of structural and anatomical defects is a well established technology with much literature (cf. the reviews [3,13]). Usually the reference to defects of furniture are only mentioned alongside with other domains of application, like for example in [8], or the quality is related more to the raw material rather than to the final product, like in [14]. The orange skin as a surface defect is considered in two papers. In [9] the images of *orange peel* (another name for *orange skin*) are generated and the visibility of this defect for humans in various conditions is investigated. The paper [2] deserves particular attention. A system which evokes what the authors call the *defect augmentation phenomena* with a complex, moving system of lighting synchronized with cameras, is used to substantially improve the results of quality inspection of painted surfaces in car industry. The system is aimed at a broad class of surface defects, including *orange peel*. The lighting makes it possible to use local adaptive thresholding as the detection method.

The motivation for this paper is as follows. In [5] we have shown that the *orange skin* surface defect can be successfully detected with the use of a set of relatively simple image processing techniques. We have used the gradient modulus as a single feature of the surface and we have shown that it is possible to attain a very good classification result by properly selecting a threshold on this feature. A 100% classification accuracy was attained for a broad range of thresholds. This result was obtained under an assumption that the entire tested furniture element is treated as the classified object. Therefore, a single sign of defectiveness was enough to reject an element. In this paper we try to do three things. First, we look more carefully at the fragments of the surface and try to classify them separately to see what are the limitations of classification accuracy of surface details, because the result obtained in our cited previous paper seems excessively optimistic. Second, we try to go forward in using the gradient as a feature and we apply one of the most successful while still relatively simple gradient-based feature generating method, namely the *Histogram of Oriented Gradients* (*HOG*) [7]. Third, to go beyond simple thresholding, but mainly because now we have numerous features, we apply the Support Vector Machine (SVM) [1,6] as a classifier. In this way, we show the possibilities and constraints of the method in a more reliable way and we test its limits with respect to classification accuracy of the local state of the object surface. We attempt not to use any special lighting and camera system as in [2] because of several reasons. First, the system of [2] is patented. Second, its complexity and cost would prevent it from being applied in the furniture industry, in which the cost of products is importantly lower than that in the car industry. Third, in our opinion the *orange skin* defect can be reliably detected with standard cameras and lighting.

Our results will not be compared to any reference ones. As already said, to our best knowledge there were no literature reports on automatic detection of *orange skin* in furniture elements up till now.

This paper is organized as follows. In the following Section we shall briefly characterize the defect considered. In the next Section we shall present the way we have prepared the data for finding the features in the images and for teaching and testing the classifier. We shall also give the details on the parameters of the methods we have used. In the following Section we shall report on the results obtained. The paper will be concluded in the final part.

2   The Defect: *Orange Skin*

Orange skin or *orange peel* is a common term used to describe a surface with small, shallow hollows. In the context of furniture manufacturing it is one of the surface defects with emerges during finishing the surface by lacquering. It manifests itself with uneven structure of the hardened surface. There are various reasons for this defect: insufficient quantity or bad quality of dilutent, excessive temperature difference between the lacquer and the surface, bad pressure or distance of spraying, excessive air circulation during spraying or drying, and insufficient air humidity. The surface processed is flat and fully covered with lacquer, so *orange skin* can be treated as the only reason for surface unevenness.

Several observations concerning this defect can be made.

1. The reasons for which a surface is considered good or defective are of esthetic nature. There are treatments and surface types for which the deviations from planarity are considered beneficial.
2. It is practically impossible to find a separate *defect* in the surface. The out-of-plane deviations or simply the small valleys and holes in the surface gradually pass to the regions of even surface. These *good* regions are smaller or larger, but irrespective of their size the whole surface should be classified as *bad*.
3. For a human eye, the smallest regions which can still be recognized as *good* or *bad* span the regions from about 0.5 × 0.5 to 1 × 1 mm.
4. The *good* surface is not free from deviations which can easily be seen in enlarged images. However, these deviations are small enough for the human to classify the whole surface as *good*.

We shall try to take these observation into account while explaining the surface classification method we have used.

3   The Data and the Methods

Images of elements with good surface and with *orange skin* were taken with a good quality, general-purpose color camera, at a resolution of 2.5 M pixels (1288 × 1936 pix). The previous study [5] indicated that the line from the light source to the center of the imaged surface should be far from perpendicular to the surface normal. In our case the light was about 70° from the normal, so the light we have used can be considered as close to tangential. A fragment of the image is shown in Fig. 1a. The resolution at the object surface was close

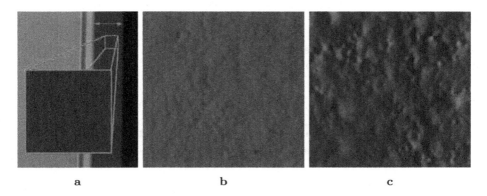

a   b   c

Fig. 1. (a) The way the training images were cut from the original; the length marked by an arrow is about 12 mm and corresponds to 400 pixels. (b) Example of a *good* region. (c) Example of a *bad* region. Contrast and brightness of images **b** and **c** enhanced for better visibility in print

to 0.03 mm per pixel, so the width of the flat surface marked with an arrow corresponding to 12 mm is slightly more than 400 pixels.

From these images, the windows of size 200 × 200 pixels containing good and bad surfaces were cut. The width of the element and the resolution allowed for cutting two windows from the width of the element, not all the windows cut were close to the element boundary. The windows were cut at random. Each window will be treated as one object by the teaching and classification processes. The numbers of these objects will be given later on in Sect. 4.

In the histograms of oriented gradients according to [7] the following parameters were used: cell size 2 × 2, 4 × 4 and 8 × 8 pixels. The remaining parameters were not varied: block size 2 × 2, block overlap 1 × 1 pixels and number of bins in the histograms 8. The classification was performed with the SVM classifier according to [1,6]. The classification accuracy was measured as the cumulated posterior probability: the number of true positive classifications versus the number of all classified objects. The calculations were made in series: for each cell size the features were calculated from the original color image, grey level image, and additionally from the binarized image. Grey images were formed by calculating the brightness as $0.299R + 0.587G + 0.114B$. Binarization was performed with the Otsu method [15] from the grey images.

The teaching and testing were preformed in two settings. In the setting denoted as *separate* the testing objects (that is, image windows) were cut from a different set of images of furniture elements than the teaching objects were. In the setting denoted *common* the teaching and testing objects were cut from the same set of images. Obviously, in each setting, no teaching object was used in the testing, and vice versa.

4 Results

The numbers of objects in the teaching and testing sets in the settings *separate* and *common* are given in Table 1.

Table 1. Numbers of objects in the teaching and testing sets in the two settings used.

Setting	No. teaching objects			No. testing objects			Total
	Good	Bad	Subtotal	Good	Bad	Subtotal	
Separate	142	142	284	44	65	109	393
Common	157	157	314	50	50	100	414

The results of teaching the classifier and testing its accuracy are summarized in Table 2. The cumulated times of the teaching and testing phases were given only for the general information purpose. They reflect rather the teaching time, as the classification time which is more important in the case of an industrial task would be negligible with the use of suitable equipment. The calculations were carried out in Matlab® environment on a four-processor, 3.6 GHz, 64 bit personal computer.

Table 2. Accuracy and time of teaching and testing phases together.

Setting	Cell size	Color		Grey		Binary	
		Accuracy [%]	Time [s]	Accuracy [%]	Time [s]	Accuracy [%]	Time [s]
Separate	2 × 2	**99.17**	13.04	97.25	12.66	**100.00**	9.69
	4 × 4	97.25	4.53	**99.08**	4.04	99.08	3.40
	8 × 8	94.50	2.86	97.25	2.04	97.25	2.10
Common	2 × 2	83.00	13.35	81.00	12.01	**99.00**	9.71
	4 × 4	**90.00**	4.60	**91.00**	3.80	**99.00**	3.46
	8 × 8	**90.00**	3.03	88.00	2.95	98.00	2.40

In Figs. 2 and 3 we give the examples of a good and defective training object, its binary image and the visualization of HOG features for the 4 × 4 pixel size. In Figs. 4 and 5 we give examples of negative classification results. The visualizations of a large number of features is not very informative but by showing them in this way we make an attempt to look more precisely at the classification process. The close analysis of these images indicates that for *bad* objects the HOG features for the binary image exhibit some regions of small and some regions of large values. This is due to the lack of homogeneity of the surface of such objects. *Good* objects are more homogeneous and do not exhibit such a behavior. This can be a candidate phenomenon for explaining the success of the classifier for binary objects.

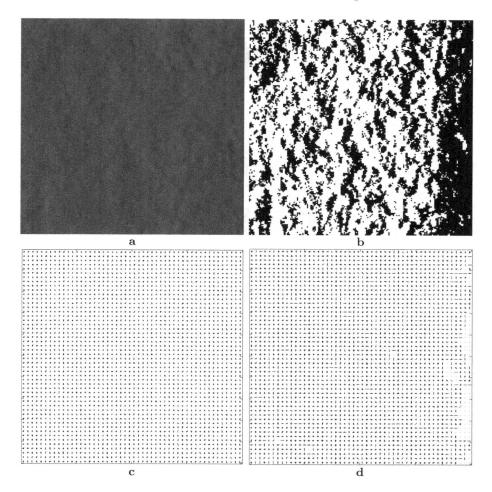

Fig. 2. An example of a *good* training object. (**a**) Grey image; (**b**) binarized **a**; (**c**) visualization of HOG features for **a** in 4 × 4 cell; (**d**) the same for **b**

The resolution of the images indicates that the windows mentioned in observation 3, Sect. 2, corresponds to from 16 × 16 to 32 × 32 pixels. It can be disappointing that in the experiments, the errors were the smallest for windows of lesser dimensions than these, and that the error increased with the increase of the window dimension, which suggests that for larger windows the errors will not be smaller. This indicates that the way the human recognizes the defect differs much from machine classification. The best results were obtained for the cell size not more than 4 × 4 pixels which corresponds to 0.12 by 0.12 mm on the object surface.

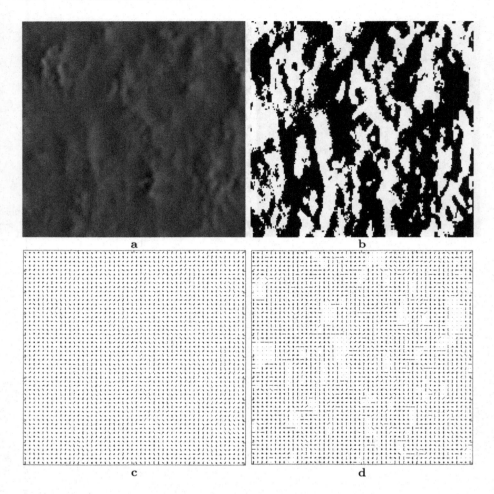

Fig. 3. An example of a *bad* training object. (**a**) Grey image; (**b**) binarized **a**; (**c**) visualization of HOG features for **a** in 4 × 4 cell; (**d**) the same for **b**

It seems reasonable that good results were received for grey images as the color information is not significant in the image we analyzed. However, it is disappointing that the best good results were obtained for binary images. This could indicate that the classical Otsu method appears to extract significant information from the images of our interest. It is also interesting that the results for the testing objects selected from the same physical objects are worse than for those from different objects.

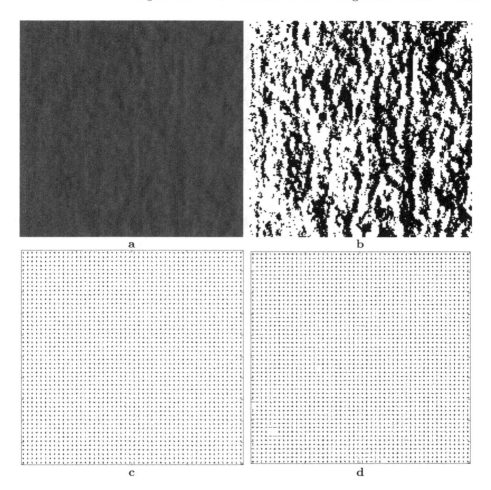

Fig. 4. An example of a false negative error: *good* object classified as *bad*. (**a**) Grey image; (**b**) binarized **a**; (**c**) HOG features for **a** in 4 × 4 cell; (**d**) the same for **b**

The results suggest that more work should be done on choosing the features for classification, including a deeper insight into the processing of the binary images. The differences between the results for the two settings indicate that the results are not stable yet and more attention should be paid to the preparation of the training and testing sets. The possible sources of errors, like the light conditions, and the resolution and focusing of the images, should be taken into account with more care.

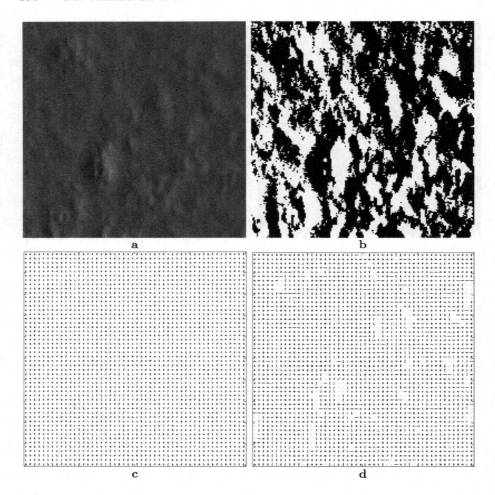

Fig. 5. An example of a false positive error: *bad* object classified as *good*. (a) Grey image; (b) binarized **a**; (c) HOG features for **a** in 4 × 4 cell; (d) the same for **b**

5   Conclusion

In this report of the work in progress it has been demonstrated that the detection of the surface defect of the type *orange skin*, although in principle realizable with relatively simple image processing methods, is a demanding task in general. The well established methods of feature extraction and classification, like the histogram of oriented gradients and the support vector machine, give positive results in the majority of cases, but can fail for a significant number of images, for which a human inspector would make no error. One of the results was that the images binarized with the Otsu method carry the information on the defect of the tested type very well. Consequently, for binary images the classification accuracy attained 99%. This result indicates that the detection of the orange skin defect is tractable with the use of a standard camera and lighting system.

Having this result in mind it can be safely stated that better controlling the lighting and imaging processes and using a broader range of image features should make it possible to attain a technically acceptable level of accuracy in the task considered. The research will be continued in these directions.

References

1. Allwein, E.L., Schapire, R.E., Singer, Y.: Reducing multiclass to binary: a unifying approach for margin classifiers. J. Mach. Learn. Res. **1**, 113–141 (2000). http://www.jmlr.org/papers/v1/allwein00a
2. Armesto, L., Tornero, J., Herraez, A., Asensio, J.: Inspection system based on artificial vision for paint defects detection on cars bodies. In: 2011 IEEE International Conference on Robotics and Automation, pp. 1–4, May 2011. doi:10.1109/ICRA. 2011.5980570
3. Bucur, V.: Techniques for high resolution imaging of wood structure: a review. Measur. Sci. Technol. **14**(12), R91 (2003). doi:10.1088/0957-0233/14/12/R01
4. Chmielewski, L.J., Laszewicz-Śmietańska, K., Mitas, P., Orłowski, A., et al.: Defect detection in furniture elements with the Hough transform applied to 3D data. In: Burduk, R., Jackowski, K., Kurzyński, M., Woźniak, M., Żołnierek, A. (eds.) Proceedings of the 9th International Conference on Computer Recognition Systems CORES 2015. AISC, vol. 403, pp. 631–640. Springer, Cham (2016). doi:10.1007/ 978-3-319-26227-7_59
5. Chmielewski, L.J., Orłowski, A., Śmietańska, K., Górski, J., et al.: Detection of surface defects of type 'orange skin' in furniture elements with conventional image processing methods. In: Huang, F., Sugimoto, A. (eds.) PSIVT 2015. LNCS, vol. 9555, pp. 26–37. Springer, Heidelberg (2016). doi:10.1007/978-3-319-30285-0_3
6. Cortes, C., Vapnik, V.: Support-vector networks. Mach. Learn. **20**(3), 273–297 (1995). doi:10.1007/BF00994018
7. Dalal, N., Triggs, B.: Histograms of oriented gradients for human detection. In: 2005 IEEE Computer Society Conference on Computer Vision and Pattern Recognition (CVPR 2005). vol. 1, pp. 886–893, June 2005. doi:10.1109/CVPR.2005.177
8. Karras, D.A.: Improved defect detection using support vector machines and wavelet feature extraction based on vector quantization and SVD techniques. In: Proceedings of International Joint Conference on Neural Networks, vol. 3, pp. 2322–2327, July 2003. doi:10.1109/IJCNN.2003.1223774
9. Konieczny, J., Meyer, G.: Computer rendering and visual detection of orange peel. J. Coat. Technol. Res. **9**(3), 297–307 (2012). doi:10.1007/s11998-011-9378-2
10. Laszewicz, K., Górski, J.: Control charts as a tool for the management of dimensional accuracy of mechanical wood processing (in Russian). Annals of Warsaw University of Life Sciences-SGGW, Forestry and Wood Technology **65**, 88–92 (2008)
11. Laszewicz, K., Górski, J., Wilkowski, J.: Long-term accuracy of MDF milling process-development of adaptive control system corresponding to progression of tool wear. Eur. J. Wood Wood Prod. **71**(3), 383–385 (2013). doi:10.1007/ s00107-013-0679-2
12. Laszewicz, K., Górski, J., Wilkowski, J., Czarniak, P.: Analysis of dimensional accuracy of milling process. Wood Res. **58**(3), 451–463 (2013)

13. Longuetaud, F., Mothe, F., Kerautret, B., et al.: Automatic knot detection and measurements from X-ray CT images of wood: a review and validation of an improved algorithm on softwood samples. Comput. Electron. Agric. **85**, 77–89 (2012). doi:10.1016/j.compag.2012.03.013
14. Musat, E.C., Salca, E.A., Dinulica, F., et al.: Evaluation of color variability of oak veneers for sorting. BioResources **11**(1), 573–584 (2015). doi:10.15376/biores.11.1. 573-584
15. Otsu, M.: A threshold selection method from gray-level histograms. IEEE Trans. Syst. Man Cybern. **1**, 62–66 (1979). doi:10.1109/TSMC.1979.4310076

Intelligent Data Analysis, Applications and Technologies for Internet of Things

Reaching Safety Vehicular Ad Hoc Network of IoT

Shu-Ching Wang, Shih-Chi Tseng, Shun-Sheng Wang[(✉)],
and Kuo-Qin Yan[(✉)]

Chaoyang University of Technology, Taichung City, Taiwan, ROC
{scwang,sl0314903,sswang,kqyan}@cyut.edu.tw

Abstract. The Internet of Things (IoT) is the interconnection between things in a network. The sensors are embedded into an equipment to support the transformation of sensor information between the carriers by the communication technology, then the smart automatic control and applications can be provided. However, Vehicular Ad hoc NETwork (VANET) is one of the applications of IoT that allowing vehicles within the network to communicate effectively with each another. It is important that VANETs are applied within a safe and reliable network topology. Therefore, in this study, a Reliable VANET Agreement Protocol (RVAP) is proposed. RVAP allows all fault-free nodes to reach safety agreement with minimal rounds of data gatherings, and tolerates the maximal number of allowable components in the VANET.

Keywords: Internet of Things · Vehicular Ad hoc Network · Intelligent Transportation Systems · Fault tolerance

1 Introduction

Due to the development of the Internet of Things (IoT), the smart objects can communicate with each other in the real world. Through the internet and combination of different types of sensors, IoT can promote the development of applications, such as intelligent city, intelligent transportation and smart home. Vehicular Ad Hoc Network (VANET) is one of the applications of IoT that has attracted a growing amount of interest and research in recent years, since they offer enhanced safety and improved travel comfort [8]. VANETs are an emerging type of network which facilitates communication between vehicles on roads, improving driving safety.

VANETs are also known as vehicular sensor networks by which driving safety is enhanced through inter-vehicle communications or communications with roadside infrastructure [5]. They are therefore an important element of Intelligent Transportation Systems (ITSs). VANET's consist of vehicles and roadside equipment that are able to communicate between each other by wireless and multi-hop communication. VANET's are prone to interference and propagation issues, as well as different types of attacks and intrusions that can harm ITS services [9]. These networks characteristically have highly mobile nodes, rapid and significant network topology changes, wireless links subject to interference and fading due to multipath propagation [3]. The absence of central entities increases the complexity of security management operations, in

© Springer International Publishing AG 2017
N.T. Nguyen et al. (Eds.): ACIIDS 2017, Part II, LNAI 10192, pp. 289–298, 2017.
DOI: 10.1007/978-3-319-54430-4_28

particular access control, node authentication and cryptographic key distribution, resulting in vulnerability to misbehaving (malicious or selfish) nodes in the network and posing nontrivial challenges to security design.

It is crucial that a VANET have a safe and reliable network topology that provides a good environment for data transmission, and this has become an important research topic. To achieve high reliability in a VANET, a protocol that allows a set of nodes to reach a common safety agreement, even in the presence of faulty nodes, is needed. Such an agreement problem was first introduced by Pease *et al.* in 1980 [7], and the problem is called the Byzantine Agreement (BA) problem [6].

In the definition of BA, there are n ($n \geq 4$) nodes in a distributed system, and an initial value is set in a source node so that the source node can send its initial data to the other nodes through a reliable network. When the other nodes receive the initial data from the source node, they communicate with each other in order to obtain enough data. One or more of the nodes may fail, so a faulty node may send incorrect data(s) to other nodes. After several rounds of data gathering, a common agreement can be reached among all the fault-free nodes if and only if the maximum number of faulty nodes allowed, t, is smaller than one third of the total number of nodes in the network ($t \leq \lfloor (n-1)/3 \rfloor$). Here, the number of rounds of data gathering is $t + 1$ [6, 7].

In many cases of distributed data processing, common agreement must be reached among all fault-free nodes before executing some special missions. Protocols designed to deal with the BA problem should thus satisfy the following requirements:

(Agreement):   All fault-free nodes in the network must reach the common agreement value;

(Validity):   If the source node is fault-free, then all fault-free nodes in the network should agree on the source node's initial value.

The classical BA problem is considered for a synchronous fixed network in which the bounds on processing and communication delays of fault-free components are fine [4]. In 1985, Fischer *et al.* indicated that agreement in an asynchronous network is impossible if just a single faulty node crashes [4]. However, most VANETs are asynchronous networks. Therefore, the previous results for the BA problem cannot solve the BA problem in asynchronous VANETs, if there is just a single faulty node in the VANET. To cope with asynchrony, Chandra and Toueg proposed a failure detector that can use to detect the non-responsive nodes in asynchronous VANETs [2].

Therefore, to ensure a reliable network, the identity-based cryptosystem (IDCrypto) is used in this study when data are transmitted, since IDCrypto facilitates further design of efficient communication and storage schemes [10]. Based on security and efficiency analyses, the system is shown to satisfy the security objectives, including preserving user privacy and enabling traceability. In this study, the BA problem is re-examined in an asynchronous VANET in order to ensure fault-tolerance and reliable distributed computing in the VANET. The protocol designed to solve the BA problem in a VANET is called the Reliable VANET Agreement Protocol (RVAP). RVAP allows fault-free nodes to reach a common agreement value for solving the BA problem by using the minimum number of data gatherings, and can tolerate a maximum number of tolerable faulty nodes.

2   The Security of a VANET

Safety in VANETs is of special concern because human lives are constantly at stake, whereas in traditional networks the major security concerns include confidentiality, integrity and availability, none of which are primarily involved in the safety of lives. It is therefore crucial that, in a VANET, it is impossible for an attacker to modify or delete vital information [10]. VANET security also includes the ability to determine driver responsibility, while maintaining driver privacy. Information about vehicles and their drivers must be exchanged securely and, more importantly, rapidly, since data delays may result in catastrophic consequences, such vehicle collisions.

The deployment of a comprehensive security system for VANETs is very important. A security breach of a VANET is often critical and hazardous. Moreover, vehicular networks are highly dynamic, with frequent and instantaneous arrivals and departures of vehicles, as well as short connection durations. In addition to its dynamic nature and high mobility, the use of wireless media also makes VANETs vulnerable to attacks that exploit the open and broadcast nature of wireless communication. VANETs are exposed to various threats and attacks [10].

An identity based security system for VANETs was proposed by Sun et al. that can effectively solve the conflicts between privacy and traceability [9]. Figure 1 depicts the entities and their interactions in the identity-based cryptosystem. It is worth noting that vehicle users are further divided into members and access group owners, because only group owners can access Road Side Units (RSUs). The identity-based cryptosystem facilitates further design of efficient communication and storage schemes. Through security and efficiency analysis, the system is shown to satisfy the security objectives, including preserving user privacy and enabling traceability.

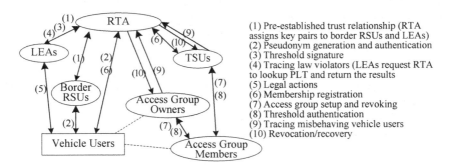

Fig. 1.   Interactions of the identity-based cryptosystem [9].

Since the proposed identity-based cryptosystem proposed by Sun et al. does not require certificates for authentication [9], it is used in this study when a data is transmitted. In this study, a VANET whose nodes are fallible during the BA execution is considered. In this scenario, nodes may be considered faulty due to interference from some noise or a hijacker, and result in exchanged data that can exhibit arbitrary behavior. However, the proposed RVAP is used to address the BA problem with malicious faults in a VANET. When all nodes reach agreement, the fault-tolerance capacity is enhanced, even if there are faults between the nodes.

3   The Proposed Reliable VANET Agreement Protocol (RVAP)

RVAP includes two parts: *data gathering* and *agreement getting*. In order for all fault-free nodes to reach agreement, each node must collect enough exchanged data from all other nodes. The gathered data are stored in a tree structure called the data storage tree (ds-tree), which is similar to that proposed by Bar-Noy *et al.* [1]. In order to make all fault-free nodes in the VANET reach an agreement, each fault-free node will execute RVAP. In the part of *data gathering*, each fault-free node communicates with other nodes and itself via radio waves by identity-based cryptosystem (IDCrypto). Furthermore, each node has high mobility, and the nodes may join into the network or depart from the network at any time. If new nodes join into the network, RVAP will execute the *migration* function to obtain the data from other nodes. Furthermore, if nodes depart from the network, then RVAP will execute the *departure* function to reconstruct the ds-tree. Because of the mobility, the required rounds (a round denotes the interval of a data gathering) of data gathering will not be an inherent value in the beginning, but it is expectable and required that at most $(t + 1)$ rounds of data gathering must be performed to reach a common value, as proposed by Fischer *et al.* at any time [4]. Each node must take care of the total number of nodes in the VANET to decide the required number of rounds of data gatherings. Thus, the protocol will use Required Round (RR) to represent the required rounds of data gatherings. After RRs, the collected data are stored in the ds-tree. In the *agreement getting* part, each fault-free node computes a common value by applying the majority voting function to the data, collected by the *data gathering* and stored on a node's ds-tree to reach an agreement.

In this study, the BA problem is considered in an asynchronous VANET with malicious faulty nodes. The assumptions and parameters of the proposed RVAP are listed as follows:

- There is only one source node that transmits the message at the first round in the BA problem.
- Let f_m be the maximum number of malicious faulty nodes.
- Let f_a be the maximum number of absent nodes.

The discussions of RVAP are given:

(1) **The number of required rounds for RVAP is expectable in a VANET:** Fischer *et al.* proved that $t + 1$ are the necessary and sufficient rounds of data gathering to solve a BA problem where $t = \lfloor (n - 1)/3 \rfloor$ and n is the number of nodes in the underlying network [4]. However, the total number of nodes in the mobile VANET may change at any time, so the number of RRs is not inherent with $(t + 1)$ in the beginning.

(2) **RVAP can tolerate f_m malicious faulty nodes and f_a absent nodes, where $n > 3f_m + f_a$ at any time:** Because each fault-free node can reach a common

agreement value if $n > 3 f_m + f_a$. Therefore, at least $n - \lfloor (n - 1 - f_a)/3 \rfloor - f_a$ nodes are fault-free and have the same agreement value. That is, in the worst case, a return node can receive $n - \lfloor (n - 1 - f_a)/3 \rfloor - f_a$ copies of the same data larger than $\lfloor (n - 1 - f_a)/3 \rfloor$, so a return node can determine the agreement value by the VOTE function.

(3) **The ds-tree and the majority voting function used by RVAP:** Finally, in using the VOTE function to root the value s for each node's ds-tree, a common value is obtained. The definition of the VOTE function is as follows. Since VOTE(s) is a common value, each fault-free node can agree on the value, and the safety agreement is reached.

In executing RVAP, some nodes may join into the network or depart from the network, but each fault-free node predetermines the number of RRs in the *data gathering*, and then collects that number of rounds' data by exchanging each node's received data. Finally, an agreement can be reached if majority voting is applied to the collected data in each fault-free node. In short, the BA protocol makes each fault-free node agree on a common value transmitted by the source node. Therefore, there are two parts in RVAP, they are the *data gathering* and the *agreement getting*. The goal of the

Reliable <u>V</u>ANET <u>A</u>greement <u>P</u>rotocol (RVAP)
RR computing: To determine the total number of RRs ($\delta = \lfloor (n-1)/3 \rfloor + 1$).

Data gathering:
 For $r = 1$ do:
     The source broadcasts its initial value v_s to other nodes and itself using IDCrypto. Each node stores v_s in the root of its ds-tree;
 For $r = 2$ to δ do:
     If a new node joins into the network, the **migration** is executed. If a node departs from the network, the **departure** is executed. The **RR computing** is executed to check the RRs.
 Each node broadcasts the value at level $(r\text{-}1)$th of its ds-tree to other nodes and itself using IDCrypto.
 Each node stores the received values at level r of its ds-tree.

Agreement getting: If a new node has joined into the network, the **migration** is executed   If a node has departed from the network, the **departure** is executed   Delete the repeatable vertices in the ds-tree.   Each node determines a common value using the VOTE function.
Function migration: 1) Each node in the original network sends its data received in the $(r\text{-}1)$th round to the new node using IDCrypto.
2) The new node obtains the majority value from the data received from other nodes.   3) The new node stores the majority value at level $r\text{-}1$ of its ds-tree.   **Function departure:**   Each node deletes the data received from the absent node and reconstructs its ds-tree.

VOTE(α) = val(α), if α is a leaf.
     The majority value in the set of {VOTE(αi)|$1 \leq i \leq n$, and vertex αi is a child of vertex α}, if such a majority value exists.
     A default value φ is chosen, otherwise.

Fig. 2. The RVAP protocol.

data gathering is to collect the data. In the VANET, each node has common knowledge of the entire or partial graphic information of the underlying VANET, and each node can transmit the data to other nodes in the VANET. The goal of the *agreement getting* is to compute a common agreement value for the BA problem. Figure 2 describes the RVAP protocol. Furthermore, each node has mobility in the VANET, and it may join into the network or depart from the network. The procedures in these two cases are as follows:

Case 1: New nodes join into the network. If a new node has joined into the network, the nodes in the original network must send the values in the $(r-1)$th level of their ds-trees to the new node. The new node must obtain the majority value from the values received from other nodes in the network and then continue the *data gathering* or *agreement getting*.

Case 2: Nodes depart from the network. If some nodes have departed from the network, the nodes that are still in the network must delete the data received from the absent nodes.

4 Example of Executing RVAP

There are two cases are described in this section. Figure 3 shows how RVAP enables each fault-free node to reach an agreement in a mobile VANET when some nodes have join intoed the network. Figure 4 shows how RVAP enables each fault-free node to reach an agreement in a mobile VANET when some nodes depart from the network.

In Fig. 3, there are five nodes in the network originally, and RVAP requires RR = $\lfloor (5-1)/3 \rfloor + 1 = 2$ rounds in the *data gathering*. For illustration, let node b (the source) is in malicious fault. Figure 3(a) is the original environment of the VANET. After the *data gathering* is executed, a two-level ds-tree is constructed. During the *agreement getting*, the VOTE function is applied to each fault-free node's ds-tree to compute a common VOTE(s) value. Figure 3(c) shows that a new node f has join intoed this network. Now, each node in the original network must send the value that it received in the first round to the new node, and the new node f must take the majority value of the received values as the value received in the first round (Fig. 3(d)). Figure 3 (e) shows that node a stores $(1,0,1,0,1)$ into its ds-tree. RVAP requires RR $= t+1 = \lfloor (n-1)/3 \rfloor + 1 = \lfloor (6-1)/3 \rfloor + 1 = 2$ rounds, thus, the *data gathering* is stopped at the end of the 2nd round. Figure 3 shows the complete steps to execute RVAP on fault-free node a when some nodes have join intoed the network. The steps for other fault-free nodes are the same as those for node a. The value that a faulty node agrees on is irrelevant.

Furthermore, a node may depart from the network. When some nodes depart from the network, they will affect the agreement of other fault-free nodes. How the nodes reach a common value in this condition is also important. In Fig. 4, there are seven nodes in the original network, and this environment requires RR $= \lfloor (7-1)/3 \rfloor + 1 = 3$ rounds in the *data gathering*. For illustration, let node b (the source) and node g be in malicious fault. The original VANET is shown in Fig. 4(a). In this case, RVAP requires 3 (RR $= t+1 = \lfloor (7-1)/3 \rfloor + 1 = 3$) rounds of data gatherings. Thus, each

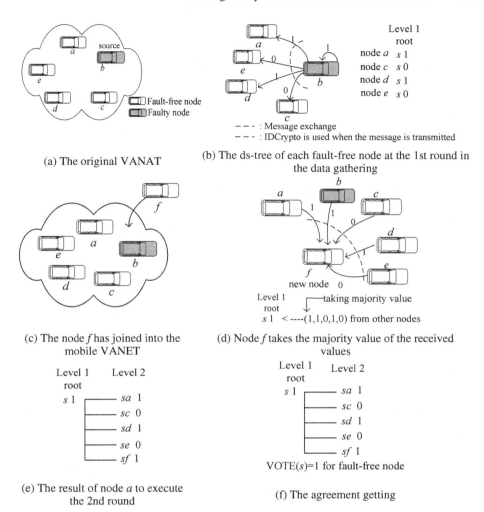

(a) The original VANAT

(b) The ds-tree of each fault-free node at the 1st round in the data gathering

(c) The node f has joined into the mobile VANET

(d) Node f takes the majority value of the received values

(e) The result of node a to execute the 2nd round

(f) The agreement getting

Fig. 3. An example of five nodes to execute RVAP

node will construct a three-level ds-tree at the *data gathering*. During the *agreement getting*, the VOTE function is applied to each fault-free node's ds-tree to compute a common value VOTE(s). Figure. 4(d) shows that node g departs from the network. Other nodes will not receive any data from node g, and they will see node g as dormant faulty. Because node g departs from the network, the total number of nodes in the VANET will be six. At this time, the required rounds of data gathering will change to $\lfloor (6-1)/3 \rfloor + 1 = 2$, so the *data gathering* is stopped at the end of the 2nd round. In order to avoid the effect of a faulty node repeating, the ds-tree is un-repeatable, thus val(sb) is omitted. Figure 4(e) shows that node a stores $(1,0,1,0,1)$ into its ds-tree. Figure 4 shows the complete steps to execute RVAP on fault-free node a when some nodes have left the network.

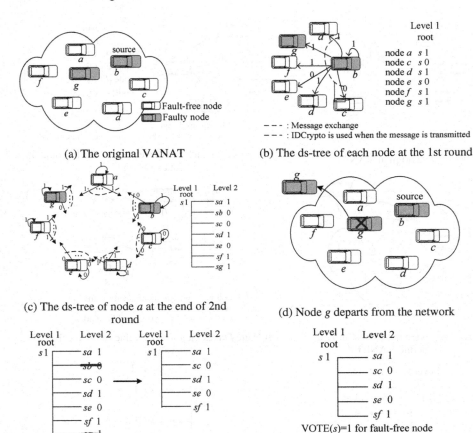

(a) The original VANAT

(b) The ds-tree of each node at the 1st round

– – – : Message exchange
– – – : IDCrypto is used when the message is transmitted

(c) The ds-tree of node *a* at the end of 2nd round

(d) Node *g* departs from the network

(e) The result of node *a* to execute the 2nd round

(f) The agreement getting

Fig. 4. An example of seven nodes to execute RVAP

5   The Correctness and Complexity of RVAP

To prove the correctness and complexity of the proposed protocol, the constraints, (Agreement) and (Validity), can be rewritten as: (Agreement'): Root s is common, and (Validity'): VOTE(s) = v_s for each fault-free node, if the source node is fault-free.

Theorem 1. RVAP can solve the BA problem in a mobile network.

Proof. To prove the theorem, it must be shown that the RVAP algorithm meets agreements (BA$_1$') and (BA$_2$'). (BA$_1$'): Root s is common. The (BA$_1$') is satisfied. (BA$_2$'): VOTE(s) = v_s for all fault-free nodes, if the source is fault-free. If the source is fault-free, then it broadcasts the same initial value v_s to all nodes. The value of correct vertices for all fault-free nodes' ds-tree is v_s. Thus, each correct vertex of the ds-tree is common, and its value is v_s. Since the source is fault-free, the root of the ds-tree is also a correct vertex. And, this root is common. The computed value VOTE(s) = v_s is stored in the root for all fault-free nodes. Thus, (BA$_2$') is satisfied.

Theorem 2. RVAP requires RR (t + 1) data gatherings, and can tolerate f_m malicious faulty nodes and f_a absent nodes, where $t \leq \lfloor (n-1)/3 \rfloor$.

Proof.

(1) Some nodes may join into the network or depart from the network, and the total number of nodes may change at any time. The required rounds must be re-estimated at any time. Although the number of required rounds may be different at any time, it is always expectable. RR (expectable rounds) is used to represent the required rounds of RVAP. Note that RR always follows the result of Fischer *et al.* at any time [4], and RR = $t+1 = \lfloor (n-1)/3 \rfloor + 1$.

(2) RVAP can tolerate f_m malicious faulty nodes and f_a absent nodes at any time, where $n > 3 f_m + f_a$. When the total number of faulty nodes exceeds the limit, then the fault-free nodes cannot reach agreement. Hence, the theorem is proved.

Theorem 3. RVAP solves the BA problem by using the expectable number of data gatherings and it is the minimum.

Proof. Fischer *et al.* noted that $t + 1$ rounds are the minimum number of rounds required to obtain enough data to achieve BA [4]. The unit of Fischer *et al.* is a node, and it is the same with RVAP [4]. Thus, the number of required rounds of data gatherings in RVAP is RR = $t + 1$ rounds at any time and this number is the minimum.

Theorem 4. The number of allowable f_m malicious faulty nodes and f_a absent nodes, where $n > 3 f_m + f_a$ in RVAP is the maximum.

Proof. If the total number of faulty nodes exceeds the limit, then there will not be sufficient data for the fault-free nodes to remove the influence caused by the faulty nodes. This is because each fault-free node can reach a common agreement value if $n > 3 f_m + f_a$. Thus, at least $n - \lfloor (n-1-f_a)/3 \rfloor - f_a$ nodes are fault-free and have the same agreement value. That is, in the worst case, a return node can receive $n - \lfloor (n-1-f_a)/3 \rfloor - f_a$ copies of the same values, which is larger than $\lfloor (n-1-f_a)/3 \rfloor$, so a return node can determine the agreement value by the VOTE function. At this point, the data will still be influenced by the faulty nodes in the *agreement getting*. For instance, there are four nodes in the network (node a, b, c, and d), in which nodes a and b are malicious. Assume that node a is a source. In the first round, node a sends 1 to nodes b and c, and 0 to nodes d and itself. In the second round, node a sends 1 to nodes b, d and itself, and 0 to node c. Node b sends 1 to nodes a, d and itself, and 0 to node c. Node c sends 1 to others and itself. Node d sends 0 to others and itself. In the *agreement getting*, node c will remove the repeatable value and take the majority value with 0 (0,1,0). Node d will take the majority of 1 (1,1,0). Finally, the fault-free nodes will not reach agreement. Therefore, the number of allowable faulty nodes is $\lfloor (n-1)/3 \rfloor$, and it is the maximum.

6 Conclusion

The future Internet, designed as an IoT is foreseen to be "a world-wide network of interconnected objects uniquely addressable, based on standard communication protocols" [12]. Identified by a unique address, any object including computers, sensors, RFID tags or mobile phones will be able to dynamically join the network, collaborate and cooperate efficiently to achieve different tasks.

VANET is one of the IoT applications. Due to the mobility of nodes in a VANET, nodes may join into or depart from the network at any time. Furthermore, some nodes in the network may be fallible, resulting in the network becoming unstable [11]. In this study, RVAP is proposed to solve the BA problem in a mobile VANET. The RVAP ensures that all fault-free nodes in the network can reach a common value to cope with the influences of the faulty nodes by using the minimum number of data gatherings, and tolerate the maximum number of faulty nodes.

References

1. Bar-Noy, A., Dolev, D., Dwork, C., Strong, R.: Shifting gears: changing algorithms on the fly to expedite Byzantine agreement. Inf. Comput. 97(2), 205–233 (1992)
2. Chandra, T., Toueg, S.: Unreliable failure detectors for reliable distributed systems. J. ACM 43(2), 245–267 (1996)
3. Dixit, A., Singh, S., Gupta, K.: Comparative study of P-AODV and improved AODV in VANET. Int. J. Adv. Res. Comput. Sci. Manag. Stud. 3(1), 270–275 (2015)
4. Fischer, M., Lynch, N.: A lower bound for the assure interactive consistency. Inf. Process. Lett. 14(4), 183–186 (1982)
5. Hsieh, Y.L., Wang, K.: Dynamic overlay multicast for live multimedia streaming in urban VANETs. Comput. Netw. 56, 3609–3628 (2012)
6. Lamport, L., Shostak, R., Pease, M.: The Byzantine general problem. ACM Trans. Program. Lang. Syst. 4(3), 382–401 (1982)
7. Pease, M., Shostak, R., Lamport, L.: Reaching agreement in the presence of faults. J. ACM 27(2), 228–234 (1980)
8. Rawat, D.B., Reddy, S., Sharma, N., Shetty, S.: Cloud-assisted dynamic spectrum access for VANET in transportation cyber-physical systems. In: 2014 IEEE International Conference on Performance Computing and Communications, 5–7 December 2010, pp. 1–2. IEEE Press, Texas (2014)
9. Sun, J., Zhang, C., Zhang, Y., Fang, Y.: An identity based security system for user privacy in vehicular ad hoc networks. IEEE Trans. Parallel Distrib. Syst. 21(9), 1227–1239 (2010)
10. Sumra, I.A., Hasbullah, H., Manan, J.A.: VANET security research and development ecosystem. In: National Postgraduate Conference, pp. 1–4. IEEE Press, Malaysia (2011)
11. Wang, S.S., Wang, S.C., Yan, K.Q.: Reaching trusted Byzantine agreement in a cluster-based wireless sensor network. Wireless Pers. Commun. 78(2), 1079–1094 (2014)
12. Internet of Things in 2020: Roadmap for the Future (2008). http://www.smart-systems-integration.org/public/internet-of-things

Short-Term Load Forecasting in Smart Meters with Sliding Window-Based ARIMA Algorithms

Dima Alberg[1] and Mark Last[2(✉)]

[1] Department of Industrial Engineering and Management, SCE - Shamoon College of Engineering, Bialik St., Beer-Sheva, Israel
albergd@gmail.com
[2] Department of Software and Information Systems Engineering, Ben-Gurion University of the Negev, 84105 Beer-Sheva, Israel
mlast@bgu.ac.il

Abstract. Forecasting of electricity consumption for residential and industrial customers is an important task providing intelligence to the smart grid. Accurate forecasting should allow a utility provider to plan the resources as well as to take control actions to balance the supply and the demand of electricity. This paper presents two non - seasonal and two seasonal sliding window-based ARIMA (Auto Regressive Integrated Moving Average) algorithms. These algorithms are developed for short-term forecasting of hourly electricity load. The algorithms integrate non - seasonal and seasonal ARIMA models with the OLIN (Online Information Network) methodology. To evaluate our approach, we use a real hourly consumption data stream recorded by six smart meters during a 16-month period.

Keywords: Internet of Things · Smart grid · Short-term forecasting · Incremental learning · Online Information Network · Sliding window · ARIMA

1 Introduction

Smart grid is a popular application of the Internet of Things (IoT). The smart grid components include smart meters, which are aimed at monitoring and controlling household energy consumption in real time [1]. The massive amounts of measurement data collected by smart meters can be used for customers' load prediction. However, power consumption patterns in both residential and non-residential buildings may change over time due to multiple reasons including variability of human behavior, changes in the number of people populating the buildings at various times, introduction of new electric appliances, etc. Hence, short-term load forecasting (STLF) algorithms should be responsive to these changes by quickly learning new consumption patterns and modifying the forecasting models accordingly. The problem of a gradual "drift" in the target concept is handled by incremental learning systems via forgetting outdated data and adapting to the most recent phenomena [2]. However, traditional ARIMA (Auto Regressive Integrated Moving Average) algorithms, which are commonly used for short-term load forecasting, are lacking such an incremental learning mechanism:

© Springer International Publishing AG 2017
N.T. Nguyen et al. (Eds.): ACIIDS 2017, Part II, LNAI 10192, pp. 299–307, 2017.
DOI: 10.1007/978-3-319-54430-4_29

they learn the parameters of a given ARIMA model only once using a fixed training set and then apply that model to all future incoming data.

In this work, we integrate non-seasonal and seasonal ARIMA modeling with the OLIN (On Line Information Network) incremental learning methodology previously developed by Last [4] for classification tasks in the presence of a concept drift. The proposed Incremental ARIMA system is aimed at continuously processing an infinite stream of incoming data such as a series of load measurements at an hourly or daily resolution. It periodically rebuilds the predictive model using the sliding window approach. We implement two non - seasonal and two seasonal sliding window-based ARIMA algorithms and evaluate them on an hourly consumption data stream recorded by six smart meters during a 16-month period.

2   Related Work

Penya et al. [5] present short-term load forecasting models for non-residential buildings. According to the authors, this special domain presents different characteristics: there is no consumption at night, or it is negligible, and anyway there exists a notable gap between idle and activity times. Another critical aspect is that usually there is scarce (if any) historical data on hourly load and the load profile is sure to vary and evolve over the time. The forecasting results presented for the consumption data from a university campus show that autoregressive models, being computationally simple, accurate, fast, and not requiring any trial-and-error customization or external data (e.g. temperature), are sufficient for providing acceptable prediction accuracy up to six days ahead (MAPE, Mean Absolute Percentage Error, between 5% and 11%). Other evaluated models, including linear and nonlinear regression, a Neural Network, a Support Vector Machine and a Bayesian Network, have provided higher values of MAPE.

Gerwig [3] evaluates five state-of-the-art approaches to short-term load forecasting on three publicly available datasets of power consumption in residential buildings. The following forecasting methods are chosen for evaluation: an autoregressive model (AR), k-nearest neighbor regression (KNN), Decision Trees (DT), Random Forest Regression (RF), and kernel ridge regression (KRR). In addition, two simple benchmarks are used: a persistent forecast (PER), where the predicted values are equal to the last observation, and an averaging method (AVG), where the predicted values are the average of the training data for the specific time of day. Compared to other methods, the autoregressive model and the KNN method achieve the best results (MAPE of about 30% in a 24-h forecast for a single household).

3   Proposed Methodology

3.1   The Incremental ARIMA Paradigm

The proposed paradigm, called "Incremental ARIMA", is presented in Fig. 1. Our Incremental ARIMA system is composed of the following four components:

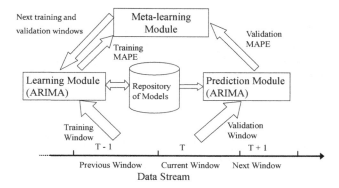

Fig. 1. Incremental learning with ARIMA

- *Learning Module*: it takes as input a sliding training window of a given size (in terms of the number of observations) and calculates the parameters for a given set of ARIMA models (seasonal or non-seasonal) as well as the Training MAPE (Mean Absolute Percentage Error) of each induced model.
- *Repository of Models*: it serves for storing the ARIMA models induced from the latest training window.
- *Prediction Module*: it takes as input a sliding validation window of a given size (a "prediction horizon" such as the next 24 h) and calculates the Validation MAPE (Mean Absolute Percentage Error) for each ARIMA model stored in the Repository of Models.
- *Meta-learning Module*: it takes as input the Training and the Validation MAPE of each ARIMA model and chooses the most accurate model, which has the lowest value of validation MAPE. It also computes the start and the end-points of the next training and validation windows so that both the end-point of a new training window and the start point of a new validation window are set to the end-point of the previous validation window.

3.2   The Evaluated Algorithms

We have used the paradigm described in the previous section to evaluate four incremental non-seasonal and seasonal (S)ARIMA algorithms: Sliding Window Hourly ARIMA Algorithm (SWH2A), Sliding Window Hourly Seasonal ARIMA Algorithm (SWHSA), Sliding Window Daily Profile ARIMA Algorithm (SWDP2A) and Window Daily Profile Seasonal ARIMA Algorithm (SWDPSA). The SWH2A and SWHSA algorithms utilize hourly consumption records in the training window for calculating the parameters of hourly ARIMA and SARIMA models. Those models are applied recursively to each hour in the validation window for predicting the hourly consumption. On the other hand, the SWDP2A and SWDPSA utilize daily consumption records in the training window for calculating the parameters of daily ARIMA and SARIMA models. In addition, the hourly consumption records in the training window are used by SWDP2A and SWDPSA for calculating the 24-h average daily profile of hourly consumption. The daily ARIMA and SARIMA models are applied recursively to each day in the validation window for predicting the overall daily consumption and then combined with the average 24-h profile for predicting the consumption during each hour.

The Training/Validation MAPE (Mean Absolute Percentage Error) of a given forecasting model is calculated on N consecutive load measurements as follows:

$$MAPE = \frac{\sum_{i=1}^{N} \left| \frac{\hat{Y}_i - Y_i}{Y_i} \right| \times 100}{N} \tag{1}$$

Where $\hat{Y}_i$ and Y_i stand for the predicted and the actual load, respectively.

The 24-h load profile is calculated from the hourly data in the training window by applying the following equation to each hour of the day:

$$P_h = \frac{\sum_{d=1}^{D} Y_{dh}}{D} \tag{2}$$

Where P_h is the average hourly load during the hour h ($h \in [1, 24]$), Y_{dh} is the actual load during the hour h of day d, and D is the number of days in the training window. The average daily load P_D can be found by summing up the values of P_h over all hours of the day:

$$P_D = \sum_{h=1}^{24} P_h \tag{3}$$

The Sliding Window Daily Profile (S)ARIMA SWDP2A and SWDPSA algorithms use the following equation to predict the hourly load during the hour h of the day d:

$$\hat{Y}_{dh} = \frac{\hat{Y}_d}{P_D} \times P_h \tag{4}$$

Where $\hat{Y}_d$ is the predicted daily load for the day d calculated by a daily (S)ARIMA models, PD is the average daily load calculated by Eq. 3 above, and Ph is the average hourly load for the hour h calculated by Eq. 2 above.

In our experiments, we have also evaluated two "naïve" models: the "naïve hourly" based on the hourly consumption during the previous day and "naïve daily profile" based on the daily consumption during the previous day and the average daily consumption profile during the training period. The "naïve hourly" model calculates the forecasted hourly load during the hour h of the day d by the following equation:

$$\hat{Y}_{dh} = Y_{d-1,h} \tag{5}$$

Where $Y_{d-1,h}$ is the actual load measured during the same hour h on the previous day $d - 1$. The "naïve daily profile" model calculates the forecasted load during the day d by the following equation:

$$\hat{Y}_d = Y_{d-1} \tag{6}$$

Then it estimates the hourly load during each hour of the day d using Eq. 4 above.

4 Evaluation Experiments

4.1 Design of Experiments

Our evaluation experiments were based on the electricity hourly load data recorded by six Powercom (www.powercom.co.il) meters during a 16-month period between 01/12/2012 and 31/03/2014. The meters were installed in different districts of a major Israeli city. The total number of recorded hourly observations was 61,646. The experiments with four algorithms (SWH2A, SWHSA, SWDP2A and SWDPSA) included nine non-seasonal hourly and nine non-seasonal daily ARIMA models[1], nine seasonal hourly SARIMA models[2] with the period parameter of 24 h, nine seasonal daily SARIMA models (see footnote 2) with the period parameter of 7 days, and two baseline models: the "naïve hourly" model and the "naïve daily profile" model. The experiments with the SWH2A, SWHSA and the "naïve hourly" models included three sizes of the training window (24, 48, and 96 days) and four sizes of the validation window (24, 48, 72, and 96 h). The experiments with the SWDP2A, SWDPSA and the "naïve daily profile" models included three sizes of the training window (24, 48, and 96 days) and three sizes of the validation window (one, two, and three days). The total number of models evaluated with each algorithm was 4 * 38 * 3 * 3 = 540.

We have explored the seasonality behavior of all meters by building the plots of daily and hourly consumption cycles of collected data.

The plots represented in Figs. 2 and 3 exhibit seasonality patterns in daily and hourly profiles, respectively. Figure 2 demonstrates a weak daily data seasonality characterized by electricity load decrease in the last three days of the week (Thursday, Friday and Saturday). This result is obvious because Friday and Saturday are official weekend days in Israel and most organizations and companies do not work and consequently consume less electricity on these days. In contrast, Fig. 3 demonstrates a strong hourly seasonality pattern expressed by the electricity load increasing from 04:00 to 21:00 and decreasing from 21:00 to 04:00, particularly during the workdays (Sunday–Thursday). This result may be explained by many people returning home from work and starting to use the electrical appliances at their homes while at the same time many companies are starting their afternoon shifts.

---

[1] ARIMA models: (000, 001, 100, 101, 010, 011, 111, 221, 222).

[2] SARIMA models: (000, 001, 100, 101, 010, 011, 111, 221, 222) (0, 1, 1).

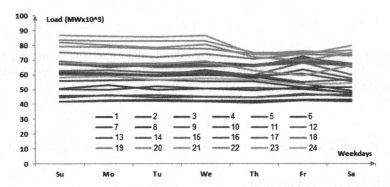

Fig. 2. Meters daily consumption cycle

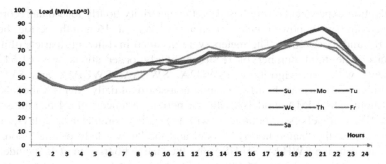

Fig. 3. Meters hourly consumption cycle

4.2   Results

Table 1 compares the average performance of four (S)ARIMA and the "naïve" models across all meters and training/validation window sizes. The Sliding Window Hourly ARIMA (SWH2A) models performed significantly worse than the other models: their average validation MAPE values are about two times higher than the "naïve hourly" MAPE (11.804%). Out of the Sliding Window Daily Profile ARIMA models, the best result (MAPE = 10.05%) is obtained with the non-seasonal SWDP2A ARIMA (101) model. Similarly, out of the Sliding Window Hourly Profile ARIMA models, the best result (MAPE = 9.19%) is obtained with the seasonal SARIMA (001) model.

Table 2 compares the average validation MAPE for each meter across various models and training/validation window sizes. In terms of the average MAPE, the daily profile model SWDP2A strongly outperforms the non - seasonal SWH2A hourly model (10.409% vs. 21.482%), slightly outperforms seasonal SWDPSA hourly model (10.409% vs. 11.451%) and has a similar performance to the SWHSA seasonal hourly model according to the paired sample t-test at the 99% significance level.

Table 1. Comparison of ARIMA models in terms of Avg. MAPE

Model	SWDP2A	SWDPSA	SWH2A	SWHSA	Average
[000]	11.592	11.457	22.826	9.689	**16.454**
[001]	10.909	11.038	21.870	9.191	**15.284**
[010]	10.329	16.708	23.631	11.600	**14.527**
[011]	10.261	11.444	23.715	11.221	**14.280**
[100]	10.363	10.241	21.640	9.246	**14.643**
[101]	10.050	10.218	21.790	9.238	**14.704**
[111]	10.155	15.045	23.864	9.543	**13.910**
[221]	10.393	15.153	22.795	11.652	**13.887**
[222]	10.358	11.331	20.999	10.432	**13.219**
Naïve daily	10.779				
Naïve hourly			11.804		

Table 2. Comparison of meters (hours)

Meter	SWDP2A	SWDPSA	SWH2A	SWHSA	Average
2478	9.627	11.331	20.083	9.740	**13.215**
4364	7.748	7.613	21.743	8.478	**12.891**
4429	14.618	15.675	21.231	15.293	**16.902**
4470	7.877	7.680	16.584	9.139	**11.170**
4740	11.042	11.643	30.980	11.559	**17.260**
5521	10.823	11.020	20.217	9.585	**13.944**
Average	**10.409**	**11.451**	**21.482**	**10.622**	**14.230**

Table 3 also shows the results of one-way ANOVA testing for statistical significance of the difference between the meters. The conclusion of one-way ANOVA is that the difference is not significant implying that we can safely refer to consolidated results of all six meters.

Table 3. ANOVA comparison of meters

Source of variation	SS	df	MS	F	P-value	F crit
Between groups	134.41	5	26.88	0.73	0.61	2.77
Within groups	658.94	18	36.61			
Total	793.34	23				

Table 4 compares the average validation MAPE for different sizes of the training window across various models, meters, and validation window sizes. It shows that in general, increasing the training window size improves the forecasting performance (reduces MAPE), which indicates a relatively stable behavior of most meters during the period of at least 96 days (about three months). However, there was one meter (4429) with the best training window size (providing the lowest value of MAPE) of 48 days

Table 4. Comparison of training window sizes (hours)

Train. Window	SWDP2A	SWDPSA	SWH2A	SWHSA	Average
576	10.729	17.232	22.238	11.642	**15.035**
1152	10.325	15.136	21.533	11.011	**14.573**
2304	10.170	11.163	20.688	9.777	**13.296**
Average	**10.409**	**11.451**	**21.482**	**10.622**	**14.230**

and another meter (5521) with the best training window size of 576 h (24 days) only. Apparently, these two meters were exposed to a faster concept drift than the other four ones.

Table 5 compares the average validation MAPE for different sizes of the validation window across various models, meters, and training window sizes. It shows that on average, extending the prediction horizon reduces the performance of the forecasting models (increases MAPE). Apparently, the rate of MAPE increase is going down for the SWH2A algorithm between the window sizes of 72 and 96 h. These results confirm the common knowledge that it is more difficult to predict a more distant future.

Table 5. Comparison of validation window sizes (hours)

Val. Window	SWDP2A	SWDPSA	SWH2A	SWHSA	Average
24	10.009	11.211	20.533	9.532	**13.052**
48	10.443	11.515	21.334	10.324	**13.665**
72	10.787	11.642	21.957	11.213	**14.177**
96	9.047	11.331	22.103	11.428	**18.036**
Average	**10.409**	**11.451**	**21.482**	**10.622**	**14.230**

Finally, Table 6 shows the best configuration of algorithm, model and training window size across all six meters for each size of the validation window. The conclusion is that the SWHSA algorithm works best for the 24 and 48 h validation window sizes. For the two larger windows (48 and 72 h), SWDPA induces the most accurate forecasting models. The maximum training window size of 96 days (2304 h) is the best one for the first three configurations. In case of the 96 h validation window (the fourth configuration), the size of the training window is 48 days (1152 h).

Table 6. The best configuration for each prediction horizon (hours)

Val. Window	Min. Avg. MAPE	Algorithm	Model	Train. Window
24	9.532	SWHSA	SARIMA(1, 0, 1)	2304
48	10.324	SWHSA	SARIMA(0, 0, 1)	2304
72	10.787	SWDP2A	ARIMA(1, 0, 1)	2304
96	9.047	SWDP2A	ARIMA(0, 1, 1)	1152

5   Discussion and Conclusions

The main contribution of this paper is the introduction of sliding window-based forecasting algorithms (SWDP2A, SWDPSA, SWH2A, SWHSA and SWHSA) for electricity load prediction in smart meters. These algorithms integrate non-seasonal and seasonal time series (S)ARIMA models with the OLIN (On Line Information Network) incremental learning methodology. The main difference between the presented algorithms concludes in seasonality adjustment and the model construction phase. The non - seasonal SWH2A and seasonal SWHSA algorithms utilize hourly consumption records in the training window, whereas non - seasonal SWDP2A and seasonal SWDPSA algorithms utilize aggregated daily consumptions and average daily profiles of hourly consumptions in order to obtain the parameters of induced (S)ARIMA models.

The experimental dataset was recorded online by state-of-the-art smart metering technology and after thorough preprocessing, was approved for use in the corresponding research experiments. The conducted experiments showed that the SWDP2A algorithm outperforms the SW2SA algorithm, performs similarly to the seasonal SWHSA algorithm and has more stable MAPE performance than the SWHSA algorithm. This remarkable contribution indicates that the hourly prediction task does not require collecting massive hourly data in the training phase of model induction. It is sufficient to use daily consumption data and aggregated hourly coefficients of daily profiles for obtaining accurate hourly predictions of electricity load.

Acknowledgments. This work was partially supported by the Israel Smart Grid (ISG) Consortium under the MAGNET Program, in the office of the Chief Scientist of the Ministry of Economics in Israel.

References

1. Depuru, S.S., Wang, L., Devabhaktuni, V.: Smart meters for power grid: challenges, issues, advantages and status. Renew. Sustain. Energy Rev. **15**(6), 2736–2742 (2011)
2. Gama, J.: Knowledge Discovery from Data Streams. CRC Press, Boca Raton (2010)
3. Gerwig, C.: Short term load forecasting for residential buildings. In: Hu, J., Leung, V.C.M., Yang, K., Zhang, Y., Gao, J., Yang, S. (eds.) SmartGIFT 2016. LNICSSITE, vol. 175, pp. 69–78. Springer, Heidelberg (2017). doi:10.1007/978-3-319-47729-9_8
4. Last, M.: Online classification of nonstationary data streams. Intell. Data Anal. **6**(2), 129–147 (2002)
5. Penya, Y., Borges, C., Fernandez, I.: Short-term load forecasting in non-residential buildings. In: AFRICON 2011, vol. 9, no. 3, pp. 1–6. Livingstone (2011)

Bus Drivers Fatigue Measurement Based on Monopolar EEG

Chin-Ling Chen[1], Chong-Yan Liao[1], Rung-Ching Chen[1(✉)],
Yung-Wen Tang[2], and Tzay-Farn Shih[1]

[1] Chaoyang University of Technology, Taichung, Taiwan
{clc, s10514901, crching, tfshih}@cyut.edu.tw
[2] Chung Shan Medical University, Taichung, Taiwan
tangyw@csmu.edu.tw

Abstract. When people are tired, their conscious activities are sluggish and slow brainwaves predominant in their brains. More specifically speaking, the health of a human body may be affected, and chances of accident may arise due to the lowering of ones' alertness as a result of the reduction of activity in cerebral cortex. In this study, by means of a portable electroencephalograph machine and brain-computer interface, the brainwave of bus drivers were measured for their fatigue quality. The results indicated that through the empirical formula proposed, we were able to measure people's fatigue state with a result comparable to sophisticated medical instrument.

Keywords: Brainwave · Fatigue · Brain-computer interface · EEG

1 Introduction

The brainwave signals were collected by the wireless monopolar EEG technique. By means of ThinkGear chips, brainwave signals were filtered, amplified and then transmitted through Bluetooth pathway to the software of PC terminal. Four types of brainwave signals, i.e., α, β, θ and δ were calculated statistically by the eSense algorithm [1] to distinguish the characteristics of concentration and relaxation. Some of the products have the effects of reducing stress, relieving tension and eliminating fatigue on the market. However, these products cannot get to the point of users' subjective symptoms. Instead, their effects need to be quantified by objective and accurate scientific measurements. Electroencephalogram (EEG) is designed to record the continuous and irregular electric potential on the surface of epicranium. The brainwave pictures reflect the fluctuation of a person's conscious activity, which can be used for the evaluation of spiritual status. In this study, a formula was designed to calculate the fatigue index based on the analysis of brainwave data. One of the common phenomena encountered in modern society is the brain fatigue caused by over-pressure among white collar workers or overburden of school work among students. In this competitive world, cases of overwork and over-stress are profoundly seen. Not only one's health is affected but also fatigue causes the rising of accident events. It is reported that among traffic accidents caused by fatigue drivers, 40% of the drivers are professional ones [1]. This study attempted the analysis of the data obtained from

© Springer International Publishing AG 2017
N.T. Nguyen et al. (Eds.): ACIIDS 2017, Part II, LNAI 10192, pp. 308–317, 2017.
DOI: 10.1007/978-3-319-54430-4_30

brainwave measurement with the hope to find some characteristic values for evaluating the spiritual status of human body.

2 Proposed Scheme

In this experiment, some long-term professional drivers are invited to participate in the test. The procedures are as following:

1. He age of all personnel selected (the participants) to do the test were in the range between 20 to 50 years old.
2. The participants were required to fill up a fatigue questionnaire which contained seven subjects regarding to their last month's sleep profile. The questionnaire was scored according to a proposed scheme.
3. The participants were asked not to drink alcoholic drinks before the test. Drinking of reinvigorating beverage or caffeinated beverage must be prior to noon time. Intense exercise should be avoided 5 h before the test and everyone kept daily schedule. Person who failed to follow the above restrictions were postponed his brainwave test to avoid the interference of the accuracy of the measurement. Participants then were asked to wear the Mind Wave Mobile and underwent a 20-min brainwave measurement. The fatigue indexing points were then calculated by the algorithm method. The scores of the questionnaire (fatigue point) and brainwave measurement (brain fatigue index) were recorded manually.
4. After all participants finished their tests, checked to see if positive correlation existed in the results between brainwave measurement and fatigue questionnaire. Data were analyzed by Pearson Correlation (or coefficient of product-moment) method. The flow chart of the procedure is shown in Fig. 1.

Experimental Environment. Participants could take the fatigue test at any place or environment in which they were familiar with, such as in a drivers' lounge or lobby. They were asked to relax themselves to avoid the interference of signals due to nerve or intensity. The hardware used includes Mind Wave.

Brainwave Measurement Instrument - Mind Wave Mobile Instrument used for the measurements of brainwave include electroencephalography (EEG), functional Magnetic Resonance Imaging (fMRI), Positron Emission Tomography (PET), and magnetoencephalography, (MEG). Both fMRI and PET associating with the blood circulation, which requires prolong observation time for the analysis of brain data, are not suitable for real-time evaluation. MEG requires a room to protect magnetic interference [2]. EEG is non-intrusive, non-radioactive, capable of doing both real-time analysis and long-term observation, which makes it suitable for conducting research.

The EEG machine that we used in this study is the Mind Wave Mobile model developed by Neuro Sky Company [3] with very high precision [4]. This brainwave detector collects the bio-electronic signals generated from human brain by dry-state electrode transducer. The collected signals are then processed by ThinkGearTM chips of Neuro Sky. Brainwaves including α, β, θ and δ waves can then be recognized.

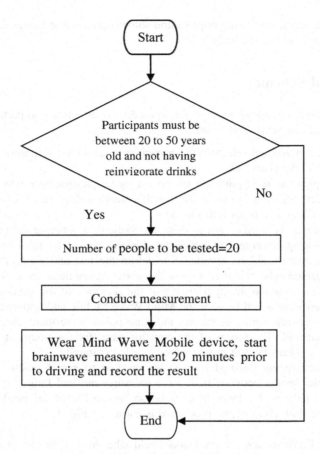

Fig. 1. The flow chart of measuring procedure

The sensor attached to epicranium and forehead can detect the bio-electronic patterns and frequency generated from human brain neuron. The position of the sensor is at Fp1 follows the international 10–20 system suggested by Jasper [5]. The reference point is at A1 as shown in Fig. 2.

In order to assure the reliability and feasibility of the Mind Wave Mobile measurements, Neuro Sky practiced a test on Birdwatching training game developed by a US online brain training company, the Lumosity and the results are satisfactory. Hence, Mind Wave mobile model was taken as measuring instrument for this study [6].

Algorithm.

1. The eSense concentration index denotes the extent of concentration level or attention level which a user pays his attention to. For example, when the user is in a state of high concentration level and able to stably control his psychological activity, the value of his concentration index will be high. When a user is distraught, absentminded, unfocused and anxious, his concentration index will be lowered. The

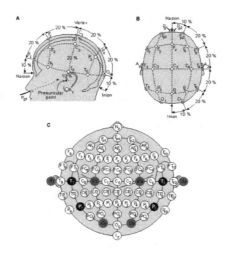

Fig. 2. The international brainwave allocation map for electrodes.

relaxation index of eSense represents the peaceful level or relaxation level of a user's spirit. It is noted that the relaxation index reflects a user's spiritual status instead of his physical status [7]. Table 1 listed all variables contained in this experiment and the meaning these variables.

2. The frequency that Mind Wave Mobile brainwave machine collects quantified brainwave value is one second per time. The values of Alpha, Beta and Theta waves measured each second were recorded and summed for the calculation through the operation of algorithm.

Among numerous algorithms proposed for the calculation of fatigue index, algorithms proposed by Larue et al. [8] and Xiaoli et al. [9] were adopted in this study. Through many practical tests, we found that taking fatigue index as Fatigue = (Alpha + Theta)/Beta in algorithm yielded most reliable data. However, the frequency ranges for each brainwave form mentioned in [8, 9] are: Alpha: 8–13 Hz; Beta:

Table 1. Variables and their meaning

Variables	Meaning
Low Alpha [k]	Quantified LowAlpha value read per second by brainwave machine
High Alpha [k]	Quantified HighAlpha value read per second by brainwave machine
Theta [k]	Quantified Theta value read per second by brainwave machine
Low Beta [k]	Quantified LowBeta value read per second by brainwave machine
High Beta [k]	Quantified HighBeta value read per second by brainwave machine
sum_Alpha	Summation of Alpha values
sum_Theta	Summation of Theta values
sum_Beta	Summation of Beta values
Fatigue	The fatigue index measured in this study

13–30 Hz; Delta: 0.5–4 Hz; Theta: 4–7 Hz. On the other hand, the frequency ranges adopted in Mind wave Mobile brainwave machine are [10]: Low Alpha: 7.5–9.25 Hz; High Alpha: 10–11.75 Hz; Low Beta: 13–16.75 Hz; High Beta: 18–29.75 Hz; Delta: 0.5–2.75 Hz; Theta: 3.5–6.75 Hz.

Therefore, in the calculation of Fatigue = (Alpha + Theta)/Beta, the frequency range of the Alpha wave must combine the frequency range of both Low Alpha and High Alpha. Likewise, the frequency range of Beta wave needs to combine the frequency range of both Low Beta and High Beta. Considering 5 conditions required for the algorithm, i.e., input, output, well-defined, finite, and efficient [11], the following formulae (1)–(3) are used:

$$sum\_Alpha = \sum_{k=1}^{n} (LowAlpha[k] + HighAlpha[k]) \tag{1}$$

$$sum\_Theta = \sum_{k=1}^{n} (Theta[k]) \tag{2}$$

$$sum\_Beta = \sum_{k=1}^{n} (LowBeta[k] + HighBeta[k]) \tag{3}$$

After many experimenting on Mind wave Mobile machine, we found formula (4) yields most accurate data for fatigue index.

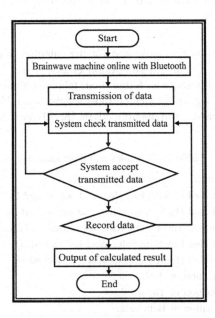

Fig. 3. Portable brainwave machine rapid detection of fatigue index

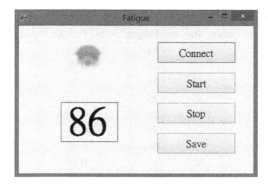

Fig. 4. Rapid detection of fatigue index by portable brainwave machine

$$Fatigue = \frac{Sum\_Alpha + Sum\_Theta}{Sum\_Beta} \tag{4}$$

Once the brainwave signals were collected, the quality of data would be determined. A small value in PoorSignal signified that the quality of the signal was acceptable. The data would then be stored as GetEGG(e). The algorism for the detection of fatigue index by the Mind Wave Mobile brainwave machine is shown in Fig. 3.

Press "save" button, and run Calc(), the system would sum up the brainwave data recorded in the test and calculated the fatigue index according to the algorism, as shown in Fig. 4.

3 Analyses

The participants were all professional bus drivers. They were proven qualified for taking the test after our preliminary screen process. Total number of participants was 20. They are 12 male and 8 female. Experimental data are for the Correlation between Fatigue Point and (Brain) Fatigue Index.

By means of the wireless brainwave machine, the fatigue indices of the participants could be rapidly determined. The brainwave signals collected were filtered and amplified and then transmitted by Bluetooth to personal computer. The results were shown by brain-computer interface after suitable processing procedure. Four brainwaves, α, β, θ, δ were analyzed in this study by the patented eSense™ algorithm developed by Neuro Sky Company to calculate the eSense™ parameters that reflected the participants in situ spiritual status. Two characters were recognized, i.e., concentration level and relaxation level [12]. These two characters are associated with Low α, High α, Low β and High β brainwaves. The result on the fatigue index acquired from the formulae is highly corresponded with that from the subjective questionnaire. We observed that younger drivers' fatigue point fell on the mean value, but their brain fatigue indices were higher than average due to their daily routine behavior, for instance, stayed up before taking the test (Table 2).

Table 2. Correlation between fatigue point and brain fatigue index

Participants	Stay up	Fatigue point	Fatigue (brain)
01	O	54	0.79
02	O	96	0.86
03	O	91	0.77
04	X	25	0.52
05	X	29	0.53
06	O	88	0.73
07	O	66	0.70
08	X	17	0.48
09	X	75	0.66
10	X	50	0.55
11	X	21	0.51
12	O	38	0.60
13	O	42	0.61
14	X	13	0.47
15	O	71	0.83
16	O	83	0.85
17	O	79	0.80
18	X	8	0.47
19	O	63	0.62

Note: O: denotes that the participant experienced stay-up within 3 days prior to taking the test; X: no staying up

Our analysis takes into consideration the following factors: (1) the spiritual status of the participants when they were on the road; (2) subjective judgment by the participants and objective observation by the machine; (3) Three days are prior to taking the test that whether the participants are experienced staying up or not? After Pearson correlation analysis, the fatigue points as obtained from the questionnaire of the participants and the fatigue index as obtained from the measurement show a high degree of correlation (Tables 3 and 4).

In Table 3, two of the participants who show high fatigue point has high correlation coefficient of 0.910***(obvious). In Table 4, nine participants who did not stay up

Table 3. Pearson correlation analysis (fatigue)

		Fatigue point	Fatigue
Fatigue point	Pearson correlation	1	0.910**
	Sig. (2-tailed)		0.000
	N	20	20
Fatigue	Pearson correlation	0.910**	1
	Sig. (2-tailed)	0.000	
	N	20	20

Table 4. Pearson correlation analysis (stay up)

Stay up all night			Fatigue point	Fatigue
0 did not stay up	Fatigue index point	Pearson correlation	1	0.976**
		Sig.(2-tailed)		0.000
		N	9	9
	Fatigue	Pearson correlation	0.976**	1
		Sig.(2-tailed)	0.000	
		N	9	9
1 did stay up	Fatigue index point	Pearson correlation	1	0.735**
		Sig.(2-tailed)		0.010
		N	11	11
	Fatigue	Pearson correlation	0.735**	1
		Sig.(2-tailed)	0.010	
		N	11	11

**Correlation is significant at the 0.01 level (2-tailed)

before the test show high correlation coefficient of 0.976****(obvious) while those who did stay up (11 participants) have correlation coefficient of 0.735****(obvious), indicating positive correlation between fatigue point and brain fatigue index.

The correlation between fatigue points and brain fatigue index is shown in Fig. 5. It is seen that when the participants recognize themselves as been tired, their fatigue index tend to be high.

In the algorism, the fatigue index was calculated by the quantified brainwaves Alpha, Theta and Beta. We found that the participants who have experienced staying up 3 days prior to the test have relatively high value in their Alpha and Theta brainwave readings. Eight of them have high fatigue point as well as high brain fatigue index. Those who did not stay up prior to the test yield a relatively low fatigue point and low brain fatigue index as well.

The calculated fatigue index reflects better the actual status of the participants as seen in the case of 3 participants who experienced staying up. Even though they filed

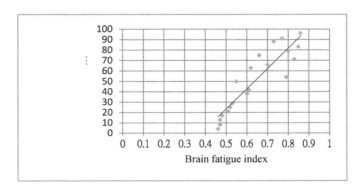

Fig. 5. Correlation between fatigue points and (brain) fatigue index

the questionnaire which yielded low fatigue point, their spiritual status was actually not good according to their high brain fatigue indices. Traffic accidents might have been resulted from drivers who considered themselves in a good spiritual status, which may be a deadly mistake. The fatigue index serves a good reminder for the professional drivers regarding to how their spiritual status is before they drive. It is wise to take proper rest to avoid the traffic accident when the brain fatigue index is high.

4 Discussions

The purpose of this study is to propose an objective number that can actually reflect the brainwave data and can also correlate with subjective information, which simplifies the process of quantitative measurement. In the past, the conventional way of measuring fatigue index is based on the questionnaire filed by the person to be tested. The result may be very subjective because everyone has his own criteria of cognition as to the status of fatigue. Therefore, even with the same working hours, some other variables may cause the fluctuation in reported fatigue point as seen in the questionnaire filed by 20 participants in this study. In order to compensate for the shortcoming of the sub-jective measurement, we propose in this research an objective measuring mechanism. The contribution of this study as follows.

In the past, when taking a fatigue test, the person to be tested would give a subjective and biased report on the spiritual status and degree of fatigue, based on variables such as age, gender and habit in daily life. It is likely that for younger, stronger people, even though they don't feel physically tired after a few nights of sleeplessness, their brains actually are in a state of fatigue. This research offers a reliable and objective brain fatigue index for people to better understand their spiritual status.

The questionnaire in this research contains topic of habit in daily life, so that the correlation between brain fatigue and daily life can be explored. According to brain fatigue index, although a person who has experienced staying up considers himself competent in driving, his brain is actually in a fatigue state and may not be save to drive. From this study, a person's recognition of fatigue may differ from the fatigue index measured by machine, and the degree of difference varies from one to another (Table 3).

The results based on participants' subjective judgment and objective observation from instrument indicated that even the participants' fatigue points were low, their brain fatigue index might not be low and have higher concentration level. Long-term habit in daily life may also have some effects on the brainwave index. Even if a person feels physically energetic, brainwave measurement shows that α wave relatively higher than β wave. In fact, that reminds the one of his brain fatigue index may be higher.

Acknowledgments. This research was supported by the Ministry of Science and Technology, Taiwan, R.O.C., under contract number MOST 104-2221-E-324-012 and MOST103-2632-E-324-001-MY3.

References

1. Jiang, S.S., Siao, G.L., Wu, Z.W.: Based on EEG characteristics attention diagnosis and training system. National Taichung University of Science and Technolog, MIS (2013)
2. Chang, F.Y.: Fourier transformation on the brain machine interface cursor control. National University of Tainan, Department of Information and Learning Technology (2007)
3. NeuroSky Official Website. http://neurosky.com/zh-Hant/. Accessed 20 Dec 2015
4. Chi, M.L., Wang, H.S., Shen, H.: The study of architecture design about multi-player brain computer interface game system. National Taichung University of Education, Department of Digital Content and Technology (2010)
5. Jasper, H.H.: The ten-twenty electrode systems of the international federation. Electroencephalogr. Clin. Neurophysiol. **10**, 371–375 (1958)
6. Lin, Y.J.: Assessing the effects of different text presentation type on attention, reading comprehension and cognitive load for mobile E-reading. National Chengchi University Graduate Institute of Library, Information and Archival Studies (2011)
7. TELDAP e-Newsletter. http://newsletter.teldap.tw/news/InsightReportContent.php?nid=6121&lid=706. Accessed 5 Jan 2016
8. Larue, G.S., Rakotonirainy, A., Anthony, N.P.: Predicting driver's hypovigilance on monotonous roads literature review. In: 1st International Conference on Driver Distraction and Inattention, Gothenburg, Sweden (2010)
9. Xiaoli, F., Qianxiang, Z., Zhongqi, L., Fang, X.: Electroen cephalogram assessment of mental fatigue in visual search. Biomed. Mater. Eng. **26**, S1455–S1463 (2015)
10. Neuro Sky Official Website. http://developer.neurosky.com/docs/doku.php?id=thinkgear_communications_protocol. Accessed 3 Jan 2016
11. Wikipedia. http://zh.wikipedia.org/zh-tw/%E7%AE%97%E6%B3%95. Accessed 3 Jan 2016
12. NeuroSky Official Website. http://www.neurosky.com.tw/products-markets/eeg-biosensors/algorithms/. Accessed 3 Jan 2016

An Improved Sleep Posture Recognition Based on Force Sensing Resistors

Yung-Fa Huang[1](✉), Yi-Hsiang Hsu[1], Chia-Chi Chang[1],
Shing-Hong Liu[2](✉), Ching-Chuan Wei[1], Tsung-Yu Yao[1],
and Chuan-Bi Lin[1](✉)

[1] Department of Information and Communication Engineering,
Chaoyang University of Technology, Taichung 41349, Taiwan (R.O.C.)
yfahuang@mail.cyut.edu.tw, {ccwei, cblin}@cyut.edu.tw
[2] Department of Computer Science and Information Engineering,
Chaoyang University of Technology, Taichung 41349, Taiwan (R.O.C.)
shliu@cyut.edu.tw

Abstract. In this paper, we applied six force sensing-resistor sensors (FSR Sensors) to perform sleep posture recognition. The analog-to-digital converter (ADC) is used to extract the resistance signals of FSRs. The recorded FSR signals are averaged as reference pattern of six values. The reference patterns and test patterns of the postures are performed pattern matching with the mean squared error (MSE) method. With a scale adjusting method, the recognition accuracy is obtained by 87%. Moreover, after the moving average windows are adopted to remove the high ripple, the recognition accuracy can be improved to 96% with window length $L = 7$.

Keywords: Force sensing-resistor sensor · Pattern matching · Moving average window · Sleep posture · Posture recognition

1 Introduction

The information communication technology and micro-controller technology are rapidly developing in recent years. Thus the smart home technologies become an important topic [1]. The smart home topics are emerging from home automation. The smart home includes the topics on Security Monitoring [2–4], Family Entertainment [3], Smart Appliance [5], and Home Care [2]. Moreover, in the smart home environment, the quality of life is important for human. For the persons, they can judge the quality of their current physical condition with the body sense organs. However, sleep time cost of living one-third, process quality through polysomnography of sleep are some physiological parameters in [6, 7].

To analyze user's physiological status, the sensors is put on the pillow to receive data [8]. This study uses force sensing-resistor sensors (FSR Sensor) placed on the pillow, to detect user gestures on the bed. The detection of sleep postures is very important for every day. Moreover, the gesture recognition study based on pattern matching had been investigated in previous works [9, 10]. Therefore, in this paper, the

© Springer International Publishing AG 2017
N.T. Nguyen et al. (Eds.): ACIIDS 2017, Part II, LNAI 10192, pp. 318–327, 2017.
DOI: 10.1007/978-3-319-54430-4_31

values of FSR sensors are collected, the users' current sleep postures are further investigated by pattern recognition.

2 System Model

In this research, the non-contact type signal acquisition method is adopted as shown in Fig. 1. In the signal acquisition, the testing user lies with the head on the pillow. Then the Arduino Yún interface board convert the sensing values of FSR to resistance values. The interface circuit is shown as Fig. 2. The nth input data for the mth posture can be expressed by

$$y_i^m(n) = R\left(\frac{1023}{s_i^m(n)} - 1\right) \tag{1}$$

where $R = 1$ KΩ is the input resistor, $s_i^m(n)$ is the nth sensing signal of the ith FSR.

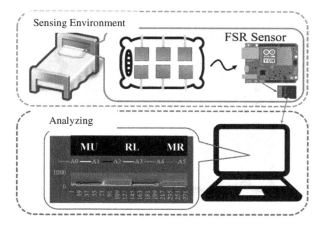

Fig. 1. System structure for sleep posture recognition based on FSRs.

In the experiments, a new pillow for sleep posture recognition where six FSR sensors were placed on, is shown in Fig. 3. The six sensors are respectively denoted by S1, S2, S3, S4, S5 and S6. The sensing values of these FSR sensors varies due to the users' sleep posture on the pillow. With the interface board Arduino and analog-digital converters, the sensing values are converted to numeric values and stored on the Micro SD Card.

The experiments are based on nine sleep postures of middle position upright lying (MU), right position upright lying (RU), left position upright lying (LU), middle position right lying (MR), left position right lying (LR), right position right lying (RR), middle position left lying (ML), right position left lying (RL), left position left lying (LL), for sleep postures recognition.

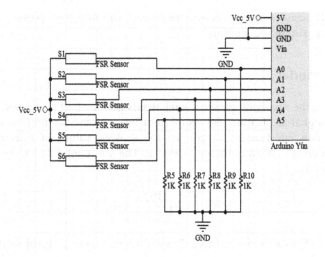

Fig. 2. The interface circuit with the Arduino Yún board.

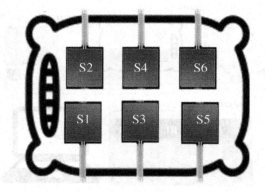

Fig. 3. The layout of the six FSRs on a pillow.

To find the received data of different posture, the received signal of MU posture are shown in Fig. 4. From Fig. 4, it is seen that the received data from S1, S2, S5, and S6 are very high about 10^6 Ω and varying between 3×10^5 to 10^6 Ω due to not forced by head. The variation between 3×10^5 to 10^6 Ω is high enough to make error propagation than the variation of S4 between 2×10^3 to 10^4 Ω. So we make a scale adjusting method by

$$y_i^m(n) = \begin{cases} R\left(\frac{1023}{s_i^m(n)} - 1\right), & s_i^m(n) > S_{th} \\ 30 \times 10^3, & s_i^m(n) \leq S_{th} \end{cases} \tag{2}$$

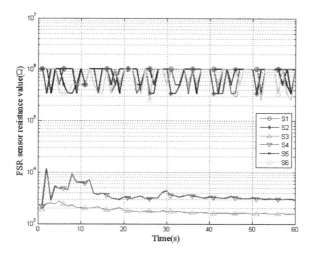

Fig. 4. A sampling example of the received data of six FSRs of MU posture.

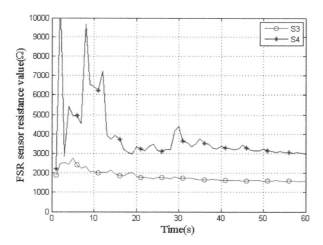

Fig. 5. A sampling example of the received data of S3 and S4 FSRs of MU posture.

where the threshold S_{th} = 33 is used to avoid the unforced fault propagation. Then we obtain a new received data of S3 and S4 of MU posture as shown in Fig. 5. And the sensing values of unforced sensors will be the same as 30×10^3 Ω.

Moreover, to investigate the difference of with upright lying, left side lying and right lying, we would like to compare the received data shown in Figs. 5, 6 and 7. On the comparison in Figs. 5, 6 and 7, it is seen that the data of S4 are almost the same but those of S3 are obviously different. Therefore, the data of S3 can be a significant feature for the middle postures.

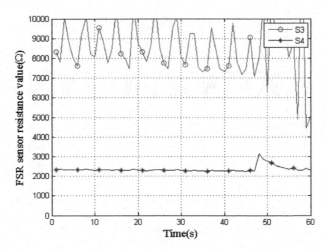

Fig. 6. A sampling example of the received data of S3 and S4 FSRs of MR posture.

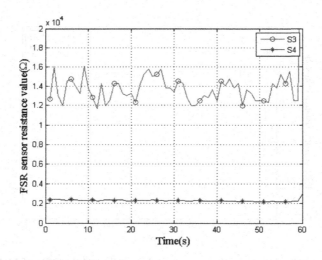

Fig. 7. A sampling example of the received data of S3 and S4 FSRs of ML posture.

3   Posture Recognition Experiments

In this section, we perform some experiments for the postures recognition. Each sleeping postures was done six times. Each sleeping posture continues for one minute. The sensing value is sampling in one second. Experimental parameters are shown in Table 1.

The pattern recognition system for posture recognition is shown in Fig. 8. In Fig. 8, in training procedure we construct the reference pattern with three steps:

Table 1. Experimental parameters

Parameters	Value
Types of postures	Nine postures of MU, RU, LU, MR, LR, RR, ML, RL, LL
Sampling rate	1 samples/second
Unit of signals	Kilo Ohms
Environment	Beyond a bed
Measurements	6 times of each posture
Number of samples each measurement	60 samples
Total no. of training samples	300 samples each postures
Number of testing samples	60 samples each postures

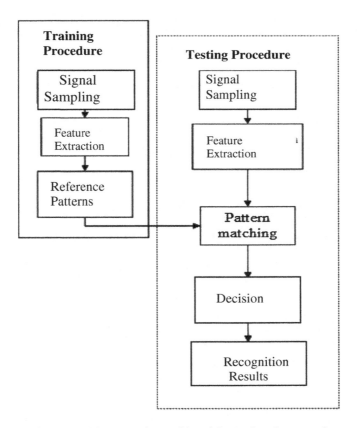

Fig. 8. Recognition procedures with training and testing procedures.

A. Signal Sampling: Via a voltage dividing circuit in Fig. 3, the sampling signal of FSR Sensor $s_i^m(n)$ can be obtained for the ith sensors Si of the mth posture.

B. Feature Extraction: By the ADC of Arduino Yún interface card, the nth sampling data of the ith FSR sensors Si, $y_i^m(n)$ can be obtained for of the mth posture as Eq. (2).

C. Reference Patterns: After we perform a minute for each posture, we collect the sampling data and obtain the average value as the reference pattern of the mth posture by

$$a^m = \{a_1^m, \cdots, a_i^m, \cdots, a_6^m\} \tag{3}$$

where the reference value of the ith FSR sensor Si can be expressed by

$$a_i^m = \frac{1}{N} \sum_{n=1}^{N} y_i^m(n) \tag{4}$$

where $N = 300$ is the total number of sampling data in training procedure of the ith FSR sensor Si for the mth posture. The reference patterns can be constructed from Table 2.

Table 2. The average values of nine postures, a_i^m.

FSR Postures $a_i^m(k\Omega)$	S1	S2	S3	S4	S5	S6
MU	30	30	2.26	2.92	30	30
MR	30	30	6.82	2.04	30	30
ML	30	30	14.43	2.54	30	30
RU	30	30	30	30	2.77	2.19
RR	30	30	30	30	2.85	2.63
RL	30	30	30	30	9.27	1.76
LU	1.79	2.50	30	30	30	30
LR	5.32	1.39	30	30	30	30
LL	2.04	6.25	30	30	30	30

After the reference patterns are constructed, the testing procedure can be proceed by five steps as shown in Fig. 8. After the first and the second step, we obtain the nth testing data $\{x_1(n), \cdots, x_i(n), \cdots, x_6(n)\}$. At the third step, the testing data and the reference patterns are pattern matching by the mean square error (MSE) for the mth posture as

$$MSE_m(n) = \frac{1}{6} \sum_{i=1}^{6} (x_i(n) - a_i^m)^2 \tag{5}$$

In decision making procedure, the recognized posture is make decision by the minimum MSE as

$$P_m(n) = \arg\min_m MSE_m(n) \tag{6}$$

After the testing results for nine postures are collected, the recognition accuracy rate for each posture can be shown in Table 3. From Table 3, the recognition rates almost are higher than 91% excepting both MR and RU postures.

Table 3. The recognition results of testing 60 sampling data of each posture.

Recognition results	MU	MR	ML	RU	RR	RL	LU	LR	LL	Accuracy
MU	59	1	0	0	0	0	0	0	0	98.3%
MR	21	39	0	0	0	0	0	0	0	65%
ML	0	5	55	0	0	0	0	0	0	91.7%
RU	0	0	0	28	30	2	0	0	0	46.7%
RR	0	0	0	2	57	1	0	0	0	95%
RL	0	0	0	0	1	59	0	0	0	98.3%
LU	0	0	0	0	0	2	58	0	0	96.7%
LR	0	0	0	0	0	0	0	59	1	98.3%
LL	0	0	0	0	0	0	3	1	56	93.3%
Accuracy rate										87%

4 Moving Average Window Methods

From Figs. 5, 6 and 7, it is seen that the signal variation of S3 is high to degrade the recognition rate of MR and RU postures to 65% and 47% respectively. Therefore, a moving average window is adopted to remove the variation. The sampling data after the proposed moving window can be expressed by

$$x_{i,L}^m(n) = \frac{1}{L}\sum_{t=n}^{n+L} x_i^m(t) \tag{7}$$

where L is the window length. We perform the experiments to improve the performance of posture recognition with moving windows $L = 3$ and $L = 7$. The improved recognition rate for $L = 3$ and $L = 7$ are shown in Tables 4 and 5, respectively. From Table 4 with $L = 3$, it is observed that the accuracy rates for eight postures exception MR posture are largely improved to be higher than 91%. Moreover, the accuracy rate for RU posture is improved from 47% to 100%. With window length $L = 7$, from Table 5, it is observed that the accuracy rate for MR posture is further improved from 63% to 71%.

Table 4. The recognition results of testing 60 sampling data of each posture with moving window, $L = 3$.

Recognition results	MU	MR	ML	RU	RR	RL	LU	LR	LL	Accuracy
MU	56	1	0	0	0	0	0	0	0	98.2%
MR	21	36	0	0	0	0	0	0	0	63.1%
ML	0	2	55	0	0	0	0	0	0	91.7%
RU	0	0	0	57	0	0	0	0	0	100%
RR	0	0	0	0	57	0	0	0	0	100%
RL	0	0	0	0	1	56	0	0	0	98.2%
LU	0	0	0	0	0	0	57	0	0	100%
LR	0	0	0	0	0	0	0	56	1	98.2%
LL	0	0	0	0	0	0	0	3	54	94.7%
Average accuracy rate										93.8%

Table 5. The recognition results of testing 60 sampling data of each posture with moving window, $L = 7$.

Recognition results	MU	MR	ML	RU	RR	RL	LU	LR	LL	Accuracy
MU	53	0	0	0	0	0	0	0	0	100%
MR	15	38	0	0	0	0	0	0	0	71.7%
ML	0	0	53	0	0	0	0	0	0	100%
RU	0	0	0	53	0	0	0	0	0	100%
RR	0	0	0	0	53	0	0	0	0	100%
RL	0	0	0	0	0	53	0	0	0	100%
LU	0	0	0	0	0	0	53	0	0	100%
LR	0	0	0	0	0	0	0	52	1	98%
LL	0	0	0	0	0	0	0	2	51	96%
Average accuracy rate										96.1%

5   Conclusions

In this paper, we applied six force sensing resistor sensors (FSR Sensors) to perform sleep posture recognition. The reference patterns and test patterns of the postures are performed pattern matching with the mean squared error (MSE) method. With a scale adjusting, the recognition accuracy is 87%. However, after the moving average windows are adopted to remove the high ripple, the recognition accuracy can be improved to 96% with window length L = 7.

Acknowledgments. This work was funded in part by Ministry of Science and Technology of Taiwan under Grant MOST 105-2221-E-324-019 and MOST 103-2632-E-324-001-MY3.

References

1. Aldrich, F.: Inside the Smart Home, pp. 17–39. Springer, Berlin (2013)
2. Leong, C.-Y., Ramli, A.-R., Perumal, T.: Rule-based framework for heterogeneous subsystems management in smart home environment. IEEE Trans. Consum. Electron. **55** (3), 1208–1213 (2009)
3. Han, D.-M., Lin, J.-H.: Smart home energy management system using IEEE 802.15.4 and ZigBee. IEEE Trans. Consum. Electron. **56**(3), 1403–1410 (2010)
4. Lee, H.-N., Lim, S.-H., Kim, J.-H.: UMONS: ubiquitous monitoring system in smart space. IEEE Trans. Consum. Electron. **55**(3), 1056–1064 (2009)
5. Suh, C., Ko, Y.-B.: Design and implementation of intelligent home control system based on active sensor networks. IEEE Trans. Consum. Electron. **54**(3), 1177–1184 (2008)
6. Pino, E.J., Morán, A.A., Paz, A.-D.-D.-l., Aqueveque, P.: Validation of non-invasive monitoring device to evaluate sleep quality. In: 37th Annual International Conference of the IEEE Engineering in Medicine and Biology Society (EMBC 2015), pp. 7974–7977, Milan (2015)
7. Pino, E.-J., Paz, A.-D.-D.-l., Aqueveque, P., Chávez, J.-A.-P., Morán, A.-A.: Contact pressure monitoring device for sleep studies. In: 35th Annual International Conference of the IEEE Engineering in Medicine and Biology Society (EMBC 2013), pp. 4160–4163, Osaka (2013)
8. Lokavee, S., Puntheeranurak, T., Kerdcharoen, T., Watthanwisuth N., Tuantranont, A.: Sensor pillow and bed sheet system: unconstrained monitoring of respiration rate and posture movements during sleep. In: IEEE International Conference on Systems, Man, and Cybernetics (SMC), pp. 1564–1568, Seoul (2012)
9. Huang, Y.-F., Yao, T.-Y., Yang, H.-J.: Performance of hand gesture recognition based on received signal strength with weighting signaling in wireless communications. In: 18-th International Conference on Network-Based Information Systems (NBiS 2015), pp. 596–600, Taipei (2015)
10. Huang, Y.-F., Yang, H.-J., Tan, T.-H.: A study of hand gesture recognition with wireless channel modeling by using wearable devices. In: 2015 International Conference on Machine Learning and Cybernetics (ICMLC), pp. 484–487, Guangzhou (2015)

Remaining Useful Life Estimation-A Case Study on Soil Moisture Sensors

Fang-Chien Chai[1], Chun-Chih Lo[1(✉)], Mong-Fong Horng[2], and Yau-Hwang Kuo[1]

[1] Center for Research of E-life DIgital Technology (CREDIT),
Department of Computer Science and Information Engineering,
National Cheng Kung University, Tainan, Taiwan
{deds, cobrageo, kuoyh}@cad.csie.ncku.edu.tw
[2] Center for Research of E-life DIgital Technology (CREDIT),
Department of Electronics Engineering,
National Kaohsiung University of Applied Sciences, Kaohsiung, Taiwan
mfhorng@cc.kuas.edu.tw

Abstract. This paper presents an approach to estimate the remaining useful life of sensors. First, a system state machine is defined to divide the sampled data received from the sensors into different categories. Then, the sampled data sets are sent to the fault model to detect whether a fault has occurred. The time of occurrence for each type of fault is recorded and weighted with different coefficient. The weighted values are accumulated to form a trend data graph. An exponential curve fitting is then used to approximate the trend of data to determine the remaining useful life function and threshold is also generated from the cumulative faults value. The experimental results shows the proposed model has a precision of 66.67% and recall rate near 100% within 10-h timespan. Thus, the proposed model may not only prolong the life span of sensors, but may also reduce the cost to replace them.

Keywords: Sensor · Remaining useful life

1 Introduction

In recent years, Internet of things (IoT) has become an emerging topic around the world. In many research fields, applications and products are being developed with IoT devices to provide useful applications to address issues in our daily life. One of the characteristic of IoT is making things smart [1]. In agriculture, precise sensor readings are essential to analyze and adjust the parameters used in the agricultural process. If part of the sensors in the environment are not functioning properly, the received data will not be trustworthy and could pose problems if these incorrect data are being used. To alleviate this problem, erroneous sensors have to be replaced in order avoid disastrous result during agricultural processes.

Sensors have a period of working lifespan specified in the sensor's specifications, but sometimes it breaks before the specified working lifespan. Moreover, some of the sensors may still works when it pass the working lifespan specified in the specifications,

N.T. Nguyen et al. (Eds.): ACIIDS 2017, Part II, LNAI 10192, pp. 328–338, 2017.
DOI: 10.1007/978-3-319-54430-4_32

but usually it will be replaced in order to prevent erroneous sensor readings being used during agricultural processes. In this case, unnecessary hardware cost emerges and the production cost of agricultural products also increases. Thus, it is much more cost-effective to estimate the remaining useful life (RUL) of sensors and replace them before they break down completely. Due to the above reasons, the primary focus of this paper is to investigate the possibility of estimating the remaining useful life of sensors and perform experiments to evaluate the feasibility of such concept.

2   Related Works

This section reviews some of the related literatures about erroneous sensors and estimating the remaining useful life of sensors. In some studies, a sufficient number of failure data can be used to train sensor degradation prediction models. This is widely accepted that the more failure histories are used to train the models, the more accurate results can be achieved. However, limited amount of data may decrease the accuracy of such prediction model. [2] proposed a model using the characteristic of piezoelectric sensors and baseline data to validate the health of the sensors. However, this model has drawbacks. It cannot directly be used to predict remaining useful life of sensors. The authors in [8, 9, 10] defined different types of possible faults in sensor nodes. Even though, the definitions of these sensor faults are complete, they still need to be redefined in mathematical form once we applied these fault type definitions in our model.

[3] proposed a model to predict the maintenance time of a tool in PECVD process using baseline data based on bathtub curve theory [4]. The result of prediction is rather accurate, but the principle of this prediction model may not be suitable for sensors. Furthermore, [5] suggested that not every tool follows this bathtub curve and correspond to the exponential curve. The Bathtub Curve Theory states that there are three stages in the life of a tool: Decreasing Failure Rate, Constant Failure Rate, and Increasing Failure Rate respectively. The error rate in the last stage is expected to have a property of exponential curve. [6] developed an integrated remaining useful life prediction method using particle filtering, then applied it to gear life prediction. [7] developed both a linear and a quadratic regression technique to predict the remaining useful life of gas turbine engines.

3   System Overview

We consider the scenario where multi-sensor data is available to the system till sensors reaches its working lifespan. The data sampled from sensors are used as an input in the Sensor Data and System State Collection phase shown in Fig. 1. In this phase, user needs to define possible events that might occur in the monitored environments, including user intended events and predictable events, to help to detect whether a fault has occurred. The data sampled in this phase and the current state of the system is recorded. Before entering the next phase, the size of sensor data set is defined first by the system and the data set is used as an input in the Fault Detection and Classification phase. This phase contains a predefined sensor fault model and uses the sampled data

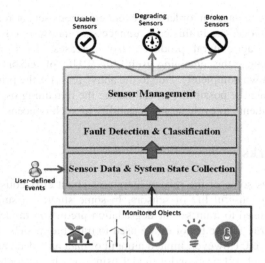

Fig. 1. System architecture

sets to determine whether the predefined faults has occur or not. Lastly, different measures will be taken in the Sensor Management phase when a fault is detected. Each type of faults are given different weights and accumulated to form a trend line. The estimation model uses this trend line to determine the approximated curve to estimate when the sensor will reach the end of its serviceable lifespan. In the end, sensors are classified into three groups, usable, degrading and broken respectively.

3.1   Sensor Data Collection and System State Recording

A set of healthy sensors are used in the system to collect data from the environment. The collected sensor data are integrated with the system state and stored into a database for fault detection and maintenance prediction. A finite state machine is set by the user-defined events and states. The state machine stores the current environment condition and uses the data collected by sensors in the fault detection model to determine the sensor faults. The state machine is defined by a 4-tuple (S, S_0, E, T_S). S is a set of states, and S_0 is the initial state. E is a set of involved events during the operation of the system. T_S is the state transition function. The transition function can be described as $(S_i, E_k) \rightarrow (S_j), S_i, S_j \in S, E_k \in E$. The detailed information of the system state machine is given in the following paragraph.

3.2   State Assisted Fault Model

The status of a sensor is examined in this phase through a predefined faults model applied with user-defined sensing status of the sensor. Series of data stored in the database are divided into smaller data sets. The size of each data set is defined by finding the maximum value of kurtosis under different sample sizes. This step is performed only once in the entire lifespan of a sensor. Thus, when a new sensor is deployed, the sample size of the sensor is redefined. Furthermore, the sampled data sets are categorized into long-term data feature and short-term data feature. The long-term data feature is all the data points with the same state in the database and transferred it to the fault model to determine whether the data contains a fault. The short-term data feature is the sampled data set itself, and it is also used to determine whether a fault has occurred. In cases where no fault is detected, the system determines that the sensor are still in working condition and goes back to the sensor data collection phase to resume sampling data. However, if a fault has occurred, corresponding measures will be taken depending on the type of fault. In this case, a bias fault means that it is highly possible that the sensor may no longer sending any meaningful data back to the system. An immediate repair or replacement of sensors is needed. If a malfunction or noise fault occurs, the condition of the sensor is likely degrading. The sensor then needs to be replaced before it is completely broken. Thus, an estimation of remaining useful life is helpful in this situation. A drift fault means that there may be ambient changes to the sensor or surrounding conditions, but the sensor is still in working condition. Hence, a software parameter correction is needed and the sensor does not need to be repaired since the sampled data is still reflecting the condition of the environment.

3.3   Remaining Useful Life Estimation

This model works with the fault detection model and estimates the remaining useful life of a sensor based on the detected faults. The detected faults in the state assisted fault model are used in this model. For drift fault, it can be corrected via software technique. Bias fault means that it is highly possible that the sensor is no longer working, an immediate repair is needed. In noise fault and malfunction fault, different weights are given based on the fault type. These weighted faults are accumulated to form a trend line and an exponential curve is used to approximate the trend of weighted faults to find out the estimation function. The threshold of the estimation function is also determined by the detected faults. With the function and the threshold value derived from the detected faults, it is possible to estimate the remaining useful life of a sensor and replace it before it breaks down completely.

4   Algorithm

In this section, the definitions of faults are explained in a mathematical form and a line chart. Then, the optimal size of data sets based on the raw data collected are defined and uses the predefined size to determine the maximum allowable faulty data points in a set. This step helps to define the weight of each detected fault and the trend line data constructed based on the detected cumulative faults. Moreover, the estimation model is introduced using the aforementioned trend line data to approximate the maximum allowable faulty data points and find the estimation function to estimate the RUL of a sensor.

4.1   Fault Definition

The defined faults used in our model are presented in this section along with the details and mathematical representation of each fault. Based on the predefined faults listed below and the recorded system states, possible faults can be detected and filters false judgements to determine whether the sensor is degrading or not.

(1) **Drift:** The sensor itself is in good working condition, but the received data slightly or greatly deviates from the true value. Since the sensor is not broken, the fault can be corrected with software settings. Following formula describes this phenomenon.

$$\left\{ \left| \mu_{D_{S_{set}}^{current}} - \frac{\sum_{i=1}^{m} \mu_{D_{S_{set}}^{history}}}{m} \right| > \frac{y_{spec}}{k_t} \right\} \tag{1}$$

(2) **Bias:** Sometimes the sensor is still in working condition, but the value of sampled data approaches to zero, upper limits or the preset value of the sensor. According to [11], it is possible that the wire connection is loose or other kinds of hardware error, but we only focus on the fault presented by the sensor itself other than hardware error.

$$S_{set}^{bias} > S_{set} * PSE \tag{2}$$

(3) **Malfunction:** For any consecutive data points, if the difference between each point is greater than the standard deviation of the sampled data set, then a malfunction fault is detected. These detected faults are used in RUL estimation of sensors.

$$\left\{ \forall D_i \in D_{S_{set}} : |D_i - D_{i+1}| > k_t * \sigma_{infant} \right\} \tag{3}$$

(4) **Noise:** Ruling out the possibility of low battery or other hardware failure presented in [11], our experiment results shows that the probable cause of this fault is the degradation of sensors.

$$k_t * \sigma_{infant} \tag{4}$$

4.2   Remaining Useful Life Estimation Method

A common specification of any kind of sensor describes its purpose, performance, recommended working environment and so on. There should be one item which specifies the natural error rate of the data retrieved from the sensor. It can be described as $D_{spec} = x \pm y_{spec}$. The notation x is the received data point at any given time and y_{spec} denotes the gradient between any two continual received data point. Normally, a small range of variation between continual data points would occur, but those points do not significantly drift from the actual value. This is a quite common phenomenon, especially in analog to digital converter. Even under the same situation, the received data points drift slightly, but do not affect the reliability of sensor data.

4.3   Defining the Threshold of RUL Estimation

When a sensor is in good working condition, especially at early stage to middle stage of its lifespan, the sensor may be degrading, but the detected faults do not affect the performance of the sensor itself. However, when the sensor reaches to end of its lifespan, it is possible that the detected faults will severely affect the sensor reading and it may break down completely in the near future. Hence, it is important to distinguish the difference between each fault as shown in Fig. 2, in order to present a precise estimation on the remaining useful life of the sensor (Figs. 3 and 4).

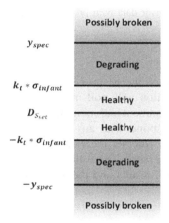

Fig. 2. Possible range of detected faults

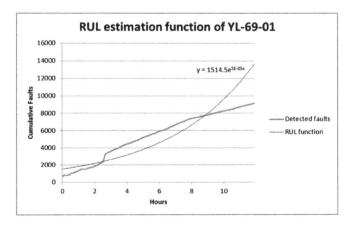

Fig. 3. Cumulative faults and an approximated function

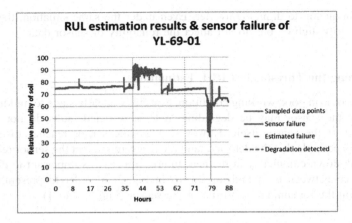

Fig. 4. Sensor data with predicted breakdown time

5 Experiment

To validate the performance of the proposed model, an experiment is conducted to emulate the real world condition with soil moisture sensors. The sensors are placed under a relatively stable environment and monitor the status of the soil. The sensors are used until they no longer respond with any meaningful data. The lifespan of each sensor is recorded to validate whether the estimation results are similar to the actual remaining useful life of the sensors. The following subsection explains the details of the experiment, the estimation results and the difference between the estimations and actual lifespan of the sensors.

5.1 Experiment Settings

The instruments used in this experiment are listed as follows: Raspberry Pi Model B, which is used as a data base. Arduino Mega 2560, which is used as an analog-digital converter and a power supply of the sensors. Three types of Soil moisture sensor are used in the experiment, soil moisture sensor (Funduino), soil moisture sensor (YL-69, with zinc plating) and soil moisture sensor (XD-28, with gold plating).

The sensors are divided into two groups as shown in Table 1. One of them is supplied with tap water and the other is supplied with tap water with liquid fertilizer. Each group has three sensors for 3-fold cross validation. The fertilizer is blended according to the receipt on the product. Each group has three different brands of soil moisture sensor and each brand has three sensors in one group. The watering time interval and the amount of water added to the soil is fixed. The sample rate of the data is one point per second. The input voltage of the sensors is 5 volts, and it is converted into digital form of 1024 bits. The calculation of relative humidity is: $\left(\frac{converted\ voltage\ in\ 1024\ bits}{1024}\right) * 100\%$. RUL estimations are performed on a Lenovo T530

Table 1. Sensor ID and corresponding environmental condition

Environment	Sensor ID		
	Funduino	XD-28	YL-69
Fertilized	Funduino-01	XD-28-01	YL-69-01
	Funduino-02	XD-28-02	YL-69-02
	Funduino-03	XD-28-03	YL-69-03
Tap water	Funduino-04	XD-28-04	YL-69-04
	Funduino-05	XD-28-05	YL-69-05
	Funduino-06	XD-28-06	YL-69-06

laptop. Table 1 records the ID number of each sensor and their corresponding brands and environmental conditions.

5.2 Experiment Results

Following formulas are defined to assess the performance of our estimation model. For derivation, we define the sensor failure time as SF, degradation detected time as DD and estimated failure time as EF. Also the success estimations are denoted as SE and all estimations are denoted as AE.

$$Estimation\ error = SF - EF \tag{5}$$

$$Precision = \frac{SE}{AE} \tag{6}$$

A $n\_hour\ Recall$ rate is defined to evaluate how many successful estimations are within a certain time interval before the sensor is considered to be broken.

$$n\_hour\ Recall = \frac{SE\ within\ n\ hours}{SE} \tag{7}$$

In Table 2, the estimation error is not very stable, it ranges from an hour to nearly 9 h. However, most of the results shows the proposed model is able to estimate the occurrence of error before the sensor breaks down completely. Except for YL-69-03

Table 2. Details of RUL estimation with Sset = 140

Sensor ID	Degradation detected time (h)	Sensor failure time (h)	Estimated failure time (h)	Estimation error (h)
YL-69-01	87.931	125.978	120.713	5.265
YL-69-02	72.704	112.822	109.839	2.983
YL-69-03	131.061	132.2	132.796	−0.596
Funduino-01	65.113	116.45	114.694	1.756
Funduino-02	77.659	116.728	116.099	0.629
Funduino-03	79.694	82.283	84.183	−1.9

and Funduino-03, the estimated results shows the error is only determined after the sensors breaks down, but the estimation errors are relatively small while the others are more than 5 h. The average estimation error is 2.507 h and the standard deviation is 1.568 h.

In Table 3, the precision rate of the estimations is 66.67% with 1-h recall rate. The precision of the estimations is admissible since the precision is neither very high nor too low. The recall rates of different time interval differ significantly from each other, but they can reach a rate of 100% if the time interval is set to 10 h. From the result in Table 3, the estimation results occurred after the end of the serviceable lifespan of a sensor. Although the estimation is quite late, but the accuracy of RUL estimation is still promising. Under most of these cases, it may not only prolong the usable time of sensors, but also allow users to replace sensors before it breaks down completely. Furthermore, the recall rate of the RUL estimations with optimal data set size reaches 100% within 10-h range, while others have an average recall rate below 40% within the same time interval. Although the precision of the RUL estimations with optimal data set size is a bit low compared with the others, it still shows a promising results based on the overall information.

Table 3. Precision and Recall rate of different size of a data set

Data set size	Precision	1-h Recall	6-h Recall	10-h Recall	24-h Recall
140	66.67%	20%	40%	100%	100%
60	83.33%	20%	40%	40%	60%
300	83.33%	0%	20%	40%	80%
600	50%	0%	0%	0%	66.67%

Table 4 lists the average estimation error and its standard deviations. The calculated optimal data set size has the lowest average estimation error and standard deviation. EE_{SE} represents the estimation errors of successful estimations.

$$Average\ estimation\ error = \frac{\sum EE_{SE}}{SE} \tag{8}$$

$$Standard\ deviation\ of\ estimation\ error = StandardDeviation(EE_{SE}) \tag{9}$$

Based on all experiment results, we can see that the proposed model shows a good forecasting performance and may prolongs the usage of sensors.

Table 4. Average estimation error and standard deviation under different size of a data set

Data set size	Average estimation error (h)	Standard deviation of estimation error (h)
140	2.658	1.72
60	14.116	11.07
300	14.899	8.58
600	17.831	4.772

6   Conclusion

The proposed RUL Estimation model for sensors shows a promising results when forecasting the sensors failure time and allows users to replace the faulty sensors before they breaks down completely. The size of a data set is determined at the beginning of the lifespan of a sensor. Based on the RUL estimation results in the experiment, it shows that the calculated data set size has outperformed others. In the experiment, the overall precision of the RUL estimations have an average rate of 66.67% and the recall rate can reach 100% within 10 h. Although by selecting different data set size will increase the precision to 83%, but this significantly decreases recall rate within 10 h. Moreover, the optimal data set size of each sensor brand can be used among the same brand and type of sensor, but will not work with different sensor brands. Furthermore, by observing the experiment results we can see that the trend lines do not necessarily follow the property of exponential curve. A small amount of the sensors corrupt very fast when they reaches the end of its serviceable lifespan, while others have exponential-like trend lines and their RUL can be successfully estimated. Thus, the error rate of the sensors can be approximated with exponential curve which exists in the wear-out stage of bathtub curve theory, but this does not mean that the lifespan of a sensor is always remained consistent with this theory.

Acknowledgement. The authors would like to thank the Ministry of Science and Technology for supporting this research, which is part of the project numbered 103-2221-E-006-257-MY3, 104-2221-E-151-007-, 105-2221-E-151-034-MY2, and 105-2221-E-006-138-MY2.

References

1. Ghasemi, A., Zahediasl, S.: Normality tests for statistical analysis: a guide for non-statisticians. Int. J. Clin. Endocrinol. Metab. **10**(2), 486–489 (2012)
2. Smith, Z., Wells, C.: Central limit theorem and sample size. In: Annual meeting of the Northeastern Educational Research Association (2006)
3. Jarque, C., Bera, A.: Efficient tests for normality, homoscedasticity and serial independence of regression residuals. Econ. Lett. **6**(3), 255–259 (1980)
4. Li, S.: Application of the internet of things technology in precision agriculture irrigation systems. In: International Conference on Computer Science and Service System (2012)
5. Cheng, T.: A critical discussion on bath-tub curve. In: Chinese Society for Quality (2006)
6. He, D., Bechhoefer, E., Ma, J., Li, R.: Particle filtering based gear prognostics using one-dimensional health index. In: 2011 Annual Conference of the Prognostics and Health Management Society (2011)
7. Li, Y.G., Nilkitsaranont, P.: Gas turbine performance prognostic for condition-based maintenance. Appl. Energy **86**, 2152–2161 (2009)
8. Baljak, V., Tei, K., Honiden, S.: Fault classification and model learning from sensory readings—framework for fault tolerance in wireless sensor networks. In: Intelligent Sensors, IEEE Eighth International Conference on Sensor Networks and Information, pp. 408–413 (2013)

9. Wei, M., Chen, M., Zhou, D., Wan, W.: Remaining useful life prediction using a stochastic filtering model with multi-sensor information fusion. In: Prognostics and System Health Management Conference, pp. 1–6 (2011)
10. Overly, T., Park, G., Farinholt, K., Farrar, C.: Piezoelectric active-sensor diagnostics and validation using instantaneous baseline data. IEEE Sens. J. 9(11), 1414–1421 (2009)
11. Sharma, A., Golubchik, L., Govindan, R.: Sensor faults: detection methods and prevalence in real-world datasets. ACM Trans. Sens. Netw. 6(3), 23 (2010)

Intelligent Algorithms and Brain Functions

Neurofeedback System for Training Attentiveness

Khuan Y. Lee[1,2,3(✉)], Emir Eiqram Hidzir[1],
and Muhd Redzuan Haron[1]

[1] Faculty of Electrical Engineering, Universiti Teknologi MARA,
40450 Shah Alam, Selangor DE, Malaysia
leeyootkhuan@salam.uitm.edu.my
[2] Computational Intelligence Detection RIG, Universiti Teknologi MARA,
40450 Shah Alam, Selangor DE, Malaysia
[3] Pharmaceutical and Life Sciences Community of Research,
Universiti Teknologi MARA, 40450 Shah Alam, Selangor DE, Malaysia

Abstract. Attention Deficit Disorder (ADD) has long been recognized as a public health concern amongst children, where its symptoms include impulsiveness, inattentiveness and unfocused. The consequence is children with poor academic performance and discipline that has negative impact on their future. Current treatment for ADD uses powerful psycho-stimulant drugs, to reduce aggression and enhance concentration. However, there are always risk factors and adverse effects with these drugs. Moreover, drugs do not alter the dysfunctional condition. Forefront research in biomedical engineering unveils neurofeedback, which presents an exciting alternative approach to neural related disorders. Our ultimate goal is to develop a neurofeedback system to enable anyone with attention deficit to practice regulating their brain to reach an attentive state of mind, with reduced dependency on drug related intervention. Relying on neuroplasticity, neurofeedback focuses on the training of brain through activities to circumvent the dysfunctional condition. In this paper, such a system has been developed and applied on normal healthy subjects, to establish the protocol on EEG subband and electrode placement as well as system functional testing. It consists of a wireless EEG acquisition module, a feature extraction module, an IoT database module, an Intel Edison microcontroller board and a feedback activity center, the humanoid robot. The protocol on subband and electrode placement is established with short time Fourier transform (STFT) and fast Fourier transform (FFT). The system rewards the subject if the root mean square voltage of his beta subband at Fp1 exceeds the target voltage, when he is attentive.

Keywords: Attention · Deficit · Disorder · Neurofeedback · Electroencephalogram · Short time Fourier transform · Fast Fourier transform

1 Introduction

Attention Deficit Disorder (ADD) is a neural impairment with long-term problematic concerns, which is estimated to affect 5–15% of school age children worldwide [1–3]. Children with ADD symptoms have low metabolism of catecholamine, low of

© Springer International Publishing AG 2017
N.T. Nguyen et al. (Eds.): ACIIDS 2017, Part II, LNAI 10192, pp. 341–350, 2017.
DOI: 10.1007/978-3-319-54430-4_33

epinephrine and noradrenalin. This causes the children to have reduced attention, behave inappropriately, be distracted easily and become impulsive [1, 4], that impedes their social-emotional development. In schools, ADD children are faced with difficulties in paying attention, following instructions from teachers, focusing on their studies and completing homework and assignments [5]. The unwarranted ridicule leaves profound negative impact to their life, in workplace, family and society subsequently.

Currently, drug treatment is the most common method to control the behavior of children with ADD symptoms. However, there are side-effects in the case of drug overdose or drug consumption over a long period of time [6, 7].

Neurofeedback is an electroencephalogram (EEG) based technology to measure and interpret real-time brain activity of patients for self-regulation [8]. Through neurofeedback system, the brainwave captured by EEG is instantaneously displayed on the computer as feedback. It is recognized as a potentially useful intervention for neural related conditions.

Human brain exhibits four types of brainwave pattern, i.e. delta wave, theta waves, alpha waves, and beta waves [1]. Theta waves (4–8 Hz) are associated with daydreaming and mental inefficiency. Alpha waves (8–12 Hz), slower and larger, are often associated with relaxation. Beta waves (>13 Hz) with its fast changing pattern are related to mental activity and outwardly focused attention. Children with ADD symptoms are found to have excessive theta wave and minimal beta wave, or high ratio of theta to beta brain wave. Deriving from this, it is hypothesized that beta wave can be an objective marker for diagnosis of ADD, while beta brainwave as well as ratio of beta-theta brainwave may be considered for basis of training [9–11].

Past and present controlled studies have identified the causes of ADD disorder to interplay between psychosocial, biological and neurological factors [2]. In general, there are three approaches to treat the disease: drug therapy, behavioral therapy and combination of drug and behavioral therapy. The most common treatment is drug therapy through oral medicine such as, methylphenidate (Ritalin), Amphetamines (Dexedrine), Pemoline (Cylert) and d-Amphetamine (Adderall) [2]. Through medication, the brain chemistry is altered that change the behavior of ADD children. Even though frequent medication reduces their impulsivity and improves their attention, the stimulant effect from drug can cause insomnia, dizziness, irritability, headache, stomach ache, sleep disorder and stunted growth [10]. However, if left untreated, ADD causes academic failure and severe intellectual disability that will lead to poor self-esteem and socially maladaptive behavior, with serious psycho-pathological consequences in later life [12].

Neurofeedback presents an alternative approach to prescribed medication for ADD children. Examples of neurofeedback systems in the market are such as SMART Mind Pro by BrainTrain for testing and training cognitive abilities [6]; Slow Cortical Potential Training for diagnosis of migraine [11]; Live Z-Score Neurofeedback Training for insomnia [13]; LORETA Neurofeedback Training using low resolution electromagnetic tomography for cognitive ability [9]. However, a neurofeedback training system for ADD has yet to be explored. In addition, all the above systems are using computer game as feedback activities, which may not be appealing to children of Z-generation, birth years ranging from the mid-1990 s to early-2000.

The presented neurofeedback training system offers a lesser drug dependent, significant and lasting improvement in attention and focus of ADD subjects. With Intel Edison, it can be made affordable and home-based, which is estimated to have a market size of 135 k–405 k sets alone (i.e. 5–15% of total primary students, based on statistic from the Ministry of Education (MOE) Malaysia, 18 May 2016 [14]). While being home-based, its effectiveness will increase with the frequency of training. In the case of ADD children in schools, it can complement current intervention programs, aligned with the National Key Result Areas (NKRAs), MOE Malaysia.

Work here intends to develop a neurofeedback system on Intel Edison IoT platform for training attentiveness. It is aimed to establish protocol on subband and electrode placement for attentiveness. Functionality of the system will also be tested. Section 2 describes architecture of the presented neurofeedback training system for ADD. Section 3 presents methods engaged to acquire, pre-process and process the EEG data. Section 4 details results on protocol for subband and electrode placement with short time Fourier transform (STFT) and fast Fourier transform (FFT).

2 Architecture

The neurofeedback system comprises of four sub-systems: the EEG data acquisition module (DAC), EEG feature extraction module (FE), the IoT database module and feedback activity module (FA) driven by Intel Edison, in an architecture as shown in Fig. 1.

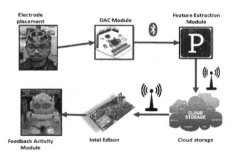

Fig. 1. Architecture of neurofeedback system for training attentiveness

The EEG DAC module consists of three gold electrodes set in a head gear designed for comfort in place of the cap, bio-potential input channels, high powered analog front-end, accelerometer, local SD storage and wireless communication interfacing circuit. The FE module enables programming, processing and embedding of filtering and feature extraction algorithms as required by our application. With our FA module, the feedback activity is not just computer games, but a humanoid robot, which can be programmed to reward attentiveness, tailored to the subject, a uniqueness of our neurofeedback system. The Intel Edison microcontroller board is based on the dual-core Intel® Atom CPU at 500 MHz. It is a tiny computer-on-module of Intel®

architecture designed for application of IoT device, providing good communication from device-to-device and device-to-cloud communication. It is able to collect, store and process data in the cloud and supports Linux,Wi-Fi and Bluetooth applications.

3   Methodology

To establish the protocol for attentiveness and test the system function, a total of 7 normal healthy adults, between 16–23 years old were recruited.

The subjects were given three tasks. Task 1 requires the subjects to relax with eyes closed. Task 2 requires the subjects to relax with eyes opened. Task 3 requires the subjects to count backward in silence. The task of counting backward demands attentiveness of the subjects. Hence it was chosen as the activity to train attentiveness [15]. Figure 2 shows the timeline for these tasks.

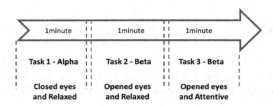

Fig. 2. Timeline of tasks

The DAC process started with agreement to sign on the patient consent form. Next the subjects were sit comfortably and relaxingly with both hands on thigh, while procedure for Task 1, Task 2 and Task 3 were explained to the subject. Then, electrode was placed at Fp1, Fp2, Fz, FC5, FC6, Cz, C3, C4, Pz, P4, CPz and FCz on the scalp, with reference to the International 10-20 System, chosen from our literature survey [7]. The reference point was taken from both ear lobes, indicated as A1 and A2. When the DAC module was first switched ON, an impedance check was conducted to ensure proper contact of electrode. Then, Task 1, Task 2, and Task 3 were carried out in sequence, with recording at one electrode each time. Subject was given time to rest between tasks.

This was then ensued with experimental procedure deliberated in Fig. 3. The raw EEG signals were first filtered by a high-pass filter with cut-off at 0.5 Hz to remove the baseline drift, EOG artefacts, low frequency noise and a Notch filter at 50 Hz to remove the harmonics of power line. To determine the protocol for electrode subband, the time-frequency features of the high-passed EEG signals were extracted with STFT of window size and number of FFT points of 64, overlap of 90%. To minimize the leakage effect so as to maintain the continuity of the first and the last points in the frame, a Hamming window was used (see 3a of Fig. 3). To establish the protocol for electrode placement, the high-passed EEG signals were processed with FFT [16] (see 3b of

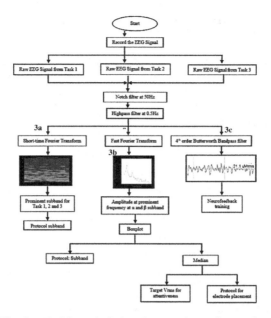

Fig. 3. Workflow depicting the experimental procedure

Fig. 3). The amplitude and frequency content were computed to produce the median, served as a guide to select the target RMS value (Vrms) for training, defined as follow,

$$\text{Vrms} = \sqrt{\frac{1}{T} \int_0^T V^2(t) dt} \tag{1}$$

For system functional test, the high-passed signals were filtered with a 4th order Butterworth filter, of frequency between 16–25 Hz, to obtain the beta subband (see 3c of Fig. 3) [17, 18]. The procedure to determine protocol (3a and 3b of Fig. 3) took precedence before neurofeedback training (3c of Fig. 3).

4   Results and Discussions

The three tasks were conducted to serve two purposes, to establish the neurofeedback protocol on EEG subband and electrode placement as well as to run a functional test on the neurofeedback system, for attentiveness. Sections 4.1 and 4.2 in the following describe the results for the former while Sect. 4.3 for the latter. Results only displayed electrodes responsive to 10% or more of study population.

4.1 Protocol: Electrode Placement for Attentiveness

Figures 4, 5 and 6 displayed the percentage of subject population with prominent frequency in alpha and beta subband for the three tasks from all the electrodes. For Task 1,
the prominent frequency in alpha subband was shown in 100% of all subjects at Fp1. For
Task 2, the prominent frequency in beta subband was shown in 72% of all subjects, at
Fp1 and Pz. And for Task 3, the prominent frequency in beta subband was shown in
72% of all subjects, at Fp1, Fp2 and Fc5. It was observed that the Fp1 of major
population displayed prominent frequency consistently for the three different tasks.

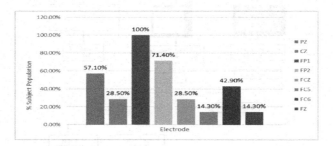

Fig. 4 Percent of subject population respond with high neural activity to Task 1 at different
electrode placements

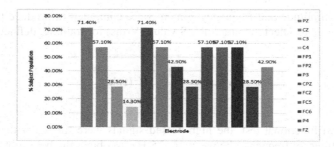

Fig. 5 Percent of subject population respond with high neural activity to Task 2 at different
electrode placements

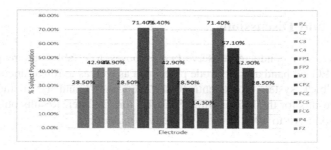

Fig. 6. Percent of subject population respond with high neural activity to Task 3 at different
electrode placements

In addition, the median of amplitude for the study population during the three tasks at Fp1 were further examined. As shown in Fig. 7, on an average, the median for the prominent frequency for Task 1, Task 2 and Task 3 were 10 μvolt (alpha subband), 1 μvolt (beta subband) and 1.5 μvolt (beta subband) respectively. Of significance, a difference of 0.5 μvolt in the amplitude of beta subband was observed at state of relaxation and attentiveness.

From the above, Fp1 was found consistently responsive to the three tasks and show difference in magnitude and frequency between the three tasks. This also concurred with literature on anatomical function of brain [19] that Fp1 is related to attention. As a consequence, Fp1 is established as the protocol for electrode placement for attentiveness.

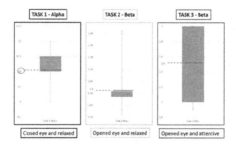

Fig. 7. Amplitude of EEG Subband for different tasks at Fp1

4.2   Protocol: EEG Subband for Attentiveness

Two representative spectrograms obtained from time-frequency analysis of EEG recording from two different subjects, during execution of the three tasks, are shown in Fig. 8.

From the frequency aspect, it was found that the alpha subband stood out for Task 1 and 2 (10 Hz), while beta subband for Task 3 (28 Hz), indicating there was a difference

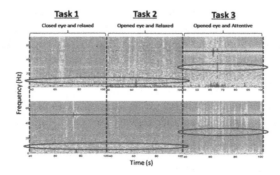

Fig. 8. Spectrogram showing high amplitude neural activity for different tasks at Fp1

in subband when the mental state was different. Furthermore, from the magnitude aspect, from FFT in Sect. 4.1, the magnitude of the beta subband was found to increase as the mental state changes from relaxation to attentiveness. With these, beta subband was established as the protocol for subband for attentiveness.

4.3   System Function Test

Figure 9 showed the assembly of the four sub-systems for our neurofeedback system for training attentiveness.

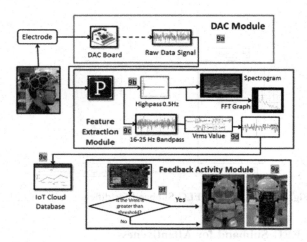

Fig. 9. Assembly to establish protocol and neurofeedback for training attentiveness

The DAC module recorded EEG from the scalp of the subject at Fp1 (see 9a of Fig. 9). The raw EEG data was then transmitted to the FE module via bluetooth.

To select protocol for electrode placement and subband, it was first filtered with cut-off at 0.5 Hz, to remove the baseline drift, EOG artefacts, low frequency noise (see 9b of Fig. 9).

For neurofeedback configuration, the raw EEG signal was further bandpassed (16–25 Hz) with a 4th order Butterworth filter, to extract the beta subband (see 9c of Fig. 9). The bandpassed signal was then converted to Vrms using (1) and transmitted to the IoT cloud for storage (see 9e of Fig. 9). Once a request from Intel Edison was initiated, the Vrms was fetched from the register in the IoT cloud, given the WIFI space and the IoT cloud address were declared. The Intel Edison then compared the Vrms fetched from IoT cloud and stored in one of its registers, with the pre-determined target Vrms for attentiveness, stored in another register (see 9f of Fig. 9). If the Vrms from the IoT cloud was equal or higher than the target Vrms, the Intel Edison yielded an output signal to be relayed to the microcontroller of the humanoid robot via its I2C bus. This output signal set the servomotor on the legs of the humanoid robot in motion, resulting in humanoid robot stepping forward with a bar of chocolate to reward subject for being attentive (see 9g of Fig. 9). Else, no output signal would be produced and the humanoid robot stayed idle, as penalty to the subject. Each time, as the subject becomes more attentive with training, it is necessary to reset and raise the value of target Vrms gradually.

5 Conclusion

A neurofeedback system on Intel Edison IoT platform for training attentiveness has been developed and functionally tested. It is applied on normal healthy subjects to establish the protocol for training attentiveness, which is found to be beta subband, from the spectrogram and Fp1 from FFT. This also concurs with findings on anatomical function of brain from literature. The median for the three tasks, relaxation with eye closed, relaxation with eye opened and attentiveness with eye opened for normal healthy population are 11 μvolt (alpha subband), 1 μvolt (beta subband), 1.5 μvolt (beta subband), respectively. The difference in median serves as indicator for target Vrms for the neurofeedback system. The IoT feature of Intel Edison facilitates pooling and sharing of data between personal caregivers, psychologists and neurofeedback trainers anytime, anywhere. To the subjects with attention deficit, our neurofeedback system will help to transform from his destructive behavior and negative outlook of life to a constructive and positive ones by circumventing the neural dysfunctional condition with his favorite feedback activity, minimal drug related intervention and a more effective intervention with IoT. In addition, instead of having a group with high liability to the national development, the nation will now benefit from a group that will contribute to the growth of the nation.

Acknowledgement. The author would like to thank the Ministry of Education (MOE), Malaysia, for providing the research funding 600-RMI/FRGS 5/3(85/2014); the Research Management Institute and the Faculty of Electrical Engineering, Universiti Teknologi MARA for the support and assistance given to the authors in carrying out this research. Procedure involving human subjects described in this paper has been approved by the Ethic Committee, Faculty of Electrical Engineering, Universiti Teknologi MARA.

References

1. Alhambra, M.A., Fowler, T.P., Alhammbra, A.: EEG biofeedback: a new treatment option for ADD/ADHD. J. Neurother. **1**, 39–43 (1993)
2. Fewell, R., Deutscher, B.: Attention deficit hyperactivity disorder in very young children: early signs and interventions. Infants Young Child. **675**, 24–32 (2002)
3. Doggett, A.M.: ADHD and drug therapy: is it still a valid treatment? J. Child Health Care **1**, 69–81 (2004)
4. Kaiser, D.A.: Efficacy of neurofeedback on adults with attentional deficit and related disorders. EEG Spectr. 1–6 (1997)

5. U.S. Department of Education: Identifying and Treating Attention Deficit Hyperactivity Disorder: A Resource for School and Home. Research to Practice Division, Washington (2003)
6. Lojek, E., Bolewska, A.: The effectiveness of computer assisted cognitive rehabilitation in brain damaged patients. Polish Psychol. Bull. **44**, 31–39 (2013)
7. Garcia, R.: A parent's guide to understand neurofeedback. J. Chem. Inf. Model. **53**, 1689–1699 (2013)
8. Liu, Y., Sourina, O., Hou, X.: Neurofeedback games to improve cognitive abilities. In: International Conference Cyberworlds, pp. 161–168 (2014)
9. Loo, S.K., Makeig, S.: Clinical utility of EEG in attention-deficit/hyperactivity disorder: a research update. Neurotherapeutics **9**, 569–587 (2012)
10. Yardley, K.: Attention deficit disorder/hyperactive disorder (ADHD) (2004). www.katolenyardley.com/Attention%20Deficit%20Hyperactive%20Disorder.pdf
11. Collura, T.F., et al.: EEG Biofeedback Training Using Live Z-Scores and a Normative Database, pp. 1–43. Press Page, Amsterdam (2009)
12. Hanna, N.: Attention deficit disorder (ADD) attention deficit hyperactive disorder (ADHD) is it a product of our modern lifestyles? Am. J. Clin. Med. **6**, 22–28 (2009)
13. Hammond, D.C.: What is neurofeedback: an update. J. Neurother. **15**, 305–336 (2011)
14. Statistics Number of Students (Enrolment). http://www.moe.gov.my
15. Detecting Concentration. http://eeghacker.blogspot.my
16. Gough, B.: FFT algorithms. In: GNU Scientific Library, pp. 1–37 (1997)
17. Lutterveld, R.V., Houlihan, S.D., Pal, P., Sacchet, M.D., McFarlane-Blake, C., Patel, P.R., Sullivan, J.S., Ossadtchi, A., Druker, S., Bauer, C., Brewer, J.A.: Source-space EEG neurofeedback links subjective experience with brain activity during effortless awareness meditation. Neuroimage 1–11 (2016)
18. Lagos, J.S., Hernández, C.A., Pinto, W.: Neurofeedback prototype with electroencephalography for stimulation and control, using the alpha wave. J. Med. Bioeng. **4**, 371–375 (2015)
19. Teplan, M.: Fundamentals of EEG measurement. Measur. Sci. Rev. **2**, 1–11 (2002)

Building Classifiers for Parkinson's Disease Using New Eye Tribe Tracking Method

Artur Szymański[1(✉)], Stanisław Szlufik[2], Dariusz M. Koziorowski[2], and Andrzej W. Przybyszewski[1]

[1] Polish-Japanese Academy of Information Technology,
Koszykowa 86, 02-008 Warsaw, Poland
{artur.szymanski,przy}@pja.edu.pl
[2] Faculty of Health Science, Department of Neurology,
Medical University of Warsaw, Warsaw, Poland
stanislaw.szlufik@gmail.com, dkoziorowski@esculap.pl

Abstract. Parkinson Disease (PD) is the second major neurodegenerative disease, which causes severe complications for patients' daily life. PD remains unspecified in many aspects including best treatment, prediction of its progression and precise diagnosis. In our study we have built machine learning (ML) models, which address some of those issues by helping to improve symptom evaluation precision by using advanced biomarkers such as fast eye movements. We have built and compared model accuracy relaying on data from two systems for recording eye movements: one is saccadometer (Ober Consulting), and another is based on the Eye Tribe (ET1000). We have reached 85% accuracy in prediction of neurologic attributes based on ET and 82% accuracy with saccadometer with help of rough set theory. The purpose of this study was to compare ET with clinically approved eye movement measurements saccadometer of Ober. We have demonstrated in 8 PD patients that both systems gave comparable results based on neurological and eye movement measurements attributes.

Keywords: Data mining · Eye tracking · Parkinson Disease · Rough set theory

1 Introduction

As the one of the most common neurodegenerative diseases we still obtain a lot of imprecise diagnosis of Parkinson's Disease. Currently in treatment we rely heavily on experience of neurologist. Symptoms and disease progression can vary significantly between patients and it is unclear what is exactly optimal treatment. In our study we have used ML methods to build classifiers in order to assist in objective assessment of PD patients using reflexive saccades (RS) as biomarker.

Approaches of using ML in PD assessment have already been carried out, using variety of biomarkers to improve objective evaluation. For example, in work of Tsanas et al. [1] we find examples of using speech signal processing which can detect dysphonia in PD patients with accuracy reaching 99%. Other studies [2] shows examples of using machine learning on MRI data in order to classify PD clinically diagnosed

© Springer International Publishing AG 2017
N.T. Nguyen et al. (Eds.): ACIIDS 2017, Part II, LNAI 10192, pp. 351–358, 2017.
DOI: 10.1007/978-3-319-54430-4_34

patients against control group. By analysing voxel data and processing it with classifier based on Support Vector Machines (SVM) it was possible to reach specificity and sensitivity above 90%.

Other recent studies [3] show that correlation of few biomarkers can give excellent results for prediction of early stages of PD. Also highlighting importance of such effort in treatment. Authors have shown that combination of non-motor features can provide high accuracy for predicting early PD. Using data from Parkinson's Progression Markers Initiative they were able to benchmark few classification approaches reaching 96% accuracy with SVM.

Our own efforts in building automated and doctor independent solution shows that ML approach could be extremely efficient in classification of PD and help neurologists in patient assessment. We have used different biomarkers to demonstrate their importance in PD diagnosis, including DTI imaging [4] and single-photon emission computed tomography [5]. In most recent works [6, 7] we have shown than RS data can be used for building intelligent classifiers which reach over 90% accuracy in predicting PD patient features which make them important biomarker.

In this work we present new approach of recording eye movement data using software developed by our team based on Eye Tribe (ET) framework. We have built models using data from two eye tracking systems: clinically approved saccadometer with a new in the clinic ET. We have demonstrated that a low cost ET framework can be effectively used in the clinic in order to improve prediction of PD symptoms.

2 Methods

In our study we analysed data of 8 PD patients, in 21 sessions. Every patient had from 1 to 4 sessions. For each patient standard neurological tests were recorded. Each session determined whether a particular patient has deep brain stimulation (DBS) of subthalamic nucleus or the best medical treatment (BMT) enabled. Patients tested: in session one (S1) were off DBS and off BMT; in session two (S2) were on DBS and off BMT; in session three (S3) were off DBS and on BMT; in session four (S4) were both on DBS and on BMT. Not all patients in our study have recordings for all four sessions examinations. As a qualification parameter we took quality of eye data captured for given patient, for given session. Eye data used in this study includes reflexive saccades (RS) which were recorded using two systems, one from Ober Consulting - saccadometer and another one developed by us on the bases of Eye Tribe (ET) tracker.

During procedure with Ober saccadometer, a patient sat in front of the wall with the device mounted on his/her head. After starting the procedure the patient saw a red dot in front of him/her. The dot moved randomly to the left or right, and after about a second came back to the central position. Patient's task was to follow fast moving spot, which is equivalent with performing RS. This experimental protocol was the same for both devices. By means of Ober saccadometer, data were recorded with sampling frequency of 1000 Hz.

The ET system has used infrared camera positioned in the front of the patient and under LCD monitor. Camera tracked positions of each eye separately. Before each examination patient was asked to perform calibration by performing short fixations on

7 or 9 spots displayed on the screen one by one in different locations. Following the calibration, the ET procedure were similar to those in Ober device with the difference that marker in ET set-up was displayed on the LCD screen. Data in ET were sampled with frequency of 30 Hz.

The process in ET solution is managed with Java application using ET API, providing build in functions for calibration, predefined and custom procedures for RS and pursuit eye movements, online preview of current procedure, simplified error correction and data preview module.

There were following differences in both systems: different stimulus displayed means, different sampling frequencies and different data presentation methods. From ET we are receiving signal data with help of provided framework while Ober produce aggregated static parameters like delay or latency of eye movement averaged for all saccades. ET method of displaying data is easier to access the raw signal data from which we can remove artefacts and manually process different data parameters. Signal data samples from ET are sent to our program every 1 s.

There is another important difference related to placing of these devices. Ober saccadometer is fixed on patient's head, while ET camera is positioned in front of patient under the LCD screen. These differences result in different sensitivities to artefacts related to patient's head movements.

Another, mentioned above difference is related to the light stimulus display. Ober has used a red dot with static spot location, which was jumping by 10-degree to the left or the right. In ET we use a light spot displayed on LCD screen. Movements of the spot on the screen were described in details above. Ober saccadometer has software that automatically calculate saccades attributes and is taking distance from the eye to the spot into account. In ET it was necessary to measure this distance for each subject that has strong influence max saccades speed calculations.

Data from ET solution displays simultaneously eye movement measurements and related movements of the light spot on the screen. We have analysed recorded data by rewriting algorithms from our previous studies [6] (with help of python with a standard frameworks like "numpy"). Attributes which we took into consideration included latencies, max speed and amplitude. All of them described in our previous studies [6].

Both systems had possibilities to measure movement of each eyes separately. However, in this study we did not take asymmetric eye responses into account, therefore we have averaged measurements for both eyes. Figure 1 represents an example of our saccades recording.

In addition to eye movements, we have collected standard neurological parameters for PD patients like age, sex, Unified Parkinson's Disease Rating, Hoehn and Yahr scale, Schwab England scale, PDQ (quality of life measurement) and others. Full index of used attributes is presented in the input data table in the results section.

We have processed our data with help of the machine learning and data mining software in order to build classifiers for predicting effects of different treatments (session number) and total UPDRS. We have used two data mining software: Rough Set Exploration System [8–10] (RSES) and KNIME.

In KNIME we have built workflow, which applied number of algorithms to our dataset by at first by binning selected attributes of the input data into the buckets. Size and type of buckets where specific to each model we built and is noted in the results

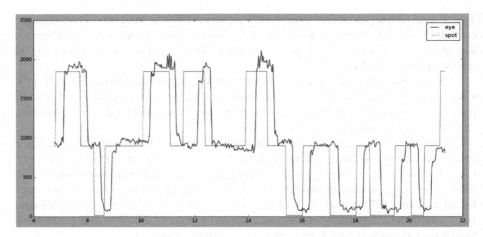

Fig. 1. This is the plot of eye position in time domain (x-axis). This recording was from the ET system. On the y-axis are position of eye gaze (ragged) and light spot (squared line)

section per classifier. KNIME provide binning methods for numeric values using two algorithms: equal bin and frequency bin method, first bin values based on minimum and maximum values to achieve numerically equal subsets while frequency bin try to perform binning so each subset have equal number of elements.

In RSES binning were implemented using built in discretization function [10], which included generating local cuts for number attributes excluding symbolic attributes.

Independently from used method, we have applied n-fold cross validation function, number of folds were specific for given model and as well as exact process parameters is noted in the results section.

3   Results

In the first part we built classifiers for predicting total UPDRS. As an input table for our model we have used data partly shown in Table 1.

We have built model for predicting total UPDRS. In order to verify predictions of our model (rules) we have used n-fold validation by dividing our set into n groups in two situations: (1) with- and (2) without-results of the eye movement measurements. In this way, it was possible to find importance of the reflexive saccades (RS) measurements on our predictions. We have binned total UPDRS into 4 groups with equal frequency algorithm, as described in the methods section. We have used RSES applying discretization with local cuts (excluding symbolic attributes) and the decision tree algorithm with 3-fold cross validation. Accuracy reached without RS data was 38.9%.

Next, we have added to our dataset RS attributes based on ET solution and applied similar process as for data in the first step. Table 2 presents information table after discretization.

Table 1. An example of data table for building classifier for total UPDRS in RSES. Legend for rows: mtre: mean delay right eye, stre: standard deviation for delay in right eye, stdredur/stdledur: standard deviation for saccade duration in right/left eye, latency_mean: mean latency for both eyes.

Patient #	'13/PD/BMT/2013'	'14/PD/POP/2010'	'55/PD/DBS/2013'	'56/PD/DBS/2013'
YearOfBirth	1948	1979	1955	1948
Sex	0	0	0	1
MonthsAfterDBS	18	48	12	12
Weight	61	58	70	88
BMT_dosage	750	400	500	MISSING
UPDRS_I	3	0	3	0
UPDRS_II	20	18	8	7
UPDRS_III	30	53	8	8
UPDRS_IV	2	2	3	2
UPDRS_TOTAL	55	73	22	17
Hoeh & Yahr scale	2.5	3	1	1
SchwabEnglandScale	70	60	90	90
PDQ39	77	34	26	49
AIMS	0	0	11	0
Epworth	9	7	6	12
ET_latency_mean	0.23	0.26	0.24	0.26
ET_latency_sd	0.06	0.09	0.08	0.08
ET_maxspeed_mean	1.73	0.41	0.53	0.97
ET_dur_mean	0.54	0.39	0.44	0.43
ET_dur_sd	0.54	1.04	0.43	0.48
Session	'S1'	'S1'	'S4'	'S4'

As states in Table 2 discretization algorithm selected only few significant attributes in order to create classification rules. Those included 5 RS attributes not only proving importance of RS saccades but also showing that ET parameters can be efficient in building classifier for PD. In contrast to model described in classification of dataset

Table 2. Discretized table for building classifier for predicting total UPDRS using ET RS data. Legend for RS attributes as in Table 1.

Patient #	'13/PD/BMT/2013'	'14/PD/POP/2010'	'14/PD/POP/2010'	'14/PD/POP/2010'
YearOfBirth	(−Inf, 1971.5)	(1971.5, Inf)	(1971.5, Inf)	(1971.5, Inf)
PDQ39	(46.5, Inf)	(30.5, 46.5)	(30.5, 46.5)	(30.5, 46.5)
AIMS	(−Inf, 5.5)	(−Inf, 5.5)	(−Inf, 5.5)	(−Inf, 5.5)
ET__mtre	(−Inf, 0.27)	(−Inf, 0.27)	(−Inf, 0.27)	(0.27, Inf)
ET__stre	(−Inf, 0.055)	(0.055, Inf)	(−Inf, 0.055)	(0.055, Inf)
ET__stdredur	(0.3349, Inf)	(−Inf, 0.3349)	(−Inf, 0.3349)	(0.3349, Inf)
ET__stdledur	(0.255, Inf)	(0.255, Inf)	(0.255, Inf)	(0.255, Inf)
ET__latency_mean	(−Inf, 0.255)	(0.255, Inf)	(−Inf, 0.255)	(0.255, Inf)
Session	'S1'	'S1'	'S2'	'S3'
UPDRS_TOTAL	(39, 64]	(64, 95]	(22, 39]	(39, 64]

without RS data we were able to achieve 72.2% accuracy, which is significantly better result.

We have performed another test using RS data from Ober saccadometer, applying the same method. The discretized table is shown below as Table 3.

Table 3. Discretized table with RS from Ober.

Patient #	'13/PD/BMT/2013'	'14/PD/POP/2010'	'14/PD/POP/2010'	'14/PD/POP/2010'
Sc_LatencyMeanLEFT	(−Inf, 217.0)	(263.0, 330.0)	(−Inf, 217.0)	(263.0, 330.0)
Sc_AmplitudeRIGHT	(10.35, 12.1)	(−Inf, 10.35)	(12.10, Inf)	(−Inf, 10.35)
Sc_PeakVelocityLEFT	(335.0, Inf)	(335.0, Inf)	(335.0, Inf)	(335.0, Inf)
Sc_LatencyMeanALL	(−Inf, 305.0)	(−Inf, 305.0)	(−Inf, 305.0)	(−Inf, 305.0)
Sc_DurationALL	(48.5, Inf)	(48.5, Inf)	(48.5, Inf)	(−Inf, 48.5)
session	'S1'	'S1'	'S2'	'S3'
UPDRS_TOTAL	(39, 64]	(64, 95]	(22, 39]	(39, 64]

As shown in Table 3 attributes significant for building classifier in our model again relay on RS, also we can note that some of the attributes like latency are used in both Ober and ET approach. Accuracy for this model was similar to model based on RS using ET and reached 66.7%.

Next we built model for predicting session number for PD patients. We have followed similar path as in building classifier for total UPDRS. Results for different datasets are shown in Table 4.

Table 4. Result for predicting session number using different classifiers.

	Accuracy
No RS data	58.30%
ET RS data	85%
Ober RS data	82.20%

There are similar trends as in case of prediction of total UPDRS, dataset with no RS had lowest accuracy. Those with RS data either from Ober or ET have similar accuracies; in case of prediction of the session number we are getting even better outcomes than for total UPDRS. Highest accuracy we have reached using combined dataset including RS from both ET and Ober systems.

We have combined list of attributes used in classification after discretization process of each dataset to show significant attributes used in building our models, those are shown in Table 5.

In the last step of our study we have run other common ML algorithms using KNIME on the datasets that proved to give best results in RSES. Additionally we have calculated other measures such as Cohen's kappa and Matthews Correlation Coefficient (MCC). As we can see in Table 5 we are able to reach high accuracy for predicting total UPDRS using standard algorithms reaching 85.7% and 71.4% respectively for WEKA decision tree algorithm and random forest (Table 6).

Table 5. Table showing significant attributes used to build model for predicting session number. Asterisk mark which columns were used while running classifier on given dataset.

	No RS data	ET RS data	OB RS data
Patient #	*	*	*
YearOfBirth	*	*	*
MonthsAfterDBS	*	*	*
UPDRS III	*	*	*
UPDRS IV	*		
ET_latency_sd		*	
OB_duration_mean			*
OB_amplitude_mean			*

Table 6. Results for building classifiers for predicting total UPDRS using common ML algorithms, including measures as Cohen's kappa and Matthews correlation coefficient.

	Accuracy	Cohen's Cappa	MCC
WEKA - decision table	85.71%	0.808	0.818
Random forest	71.43%	0.618	0.611
WEKA - random forest	57.14%	0.434	0.499
Tree ensemble	61.90%	0.488	0.421

4 Conclusions

Our study proved that we have successfully used a low cost ET eye tracker for clinically relevant eye movement measurements. We have measured parameters of the reflexive saccades (RS) and with help of the discretization process choose only relevant ones. We have built models for predicting PD patient session number as different treatments effectiveness and total UPDRS as general patient conditions. Our predictions of session number and UPDRS had a high accuracy when ET was used as well as the commercial saccadometer was utilized.

We have confirmed that fast eye movements are important biomarker for PD. Both devices, clinically approved Ober saccadometer and adapted by us to the clinical measurements - Eye Tribe can be used to improve diagnostic of PD symptoms.

Bibliography

1. Tsanas, A., Little, M.A., McSharry, P.E., Spielman, J., Ramig, L.O.: Novel speech signal processing algorithms for high-accuracy classification of Parkinson's disease. IEEE Trans. Biomed. Eng. **59**, 1264–1271 (2012)
2. Salvatore, C., Cerasa, A., Castiglioni, I., Gallivanone, F., Augimeri, A., Lopez, M., Arabia, G., Morelli, M., Gilardi, M.C., Quattrone, A.: Machine learning on brain MRI data for differential diagnosis of Parkinson's disease and progressive supranuclear palsy. J. Neurosci. Methods **222**, 230–237 (2014)

3. Prashanth, R., Dutta Roy, S., Mandal, P.K., Ghosh, S.: High-accuracy detection of early Parkinson's disease through multimodal features and machine learning. Int. J. Med. Inf. **90**, 13–21 (2016)

4. Szymański, A., Przybyszewski, A.W.: Rough set rules help to optimize parameters of deep brain stimulation in Parkinson's patients. In: Ślęzak, D., Tan, A.-H., Peters, J.F., Schwabe, L. (eds.) BIH 2014. LNCS (LNAI), vol. 8609, pp. 345–356. Springer, Heidelberg (2014). doi:10.1007/978-3-319-09891-3_32

5. Szymanski, A., Szlufik, S., Dutkiewicz, J., Koziorowski, D.M., Cacko, M., Nieniecki, M., Przybyszewski, A.W.: Data mining using SPECT can predict neurological symptom development in Parkinson's patients. In: 2015 IEEE 2nd International Conference on Cybernetics (CYBCONF), pp. 218–223 (2015)

6. Przybyszewski, A.W., Kon, M., Szlufik, S., Dutkiewicz, J., Habela, P., Koziorowski, D.M.: Data mining and machine learning on the basis from reflexive eye movements can predict symptom development in individual Parkinson's patients. In: Gelbukh, A., Espinoza, F.C., Galicia-Haro, S.N. (eds.) MICAI 2014. LNCS (LNAI), vol. 8857, pp. 499–509. Springer, Heidelberg (2014). doi:10.1007/978-3-319-13650-9_43

7. Przybyszewski, A.W., Kon, M., Szlufik, S., Szymanski, A., Habela, P., Koziorowski, D.M.: Multimodal learning and intelligent prediction of symptom development in individual Parkinson's patients. Sensors **16**, 1498 (2016)

8. Pawlak, Z.: Rough Sets. Springer, Dordrecht (1991)

9. Bazan, J.G., Nguyen, H.S., Nguyen, S.H., Synak, P., Wróblewski, J.: Rough set algorithms in classification problem. In: Polkowski, P.L., Tsumoto, P.S., Lin, P.T.Y. (eds.) Rough Set Methods and Applications, pp. 49–88. Physica-Verlag HD, Heidelberg (2000)

10. Bazan, J.G., Szczuka, M.: RSES and RSESlib - A Collection of Tools for Rough Set Computations. In: Ziarko, W., Yao, Y. (eds.) RSCTC 2000. LNCS (LNAI), vol. 2005, pp. 106–113. Springer, Heidelberg (2001). doi:10.1007/3-540-45554-X_12

Rules Found by Multimodal Learning in One Group of Patients Help to Determine Optimal Treatment to Other Group of Parkinson's Patients

Andrzej W. Przybyszewski[1(✉)], Stanislaw Szlufik[2], Piotr Habela[1], and Dariusz M. Koziorowski[2]

[1] Polish-Japanese Academy of Information Technology,
02-008 Warszaw, Poland
{przy,piotr.habela}@pja.edu.pl
[2] Department of Neurology, Faculty of Health Science,
Medical University, Warsaw, Poland
stanislaw.szlufik@gmail.com, dkoziorowski@esculap.pl

Abstract. We have already demonstrated that measurements of eye movements in Parkinson's disease (PD) are diagnostic. We have performed experimental measurements of fast reflexive saccades (RS) in PDs in order to predict effects of different therapies. We have also found rules by means of data mining and machine learning (ML) in order to classify how different doses of medication have determined motor symptoms (UPDRS III) improvements. These rules from one group of 23 patients only on medications were supplied to another group of 18 patients under medications and DBS (deep brain stimulation) therapies in order to predict motor symptoms changes. Such parameters as patient's age, neurological and saccade's parameters gave a global accuracy in the motor symptoms predictions of 76% based on the cross-validation. Our approach demonstrated that rough set rules are universal between groups of patients with different therapies that may help to predict optimal treatments for individual PDs.

Keywords: Neurodegenerative disease · Rough set · Decision rules · Granularity

1 Introduction

Our knowledge about brain's algorithms, especially about their plastic properties is still very limited. The compensatory mechanisms of the brain are many times better than in artificial NN, which gives a great advantage of the natural in comparison to artificial intelligent systems. A disadvantage of such great plastic system is that in the most cases, subject may notice first symptoms of brain dysfunctions when a big part of his/her brain is already dead. We still do not know how to recover dead cells, as it is a case in the neurodegenerative diseases (ND) such as Alzheimer (AD) or Parkinson's (PD). Neurodegenerative are related to the brain inability to further compensate lost of many neurons in the Central Nervous System. As these diseases starts many years

© Springer International Publishing AG 2017
N.T. Nguyen et al. (Eds.): ACIIDS 2017, Part II, LNAI 10192, pp. 359–367, 2017.
DOI: 10.1007/978-3-319-54430-4_35

before the first visible symptoms, their progress is complex and individually variable. Good neurologists use their experience to propose and optimize various treatments in order to improve patients' symptoms. Doctors use their knowledge and intuition based on their measurements and assumptions of typical cases that are probably based on averaging severities of symptoms. As they have limited time, they cannot see each patient very often, adjustment of the treatment is spare. We propose to improve their diagnosis and statistical analysis of PD symptoms with help of data mining and machine learning (ML) procedures such as rough set theory or C4.5 decision tree in Weka.

Better symptoms classifications may improve therapy, lead to slowing down symptoms and in the consequence improve in patients' quality of life.

Our symptom classification methods are similar to that used by our visual system in object recognition [1], as we have discussed in more details before [2]. ML algorithms by comparing large amount of data should improve prediction and indication for more precise analysis of variations in different patients' conditions.

Standard methods of the disease stage estimation used by neurologists are grounded on Hoehn and Yahr and the UPDRS (Unified Parkinson's Disease Rating) scales. Obtained results of these measurements are dependent on the doctor's experience and not always consistent between different hospitals. In other words, they are imprecise and not objective in contrast to our measurements. In addition, we also propose to process our experimental measurements and data obtained by doctors in an intelligent way. We will use data mining and ML approaches grounded on neurologist's diagnosis for learning rules that will be applied to different groups of patients in order to predict potential effectiveness of different therapies.

2 Methods

Our data are from 41 Parkinson Disease (PD) patients divided into two groups: (1) BMT-group (best medical treatment group) with 23 PD patients that were only on medications; (2) DBS-group of 18 PDs that went to the DBS (Deep Brain Stimulation) surgery in the Institute of Neurology and Psychiatry WUM that was mainly determined by fluctuations (ON-OFF effects) of their motor symptoms. All patients went through the EM tests. They have to perform horizontal RS (reflexive saccades) in four sessions. BMT-group of patients was tested only in the first and third sessions: MedOn/Off sessions (sessions with or without medication). The second DBS-group of patients was tested in all 4 sessions related to medication and stimulation combinations: (1) MedOffDBSOff; (2) MedOffDBSOn; (3) MedOnDBSOff; (4) MedOnDBSOn. Details of these procedures were described earlier [2]. Tests of different motor and non-motor tasks (UPDRS I – IV) and psychological tests were performed by doctors in agreement with the standard neurological procedures. Fast eye movements (reflexive saccades - RS) were recorded by means of the Ober saccadometer (Poland) as described in [2]. This infrared and head mounted system was practically insensitive to patients' small movements that are partly related to the specificity of the disease. Patients were sitting in a chair without an unnatural chinrest.

As described in more details before [2, 3] patient was observed a computer monitor and followed moving randomly to the right or left from the fixation point light spot [2].

In this study we have measured and analyzed only fast saccadic (RS) eye movements that means that other eye movements (pursuit or antisaccades) were rejected. As described earlier [2], we were interested in the resulting parameters such as: the latency – a time difference between the beginning of the light spot movements and beginning of the saccade; saccade's amplitude in comparison to the light spot amplitude; max velocity averaged for different saccades; duration of saccade measured as the time from the beginning to the end of the saccade.

Each PD patient had to complete 10 reflexive saccades without medications (did not obtain medication) in one series for the BMT-group and two series: with DBS off and DBS on for the DBS-group. After taking the appropriate dosage of medication, patient was waiting about half of hour before the next series of measurements. Then the identical RS measurements were completed in one series for the BMT-group and two series with DBS off and DBS on for the DBS-group.

Institutional Ethic Committee at the Warsaw Medical University approved all procedures.

2.1 Theoretical Basis

In this work we have represented our data as a table (so-called decision table) where rows represented different measurements (on the same or different patients) and columns were related to different attributes. An information system was defined after Pawlak [4]) as a pair $S = (U, A)$, where U, A are finite sets: called U – is the universe of objects; and A - the set of attributes. The value $a(u)$ is a unique element of V (where V is a value set) for $a \in A$ and $u \in U$.

Following [4] we define the *indiscernibility relation* of any subset B of A or *IND* (B) as: $(x, y) \in IND(B)$ or $xI(B)y$ iff $a(x) = a(y)$ for every $a \in B$ where the value of $a(x) \in V$. It is an equivalence relation $[u]_B$ that we understand as a B-*elementary granule*. The family of $[u]_B$ gives the partition U/B containing u will be denoted by $B(u)$.

A **lower approximation** of set $X \subseteq U$ in relation to an attribute B is defined as $\underline{B}X = \{u \in U : [u]_B \subseteq X\}$. The **upper approximation** of X *is* defined as $\overline{B}X = \{u \in U : [u]_B \cap X \neq \phi\}$. The difference of $\overline{B}X$ *and* $\underline{B}X$ is the boundary region of X that we denote as $BN_B (X)$. If $BN_B (X)$ is empty then set than X is *exact* with respect to B; otherwise if $BN_B (X)$ is not empty and X is not *rough* with respect to B. We can interpret the B-lower approximation of the set X as union of all B-*granules* that are included in X and the B-upper approximation of X is of the union of all B-*granules* that have nonempty intersection with X.

A decision table for S is the triplet: $S = (U, C, D)$ where: C, D are condition and decision attributes [5]. As we will show below, we can interpret each row of this table as a particular rule that connects condition and decision attributes (e.g. Eqs. 1 and 2 below). As there are many rows related to different patients and sessions, they gave many particular rules. Our rough set approach allows generalizing these rules into universal hypotheses that may determine optimal treatment options for an individual PD patient.

It is well known that each PD patient needs to some extend a different treatment. As a consequence, we need rules that are enough flexible e.g. they are related to the granular computation that may simulate how neurologists interact with patients. As experienced doctor knows by intuition that some symptoms are more important than others and has a feeling which treatment may work. Good neurologist may consider different approaches to the disease and its treatment, or in other words, various levels of abstraction (different granularities). Granular computing is similar to the intelligent visual bran object classifications [3], where an object is set of different attributes and each measurement gives different subjective rule. The higher areas of the visual system can recognize different parts of the object as doctor classifies different symptoms. The right object recognition is like the optimal classification of the disease.

In our previous papers [2, 3] we have divided experimental measurements into several subsets and we were testing each subset by applying rules that we have obtained from other subsets. In this work, we have changed our procedure in such a way that we used sets from different patients for training and testing. Moreover, these different groups of patients have different treatments. These approach shows that our rules are more universal and may be applied to different groups of patients.

In addition to rough set algorithms [6], we have used C4.5 (J48) classifier and visualized in a decision tree implemented in WEKA (as in [2]).

3 Results

All 41 PD patients have mean age of 56 +/− 11.7 (SD) years with mean disease duration of 8.3 +/− 3.7 years and with LEDD (L-Dopa equivalent daily doses calculated as L-Dopa and equivalent L-Dopa agonist dosage) 700 +/− 550 mg have two different treatments: BMT-group (only medication) and DBS-group (medication and with implanted electrodes in the basal ganglia STN nucleus).

In the BMT group of 23 patients with mean age of 57.8 +/− 13 (SD) years; disease duration was 7.1 +/− 3.5 years, with LEDD 1010 +/− 530 mg. The second DBS group of 18 patients with mean age of 53.7 +/− 9.3 (SD) years; disease duration was 9.7 +/− 3.5 years (statically significant longer disease duration than BMT-group: $p < 0.025$), with LEDD 300 +/− 220 mg (stat. sig. lower than group BMT-group: $p < 0.0005$).

3.1 J48 Decision Tree in WEKA

We have downloaded our data to WEKA from the DBS-group of 18 patients that in form are equivalent to the information table (Table 1).

The full table has 18 (subjects) × 4 (sessions) = 72 objects (measurements). In the Table 1 are values of eleven attributes for two patient where: P# is the patient number, age is the patient's age, LEED is L-Dopa equivalent daily dose of medications, t_dur is the duration of the PD disease, UPDRS III – scale related to movement symptoms, UPDRST – scale related to integration of many different disorders, parameters of saccades are following: SccDur is the mean duration of 10 saccades; SccLat is the mean

Table 1. Part of the information table.

P#	Age	t_dur	LEED	UPDRSIII	UPDRST	SccLat	SccDur	SccAmp	SccVel	S#
45	69	13	0	42	62	255	44	8.8	396	1
45	69	13	750	27	47	331	43	9.3	406	2
45	69	13	0	21	39	336	44	10.3	459	3
45	69	13	750	12	30	241	46	10.7	458	4
46	58	7	0	41	63	276	49	14.2	579	1
46	58	7	200	38	60	83	55	8.3	284	2
46	58	7	0	18	31	321	51	11.9	429	3
46	58	7	200	12	25	312	51	9.9	360	4

latency of saccades, SccAmp is the mean amplitude of 10 saccades, SccVel is the mean velocity of saccades, and S# - session number.

In the decision tree (Fig. 1) root is LEDD (dosage of medication) as we have two different situations when patients are off (sessions: 1, 2) and on (sessions 3, 4) – medications represented by two branches. The next two leaves are interesting as without medication important is the value of the UPDRS III whereas with medication – the value of total UPDRS is critical. In the lowest row are predictions for each node, accompanying with numbers in brackets. The first number is the total number of instances (weight of instances) reaching the leaf. The second number is the number (weight) of those instances that are misclassified. E.g. in the first leaf node from the left, assigned as instance class 3 there are 22 classified instances assigned to this leaf and 4 instances are assigned in error.

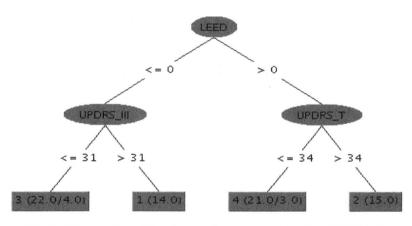

Fig. 1. Visualization of the decision tree output obtained by J48 (Weka).

As states above in Table 2, data were tested by 10-fold validation, and the global precision was 0.854, the global recall 0.847, the global accuracy 0.897, and the MCC (Matthews correlation coefficient) was 0.8.

Table 2. Confusion matrix for different session numbers (1–4), DBS-group

Predicted

		Session 1	Session 2	Session 3	Session 4	Precision
Actual	Session 1	14	0	4	0	0.93
	Session 2	0	14	0	4	0.88
	Session 3	1	0	17	0	0.81
	Session 4	0	2	0	16	0.8

3.2   Data Mining Related to Rough Set (RS) Theory and ML Methods

As mentioned in the Methods section we have used the RSES 2.2 (Rough System Exploration Program) [5] with implementation of RS rules to process our data. At first, our BMT-group data was placed in the information table as originally proposed by Pawlak [4]. There is an extract of a part of data in Table 3 where P# - number of the patient; LEED – dosage of medications; UPDRST – total UPDRS; SccLat, SccDur, SccAmp, and SccVel are saccades latency, duration, amplitude and velocity; UPDRS III – motor symptoms tested as UPDRS procedures.

In the next step, using RSES, we have completed reduction and discretization of all attributes except of the patient number (see reduct in [2, 3] and Method section). In the table below (Table 4) for the same data as Table 3 we have performed discretization:

In the first column are number of PD patients same as in Table 3 not digitized; in the second column, the patient's dosage of medication divided in two groups below and above 732.5 mg; in the third total UPDRS also divided into two ranges; and other parameters of saccades were also divided into ranges. Notice that SccVel (saccade velocity) was eliminated as significant condition attribute. The decision attribute UPDRS III was divided into three ranges: above 30.5; between 4.5 and 30.5; and below 4.5 (the last column). The first row of the Table 4 we can replace by the following rule:

('Pat'=4)&('LEED'="(-Inf,732.5)"))&('UPDRST'="(28,Inf)")&('SccLat'="("(-Inf, 231.0)")&('SccDur'="(53.35,Inf)")&('SccAmp')="(9.6,10.2)")=>('UPDRSIII'="(30.5, Inf)")

$$(1)$$

Table 3. Part of the information table.

P#	LEED	UPDRST	SccLat	SccDur	SccAmp	SccVel	UPDRSIII
4	0	44	152.3	53.7	10	340.4	35
4	1540	14	162.1	57.3	9.1	319.8	9
5	0	40	253	51.1	12.1	454.3	15
5	660	31	221.7	50.4	10.6	467.5	10
6	0	65	212	51	8	356	37
6	1500	44	457	63	9.5	298	18
7	0	44	233	62.4	10	302.4	19
7	1330	22	247	56	9.7	371	7

Table 4. Extract from the discretized-table for BMT-group

P#	LEED	UPDRST	SccLat	SccDur	SccAmp	SccVel	UPDRSIII
4	"(-Inf,732.5)"	"(28,Inf)"	"(-Inf,231.0)"	"(53.35,Inf)"	"(9.6,10.2)"	*	"(30.5,Inf)"
4	"(732.5,Inf)"	"(-Inf,28)"	"(-Inf,231.0)"	"(53.35,Inf)"	"(-Inf,9.6)"	*	"(4.5,30.5)"
5	"(-Inf,732.5)"	"(28,Inf)"	"(231,270.5)"	"(50.75,53.35)"	"(10.2,Inf)"	*	"(4.5,30.5)"
5	"(-Inf,732.5)"	"(28,Inf)"	"(-Inf,231)"	"(-Inf,50.75)"	"(10.2,Inf)"	*	"(4.5,30.5)"
6	"(-Inf,732.5)"	"(28,Inf)"	"(-Inf,231)"	"(50.75,53.35)"	"(-Inf,9.6)"	*	"(30.5,Inf)"
6	"(732.5,Inf)"	"(28.0,Inf)"	"(270.5,Inf)"	"(53.35,Inf)"	"(-Inf,9.6)"	*	"(4.5,30.5)"
7	"(-Inf,732.5)"	"(28,Inf)"	"(231,270.5)"	"(53.35,Inf)"	"(9.6,10.2)"	*	"(4.5,30.5)"
7	"(732.5,Inf)"	"(-Inf,28)"	"(231,270.5)"	"(53.35,Inf)"	"(9.6,10.2)"	*	"(4.5,30.5)"

We read this formula above (Eq. 1) as stating that each row of the table (Table 1) can be written in form of this equation (Eq. 1). It states that if we evaluate patient #4 *and* with LEED below 732,5 *and* total UPDRS above 28 *and* saccade latency below 231 *and* with saccade duration above 53.35 *and* ... *and* saccade amplitude between 9.6 and 10.2 *then* patient's UPDRS III is above 30.5. In the next step, the RSES found more general rules such as below (2):

$$('LEED'="(-Inf,732.5)")\&('SaccDur'="(-Inf,50.75)")\&('UPDRS\_T'="(-Inf,28.0)")$$
$$\&('SaccLat'="(-Inf,231.0)")=>('UPDRS\_III'="(4.5,30.5)"[5])$$

$$(2)$$

This rule states that if LEED is below 732,5 *and* saccade duration below 50.75 *and* total UPDRS below 28 *and* saccade latency below 231 *then* patient's UPDRS III is between 4.5 and 30.5 in five cases.

Such equations as Eq. 2 are parts of a data mining system bases on rough set theory [6]. We have tested our rule using the machine-learning concept. We have classified DBS-group data using rules generated by BMT-group population using a standard voting in order to generate a confusion matrix. The results are gives in Tables 3 and 5.

TPR: True positive rates for decision classes; ACC: Accuracy for decision classes: the global coverage was 0.31 and the **global accuracy was 0.77**, the coverage for decision classes: 0.39, 0.75, 0.18.

Table 5. Confusion matrix for UPDRS III of DBS-group by rules obtained from BMT-group.

Predicted		"(4.5, 30.5)"	"(-Inf, 14.0)"	"(30.5, Inf)"	ACC
Actual	"(4.5, 30.5)"	24	0	1	**0.96**
	"(-Inf, 14.0)"	3	0	0	**0.0**
	"(30.5, Inf)"	4	0	3	**0.429**
	TPR	**0.76**	**0.0**	**0.75**	

In summary, the last result has demonstrated that we can classify one group of 18 PD patients by another one of 23 PD with a high accuracy even if they have different treatments.

4 Discussion

In the actually used neurological procedures that are approved after many cases and clinical tests, we still do not know if the actual treatments are optimal. Also treatments between different centers are not always consistent in addition to differences between different neurologists and individual PD cases. We have compared standard neurological diagnostic protocols with our methods related to the data mining and machine learning algorithms. The main advantages of our methods are their precision and objectivity. They are not dependent on a particular center or neurologist and in addition they can be performed automatically. As they can be performed without doctor, they can also be effective by saving precious time of neurologist or even, in the future, performed in patient's home. Another advantage of our methods, which we have demonstrated in this work, is that they can estimate symptoms and treatments of one population on the basis of the other one. It may be used to predict effects of different treatment on an individual patient and or to compare methods of different centers. Especially the second reason is important as it may lead to unification of the treatments between different centers by objective comparison which methods are more effective in different cases. It also may lead to create large databases with thousands cases that may experimentally support any particular, new case. It also points to another question if we can treat some patients with different neurodegenerative diseases in a similar ways, as partly the mechanisms are similar. In order to response to such questions, we need databases of different diseases with similar objective measurement and intelligent comparisons between different populations as is, for a particular disease, demonstrated above. If successful such approach may, in the future, open possibilities of automatic, human independent diagnosis and treatments.

5 Conclusions

In this study that follows others our experiments [2, 3], we have compared the popular statistical approach used by majority of neurologists for PD diagnosis with data mining classifications. We have matched symptoms and treatments of individual patients belonging to the one group (DBS-group with two different treatments) with another BMT-group only on medication treatment. We have demonstrated that with help of the eye movement measurements we can predict, using intelligent methods, motor symptoms of another DBS-group, even if therapies effects are strongly patient dependent. Our intelligent approaches gave precise motor symptoms predictions on the basis of one group of patients' treatments to another group of patients with different treatments. In addition, we have also demonstrated that depends on the dosage of the medication different attributes (neurological testing) are important for symptoms classification. This last finding may simplify some neurological procedures and help doctors to find the optimal treatment.

Acknowledgements. This work was partly supported by projects Dec-2011/03/B/ST6/03816, from the Polish National Science Centre.

References

1. Przybyszewski, A.W.: Logical rules of visual brain: from anatomy through neurophysiology to cognition. Cogn. Syst. Res. **11**, 53–66 (2010)
2. Przybyszewski, A.W., Kon, M., Szlufik, S., Szymanski, A., Koziorowski, D.M.: Multimodal learning and intelligent prediction of symptom development in individual Parkinson's patients. Sensors **16**(9), 1498 (2016). doi:10.3390/s16091498
3. Przybyszewski, A.W., Kon, M., Szlufik, S., Dutkiewicz, J., Habela, P., Koziorowski, D.M.: Data mining and machine learning on the basis from reflexive eye movements can predict symptom development in individual Parkinson's patients. In: Gelbukh, A., Espinoza, F.C., Galicia-Haro, S.N. (eds.) MICAI 2014. LNCS (LNAI), vol. 8857, pp. 499–509. Springer, Heidelberg (2014). doi:10.1007/978-3-319-13650-9_43
4. Pawlak, Z.: Rough Sets: Theoretical Aspects of Reasoning About Data, pp. 499–509. Kluwer, Springer, Dordrecht (1991, 2014)
5. Bazan, J., Son Nguyen, H., Trung, T., Nguyen, Skowron A., Stepaniuk, J.: Decision rules synthesis for object classification. In: Orłowska, E. (ed.) Incomplete Information: Rough Set Analysis, pp. 23–57. Physica-Verlag, Heidelberg (1998)
6. Bazan, J.G., Szczuka, M.: RSES and RSESlib - a collection of tools for rough set computations. In: Ziarko, W., Yao, Y. (eds.) RSCTC 2000. LNCS (LNAI), vol. 2005, pp. 106–113. Springer, Heidelberg (2001). doi:10.1007/3-540-45554-X_12

Intelligent Systems and Algorithms in Information Sciences

Reasoning in Formal Systems of Extended RDF Networks

Alena Lukasová, Martin Žáček[(⊠)], and Marek Vajgl

Department of Computers and Informatics,
University of Ostrava, Ostrava, Czech Republic
{alena.lukasova,martin.zacek,marek.vajgl}@osu.cz

Abstract. It is a fact that the RDF(S) model has been declared as the ground base for implementations of further web development conception. RDF provides a common and flexible way to decompose knowledge to elementary statements that allows, as networks of indivisible knowledge atoms, to be represented by RDF triples or by RDF graph vectors.

The article presents two graph based formal systems GRDF and RDFCFL defined on the base of extended RDF model with the help of clausal form logic principle and notation. The transformation process from the first order predicate logics (FOPL) to the RDF graph notation protects language expressivity and moreover both the presented systems share the partial decidability with the FOPL. As an example it is shown a reasoning of consequents in a monotonic version of the RDFCFL system.

Keywords: Knowledge representation · Graph formal language · Clausal form logics RDF model · Semantic web · Linked data

1 Motivation

To have a high level tool of knowledge representation and manipulation on the Web site it is important to have a formal language (within a formal system) on the base of a modification of first order predicate logic (FOPL) that

- is shared by a wide community,
- uses a relative simple language syntax, allowing a machine readability and handling,
- has a high expressivity relative to natural languages, similarly to the case of the FOPL,
- has a property to be placed into some external contexts, in order to capture better its semantics,
- The formal language with the characteristics above naturally ought to provide a mechanism of automated deduction in the frame of semantically correct and complete formal system.

When the idea of the Semantic Web has appeared at the beginning of this Millennium, the file of the properties shaped above has got one requirement more:

© Springer International Publishing AG 2017
N.T. Nguyen et al. (Eds.): ACIIDS 2017, Part II, LNAI 10192, pp. 371–381, 2017.
DOI: 10.1007/978-3-319-54430-4_36

- The language of knowledge representation should be a web language that ought to be suitable to maintain knowledge stored in linked data [1]. Thus, the language ought to follow the RDF(S) (Resource Description Framework (Schema)) model.

Recently all the requirements listed above in the computer science have been working partly in some way and have brought a lot of good results. Here we present two of the tools that according to our belief can fulfil the condition listed above.

2   Two RDF Graph Based Languages Following CFL Principle

Richards [8] proposed the Clausal Form Logic (CFL) on the base of the FOPL that corresponds to the mostly used conditional "if – then" statements (rules). Generally, a conditional statement (clause) says that a consequent composed as a disjunction of some predicate atoms follows from antecedent composed as a conjunction of some predicate atoms.

Developments in the issue of a formal knowledge representation clearly show that the language of the FOPL and its clausal form specifically is the appropriate formal language that can virtually represent any assertion formulated in a natural language.

Apart from a general predicate form, a special representation of a knowledge atom is used here – a vector in the graph language, which is already very well established in associative networks [3]. The graphic form of clausal form logic CFL language became the main idea of the two RDF graph languages in formal systems GRDF and RDFCFL presented here.

In relation to the FOPL, graphs of the GRDF and RDFCFL we present here, as well as associative networks, have been built on the foundations of atomic vectors using binary predicates representing simple network statements. A graph view of both languages is an easy-to-understand method of visualization that can represent information "as a whole" consisting of knowledge atoms together with some relevant or important atoms contributing to understanding of such the composed information.

3   Syntax of GRDF and RDFCFL Languages

The GRDF graph version language built on RDF has been presented in [4] as a follower of the language of associative networks with quantifiers using formal means of knowledge representation based on concept-oriented paradigm.

RDFCFL is in fact a direct follower of the graphic CFL nested into RDF model environment.

RDF provides a common and flexible way to decompose knowledge to elementary statements that allows, as networks of indivisible atoms of knowledge, to be represented by RDF triples or by RDF graph vectors. The transformation process from the first order predicate logics (FOPL) to the RDF graph notation protects language expressivity.

As both of the formal systems GRDF and RDFCFL belong to extended RDF model and language, each of the name symbols at nodes and edges in the Fig. 1 is in fact a URI name identifier (Fig. 1).

Fig. 1. The statement "X is a citizen of London"

The statement "X is a citizen of London" represents a RDF triple

$$< URI \, beginning > /universal\#X$$
$$< URI \, beginning2 > /citizen\_of$$
$$< URI \, beginning3 > /london$$

with a graphic shape at the figure.

3.1   Knowledge Bases and Formal Reasoning

Knowledge-aimed formal systems usually are built because of formal reasoning.

They consist of knowledge bases (formulas/networks) written in a special formal language and a mechanism of inference in the form of rules.

A knowledge base, usually a special domain-oriented, is created as a set of networks representing clauses, so that it contains unconditional facts and basic general "if – then" statements (conditional clauses as special axioms) about a domain of interest, usually supported by a corresponding domain ontology that ensures proper terminological clarity.

An important thing of the both formal systems presented here is a fact that they share partial decidability property with the first order predicate logics (FOPL).

4   Semantics of GRDF and RDFCFL

4.1   Model-Theoretical and Axiomatic Defining of Language Semantics

Formal systems derived from the language of FOPL usually share with the FOPL its generally wide used model-theoretic way of defining language semantics. Predicates in this case reflect real situations with all possible concrete real contents of concepts and their relationships. But the same work can do also language formulas (axioms) expressing generally in mutual relationships each of the concepts used within the represented reality. This fact has brought [3] an idea of semantic network representation into formal tools.

Both the GRDF and the RDFCFL are built besides the RDF also on the base of semantic network. It means both the systems brought a new sight into a reality to be represented. It is something more expressive relative to the original FOPL representation with its model-theoretic semantics.

4.2 Negative Atoms in GRDF

In the case of the GRDF the denial of a vector V is labelled by the special symbol of falsity ⊗, which says that if V is true, then it is followed by the consequent false (Fig. 2).

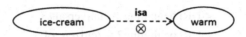

Fig. 2. The statement "Ice cream is not warm."

Unconditional networks in GRDF as well as in associative networks (ground (facts)/universal/existential) use only the logical connective & of conjunction to express that all the statements (vectors) must hold simultaneously even if some of them use in graphic vectors negative atoms with a special vector of the false.

As the conditions - ground/universal/existential of the general form <antecedent> → <consequent> in the GRDF have the meaning of the implication p_1 & p_2 & ... & p_n q the conditional networks in general give a possibility to express all logical connectives because the set of connectives {¬, &, →} has a property of the functional completeness.

The same holds in the case of RDFCFL that manipulates only with clauses of the form <antecedent> → <consequent> even if the link or the right side of the "→" can contain an empty set of atoms.

4.3 Clauses in RDFCFL

RDFCFL language does not allow using of negative atoms. The process of expression negative statements in the graph language uses in the case of RDFCFL the following "transfer rule":

"Instead of including negative ground atom into a set of antecedent/consequent of the clause, transfer it into a set of atoms consequent/antecedent, where the atom becomes a positive one."

The RDFCFL system applies a special transfer of negative atoms rule in the case of universal/existential clauses. Under this rule, the variable after the transfer becomes a new existential constant and after the transfer the existential constant it becomes a new variable.

As the conditions - ground/universal/existential of the general form <antecedent> → <consequent> in the RDFCFL have the meaning of the implication p_1 &

p_2 & ...& $p_n \rightarrow q_1, q_2, ... q_n$ the conditional networks in general give a possibility to express all logical connectives because the set of connectives $\{\&, \vee, \rightarrow\}$ has a property of the functional completeness.

4.4 Representation of Vector Edges

Conventions of two different representation types of the vector edges, GRDF or RDFCFL ensure that there is no need for an additional method of separation the antecedent from the consequent of a clause, the same as in the case of using CFL meta-symbol "$\rightarrow$". In doing so, as well as at the CFL version, antecedents are logically connected vectors by connective &, and consequents are logically connected vectors by logical connective $\vee$.

5 Reasoning in Formal Systems GRDF and RDFCFL

In previous paragraphs two RDF based formal languages GRDF and RDFCFL has been introduced. The both language formalisms mutually distinct in two things:

- in the manner how to represent negation of a knowledge atom,
- in general forms of conditions.

Inference rules of two formal systems with so distinct languages must also be different in a corresponding way. The only exception consists in the substitution rule that exploits in both systems a common property of universal quantified variables within a given universal network. The universality of some variables allows a substitution uniformly a concrete relevant FOPL term (usually constant) for a chosen universally quantified variable. This fact implies that the substitution rule holds for universal networks in both formal systems GRDF and RDFCFL.

Substitution rule in both systems GRDF/RDFCFL: A clause of a network form with the occurrence of variables can derive a new clause by a uniform substitution of a convenient language term (usually constant) for each occurrence of the same variable.

5.1 Reasoning with Modus Ponens Rule in the GRDF

Besides the substitution (unification) rule uses the formal system GRDF the transfer rule corresponding to the Modus Ponens rule.

Transfer rule in GRDF: If it is possible to make the unification of a convenient universal condition that is taken for the derivation of the form $A \rightarrow B$ with a part A by the help of substitutions then the consequent B becomes a new added vector of the main network.

A special axiom is usually a universal "if – then" condition allowing a unification with a sub-network of the main one by a substitution of a corresponding term of the Herbrand universe (see for example [7]). Then the reasoning by means of the transfer rule can be made with participating of only ground vectors and clauses of the input networks. It is the right process supported by the fact that the Herbrand theorem holds:

> *Herbrand Theorem: A set of Herbrand ground (ground) clauses is satisfiable if and only if it has a Herbrand model that satisfies it.*

Conditional clause is a tool of transferring its consequent into another clause unifiable with its antecedent.

5.2   Resolution Reasoning and Decidability in Both RDFCFL and GRDF

Both GRDF as well as RDFCFL can use the well-known resolution rule to obtain logical consequents from a knowledge base. The rule holds at the original CFL version as well as in the graph-based CFL version and as well as at in the GRDF formal system and the RDFCFL formal system.

Resolution inference rule works by the help of two partial rules - the substitution rule and the cut rule.

RDFCFL Cut rule: If the knowledge base designed for derivation contains a pair of clauses sharing the same atom (vector), one in the antecedent (with a dashed line in the graph of the atom), one in the consequent (with a full line in the graph of the atom), then, as a result a new clause is derived from this pair of clauses. This clause cuts the shared atom and connects it together (drawing one common network) with the remaining atoms of their antecedents (consequents) by conjunction (disjunction) connectivity.

As the transformation from the FOPL language to the GRDF or RDFCFL languages shares the property of satisfiability of knowledge base formulas both the graphic formal systems proposed here are partially decidable as well as the formal system of the FOPL.

6   Reasoning in GRDF/RDFCFL Graphs by Example

To explain steps of transforming statements in the FOPL into graph language GRDF or RDFCFL we use here an example. We show the way from a text in a natural language to the FOPL language and consequently to the RDFCFL graph language. The reasoning in graphs is then outlined.

Example - Representation of a statement about a suspect of a murder at a marketplace in a broad daylight:

"Who could be present at the place at the time of the murder and has no alibi is then the suspect."

1. The first-order predicate logic can e.g. be used to represent sentences with the following structure of predicates:

person(<who>), **place**(<of_what> , <where>),
time(<of_what> , <when>), **has**(<who> alibi),
is_present(<who> , <where> , <when>), **suspect**(<who>)

2. A predicate formula representing sentences as a whole has then a form

$$\exists y \, \exists z \, \forall x \, , \, (\textbf{person}(x) \ \& \ \textbf{place}(\text{murder } y) \ \& \ \textbf{time}(\text{murder}, \ z) \ \& \ \neg\textbf{has}(x, \ \text{alibi}) \ \&$$
$$\textbf{is\_present}(x, \ y, \ z) \ \rightarrow \ \textbf{suspect}(x, \ \text{murder})).$$

3. One of the possible conversions of the formula into a clausal form [10]:

person(X) & **place**(murder, loc_m) & **time**(murder, time_m) &
¬**has**(X, alibi) & **is_present**(X, place_m, time_m) → **suspect**(X, murder).

4. Clauses of the clausal form logic now provide a possibility of representation formula with the occurrence of all logical connectors, as they can use a functional complete couple (&,∨, →). Here we use the "negative atom transfer rule" to transfer negative atoms as positive ones at the opposite side of the implication. The clause would then be of the shape

person(X) & **location**(murder, loc_m) & **time**(murder, time_m) &
is_present(X, loc_m, time_m) → **has**(X, alibi) ∨ **suspect**(X, murder),

which says that a person X has either alibi, or is suspect of a murder.
5. to make a decomposition of a ternary predicate **is_ present**(X, loc_m, time_n) we use here two binary predicates **presence**(X, loc_m) in conjunction & with **in_time**(loc_m, time_n).
6. to remake unary predicates to binary predicates expressing the same statement – here we use the predicate **isa**(X, person) - X is a person.
7. to use the notation introduced in the graphic representation of an implication <antecedent> → <consequent> to display the specified formula of predicate logic converted into a clause.
8. The meaning of RDF graph participated concepts and relationships (Fig. 3) has been clarified by URI references (its beginning part) to appropriately chosen dictionaries.
9. It was therefore elected a reference to the fictional ontology concerning criminal terminology created as a *namespace* crime semantic network in the Fig. 3.

Individual identifiers of subjects, objects and predicates will then be shaped for instance crime:murder, crime:alibi, crime:suspected.

All the names of graph nodes or edges in both formal languages have to be formatted URIs. Both similarly to CFL share the ability to use universal and existential quantified statements.

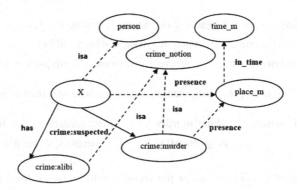

Fig. 3. RDF graph showing complete URIs of the concepts murder and alibi and the predicate suspect.

To mark the quantified variables GRDF and RDFCFL we introduced:

- The name of a variable is a hash fragments in the URI string behind the symbol # beginning with a capital letter: X, Y, People, Animals, ... in the label RDF term.
- The name of an existence term is a hash fragment in the URI string behind the symbol #beginning with the symbol @, @someone @known, ...

Established format of the designation of GRDF as well as RDFCFL graphs in the preceding paragraphs is now able to express statements such as "X is suspected from murder" (Fig. 4).

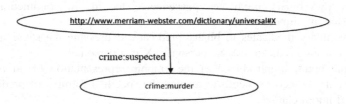

Fig. 4. RDF graph with a variable X

10. To deduce some conclusions about a participation of persons at the murder presented at the place_m in the time_m we must have further information about them.

For example Fred was a salesman presented at the place_m in the time_m because of sailing ice-cream at the marketplace. Figure 5 shows the information about Fred's activity:

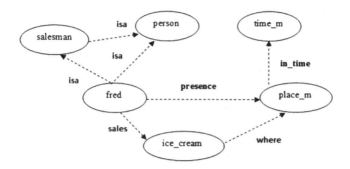

Fig. 5. Facts about an ice-cream salesman Fred.

11. To obtain a conclusion about Fred's alibi we need two general rules:

- isa(X, Y) & isa(Y, Z) → isa(X, Y) gives us a possibility to deduce isa(fred, person),
- sales(X, Y) & where(Y, place_m) & in_time(place_m, time_m), → has(X, alibi) that gives us a possibility to obtain that salesman Fred is not suspect.

The whole proof described above is representable in a graphic form.

7 Conclusions and Support the Reasoning by RDFCFL

7.1 Necessity to Obtain Conclusions Direct Within RDF Extended Model and Language

Because every formal system ought to have an inference formalism to work on knowledge bases in order to obtain new effects and relations, it is natural and beneficial to attempt to develop such systems that mainly derive additional knowledge and can manipulate structured data straightforwardly without rewriting knowledge into a language of a higher complexity such as OWL. Moreover OWL formal language is coming out from description logics with a rather doubtful decidability (see [8]).

Currently, realization of the semantic web ideas is more or less directed to the concept of the linked data through the RDF model of knowledge representation. Specifically, this means that the integration of RDF knowledge represented in large-scale considerations will naturally focus on the task of creating new interesting relationships of this knowledge.

In short, the system of knowledge in the realization of the idea of the Semantic Web based RDF model aims at:

- providing new and useful complementary knowledge from large collections of RDF structured data based on human experience,
- enhancing RDF-based language by the common interface technology of the RDF structured data of the triple/vector notation,
- derivation of answers on the base of solutions, discoveries or other results beyond the original data based on calculations and derived conclusions.

7.2   Further Tasks

RDFCFL formal system proposed here serves a tool of deriving conclusion over a data sets that has been structured as RDF triple/vector linked collections.

The realization of the idea of inference conclusions in the linked data RDF model in Semantic Web has brought minimally one further important task to solve:

1. We have at one side a giant database of linked data that does not dispose with triples (vectors) of a quantified form because all the content of the data linked by resource URIs has been created as individual facts without a necessity to take into account generalized (quantified) statements at a higher level of abstraction.
2. At the opposite side there are special rules (axioms) of a general form within ontology of a special topic that have a form of some general logical formulas with quantifiers.

Linking conditions (1) and (2) into a real tool of reasoning by the RDFCFL in a space of semantic web database is a problem to solve that becomes in early future an important task for specialists in computer science.

Acknowledgments. The research described here has been financially supported by University of Ostrava grant SGS13/PřF/2016. Any opinions, findings and conclusions or recommendations expressed in this material are those of the authors and do not reflect the views of the sponsors.

References

1. Auer, S., Lehmann, J., Ngonga Ngomo, A.-C.: Introduction to linked data and its lifecycle on the web. In: Polleres, A., d'Amato, C., Arenas, M., Handschuh, S., Kroner, P., Ossowski, S., Patel-Schneider, P. (eds.) Reasoning Web 2011. LNCS, vol. 6848, pp. 1–75. Springer, Heidelberg (2011). doi:10.1007/978-3-642-23032-5_1
2. Tim, B.L.: Giant Global Graph. MIT, Cambridge (2007). CSail: http://dig.csail.mit.edu/breadcrumbs/node/215
3. Lukasová, A.: Reprezentace znalostí v asociativních sítích. In: Proceedings of Znalosti (2001). (in Czech)
4. Lukasová, A., Vajgl, M., Žáček, M.: Reasoning in RDF graphic formal system with quantifiers. In: Proceedings of the International Multiconference on Computer Science and Information Technology, Mragowo, Poland. pp. 67–72. IEEE Computer Society (2010). ISBN 978-1-4244-6432-6
5. Lukasová, A., Žáček, M., Vajgl, M.: Building a non-monotonic default theory in GCFL graph-version of RDF. In: Silhavy, R., Senkerik, R., Oplatkova, Z.K., Silhavy, P., Prokopova, Z. (eds.) CSOC 2014. AISC, vol. 285, pp. 455–465. Springer, Heidelberg (2014). doi:10.1007/978-3-319-06740-7_38
6. Lukasová, A., Vajgl, M., Žáček, M.: Knowledge represented using RDF semantic network in the concept of semantic web. In: Proceedings of the International Conference of Numerical Analysis and Applied Mathematics 2015, (ICNAAM 2015). American Institute of Physics, USA (2016)
7. Lukasová, A., Vajg, M., Žáček, M.: Reasoning in graph-based clausal form logic. IJCSI-Int. J. Comput. Sci. **9** (1(3)) 37–43 (2012). ISSN 1694-0814

8. Lukasová, A.: Formal Logics in Artificial Intelligence (Formální logika v umělé inteligenci) Computer Press (2003). (in Czech)
9. Lukasová, A., Vajgl, M.: Genzen-like proofs in description logic DL1. In: Proceedings of the Tenth International Conference on Informatics 2009, Košice (2009)
10. Richards, T.: Clausal Form Logic an Introduction to the Logic of Computer Reasoning. Addison-Wesley, Boston (1989)
11. Žáček, M., Lukasová, A., Miarka, R. Modeling knowledge base and derivation without predefined structure by graph-based clausal form logic. In: Proceedings of the 2013 International Conference on Advanced ICT and Education, pp. s.546–s.549. Atlantis Press: AISR, France (2013). ISBN 978-90786-77-79-6
12. Žáček, M., Lukasová, A.: English grammatical rules representation by a meta-language based on RDF model and predicate clausal form. Inf.-Int. Interdisc. J. **19**, 4009–4015 (2016). ISSN 1343-4500

Big Data Filtering Through Adaptive Resonance Theory

Adam Barton, Eva Volna[✉], and Martin Kotyrba

Department of Informatics and Computers, University of Ostrava, 30. Dubna 22,
70103 Ostrava, Czech Republic
{adam.barton, eva.volna, martin.kotyrba}@osu.cz

Abstract. The aim of the article is to use Adaptive Resonance Theory (ART1) for Big Data Filtering. ART1 is used for preprocessing of the training set. This allows finding typical patterns in the full training set and thus covering the whole space of solutions. The neural network adapted by a reduced training set has a greater ability of generalization. The work also discusses the influence of vigilance parameter settings for filtering the training set. The proposed method Big Data Filtering through Adaptive Resonance Theory is experimentally verified to control the behavior of an autonomous robot in an unknown environment. All obtained results are evaluated in the conclusion.

Keywords: Adaptive Resonance Theory (ART) · Big data · Control neural network · Data Filtering

1 Introduction

Big Data has been one of the current and future research frontiers. Big data concerns large-volume, complex, growing data sets with multiple, autonomous sources. With the fast development of networking, data storage, and the data collection capacity, big data is now rapidly expanding in all science and engineering domains, including physical, biological and bio-medical sciences.

The most fundamental challenge for the big data applications is to explore the large volumes of data and extract useful information or knowledge for future [4]. In many situations, the knowledge extraction process has to be very efficient and close to real time because storing all observed data is nearly infeasible. Four things are required to mine data effectively [6]: high-quality data, the "right" data, an adequate sample size and the right tool. There are many tools available to a data mining practitioner. These include decision trees, various types of regression and neural networks. In more practical terms, neural networks are non-linear statistical data modeling tools. They can be used to model complex relationships between inputs and outputs. Using neural networks as a tool, we are able to harvest information from big data [2].

The work describes an effective way of Big Data Filtering through Adaptive Resonance Theory, which is experimentally verified to control the behavior of an autonomous robot in an unknown environment.

© Springer International Publishing AG 2017
N.T. Nguyen et al. (Eds.): ACIIDS 2017, Part II, LNAI 10192, pp. 382–391, 2017.
DOI: 10.1007/978-3-319-54430-4_37

2   Data Filtering Using Adaptive Resonance Theory

Adaptive Resonance Theory (ART1) is composed of two layers and a control neuron R, Fig. 1. Layer F_1 includes computational neurons X_i from the input layer $F_1(a)$ and neurons I_i from the interface layer $F_1(b)$. Neurons Y_i form layer F_2. Synapses between F_1 and F_2 are oriented bottom-up and synapses between F_2 and F_1 are oriented top-down. This approach enables to control the degree of similarity between weight values of synapse top-down and the input vector. Deciding is provided by neuron R (reset), which that separates states of neurons in F_2 three possibilities: active, inactive - acceptable for competition, inactive - unacceptable for another competition [5].

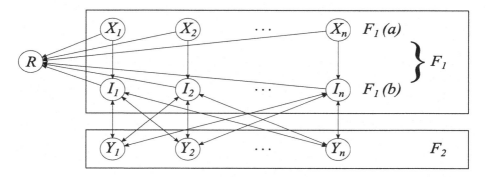

Fig. 1. ART1

Using a heuristic of self-organizing networks for preprocessing of the training set, we are able to find typical patterns and thus cover the entire space. We utilize ART1 to filter a training set, which works with a binary input. The present input is compared based on parameter vigilance $\rho \in [0, 1]$ with stored patterns in the network. If it differs significantly, this pattern is stored in the network as a new class. ART1 is able to create groups of patterns and the user has the opportunity to influence the level of relative similarity of patterns stored in the same group.

2.1   ART1 for Training Set Filtering

To create a filtered training set, the basic algorithm ART1 [5] is modified as follows:

1. Initialization of synaptic weights, bottom-up b_{ij}, top-down t_{ij}, and parameter vigilance $\rho \in [0, 1]$

$$b_{ij} = \frac{1}{(1+n)}, \quad t_{ij} = 1, \quad \rho \in [0, 1], \tag{1}$$

where n is the number of input neurons.

2. The submission of input $x = [x_1, \ldots, x_n]^T$ to input layer F_1.
3. Activation of neurons in layer F_2 and calculation the response according to:

$$y_j = \sum_{i=1}^{n} b_{ij} x_i, \tag{2}$$

where y_j is the response of the j-th neuron in F_2 and x_i is the i-th element of input vector x.

4. Selection of the neuron with the greatest response $y_{max} = \max(y_j)$.
5. Calculation of similarity δ of input pattern with back weighting factor t_{imax} regarding the neuron y_{max}:

$$\delta = \frac{\sum t_{imax} x_i}{\sum x_i}, \tag{3}$$

6. If $\delta \geq \rho$ a resonance of input with expectations occurs and the candidate y_{max} is accepted, the algorithm proceeds to step 7. Otherwise, the candidate is rejected and its output is set to -1 and other suitable candidate will be chosen and the algorithm proceeds to step 3.
7. Weights' adaptation b_{imax} a t_{imax} of neuron y_{max} according to formulas (Weber Law):

$$b_{imax} = \frac{L t_{imax} x_i}{L - 1 + \sum t_{imax} x_i}, \tag{4}$$

$$t_{imax} = t_{imax} x_i, \tag{5}$$

where L is a constant (excelling unit), its value is usually $L = 2$.

8. Test for gradual addition of input vectors into a reduced training set (RTS):
   (a) If there is no vector in the given group of RTM yet, the currently presented vector x is stored as a typical representative of the group.
   (b) In the case that the group is not empty and also the currently submitted vector x carries a greater amount of information than the vector stored, as the representative of the group, vector x will replace it.
9. Checking the termination condition. If it is satisfied, the algorithm proceeds to step 11.
10. Suppression of neurons which were inhibited and the algorithm proceeds to step 2.
11. Stopping the algorithm.

The key step in the modified algorithm ART1 is Step 8, where the reduced training set (RTM) is compiled based on the classification of the network. The user chooses a rule setting of a vector with a higher content of carried information. In this work, we accept vectors containing more critical states, i.e. elements with value 1.

2.2 Impact of Vigilance

ART1 requires parameter vigilance ρ, which defines the degree of similarity for assigning the pattern to the same class. This parameter is defined by the user. Figure 2 shows the effect of parameter ρ on pattern matching $\boldsymbol{p} = [p_1, \ldots, p_n]^T \in \{0, 1\}$ into classes c.

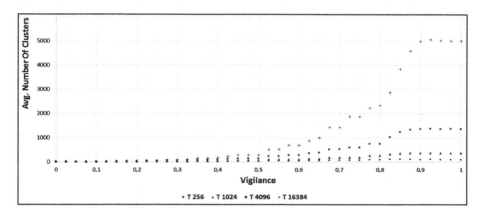

Fig. 2. Influence of the vigilance parameter on the training set clustering

If a low value of ρ is set, it can lead to the classification of different patterns in the same class and the number of classes is small. On the contrary a big value of ρ results in the formation of a large number of classes. In this work, ρ is set so that its value is from the interval $[0.5; 0.7]$. This leads to the creation of a big number of groups or to an inadequate classification of the input pattern.

3 Big Data Filtering Due to Control of Autonomous Robot Behavior

In the centralized control [3], the robot has information from sensors and the base of stored knowledge. Based on this information, it creates a plan of action. It scans potentially appropriate plans and it tries to predict. It is also the reason why this approach of "strategic thinking" requires a lot of memory and time.

The proposed model of the robot is assembled from Lego Mindstorm EV3 [1]. The robot uses LeJOS operating system. The robot uses two large engines that propel a tracked chassis on which is placed a programmable brick. A sensory head with 4 infrared sensors is attached to the chassis. The sensors emit infrared signals and detect reflections from an object which is located in front of the encoder. The strength of the reflected signal indicates the distance from the object in the range from 0 to 100 units.

As the controlling neural network was chosen a three-layer feedforward network with a sigmoidal transfer function, which has been adapted by backpropagation. Hidden and output neurons have a threshold equaled to the value "1". The input into the neural network is an input vector x $\in \{0,1\}$ and the output from the neural network is an output vector $\boldsymbol{y} \in \mathbb{R}$, which values are normalized to the interval $[0, 1]$.

The training set is automatically generated. If we consider that the model of the robot has four infrared sensors that are able to detect the distance from 0 to 100 cm, it is appropriate to decompose this input interval into several parts. We have decomposed the input interval into five parts Therefore, the training set contains 20 inputs (each of the four sensors is associated with five terms) and 4 outputs. Then the number of possibilities (input patterns) that can occur is $2^{4 \times 5} = 2^{20}$. The output vector consists of real numbers, and thus we have been able to reduce the necessary number of output neurons to four, due to the use of two motors - two neurons for each motor (forward and reverse).

We describe the principle of training set generating using an example. We consider four infrared sensors, each of which is able to detect the distance 0–100 cm. We split this distance in 5 equal parts, see Fig. 3. We express these parts using a set of terms: vc – very close (0–19 cm), c – close (20–39 cm), mf – medium far (40–59 cm), f – far (60–79 cm), vf – very far (80–100 cm).

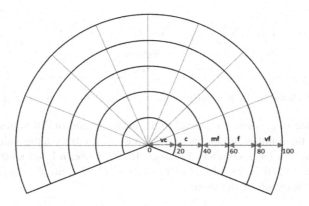

Fig. 3. Scanning area

Each sensor is associated with those five terms that represent obstacles in the sensor range for the given interval. Value "1" indicates the presence of obstacles in a given interval, and value "0" represents a blank space. The value 100 cm is considered as an obstacle in the relevant interval.

We focus only on the part of the training set, which describes inputs. The given training set, based on the robot configuration and the distribution of the scanned area into five parts, comprises 1,048,576 patterns. The training set also contains patterns that admit a possibility of the existence of obstacles in several given intervals. Therefore, it is necessary to specify a group of inputs I_i belonging to a given sensor. In the case that the group comprises more than one value of "1", such a pattern is not included in the training set. The reduced training set (RTS) comprises 1296 patterns, see Table 1.

After performing exclusion of groups with zero values, the reduced training set is reduced into 625 patterns (ERTS). Unfortunately, the resulting training set will increase depending on the number of sensors and on the distribution of the scanned areas. Therefore, we will process further filtration. The impact of vigilance setting for this training set is shown in Table 2. Characteristics of groups reflect ten filtering ERTS using ART for each vigilance setting.

Table 1. Input patterns of reduced training set

Sensor 1					Sensor 2					Sensor 3					Sensor 4				
I_1	I_2	I_3	I_4	I_5	I_1	I_2	I_3	I_4	I_5	I_1	I_2	I_3	I_4	I_5	I_1	I_2	I_3	I_4	I_5
1	0	0	0	0	0	0	0	0	0	0	1	0	0	0	0	0	1	0	0
0	0	1	0	0	0	1	0	0	0	0	0	0	1	0	0	0	0	1	0
...	...	...	...	...	...	...	...	...	...	...	...	...	...	...	...	...	...	...	...

Table 2. Number of groups ERTM depending on ρ

Vigilance ρ	Minimal value of the number of classes	Maximal value of the number of classes	The average number of groups	Mean number of groups
0	6	12	8.6	7
0.1	5	11	7.8	5
0.2	5	10	7.8	8
0.3	36	53	46.9	51
0.4	36	54	46.2	47
0.5	40	51	46.3	50
0.6	177	188	182.6	188
0.7	179	192	184.9	184
0.8	625	625	625.0	625
0.9	625	625	625.0	625
1	625	625	625.0	625

The number of groups of the filtered training set is variable due to mixing of samples before their submission to ART1. This step influences the final training set significantly. From the view of the user, it was experimentally verified that the given training set of 625 patterns and $\rho > 0.5$ achieves the best results due to the consequent need for setting output values to these patterns manually. Therefore $\rho = 0.5$ is used in this work. The resulting filtered training set contains 51 patterns.

For clarity, columns of associated intervals of individual sensors are combined into themselves, i.e. that each 5 binary inputs representing an obstacle in a given interval are replaced with only one linguistic value representing the distance for each sensor. For the adaptation of the neural network, the training set is expanded backwards. Table 3 describes outcomes of the training set, where S_1, S_2, S_3 and S_4 indicates sensors in a given port; B_F and A_F indicates the speed of motors for forward, B_B and A_B indicates the speed of motors for backward of ports A, B. Speed settings for individual run of both of engines were set experimentally with regard to limitation of maximum speed at 500 degrees/s because of skidding belts on the smooth surface at higher speeds.

Pattern number 52, which defines the default behavior for moving forward, was added subsequently at the end of the training set. By adding the pattern, the robot is prevented from going a different direction than straight if no obstacle has been detected.

Table 3. Training set for the adaptation of the neural network

Pattern	Evaluation of sensors with use of terms				Setting the engine speed for forward and reverse degrees (per second)			
	S4	S3	S2	S1	BF	BB	AF	AB
1	vb	d	vd	b	350	0	300	0
2	d	s	s	vb	300	0	350	0
3	b	d	s	vd	350	0	300	0
...	...	...	...	...	...	...	...	...
51	vb	b	vb	d	100	0	0	100
52	vd	vd	vd	vd	500	0	500	0

The life cycle of the training set is shown in Fig. 4. In this way, we are able to find typical patterns and thus cover the entire space of solutions. Filtration of the training set using ART1 adjusts the input vectors so that these patterns are mixed pseudo-randomly.

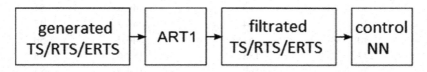

Fig. 4. Life cycle of the training set

4 Experimental Verification

The aim of the experimental study is to verify the behavior of the robot controlled by artificial neural networks. This controlling neural network is adapted by the training set that is filtered through ART1. The topology of the controlling neural network had the following configuration: 20 input, 20 hidden, and 4 output neurons. Other parameters were set as follows: learning rate $\alpha = 0.85$, momentum factor $\mu = 0.9$, and the steepness parameter $\lambda = 1$. The termination condition for adaptation has been set as $E \leq 0.0001$ due to normalization of the training set's outputs. Adaptation lasted for 109 s and included 14393 cycles.

The model of the autonomous robot was placed in a space which consists of squares 15 cm^2. Motor acceleration was set to 600 degrees/s for all experiments. Scanning of a space was carried out once per 500 ms. The following data is recorded for the purposes of evaluation of the experimental part.

- number of primary collisions (direct collisions of the robot with an obstacle, i.e. running into obstacles),
- number of secondary collisions (collision only some parts of the robot such as the sensor cables, etc.),
- number of deadlocks,
- duration of the experiment,
- trajectory of the robot's movement.

Table 4. Experimental outcomes (secondary collisions are represented by points).

Movement of the robot in the environment	Number of primary collisions	Number of secondary collisions	Number of deadlocks	Average duration of the experiment (in second)
	0	1	0	17.7
	0	1	0	28
	0	2	0	25
	0	1	0	24
	0	1	0	36.3
	0	0	0	47.3
	0	1	0	49.7
	0	0	0	67

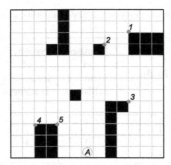

Fig. 5. Experimental environment for free movement of the robot

Experimental outcomes are shown in Table 4. The robot has completed all objectives defined in individual experiments without deadlocks as well as without a primary collision with obstacles. In some cases, it is apparent that the signal from sensors reflects the way leading to an erroneous evaluation of the measured distance, see cases of corner collision.

The last experiment represents free movement of the robot in the environment 195 × 195 cm, which contains a number of obstacles placed so that they form extra "rooms", in which the robot may become trapped or cause collision by its moving inside, Fig. 5. The experiment was conducted for 30 min, i.e. 3603 cycles of environment scanning. During the experiment, there were a total of 10 secondary collisions, no primary collisions and no deadlocks. The robot explored the whole experimental environment. There were the following collisions:

- A secondary collision marked as "1" occurred three times, always after the turn round in the space behind the barrier and coming out, when the robot touched bands on the corner of obstacles.
- A secondary collision marked as "2" occurred twice, when the robot travelled around the obstacles left.
- A secondary collision marked as "3" occurred twice, when the robot came out from the room and its sensors touched the barrier.
- A secondary collision marked as "4" occurred once, i.e. collision of a sensor with a corner, when the robot came out from the environment behind the barrier.
- A secondary collision marked as "5" occurred twice, i.e. collision of belt, when the robot traveled around the obstacles left and during rotation.

A video showing the progress of the experiment is placed on https://www.youtube.com/watch?v=r-joFp3vh1k under a title "EV3 Neural Network Controller".

5   Conclusions

The whole training set included 1048576 patterns. We obtained 51 patterns after the training set filtration using ART1. Pattern 52 was added to them, which defines the default behavior for forward movement. Vigilance parameter ρ was set to 0.5. The control neural network was adapted by this reduced training set. Experimental outcomes are the following:

- The speed of each motor was adjusted appropriately, considering the situation in which the robot is found.
- The robot moved effectively in all environments, where precision of movement was not required too much.
- The ability of the robot to move backwards was observed during the experiment. Reversing was always implemented only for a short path, e.g. when the robot had little space to turn on the spot. This ability was not a part the training set.
- It has been observed that the robot is able to move too close to obstacles without running into them.

The use of self-organizing networks heuristic (ART 1) for preprocessing of the training set allows to find typical patterns, and thus cover the whole space. There is used, based on the results of previous work [7], the ability of generalization of neural networks, to reduce the possibility of user error in order to achieve better results compared to a complete training set. It can be stated that ART1 is a good tool for Big Data Filtering.

Acknowledgments. The research described here has been financially supported by University of Ostrava grant SGS14/PřF/2016. Any opinions, findings and conclusions or recommendations expressed in this material are those of the authors and do not reflect the views of the sponsors.

References

1. Bartoň, A.: Control of autonomous robot using neural networks (in Czech). Master thesis, University of Ostrava, Czech Republic (2015)
2. Mironovova, M., Bíla, J.: Fast fourier transform for feature extraction and neural network for classification of electrocardiogram signals. In: 2015 Fourth International Conference on Future Generation Communication Technology (FGCT), pp. 1–6. IEEE (2015)
3. Dudek, G., Jenkin, M.: Computational Principles of Mobile Robotics. Cambridge University Press, Cambridge (2010)
4. Rajaraman, A., Ullman, J.: Mining of Massive Datasets. Cambridge University Press, Cambridge (2011)
5. Rojas, R.: Neutral Networks: A Systematic Introduction. Springer, Berlin (1996)
6. Singh, Y., Chauhan, A.S.: Neural networks in data mining. J. Theor. Appl. Inf. Technol. **5**(6), 36–42 (2009)
7. Volná, E., Kotyrba, M., Žáček, M., Bartoň, A.: Emergence of an autonomous robot's behavior. In: Proceedings of the 29th European Conference on Modelling and Simulation, ECMS 2015, Albena, Bulgaria, pp. 462–468 (2015)

The Effectiveness of the Simplicity in Evolutionary Computation

Michal Witold Przewozniczek[1]([✉]), Krzysztof Walkowiak[2], and Michal Aibin[2]

[1] Department of Computational Intelligence, Wroclaw University of Science and Technology, Wybrzeze Wyspianskiego 27, 50-370 Wroclaw, Poland
michal.przewozniczek@pwr.edu.pl
[2] Department of Systems and Computer Networks, Faculty of Electronics, Wroclaw University of Science and Technology, Wybrzeze Wyspianskiego 27, 50-370 Wroclaw, Poland
{krzysztof.walkowiak,michal.aibin}@pwr.edu.pl

Abstract. Current research in Evolutionary Computation concentrates on proposing more and more sophisticated methods that are supposed to be more effective than their predecessors. New mechanisms, like linkage learning (LL) that improve the overall method effectiveness, are also proposed. These research directions are promising and lead to effectiveness increase that cannot be questioned. Nevertheless, in this paper, we concentrate on a situation in which the simplification of the method leads to the improvement of its effectiveness. We show situations when primitive methods, like Random Search (RS) combined with local search, can compete with highly sophisticated and highly effective methods. The presented results were obtained for an up-to-date, practical, NP-complete problem, namely the Routing and Spectrum Allocation of Multicast and Unicast Flows (RSA/MU) in Elastic Optical Networks (EONs). None of the considered test cases is trivial. The number of solutions possible to encode by an evolutionary method is large.

Keywords: Hybrid methods · Coevolution · Strategies of dynamic subpopulation number control · Elastic optical networks

1 Introduction

One of the main research directions in the Evolutionary Computation (EC) field is the effectiveness improvement of the already known methods and searching for new method frameworks. To improve the method effectiveness, different techniques are used. Among these techniques are: the introduction of the linkage learning mechanisms [4–7, 10, 12, 13, 17], the use of coevolution [2, 6–8, 10, 12, 13, 17], the Baldwin effect incorporation [13] and other. All these general mechanisms may be mixed, and they all have their subtypes. For instance, the coevolution may include the use of the Island Model [6, 8] with different migration subject: we can migrate individuals, probabilistic models [9], or, as recently proposed, the building blocks [6]. Moreover, the coevolution may be used in a classical way – using the static number of coevolving subpopulations,

© Springer International Publishing AG 2017
N.T. Nguyen et al. (Eds.): ACIIDS 2017, Part II, LNAI 10192, pp. 392–402, 2017.
DOI: 10.1007/978-3-319-54430-4_38

or the Strategies of Dynamic Subpopulation Number Control (SDSNC). When SDSNC is used, the number of coevolving subpopulations may increase or decrease depending on the method state [2, 7, 10, 12, 13, 17]. Even the brief and general description of available possibilities according to the coevolution shows that the evolutionary methods become more and more complex to improve their effectiveness. However, sometimes the key to creating an effective method may lie in the simplification of the already proposed methods, rather than in introducing new and more sophisticated mechanisms.

It is expected that optical transport networks will evolve from the current wavelength-switched optical network (WSON) architectures built with the wavelength division multiplexing (WDM) technology towards elastic optical network (EON) architectures [21]. Flexible-grid EONs using a frequency grid with 6.25 GHz granularity, in contrast to the 50 GHz fixed grid used in traditional WSONs, better utilize the spectrum resources. Consequently, EONs can provide variable bit-rate demands to adapt to the dynamic requests from DC services. In EONs, the problem of finding unoccupied spectrum resources to establish a lightpath is called the Routing and Spectrum Allocation (RSA) problem. Considering the physical topology discussed above, our goal is to find a route, and allocate spectrum for each demand. The solution to the problem is subject to the following constraints: spectrum contiguity, spectrum continuity and slice opacity. In EONs, continuous spectrum resources to the specific demand must be assigned. In other words, once chosen set of slices have to be used over the entire end-to-end optical path. Moreover, if one frequency slice has been allocated to an existing request, this slice cannot be assigned to another request.

In this paper, we show the effectiveness of algorithms simplification phenomenon, on the base of research performed for NP-hard problem, namely the Routing and Spectrum Allocation of Multicast and Unicast Flows (RSA/MU) in Elastic Optical Networks (EONs). For all considered test cases, if non-simplified problem decoding way is used, the number of available solutions, possible to encode is in the range from $3.62 \cdot 10^{16}$ up to $5.9 \cdot 10^{440}$. Therefore, all considered tests cases are non-trivial, and the use of evolutionary methods seems to be justified.

This paper is organized as follows: Sect. 2 presents the RSA/M problem, the third section presents considered Solution Construction Algorithms, the competing methods are presented in Sect. 4, the fifth section presents the considered test cases, method settings and obtained results, finally the last section concludes this work.

2 Problem Formulation

In this section, we describe the optimization problem, namely Routing and Spectrum Allocation of multicast and unicast flows (RSA/MU). We consider static multicast and unicast demands, i.e., all requests to be established in EON are known *a priori*, and they are not subject to blocking. The EON is defined as a directed graph $G = (V, E)$ where V is a set of nodes and E is a set of links. All network nodes are equipped with *multicasting capable* (MC) optical cross-connects that can replicate the input data stream to multiple outputs. The EON operates within a flexible frequency grid, where the entire optical spectrum is divided into narrow frequency *slices*.

A transmission channel is created by grouping an even number of slices, adjusted to the demand volume. Let S denote a set of all slices available on each network link $e \in E$.

Set D includes both multicast and unicast demands. Each multicast demand $d \in D$ is denoted by a source node (root) $s(d)$, a set of receivers $R(d)$ and bit-rate h_d. To provision a multicast demand, a *light-tree* defined as a point-to-multipoint connection is established in the EON using the MC nodes. To model routing of multicast demands, we use the *candidate tree* (CT) approach described in [19, 20]. More precisely, for each multicast demand $d \in D$ (source and set of receivers), there is a set of t candidate trees $P(d)$ that originate at the source node $s(d)$ and include all receivers $R(d)$. To establish a multicast demand d in the EON, one of the candidate trees from set $P(d)$ is selected. To model unicast routing, we apply the link-path approach, i.e., for each demand d we are given set $P(d)$ of k candidate paths. As in [20], the spectrum requirement for a particular multicast or unicast demand is determined according to a *distance-adaptive transmission* (DAT) rule. Let n_{dp} denote the requested number of slices for demand d realized on tree p according to the SAT rule and the requested bit-rate of demand d.

To model spectrum usage in EONs, we use the concept of spectrum channels [16]. Channel c is defined as a contiguous (adjacent) subset of slices in ordered set S, i.e., $c \subseteq S$. Let $C(d, p)$ denote a set of channels accessible for demand d on candidate tree/path p. Each channel $c \in C(d, p)$ includes n_{dp} adjacent slices to support traffic volume of demand d (given by h_d) according to the modulation format selected using the DAT rule. The RSA/MU optimization problem consists in the selection of the routing structure (tree or path) and spectrum channel for each demand. The goal of optimization is to minimize the width of spectrum required in the network, i.e., the indexed of the highest allocated slice in the whole network required to provision all demands included in set D. Note that the considered optimization problem RSA/MU is an offline NP-complete problem and according to our previous works, can be solved in optimal way only for small problem instances [19]. Due to the lack of space, we do not present a formal Integer Linear Programming (ILP) model. However, for more details on the optimization model as well as the ILP formulation refer to [19, 20].

3 Solution Construction Algorithms

The methods presented in this paper use Solution Construction Algorithm (SCA) to translate the genotype into the final solution. Recently, the SCA-Baldwin was proposed [13] for the RSA/M problem. It was shown that SCA-Baldwin improves the performance of evolutionary methods. In this paper, another SCA, called SCA-Limited is considered. SCA-Limited is a simplified version of SCA-Baldwin and was proposed in [1]. In this section, both SCAs are presented and discussed.

When SCA-Baldwin is used, the solution to the RSA/M problem is a list of triples: [d, p, w], where d is the demand number, p is the either the candidate tree or path required for the demand d and w is the weight ($w \in <0, 1>$). The pseudocode of the SCA- Baldwin is presented in Fig. 1.

If the SCA-Baldwin is used, any solution is feasible. Each solution is also optimized in a greedy way. If a better path/tree is available than the one specified in the genotype, the better path/tree is used instead. On the other hand, if the group of the best

```
1: sort all [d, p, w] triples with respect to the weight w;
2: model ← empty;
3: for each demand d in the [d, p, w] Set:
4:        BestTreePath ← null;
5:        for each TreePath appropriate for demand d:
6:                MaxEndSlice, NumOfSlices ← GetMaxEndSlice(TreePath, model);
7:                if (
8:                    (BestTreePath = null) or
9:                    (BestMaxEndSlice > MaxEndSlice) or
10:                   (BestMaxEndSlice = MaxEndSlice and BestNumOfSlices > NumOfSlices)
                      or
                      (BestMaxEndSlice = MaxEndSlice and BestNumOfSlices = NumOfSlices
                      and TreePath = p)
11:                ):
12:                    BestNumOfSlices ← NumOfSlices;
13:                    BestMaxEndSlice ← MaxEndSlice;
14:                    BestTreePath ← TreePath;
15:       SetInModel(TreePath, model);
16: return model;
```

Fig. 1. SCA-Baldwin pseudocode

paths is larger than the path specified in the genotype, the genotype's path is preferred. Let us consider an example with two demands and two candidate trees/paths available for each of the demand. We consider the solution: [(1, 2, 0.4) (2, 4, 0.1)]. The way the solution is constructed for such genotype is presented in Fig. 2.

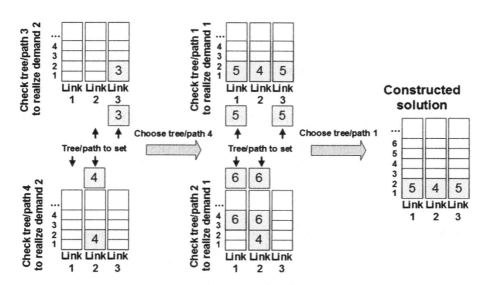

Fig. 2. SCA-Baldwin solution construction example

In the example presented in Fig. 2, the network state after setting particular paths/trees is presented. Note, that both, trees and paths, can be defined as a set of links. Therefore, from the network model point of view, there is no difference between them. In the proposed example, the demand number 2 is the first one considered, because, in the proposed genotype, it has the lowest weight. Both available paths/trees are equally useful, but path/tree number 4 is chosen, because it is the one specified in the genotype. For the demand number 1, the fifth path/tree is chosen, because maximum end slice is higher for the path/tree number 6. Note, that in this case the fact that path/tree number 6 is directly specified in the genotype, does not have any influence on the final path/tree choice.

The SCA-Limited is a simplified version of SCA-Baldwin. It works in the same way but does not require specifying the preferred tree/path in the genotype. Therefore, if the SCA-Limited is used, the solution is specified as a list of pairs: $[d, w]$. The SCA-Limited is less precise than SCA-Baldwin. In [13] methods that used SCA-Baldwin have significantly outperformed simulated annealing (SA) method that used SCA-Limited. However, in [13] the pros of SCA-Limited are also pointed – this encoding limits the number of solutions possible to encode. Thus it allows for faster convergence.

In [13] the considered test cases used only multicast demands with very high number of trees available for a particular demand (up to 1000 trees per single demand). The demand number was in the range from 17 up to 23. Therefore, the genotypes were fairly short (no more than 23 genes), but the number of gene values available was very high (up to 1000). In this paper, we also consider another type of test cases - the mixed test cases which include both: multicast and unicast demands. For the unicast demands, the maximum number of paths available is only 30, but the number of such demands is up to 111. We use the competing methods proposed in [13]. For the most promising method, the MuPPetS-EON, we use both, SCA-Baldwin and SCA-Limited encoding. The dedicated SA uses SCA-Limited. The rest of the methods (standard GA and Random Search) use SCA-Baldwin.

4   The Competing Methods

This section presents the considered competing methods: Multi Population Pattern Searching Algorithm for Elastic Optical Networks (MuPPetS-EON) [10, 13], Simulated Annealing (SA) [1, 13], Standard Genetic Algorithm for Elastic Optical Networks (sGA-EON) [10, 13] and dedicated Random Search (RS-EON). The competing methods choice is based on the literature review for the considered practical problem.

4.1   MuPPetS-EON

MuPPetS-EON was originally proposed in [10]. Recently, it was developed by adding the Baldwin effect which has significantly increased its effectiveness in solving RSA/M problem test cases [13]. If SCA-Baldwin is used, MuPPetS-EON encodes the solution as a list of triples: $[d, path, w]$, where d is the demand number *path* is the candidate path proposed for the demand d and w is the weight ($w \in$ <0, 1>). When the method is using SCA-Limited, then the method works in the same way, but the *path* value is ignored during the solution construction process.

MuPPetS-EON consists a set of coevolving virus populations. Viruses are messy-coded individuals proposed in [7]. Each virus is a list of pairs [(*pos, val*), (*pos, val*), (*pos, val*)], where *pos* is a position of a gene in a standard GA-individual genotype, and *val* is its value. In the encoding employed by MuPPetS-EON, the demand number d is a gene position, while the combination of *path* and *weight* is its value. Every virus subpopulation is assigned to a standard (GA-coded) complete solution called Competitive Template (CT). The population of viruses is used to update and optimize the CT. The detailed information about viruses, messy-coding and messy operators can be found in [7, 10, 12, 13].

MuPPetS-EON uses a set of coevolving virus populations. Every virus population is assigned to a single CT (the typical GA-like individual using typical GA-like encoding). The number of CTs is always equal to the number of coevolving virus populations. The method starts with a single virus population. Later on, during the method run, their number may increase or decrease, depending on the methods state. In general, if the method is stuck the number of coevolving virus subpopulations increases, otherwise it may decrease. Such Strategy of Dynamic Subpopulation Number Control (SDSNC) allows the method to remain flexible and avoid being stuck in local optima. More information about SDSNC, as well as about pros and cons of available strategies can be found in [11, 12].

MuPPetS-EON is a linkage learning method, which means that it collects the information about possible gene dependencies during the method run [4–7, 10, 12, 13, 17]. The linkage is represented by structures called *gene patterns*. Gene pattern is represented as a list of genotype positions without values: [(*d*), (*d*), (*d*), (*d*), (*d*)], where d is the demand number (gene position). For example, gene pattern: [(5) (2) (2)] marks genes 5 and 2. As shown the gene pattern may point the same position many times. The knowledge about linkage is used to exchange the data between CTs and during virus initialization process. More information about linkage learning in MuPPetS-EON can be found in [10, 13].

4.2 Other Competing Methods

The other competing methods are Standard Genetic Algorithm for Elastic Optical Networks (sGA-EON) [10, 13], dedicated Random Search (RS-EON) and Vector Simulated Annealing (VSA) [1, 13]. Both, sGA-EON and RS-EON use SCA-Baldwin, while VSA uses SCA-Limited. sGA-EON uses a classical GA framework. The mutation operator was adjusted to the solved problem. When an individual is mutated in 50% of the cases gene weight is randomly chosen from the available range. In the rest of the cases, all available trees/paths are rated with the use of the dedicated quality measure [13], and the new tree/path is randomly chosen with the use of the roulette wheel method and the dedicated rating. RS-EON is a classical Random Search, but it uses the SCA-Baldwin. The results of the above methods are also compared to the simple constructive heuristic algorithm AFA proposed in [18]. Finally, the Vector

Simulated Annealing (VSA) [1] is a classical SA, but it incorporates the automatic tuning process. Therefore, the VSA is parameterless [4] which is a rare and important benefit. The VSA starts its search from the automatically generated AFA algorithm solution.

5   The Results

In this section, the results of experiments are presented. The competing methods were chosen on the base of current research state: MuPPetS-EON [10, 13], sGA-EON [10, 13], problem dedicated Simulated Annealing (SA) and Random Search for EONs (RS-EON). The sGA-EON and RS-EON used SCA-Baldwin solution construction in the way it was proposed in [13]. SA used SCA-Limited as it is unsuitable for using SCA-Baldwin. Finally, MuPPetS-EON was tested with both available SCAs: SCA-Baldwin and SCA-Limited. In the latter case, MuPPetS-EON simply ignored the path suggested in the genotype.

All the source codes, detailed results, and experiments setup files are available at: http://www.mp2.pl/download/ai/20161016_eon_simplicity_eff.zip. All methods were coded in C++ and shared all possible source codes parts. All the experiments were conducted on HP Elite Desk800 3.4 GHz 8 GB RAM server with Intel Core i7-4770 CPU and Windows 7 64-bit installed. All test runs were executed in a repeatable environment state with no other resource consuming processes running. Since the environment and the time-constraint were the same as in [13], the method settings were adopted from [13].

The experiments were performed for the pan-European backbone network, named Euro28 (28 nodes, 82 directed links). The experiments were divided into four groups. Two different criteria were considered: the number of TBps (10TBps and 20TBps – influences the number of multicast demands), demands type used (only multicast demands and "mixed demands", using both: unicast and multicast). Among every group, the experiments were divided into 7 subgroups depending on the number of paths and trees available per single demand. Every subgroup included ten different test cases. Therefore, the total number of considered test cases was: $4 \cdot 7 \cdot 10 = 280$. The details of considered test case groups including the demand number and the of solutions possible to encode are given in Table 1.

As shown in Table 1 the T10 and T20 test case groups differ only in the existence of the unicast demands. This difference is important since it directly influences the number of genes necessary to encode the full problem solution. The number of solutions possible to encode is significantly different for the group of experiments were only multicast demands are considered (10^{16} up to 10^{91} of possible encodings for SCA-Baldwin) and the group where both demand types are mixed (10^{201} up to 10^{440} for SCA-Baldwin). Note, that when SCA-Limited is used the number of solutions possible to encode is always equal to: $(DemNum)!$, where $DemNum$ is a total number of demands. The number of solutions possible to encode for SCA-Limited is given in Table 2.

The measure of results quality is a ranking – the method that has proposed the best solution receives the number of points equal to the number of competing methods. The

Table 1. The details of considered test case groups.

	Demands number		The number of solutions possible to encode for SCA-Baldwin [10^x]						
	Unicast	Multicast	k = 2; t = 10	k = 4; t = 50	k = 6; t = 100	k = 8; t = 200	k = 10; t = 300	k = 20; t = 500	k = 30; t = 1000
	Min Max	Min Max	Min Max	Min Max	Min Max	Min Max	Min Max	Min Max	Min Max
T10 only multicast	0 0	10 10	16 16	23 23	26 26	29 29	31 31	33 33	36 36
T10 unicast and multicast	93 111	10 10	201 244	236 284	256 307	270 324	281 336	311 372	331 394
T20 only multicast	0 0	17 23	31 45	43 61	48 68	53 75	56 79	60 84	65 91
T20 unicast and multicast	93 111	17 23	229 272	270 318	292 344	310 363	322 377	354 415	376 440

Table 2. The number of solution possible to encode with SCA-limited.

T10 only multicast	T20 only multicast	T10 unicast and multicast		T20 unicast and multicast	
		Min	Max	Min	Max
$3.6 \cdot 10^6$	$2.4 \cdot 10^{18}$	$9.9 \cdot 10^{163}$	$8.9 \cdot 10^{200}$	$1.6 \cdot 10^{178}$	$2.0 \cdot 10^{228}$

second method receives one point less than a number of competing methods and so on. If more than one method proposes the solution of the same quality, then all these methods receive the number of points the same as if the other methods that have proposed the same solution did not exist. For instance, let us assume 4 competing methods, method A has proposed the result of quality 100 (the best one), methods B and C has proposed the result of quality 150 and method D has proposed the result of quality 200. In such situation, method A receives four points, methods B and C receive three points, and method D receives one point. The results for 10TBps and 20 TBps are given in Tables 3 and 4.

As shown in Tables 3 and 4 MuPPetS-EON using SCA-Limited significantly outperforms all competing methods in "mixed demands" experiment groups. The reason of this is simple – SCA-Limited limits the number of solutions possible to encode. Therefore, the method considers significantly smaller solution spaces. If the number of possible solutions is large, then the method using SCA-Baldwin may be unable to effectively process the problem in the search of valuable solutions. In the "mixed demands" test case group the number of available solutions when SCA-Baldwin is used is large (from 10^{201} up to 10^{440}). SCA-Limited significantly limits this number to the range starting from 10^{163} up to 10^{200}. Method employing SCA-Limited gains the possibility of fast convergence to the most promising regions of solutions space. Of course, this positive feature has its price – some solutions are impossible to encode. Therefore, in "Only multicast" test cases group, significantly better MuPPetS-EON version is the one using SCA-Baldwin.

Table 3. Summarized results for 10TBps experiments.

k/t	Unicast and multicast						Only multicast					
	AFA	SA	MuPPetS-EON		sGA-EON Bald.	RS-EON Bald.	AFA	SA	MuPPetS-EON		sGA-EON Bald.	RS-EON Bald.
			Limited	Bald.					Limited	Bald.		
2/10	1.1	2.0	3.6	**3.9**	3.1	2.4	1.0	1.9	2.0	**3.2**	**3.2**	**3.2**
4/50	1.5	2.0	**6.0**	4.5	3.8	2.6	1.0	2.2	6.0	**4.1**	3.0	2.8
6/100	1.3	2.9	**5.8**	4.9	3.1	2.3	1.0	2.0	2.2	**3.6**	2.9	**3.6**
8/200	1.2	3.3	**6.0**	4.7	3.0	1.9	1.0	2.0	2.4	**3.5**	3.2	3.2
10/300	1.3	3.1	**5.9**	4.0	4.0	1.7	1.0	2.1	2.5	**3.4**	3.3	3.2
20/500	1.8	4.0	**6.0**	3.9	2.7	1.8	1.0	2.1	2.2	**3.7**	2.8	3.4
30/1000	1.9	4.5	**6.0**	3.2	3.3	1.6	1.0	2.0	2.5	**3.1**	3.0	2.9

Table 4. Summarized results for 20TBps experiments.

k/t	Unicast and multicast						Only multicast					
	AFA	SA	MuPPetS-EON		sGA-EON Bald.	RS-EON Bald.	AFA	SA	MuPPetS-EON		sGA-EON Bald.	RS-EON Bald.
			Limited	Bald.					Limited	Bald.		
2/10	1.0	2.0	**5.2**	5.0	4.0	3.0	1.0	2.0	3.0	**4.3**	3.6	3.5
4/50	1.5	2.0	**5.7**	4.6	3.9	2.6	1.0	2.2	4.1	**4.6**	3.3	3.1
6/100	1.1	3.0	**5.7**	4.9	3.2	2.0	1.0	2.1	3.8	**5.2**	3.8	2.9
8/200	1.4	3.3	**5.8**	4.5	3.5	1.6	1.0	2.1	**4.7**	**4.7**	3.5	2.8
10/300	2.0	4.1	**5.9**	4.0	3.0	1.3	1.0	2.4	4.4	**5.4**	3.9	2.8
20/500	2.1	4.4	**6.0**	3.9	2.5	1.4	1.0	2.4	4.2	**5.5**	3.9	2.5
30/1000	2.9	4.6	**6.0**	3.1	2.6	1.4	1.0	2.8	**5.0**	4.4	3.4	3.1

Another interesting observation is that in the T10 experiment group with only multicast demands, the simplest RS-EON-Baldwin method is almost equally effective as highly sophisticated MuPPetS-EON-Baldwin. The reason of this phenomenon is quite simple – the SCA-Baldwin supports quite an effective optimization of any randomly generated solution. In T10 group, the number of all solutions possible to encode

Table 5. MuPPetS-EON with SCA-Baldwin and SCA-Simple comparison on the base of p-values returned by Sign test.

k/t	T10						T20					
	Unicast and multicast			Only multicast			Unicast and multicast			Only multicast		
	Equal	SCA-limited no worse	SCA-Baldwin no worse	Equal	SCA-limited no worse	SCA-Baldwin no worse	Equal	SCA-limited no worse	SCA-Baldwin no worse	Equal	SCA-limited no worse	SCA-Baldwin no worse
2/10	0.62	0.31	**0.94**	0.03	0.02	**1.00**	**1.00**	0.64	0.64	0.02	0.01	**1.00**
4/50	0.00	**1.00**	0.00	0.00	**1.00**	0.00	0.04	**0.99**	0.02	0.45	0.23	**0.94**
6/100	0.01	**1.00**	0.00	0.03	0.02	**1.00**	0.18	**0.98**	0.09	0.02	0.01	**1.00**
8/200	0.00	**1.00**	0.00	0.03	0.02	**1.00**	0.00	**1.00**	0.00	**1.00**	0.77	0.50
10/300	0.02	**0.99**	0.01	0.12	0.06	**1.00**	0.02	**0.99**	0.01	0.12	0.06	**0.99**
20/500	0.00	**1.00**	0.00	0.06	0.03	**1.00**	0.00	**1.00**	0.00	0.18	0.09	**0.98**
30/1000	0.00	**1.00**	0.00	0.50	0.25	**1.00**	0.00	**1.00**	0.00	0.51	**0.91**	0.25

is lower than in T20 group. Also, the number of genes is low (only 10 genes). Therefore, the main key to proposing high-quality results is the diversity maintenance [2, 6–10, 12, 13, 17]. The above observations were verified with the use of proper statistical tests. Their results are listed in Table 5.

6 Summary

In this paper, we have shown, that the simplification of the method parts, may lead to the significant increase of the method effectiveness. When SCA-Limited is used, some valuable solutions are impossible to encode and thus will never be found. However, thanks to the limitation of available solutions number the method that uses the SCA-Limited can converge faster to the promising regions of solution space. Another important phenomenon is that the Random Search method properly hybridized with a local search may successfully compete with sophisticated methods [14]. Despite its framework is the one of the most primitive. Note, that the test case group, for which RS-EON is effective, is non-trivial. The key to the method effectiveness for this test case subgroup is a high level of population diversity preservation. The only other method that was also successful for this subgroup is MuPPetS-EON. The reason is that MuPPetS-EON uses SDSNC, which properly increases the number of coevolving subpopulations when the method is stuck.

The next research steps will concentrate on mixing the advantages of SCA-Baldwin and SCA-Limited. Such SCA will integrate the capability of fast convergence offered by the SCA-Limited and the precision in finding high-quality results given by SCA-Baldwin. Such SCA should significantly increase the effectiveness of methods used to solve RSA/M problems. Finally, another important observation is that even the most sophisticated methods should always be compared with the simple ones. Only such comparison can give the answer if the advanced mechanisms are beneficial and how significant are these benefits.

Acknowledgements. This work was supported in by the Polish National Science Centre (NCN) under Grant 2015/19/D/ST6/03115.

References

1. Aibin M., Walkowiak K.: Simulated annealing algorithm for optimization of elastic optical networks with unicast and anycast traffic. In: Proceedings of 16th International Conference on Transparent Optical Networks ICTON (2014)
2. Fieldsend, J.E.: Running up those hills: multi-modal search with the niching migratory multi-swarm optimizer. In: IEEE Congress on Evolutionary Computation, pp. 2593–2600 (2014)
3. Goldberg, D.E., et al.: Rapid, accurate optimization of difficult problems using fast messy genetic algorithms. In: Proceedings of 5th International Conference on Genetic Algorithms, pp. 55–64 (1993)
4. Goldman, B.W., Punch, W.F.: Parameter-less population pyramid. In: Proceedings of 2014 Annual Conference on Genetic and Evolutionary Computation (GECCO 2014), pp. 785–792 (2014)

5. Hsu, S.-H., Yu, T.-L.: Optimization by pairwise linkage detection, incremental linkage set, and restricted/back mixing: DSMGA-II. In: Proceedings of 2015 Annual Conference on Genetic and Evolutionary Computation (GECCO 2015), pp. 519–526 (2015)

6. Komarnicki, M.M., Przewozniczek, M.W.: Linked genes migration in island models. In: Proceedings of International Conference on Evolutionary Computation Theory and Applications (ECTA 2016) (2016, in press)

7. Kwasnicka, H., Przewozniczek, M.: Multi population pattern searching algorithm: a new evolutionary method based on the idea of messy genetic algorithm. IEEE Trans. Evol. Comput. **15**(5), 715–734 (2011)

8. Kurdi, M.: An effective new island model genetic algorithm for job shop scheduling problem. Comput. Oper. Res. **67**, 132–142 (2016)

9. Muelas, S., Mendiburu, A., LaTorre, A., Peña, J.-M.: Distributed estimation of distribution algorithms for continuous optimization: how does the exchanged information influence their behavior? Inf. Sci. **268**, 231–254 (2014)

10. Przewozniczek, M., Goscien, R., Walkowiak, K., Klinkowski, M.: Towards solving practical problems of large solution space using a novel pattern searching hybrid evolutionary algorithm - an elastic optical network optimization case study. Expert Syst. Appl. **42**, 7781–7796 (2015)

11. Przewozniczek, M.: Dynamic subpopulation number control for solving routing and spectrum allocation problems in elastic optical networks. In: Proceeding of 3rd European Network Intelligence Conference (2016)

12. Przewozniczek, M.: Active multi population pattern searching algorithm for flow optimization in computer networks – the novel coevolution schema combined with linkage learning. Inf. Sci. **355–356**, 15–36 (2016)

13. Przewozniczek, M., Walkowiak, K., Aibin, M.: The evolutionary cost of Baldwin effect in the routing and spectrum allocation problem in elastic optical networks. Appl. Soft Comput. (2016, in press)

14. Resende, M.G.C., Ribeiro, C.C.: Optimization by GRASP: Greedy Randomized Adaptive Search Procedures. Springer, Heidelberg (2016). ISBN-10: 149396528X ISBN-13: 978-1493965281

15. Simmons, J.M.: Optical Network Design and Planning, 2nd edn. Springer, Heidelberg (2014)

16. Velasco, L., Klinkowski, M., Ruiz, M., Comellas, J.: Modeling the routing and spectrum allocation problem for flexgrid optical networks. Photon Netw. Commun. **24**(3), 177–186 (2012)

17. Walkowiak, K., Przewozniczek, M., Pajak, K.: Heuristic algorithms for survivable P2P multicasting. Appl. Artif. Intell. **27**(4), 278–303 (2013)

18. Walkowiak, K., Goścień, R., Klinkowski, M., Woźniak, M.: Optimization of multicast traffic in elastic optical networks with distance-adaptive transmission. IEEE Commun. Lett. **18**, 2117–2120 (2014)

19. Walkowiak, K., Goścień, R., Woźniak, M., Klinkowski, M.: Joint optimization of multicast and unicast flows in elastic optical networks. In: Proceedings of IEEE International Conference on Communications ICC, pp. 5186–5191 (2015)

20. Walkowiak, K.: Modeling and Optimization of Cloud-Ready and Content-Oriented Networks. Springer, Heidelberg (2016)

21. Yu, X., Tornatore, M., Xia, M., Wang, J., Zhang, J., Zhao, Y., Zhang, J., Mukherjee, B.: Migration from fixed grid to flexible grid in optical networks. IEEE Commun. Mag. **53**(2), 34–43 (2015)

Genetic Algorithm for Self-Test Path and Circular Self-Test Path Design

Miłosław Chodacki$^{(\boxtimes)}$

Institute of Computer Science, Silesian University in Katowice, Katowice, Poland
miloslaw.chodacki@us.edu.pl

Abstract. The article presents the use of Genetic Algorithm to search for non-linear Autonomous Test Structures (ATS) in Built-In Testing approach. Such structures can include essentially STP and CSTP and their modifications. Nonlinear structures are more difficult to analyze than the widely used structures like independent Test Pattern Generator and the Test Response Compactor realized by Linear Feedback Shift Register. To reduce time-consuming test simulation of sequential circuit it was used an approach based on the stochastic model of pseudo-random testing. The use of stochastic model significantly affects the time effectiveness of the search for evolutionary autonomous structures. In test simulation procedure the block of sequential circuit memory is not disconnected. This approach does not require a special selection of memory registers like BILBOs. A series of studies to test circuits set ISCAS'89 are made. The results of the study are very promising.

1 Introduction

Digital systems should provide services according to the specifications reliably. Impairments of dependability are associated with a large class of faults, errors and failures [1]. These impairments may be caused by design, produce or rarely operational imperfections and improper use. There are lots of possible circuit failures: single stuck at 0 or 1 faults, delay and synchronization faults, bridging and open faults, in Metal Oxide Semiconductor technique (MOS) these faults consist in transistor stuck on or stuck off in a logical gates [2]. Some faults cannot be logically represented. Other class of faults can be connected with operational timing frequency and they are related to change impedance parameters, but in that case the built-in testing is one of the most resistant techniques because of common silicon space. Faults that are stimulated may manifest itself as an error. For do that the fault have to be stimulated and propagated to one of internal (to memory module of sequential circuit) or external (primary) circuit output. The error, that is accessible from circuit output, is an information on detected fault and indicates that functional specification of circuit is violated. There is therefore a need for hardware testing.

For Very Large Scale Integration circuits (VLSI) the Built-In Self-Testing (BIST) concept is well used. Embedding the whole or major part of the tester into the circuit is considered as BIST. Production standard involves the use of Linear

© Springer International Publishing AG 2017
N.T. Nguyen et al. (Eds.): ACIIDS 2017, Part II, LNAI 10192, pp. 403–412, 2017.
DOI: 10.1007/978-3-319-54430-4_39

Feedback Shift Registers (LFSR) that are used as a Test Pattern Generators (TPG) and Test Response Compactors (TRC) in a signature analysis [3,4]. These LFSR registers can generate pseudo-random test vectors that may cover many faults. For the evaluation of the effectiveness of coating defects by sequence of test vectors the Fault Coverage (FC) is applied. As TRC registers are mainly used Multi-Input Signature Registers (MISR) and Single-Input Signature Registers (SISR) [5]. These Compactors perform data compression generally lossy, but are known lossless Zero-Aliasing ones without faults masking phenomena [6].

There are also a nonlinear technique with Self-Test Path (STP) or Circular STP (CSTP) [7]. Some modifications of these self-testing techniques are also known e.g. Circular CSTP (C2STP) [8]. Contrary to linear technique, the Circuit Under Test (CUT) in nonlinear technique is a feedback of STP or CSTP, thus posing a problem with parameter selection for these structures. These structures can be implemented in Field Programmable Gate Arrays (FPGA) [9], Application Specific Integrated Circuit (ASIC), System-on-Chip (SOC), which consist of many virtual Intellectual Property modules (IP Core). For SOC the STP structures can also link IP modules [10].

Nevertheless, simulations presented in this paper, show that it is possible to design such BIST, modeled as NLFSR, that achieve higher effectiveness than those of solutions reported in the literature and often are minimized. The minimization consists in the concept of external self-testing, where internal Memory Module (MM) of the circuit is disconnected during test, thus no additional conditions are imposed on its operation. It should be noted that both in linear and nonlinear testing techniques, the circuit MM is typically included into self-testing structure registers, as results from ability to improve testability and application of Design for Testability (DFT). An important observations was made in [11].

The properties of evolutionary algorithms make possible to solve a non-linear structures designing problem. The modeling of NLFSRs is important for the sake of correct representation of it. In this paper many evolutionary models of non-linear register for evolutionary computer simulations are shown. Also many different design methods, often with the use of Genetic Algorithms (GATTO, GATTO+, GATTO*, SELFISH GENE GA [12–14]) and deterministic systems based, among other things, on Automatic Test Pattern Generation (ATPG) [15], Cellular Automata (CA) [16], Finite State Machines (FSM), and Binary Decision Diagrams (BDD) are used to built-in testing design. There are some solutions that can create a sequence of test vectors. In this paper it is searching for not only a selection of sequences, but for structures that can generate these sequences with good FC parameter.

The paper is organized as follows. In Sect. 2 basic information on NLFSR as ATS, essentially STP, evolutionary model are presented. Section 3 includes some description of the Genetic Algorithm and its using to create ATS structures. Next, in Sect. 4 some results of evolutionary searching for STP/CSTP structures are presented, and finally Sect. 5 concludes the paper.

2 Non-Linear Feedback Shift Register as STP and CSTP Model

Self-Test Path, Circular Self-Test Path and Condensed Circular Self-Test Path (C2STP) and in general NLFSRs can be seen as realization of Autonomous Test Structure.

In Fig. 1 a schema of autonomous structure STP, that works accordingly to (1) is presented.

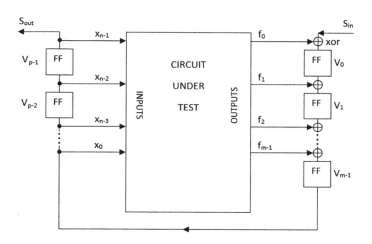

Fig. 1. Self-Test Path

In Fig. 1 V_i is an element of STP and it is mainly D-type flip-flop, t - is a discrete time (clock time) and $p = n + m - 1$ is a length of STP (number of flip-flops).

$$\begin{vmatrix} V_0(t+1) \\ V_1(t+1) \\ \vdots \\ V_{m-1}(t+1) \\ \vdots \\ V_{p-2}(t+1) \\ V_{p-1}(t+1) \end{vmatrix} = T * \begin{vmatrix} V_0(t) \\ V_1(t) \\ \vdots \\ V_{m-1}(t) \\ \vdots \\ V_{p-2}(t) \\ V_{p-1}(t) \end{vmatrix} \oplus \begin{vmatrix} f_0(V_{m-1}(t), \ldots, V_{m-n-2}(t), V_{m-n-1}(t)) \oplus S_{in} \\ f_1(V_{m-1}(t), \ldots, V_{m-n-2}(t), V_{m-n-1}(t)) \\ \vdots \\ f_{m-1}(V_{m-1}(t), \ldots, V_{m-n-2}(t), V_{m-n-1}(t)) \\ \vdots \\ 0 \\ 0 \end{vmatrix}$$

$$(1)$$

where $\oplus$ is an addition operator over $GF(2)$ and

$$T = \begin{vmatrix} 0\,0\,\ldots\,0\,\ldots\,0\,0 \\ 1\,0\,\ldots\,0\,\ldots\,0\,0 \\ \vdots\,\vdots\,\ddots\,\vdots\,\ldots\,\vdots\,\vdots \\ 0\,0\,\ldots\,1\,\ldots\,0\,0 \\ \vdots\,\vdots\,\ldots\,\vdots\,\ddots\,\vdots\,\vdots \\ 0\,0\,\ldots\,0\,\ldots\,1\,0 \end{vmatrix}, \tag{2}$$

is the Connection Matrix for D-type flip-flops which are used to create STP register. If some i^{th} flip-flop is a T-type one then $[T_{i,i}] = 1$.

By taking into account the type of additional STP linear feedback it can be distinguished different connection matrix forms: external ExOR (3), internal ExOR (4) and external-internal ExOR (5).

$$T_1 = \begin{vmatrix} g_0 & g_1 & \cdots & g_{m-1} & \cdots & g_{m+n-2} & g_{m+n-1} = 1 \\ 1 & 0 & \cdots & 0 & \cdots & 0 & 0 \\ \vdots & \vdots & \ddots & \vdots & \cdots & \vdots & \vdots \\ 0 & 0 & \cdots & 1 & \cdots & 0 & 0 \\ \vdots & \vdots & \cdots & \vdots & \ddots & \vdots & \vdots \\ 0 & 0 & \cdots & 0 & \cdots & 1 & 0 \end{vmatrix}, \tag{3}$$

$$T_2 = \begin{vmatrix} 0\,0\,\ldots\,0\,\ldots\,0 & h_0 = 1 \\ 1\,0\,\ldots\,0\,\ldots\,0 & h_1 \\ \vdots\,\vdots\,\ddots\,\vdots\,\ldots\,\vdots & \vdots \\ 0\,0\,\ldots\,1\,\ldots\,0 & h_{m-1} \\ \vdots\,\vdots\,\ldots\,\vdots\,\ddots\,\vdots & \vdots \\ 0\,0\,\ldots\,0\,\ldots\,1 & h_{m+n-1} \end{vmatrix}, \tag{4}$$

$$T_3 = \begin{vmatrix} g_0 & g_1 & \cdots & g_{m-1} & \cdots & g_{m+n-2} & g_{m+n-1} = h_0 = 1 \\ 1 & a_{1,1} & \cdots & a_{1,m-1} & \cdots & a_{1,m+n-2} & h_1 \\ \vdots & \vdots & \ddots & \vdots & \cdots & \vdots & \vdots \\ 0 & 0 & \cdots & 1 & \cdots & a_{m-1,m+n-2} & h_{m-1} \\ \vdots & \vdots & \cdots & \vdots & \ddots & \vdots & \vdots \\ 0 & 0 & \cdots & 0 & \cdots & 1 & h_{m+n-1} \end{vmatrix}, \tag{5}$$

where

$$a_{i,j} = \begin{cases} 1 \text{ if } h_i = g_j = 1, \\ 0 \text{ if } h_i \neq g_j \text{ or } h_i = g_j = 0, \end{cases} \tag{6}$$

and $g_i, h_j, a_{i,j} \in GF(2)$.

NLFSR register can be connected to CUT in different ways. Equation (1) can be expressed simply by (7).

$$V(t+1) = T * V(t) \oplus F(V(t)) \tag{7}$$

It is possible to change connection schema from CUT to STP/CSTP register by using Output Matrices (OM) in a few modes (9) according to (8).

$$V(t+1) = T * V(t) \oplus OM * F(V(t)) \tag{8}$$

$$OM_E = \begin{vmatrix} 1\,0 \ldots 0 \ldots 0\,0 \\ 0\,1 \ldots 0 \ldots 0\,0 \\ \vdots\,\vdots\,\ddots\,\vdots\,\ldots\,\vdots\,\vdots \\ 0\,0 \ldots 1 \ldots 0\,0 \\ \vdots\,\vdots\,\ldots\,\vdots\,\ddots\,\vdots\,\vdots \\ 0\,0 \ldots 0 \ldots 0\,0 \end{vmatrix}, OM_1 = \begin{vmatrix} 0\,1 \ldots 0 \ldots 0\,0 \\ 0\,0 \ldots 1 \ldots 0\,0 \\ \vdots\,\vdots\,\ddots\,\vdots\,\ldots\,\vdots\,\vdots \\ 1\,0 \ldots 0 \ldots 0\,0 \\ \vdots\,\vdots\,\ldots\,\vdots\,\ddots\,\vdots\,\vdots \\ 0\,0 \ldots 0 \ldots 0\,0 \end{vmatrix}, OM_{FREE} = \begin{vmatrix} 1\,1 \ldots 0 \ldots 0\,0 \\ 0\,1 \ldots 1 \ldots 0\,0 \\ \vdots\,\vdots\,\ddots\,\vdots\,\ldots\,\vdots\,\vdots \\ 1\,1 \ldots 1 \ldots 0\,0 \\ \vdots\,\vdots\,\ldots\,\vdots\,\ddots\,\vdots\,\vdots \\ 0\,0 \ldots 0 \ldots 0\,0 \end{vmatrix} \tag{9}$$

The OM_E matrix is identity matrix, but OM_1 matrix must meet the following condition (10):

$$\forall_i \exists_{1 \leq j \leq m}! [OM_{i,j}] = 1 \quad and \quad \forall_j \exists_{1 \leq i \leq m}! [OM_{i,j}] = 1, \tag{10}$$

where i represents rows, j columns of OM_1 matrix respectively and m is a number of circuit outputs. Matrix OM_{FREE} must meet following condition (11):

$$\forall_i \exists_{1 \leq i \leq m}! [OM_{i,j}] = 1. \tag{11}$$

Depending on the contents of the minor the three types of output connection matrices can be distinguished Output Matrix E, Output Matrix 1 and Output Matrix Free. The same notation can be applied to input connection matrices of INPUT MATRIX type [11].

The following linear feedback types can be chosen when configuring the ATS model:

- AIJ TOP-BOTTOM LFSR (1–1500), additional external and internal linear feedbacks are possible,
- BOTTOM LFSR (1500–3000), additional internal linear feedback is possible,
- SHIFT REGISTER (3000–4500), no additional linear feedback,
- TOP LFSR (4500–6000), additional external linear feedback is possible,
- TOP-BOTTOM LFSR (6000–7500), additional external and internal linear feedbacks (other than AIJ TOP-BOTTOM LFSR) are possible.

To configure the STP register connections with the tested circuit, the following connection diagram types were distinguished: for circuit inputs:

- INPUT MATRIX 1 (1–300), complex connections available to the part of the STP register that controls inputs of the tested circuit,
- INPUT MATRIX 1 LONG (300–600), complex connection, while allowing connections with any component of the STP register,
- INPUT MATRIX E (600–900), simple connections (as shown in Fig. 1),
- INPUT MATRIX FREE (900–1200), connections through XOR matrices, but only with those STP register components that control inputs of the tested circuit,

– INPUT MATRIX FREE LONG (1200–1500), connection through XOR matrices with any STP register components.

For circuit outputs:

– OUTPUT MATRIX 1 (1–100), complex connections, available for those components of STP register that are responsible circuit response.
– OUTPUT MATRIX E (100–200), simple connections (as shown in Fig. 1),
– OUTPUT MATRIX FREE (200–300), connections through XOR matrices, but only with those STP components that are responsible for circuit response receiving.

The matrix names listed above contain the type of linear feedback and connection matrix. In brackets there are identifiers being useful in analysis of simulation graphs presented in Figs. 3 and 4.

3 Genetic Algorithm as NLFSR Design Method

Genetic Algorithm (GA) has some useful features, such as the ability to deliver multiple point solutions, and so the lack of concentration of solutions around a certain subclass of STP/CSTP structure and configuration. The algorithm mimics natural evolutionary processes, and therefore there exists the possibility of self-control calculations in such a way that a solution better adapted to a greater extent affects the entire population of solutions (Selective Pressure).

The GA directs the search in the space of feasible solutions by environmental evaluation of the fitness function of each solution (Individual). The course of the algorithm is presented in Fig. 2.

The process of STP/CSTP design creation is a complicated one, especially due to the difficulty of non-linear circuit feedback and BIST simulation time. Every Individual have to be simulated and this process is a great time-consuming task. In [11] the stochastic model of pseudo-random testing, which significantly reduces this problem, was introduced. By using the stochastic model the simulation of each solution is well reduced due to the conversion of exploration FC in search of a suitable length of sequence. Fitness function can be described as some optimization problem in which one is looking for such $x^* \in V(p)$ that maximizes the following formula (12):

$$f(x^*) = \max_{x \in V(p)} f(x), \tag{12}$$

where $V(p)$ is a multidimensional vector of parameters. The fitness function is defined as follows (13):

$$Fitness(x \in V(p)) = w_0 x_0 + w_1(x_{1max} - x_1) + w_2(x_{2max} - x_2), \tag{13}$$

where x_0 is a length of sequence, x_1 is a number of ExORs used to create additional linear feedback, x_2 is a number of T-type flip-flops used to design NLFSR and $\sum_{i=0}^{n=2} w_i = 1$.

The linear code of solution is a binary array of bit values that stores:

- Initial state of register;
- Initial state of circuit memory module (MM);
- Type of register flip-flops e.g. D, T;
- Schema of additional external/internal linear feedback;
- Type and values of Input Matrix IM;
- Type and values of Output Matrix OM;
- Number of included circuit memory elements MM as part of ATS.

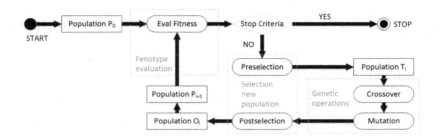

Fig. 2. Evolution in genetic algorithm

In the initial stage of the genetic algorithm essentially random P_0 population base is created. The population is further assessed by the environment (Fitness Function). Based on the adaptation of individuals their reproduction to temporary populations T_i is made. Then from T_i using genetic operators crossover and mutation with some probabilities the descendant population (Offspring) O_i is created. Next evaluation of the newly established offspring population P_{i+1} takes place iteratively until the stopping criteria are fulfilled. This evolutionary process is common to all Genetic Algorithms.

4   Results

In Fig. 3b, one can notice a specific repeatability of FC 0.6–0.8 for STP/CSTP resulting from the presence of type 1 LONG connection matrix (e.g. Seq. Id. 300–600, 1800–2100 and so on). This type of matrix can actually reduce the length of the ATS structure and thereby reduce the length of the test sequence (Fig. 3a). For other connection matrices FC reaches the value of 1. However, the described phenomenon occurs for the s349, only.

Figure 3a shows that for the s382 within certain ATS structures the focused values FC of small discrepancies are obtained. Figure 3b for the same circuit can be noted that only a few configurations of ATS structure can be able to obtain the FC at around 0.9 (Seq. Id 2100–2300 and 9700–9800 for STP, 12800–900 for CSTP). In these structures matrices like INPUT MATRIX FREE and OUTPUT MATRIX FREE are used. An interesting area is that identified by Id Seq. 10500 and 11700, there is ATS structure realized by a simple Shift Register

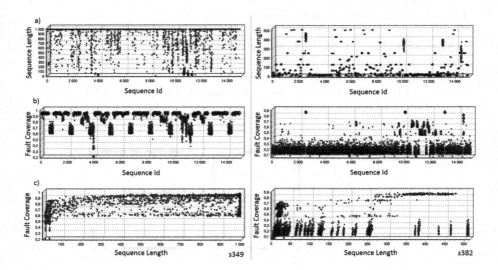

Fig. 3. Simulation graph (from left) s349 and s382 (a) sequence length vs. sequence id, (b) FC vs. sequence id and (c) FC vs. sequence length. STP (Id 1–7500), CSTP (Id 7501–1500)

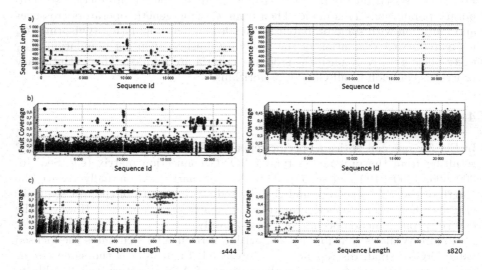

Fig. 4. Simulation graph (from left) s444 and s820 (a) sequence length vs. sequence id, (b) FC vs. sequence id and (c) FC vs. sequence length. STP (Id 1–7500), CSTP (Id 7501–1500), CSTP with additional DFF (Id 1501–2250)

(SR) generating short sequences, which however allow for the acquisition of a relatively high FC value. In this area, there is no additional CSTP linear feedback.

The charts shown in Fig. 4 for the s444 are in some ways similar to the graph in Fig. 3 for the s382. Both test circuits are traffic light controllers and have the BCD counters (timers).

In Fig. 4 for the s820 it can be seen that almost independently of the ATS structure the FC is recovered in the range of 0.3 to 0.45. The best achieved result was FC = 0.598. The s820 circuit has only a few flip-flops but some portion of the state space which should be taken into account is faults dependent.

Statistical analysis of the results showed a correlation from low (below than 0.1) to strong (greater than 0.9), between the length of the sequences and ATS structures and between FC and ATS structures.

In Table 1 the FC values obtained for a subset of ISCAS'89 are presented. In test simulation procedure the block of sequential circuit memory is not disconnected. In other case the greatest value of FC = 0.997 was calculated for the structure of the CSTP including all s298 circuit flip-flops (MM).

Table 1. ISCAS'89 benchmark Fault Coverage Results. Highlighted GA-ATS is the approach presented in this paper.

ISCAS'89	s208.1	s298	s349	s382	s444	s641	s713	s820	s953	s1196	s1238	s1423
GA-ATS	0.995	**0.913**	**0.991**	0.891	**0.924**	**0.921**	**0.877**	0.598	0.983	0.894	0.812	0.530
GATTO	0.679	0.886	NA	0.917	0.890	0.873	0.826	0.918	NA	0.995	0.946	0.963
CA2	0.673	0.876	0.973	0.877	0.863	0.873	0.826	0.598	0.983	0.832	0.812	0.882
ATPG	0.677	0.876	0.978	0.949	0.926	0.873	0.826	0.949	0.990	0.997	0.945	0.896
ATPG-LP	1.000	0.877	0.984	0.927	0.924	0.874	0.877	0.529	0.991	0.995	0.960	0.973
GATTO+	0.697	0.886	0.978	0.947	0.924	0.873	0.826	0.941	0.991	0.995	0.944	0.967
CSTP	0.748	0.886	0.833	0.883	0.831	0.834	0.841	NA	NA	0.641	0.622	NA
FSM-ATPG	0.976	0.913	0.954	0.286	0.317	0.887	0.848	0.965	0.995	0.999	0.971	0.445
CA-GA	1.000	0.893	0.959	0.943	0.924	0.886	0.846	0.528	0.993	0.894	0.954	0.445
HITEC	NA	0.860	0.954	0.754	0.787	NA	NA	0.956	NA	NA	NA	0.518
HITEC-BDD	NA	0.860	0.957	0.779	0.820	NA	NA	0.956	NA	NA	NA	0.564
CCPS	1.000	0.893	0.968	0.943	0.924	0.886	0.846	0.528	0.993	0.894	0.854	0.866
CA 90/150	0.948	0.238	0.610	0.165	0.138	0.886	0.847	0.456	0.994	0.942	0.915	0.635
SELFISH GA	1.000	0.895	0.806	0.942	0.923	0.887	0.847	0.479	0.994	0.953	0.919	0.876

5 Conclusions

Genetic algorithm was able to find appropriate solutions, i.e. the structure of ATS, which were able to generate adequate quality test sequences and the high value of FC was possible to obtain. The conducted experiments show that it is possible to identify ATS structures with a high correlation coefficient between the sequence length and FC. Finding a suitable ATS structure evolutionary with those properties requires the circuit test simulation without faults, and therefore significantly affects the efficiency of the search (exploration) of the solutions space. Next, the diagnostic efficacy of ATS structure have to be finally confirmed by simulation circuits with faults. It should be noted that memory block of circuit operate in accordance with the specification and has not been disconnected, and so the same process of testing simulation was significantly complicated.

References

1. EDCC-2 Companion Workshop on Dependable Computing, Workshop Proceedings. Silesian Technical University, Department of Automatic Control, Electronics and Computer Science, Gliwice, 15 May 1996 (1997). ISBN 83-906582-1-6
2. Agarwal, V.K., Fung, A.S.: Multiple fault testing of logic circuits by single fault test sets. IEEE Trans. Comput. **C-30**(11), 855–865 (1981)
3. Koeter, J.: What's an LFSR? Texas Instruments (1996)
4. Gościniak, I.: Extended class of linear feedback shift registers. In: ICSES, International Conference on Signals and Electronic Systems, Poland (2016)
5. Novak, O., Pliva, Z., Nosek, J., Hławiczka, A., Garbolino, T., Gucwa, K.: Test-per-clock logic BIST with semi-deterministic test patterns and zero-aliasing compactor. J. Electron. Test. Theory Appl. **20**, 109–122 (2004)
6. Chakrabarty, K., Hayes, J.P.: Zero-aliasing space compaction of test responses using multiple parity signatures. IEEE Trans. Very Large Scale Integr. VLSI Syst. **6**(2), 309–313 (1998)
7. Njinda, C., Srinivasan, R., Breuer, A.: On applying circular self-test path (CSTP) technique to circuit. In: IEEE Custom Integrated Circuits Conference (1991)
8. Hławiczka, A., Badura, D.: Condensed circular self-test path: a low cost circular BIST. In: IEEE European Test Workshop - ETW 1996, France (1996)
9. Stroud, C., Lee, E., Chen, P., Abramovici, M.: Built-in self-test of logic blocks in FPGAs. In: 14th VLSI Test Symposium (1996)
10. Stroud, C., Sunwoo, J., Garimella, S., Harris, J.: Built-in self-test for system-on-chip: a case study. In: IEEE ITC International Test Conference (2004)
11. Chodacki, M., Badura, D.: Autonomous test structures for synchronous sequential circuits. In: 13th International Carpathian Control Conference, Czech Republic (2012)
12. Corno, F., Prinetto, P., Sonza Reorda, M.: A genetic algorithm for automatic generation of test logic for digital circuits. In: IEEE International Conference on Tools with Artificial Intelligence, Toulouse, France (1996)
13. Corno, F., Prinetto, P., Rebaudengo, M., Sonza Reorda, M.: A parallel genetic algorithm for automatic generation of test sequences for digital circuits. In: International Conference on High-Performance Computing and Network, Belgium (1996)
14. Corno, F., Prinetto, P., Sonza Reorda, M., Mosca, R.: Advanced techniques for GA-based sequential ATGs. In: IEEE Design & Test Conference, France (1996)
15. Corno, F., Patel, H.J., Rudnicki, E.M., Sonza Reorda, M., Vietti, R.: Enhancing topological ATPG with high-level information and symbolic techniques. In: ICCD 1998, International Conference on Circuit Design, USA (1998)
16. Corno, F., Reorda, M.S., Squillero, G.: Evolving cellular automata for self-testing hardware. In: Miller, J., Thompson, A., Thomson, P., Fogarty, T.C. (eds.) ICES 2000. LNCS, vol. 1801, pp. 31–40. Springer, Heidelberg (2000). doi:10.1007/3-540-46406-9_4

IT in Biomedicine

The Use of Tuned Shape Window for the Improvement of Scars Imaging in Static Renal Scintigraphy in Children

Janusz Pawel Kowalski[1(✉)], Bozena Birkenfeld[2], Piotr Zorga[2], Jakub Peksinski[3], and Grzegorz Mikolajczak[3]

[1] Department of Computer Science in Medicine,
Pomeranian Medical University, Zolnierska 54, Szczecin, Poland
janus@pum.edu.pl
[2] Department of Nuclear Medicine, Pomeranian Medical University,
Unii Lubelskiej 1, Szczecin, Poland
{birka,pzorga}@pum.edu.pl
[3] Faculty of Electrical Engineering, West Pomeranian University of Technology,
Sikorskiego 37, Szczecin, Poland
{jakub.peksinski,grzegorz.mikolajczak}@zut.edu.pl

Abstract. Physiological renal processes are evaluated based on planar images registered by a gamma camera, SPECT or PET. However, detection of small disorders in standard planar scintigraphic imaging is difficult and sometimes impossible. The aim of the research conducted at Pomeranian Medical University is to increase the sensitivity of the method for detecting the areas of renal functional disorders called scarring. The image recorded as a result of the test was subject to digital processing. In that purpose, a novel window function was used for filter designing. Standard and processed images were presented to three independent experts. The diagnosis results were subject to statistical analysis. As a result, a large share of changes in diagnosis was reported in the total number of conducted tests, a strong correlation between positive evaluation of the role of processed images and a change in diagnosis as well as an increased possibility of kidney condition evaluation.

Keywords: Digital image processing · Radioisotope imaging · Renal scarring · Static renal scintigraphy

1 Introduction

Nuclear medicine techniques allow for functional organism evaluation based on biodistribution of the radioisotope index, unavailable for the other imaging diagnostic methods. Scintigraphic images, however, have lower resolution and contain fewer anatomic details than the pictures obtained as a result of using other techniques: planar radiology, transmission CT or MRI. Three-dimensional visualization of image test results allows for bringing out the existing, but not very distinctive information. The role of such imaging has been emphasised by many authors, e.g. Farrell *et al.* [1, 2]. Complex technical diagnostic systems in use provide a real three-dimensional image

© Springer International Publishing AG 2017
N.T. Nguyen et al. (Eds.): ACIIDS 2017, Part II, LNAI 10192, pp. 415–423, 2017.
DOI: 10.1007/978-3-319-54430-4_40

consisting of many planar images with the use of back propagation method or OSEM [3–6]. Spatial visualization limited to one scanning direction can also be used for planar images [7]. Then, higher readability of the analysed scintigraphic images is obtained. This is achievable through digital processing. The results of using chosen digital filtration methods for this purpose have been presented in previous papers of Kowalski [8, 9].

The aim of the research is to increase the sensitivity of the method for detecting the areas of renal functional disorders called scarring [10]. Scintigraphic examination with ^{99m}Tc DMSA is used as a standard one [11, 12]. Nevertheless, this method has finite sensitivity. Detection of small disorders in standard planar scintigraphic imaging is difficult and sometimes impossible. In order to achieve higher sensitivity of this diagnostic method, the algorithm of digital filtration based on new tuned shape window, described in [13, 14], was used.

2 Material and Methodology

As a result of the examination aiming at detection of renal functional disorders (scars), patients are divided into three groups: patients with proper renal image, patients with scar renal changes and patients with possibility of such changes.

From the therapeutic point of view, qualification into the group of patients with a confirmed scar is of importance.

Qualification into the group of patients with possible scars results in providing such people with medical control on account of high risk of pathology occurrence and its earlier detection [10].

The material used in the research presented in this article included scintigraphic renal images registered with the use of a single head rotational gamma camera Mediso Nucline X-Ring. 128 × 128 pixel matrix was used with 16-bit impulse coding depth. Images in the DICOM format with 16-bit coding depth were used for processing. The images were processed in Mathematica 10.0 environment with the use of algorithms developed by Kowalski [9, 13].

The algorithm increasing standard scintigraphic image readability involves processing of the source image through a system of two-dimensional digital filters. This is presented in the schema below (Fig. 1).

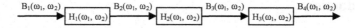

Fig. 1. Block diagram of the filtration process

Transmittance of the second, high-pass, filter $H_2(\omega_1, \omega_2)$ is the inverse of the transmittance of the first, low-pass, filter $H_1(\omega_1, \omega_2)$. As roll-off of the frequency response of both filters in the transition region is of finite value, the use of the cascaded filters results in highlighting the desired features of the processed image. The effect of this filtration was described by Kowalski [8].

The third filter $H_3(\omega_1, \omega_2)$ in the cascade is of low-band nature. The novel window function (Tuned Shape Window [14]) used for designing this filter has the following form:

$$
w(\tau) = \begin{cases} \dfrac{1 - \left(\frac{|\tau|}{\tau_M}\right)^{\beta}}{1 + \left(\frac{\alpha \cdot |\tau|}{\tau_M}\right)^{\beta}} & if\ |\tau| \leq \tau_M \\[2ex] 0 & otherwise \end{cases} \tag{1}
$$

where α, β are non-negative real numbers that determine the shape of the window and τ_m is a half of the length of the window.

The above window function complies with the requirements:

– symmetry - first derivative is equal 0 for $\tau = 0$:

$$
\frac{d}{d\tau} w(\tau) = \frac{-\frac{\beta}{\tau_M^{\beta}} \cdot \left(1 + \alpha^{\beta}\right) \cdot sgn(\tau) \cdot |\tau|^{\beta-1}}{\left(1 + \left(\frac{\alpha \cdot |\tau|}{\tau_M}\right)^{\beta}\right)^2} = 0 \quad for\ \tau = 0 \tag{2}
$$

– maximum value for $\tau = 0$: $w(0) = 1$,
– minimum value for $|\tau| = \tau_M$:$w(-\tau_M) = w(\tau_M) = 0$.

Its second derivative is equal to:

$$
\frac{d^2}{d\tau^2} w(\tau) = \frac{-\frac{\beta}{\tau_M^{\beta}} \cdot \left(1 + \alpha^{\beta}\right) \cdot |\tau|^{\beta-1}}{\left(1 + \left(\frac{\alpha \cdot |\tau|}{\tau_M}\right)^{\beta}\right)^3}
$$
$$
\cdot \left\{ \left(1 + \left(\frac{\alpha \cdot |\tau|}{\tau_M}\right)^{\beta}\right) \cdot \left(2 \cdot \delta(0) + (\beta - 1) \cdot |\tau|^{-1}\right) - 2 \cdot \beta \cdot \left(\frac{\alpha}{\tau_M}\right)^{\beta} \cdot |\tau|^{\beta-1} \right\} \tag{3}
$$

Dirac delta function appears in the second derivative. Hence, side lobes fall off at $|\omega^{-2}|$ or 12 dB per octave [4]. Smooth order of the tuned shape window is 2. Using Landau symbol:

$$
W(\omega) = o\left(|\omega^{-2}|\right) \tag{4}
$$

which means that $W(\omega)$ falls faster then $|\omega^{-2}|$. $W(\omega)$ is Fourier transform of window $w(\tau)$.

Parameters α, β determine the desired filter features in the time and frequency domain. Properties of the proposed window function (1) were investigated and described by Kowalski in [13, 14].

The choice of the TSW was determined by its advantage in the above application, in comparison with popular windows, e.g. Hamming, Blackman, Kaiser etc. Comparative analysis of the TSW and the other windows is in [13, 14].

The main lobe width and the intensity of the side lobes decrease were assumed as criterions for choosing the parameters α, β of the TSW function used for filtration of the scintigraphic images. The function is characterized by the following parameters:

Side Lobe Drop Rate with the value of 12 dB per octave, Normalized Equivalent Noise Bandwidth with the value of 1.16819 bins, 50% Overlap Correlation of 38.9998%, Coherent Gain 0.768235, the first zero in 1.28492 bins, the Highest Side Lobe Peak Level −15.78 dB for 1.77002 bins, Scalloping Loss equal to 2.43379 dB, 3 dB bandwidth with 0.551732 bins width and 6 dB bandwidth with 0.754536 bins width.

The scintigrams obtained as a result of the filtration were still characterised by background radiation of a varied intensity level. This made it more difficult to determine the border between the examined organ and the background. In order to obtain a clear kidney outline, the scintigraphic image background was smoothened with the use of an adaptively chosen cut-off level.

An increase in the readability of the processed image was influenced by the introduction of a colour scale. In this way, diversification was obtained of the levels of radiation coming from different areas of a kidney under examination. The effect of three-dimensional visualization was obtained as a result of the whole scintigram processing. The results of the performed operations are presented in Fig. 2.

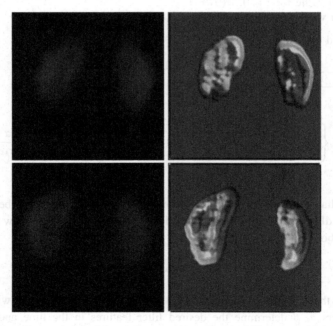

Fig. 2. Kidney image: original from gamma camera (*on the left*), processed (*on the right*)

The aim of the study was to develop a real tool that increases the sensitivity of the diagnostic method. Evaluation of the quality of images based on theoretical criteria is not reflected in their real usefulness. Therefore, the effect of image processing was evaluated from the practical point of view, in regard to the clinical application. The standard images and the processed images were presented to three independent experts, nuclear medicine doctors. The experts made independent diagnoses based on the standard images and, then, the processed images and evaluated the usefulness of the processed images with regard to the aspect of an improved diagnostic process of the right and left kidney. The obtained diagnosis results were subject to statistical analysis.

3 Results

The influence of the processed image form on the change in diagnosis from a quantitative perspective was stated. In order to verify the null hypothesis about the compliance of experts' opinions concerning the change of diagnostic decision, the McNemar test was used.

The test was performed separately for examinations of the right and left kidney. The assumed significance level equals 0.05. The critical value of chi-square for the McNemar test and the assumed significance level amounts to 3.841. Tables 1, 2, 3, 4, 5 and 6 list the results of the investigation concerning the change of diagnostic decision.

The resulting value (0.125) is less than the critical value of the chi-square distribution, which confirms the compliance of the opinion of experts 1 and 2 concerning the change of diagnostic decision with respect to the right kidney.

Table 1. Contingency table for expert 1 vs. expert 2. Right kidney.

	Positive	Negative
Positive	1	3
Negative	5	35

Table 2. Contingency table for expert 1 vs. expert 3. Right kidney.

	Positive	Negative
Positive	1	3
Negative	1	39

Table 3. Contingency table for expert 2 vs. expert 3. Right kidney.

	Positive	Negative
Positive	0	6
Negative	2	36

Table 4. Contingency table for expert 1 vs. expert 2. Left kidney.

	Positive	Negative
Positive	8	3
Negative	17	16

Table 5. Contingency table for expert 1 vs. expert 3. Left kidney.

	Positive	Negative
Positive	4	7
Negative	2	31

Table 6. Contingency table for expert 2 vs. expert 3. Left kidney.

	Positive	Negative
Positive	4	21
Negative	2	17

The resulting value (0.250) is less than the critical value of the chi-square distribution, which confirms the compliance of the opinion of experts 1 and 3 concerning the change of diagnostic decision with respect to the right kidney.

The resulting value (1.125) is less than the critical value of the chi-square distribution, which confirms the compliance of the opinion of experts 2 and 3 concerning the change of diagnostic decision with respect to the right kidney.

The resulting value (8.450) is greater than the critical value of the chi square, which indicates the lack of compliance of the opinions of experts 1 and 2 concerning the change of diagnostic decision with respect to the left kidney.

The resulting value (1.778) is less than the critical value of the chi-square distribution, which confirms the compliance of the opinion of experts 1 and 3 concerning the change of diagnostic decision with respect to the left kidney.

The resulting value (14.087) is greater than the critical value of the chi square, which indicates the lack of compliance of the opinions of experts 2 and 3 concerning the change of diagnostic decision with respect to the left kidney.

The strength of correlation between the change of diagnosis based on the processed image and the indication of an improved examination quality was determined based on Yule's coefficient. The obtained results are presented in Table 7.

Table 7. Correlation between the change in diagnosis and the indication of an improved examination quality.

Expert	Kidney	Yule's coefficient
Expert 1	Right	1
	left	1
Expert 2	Right	1
	Left	1
Expert 3	Right	1
	Left	0.88

The Yule's coefficient values explicitly indicate an increased possibility of the evaluation of kidney condition in children as a result of using the algorithm. Compilation of the improved diagnosis quality in percentages is presented in Table 8.

Table 8. Improved diagnosis quality.

Expert	Kidney	Reported quality improvement in %
Expert 1	Right	13.64
	Left	36.36
Expert 2	Right	18.18
	Left	63.63
Expert 3	Right	6.82
	Left	31.82

The processed images improved the possibility of diagnostic evaluation of the left kidney to a larger extent than of the right one. From a global perspective, for the diagnoses of all the experts, the quality improvement was reported in 12.88% cases for the right kidney compared to 43.94% in the case of the left kidney.

The obtained highest reliability estimator of the diagnosis change probability was $p_{zp} = 0.111$ for the left kidney and $p_{zl} = 0.318$ for the right kidney, while the highest reliability estimator of the improved scintigram quality evaluation probability was $p_{pp} = 0.129$ for the right kidney and $p_{pl} = 0.500$ for the left kidney.

Such a distribution asymmetry of pathological changes with predominance of the left kidney pathology has also been observed by other researchers [15]. Nevertheless, in the literature the cause of this phenomenon is not explained. This follows from pragmatism. From the point of view of clinicians, neither the number of scars nor their location matters, because systemic treatment is used.

4   Discussion and Conclusions

Scintigraphic examination provides a dynamic image of the organism. Various techniques and radiopharmaceuticals are used depending on the diagnostic indications. The evaluation of physiological renal processes is performed based on planar images recorded by a gamma camera, SPECT or PET. The positrons used in PET imaging, as a result of annihilation with the electrons in a patient's body, create a pair of gamma rays propagating at the angle of 180° relative to each other. The rays are recorded by a chosen pair of detectors located on the same line. The use of this phenomenon allows for obtaining higher sensitivity and resolution than in the case of a planar technique or SPECT using a single gamma ray incident on a single detector [16, 17]. Therefore, it is assumed that the obtained PET results are referential in relation to other methods. This does not mean supporting planar techniques or SPECT. The number of SPECT diagnostic devices is much higher than in the case of PET devices, which are only available in large medical centres. Preparation of the markers used in SPECT examination is cheaper than cyclotron-radiochemical production of PET markers on the premises. Longer half-life of SPECT markers makes them more useful in comparison to PET [18, 19].

Other methods of scar detection rely on the application of the same radiopharmacy and performing a SPECT or SPECT/CT examination. Both methods are more precise, but the problem is the need for child sedation. Children are not able to stay still alone for 15–20 min. The planar techniques, used for the study, allow the presence of a parent with a child. Thanks to this parents, if necessary, can hold a child.

The above mentioned properties, especially taking into consideration wider access to planar gamma cameras as well as diagnostic and economic conditions [20, 21], support the necessity for research on the improvement of this diagnostic method.

The scintigram quality resulting from the imaging technology cannot be compared to images created with other diagnostic methods characterized by high level of detail. However, as scintigraphic images are the basic source of information on the regularity of physiological processes, many centres have been conducting research on improving their readability [22, 23].

As the radioisotopic image obtained during static renal scintigraphy in children does not contain any details by nature, low-pass filtration can be used. This does not influence the reconstruction of the scars' shape. The applied new algorithm highlights the low-frequency image components carrying vital diagnostic information.

The background in the scintigram does not contain any information of diagnostic significance. The problem is explicit determination of the border between the background and the examined organ [10, 15, 24–26]. Definitely, interpretation of images is facilitated by smoothening of the background level, which requires determination of its features.

The proposed method of scintigraphic image processing increases its readability. As the applied algorithm is supposed to support the diagnostic decision, it is meant for the users who are not experts. Therefore, a relatively high share of changes in diagnosis in the total number of examinations, strongly correlated with positive evaluation of the role of processed images indicates an increased possibility of kidney condition evaluation.

References

1. Farrell, E.J., Gorniak, R.J., Kramer, E.L., Noz, M.E., Maguire Jr., G.Q., Reddy, D.P.: Graphical 3D medical image registration and quantification. J. Med. Syst. **21**(3), 155–172 (1997)
2. Farrell, E.J., Zappulla, R.A.: Three-dimensional data visualization and biomedical applications. Crit. Rev. Biomed. Eng. **16**(4), 323–363 (1989)
3. Wiant, D., Gersh, J.A., Bennett, M., Bourland, J.D.: Evaluation of the spatial dependence of the point spread function in 2D PET image reconstruction using LOR-OSEM. Med. Phys. **37** (3), 1169–1182 (2010)
4. Tong, S., Alessio, A.M., Kinahan, P.E.: Noise and signal properties in PSF-based fully 3D PET image reconstruction: an experimental evaluation. Phys. Med. Biol. **55**(5), 1453–1473 (2010)
5. Leong, L.K., Kruger, R.L., O'Connor, M.K.: A comparison of the uniformity requirements for SPECT image reconstruction using FBP and OSEM techniques. J. Nucl. Med. Technol. **29**(2), 79–83 (2001)
6. van Velden, F.H., Kloet, R.W., van Berckel, B.N., Lammertsma, A.A., Boellaard, R.: Accuracy of 3-dimensional reconstruction algorithms for the high-resolution research tomography. J. Nucl. Med. **50**(1), 72–80 (2009)

7. Cook, G.J., Lewis, M.K., Clarke, S.E.: An evaluation of 99Tcm-DMSA SPET with three-dimensional reconstruction in 68 patients with varied renal pathology. Nucl. Med. Commun. **16**(11), 958–967 (1995)
8. Kowalski, J.P.: Inverse filtering and readability of radioisotopic images. Acta Bio-Opt. Inform. Med. **10**(3–4), 83–86 (2004). (in Polish)
9. Kowalski, J.P.: Visualization of pseudo 3D radioisotopic images and chosen static features. Methods Appl. Inf. Technol. **12**(2), 55–60 (2007). Szczecin (in Polish)
10. Hitzel, A., Liard, A., Dacher, J.N., Gardin, I., Ménard, J.F., Manrique, A., Véra, P.: Quantitative analysis of 99mTc-DMSA during acute pyelonephritis for prediction of long-term renal scarring. J. Nucl. Med. **45**(2), 285–289 (2004)
11. Piepsz, A., Blaufox, M.D., Gordon, I., Granerus, G., Majd, M., O'Reilly, P., Rosenberg, A.R., Rossleigh, M.A., Sixt, R.: Consensus on renal cortical scintigraphy in children with urinary tract infection. Scientific committee of radionuclides in nephrourology. Semin. Nucl. Med. **29** (2), 160–174 (1999)
12. Celik, T., Yalcin, H., Gunay, E.C., Ozen, A., Ozer, C.: Comparison of the relative renal function calculated with 99mTc-diethylenetriaminepentaacetic acid and 99mTc-dimercaptosuccinic acid in children. World J. Nucl. Med. **13**(3), 149–153 (2014)
13. Kowalski, J.P.: Tuned shape window. In: Advances in System Science, pp. 53–66. Academic Publishing House EXIT, Warsaw (2010)
14. Kowalski, J.P.: Resolution ability of the tuned shape window. In: Advances in System Science, pp. 67–74. Academic Publishing House EXIT, Warsaw (2010)
15. Khan, J., Charron, M., Hickeson, M.P., Accorsi, R., Qureshi, S., Canning, D.: Supranormal renal function in unilateral hydronephrotic kidney can be avoided. Clin. Nucl. Med. **29**(7), 410–414 (2004)
16. Peters, A.M.: Scintigraphic imaging of renal function. Exp. Nephrol. **6**, 391–397 (1998)
17. Willkomm, P., Bangard, M., Guhlke, S., Sartor, J., Bender, H., Gallkowski, U., Rexchmann, K., Biersack, H.J.: Comparison of [18F]FDG-PET and L-3[123I]-iodo-α-methyl tyrosine (I-123 IMT)-SPECT in primary lung cancer. Ann. Nucl. Med. **16**(7), 503–506 (2002)
18. Mitsutaka, F.: Single-photon agents for tumor imaging: 2-99mTc-MIBI, and 99mTc-tetrofosmin. Ann. Nucl. Med. **18**(2), 79–95 (2004)
19. van Cauter, S.C.: The advantages of SPECT vs PET. Next Generation Pharmaceutical. http://www.ngpharma.com/article/The-Advantages-of-SPECT-vs-PET/
20. Saleh, F., Hussein, R., Mohamed, S., Mohamed, H.: Technetium-99m dimercaptosuccinic acid scan in evaluation of renal cortical scarring: is it mandatory to do single photon emission computerized tomography? Indian J. Nucl. Med. **30**(1), 26–30 (2015)
21. Nirmal, T.J., Kekre, N.S.: Management of urological malignancies: has positron emission tomography/computed tomography made a difference? Indian J. Urol. **31**(1), 22–27 (2015)
22. Chen, S., Feng, D.: Noninvasive quantification of the differential portal and arterial contribution to the liver blood supply from PET measurements using the 11C-acetate kinetic model. IEEE Trans. Biomed. Eng. **51**(9), 1579–1585 (2004)
23. Wen, L., Eberl, S., Fulham, M.J., Feng, D., Bai, J.: Constructing reliable parametric images using enhanced GLLS for dynamic SPECT. IEEE Trans. Biomed. Eng. **56**(4), 1117–1126 (2009)
24. Fleming, J.S.: A technique for analysis of geometric mean renography. Nucl. Med. Commun. **27**(9), 701–708 (2006)
25. Garcia, E.V., Folks, R., Pak, S., Taylor, A.: Totally automatic definition of renal regions of interest from 99mTc-MAG3 renograms: validation in patients with normal kidneys and in patients with suspected renal obstruction. Nucl. Med. Commun. **31**(5), 366–374 (2010)
26. Buijs, W.C., Siegel, J.A., Boerman, O.C., Corstens, F.H.: Absolute organ activity estimated by five different methods of background correction. J. Nucl. Med. **39**(12), 2167–2171 (1998)

PCA-SCG-ANN for Detection
of Non-structural Protein 1 from SERS
Salivary Spectra

N.H. Othman[1], Khuan Y. Lee[1,2(✉)], A.R.M. Radzol[1,2],
and W. Mansor[1,2]

[1] Faculty of Electrical Engineering, Universiti Teknologi MARA,
Shah Alam, Selangor DE, Malaysia
leeyootkhuan@salam.uitm.edu.my
[2] Computational Intelligence Detection RIG,
Pharmaceutical and Lifesciences Communities of Research,
Universiti Teknologi MARA, Shah Alam, Selangor DE, Malaysia

Abstract. With non-structural protein (NS1) being acknowledged as biomarker
for Dengue fever, the need to automate detection of NS1 from salivary surface
enhanced Raman spectroscopic (SERS) spectra, with claim of sensitivity up to a
single molecule thus become eminent. Choice for Principal Component Analysis
(PCA) termination criterion and artificial neural network (ANN) topology crit-
ically affect the performance and efficiency of PCA-SCG-ANN classifier. This
paper aims to explore the effect of number of hidden node for the ANN topology
and PCA termination criterion on the performance of the PCA-SCG-ANN
classifier for detection of NS1 from SERS spectra of saliva of subjects. The
Eigenvalue-One-Criterion (EOC), Cumulative Percentage Variance (CPV) and
Scree criteria, integrated with ANN topology containing hidden nodes from 3 to
100 are investigated. Performance of a total of 42 classifier models are examined
and compared in terms of accuracy, precision, sensitivity. From experiments, it
is found that EOC criterion paired with ANN topology of 13 hidden node
outperforms the other models, with a performance of [Accuracy 91%, Precision
94%, Sensitivity 94%, Specificity 96%].

Keywords: NS1 · SERS · PCA · ANN

1 Introduction

Dengue fever is one of the major public health problems in the tropical and subtropical
regions [1, 2]. In Malaysia, outbreak of dengue cases has been made as a national major
agenda. Dengue virus is classified into four serotypes namely, DENV 1, DENV 2,
DENV 3 and DENV 4 [3]. All the four serotypes of dengue virus is an arboviral disease
that is transmitted by mosquito genus Aedes. The Aedes aegepti and Aedes albopictus
[4] are the primary mosquito vector and human are the primary hosts [5]. There are two
types of dengue infection, primary infection and secondary infection [1]. The primary
infection occurs in febrile phase, that is known as dengue fever (DF), while the secondary infection in critical phase which could result in hemorrhagic fever (DHF) and

© Springer International Publishing AG 2017
N.T. Nguyen et al. (Eds.): ACIIDS 2017, Part II, LNAI 10192, pp. 424–433, 2017.
DOI: 10.1007/978-3-319-54430-4_41

dengue shock syndrome (DSS) with fatal consequences [1]. It is believed that the global increase in dengue cases is owing to increase in international travel and urbanization [2, 5].

NS1 is Non-structural protein 1 that can cause diseases such as Dengue Fever (DF), Murray Valley Encephalitis (MVE), Yellow Fever (YF), Tick-borne Encephalitis, West Nile Encephalitis (WNE) and Japanese Encephalitis (JE). NS1 is one of the non-structural proteins that come from flavivirus genome. The flavivirus genome consists of seven non-structural proteins: replication pathogenicity (NS1), NS2A, cofactor (NS2B), protease, helicase, RNA-triphosphatase (NS3), NS4A, NS4B, RNA polymerase, methyltranferase (NS5) and three structural proteins, capsid (C), membrane precursof (prM) and envelope (E) [6]. Of these, NS1 plays an important role in viral replication and pathogenesis. As such, recently NS1 has been acknowledged as a biomarker for flavivirus infected diseases. Besides, it has the advantage of being present in the blood of infected person from or before the onset of the symptom [7].

Raman spectroscopy is a spectroscopic technique that has been acknowledged as one of the technique that can be used to obtain spectra from gases, solid, liquid, slurries, gel and powder. It is founded by C.V. Raman and K.S. Krisnan since 1928 [8]. It is a non-destructive analysis technique that provides information on molecular interaction with structure of the sample, through molecular vibration. It detects biochemical and structural changes in nucleic acids and proteins of the diseases transformation [9]. This makes it a comprehensive investigation diagnosis technique to detect diseases such as cancer [10], atherosclerosis [11], thyroid [12], and gastric [13]. Since the Raman scattered photon is scarse, the signal is a weak signal. However, Surface Enhanced Raman Spectroscopy (SERS) is found capable to amplify the Raman signal by 10^3 to 10^7 times [14]. Previous works from our research group have established the fingerprint Raman spectrum of NS1 [15] and conducted some preliminary analysis and classification using adulterated saliva set [16, 17].

In this study, we are investigating the performance of a new classifier, PCA-SCG-ANN, in order to find the optimal number of hidden nodes for ANN network topology and optimal PCA stopping criterion for detecting NS1 fingerprint features from SERS spectra of saliva. Section 2 elaborates on the theoretical background of PCA and ANN. Section 3 describes the dataset, PCA algorithm and ANN algorithm. Section 4 presents and compares performance from the different models of PCA-SCG-ANN.

2   Theoretical Background

2.1   Principal Component Analysis

Principal Component Analysis (PCA) is a dimension reduction tool that can be used to reduce the dimensionality of a dataset with large number of correlated variables, while keeping as much as possible of the variation presents in the data set [18]. This reduction is achieved by transforming the original data set (correlated variables) to a new data set (uncorrelated variables) known as principal components (PCs) that emphasize variation and bring out strong patterns in the dataset. In other words, PCA compresses the data with less information lost by extracting features with significant variation and leaving out the redundant features [18]. Mathematically, PCA algorithm involves the calculation of variance, covariance, eigenvalues and eigenvector. PCA outcomes are generally in terms of principle component scores (the transformed variable values corresponding to a particular data point), and loadings (the weight by which each standardized original variable should be multiplied to get the component score). The PCs are ranked subsequently according the corresponding eigenvalues. Generally, PC1 contains the largest possible variance and PC2 has the second largest possible variance. The variance will reduce as the principle component increase. The PCs may then be used as predictor [19] or criterion variables, classification, data discrimination [20–22] as well as description and interpretation in following analysis.

In the implementation of PCA, researchers need to choose the number of meaningful PCs to be retained for interpretation. There are several criteria available that may be used in making this decision. Eigenvalue-One-Criterion (EOC), Scree test and Cumulative Percent of Variance (CPV) are amongst the commonly used. EOC, also known as Kaiser criterion retains PCs with eigenvalues more than one and discards PCs with eigenvalues less than one [23]. Implicitly, the criterion only chooses principal components that extract as much as the equivalent of one original variable, at the least. This criterion is probably the most widely used. Scree test criterion chooses its meaningful PCs by observing the Scree plot, which is a plot of eigenvalues versus the number of PCs. The number of PCs is determined by the point where the slope of the scree plot changes, also known as the elbow or inflection point [24]. Meanwhile, Cumulative Percent of total Variance (CPV) regulates the number of PCs based on a pre-selected cumulative variance threshold, normally between 70–99%. This criterion retains PCs that build up a CPV equal to or more than the designated threshold [18].

2.2 Artificial Neural Network (ANN)

Artificial neural network (ANN) is a highly efficient mathematical algorithmic model that can be used to solve classification problems. ANN is inspired by human brain that contains biological neural networks [25]. The invention of back propagation algorithm [26] has revolutionaries the popularity of ANN in machine learning. It comprises of information processing neurons of input layer, hidden layer and output layer, networked to each other by weighted link as illustrated in Fig. 1. The algorithm forwardly propagates the inputs from input layer to output layer. Then input is compared to a desired output using an activation function. The difference or error at output layer is propagated backward to update weight of node [26].

The ability of ANN to imitate the human brain learning process and intuition is acquired through learning algorithms during training sessions, which require database of experiential and practical knowledge. There are many types of learning algorithms for ANN. The dependent factors for the speed of learning include the number of data, the complexity of the problem and the type of problem, i.e. pattern recognition or function approximation. The choice of network topology, learning algorithm and parameterization of ANN architecture is critical for accurate and efficient solution [27] [25]. From previous studies, it has been established that one hidden layer is enough to

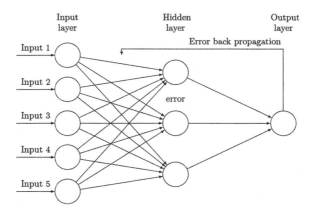

Fig. 1. Network structure of ANN back propagation algorithm.

produce satisfactory results [27–29]. As the number of hidden layer is increased, the computational time and probability of over fitting increases [30]. The number of neurons for hidden layer depends on the type of data and the type of problem. It is usually determined by brute force or some heuristic techniques and found to fall in a range from 3 to 100, from literature review [28, 29, 31].

3 Methodology

3.1 Salivary Raman Spectra Dataset

NS1 ELISA salivary Raman dataset from UiTM-NMRR 12868-NS1-DENV database was used in this study. The dataset contains Raman spectra of saliva samples collected from suspected dengue patients and healthy volunteers benchmarked with NS1-ELISA serological test result. It consists of Raman spectra of 142 positive dengue, 82 dengue negative and 60 control group samples. The spectra was acquired using PeekSeeker Pro with a resolution of 6 cm^{-1}. Being biological, the salivary samples would be damaged if laser wavelength, power and exposure time was not carefully selected. Laser of 785 nm wavelength at a power of 300 mW was used as the excitation source. The exposure time and objective lens were selected at 20 s and 50X, respectively. Raman spectra was captured for Raman shift between 200 to 2000 cm^{-1} comprises of 1801 Raman shift features. Overall, the dataset has 284 spectra with 1801 features. Prior to Raman analysis, 10 µl saliva samples supernatant was dried on gold coated slide, the SERS substrate that amplify the peak intensity.

The saliva samples of dengue suspects were collected from Hospital Pulau Pinang, Hospital Seberang Jaya, Perai and six public polyclinics around Petaling district, Selangor. Saliva samples were collected in accordance to standard unstimulated protocol [31] from patients that fulfilled the WHO 2009 definition of probable dengue and inclusion criteria, i.e. age between 15 and 40; day of fever less than or equal to 5 days and body temperature more than or equal to 38 °C. The exclusion criteria are patients with oral diseases; patients experiencing mouth and gum bleeding; patients unable to provide adequate sample and patients suffering from diseases such as diabetes, high blood pressure, cancer and HIV. Saliva samples from patients that fulfilled both the inclusion criteria and exclusion criteria were not collected.

3.2   Feature Extraction by Principal Component Analysis

Every Raman spectrum in the dataset has 1801 Raman shift features. However, most of the features are redundant or insignificant for ANN classification. Removing the features will help in reducing the complexity of computation. In this study, PCA is employed to reduce the dimension of data prior to ANN classification. It removes the less significant and redundant features while retaining only the most significant features with maximum variance to differentiate between positive and negative dengue cases. The algorithm was implemented using *princomp* command in the MATLAB environment. The algorithm produces the eigenvalues and eigenvectors, which in turn were used to derive the ranked PCs. Features with maximum variance are relocated to the top ranked PCs. Prior to ANN, another critical step in PCA is to determine the number non-trivial PCs for ANN inputs. Scree test, CPV and EOC termination criteria are used and compared in deciding the optimal criterion for this study. The number of retained PCs and its cumulative variance, in accordance to the different criteria are tabulated in Table 1. Scree test retains four PCs that carry 34.57% of the total cumulative variance. CPV criterion with threshold of 90% retained 95 PCs while EOC retains 126 PCs with 94.49% of the total cumulative variance.

Table 1. Number of retained PCs and its cumulative variance for the different PCA termination criteria

Criteria	No of retained PCs	% of cumulative variance
EOC	126	94.49
CPV	95	90.18
Scree test	4	34.57

3.3   Classification by Artificial Neural Network

PCs retained by the PCA termination criteria were served as input vectors to ANN classifier. ANN was used to classify the healthy sample spectra from the NS1 infected sample spectra. Figure 2 shows the procedural flowchart to develop the ANN classifier architecture and the classification process. The algorithm started by uploading the inputs, or the PCs, to the ANN as according to the respective criterion. The PCs were randomized using *RandStream* function to fix the network configuration. Subsequently, the number of PCs, ANN parameters such as type of learning algorithms, number of hidden nodes, number of hidden layers and data divide ratio were set.

For this work, backpropagation network with sigmoid activation function was used. The number of hidden layer was one and the hidden node vector was of the range [3, 5, 8, 10, 13, 15, 17, 18, 20, 25, 30, 40, 50, and 100]. The data was divided into training, test and validate subsets in the ratio of [70:15:15]. The learning algorithm was Scaled Conjugated Gradient (SCG). For every combination of PCA termination criterion and ANN network topology, performance indicator vector [accuracy, sensitivity, specificity and precision] was adopted to evaluate and compare between the classifier models.

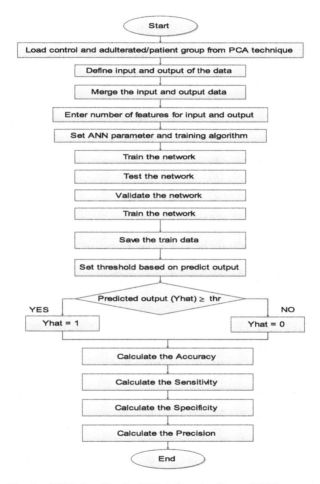

Fig. 2. ANN classifier for NS1 infected salivary SERS spectra

4   Result and Discussion

This section presents results from classification of NS1 infected salivary SERS spectra by using PCA termination criteria integrated with ANN classifier. Table 2 to Table 5 show the performance of 14 different neural network topologies integrated with three different PCA termination criteria, i.e. 42 classifier models.

Table 2. Accuracy for different PCA-SCG-ANN models.

Hidden nodes	PCA termination criterion		
	EOC	CPV	SCREE
3	74	74	70
5	77	77	81
8	74	88*	67
10	81	84	79
13	91*	74	88*
15	74	72	61
17	63	84	67
18	74	81	77
20	70	77	77
25	84	86	67
30	74	81	70
40	81	81	61
50	84	77	67
100	67	74	79

Table 3. Precision for different PCA-SCG-ANN models.

Hidden nodes	PCA termination criterion		
	EOC	CPV	SCREE
3	89	73	70
5	67	72	83
8	65	87	79
10	91	81	78
13	94*	68	78
15	86	67	63
17	52	88	75
18	82	82	79
20	60	78	84
25	83	86	59
30	65	90*	88*
40	90	71	41
50	73	87	55
100	58	83	82

Table 4. Sensitivity for different PCA-SCG-ANN models.

Hidden nodes	PCA termination criterion		
	EOC	CPV	SCREE
3	64	76	73
5	82	86	83
8	83	91*	60
10	76	85	74
13	83	79	93*
15	57	80	55
17	65	75	63
18	78	82	71
20	83	70	70
25	87	86	59
30	77	75	58
40	74	88	69
50	94*	74	75
100	78	73	85

Table 5. Specificity for different PCA-SCG-ANN models.

Hidden nodes	PCA termination criterion		
	EOC	CPV	SCREE
3	89	73	67
5	73	68	79
8	68	86	78
10	89	83	83
13	96*	71	86
15	91	65	67
17	62	91*	74
18	72	81	82
20	60	83	85
25	80	86	73
30	73	90	88*
40	90	77	57
50	77	81	63
100	60	77	71

Table 2 shows the performance in accuracy of classification. It is found that, for all PCA termination criteria and all hidden nodes, the accuracy ranges from 60–100%. The highest accuracy reported is 91% and 88%, based on EOC and Scree criterion respectively, with the use of 13 hidden nodes. The highest accuracy attained with CPV criterion is 88% with 8 hidden nodes. With reference to Table 3, with hidden nodes of 30, CPV and Scree criterion are found to attain the highest precision of 90% and 88% respectively. With EOC criterion, the highest precision is 94% at hidden node of 13. Table 4 records highest sensitivity of 94% for EOC criterion by using 50 hidden nodes. With CPV criterion, at hidden node of 8, it is found that the highest sensitivity achieved is 91%, whereas the highest sensitivity for Scree criterion is 93% with the use of 13 hidden nodes. The highest specificity of 96% as shown in Table 5 is from EOC criterion, with the use of 13 hidden nodes. With CPV criterion, the highest specificity attained is 91% for hidden node of 17, while for Scree criterion the highest specificity is 88% by using 30 hidden nodes.

From the results, of all the three criteria, EOC scores the highest in accuracy, precision, sensitivity and specificity, at hidden node of 13, except for sensitivity. Scree give the worst performance of 88%, except for sensitivity, with the use of higher number of hidden nodes. An explanation for this is EOC retains the most PCs, which is 126. These retained PCs embody sufficiently important attributes of the original signal for classification of infected salivary SERS spectra, as compared to CPV and Scree which have retained PCs of 95 and 4. It can be observed that the more the number of PCs retained, the better is the performance of the ANN classifier, since neural network is known to be *greedy* for input data [28]. However, it is also noticed that for a difference of 97% in the number of PCs retained, the improvement in accuracy, precision, sensitivity and specificity is only [3%, 6%, 3%, 6%].

In addition, it can also be observed highest performance in accuracy, precision, sensitivity and specificity is achieved with hidden nodes in the range of 8 to 50, and hence deemed sufficient to classify the NS1 infected salivary SERS spectra. Less hidden nodes yields lower performance, however too many hidden nodes lead to over fitting and excessive computational load [32].

5   Conclusion

This paper intends to determine an optimal PCA-SCG-ANN classifier model for detection of NS1 molecule in salivary SERS spectra of human subjects. It is observed that the performance of the classifier depends on the number of hidden nodes of SCG-ANN and criterion of PCA. Results here identified EOC criterion surpasses the CPV and Scree criterion. In addition, hidden nodes in the range of 8 to 50 is sufficient to classify the NS1 infected salivary SERS spectra from the normal spectra. Highest performance in accuracy, precision, sensitivity and specificity is achieved PCA-SCG-ANN classifier with topology of 13 hidden nodes and EOC criterion.

Acknowledgment. The author would like to thank the Ministry of Science and Technology (MOSTI), Malaysia, for providing the research funding 100-RMI/SF 16/6/2 (14/2015); the Research Management Institute, Universiti Teknologi MARA, Malaysia and the Faculty of

Electrical Engineering, Universiti Teknologi MARA, Malaysia, for the support and assistance given to the authors in carrying out this research; Ministry of Health, Selangor (JKNS) for their assistance and permission in providing subjects and advice. The experimental procedures involving human subjects described in this paper were approved by the Medical Research & Ethics Committee, National Medical Research Register (NMRR 12868), Malaysia.

References

1. Idrees, S., Ashfaq, U.A.: RNAi: antiviral therapy against dengue virus. J. Trop. Biomed. **3**, 232–236 (2013)
2. Moi, M.L., Omatsu, T., Tajima, S., Lim, C.K., Kotaki, A., Ikeda, M., Harada, F., Ito, M., Saijo, M., Kurane, I., Takasaki, T.: Detection of dengue virus nonstructural protein 1 (NS1) by using ELISA as a useful laboratory diagnostic method for dengue virus infection of international travelers. J. Travel Med. **20**, 185–193 (2013)
3. Shrivastava, A., Dash, P.K., Tripathi, N.K., Sahni, A.K., Gopalan, N., Rao, P.V.L.: Evaluation of a commercial dengue NS1 enzyme-linked immunosorbent assay for early diagnosis of dengue infection. Indian J. Med. Microbiol. **29**, 51–55 (2011)
4. Da Moura, A.J.F., De Melo Santos, M.A.V., Oliveira, C.M.F., Guedes, D.R., De Carvalho Leandro, D.D., Da Cruz Brito, M.L., Rocha, H.D.R., Gómez, L.F., Ayres, C.F.J.: Vector competence of the aedes aegypti population from Santiago Island, cape verde, to different serotypes of dengue virus. J. Parasit. Vectors **8**, 1–9 (2015)
5. Chawla, P., Yadav, A., Chawla, V.: Clinical implications and treatment of dengue. J. Trop. Med. **7**, 169–178 (2014)
6. Lindenbach, B.D., Rice, C.M.: Molecular biology of flavivirus. Adv. Virus Res. **59**, 23–61 (2003)
7. Muller, D.A., Young, P.R.: The flavivirus NS1 protein: molecular and structural biology, immunology, role in pathogenesis and application as a diagnostic biomarker. Antiviral Res. **98**, 192–208 (2013)
8. Raman, C.V.: A change of wave-length in light scattering. Nature **121**, 619 (1928)
9. Huang, S., Lin, D., Chen, G., Xu, Y., Li, Y., Huang, Z., Pan, J., Chen, R., Zeng, H.: Surface-enhanced Raman spectroscopy of saliva proteins for the noninvasive differentiation of benign and malignant breast tumors. Int. J. Nanomed. **10**, 537–547 (2015)
10. Sigurdur, S., Peter, A.P., Lars, K.H., Jan, L., Monika, G., Hans, C.W.: Detection of skin cancer by classification of Raman spectra. IEEE Trans. Biomed. Eng. **51**, 1784–1793 (2004)
11. Alderico, R.D.P.J., Sokki, S.: Raman spectral classification of atherosclerosis using neural networks and discriminant analysis. In: Fourth IEEE International Caracas Conference on Devices, Circuits and Systems, pp. 1–6 (2002)
12. Andrew, T.H., Manjree, G., Xuebin, B.Y., Sheila, E.F., Jennifer, K., Alastair, S., Dominic, P.M.H., Alec, S.H.: Raman spectroscopy and advanced mathematical modelling in the discrimination of human thyroid cell lines. J. Head Neck Oncol. **1**, 1–6 (2009)
13. Feng, S., Chen, R., Lin, J., Pan, J., Wu, Y., Li, Y., Chen, J., Zeng, H.: Gastric cancer detection based on blood plasma surface-enhanced Raman spectroscopy excited by polarized laser light. J. Biosens. Bioelectron. **26**, 3167–3174 (2011)
14. Kleinman, S.L., Frontiera, R.R., Henry, A.I., Dieringer, J.A., Van Duyne, R.P.: Creating, characterizing, and controlling chemistry with SERS hot spots. J. Phys. Chem. **15**, 21–36 (2012)

15. Radzol, A.R.M., Lee, Y.K., Mansor, W.: Raman molecular fingerprint of non-structural protein 1 in phosphate buffer saline with gold substrate. In: Proceedings of International Annual Conference on IEEE Engineering Medical Biology Society, pp. 1438–1441 (2013)
16. Twon Tawi, F.M., Lee, K.Y., Mansor, W., Radzol, A.R.M.: Automatic non-structural protien 1 recognition based on LDA classifier. In: IEEE International Conference on Control System Computing Engineering, pp. 340–343 (2013)
17. Radzol, A.R.M., Lee, K.Y., Mansor, W.: Nonstructural protein 1 characteristic peak from NS1-saliva mixture with surface-enhanced Raman spectroscopy. In. Proceedings of Annual International Conference on IEEE Engineering Medical Biology Society, pp. 2396–2399 (2013)
18. Joliffer, I.T.: Principal Component Analysis. Springer Series of Statistic, 2nd edn. Springer, Heidelberg (2002)
19. Grimbergen, M.C.M., van Swol, C.F.P., van Moorselaar, R.J.A., Uff, J., Mahadevan-Jansen, A., Stone, N.: Raman spectroscopy of bladder tissue in the presence of 5-aminolevulinic acid. J. Photochem. Photobiol. B **95**(3), 170–176 (2009)
20. Krishna, C.M., Prathima, N.B., Malini, R., Vadhiraja, B.M., Bhatt, R.A., Fernandes, D.J., Kartha, V.B.: Raman spectroscopy studies for diagnosis of cancers in human uterine cervix. Vib. Spectrosc. **41**(1), 136–141 (2006)
21. Li, X., Guo, X., Wang, D., Wang, Y., Li, X., Zhang, X., Lin, J.: Spectral analysis for diagnosis of rectum cancer using fluorescence and Raman spectroscopy of serum. In: Proceedings :Annual International Conference of the IEEE Engineering in Medicine and Biology Society, vol. 5, pp. 5449–5452 (2005)
22. Abramczyk, H., Surmacki, J., Brozek-Płuska, B., Morawiec, Z., Tazbir, M.: The hallmarks of breast cancer by Raman spectroscopy. J. Mol. Struct. **924–926**, 175–182 (2008)
23. Kaiser, H.F.: The application of electronic computers to factor analysis. Educ. Psychol. Measur. **20**(1), 141–151 (1960)
24. Cattell, R.B.: The scree test for the number of factors. Multivar. Behav. Res. **1**(2), 245–276 (1966)
25. McCulloch, W., Pitts, W.: A logical calculus of ideas immanent in nervous activity. Bull. Math. Biophys. **5**(4), 115–133 (1943)
26. Werbos, P.: Beyond regression: new tools for prediction and analysis in the behavioral sciences. Ph.D. dissertation, Harvard University, Cambridge (1974)
27. Ceke, D., Kunosic, S., Kopric, M., Lincender, L.: Using neural network algorithms in prediction of mean glandular dose based on the measurable parameters in mammography. Acta Inform. Medica. **17**, 194–197 (2009)
28. Slabbinck, B., Baets, B., Dawyndt, P., De Vos, P.: Genus-wide bacillus species identification through proper artificial neural network experiments on fatty acid profiles. J. Antonie Van Leeuwenhoek **94**, 187–198 (2008)
29. Akbulut, F.P., Akkur, E., Akan, A., Yarman, B.S.: A decision support system to determine optimal ventilator settings. J. BMC Med. Inform. Decis. Making **14**, 1–12 (2014)
30. Beale, M.H., Hagan, M.T., Demuth, H.B.: Neural Network Toolbox ^{TM} User's Guide R 2014 b (2014)
31. Khare, V.: Performance comparison of neural network training methods based on wavelet packet transform for classification of five mental tasks. J. Biomed. Sci. Eng. **03**, 612–617 (2010)
32. Autio, L., Juhola, M., Laurikkala, J.: On the neural network classification of medical data and an endeavour to balance non-uniform data sets with artificial data extension. J. Comput. Biol. Med. **37**, 388–397 (2007)

Prediction of Arterial Blood Gases Values in Premature Infants with Respiratory Disorders

Wiesław Wajs[1], Hubert Wojtowicz[1(✉)], Piotr Wais[2], and Marcin Ochab[3]

[1] The University of Rzeszów, 16c Al. Rejtana, 35-959 Rzeszów, Poland
wwa@agh.edu.pl, hubert.wojtowicz@gmail.com
[2] State Higher Vocational School in Krosno, Rynek 1, 38-400 Krosno, Poland
waisp@poczta.onet.pl
[3] AGH University of Science and Technology,
30 Mickiewicza, 30-059 Kraków, Poland
mj.ochab@labor.rzeszow.pl

Abstract. Arterial blood gases sampling (ABG) is a method for acquiring neonatal patients' acid-base status. Variations of blood gasometry parameters values over time can be modelled using multi-layer artificial neural networks (ANNs). Accurate predictions of future levels of blood gases can be useful in supporting therapeutic decision making. In the paper several models of ANN are trained using growing numbers of feature vectors and assessment is made about the influence of input matrix size on the accuracy of ANNs' prediction capabilities.

1 Introduction

Measurements of blood gases are used to ascertain characteristics of respiratory function, levels of oxygen and carbon dioxide, as well as neonatal acid-base balance. Together with other complementary tests they give information crucial for the assessment of patient's state and making therapeutic decisions [3]. The chemical assessment of blood gases provides complementary information for the clinical assessment of a respiratory disease [2]. The objective goal of acquiring blood gases measurements is the establishment of facts about ventilation and perfusion of a neonate. Values of blood gases are also used to check the adequacy of oxygenation. Frequency of blood gases sampling is dependant on the neonate's response to the treatment. The ABD test is both expensive to carry out and stressful to the subject therefore the frequency of sampling should be optimized based on previous test' results. Therefore the importance of fast and accurate interpretation of blood test results cannot be overstated. With the advancements in technology in areas such as synthetic surfactants and high-frequency pulmonary ventilation, the requirements for timely response to changes in clinical conditions have increased. Currently concerns associated with excessive blood loss in neonates are mitigated by the use of in-line gas monitoring

© Springer International Publishing AG 2017
N.T. Nguyen et al. (Eds.): ACIIDS 2017, Part II, LNAI 10192, pp. 434–444, 2017.
DOI: 10.1007/978-3-319-54430-4_42

equipment and indwelling probes. An important issue in interpretation of blood gases dynamics is keeping regard to the set of their normal values. Presence of fetal hemoglobin in blood of neonates results in differences between values of infants and adults. Moreover the ascertained normal values differ depending on the institution providing hospitalization services. Four major components of the arterial blood gas are $pH, PaCO_2$, bicarbonate HCO_3 or base excess, and PaO_2. For these parameters their established normal values are as follows: pH 7.35–7.45, $PaCO_2$ 35–45 mm Hg, PaO_2 50–70 mm Hg (term infant), 45–65 mm Hg (preterm infant), HCO_3 22–26 mEq/liter. Range of base excess is -2 to $+2$ mEq/liter and O_2 saturation is 92–94. The neonate's respiratory system and kidneys are responsible for maintaining the acid-balance within the narrow limits. Oxygen is moved by the difference in pressures between alveolus and the blood is diffused across the alveolar-capillary membrane. It is then dissolved in the plasma where it binds to hemoglobin.

2   Arterial Blood Gas Data

Measurements of blood gas values are acquired from earlobe blood samples. The same information can be obtained alternatively from pulse oximetry and transcutaneous carbon dioxide measurements as well. Four parameters need to be considered in evaluating blood gases: pH, partial pressure of oxygen PaO_2, partial pressure of carbon dioxide $PaCO_2$ and standard bicarbonate HCO_3. pH is a measure of the hydrogen ion concentration in the blood and indicates acidity or alkalinity. PaO_2 is a measure of the partial pressure of oxygen dissolved in the blood. The unit of measurement is the kilopascal (kPa). $PaCO_2$ is a measure of the partial pressure of carbon dioxide dissolved in the blood. The unit of measurement is the kilopascal (kPa). Arterial blood has a normal $PaCO_2$ of 4–6 kPa. Standard Bicarbonate (St HCO_3) is a measure of the amount of bicarbonate (HCO_3) in the blood. The unit of measurement is millmoles/Litre (mml/L). Arterial blood has a normal value of 18. Interdependencies exist between the studied parameters. The $PaCO_2$ parameter shall decrease by roughly $1\,mmHg$ for the simultaneous decrease of HCO_3^- by $1\,mEq/L$ below $24\,mEq/L$ value. A shift of HCO_3^- value by $10\,mEq/L$ causes a shift in the same direction of pH parameter by a value of about $0.15\,pH$ units. The aforementioned interrelations are considered as standard reference ranges, however in some laboratory analysis equipment different ranges may be employed [5]. The pH or H^+ is used to test whether a neonate is acidemic ($pH < 7.35$; $H+ > 45$) or alkalemic ($pH > 7.45$; $H+ < 35$). A low value of $PaCO_2$ shows improper oxygenation of the hypoxemic patient. It must be mentioned that the low value of $PaCO_2$ is not a necessary condition for the neonate to be hypoxic. As the level of PaO_2 falls below $60\,mmHg$ it becomes necessary to administer supplemental oxygen. The value of PaO_2 lower than $26\,mmHg$ means there is a significant risk of death for the patient, which necessitates immediate oxygenation. The partial pressure of carbon dioxide ($PaCO_2$) is used to assess production and elimination processes of CO_2. With the metabolism rate constant, the level of $PaCO_2$

is assessed through its elimination via pulmonary ventilation. A high level of $PaCO_2$ implies that the neonate is hypoventilating, a low level of this parameter indicates hyperventilation. A HCO_3^- is used to check if a metabolic state of ketoacidosis is present. A low level of hydrogen carbonate is an indicator of metabolic acidosis, a high level of this ion is an indicator of metabolic alkalosis. The level of hydrogen carbonate is usually calculated by the analyzing equipment together with other blood gases results so it's a good practice to check the correlation of this value with directly measured levels of CO_2. To estimate the metabolic component of acid-base disorder a parameter known as base excess is used. Using base excess it can be assessed if the neonate suffers from acidosis or alkalosis. In contrast with HCO_3^- levels, the value of base excess is used to separate the non-respiratory unit of the pH change. This value is total amount of CO_2 and it constitutes the sum of CO_2 and bicarbonate as derived by the following formula: $tCO_2 = [HCO_3^-] + \alpha * PaCO_2$, where $\alpha = 0.226\,mM/kPa$. Bicarbonate value is given as a concentration in millimoles (mmol/l) and $PaCO_2$ value is given in kPa. A care should be taken to avoid sample contamination with the room air as it may give false results of abnormally low levels of carbon dioxide and high levels of oxygen and pH. Contamination of the sample with room air will result in abnormally low carbon dioxide and possibly elevated oxygen levels, and a concurrent elevation in pH. Similar issues may arise if the analysis is performed with delay and the samples were not chilled, which may give inaccurate values of high levels of carbon dioxide and low levels of oxygen. The pH has a normal range between 7.35 and 7.45. The decrease of pH below the lower range indicates acidocis, while the increase above the upper range indicates alkalosis.

The most frequently encountered disorder in blood gases analysis is a respiratory failure also known as respiratory acidosis. The CO_2 dissolves in blood as a weak carbonic acid but if present in large concentrations it can have a huge influence on the value of pH. If neonate's lungs are not able to remove carbon dioxide then its levels in the blood are rising. This in turn results in a rise of carbonic acid concentration and lower value of pH. Plasma proteins act as a buffer and help in maintaining pH level by accepting hydrogen ions. With the increase of CO_2 concentrations a respiratory acidosis is starting to occur. The organism attempts to uphold homeostasis through the increase of respiratory rate. It enables the escape of greater amounts of carbon dioxide through the lungs, which results in rising of pH. In case of neonate being in a critical state and intubated, the frequency of breaths must be increased artificially. Conversely, when the concentration of carbon dioxide is too low a respiratory alkalosis occurs. This may happen because of hyperventilation or improper setting of mechanical ventilation frequency.

3   Artificial Neural Network Method

The multi-layer perceptron neural network architecture trained with gradient descent algorithm was chosen to build prediction models [7]. A simple neural

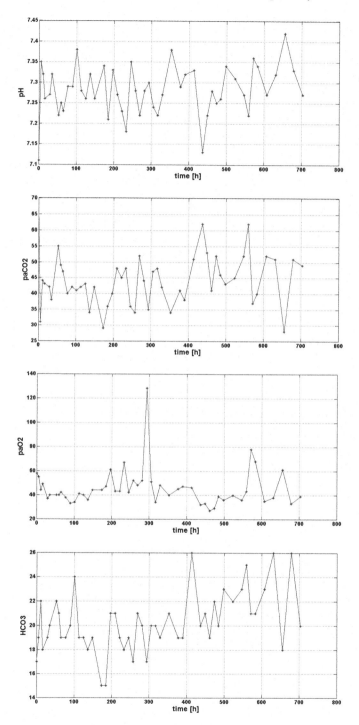

Fig. 1. Dynamics of blood gasometry parameters

network consisting of two layers with sigmoid activation functions in both layers can act as a universal approximator for any function. The gradient descent algorithm is a standard tool for solving non-linear least square problems [1, 4, 10]. These problems are concerned with fitting a parameterized function to a set of vectors through minimization of errors between values of data vectors and function outputs [6, 8]. The steepest gradient descent algorithm is a general minimization technique, which involves calculation of updated values of function parameters in the direction opposite to the estimated objective function gradient. This optimization algorithm is capable of being convergent to a good solution for models with relatively simple objective functions. It is also very efficient for solving problems with great numbers of parameters. An artificial neural network used in our experiments is comprised of two layers. With each layer a weight matrix denoted as W is associated (Fig. 1).

Input and output data for learning process if $R5 = 5$ has a form:

$$R5 = \begin{bmatrix} r11 & r12 & r13 & r14 & r15 \\ r21 & r22 & r23 & r24 & r25 \\ r31 & r32 & r33 & r34 & r35 \\ r41 & r42 & r43 & r44 & r45 \end{bmatrix} =$$

$$\begin{bmatrix} pH_{t-5} & pH_{t-4} & pH_{t-3} & pH_{t-2} & pH_{t-1} \\ PaO2_{t-5} & PaO2_{t-4} & PaO2_{t-3} & PaO2_{t-2} & PaO2_{t-1} \\ PaCO2_{t-5} & PaCO2_{t-4} & PaCO2_{t-3} & PaCO2_{t-2} & PaCO2_{t-1} \\ HCO3_{t-5} & HCO3_{t-4} & HCO3_{t-3} & HCO3_{t-2} & HCO3_{t-1} \end{bmatrix}$$

3.1   Input and Output Data for Testing Process

The neuron input has a following form:

$$I = \begin{bmatrix} a1 & a2 & a3 & a4 & a5 \end{bmatrix} = \begin{bmatrix} pH_{t-4} & pH_{t-3} & pH_{t-2} & pH_{t-1} & pH_t \end{bmatrix} \tag{1}$$

The neuron output is given below:

$$O = \begin{bmatrix} a1 & a2 & a3 & a4 & a5 \end{bmatrix} = \begin{bmatrix} pH_{t-3} & pH_{t-2} & pH_{t-1} & pH_t & pH_{t+1} \end{bmatrix} \tag{2}$$

The layers of a multi-layer neural network play different roles. A layer that produces the network output is called an output layer. Multi-layer networks are quite powerful. A network consisting of two layers, where the first layer is sigmoid and the second layer is also sigmoid, can be trained to approximate any function. We consider five input arrays: $R1, R2, R3, R4$ and $R5$ for tests. We calculate an error e as the difference between the output vector s and a desired target vectors $pH_{t-3}, pH_{t-2}, ph_{t-1}, pH_t, pH_{t+1}$ in the form (Fig. 2):

$$e = 1/2 * ((pH_{t-3} - s1)^2 + pH_{t-2} - s2)^2 + (pH_{t-1} - s3)^2 + (pH_t - s4)^2 + (pH_{t+1} - s5)^2). \tag{3}$$

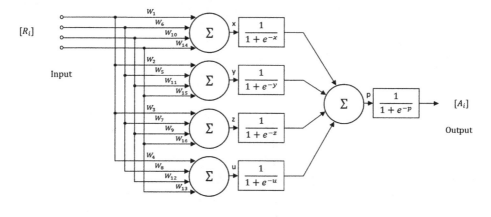

Learning Value $pH(t)$

Fig. 2. Artificial neural network structure and data for learning and testing processes

4   Neural Network Models Training and Testing Results

The prediction of continuous values of blood gases is a nonlinear regression problem. A neural network model after supervised learning should be capable to accurately predict future values of one or more of the selected blood gases parameters. The single input vector for the training of neural network is comprised of values of four blood gases parameters pH, PaO_2, $PaCO_2$ and HCO_3 measured in time. The output vector used for the training is composed of values of pH parameter. In the testing phase the time window for input vectors is shifted forward by one value, which allows for prediction of one future value of the pH parameter. In the experiment a problem of the quantity of input data vectors is considered in terms of its impact on the prediction accuracy. For this purpose a group of neural network models is trained with a growing number of input vectors and for each of these models accuracy of predictions is analyzed. We obtained results for five different input cases: $R1$, $R2$, $R3$, $R4$ and $R5$ array (Figs. 3, 4, 5, 6 and 7).

Case $R1$:

$$R1 = \begin{bmatrix} r15 \\ r25 \\ r35 \\ r45 \end{bmatrix} = \begin{bmatrix} pH_{t-1} \\ PaO2_{t-1} \\ PaCO2_{t-1} \\ HCO3_{t-1} \end{bmatrix}$$

Case $R2$:

$$R2 = \begin{bmatrix} r14 \ r15 \\ r24 \ r25 \\ r34 \ r35 \\ r44 \ r45 \end{bmatrix} = \begin{bmatrix} pH_{t-2} & pH_{t-1} \\ PaO2_{t-2} & PaO2_{t-1} \\ PaCO2_{t-2} & PaCO2_{t-1} \\ HCO3_{t-2} & HCO3_{t-1} \end{bmatrix}$$

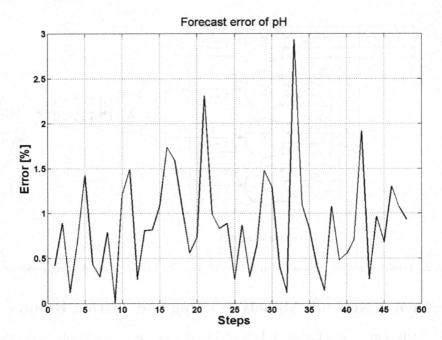

Fig. 3. Error of pH in time [h] for R1

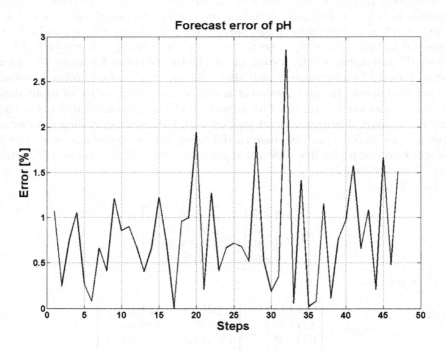

Fig. 4. Error of pH in time [h] for R2

Case $R3$:

$$R3 = \begin{bmatrix} r11 & r12 & r13 \\ r21 & r22 & r23 \\ r31 & r32 & r33 \\ r41 & r42 & r43 \end{bmatrix} = \begin{bmatrix} pH_{t-3} & pH_{t-2} & pH_{t-1} \\ PaO2_{t-3} & PaO2_{t-2} & PaO2_{t-1} \\ PaCO2_{t-3} & PaCO2_{t-2} & PaCO2_{t-1} \\ HCO3_{t-3} & HCO3_{t-2} & HCO3_{t-1} \end{bmatrix}$$

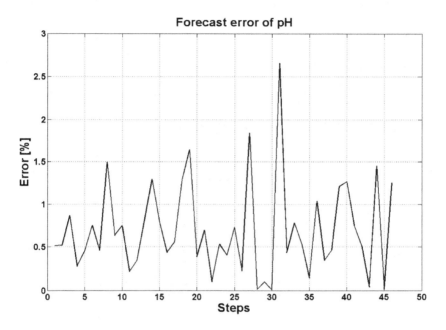

Fig. 5. Error of pH in time [h] for R3

Case $R4$:

$$R4 = \begin{bmatrix} r11 & r12 & r13 & r14 \\ r21 & r22 & r23 & r24 \\ r31 & r32 & r33 & r34 \\ r41 & r42 & r43 & r44 \end{bmatrix} =$$

$$\begin{bmatrix} pH_{t-4} & pH_{t-3} & pH_{t-2} & pH_{t-1} \\ PaO2_{t-4} & PaO2_{t-3} & PaO2_{t-2} & PaO2_{t-1} \\ PaCO2_{t-4} & PaCO2_{t-3} & PaCO2_{t-2} & PaCO2_{t-1} \\ HCO3_{t-4} & HCO3_{t-3} & HCO3_{t-2} & HCO3_{t-1} \end{bmatrix}$$

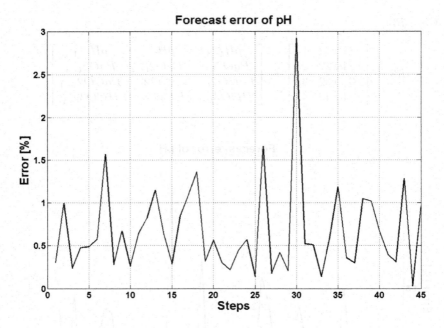

Fig. 6. Error of pH in time [h] for R4

Case $R5$:

$$R5 = \begin{bmatrix} r11 \ r12 \ r13 \ r14 \ r15 \\ r21 \ r22 \ r23 \ r24 \ r25 \\ r31 \ r32 \ r33 \ r34 \ r35 \\ r41 \ r42 \ r43 \ r44 \ r45 \end{bmatrix} =$$

$$\begin{bmatrix} pH_{t-5} & pH_{t-4} & pH_{t-3} & pH_{t-2} & pH_{t-1} \\ PaO2_{t-5} & PaO2_{t-4} & PaO2_{t-3} & PaO2_{t-2} & PaO2_{t-1} \\ PaCO2_{t-5} & PaCO2_{t-4} & PaCO2_{t-3} & PaCO2_{t-2} & PaCO2_{t-1} \\ HCO3_{t-5} & HCO3_{t-4} & HCO3_{t-3} & HCO3_{t-2} & HCO3_{t-1} \end{bmatrix}$$

Table 1. Values of prediction errors for neural network models.

Input array	Prediction error %
R1	0.8780
R2	0.7887
R3	0.6955
R4	0.6647
R5	0.7030

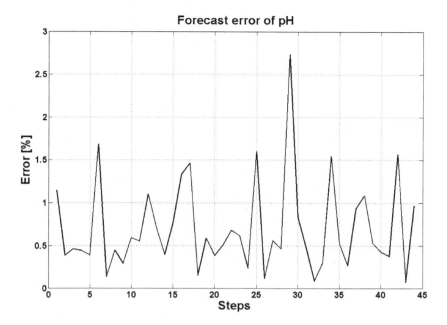

Fig. 7. Error of pH in time [h] for R5

Table 1 contains values of errors calculated between real and predicted values of the pH parameter. The best result was obtained for array $R4$. Testing error for array $R5$ containing the highest number of input vectors is 0.7030%.

Biochemical processes for each of the newborns have their own unique dynamic that is also greatly affected by the process of treatment. Developing a reliable model, which describes future values of blood gases is a complex task. The results show that the neural network model is capable of modeling the dynamics of blood parameters. To account for the changes of the model happening over the course of treatment the neural network model can be retrained in every time step using only last 72 h of historical data.

5   Summary

The results obtained in the course of the research indicate the ability of neural networks to predict future values of selected blood gas parameters with high accuracy in the analyzed time window. Testing error for neural network models decreases with the increase of the number of input vectors presented to the neural network in the training phase. Further research will focus on the possibility of predicting future values over a longer time window. Another issue is finding a compromise between the increasing amount of data, the complexity of network structure and the amount of time needed to retrain the network, while maintaining high demands on the reliability of the prediction and the shortest possible response time of the trained neural network model.

References

1. Antoniou, A., Lu, W.: Practical Optimization: Algorithms and Engineering Applications. Springer, Heidelberg (2007)
2. Aaron, S.D., Vandemheen, K.L., Naftel, S.A., Lewis, M.J., Rodger, M.A.: Topical tetracaine prior to arterial puncture: a randomized, placebo-controlled clinical trial. Respir. Med. **97**(11), 1195–1199 (2003). PMID 14635973
3. Brouillette, R.T., Waxman, D.H.: Evaluation of the newborns blood gas status. Clin. Chem. **43**(1), 215–221 (1997). AACC
4. Kelley, C.T.: Iterative Methods for Optimization. North Carolina State University/SIAM, Raleigh/Philadelphia (1999)
5. Kofstad, J.: Blood gases and hypothermia: some theoretical and practical considerations. Scand. J. Clin. Lab. Invest. **224**(Suppl.), 21–26 (1996). PMID 8865418
6. Levenberg, K.: A method for the solution of certain problems in least squares. Quart. Appl. Math. **2**, 164–168 (1944)
7. Lippman, R.P.: An introduction to computing with neural nets. IEEE ASSP Mag. **4**, 4–22 (1987)
8. Lourakis, M.I.A.: A brief description of the Levenberg-Marquardt algorithm implemented by levmar. Technical report, Institute of Computer Science, Foundation for Research and Technology - Hellas (2005)
9. Transtrum, M.K., Machta, B.B., Sethna, J.P.: Why are nonlinear fits to data so challenging? Phys. Rev. Lett. **104**, 060201 (2010)
10. Marquardt, D.: An algorithm for least-squares estimation of nonlinear parameters. SIAM J. Appl. Math. **11**, 431–441 (1963)
11. Raoufy, M.R., Eftekhari, P., Gharibzadeh, S., Masjedi, M.R.: Predicting arterial blood gas values from venous samples in patients with acute exacerbation chronic obstructive pulmonary disease using artificial neural network. J. Med. Syst. **35**(4), 483–488 (2011)

Extraction of Optical Disc Geometrical Parameters with Using of Active Snake Model with Gradient Directional Information

Jan Kubicek[1(✉)], Juraj Timkovic[2], Marek Penhaker[1],
Martin Augustynek[1], Iveta Bryjova[1], and Vladimir Kasik[1]

[1] VSB–Technical University of Ostrava, FEECS, K450,
17. listopadu 15, 708 33 Ostrava–Poruba, Czech Republic
{jan.kubicek,marek.penhaker,martin.augustynek,
iveta.bryjova,vladimir.kasik}@vsb.cz
[2] Clinic of Ophthalmology,
University Hospital Ostrava, Ostrava, Czech Republic
timkovic.j@bluepoint.sk

Abstract. An analysis of the optical disc is challenging task in the field of clinical ophthalmology. Optical disc (OD) is frequently utilized as reference parameter for time evolution of retinal changes therefore, their analysis is significantly important. In the clinical practice, there are especially problem with lower quality of retinal records acquired by retinal probe of RetCam 3, and worse observation of OD area. Therefore, many algorithms are unable to precisely approximate of OD area. We propose a method based on the active snake model carrying out automatic extraction of retinal disc area even in the spots where an OD is not clearly observable, or image edges completely missing. Furthermore, the proposed solution calculates OD centroid, and respective area for further comparison of OD with retinal lesions.

Keywords: Retinal image · RetCam 3 · Image segmentation · Optical disc · Active contour · Snake

1 Retinal Image Processing

Acquiring of fundus images is relatively complicated process. Fundus camera is composed from a low power microscope. The main their task is capturing retinal image of the posterior retinal area and the whole retina as well [1, 2].

Retinal imaging allows of the three types of clinical examination:

- Color, where white lights is on the retina to examine in full color.
- Red-free in which the contrast among blood vessels and other retinal structures is improved by removing the red color through filtering the imaging light.
- Angiography in which the contrast of vessels is improved by intravenous injection of a fluorescent dye [3–5] (Fig. 1).

OD is one of the essential parts of retinal area. OD is perceived as reference component in many methods leading to automatic image segmentation of retinal

© Springer International Publishing AG 2017
N.T. Nguyen et al. (Eds.): ACIIDS 2017, Part II, LNAI 10192, pp. 445–454, 2017.
DOI: 10.1007/978-3-319-54430-4_43

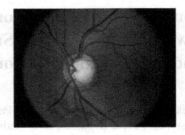

Fig. 1. Optical disc acquired by fundus camera [1] (Color figure online)

structures. From the view of an optical disc morphological structure, OD has vertical oval (elliptical) shape, and OD is separated into two zones: the central zone and peripheral zone also called as neuro-retinal rim.

Ophthalmologic pathologies are indicated on the base of color change, shape or depth of an optical disc. Therefore, OD extraction and geometrical parameters measurement have significant diagnostic value. Furthermore, correct segmentation of OD requires accurate detection of the boundary between the retina and the rim. OD segmentation is also important for estimation of blood vessel density. Accurate disc segmentation is very important in all pathological cases since errors in disc segmentation may mislead the physicians and hence affect their diagnosis.

The outcome of optic disc segmentation process is pixel based. We recognize the three essential retinal areas (Fig. 2):

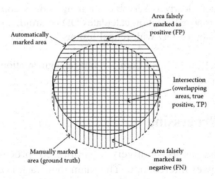

Fig. 2. Similarities between the ground truth and automatically extracted area [13]

- the true positive area representing the overlapping area between the manually marked (ground truth) and automatically marked (segmented image) areas,
- the false negative area where a pixel is classified only in the manually marked area, and
- the false positive area where a pixel is classified only in the automatically segmented area.

On the other hand, there are other several parameters indicating whether retina is physiological, or containing pathological signs. Parameter cup to disc represents ratio of vertical distances between pixels on the highest and the lowest vertical positions inside the cup and OD area (Fig. 3) [6–8, 13, 18].

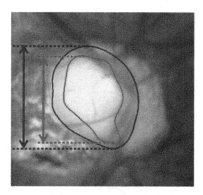

Fig. 3. Representation of the cup to disc ratio for a tilted optical disc [13]

2 System RetCam 3

RetCam 3 is a very useful tool for screening of retinopathy of prematurity (ROP). RetCam stands for the abbreviation for retinal camera, a wide field digital imaging equipment intended for the pediatric fundus examination. While indirect ophthalmoscopy needs expertise, the RetCam 3 allows for ROP screening to be performed by anyone with suitable training and get reproducible results quickly. In poorly cooperative children, the RetCam 3 may be a good alternative for examination under anesthesia or sedation in selected cases.

While it is an excellent tool for peripheral screening in ROP, it has also been used extensively for examination in other diseases like retinoblastoma, and is also useful for anterior segment imaging and gonio-imaging in cases of glaucoma and iris lesions.

RetCam 3 is a useful tool to snap digital images of both the anterior and posterior segment of the eye. All images are stored in the local database allowing for image access over the time. User is allowed to perform fluorescent angiography as well.

Multiple lenses equipped by different magnification are available for attaching to the handheld camera. The most clinically used lens has of 130°, and can easily visualize the retinal periphery in ROP eyes.

Examination by the system RetCam3 goes by similar way as the indirect ophthalmoscopy. A Retina is illuminated by very intensive light which is placed directly on the apex of the camera. Consequently, images or video records are captured, and stored in the hard drive. Since the device has smaller size it captures images relatively in lower resolution. Physician can browse individual images in programme menu, emphasize important artefacts, and useless images can be deleted (Figs. 4 and 5).

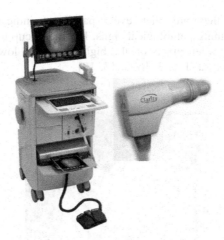

Fig. 4. RetCam 3 (left) and retinal probe (right) [10]

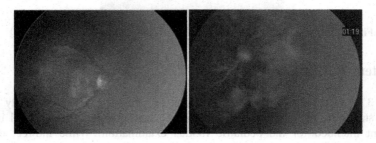

Fig. 5. RetCam 3 image of right eye indicating zone1: aggressive posterior retinopathy of prematurity (left) and RetCam 3 fundus fluorescein angiography indicating clear demarcation of vascular-avascular retina and flat neovascularization [9]

3 Active Snake Model

Active snake represents energy-minimizing spline controlled by internal constrained and external image forces detecting of optical disc features as lines and edges. Snake is defined on the base of conventional energy function in discrete form:

$$E_{snake} = \sum_{i=1}^{N}(E_{int}(i) + E_{ext}(i)) \tag{1}$$

where $E_{int}(i)$ and $E_{ext}(i)$ represent internal spline and external image energy of i^{it} contour position. Internal and external energies are defined by:

$$E_{int}(i) = \alpha_i |s_i - s_{i-1}|^2 + \beta_i |s_{i-1} - 2s_i + s_{i+1}|^2 \tag{2}$$

$$E_{ext}(i) = -|\nabla f(s_i)| \tag{3}$$

where $f(s_i)$ represents the original pixel value at the point s_i. N represents number of snaxels (distance between adjacent snake points). Parameters α_i and β_i represent the weighting factors and control relative importance of the membrane. The external energy is set, so that image gradient has negative magnitude as it is given in Eq. 3.

The optimum locations of the snaxels are searched, so that snake energy E_{snake} can be minimized. There are more variable ways for obtaining new snaxel positions having lower snake energy, for example gradient descent methods and dynamic programming methods can be used. The gradient descent method performs minimization of snake energy on the base of finding of new snaxel position iteratively.

$$\overrightarrow{x_t} = (A - \gamma I)^{-1}(\gamma \overrightarrow{x_{t-1}} - \overrightarrow{h_x}(\overrightarrow{x_{t-1}}, \overrightarrow{y_{t-1}})) \tag{4}$$

$$\overrightarrow{y_t} = (A - \gamma I)^{-1}(\gamma \overrightarrow{y_{t-1}} - \overrightarrow{h_y}(\overrightarrow{x_{t-1}}, \overrightarrow{y_{t-1}})) \tag{5}$$

where A represents the pentadiagonal matrix containing coefficients of descending gradients of the internal energy E_{int}. Parameter γ represents step size, and column vectors $\overrightarrow{x_t}$, $\overrightarrow{y_t}$, $\overrightarrow{h_x}(\overrightarrow{x_{t-1}}, \overrightarrow{y_{t-1}})$ and $\overrightarrow{h_y}(\overrightarrow{x_{t-1}}, \overrightarrow{y_{t-1}})$ are given by the following expressions:

$$\overrightarrow{x_t} = (x_1, x_2, x_3, \ldots, x_{n-1}, x_n)^T \text{ at time t}$$
$$\overrightarrow{y_t} = (y_1, y_2, y_3, \ldots, y_{n-1}, y_n)^T \text{ at time t}$$
$$\overrightarrow{h_x}(\overrightarrow{x_{t-1}}, \overrightarrow{y_{t-1}}) = \left(\frac{\partial E_{ext}(1)}{\partial x_1} \frac{\partial E_{ext}(2)}{\partial x_2} \frac{\partial E_{ext}(3)}{\partial x_3} \cdots \frac{\partial E_{ext}(N-1)}{\partial x_{N-1}} \frac{\partial E_{ext}(N)}{\partial x_N} \right)^T \text{ at time t}-1$$
$$\overrightarrow{h_y}(\overrightarrow{x_{t-1}}, \overrightarrow{y_{t-1}}) = \left(\frac{\partial E_{ext}(1)}{\partial y_1} \frac{\partial E_{ext}(2)}{\partial y_2} \frac{\partial E_{ext}(3)}{\partial y_3} \cdots \frac{\partial E_{ext}(N-1)}{\partial y_{N-1}} \frac{\partial E_{ext}(N)}{\partial y_N} \right)^T \text{ at time t}-1$$

where x^T represents transpose matrix of x. I is $N \times N$ identity matrix. As shown in these equations, the external energy is defined from the magnitude of the gradient image. When the gradient magnitude is computed, the directional information of the gradient can be easily obtained. This additional information can be used in improving the segmentation performance of the active contour model [12, 14–16].

4   Gradient Directional Information

Gradient image is generated on the base of gradient operator searching for image edges. For practical purposes can be used of Sobel, Prewitt, or derivative of Gaussian operator (DOG). It is used of the image convolution to obtain gradient image. Parameters Gradient magnitude $|\nabla f(x, y)|$ and edge direction $\theta_G(x, y)$ are obtained on the base of the expressions:

$$|\nabla f(x,y)| = \sqrt{G_x(x,y)^2 + G_y(x,y)^2} \tag{6}$$

$$\theta_G(x,y) = arctan\left(\frac{G_y(x,y)}{G_x(x,y)}\right) \tag{7}$$

where:

$$G_x(x,y) = f(x,y) * K_x(x,y)$$
$$G_y(x,y) = f(x,y) * K_y(x,y)$$

2D discrete convolution is denoted by $*$ symbol, and operators $K_x(x,y)$ and $K_y(x,y)$ represent horizontal, respective vertical gradient kernels.

In the context of commonly used snakes, snake energy is formed just by gradient magnitude. A conventional snake fails to converge to either outer or inner boundaries. Using of information about gradient direction is beneficial in the context of snake convergence on either outer or inner boundaries.

External energy is reformulated, so that we could consider directional information of gradient:

$$E_{ext}(i) = \begin{cases} -|\nabla f(s_i)|, & if \ |\theta_G(s_i) - \theta_N(s_i)| \leq \frac{\pi}{2} \\ 0, otherwise \end{cases} \tag{8}$$

$\theta_N(s_i)$ is the contour normal direction in the ith snaxel:

$$\theta_N(s_i) = \theta_C(s_i) + \lambda_i \frac{\pi}{2} \tag{9}$$

where $\theta_C(s_i)$ represents counter clockwise contour direction in s_i, and λ_i denotes to the directional parameter of the ith snaxel defined as 1 or -1 in the dependence of the predefined gradient direction of the contour. Parameter θ_C is given by the following expression:

$$\theta_C(s_i) = \frac{1}{2}\left(arctan\left(\frac{y_i - y_{i-1}}{x_i - x_{i-1}}\right) arctan\left(\frac{y_{i+1} - y_i}{x_{i+1} - x_i}\right)\right) \tag{10}$$

Snaxels s_{i-1} and s_{i+1} are represented by coordinates $(x_{i-1}, y_{i-1}), (x_i, y_i)$ and (x_{i+1}, y_{i+1}). In the case when contour normal direction is opposite to the gradient direction, the external energy reaches maximum. Only edges having correct gradient direction participate in the context of a snake minimization process. The directional parameter λ_i is defined by the following way [11, 17]:

$$\lambda_i = \begin{cases} 1, & if \ |\theta_G(s_i) - \theta_C(s_i)| \leq \frac{\pi}{2} \\ -1, otherwise \end{cases} \tag{11}$$

5   Optical Disc Segmentation

The mathematical model of OD shall reflect area of optical disc localization, and also allow for geometrical parameters extraction. The output is in the form of binary image. By binary representation, pixels are classified into two classes. The image background is represented by black color contrarily the area of optical nerve is indicated by white color.

Firstly, the initial contour is placed inside of the optical disc. This contour roughly approximates of optical disc area. The initial contour is shown on the Fig. 6.

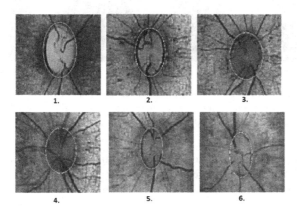

Fig. 6. Representation of initial contour of the OD. Extract represents six patient's records ((1)–(6)), where RoI containing OD is extracted

The model of the optical disc is formed within consecutive steps on the base of the evolution of the active snake contour. Final shape is especially affected by number of iterations. On the base of experiments, using of 60 iterations appears itself as best compromise for reaching as relevant approximation of optical nerve as possible. Time evolution of optical nerve segmentation is depicted on the Fig. 7.

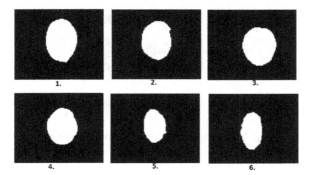

Fig. 7. Final mathematical model of the OD area. Steps (1)–(6) describe the evolution process of the OD model. The final model of the OD is formed on the base of the 60 iterations

452   J. Kubicek et al.

The key fact influencing of the whole process of segmentation is the number of iterations. In the case of using lower number of iterations, initialization contour should not reach the observed object, contrarily if we used higher number of iterations contour would have tendency to spread itself out of the OD. The comparison of individual iterations is depicted on the Fig. 8.

5   60   300

Fig. 8. Underestimated model of the optical disc, number of iterations = 5 (left), ideal approximation of the optical disc, number of iterations = 60 (right) and overestimated model with number of iterations = 300 (right)

The important part of any segmentation algorithm is certain form of feedback which corresponds with the algorithm effectivity.

The disadvantage of the optical nerve processing is absence of real measurable geometrical parameters of optical nerve from the native data, so that there is no any reference level for the assessing objectivity of the proposed method. This fact can be at least partially compensated by the image fusion algorithm serving for the overlaying of the native records and the segmentation model. On the base of the fusion algorithm, we can state that the segmentation model reliably approximates area of optical nerve. The process of image fusion is depicted on the Fig. 9.

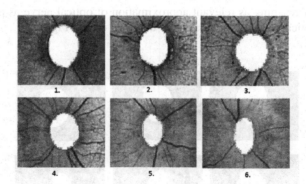

Fig. 9. Output of the image fusion algorithm

6 Conclusion

In the clinical ophthalmology, an optical disc is one of the most important structures characterizing of physiological state of a retinal system. OD is commonly clinically perceived as reference indicator for retinal lesion detection, and their time comparison. There are several approaches attempting to approximate of an OD by symmetric shape especially by ellipse. Unfortunately, it is not correct approximation, the true is that is some special cases optical disc appears itself as symmetric object reminding of an ellipse, but it is not universally applicable rule. Especially for the aforementioned reasons, we are focusing to the task of optical disc segmentation and mathematical modeling. We proposed method based on the active snake model reliably approximating an OD area by time evolving closed curve without regards whether a respective optical disc exhibits itself of elliptical features. Furthermore, in many cases, optical disc does not have clearly recognizable boundaries which can be disconnected, or they completely missing. Snake method appears as suitable alternative even for such situations. In the result of the segmentation process, we obtain mathematical model of an optical disc in the binary form where optical disc is highlighted, and other structures are suppressed. The model also allows for computing optical disc area, and optical disc centroid. These parameters will be used in our future research where we are going to focus on retinal lesion time comparison in the dependence on OD.

Acknowledgment. The work and the contributions were supported by the project SV4506631/2101 'Biomedicínské inženýrské systémy XII'. This article has been supported by financial support of TA ČR, PRE SEED Fund of VSB-Technical university of Ostrava/TG01010137.

References

1. Kumar, S.N., Fred, A.L., Kumar, H.A., Miriam, L.R.J., Asha, M.R.: Retinal blood vessel extraction using wavelet transform and morphological operations. Res. J. Pharm. Biol. Chem. Sci. **7**(5), 1479–1487 (2016)
2. Ko, M.W.L., Leung, C.K.S., Yuen, T.Y.P.: Automated segmentation of optic nerve head for the topological assessment. In: Global Medical Engineering Physics Exchanges/Pan American Health Care Exchanges, GMEPE/PAHCE 2016 (2016). Article no. 7504611
3. Reza, M.N., Ahmad, M.: Automatic detection of optic disc in fundus images by curve operator. In: 2nd International Conference on Electrical Information and Communication Technologies, EICT 2015, pp. 143–147 (2015). Article no. 7391936
4. Gopinath, K., Sivaswamy, J., Mansoori, T.: Automatic glaucoma assessment from angio-OCT images. In: Proceedings - International Symposium on Biomedical Imaging, June 2016, pp. 193–196 (2016). Article no. 7493242
5. Karimi, S., Pourghassem, H.: Optical disc detection in retinal image based on spatial density of grayscale pixels. Int. J. Imaging Robot. **16**(2), 105–117 (2016)
6. Claro, M., Santos, L., Silva, W., Araújo, F., Santana, A.D.A.: Automatic detection of glaucoma using disc optic segmentation and feature extraction In: Proceedings - 2015 41st Latin American Computing Conference, CLEI 2015 (2015). Article no. 7360047

7. Issac, A., Sarathi, M.P., Dutta, M.K.: An adaptive threshold based image processing technique for improved glaucoma detection and classification. Comput. Methods Programs Biomed. **122**(2), 229–244 (2015)
8. Park, H., Schoepflin, T., Kim, Y.: Active contour model with gradient directional information: directional snake. IEEE Trans. Circ. Syst. Video Technol. **11**(2), 252–256 (2001)
9. Zhu, S., Zhou, Q., Gao, R.: A novel snake model using new multi-step decision model for complex image segmentation. Comput. Electr. Eng. **51**, 58–73 (2016)
10. Kubicek, J., Timkovic, J., Augustynek, M., Penhaker, M., Pokrývková, M.: Optical nerve disc segmentation using circual integro differencial operator. In: Sulaiman, H.A., Othman, M.A., Othman, M.F.I., Rahim, Y.A., Pee, N.C. (eds.) Advanced Computer and Communication Engineering Technology. LNEE, vol. 362, pp. 387–396. Springer, Heidelberg (2016). doi:10.1007/978-3-319-24584-3_32
11. Kubicek, J., Penhaker, M., Bryjova, I., Kodaj, M.: Articular cartilage defect detection based on image segmentation with colour mapping. In: Hwang, D., Jung, J.J., Nguyen, N.-T. (eds.) ICCCI 2014. LNCS (LNAI), vol. 8733, pp. 214–222. Springer, Heidelberg (2014). doi:10.1007/978-3-319-11289-3_22
12. Kubicek, J., Penhaker, M., Bryjova, I., Augustynek, M.: Classification method for macular lesions using fuzzy thresholding method. In: Kyriacou, E., Christofides, S., Pattichis, Constantinos, S. (eds.) XIV Mediterranean Conference on Medical and Biological Engineering and Computing 2016. IP, vol. 57, pp. 239–244. Springer, Heidelberg (2016). doi:10.1007/978-3-319-32703-7_48
13. Almazroa, A., Burman, R., Raahemifar, K., Lakshminarayanan, V.: Optic disc and optic cup segmentation methodologies for glaucoma image detection: a survey 2015. J. Ophthalmol. (2015). Article no. 180972
14. Augustynek, M., Penhaker, M.: Finger plethysmography classification by orthogonal transformatios. In: 2010 Second International Conference on Computer Engineering and Applications (ICCEA), vol. 2, pp. 173–177. IEEE (2010)
15. Penhaker, M., Stula, T., Cerny, M.: Automatic ranking of eye movement in electrooculographic records, pp. 456–460 (2010 edn.)
16. Partila, P., Voznak, M., Peterek, T., Penhaker, M., Novak, V., Tovarek, J., Mehic, M., Vojtech, L.: Impact of human emotions on physiological characteristics. In: SPIE (2014 edn.)
17. Augustynek, M., Penhaker, M.: Non invasive measurement and visualizations of blood pressure. Elektronika Ir Elektrotechnika **10**, 55–58 (2011)
18. Krawiec, J., Penhaker, M., Krejcar, O., Novak, V., Bridzik, R.: System for storage and exchange of electrophysiological data. In: Proceedings of 5th International Conference on Systems, ICONS, pp. 11–16 (2010)

Segmentation of Vascular Calcifications and Statistical Analysis of Calcium Score

Jan Kubicek$^{(\boxtimes)}$, Iveta Bryjova, Jan Valosek, Marek Penhaker, Martin Augustynek, Martin Cerny, and Vladimir Kasik

VSB–Technical University of Ostrava, FEECS,
K450 17. Listopadu 15, 708 33 Ostrava–Poruba, Czech Republic
{jan.kubicek,iveta.bryjova,jan.valosek.st,
marek.penhaker,martin.augustynek,martin.cerny,
vladimir.kasik}@vsb.cz

Abstract. Assessment of blood vessels is current task in the field of clinical angiography. Calcification spots are usually observable well from native CT angiography records nevertheless, it is absence of methods precisely calculating calcium score (CS) serving as indicator of blood vessel deterioration by calcification process. The paper deals with the segmentation method for segmentation, extraction and differentiation of vascular calcifications based on the multilevel Otsu thresholding. The method generates mathematical model of physiological and calcification part of blood vessels and, it consequently allows for CS calculation as calcified blood vessel to physiological ratio. We performed analysis of blood vessels modeling and CS calculation on the base of wide dataset 90 patient's records. We compared our CS results with three independent clinical experts. On the base of this comparative analysis, we propose estimated intervals for CS serving for objective analysis of this vascular parameter in the dependence of a calcification level. The main clinically applicable result is SW *VesselsCalc* application for complex analysis of blood vessels, modeling of calcification areas and calculation of calcium score.

Keywords: Blood vessels · Image segmentation · Vascular calcification · Calcium score · CT angiography

1 Introduction

Ischemic diseases like it is ischemic heart disease (ICHS), chronical peripheral arterial disease (ICHDK), peripheral vascular disease and ischemic stroke (CMP) are one of the main causes of adult population morbidity and death rate across the world. All these diseases create themselves on the base of atherosclerosis. It is long-term process which can go without greater clinical signs by many years. This process causes blood vessel deterioration by storing of fat in the blood and calcium salt. These sclerotic plaques narrow of vascular lumen, restrict blood flow, and can be basis for thrombosis. On the base of the aforementioned facts, it is obvious that early diagnosis and prediction of ischemic diseases are necessary for clinical practice. [1–3, 10, 11]

© Springer International Publishing AG 2017
N.T. Nguyen et al. (Eds.): ACIIDS 2017, Part II, LNAI 10192, pp. 455–464, 2017.
DOI: 10.1007/978-3-319-54430-4_44

The first mutual sign of ischemic diseases is occurrence of calcium plaques in a blood stream. Their amount can be found and assessed by Computed Tomography (CT). On the base of the CT data, it is needed to determine of the Calcium score (CS). This parameter allows for assessing of calcium plaques occurrence in coronal arteries. One expression of CS is related to the density of calcifications in given artery calculated in Hounsfield units (HU) multiplied by size of given area (mm^2). Implicitly, it is calculated with only pixels having HU bigger than 130, creating area ≥ 1 mm^2. CS values are from the range: 0–400, and express deterioration level of calcium plaques, and probability of coronal arteries deterioration (Table 1) [4, 7, 12, 13].

Table 1. Calcium score values [5, 6]

Calcium score	Calcium deterioration	Deterioration probability of coronal arteries
0	Without plaque	Very low
1–10	Very little of plaque	Very probably
11–100	Light deterioration	Low probability of significant stenosis
101–400	Middle deterioration	Middle probability of significant stenosis
≥ 400	Wide deterioration	High probability of significant stenosis

2 Multilevel Thresholding Algorithm for Vascular Calcifications

Segmentation using only one thresholding it is not appropriate for blood vessel segmentation due to the fact that image data often contain image noise which cannot be properly classified. We use optimized Otsu method utilizing of multi-thresholding increasing sensitivity of blood vessel segmentation, and extraction of vascular calcifications [14, 15].

The core of the method is finding a specific intensity level on the base of the histogram distribution into evenly large areas. Specific thresholding level is used for each area. The analyzed image is consequently segmented according to thresholding levels. Pixels having different shade levels are labeled by parameter L from interval: $[0, 1, 2, \ldots, L]$. Number of levels is indicated by p. A size of respective segmented area is given by the following equation:

$$a = \frac{L}{p} \tag{1}$$

The between class variance σ^2 is calculated similarly as in the Otsu method:

$$\sigma^2 = W_0 * W_1 * (\mu_0 - \mu_1)^2 \tag{2}$$

Parameter W represents weight, and average intensity is represented by μ. The number of separated histogram image regions is identical to the number of thresholding levels p. Optimal number of thresholding levels is calculated as:

$$P_p = max_p\left(\sigma^2\right) \tag{3}$$

It is necessary to number of pixels in different shades of gray L would be equal to 256 * j. Parameter p must belong to the range: $[2*j, 4*j, 8*j]$, where j belongs to the range: $1, 2, \ldots, \infty]$. The overall structure of the multilevel segmentation approach is depicted on the Fig. 1 [8, 9].

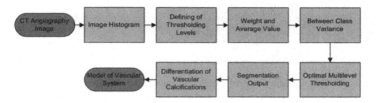

Fig. 1. Flow chart of multilevel segmentation method for vascular calcification modeling

3 Modelling of Vascular Calcifications

The algorithm was tested on the sample of the 90 patient's records including calcification plaques. Data were acquired from CT angiography from Hospital in Trinec. Example of analysed records is depicted on the Fig. 2.

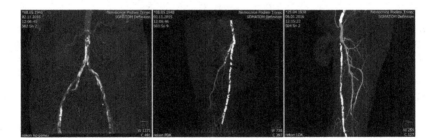

Fig. 2. Extract of analyzed images acquired by CT angiography

In common physiological situation, blood vessel system is represented by shade intensity without more significant intensity changes. Presence of calcifications is represented by white color. Calcification changes are obviously observable by naked eye. In the case of the CT angiography, calcification areas are usually represented by bright white color spectrum, while physiological vessels are imagined in gray. Nevertheless, important issue is quantification of calcification area in the form of Calcium score. We approached to extract of this parameter on the base of mathematical model separating physiological area of respective blood vessel, and vascular calcification (Fig. 3). The vascular model consequently allows for computing of respective blood vessel area and calcification area (Fig. 4).

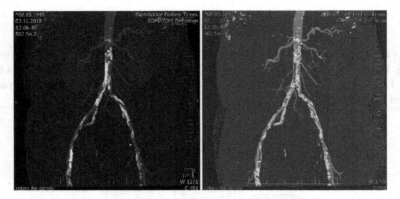

Fig. 3. Native CT angiography record (left), output of the segmentation model, where early calcification plaque is indicated by yellow, and advanced calcification is indicated by red (right) (Color figure online)

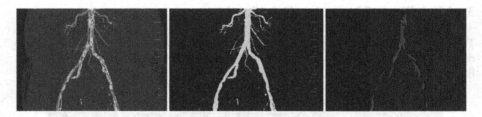

Fig. 4. Overall segmentation model of vascular system with color calcification differentiation (left), model of overall vascular system (middle) and calcification model (right) (Color figure online)

4   Statistical Evaluation of Calcium Score

From the view of the clinical practice, there is an absence of clinical instrument automatically calculating of Calcium score from native images, thus it would perform of diagnosis of blood vessel deterioration level by calcification. The multilevel segmentation algorithm separates individual areas of blood system into individual classes reflecting state whether in the particular area it is calcification, or not. On the base of this model, we can calculate of calcification score:

$$CS = \frac{R_c}{R_o} [\%] \qquad (4)$$

Where R_c corresponds with area of calcification, and R_o corresponds with overall blood vessel area.

4.1   Testing of Statistical Significance of Calcium Score

Calcium score has greater ambitions to be used in the field of clinical radiology and angiography. Physician performs of diagnosis, usually, on the base of the subjective opinion and own experience. For one angiographic examination MR commonly generates 150–200 patient records of region of interest. On the base of the clinical evaluation of three clinical experts, patients are clinically classified into three groups:

- **Calcified blood vessels (CBV)** – blood vessels are completely calcified
- **Partially calcified blood vessels (PCBV)** – blood vessels are partially deteriorated by calcification
- **Early calcification (EC)** – blood vessels are not visually impaired but there is a predisposition of calcification

Descriptive statistics for all patient groups are summarized in Table 2. Calcification score was calculated on the base of the 90 patients. 30 patients were used for every group ($n_{CBV} = 30$, $n_{PCBV} = 30$ and $n_{EC} = 30$).

Table 2. Descriptive statistics of calcification score

	CBV	PCBV	EC
Median	89.5%	48.5%	23.5%
Average value	86%	41%	19%

On the base of the observation, it is evident that median and average value are significantly different for individual groups. On the base of the physician opinions, expected values of calcification score should belong to the following intervals:

- CBV: <85–100> [%]
- PCBV: <35–60> [%]
- EC: <10–30> [%]

Robust Interval Estimation of Average Value. For each type of the calcification deterioration (CBV, PCBV, EC), there are subjectively defined intervals by physicians where outputs of CS calculated by multilevel segmentation are expected. These subjectively defined intervals should be verified on the base of the median which is robust against outlying observations. The interval estimation of median is proved on the base of the Gastwirth median estimation. Interval median estimation with reliability 95% is estimated from the interquartile range:

$$\left\langle x_{0.5} - 1.57 \frac{(x_{0.75} - x_{0.25})}{\sqrt{n}} ; x_{0.5} + 1.57 \frac{(x_{0.75} - x_{0.25})}{\sqrt{n}} \right\rangle \tag{5}$$

where x_p represent 100p% selection quartiles.

Median Estimation for Calcified Blood Vessels (CBV). Defining of parameters for median estimation is summarized into Table 3.

Table 3. CBV quantiles for median estimation

$CBV_{0.5}$ [%]	$CBV_{0.75}$ [%]	$CBV_{0.25}$ [%]
89.5	93	87

Median estimation for calcified blood vessels is calculated:

$$\left\langle 89.5 - 1.57\frac{93-87}{\sqrt{30}} ; 89.5 + 1.57\frac{93-87}{\sqrt{30}} \right\rangle$$

$$\langle 87.7; 91.2 \rangle$$

Median Estimation for Partially Calcified Blood Vessels (PCBV). Defining of parameters for median estimation is summarized into Table 4.

Table 4. PCBV quantiles for median estimation

$PCBV_{0.5}$ [%]	$PCBV_{0.75}$ [%]	$PCBV_{0.25}$ [%]
48.5	56	42

Median estimation for partially calcified blood vessels is calculated:

$$\left\langle 48.5 - 1.57\frac{56-42}{\sqrt{30}} ; 48.5 + 1.57\frac{56-42}{\sqrt{30}} \right\rangle$$

$$\langle 44.4; 52.5 \rangle$$

Median Estimation for Early Calcification (EC). Defining of parameters for median estimation is summarized into Table 5.

Table 5. EC quantiles for median estimation

$EC_{0.5}$ [%]	$EC_{0.75}$ [%]	$EC_{0.25}$ [%]
23.5	28	15

Median estimation for partially early calcification is calculated:

$$\left\langle 23.5 - 1.57\frac{28-15}{\sqrt{30}} ; 23.5 + 1.57\frac{28-15}{\sqrt{30}} \right\rangle$$

$$\langle 19.7; 27.2 \rangle$$

Gastwirth Median Estimation. Gastwirth median also belongs among of robust average value estimation. This estimation is determined by the sample median, lower ($x_{0.33}$) and upper ($x_{0.67}$) decile. Gastwirth median estimation is given by the expression:

$$\left\langle x_{GST} - 1.57\frac{x_{0.75} - x_{0.25}}{\sqrt{n}}; x_{GST} + 1.57\frac{x_{0.75} - x_{0.25}}{\sqrt{n}} \right\rangle \qquad (6)$$

where:

$$x_{GST} = 0.4.x_{0.5} + 0.3.(x_{0.33} + x_{0.67}) \qquad (7)$$

Gastwirth Median Estimation for Calcified Blood Vessels (CBV). Defining of parameters for Gastwirth median estimation is summarized into Table 6.

Table 6. CBV quantiles for Gastwirth median

$CBV_{0.5}$ [%]	$CBV_{0.75}$ [%]	$CBV_{0.25}$ [%]	$CBV_{0.33}$ [%]	$CBV_{0.67}$ [%]	CBV_{GST} [%]
89.5	93	87	87.4	91.6	89.5

Gastwirth median estimation for calcified blood vessels is calculated:

$$\left\langle 89.5 - 1.57\frac{93 - 87}{\sqrt{30}}; 89.5 + 1.57\frac{93 - 87}{\sqrt{30}} \right\rangle$$

$$\langle 87.7; 91.2\rangle [\%]$$

Gastwirth Median Estimation for Partially Calcification (PCBV). Defining of parameters for Gastwirth median estimation is summarized into Table 7.

Table 7. PCBV quantiles for Gastwirth median

$PCBV_{0.5}$ [%]	$PCBV_{0.75}$ [%]	$PCBV_{0.25}$ [%]	$PCBV_{0.33}$ [%]	$PCBV_{0.67}$ [%]	$PCBV_{GST}$ [%]
48.5	56	42	45.4	54.2	49.2

Gastwirth median estimation for partially calcified blood vessels is calculated:

$$\left\langle 49.2 - 1.57\frac{56 - 42}{\sqrt{30}}; 49.2 + 1.57\frac{56 - 42}{\sqrt{30}} \right\rangle$$

$$\langle 45.2; 53.2\rangle [\%]$$

Gastwirth Median Estimation for Early Calcification (EC) Defining of parameters for Gastwirth median estimation is summarized into Table 8.

Table 8. EC quantiles for Gastwirth median

EC$_{0.5}$ [%]	EC $_{0.75}$ [%]	EC $_{0.25}$ [%]	EC $_{0.33}$ [%]	EC $_{0.67}$ [%]	EC $_{GST}$ [%]
23.5	28	15	17.4	25.6	22.3

Gastwirth median estimation for early calcification is calculated:

$$\left\langle 22.3 - 1.57\frac{28 - 15}{\sqrt{30}}; 22.3 + 1.57\frac{28 - 15}{\sqrt{30}} \right\rangle$$

$$\langle 18.5; 26 \rangle [\%]$$

The following charts (Tables 9 and 10) summarize median interval estimation for calcium score of three tested groups of patients (CBV, PCBV, EC).

Table 9. Comparison of CS interval estimation

	CBV [%]	PCBV [%]	EC [%]
Median estimation	$\langle$**87.7; 91.2**$\rangle$	$\langle 44.4; 52.5 \rangle$	$\langle 19.7; 27.2 \rangle$
Gastwirth median	$\langle 87.7; 91.2 \rangle$	$\langle 45.2; 53.2 \rangle$	$\langle 18.5; 26 \rangle$
Expected values of CS	$\langle 85; 100 \rangle$	$\langle 35; 60 \rangle$	$\langle 10; 30 \rangle$

Table 10. Comparison on CS interval lengths

	CBV [%]	PCBV [%]	EC [%]
Median estimation	3.5	8.1	7.5
Gastwirth median	3.5	8	7.5
Expected values of CS	15	25	20

The important fact from the view of Calcium score is confidence interval of median for the individual analyzed groups of patients. In this context it is important, so that these interval estimations would correspond with estimated range of values proposed by physicians from clinical practice. The individual interval estimations lay inside of expected intervals of CS. Interesting fact is comparison of individual interval lengths. Calcified blood vessels have the narrowest interval of Calcium score. The next important fact is also comparison between length of estimated intervals and average value estimation. These parameters are directly proportional. Gastwirth median exhibits minimal differences in comparison with median estimation. In the case of the CBV group, the Gastwirth median is identical to median estimation, in the other cases observed differences are not perceived as statistically significant.

5   Conclusion

We proposed method performing modelling blood vessel system, consequently allowing for calculation of calcium score. The model is generated on the base of the multilevel thresholding method separating physiological and calcification part of a respective blood vessel. Furthermore, the proposed model utilizes color mapping of originally monochromatic areas where blood vessels are clearly recognizable from calcification plaques. In the context of the testing, patients were divided into three groups (CBV, PCBV, EC) according to the clinical expert's opinion. For prediction of method utilization in the clinical practice, it is important distribution of Calcium score for the individual groups regards to expected intervals given by physicians. Across of the groups, there are not outlying values. Interval estimations shall demonstrate of reliability of CS measurement. It is needed, so that these interval estimations correspond with estimated value ranges. This assumption is fulfilled for each group of patients. The interval estimations are calculated by robust interval estimation of average value and by estimation of Gastwirth median.

Acknowledgment. This article has been supported by financial support of TA ČR, PRE SEED Fund of VSB-Technical univerzity of Ostrava/TG01010137. The work and the contributions were supported by the project SV4506631/2101 'Biomedicínské inženýrské systémy XII'.

References

1. Wilson, P.W.F., D'agostino, R.B., Levy, D., Belanger, A.M., Silbershatz, H., Kannel, W.B.: Prediction of coronary heart disease using risk factor categories. Circulation **97**(18), 1837–1847 (1998). doi:10.1161/01.CIR.97.18.1837. ISSN 0009-7322
2. Agatston, A.S., Janowitz, W.R., Hildner, F.J., Zusmer, N.R., Viamonte, M., Detrano, R.: Quantification of coronary artery calcium using ultrafast computed tomography. J. Am. Coll. Cardiol. **15**(4), 827 (1990). doi:10.1016/0735-1097(90)90282-T. ISSN 07351097
3. Detrano, R., Guerci, A.D., Carr, J.J., et al.: Coronary calcium as a predictor of coronary events in four racial or ethnic groups. N. Engl. J. Med. **358**(13), 1336–1345 (2008). doi:10.1056/NEJMoa072100. ISSN 0028-4793
4. Carr, J.J., Nelson, J.C., Wong, N.D., et al.: Calcified coronary artery plaque measurement with cardiac CT in population-based studies: standardized protocol of multi-ethnic study of atherosclerosis (MESA) and coronary artery risk development in young adults (CARDIA) study1. Radiology **234**(1), 35–43 (2005). doi:10.1148/radiol.2341040439. ISSN 0033-8419, [cit. 2016-05-23]
5. Cook, N.R., Paynter, N.P., Eaton, C.B., et al.: Comparison of the Framingham and Reynolds Risk scores for global cardiovascular risk prediction in the multiethnic Women's Health Initiative. Circulation **125**(14), 1748–1756 (2012). doi:10.1161/CIRCULATIONAHA.111.075929. ISSN 0009-7322
6. Van Gils, M.J., Bodde, M.C., Cremers, L.G.M., Dippel, D.W.J., Van Der Lugt, D.W.J.: Determinants of calcification growth in atherosclerotic carotid arteries; a serial multi-detector CT angiography study. Atherosclerosis **227**(1), 95–99 (2013). doi:10.1016/j.atherosclerosis.2012.12.017
7. Huang, Ch.L., Wu, I.H., Wu., Y.W.: Association of lower extremity arterial calcification with amputation and mortality in patients with symptomatic peripheral artery disease (2014). doi:10.1371/journal.pone.0090201
8. Kubicek, J., Valosek, J., Penhaker, M., Bryjova, I., Grepl, J.: Extraction of blood vessels using multilevel thresholding with color coding. In: Sulaiman, H.A., Othman, M.A., Othman, M.F.I., Rahim, Y.A., Pee, N.C. (eds.). LNEE, vol. 362, pp. 397–406. Springer, Heidelberg (2016). doi:10.1007/978-3-319-24584-3_33

9. Kubicek, J., Valosek, J., Penhaker, M., Bryjova, I.: Extraction of chondromalacia knee cartilage using multi slice thresholding method. In: Vinh, P.C., Alagar, V. (eds.) ICCASA 2015. LNICSSITE, vol. 165, pp. 395–403. Springer, Heidelberg (2016). doi:10.1007/978-3-319-29236-6_37

10. Meershoek, A., van Dijk, R.A., Verhage, S., Hamming, J.F., van den Bogaerdt, A.J., Bogers, A.J.J.C., Schaapherder, A.F., Lindeman, J.H.: Histological evaluation disqualifies IMT and calcification scores as surrogates for grading coronary and aortic atherosclerosis. Int. J. Cardiol. **224**, 328–334 (2016)

11. Lee, W.-C., Fang, H.-Y., Wu, C.-J.: Coronary artery perforation and acute scaffold thrombosis after bioresorbable vascular scaffold implantation for a calcified lesion. Int. J. Cardiol. **222**, 620–621 (2016)

12. Burgers, L.T., Redekop, W.K., Al, M.J., Lhachimi, S.K., Armstrong, N., Walker, S., Rothery, C., Westwood, M., Severens, J.L.: Cost-effectiveness analysis of new generation coronary CT scanners for difficult-to-image patients. Eur. J. Health Econ., 1–12 (2016)

13. Penhaker, M., Stula, T., Cerny, M.: Automatic ranking of eye movement in electrooculographic records, pp. 456–460 (2010)

14. Kubicek, J., Penhaker, M., Bryjova, I., Augustynek, M.: Classification method for macular lesions using fuzzy thresholding method. In: Kyriacou, E., Christofides, S., Pattichis, Constantinos, S. (eds.) MEDICON 2016. IP, vol. 57, pp. 239–244. Springer, Heidelberg (2016). doi:10.1007/978-3-319-32703-7_48

15. Peterek, T., Krohova, J., Smondrk, M., Penhaker, M.: Principal component analysis and fuzzy clustering of SA HRV during the Orthostatic challenge, pp. 596–599 (2012)

Rough Hypercuboid and Modified Kulczynski Coefficient for Disease Gene Identification

Ekta Shah and Pradipta Maji[(✉)]

Biomedical Imaging and Bioinformatics Lab, Machine Intelligence Unit,
Indian Statistical Institute, Kolkata, India
{ekta_r,pmaji}@isical.ac.in

Abstract. The most important objective of human genetics research is the discovery of genes associated to a disease. In this respect, a new algorithm for gene selection is presented, which integrates wisely the information from expression profiles of genes and protein-protein interaction networks. The rough hypercuboid approach is used for identifying differentially expressed genes from the microarray, while a new measure of similarity is proposed to exploit the interaction network of proteins and therefore, determine the pairwise functional similarity of proteins. The proposed algorithm aims to maximize the relevance and functional similarity, and utilizes it as an objective function for the identification of a subset of genes that it predicts as disease genes. The performance of the proposed algorithm is compared with other related methods using some cancer associated data sets.

1 Introduction

Abnormalities in the functioning of genes are causals of several diseases, which may be inherited over generations. The inherited genetic condition is an area of active research, which reports that the dysfunctioning of genes is due to mutations in the DNA, and such genes are known as disease genes [22]. Identification of disease genes may play a pivotal role in improving the existing diagnostic tools and thereby, assisting in better prognosis and boosting the survial chances of a patient. It has been observed that there exists a difference in expression pattern of the genes, across infected and uninfected patients, which is known to represent the propensity level of a disease [22]. Therefore, microarray technology has been used to capture this changing expression patterns over different substrates. Various approaches for feature selection have been developed for the task of discovering disease genes from the microarray [2,21].

Some recent studies have reported that genes related to similar disorders have an affinity for sharing similar functional features, thereby, validating the affinity of their protein products for interacting with each other [13]. Thus, protein-protein interaction (PPI) networks have been intensely used for the purpose of disease gene identification [12,14,17,24]. The need for data integration arises due to the limitations associated with the chances of finding novel disease genes, when using PPI network data or microarray data individually. Data integration

© Springer International Publishing AG 2017
N.T. Nguyen et al. (Eds.): ACIIDS 2017, Part II, LNAI 10192, pp. 465–474, 2017.
DOI: 10.1007/978-3-319-54430-4_45

methods have gained a huge popularity in identifying pleiotropic genes involved in the physiological cellular processes of many diseases [1,3,15,25].

The dense association between the disease related proteins and differentially expressed genes is the basic assumption of the integrated approaches. Recently, Wu et al. developed a method that ranks the genes related to cancer, using their expression profiles and interaction networks [6]. Li et al. [3] and Maji et al. [25] proposed two algorithms, using the mRMR [2] and MRMS [21] criteria, respectively, to extract a set of candidate genes from the expression data and later used them as inputs to the PPI network to extract the connector genes linking two candidate genes. Nonetheless, most of the methods reported above have considered expression profiles and PPIN singly for the selection of potential disease genes. Maji et al. have recently proposed two gene selection algorithms that utilize both the expression and interaction data together for disease gene identification [16,23]. However, the field of data integration with simultaneous use of different forms of data remains relatively unexplored.

The paper, therefore, addresses the task of identifying disease genes by developing a novel gene selection algorithm. The proposed algorithm makes use of the rough hypercuboid based approach for curation of a subset of differentially expressed genes from the microarray. A new measure of similarity has also been proposed, for quantifying the pairwise functional similarity between genes. The algorithm tries to efficiently integrate two sources of information, while augmenting the relevance of the genes to the class labels and the functional similarity between them. Disease Ontology, Gene Ontology and KEGG pathway based enrichment is used to compare the performance of the proposed algorithm with some existing integrated approaches of gene selection. The intensive analysis demonstrates higher efficiency of the proposed algorithm in extracting a subset of genes, having a high relevance and high functional similarity, from the microarray data and PPIN. It is clearly reflected from the results that the proposed method may become a useful tool for identifying disease genes.

2   Proposed Algorithm

The technological advancement in the field of biology has enabled the production of huge volume of data, which is being utilized to study the phenotype of various diseases and the different processes associated to the disease. The advent of microarray technology helped in harnessing the phenotype information in the form of gene expression. A difference in the expression pattern has been observed between diseased and non-diseased samples. Moreover, it is known that genes associated with similar disorders have an affinity for sharing functional features and their protein products are known to interact with other disease gene proteins [8,13]. Such an activity can be attributed to the fact that the query protein and its interacting neighbors form a complex, which collectively participate in several pathways and perform various functions. Thus, it is essential to capture the information that can be extracted from the two data sources.

In this aspect, a new algorithm for the selection of disease genes is introduced in this paper. It blends vigilantly the benefits of gene expression data and PPI network, and utilizes it for the purpose of identifying disease associated genes. If $\mathbb{C} = \{\mathcal{A}_1, \cdots, \mathcal{A}_i, \cdots, \mathcal{A}_m\}$ be a set of m genes present in a given microarray data, then the proposed algorithm intends to escalate the relevance and functional similarity of a subset $\mathbb{S}$ of selected genes. Let us define $\mathbb{D}$ as the class label, $\gamma(\mathcal{A}_i, \mathbb{D})$ as the relevance of the gene $\mathcal{A}_i$ with respect to the class labels $\mathbb{D}$, while $\mathcal{S}(\mathcal{A}_i, \mathcal{A}_j)$ as the pairwise functional similarity between two genes $\mathcal{A}_i$ and $\mathcal{A}_j$. The total relevance of the set of selected genes is

$$J_{\text{relevance}} = \sum_{\mathcal{A}_i \in \mathbb{S}} \gamma(\mathcal{A}_i, \mathbb{D}), \tag{1}$$

while the total functional similarity among the selected genes is

$$J_{\text{similarity}} = \sum_{\mathcal{A}_i \neq \mathcal{A}_j \in \mathbb{S}} \mathcal{S}(\mathcal{A}_i, \mathcal{A}_j). \tag{2}$$

Thus, the task of selecting the subset $\mathbb{S}$ of relevant and functionally similar genes from the set $\mathbb{C}$ of m genes can be stated as being equivalent to maximize $J_{\text{relevance}}$ and $J_{\text{similarity}}$, that is, to maximize the objective function

$$J = \alpha J_{\text{relevance}} + (1 - \alpha) J_{\text{similarity}}, \tag{3}$$

where α is a weight parameter. The algorithm follows a greedy approach for the curation of the desired subset of genes, $\mathbb{S}$ from the complete set of genes, $\mathbb{C}$. In the initial stage of the algorithm, we have $\mathbb{C} = \{\mathcal{A}_1, \cdots, \mathcal{A}_i, \cdots, \mathcal{A}_j, \cdots, \mathcal{A}_m\}$ and $\mathbb{S} = \emptyset$. The relevance $\gamma(\mathcal{A}_i, \mathbb{D})$ is computed for each gene in $\mathbb{C}$, then the most relevant gene is selected. In effect, $\mathcal{A}_i \in \mathbb{S}$ and $\mathbb{C} = \mathbb{C} \setminus \mathcal{A}_i$. The process iterates till the selection of desired number of genes is accomplished. The functional similarity between the genes of $\mathbb{C}$ and that of, the already selected set $\mathbb{S}$ of genes is computed. If a gene of $\mathbb{C}$ has an insignificant functional similarity, denoted by a zero value, with respect to any one of the genes of $\mathbb{S}$, then that gene is rendered unsuable and is discarded from the set $\mathbb{C}$. From the remaining genes of $\mathbb{C}$, the gene $\mathcal{A}_j$ is selected, which maximizes the condition given in (4). As a result of which, $\mathcal{A}_j \in \mathbb{S}$ and $\mathbb{C} = \mathbb{C} \setminus \mathcal{A}_j$.

$$\alpha \gamma(\mathcal{A}_j, \mathbb{D}) + \frac{(1 - \alpha)}{|\mathbb{S}|} \sum_{\mathcal{A}_i \in \mathbb{S}} \mathcal{S}(\mathcal{A}_i, \mathcal{A}_j). \tag{4}$$

Figure 1 represents the flow chart for the proposed algorithm.

2.1   Computation of Relevance

The concept of hypercuboid equivalence partition matrix of rough hypercuboid approach [20] is used to compute the relevance of a feature. The relevance $\gamma(\mathcal{A}_k, \mathbb{D})$ of a feature or condition attribute $\mathcal{A}_k$ with respect to the class labels or decision attribute $\mathbb{D}$ is computed as follows [20]:

$$\gamma(\mathcal{A}_k, \mathbb{D}) = 1 - \frac{1}{n} \sum_{j=1}^{n} v_j(\mathcal{A}_k) \tag{5}$$

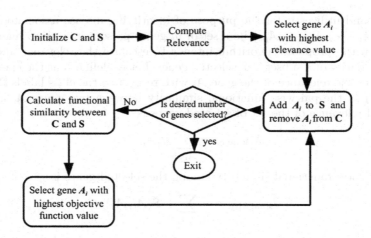

Fig. 1. Flow chart for the proposed algorithm

where n is the number of samples and

$$\mathbb{V}(\mathcal{A}_k) = [v_1(\mathcal{A}_k), \cdots, v_j(\mathcal{A}_k), \cdots, v_n(\mathcal{A}_k)] \qquad (6)$$

is termed as the confusion vector for the attribute $\mathcal{A}_k$, where

$$v_j(\mathcal{A}_k) = \min\{1, \sum_{i=1}^{c} h_{ij}(\mathcal{A}_k) - 1\}, \qquad (7)$$

c is the number of classes, and the matrix $\mathbb{H}(\mathcal{A}_k) = [h_{ij}(\mathcal{A}_k)]_{c \times n}$ is termed as hypercuboid equivalence partition matrix of the feature $\mathcal{A}_k$ [20], where

$$h_{ij}(\mathcal{A}_k) = \begin{cases} 1 \text{ if } \mathcal{L}_i \leqslant O_j(\mathcal{A}_k) \leqslant \mathcal{U}_i \\ 0 \text{ otherwise} \end{cases} \qquad (8)$$

represents the membership of object O_j in the i-th class β_i. Here $[\mathcal{L}_i, \mathcal{U}_i]$ represents the range of ith class β_i based on the decision attribute set $\mathbb{D}$. The interval $[\mathcal{L}_i, \mathcal{U}_i]$ is the range of values spanned by condition attribute $\mathcal{A}_k$ with respect to class β_i. It comprises of objects with same class label β_i. That is, the value of each object O_j with class label β_i falls within interval $[\mathcal{L}_i, \mathcal{U}_i]$. A $c \times n$ hypercuboid equivalence partition matrix $\mathbb{H}(\mathcal{A}_k)$ represents the c-hypercuboid equivalence partitions of the universe generated by an equivalence relation. Each row of the matrix $\mathbb{H}(\mathcal{A}_k)$ is a hypercuboid equivalence class.

So, the relevance $\gamma(\mathcal{A}_k, \mathbb{D}) \in [0, 1]$. If $\gamma(\mathcal{A}_k, \mathbb{D}) = 1$, $\mathbb{D}$ depends totally on $\mathcal{A}_k$, if $0 < \gamma(\mathcal{A}_k, \mathbb{D}) < 1$, $\mathbb{D}$ depends partially on $\mathcal{A}_k$, and if $\gamma(\mathcal{A}_k, \mathbb{D}) = 0$, then $\mathbb{D}$ does not depend on $\mathcal{A}_k$.

2.2   Computation of Similarity

Graphical data structures have generally been used to characterize PPI networks, where nodes depict proteins, edges depict the interactions between them and weights along the edges are depictions of the experimental and predicted evidence supporting the presence of the association. The network under consideration is undirected. Let $\mathcal{N}_i$ represent the set of nodes directly linked to the candidate gene $\mathcal{A}_i$, where every such edge is marked with a weight value $\omega_{ij} \in [0, 1]$, representing the confidence in the interaction. Thus, PPI network of proteins can be utilized to derive the set of direct neighbors or descendants, $\mathcal{N}_i$ for any gene, $\mathcal{A}_i$ and the associated weight value ω_{ij}. Any two genes $\mathcal{A}_i$ and $\mathcal{A}_k$ may have a set of common descendants, say $\mathcal{N}_{ik}$, where $\mathcal{N}_{ik} = \mathcal{N}_i \cap \mathcal{N}_k$. The pairwise functional similarity between genes $\mathcal{A}_i$ and $\mathcal{A}_k$, having $\mathcal{N}_i$ and $\mathcal{N}_k$ as set of descendant, is defined as follows:

$$\mathcal{S}(\mathcal{A}_i, \mathcal{A}_k) = \frac{1}{2} \times \sum_{\mathcal{A}_j \in \mathcal{N}_{ik}} \min\{\omega_{ij}, \omega_{kj}\} \times \left\{ \frac{1}{\sum\limits_{\mathcal{A}_j \in \mathcal{N}_i} \omega_{ij}} + \frac{1}{\sum\limits_{\mathcal{A}_j \in \mathcal{N}_k} \omega_{kj}} \right\}. \tag{9}$$

Careful study of the proposed measure reveals the fact that maximum functional similarity, for a pair of genes, is achieved when the set of neighbors and their designated edge weights are same. Conversely, a pair of candidate genes that do not share any neighbors is functionally dissimilar. In a similar manner, other characteristics of the measure can be elaborated as follows:

1. $\mathcal{S}(\mathcal{A}_i, \mathcal{A}_k) \in [0, 1]$.
2. $\mathcal{S}(\mathcal{A}_i, \mathcal{A}_k) = 1$ iff two sets $\mathcal{N}_i$ and $\mathcal{N}_k$ contain an indistinguishable set of descendants, that is, $\mathcal{N}_{ik} = \mathcal{N}_i = \mathcal{N}_k$, and weight value $\omega_{ij} = \omega_{kj}, \forall \mathcal{A}_j \in \mathcal{N}_{ik}$.
3. $\mathcal{S}(\mathcal{A}_i, \mathcal{A}_k) = 0$ if and only if $\mathcal{N}_{ik} = \emptyset$.
4. $\mathcal{S}(\mathcal{A}_i, \mathcal{A}_k) = \mathcal{S}(\mathcal{A}_k, \mathcal{A}_i)$ (symmetric).

The proposed similarity measure would condense to

$$\tilde{\mathcal{S}}(\mathcal{A}_i, \mathcal{A}_k) = \frac{1}{2} \times |\mathcal{N}_i \cap \mathcal{N}_k| \times \left\{ \frac{1}{|\mathcal{N}_i|} + \frac{1}{|\mathcal{N}_k|} \right\}; \tag{10}$$

if $\omega_{ij} \in \{0, 1\}$, which is Second Kulczynski coefficient between $\mathcal{A}_i$ and $\mathcal{A}_k$ [4].

2.3   Estimation of Optimal Weight Parameter

The variable α in (4) is a parameter that monitors the relative weightage of relevance and functional similarity, in the task of candidate gene selection. If $\alpha = 1$, then gene selection is performed based on the relevance parameter only, leading to selection of functionally dissimilar genes. To the contrary, if $\alpha = 0$, then gene selection is performed entirely on the basis of similarity. In order to consider both relevance and functional similarity for candidate gene selection, it is important to consider an optimal value of α that would lie between 0 to 1.

The optimal value of α for a given data set can be selected using area under the curve (AUC) of a receiver operating characteristic (ROC). The ROC graph is a technique for analyzing the performance of a classifier, given a particular data set. A two-by-two confusion matrix can be created given a classifier and the set of instances. Each element of the matrix denotes the number of true positive (tp), false positive (fp), true negative (tn) and false negative (fn). The true positive rate (tpr) and false positive rate (fpr) can be computed as follows:

$$\text{tpr} = \frac{\text{tp}}{\text{P}}; \quad \text{and} \quad \text{fpr} = \frac{\text{fp}}{\text{N}}; \tag{11}$$

where P and N denotes the total number of positive and negative instances, respectively. The ROC graphs are two-dimensional graphs where, tpr is plotted against fpr, depicting the relative tradeoffs between benefits (true positives) and costs (false positives). Every point in the ROC space is joined to form the ROC curve, the area under which can be computed using the trapezoidal rule. The value of α that maximizes the change in area under the ROC curve (ΔAUC) is considered to be the optimal value $\alpha^\star$.

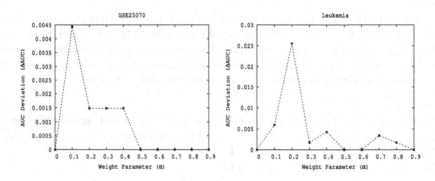

Fig. 2. Variation of ΔAUC for different values of weight parameter α

3   Experimental Results and Discussion

In this study, the expression profiles of two different forms of cancer have been considered. The first one is GSE25070, colorectal cancer (CRC) data put forward by the study of Hinoue et al. [18], consisting of 24526 genes, 26 diseased samples and 26 non-diseased samples. The other data set is an Affymetrix high density oligonucleotide array consisting of 7070 genes and 72 samples from two different classes of leukemia: 47 acute lymphoblastic leukemia and 25 acute myeloid leukemia [19]. Each of the data sets is pre-processed by standardizing each sample to zero mean and unit variance. Analysis of Fig. 2 reveals that the optimum values $\alpha^\star$ are 0.1 and 0.2 for GSE25070 and Leukemia data, respectively.

Some existing algorithms of gene selection, like CLAIM [7], mRMR+PPIN [3], MR+PPIN [22], MRMS+PPIN [25], RelSim [16] and SiFS [23] are used for performance based comparison with the proposed method. The comparative analysis of these algorithms are reported in Tables 1, 2, and 3. The different

Table 1. Biological process based results for GSE25070 and Leukemia data

Data set	Methods	Term	P-value
GSE25070	MR+PPIN	intracellular steroid hormone receptor signaling pathway	6.78E-13
	mRMR+PPIN	DNA-templated transcription, initiation	1.32E-12
	MRMS+PPIN	cellular response to lipid	2.68E-13
	CLAIM	one-carbon metabolic process	8.19E-04
	GenePEN	response to lipid	6.21E-19
	RelSim	response to lipid	2.18E-29
	SiFS	response to lipid	2.00E-23
	Proposed	response to bacterium	9.63E-33
Leukemia	MR+PPIN	platelet degranulation	6.93E-04
	mRMR+PPIN	regulation of tissue remodeling	8.08E-05
	MRMS+PPIN	chaperone-mediated autophagy	3.23E-05
	CLAIM	chaperone-mediated autophagy	7.96E-05
	GenePEN	cell-type specific apoptotic process	1.81E-19
	RelSim	leukocyte activation	1.09E-18
	SiFS	leukocyte activation	5.93E-22
	Proposed	leukocyte activation	3.89E-45

methods are evaluated using disease ontology (DO), KEGG pathway and gene ontology (GO). The ClueGO v1.8 [9] is used for enrichment analysis of the set of genes generated. The tool efficiently coalesces the KEGG/BioCarta pathways with the GO terms. The enrichment scores for the identified GO terms and pathways are computed using the hypergeometric distribution. The corrected p-value obtained for each term is used to represent the significance of the GO term and pathway to the group of genes. A higher significance being denoted by a lower p-value. Table 1 compares different algorithms using their annotated biological processes (BP) along with their respective p-values. Comparative performance analysis between the proposed and existing algorithms, using KEGG pathway analysis and DO, are reported in Tables 2 and 3, respectively. DO based evaluation is performed using the DOSE package in R [10].

A stonger association between the subset of genes selected by the proposed algorithm and the disease under study, is established by the results reported in Table 1. The proposed algorithm annotates to the biological process "response to bacterium" with the lowest p-value (9.63E-33). In [11] it has been reported that enteropathogenic *Escherichia coli* has the ability to downregulate DNA mismatch repair proteins and therefore, promote colonic tumorigenesis. This bacteria is known to possess a unique ability to produce reactive oxygen and nitrogenous species, leading to chromosomal instability and epithelial DNA damage and thereby promoting initiation of CRC. The role of microbiota in CRC growth and progression is demonstrated by the set of genes selected using the proposed algorithm. The proposed method and SiFS both annotate to the term "leukocyte activation" for Leukemia data set. However, a lower p-value is obtained for the proposed algorithm. The biological process "leukocyte activation" refers to any change in behavior or morphology of leukocytes, in response to various agents

Table 2. KEGG Pathway Based Results

	Methods	Term	P-value	AUC
GSE25070	MR+PPIN	Thyroid hormone signaling pathway	2.74E-09	0.9822
	mRMR+PPIN	Thyroid hormone signaling pathway	2.08E-13	0.9053
	MRMS+PPIN	Thyroid hormone signaling pathway	1.47E-08	0.9911
	CLAIM	Retinol metabolism	1.61E-03	0.9941
	GenePEN	Pathways in cancer	3.83E-10	0.9970
	RelSim	Cytokine-cytokine receptor interaction	5.42E-18	0.9985
	SiFS	TNF signaling pathway	1.57E-11	0.9985
	Proposed	Cytokine-cytokine receptor interaction	3.45E-22	0.9985
Leukemia	MR+PPIN	Inositol phosphate metabolism	1.08E-01	0.7923
	mRMR+PPIN	Malaria	1.04E-04	0.7940
	MRMS+PPIN	B cell receptor signaling pathway	1.16E-03	0.8102
	CLAIM	Epstein-Barr virus infection	4.36E-05	0.7047
	GenePEN	Pathways in cancer	6.83E-17	0.8638
	RelSim	Pathways in cancer	6.44E-11	0.9685
	SiFS	NF-kappa B signaling pathway	3.19E-12	0.9643
	Proposed	Hematopoietic cell lineage	8.31E-23	0.9898

Table 3. Disease ontology based results

Integrated methods	GSE25070		Leukemia	
	Term	P-value	Term	P-value
MR+PPIN	Colon carcinoma	1.01E-03	*	*
mRMR+PPIN	Coronary artery disease	1.12E-02	*	*
MRMS+PPIN	Autosomal genetic disease	*	Leukemia	1.19E-02
CLAIM	*	*	Multiple sclerosis	2.95E-02
GenePEN	Colon cancer	1.51E-10	Leukemia	2.21E-27
RelSim	Colon carcinoma	1.24E-05	Leukemia	4.87E-20
SiFS	Colorectal cancer	8.81E-08	Leukemia	3.90E-25
Proposed	Colorectal cancer	1.97E-14	Leukemia	4.21E-28

like, a specific antigen, cytokine, mitogen, cellular ligand, etc. The annotated term denotes that the set of genes selected by the proposed algorithm plays a significant role in activating different forms of leukocytes.

KEGG enrichment analysis is performed on the gene sets generated using the different algorithms, results of which are reported in Table 2. For GSE25070, the proposed method annotates to the term "Cytokine-cytokine receptor interaction" and the corresponding p-value being the lowest among the other algorithms. The importance of cytokines in colorectal cancer growth and progression has previously been elaborated. Similarly, for the Leukemia data set, proposed method annotates to "hematopoietic cell lineage", with the lowest p-value. The hematopoietic process is known to generate different blood cell lineages, which when distrupted lead to blood cell diseases [5]. Table 2 also compares the AUC values obtained using different algorithms. It can be clearly seen that the proposed method covers a larger area under the curve, in comparison with other existing algorithms of gene selection. Table 3 compares different algorithms on the basis of DO and shows that relevant terms like "colorectal cancer" and "leukemia" are annotated by the proposed method, with the lowest p-value. Tables 2 and 3 show that performance of the proposed method is comparable to that of the SiFS and RelSim for the GSE25070 data set and better than all other existing algorithms for the Leukemia data set. Thus, careful analysis of Tables 1, 2, and 3 elaborate the efficiency of the introduced algorithm.

4   Conclusion

A new algorithm for the identification of genes associated to a disease has been reported. Information extracted from the expression profiles of genes and the interaction network of proteins is judiciously merged to extract the disease associated genes. The algorithm identifies those genes as diseased, which tend to maximize the relevance and functional similarity among the set of selected genes. Information extracted from the PPI network has been utilized to define a new weighted measure of similarity. Some cancer data sets have been used for the extensive study of the proposed algorithm. They have also been utilized for performance based comaprison with other related methods. The proposed method shows improved performance over existing algorithms and identifies more disease causing genes. This establishes the underlying potential of the proposed algorithm to be a useful approach for disease gene identification.

Acknowledgements. This work is partially supported by the Department of Electronics and Information Technology, Government of India (PhD-MLA/4(90)/2015-16).

References

1. Nitsch, D., et al.: Network analysis of differential expression for the identification of disease-causing genes. PLoS One **4**(5), e5526 (2009)
2. Ding, C., Peng, H.: Minimum redundancy feature selection from microarray gene expression data. J. Bioinform. Comput. Biol. **3**(2), 185–205 (2005)
3. Li, B.-Q., et al.: Identification of colorectal cancer related genes with mRMR and shortest path in protein-protein interaction network. PLoS One **7**(4), e33393 (2012)
4. Bass, J.I.F., et al.: Using networks to measure similarity between genes: association index selection. Nat. Methods **10**(12), 1169–1176 (2013)
5. Riether, C., et al.: Regulation of hematopoietic and leukemic stem cells by the immune system. Cell Death Differ. **22**(2), 187–198 (2015)
6. Wu, C., et al.: Integrating gene expression and protein-protein interaction network to prioritize cancer-associated genes. BMC Bioinform. **13**(1), 182 (2012)
7. Santoni, D., et al.: An integrated approach (cluster analysis integration method) to combine expression data and protein-protein interaction networks in agrigenomics: application on Arabidopsis Thaliana. OMICS: J. Integr. Biol. **18**(2), 155–165 (2014)

8. Barrenas, F., et al.: Network properties of complex human disease genes identified through genome-wide association studies. PLoS One **4**(11), e8090 (2009)

9. Bindea, G., et al.: ClueGO: a cytoscape plug-in to decipher functionally grouped gene ontology and pathway annotation networks. Bioinformatics **25**(8), 1091–1093 (2009)

10. Yu, G., et al.: DOSE: an R/Bioconductor package for disease ontology semantic and enrichment analysis. Bioinformatics **31**(4), 608–609 (2015)

11. Sobhani, I., et al.: Microbial dysbiosis and colon carcinogenesis: could colon cancer be considered a bacteria-related disease? Ther. Adv. Gastroenterol. **6**(3), 215–229 (2013)

12. Chen, J., et al.: Disease candidate gene identification and prioritization using protein interaction networks. BMC Bioinform. **10**(1), 73 (2009)

13. Goh, K.-I., et al.: The human disease network. Proc. Natl. Acad. Sci. USA **104**(21), 8685–8690 (2007)

14. Oti, M., et al.: Predicting disease genes using protein-protein interactions. J. Med. Genet. **43**(8), 691–698 (2006)

15. Jia, P., et al.: dmGWAS: dense module searching for genome-wide association studies in protein-protein interaction networks. Bioinformatics **27**(1), 95–102 (2011)

16. Maji, P., et al.: RelSim: an integrated method to identify disease genes using gene expression profiles and PPIN based similarity measure. Inf. Sci. **384**, 110–125 (2017)

17. Kohler, S., et al.: Walking the interactome for prioritization of candidate disease genes. Am. J. Hum. Genet. **82**(4), 949–958 (2008)

18. Hinoue, T., et al.: Genome-scale analysis of aberrant DNA methylation in colorectal cancer. Genome Res. **22**(2), 271–282 (2012)

19. Golub, T.R., et al.: Molecular classification of cancer: class discovery and class prediction by gene expression monitoring. Science **286**(5439), 531–537 (1999)

20. Maji, P.: A rough hypercuboid approach for feature selection in approximation spaces. IEEE Trans. Knowl. Data Eng. **26**(1), 16–29 (2014)

21. Maji, P., Paul, S.: Rough set based maximum relevance-maximum significance criterion and gene selection from microarray data. Int. J. Approximate Reasoning **52**(3), 408–426 (2011)

22. Maji, P., Paul, S.: Scalable Pattern Recognition Algorithms: Applications in Computational Biology and Bioinformatics, page 304. Springer, London (2014)

23. Maji, P., Shah, E.: Significance and functional similarity for identification of disease genes. IEEE/ACM Trans. Comput. Biol. Bioinform. doi:10.1109/TCBB.2016. 2598163

24. Navlakha, S., Kingsford, C.: The power of protein interaction networks for associating genes with diseases. Bioinformatics **26**(8), 1057–1063 (2010)

25. Paul, S., Maji, P.: Gene expression and protein-protein interaction data for identification of colon cancer related genes using f-information measures. Nat. Comput. **15**(3), 449–463 (2016)

Detection of Raynaud's Phenomenon by Thermographic Testing for Finger Thermoregulation

Orcan Alpar and Ondrej Krejcar[(✉)]

Faculty of Informatics and Management, Center for Basic and Applied Research,
University of Hradec Kralove, Rokitanskeho 62,
Hradec Kralove 500 03, Czech Republic
orcanalpar@hotmail.com, ondrej@krejcar.org

Abstract. Raynaud phenomenon is a disorder of blood flow in the fingers and generally said to be a vasospastic response to cold or emotional stress due to blockage in constricted digital artery. However some cases triggered by excessive use of the hands are also reported besides the idiopathic vasospasms. During the attack, the vessels temporarily narrow down limiting the blood supply to fingers thus the small arteries may thicken overtime. The severe episodes result in numbness, color change of the affected fingers, eventually gangrene. Therefore in this paper we present a detection methodology for early diagnosis to prevent the reduced blood flow by thermal image processing based on thermoregulation in fingers. Several experiments were conducted by altering the conditions to understand the differences between the states after cooling process. The first results are greatly encouraging that reduced blood flow is mathematically identified by grid histogram matrices for red-channel conversion of higher radiation.

Keywords: Thermal image processing · Raynaud's phenomenon · Warning system · Thermoregulation · Infrared thermogram

1 Introduction

Raynaud's syndrome is a general condition affecting the blood flow to extremities of the body, mostly fingers. There mainly are two major types of this syndrome: Primary Raynaud's or Raynaud's Disease often has no clear underlying cause thus idiopathic while symptoms are usually milder. The cause could be genetic or triggered by another unknown cause so that the differential diagnosis is unknown source of the vasospasm. Secondary Raynaud's or Raynaud's phenomenon is a severe version of this syndrome which is a type of vasospasm associated with another disease or disorder. It mostly occur by an autoimmune disease such as scleroderma that hardens the skin, rheumatoid arthritis that causes joint pain, lupus that causes tiredness and skin rash or Sjogren's syndrome that makes immune system attack tear glands. Moreover it could be triggered by infections like hepatitis B or C, cancer like lymphoma, myeloma or leukemia. The other underlying conditions could be some medications, artery disorders, injuries, smoking, frostbite and exposure to chemicals. The only mechanical causes are reported

© Springer International Publishing AG 2017
N.T. Nguyen et al. (Eds.): ACIIDS 2017, Part II, LNAI 10192, pp. 475–484, 2017.
DOI: 10.1007/978-3-319-54430-4_46

as repetitive action or vibration and overuse of fingers and hands which is the main topic of this paper.

There are few researches in the literature regarding the excessive use of the hands causing Raynaud's. For instance; Atashpaz and Ghabili [1] reported that long-term playing an acoustic guitar could be a potential risk factor for Raynaud's phenomenon due to long exposure to vibration. Yassi [2] presented that strain injuries caused by repetitive and forceful motions, awkward postures, and other work-related conditions may result in Raynaud's phenomenon. In line with these papers, we focus on excessive use of computer keyboard that may cause acute, sub-acute but potentially progressive Raynaud's for prevention as an early warning system using thermoregulation methodology with infrared thermal imaging. The example of thermal vision of the phenomenon could be seen in Fig. 1 taken from the paper of Ring and Ammer [3].

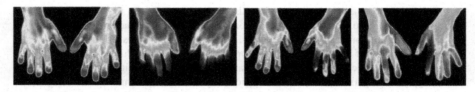

Fig. 1. Thermogram of dorsal hands: on the left the healthy hand, others suffer from several types of Raynaud's.

These images were taken to figure out the finger thermoregulation after 1 min of immersion in 20 °C water. The image on the left shows full recovery after 10 min of exposition to water while the other images show delay in recovery from different parts of the hands. Thermoregulation is an internal process to turn the affected body part to homeostasis which is a state of equilibrium, governed by hypothalamus. Lowering the internal temperature bring sweating and vasodilation on the contrary; warming up needs vasoconstriction and thermogenesis. During the period of cooling, the vessels of the hands initially narrow down decreasing the blood flow going to fingers to protect inner body temperature, which is vasoconstriction. When warming, vasodilation takes place where vessels get wider to increase the blood flow to fingers which is supposed to be near-simultaneously and symmetrically.

The thermogram methodology is one of the common differential diagnostic technique studied by researchers to identify Raynaud's disease. Dinsdale and Herrick [4] dealt with vascular diagnosis for Raynaud's and presented several scenarios for primary and secondary. They investigated rewarming process of the hands after 10 min to identify abnormality of blood flow. Since the main feature of this disease on the infrared thermal imaging is irregularity of temperature distribution, they first tested the hand of the subject suffering from Raynaud's at a room temperature of 23 °C and they observed the gradient while at 30 °C the gradient disappeared. They also researched the primary Raynaud's versus secondary caused by systemic sclerosis to distinguish the thermograms by dynamic temperature challenge.

Furthermore, Ismail et al. [5] processed the thermal images of 18 healthy subjects versus 48 Raynaud's patients, where 20 of them are suffering from primary Raynaud's and the rest have systemic sclerosis. Their data collection methodology was a bit different since they made the experiments with the subjects wearing latex gloves while in 2 min of cold stress and collected data after 20 min after the cold exposure of 10 °C. Even though it is not about thermograms, Foerster et al. [6] focused only on fingertips to identify Raynaud's by analyzing cold response using led based duosensor. They made the experiments by cooling 1 min in 16 °C water and recorded the warming pattern differences between the healthy subjects and Raynaud's patients. In the past decade, quite a few related papers were published [7–12]. Although they are not fully based on image processing of thermograms, they give valuable information about the disease itself.

Considering the differential diagnosis presented in previous researches, we firstly take the thermal images of the hand of a patient, diagnosed as having Guyon's Canal syndrome, by FLIR Infrared Thermal camera. Preserving the temperature data, the images are reconstructed and grayscaled to omit the RGB layers. The grayscale images are analyzed to identify possible temperature gradient and stages of recovery. The experiments are repeated after the overuse of the hands to recognize the differences. The purpose of this paper is presenting image processing techniques for the thermograms for early diagnosis and warning of Raynaud's. On the other hand, this research is based but not strictly dependent on Raynaud's, therefore it can shed light on the other kind of blood flow disorders on hands such as Acroparesthesia [13] or Hypothenar Hammer Syndrome [14] or recurrent nerve entrapments in Guyon's Canal or Carpal Tunnel [15].

The paper starts with presenting the infrared thermogram technology with the fundamentals of thermal image processing we used in this article. Afterwards we introduced a case study analyzed by grid histogram method to identify possible blood flow blockage or reduction with several experiments and difference matrices. The paper concludes with results and discussion sections.

2   Infrared Thermograms

Infrared thermography is an investigative methodology to reveal beyond the visible wavelengths. Thermal energy is a light from heat emitting substances and actually is a part of the electromagnetic spectrum. We cannot see this energy since the wavelengths are too long to be detected yet we perceive this light as heat. On the other hand cold substances emit heat in infrared spectrum as well, though the higher the temperature the greater the infrared radiation.

The outputs of thermography are actually RGB images called thermograms giving sensitive information about the temperature colored by a colormap. Using the infrared spectrum the colormap is chosen upon the requirements. For instance if a leakage is investigated the colormap could be gray toned with a high-pass filter that allows to higher radiation pass through the filter and turns into visible spectrum. On the other hand if the temperature distribution is a concern, then the colormap could be false-color, scaled in various ways depending on the requirements. There are some examples of various colormaps in Fig. 2.

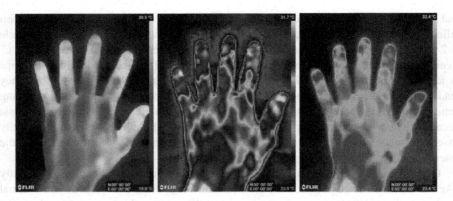

Fig. 2. Thermograms colored in different rendering. (Colour figure online)

In Fig. 2 left, the rendering is perfect for analyzing the whole temperature scale yet it is not sensitive to giving the contrast information. The image in the middle is focused on contrast however it is fully in interaction with the environment which could be used for recognizing the leakages in industry. We rendered our images using the third one on the right since it has both contrast information with the gradient and no interaction with the temperature of the environment. These colormaps are provided by built-in FLIR pseudo colorization palettes, Iron, Rainbow HC and Rainbow form left to right.

3   Thermal Image Processing

Any image taken from an infrared camera would be in RGB format that has three layers and rendered as a true-color image, unless specified otherwise. Any pixel p on an RGB image R in state of a, could be pointed as;

$$p^a_{i,j,k} \in R^a_{i,j,k}(i = [1 : w], j = [1 : h], k = [1 : 3]) \tag{1}$$

where w is the width, h is the height of the image; i is the row number, j is the column number, k is the layer of color channel and a represents rested/overuse states of before/immediately after/20 min after cooling process. For this paper, all images we analyzed are in 300×400 resolution captured by 0.96 emissivity, from 30 cm distance in an approximate 25 °C room temperature. The RGB images are converted by splitting into layers of R, G and B, namely;

$$p^a_{i,j,1} \in R^a_{i,j,k}(i = [1 : w], j = [1 : h], 1) \tag{2}$$

$$p^a_{i,j,2} \in R^a_{i,j,k}(i = [1 : w], j = [1 : h], 2) \tag{3}$$

$$p^a_{i,j,3} \in R^a_{i,j,k}(i = [1 : w], j = [1 : h], 3) \tag{4}$$

The results of conversion are presented in Fig. 3.

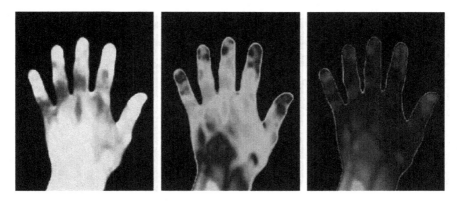

Fig. 3. Conversion into Red-Green-Blue layers from left to right.

Afterwards, the histogram matrices are calculated as 10×10 submatrices having an area of $S = 1200\,px^2$ separately for each layer. The main purpose of this reconstruction is to build histograms that have two dimensional information using the averages of submatrices. For each layer k in the state of a, the histogram matrix is computed as follows.

$$
CR_k^a = \begin{bmatrix} & \cdots & & \cdots & \left(\sum_{i=271}^{300} \sum_{j=361}^{400} p_{i,j,k}^a \right)/1200 \\ & \cdots & & \cdots & \cdots \\ \left(\sum_{i=1}^{30} \sum_{j=1}^{40} p_{i,j,k}^a \right)/1200 & \cdots & & & \end{bmatrix}_{10x10}
$$

(5)

The results of the reconstruction are in Fig. 4 below,

With the help of double sided histogram matrices, it is possible to calculate the difference matrices from state to state for identification of blood flow disorders in hand vessels. For instance for the separate RGB layers of the example image in Fig. 5, the differences matrices are calculated by subtracting the matrices from the matrices in Fig. 4.

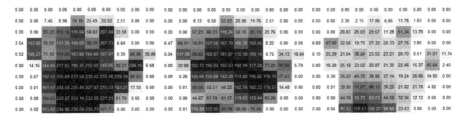

Fig. 4. Reconstruction to 10×10 histogram matrices of Red-Green-Blue layers from left to right

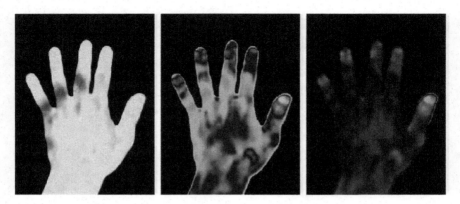

Fig. 5. Another example for RGB layer conversion

Since the layers from red to blue represent the higher radiation to lower, any increase in the histogram matrix of red channel will stand for warming up. Given the protocol of the experiment it therefore will be enough to analyze the difference matrix only in red channel to identify the temperature change in the gradient. The difference matrices are shown in Fig. 6.

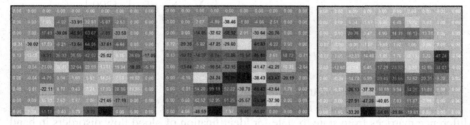

Fig. 6. Difference matrices for Red-Green and Blue channels from left to right. (Color figure online)

It is revealed from the difference matrix of red channel that there is a slight warming in the middle of the hand while some parts are cooling down. Although only the red channel conversion and reconstruction is used in this research, the other difference matrices could give valuable information as a supportive material when necessary.

4   Case Study

A 36 year old male diagnosed as Guyon's Canal syndrome 4 years ago, was suffering from numbness in his left hand. The first experiment was done in the early morning with the rested left hand. Cooling is done by 1 min of 10 °C cold water exposure and the images are recorded immediately after and 20 min of staying at room temperature.

The experiments was repeated at night after 8 h of intensive computer usage. The images presented in Fig. 7 show the thermograms after the cooling process.

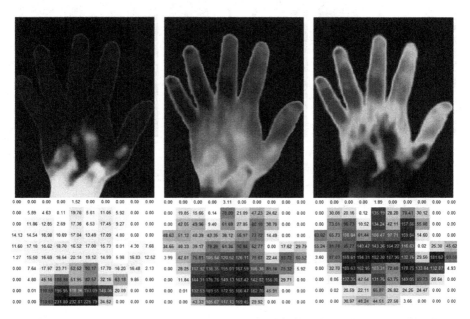

Fig. 7. Images taken and histogram matrices calculated right after the cooling process

The second image set was taken after 20 min of warming in room temperature where the hands of the subject were fully rested, as presented in Fig. 8. The final experiment was done under the same circumstances after 8 h of intensive keyboard use and the result of this experiment is presented in Fig. 9.

The images or the histogram matrices don't reveal the disorder therefore the difference matrices are calculated to find out whether there is a problem in thermoregulation. Firstly, the difference between the cooled and the rested state is calculated only for red channel conversion. Secondly, the difference between the overuse and cooled state are computed. Finally, outcomes are compared and presented in Fig. 10.

As expected, we discovered a problem in thermoregulation around the fourth and fifth finger on left hand of the patient which is also the symptom of Guyon's Canal syndrome. When compared with the outcomes mentioned in Fig. 6, the differences are more significant considering the image on the right of Fig. 10. Given these experiments, we may state that overuse of the computer causes the vasospasm and disorder in thermoregulation.

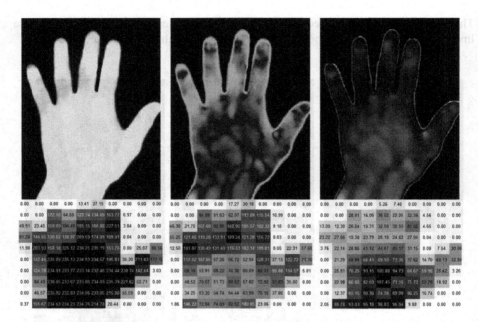

Fig. 8. Images taken and histogram matrices calculated after 20 min – rested state

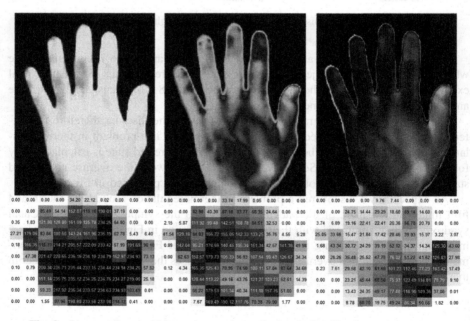

Fig. 9. Images taken and histogram matrices calculated after 20 min – overuse state

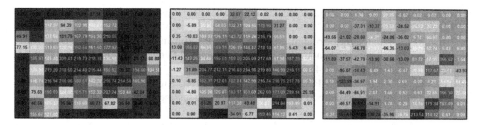

Fig. 10. Difference histograms: rested vs cooled, overuse vs cooled, overuse vs rested from left to right

5   Discussion and Conclusion

The main idea beneath this paper is analyzing the thermal images to recognize the disorder in blood flow in hand vessels. Although the Raynaud's disease could be seen by naked eye due to color change in fingers, thermograms will be helpful for early diagnosis and prevention. The results seem encouraging since the first objective of the research is accomplished. The overuse of the hands especially if there is a background of any kind of diagnosed nerve entrapment triggers disorder in blood flow. Moreover turning the RGB images into the histogram matrices enables mathematical calculation and ease the procedure of automatic diagnosis. During the protocol, there was no important assumption, drawback, difficulty or weakness that is worth mentioning. We determined the matrix size of 10×10 for better presentation however high-res histograms would give precise information about the place of blockage or reduction of blood flow in the hand vessels. In addition, for more accurate results, some stabilizers could be utilized to fix the finger and the hand positions.

As the future research topics, the dorsal hand recognition algorithms that we have presented before [16] could be implemented before the analysis, if necessary. We haven't included the training session for the images yet the Levenberg-Marquardt or Guass-Newton based artificial neural network classifiers [17–19] could be useful since the inputs are matrices. In addition, the several RGB histogram techniques could be implementable as we have introduced in one of our researches before [20].

Acknowledgment. This work and the contribution were supported by project "SP/2017 - Smart Solutions for Ubiquitous Computing Environments" from University of Hradec Kralove. We are also grateful for the support of Ph.D. students of our team (Richard Cimler and Jan Trejbal) in consultations regarding application aspects.

References

1. Atashpaz, S., Ghabili, K.: Color triad in guitarist's fingers: a probable case of Raynaud's phenomenon due to string vibration. Med. Probl. Perform. Art **23**(3), 143–144 (2008)
2. Yassi, A.: Repetitive strain injuries. Lancet **349**(9056), 943–947 (1997)
3. Ring, E.F.J., Ammer, K.: Infrared thermal imaging in medicine. Physiol. Meas. **33**(3), 33–46 (2012)

4. Dinsdale, G., Herrick, A.L.: Vascular diagnostics for Raynaud's phenomenon. J. Vasc. Diagn. **2**, 127–139 (2014)
5. Ismail, E., Orlando, G., Corradini, M.L., Amerio, P., Romani, G.L., Merla, A.: Differential diagnosis of Raynaud's phenomenon based on modeling of finger thermoregulation. Physiol. Meas. **35**(4), 703–716 (2014)
6. Foerster, J., Wittstock, S., Fleischanderl, S., Storch, A., Riemekasten, G., Hochmuth, O., Worm, M.: Infrared-monitored cold response in the assessment of Raynaud's phenomenon. Clin. Exp. Dermatol. **31**(1), 6–12 (2006)
7. Lim, M.J., Kwon, S.R., Jung, K.H., Joo, K., Park, S.G., Park, W.: Digital thermography of the fingers and toes in Raynaud's phenomenon. J. Korean Med. Sci. **29**(4), 502–506 (2014)
8. Kolesov, S.N.: Thermal-vision diagnosis of Raynaud's syndrome and its stages. J. Opt. Technol. **82**(7), 478–486 (2015)
9. Ammer, K.: The sensitivity of infrared imaging for diagnosing Raynaud's phenomenon is dependent on the method of temperature extraction from thermal images. In: Infrared Imaging, pp. 161–166. IOP Publishing (2015)
10. Herrick, A.L.: The pathogenesis, diagnosis and treatment of Raynaud phenomenon. Nat. Rev. Rheumatol. **8**(8), 469–479 (2012)
11. Pauling, J.D., Shipley, J.A., Harris, N.D., McHugh, N.J.: Use of infrared thermography as an endpoint in therapeutic trials of Raynaud's phenomenon and systemic sclerosis. Clin. Exp. Rheumatol.-Incl Suppl. **30**(2), 103–115 (2012)
12. Roustit, M., Blaise, S., Millet, C., Cracowski, J.L.: Impaired transient vasodilation and increased vasoconstriction to digital local cooling in primary Raynaud's phenomenon. Am. J. Physiol. Heart Circulatory Physiol. **301**(2), 324–330 (2011)
13. Boskovski, M.T., Thomson, J.G.: Acroparesthesia and carpal tunnel syndrome: a historical perspective. J. Hand Surg. **39**(9), 1813–1821 (2014)
14. Carter, P.M., Hollinshead, P.A., Desmond, J.S.: Hypothenar hammer syndrome: case report and review. J. Emerg. Med. **45**(1), 22–25 (2013)
15. Mosier, B.A., Hughes, T.B.: Recurrent carpal tunnel syndrome. Hand Clin. **29**(3), 427–434 (2013)
16. Alpar, O., Krejcar, O.: Dorsal hand recognition through adaptive YCbCr imaging technique. In: Nguyen, N.-T., Manolopoulos, Y., Iliadis, L., Trawiński, B. (eds.) ICCCI 2016. LNCS (LNAI), vol. 9876, pp. 262–270. Springer, Heidelberg (2016). doi:10.1007/978-3-319-45246-3_25
17. Alpar, O.: Intelligent biometric pattern password authentication systems for touchscreens. Expert Syst. Appl. **42**(17), 6286–6294 (2015)
18. Alpar, O., Krejcar, O.: Biometric swiping on touchscreens. In: Saeed, K., Homenda, W. (eds.) CISIM 2015. LNCS, vol. 9339, pp. 193–203. Springer, Heidelberg (2015). doi:10.1007/978-3-319-24369-6_16
19. Alpar, O., Krejcar, O.: Pattern password authentication based on touching location. In: Jackowski, K., Burduk, R., Walkowiak, K., Woźniak, M., Yin, H. (eds.) IDEAL 2015. LNCS, vol. 9375, pp. 395–403. Springer, Heidelberg (2015). doi:10.1007/978-3-319-24834-9_46
20. Alpar, O.: Keystroke recognition in user authentication using ANN based RGB histogram technique. Eng. Appl. Artif. Intell. **32**, 213–217 (2014)

Intelligent Technologies in the Smart Cities in the 21st Century

Enhancing Energy Efficiency of Adaptive Lighting Control

Adam Sędziwy$^{(\boxtimes)}$, Leszek Kotulski, and Artur Basiura

AGH University of Science and Technology,
al. Mickiewicza 30, 30-059 Kraków, Poland
{sedziwy,kotulski,basiura}@agh.edu.pl

Abstract. Lighting standards (generally based on the CEN/TR 13201 standard) allow assigning different lighting classes to a single roadway dependently on an actual traffic flow. Applying a less restrictive lighting class decreases energy consumption. Thanks to this setting lighting classes accordingly to actual road conditions yields up to 34% reduction of the power usage. This result may be further improved if one performs control on a well design installation. The optimization of a lighting design for one lighting class is the subject to the intensive research. In this article we show in turn that the optimization within one class is not sufficient for preparing an optimal solution. The metrics allowing the comparison of two solutions has been also introduced. The energy consumption in the presented case is reduced by 2.73% only but in the context of the global energy consumption of lighting systems (estimated for 3 GWh in the EU) it seems to be interesting.

Keywords: Lighting design · Energy efficiency · LED · Solid state lighting

1 Introduction

Light emitting diodes became the key technology in the lighting of public places today. Properties of LED-based sources make engineers and scientists develop lighting solutions aiming at the power usage reduction. Besides of the energy usage optimization they allow reducing the light pollution as well, being an issue especially in large cities. This approach is timely especially in the area of *smart cities*. There exists a broad spectrum of methods of lighting power saving ranging from such sophisticated ones as changing the reflective properties of a road surface [13] or transmitting the sunlight through special optical fiber cables [10] (both applicable for road tunnels) to preparing well suited lighting designs preventing over-illumination [16] and advanced adaptive control methods based on sensor data [21]. It should be stressed that last methods (i.e., design and control) should not be considered separately as a well tailored lighting design is necessary for an energy efficient lighting control [21].

In the article we focus on the lighting design area. The primary goal is determining configuration of lighting installations (in terms of such parameters as a

© Springer International Publishing AG 2017
N.T. Nguyen et al. (Eds.): ACIIDS 2017, Part II, LNAI 10192, pp. 487–496, 2017.
DOI: 10.1007/978-3-319-54430-4_47

fixture model, pole height, arm length, luminaire spacing, luminous flux dimming and so on) resulting in a minimized power usage and satisfied lighting standard requirements for a relevant road. Such a problem was subject to research works [16,17,20] but one important element was omitted so far. It is the influence of the traffic intensity changes on a lighting class selection being made prior to a design process.

A term "lighting class" refers to the categorization of a given public space such as a roadway, road junction, walkway and so forth, made on the basis of such factors as the dominant user type, traffic volume, ambient luminosity, navigational task, speed limit and others. Properties of an area which influence a lighting class selection are defined in the standard [4]. The standard [1] defines lighting performances required for particular classes.

A lighting class for a highway is more restrictive than the one for residential area roads. In turn, a lighting level (and thus a power usage) for more restrictive classes will be higher than for lower ones. The problem becomes more complex when traffic intensity changes during a day and several lighting classes are applicable. Then one has to consider which class is the most restrictive, which one has the highest duration time, what effective powers will be used. On this technical background one can also consider business-specific decision such as investment costs of creating/retrofitting an installation (a cost of poles, fixtures and so on), payback period or return on investment (ROI) which can influence the final decision concerning an installation setup.

The practical tests made in the city of Cracow showed that implementing a dynamic assignment of lighting classes to a roadway allows reducing the power consumption up to 34% [20]. In the article we drill down this result to report the annual savings for particular lighting classes for the normalized, 1000 m long road section. This normalization allows assessing potential savings for other roadways.

The structure of the article is following. The next section contains the state of the art overview. In Sect. 3 a basic background necessary for future considerations is presented. Section 4 we define the core problem, i.e., optimization of adaptive lighting installations, present the comprehensive case study and discuss the strictly economic aspect of the problem. In Sect. 5 the final conclusions are formulated.

2   State of the Art

Two common approaches to creating energy-efficient road lighting solutions are (i) the control-based and (ii) the design-based methods.

2.1   Control-Based Energy Usage Reduction

The control logic of main industry-standard solutions for public lighting management (LightGrid [5], CityTouch [11] or Owlet [14]) relies on performance schedules for particular lighting points (fixtures) or groups of fixtures (see Fig. 1). Such a schedule defines the level of a luminous flux emitted by the fixture, for

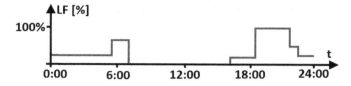

Fig. 1. The sample performance schedule for a fixture. The horizontal axis corresponds to the hour of the day. The vertical one denotes luminous flux ratio (LF), i.e., a percentage of a nominal (non-dimmed) luminous flux of a fixture

particular periods of a day. Although such a predefined schedule can be remotely uploaded to a control cabinet or even a particular fixture a control remains static in the sense that it does not follow the changes of an environment state. For that reason it is a rough approach to enhancing road lighting power efficiency. Additional issue of this method is a workload required for manual updating schedules on the area of an entire city even if it applies to groups of streetlights.

In the another, more advanced approach, fixture dimming is triggered on the basis of an analysis of an actual environment state which is continuously reported by sensors (induction loops, photo cells, cameras and so on). Thus a lighting system adapts to actual road/weather conditions. To enhance a lighting system control the concept of lighting profiles was introduced [19]. It assumes preparation of a set of dimming levels for particular luminaries (or groups of luminaries) appropriate for given environment conditions. Thus they do not have to be calculated on the fly, as that would not be feasible (due to the complexity) in an adaptive lighting control.

2.2   Design-Based Energy Usage Reduction

The outdoor lighting design process is carried out according to the rules included in relevant standards [1, 2, 4]. The designer's objective is twofold: ensuring compliance with mandatory standards, good practices and recommendations [7, 9] and optimizing a design with respect to measurable criteria related to energy-efficiency (besides that the other factors such as investment costs or payback period which can be also considered).

Due to the great number of available variants corresponding to such factors as positioning of objects or installation parameters (see the example shown in Table 1), a designer has to use, besides his expert knowledge, the trial-and-error design method. Even assuming that the experience in a domain allows reducing a number of possible scenarios, the search space size still remains too high to determine an optimal solution in a reasonable time. It has to be also mentioned that lighting situations are simplified in the practice to accelerate the design process. This simplification is achieved by averaging luminaire spacings, road widths, pole setbacks, and so forth. Anyway, to make a design process feasible one has to use a fully automated method [15, 17] supported by the well defined formal representation of objects [18].

Table 1. Variable parameters of the design process. The total number of possible configurations is 1.6×10^7

Parameter	Range	Step	Number of variants
Pole height	8 m–10 m	1 m	3
Arm length	0.5 m–2.5 m	0.5 m	5
Dimming level	0–99%	1%	100
Luminaire spacing	20 m–30 m	0.5 m	21
Fixture model*	n/a	n/a	518

* Each fixture model has a unique photometric solid hence a number of variants refers to the number of photometric solids being tested.

Well tailored lighting installation created according to the customised design [16] is shown to have the power usage reduced by up to 15% compared to installations designed using industry-standard tools like DIALux software [3].

3   Preliminaries

The core requirement which has to be addressed in a lighting design is the compliance of a project with mandatory standards for illumination, defined for various areas such as roadways, driveways, intersections, sidewalks etc. For each of them, an appropriate lighting class is assigned [4] depending on the traffic characteristics: dominant user types, traffic flow intensity (see Table 2), average speed of vehicles etc. For each lighting class, the performance requirements are assigned. For example, Table 3 shows so called ME-lighting classes which establish performance requirements for traffic routes of medium to high driving speeds as defined in the European standard EN 13201-2 [1]. Similarly, performance requirements are defined for road intersections, conflict areas and others.

Table 2. Lighting classes *vs.* flows

Flow (vehicles/day)	Lighting class
≥25,000	ME2
15,000–25,000	ME3b
15,000–7,000	ME4a
≤7000	ME5

It can be easily found that a lighting class for a given traffic route may change during a day due to the variable traffic load (Table 2). In the practice, such traffic load changes are tracked by probing an actual vehicle flow in 15 min long time windows (narrow time windows may cause the random flickering). If the flow remains stable during this period of time then an appropriate lighting class relevant to the flow can be set and, if a control system is present, luminous flux of streetlights can be adjusted accordingly (Fig. 2).

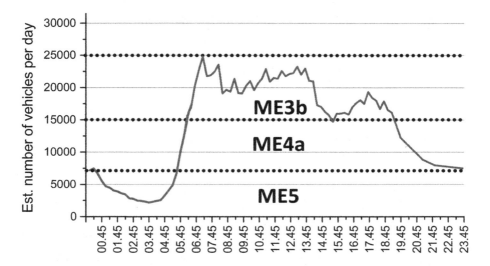

Fig. 2. The 24 h traffic load measured in 15 min time windows.

After aggregating acquired data one can also create a lighting class duration time diagram showing what is the contribution of particular classes to the total performance time. For example, the vehicle flow on one of the major streets in the city of Cracow, being measured with the help of induction loops was shown to have the 24-hour structure giving three lighting classes (percentages refer to the duration time): ME2 – 51%, ME3a – 25% and ME4a – 24%[1]. Note that environment state change events can be reported to a lighting control system either from a telemetric layer (induction loops) or from other sources such as BTS-based (Base Transceiver Station) monitoring facilities [6].

4   Problem Formulation

In this context we can formulate and illustrate the core idea of this article. Since there are multiple lighting class ascribed to a road (as shown in the above example) the question arises: which lighting class has to be selected when designing a new optimal installation (or retrofitting an existing one)? In the *conservative* approach one chooses the most restrictive one (e.g., ME3a) as the resultant installation's setup allows meeting requirements for lower classes by dimming

---

[1] The Smart Power Grids project, financed by the Polish National Fund for Environmental Protection and Water Management [12].

Table 3. ME-lighting classes (for traffic routes) according to the EN 13201-2 standard. L_{avg}—min. average luminance maintained, U_o—min. overall uniformity, U_l—min. longitudinal uniformity, TI—max. threshold increment, SR—min surround ratio.

Class	L_{avg} [cd/m^2] (min.*)	U_o (min.)	U_l (min.)	TI [%] (max.**)	SR (min.)
ME1	2.0	0.4	0.7	10	0.5
ME2	1.5	0.4	0.7	10	0.5
ME3a	1.0	0.4	0.7	15	0.5
ME3b	1.0	0.4	0.6	15	0.5
ME3c	1.0	0.4	0.5	15	0.5
ME4a	0.75	0.4	0.6	15	0.5
ME4b	0.75	0.4	0.5	15	0.5
ME5	0.5	0.35	0.4	15	0.5
ME6	0.3	0.35	0.4	15	-

*min.: minimum allowed value; **max.: maximum allowed value.

the fixtures. In this article we show that the lower total power consumption can be obtained in an optimization performed against a less restrictive class (e.g., ME4a or ME5) on condition that the resultant solution will also comply with requirements of more restrictive classes considered in a given context. The following case study presents such a situation.

Case Study

Let us suppose that the two-lane carriageway of the width $w = 4$ m and length $D = 1000$ m is given. The parameters being varied during a solution search process are shown in Table 1. Solution finding is performed using PhoCa (**Pho**tometric **Ca**lculations) software presented in [8], capable of performing massive photometric computations and enabling output data visualization. We test fixtures taken from the pool containing 518 items[2]. The lighting class duration structure is ME3a – 51%, ME4a – 25%, ME5 – 24%. Moreover we assume that the optimization is made twice: against ME3a and ME5 class respectively. The goal of an optimization for a given class is obtaining the lowest effective power for a single fixture. Besides varying a fixture model during an optimization process we can also test various spacings what means that the resultant installations may require different numbers of luminaries in each case.

The following questions arise for the above example: (i) does this method bring expected power savings? (ii) are the obtained solutions equally good? To answer them we have to define a suitable metrics:

$F(x)$ measuring an annual energy consumption of an installation (if duration structure is made for a per-year basis):

---

[2] We selected fixtures manufactured by Philips, Schreder and GE Lighting.

$$F(x) = D\frac{P_x}{S_x} \sum_{i \in \text{LClasses}} LF_x^{(i)} \cdot \Delta t_i, \tag{1}$$

where D is an installation length, S_x is luminaire spacing (the subscript x will refer to the solution x from now on) thus D/S_x is a number of luminaries in an installation, P_x is a nominal power of a single fixture, LF_x denotes luminous flux ratio; the sum iterates over all classes present in a lighting class duration schema; Δt_i is a duration time for an i-th class.

Having the function $F(x)$ defined we can answer the question put above. For two solution, say x and y, we regard x as being better than y if $F(x) < F(y)$.

We apply this algorithm to all fixtures from the test set and calculate the annual energy consumption for the base class ME3a and ME5 respectively. Next, we get the best result for each class with respect to the annual power usage. The results are shown in Table 4.

Table 4. Optimization results for two classes

Lighting class	ME3a	ME5
Mounting height	8 m	
Arm length	2 m	
Spacing	20.0 m	20.5 m
Fixture	AMPERA MAXI 5139 80 Cree XP-G2 700mA	AMPERA MAXI 5139 128 Cree XP-G2 700mA

Let us consider Tables 5 and 6 containing installation setups obtained for the ME3a-based and ME5-based solutions respectively (selected from over 3800 standard-compliant solutions).

Table 5. The ME3a-oriented solution (fixture: AMPERA MAXI 5139 80 Cree XP-G2 700 mA, mounting height: 8 m, spacing: 20.0 m, arm length: 2 m)

Lighting class	ME3a	ME4a	ME5	
LFR	0.31	0.23	0.16	
Nominal power [W]	180.00			
Effective power [W]	55.80	41.40	28.80	**Total**
Annual energy consumption [MWh/km]	6.09	2.22	1.48	**9.79**

In the rightmost cells of last rows there are placed total annual energy consumptions for both solutions which were normalized to a 1000 m-long installation section:

Table 6. The ME5-oriented solution (fixture: `AMPERA MAXI 5139 128 Cree XP-G2 700 mA`, mounting height: 8 m, spacing: 20.5 m, arm length: 2 m)

Lighting class	ME3a	ME4a	ME5	
LFR	0.20	0.15	0.10	
Nominal power [W]	279.00			
Effective power [W]	55.80	41.85	27.90	**Total**
Annual energy consumption [MWh/km]	5.95	2.19	1.40	**9.53**

$$F(\text{ME3a}) = 9.79 \, \frac{MWh}{km \cdot year}, \tag{2}$$

$$F(\text{ME5}) = 9.53 \, \frac{MWh}{km \cdot year}. \tag{3}$$

It can be seen that the ME5-oriented solution is better than the other one in spite of the fact that an effective power of a single fixture for the ME3a duration is the same for both solutions.

The above case shows that even highly power-efficient adaptive lighting control solutions can be tuned up by preparing a proper installation design. The term *proper* means that a project is optimized for a lighting class determined by the statistical analysis of traffic data and calculating appropriate quality metrics.

5   Conclusions

In this paper we present different approaches to creating optimal roadway lighting installations in terms of the power usage. Such installations have to comply with EU lighting standards and, what is also very important, generate savings of operation costs and reduce the lighting pollution. Wasting energy in the public lighting area is significant as each 1% of the power usage reduction in EU Union is equivalent to generation of 30 TWh of energy.

A proper approach to creating a lighting project is very important. There are many factors influencing it. One of them is selecting lighting classes applicable to a specific road situation. A designer usually prepares a project for the most restrictive lighting class only. We show that this approach is not effective. In some cases we are able to find a better set of installation parameters relying on weaker lighting performance requirements (i.e., a less restrictive class). We suggest the method of optimizing the energy efficiency of roadway lighting installations being adaptively controlled (i.e., a luminous flux is adjusted to actual traffic conditions). This optimization is based on two elements. The first is an analysis of statistical data describing traffic intensity of relevant roads and the second element is adjusting luminous fluxes of fixtures so that energy usage is minimized and ensures the compliance with mandatory lighting performance requirements. The obtained reduction of the energy consumption (by 2.73%) shows the potential in the field of further optimizations.

Note that in real life cases such an analysis cannot be limited to energy costs only. It has to cover economic issues as well so it should enable striking golden mean among such factors as investments and exploitation costs, return of investment, net present value, payback period and so on.

References

1. European Committee for Standardization CEN 13201-2: Road lighting - part 2: Performance requirements. Ref. No. EN 13201-2:2003 E (2003)
2. European Committee for Standardization CEN 13201-3: Road lighting - part 3: Calculation of performance. 2003. Ref. No. EN 13201-3:2003 E
3. DIAL: DIALux (2015). http://goo.gl/21XOBx. Accessed 5 Dec 2016
4. European Committee For Standardization: Road lighting - Part 1: Guidelines on selection of lighting classes, CEN/TR 13201-1:2015 (2015). Ref. No. EN 13201-1:2015 E
5. GE Lighting: LightGrid (2015). http://goo.gl/cZSBY3. Accessed 5 Dec 2016
6. Klimek, R., Rogus, G.: Modeling context-aware and agent-ready systems for the outdoor smart lighting. In: Rutkowski, L., Korytkowski, M., Scherer, R., Tadeusiewicz, R., Zadeh, L.A., Zurada, J.M. (eds.) ICAISC 2014. LNCS (LNAI), vol. 8468, pp. 257–268. Springer, Heidelberg (2014). doi:10.1007/978-3-319-07176-3_23
7. Kostic, M., Djokic, L.: Recommendations for energy efficient and visually acceptable street lighting. Energy **34**(10), 1565–1572 (2009). 11th Conference on Process Integration, Modelling and Optimisation for Energy Saving and Pollution Reduction
8. Kotulski, L., De Landtsheer, J., Penninck, S., Sędziwy, A., Wojnicki, I.: Supporting energy efficiency optimization in lighting design process. In: Proceedings of 12th European Lighting Conference Lux Europa, Krakow, 17–19 September 2013 (2013). Accessed 5 Dec 2016
9. NYSERDA: How-to guide to effective energy-efficient street lighting for planners and engineers (2015). http://goo.gl/YRzF3Q. Accessed 5 Dec 2016
10. Peña-García, A., Gil-Martín, L.M., Hernández-Montes, E.: Use of sunlight in road tunnels: an approach to the improvement of light-pipes' efficacy through heliostats. Tunn. Undergr. Space Technol. **60**, 135–140 (2016)
11. Philips: CityTouch (2015). http://goo.gl/GhCJsX. Accessed 5 Dec 2016
12. Polish National Fund for Environmental Protection and Water Management. ISE Project (Polish) (2014). Accessed 5 Dec 2015
13. Salata, F., Golasi, I., Bovenzi, S., Vollaro, E.D.L., Pagliaro, F., Cellucci, L., Coppi, M., Gugliermetti, F., Vollaro, A.D.L.: Energy optimization of road tunnel lighting systems. Sustainability **7**(7), 9664 (2015)
14. Schréder: Owlet (2015). http://goo.gl/9g53OP. Accessed 5 Dec 2016
15. Sędziwy, A.: Effective graph representation supporting multi-agent distributed computing. Int. J. Innov. Comput. Inf. Control **10**(1), 101–113 (2014)
16. Sędziwy, A.: A new approach to street lighting design. LEUKOS **12**(3), 151–162 (2016)
17. Sędziwy, A.: Sustainable street lighting design supported by hypergraph-based computational model. Sustainability **8**(1), 13 (2016)
18. Sędziwy, A., Kozień-Woźniak, M.: Computational support for optimizing street lighting design. In: Zamojski, W., et al. (eds.) Complex Systems and Dependability. AISC, vol. 170, pp. 241–255. Springer, Heidelberg (2012)

19. Szmuc, T., Kotulski, L., Wojszczyk, B., Sedziwy, A.: Green AGH Campus. In: Donnellan, B., Lopes, J.A.P., Martins, J., Filipe, J. (eds.) SMARTGREENS, pp. 159–162. SciTePress (2012)
20. Wojnicki, I., Ernst, S., Kotulski, L.: Economic impact of intelligent dynamic control in urban outdoor lighting. Energies 9(5), 314 (2016)
21. Wojnicki, I., Ernst, S., Kotulski, L., Sędziwy, A.: Advanced street lighting control. Expert Syst. Appl. Part 1 41(4), 999–1005 (2014)

Comparative Analysis of Selected Algorithms in the Process of Optimization of Traffic Lights

K. Małecki[1(✉)], P. Pietruszka[1], and S. Iwan[2]

[1] Faculty of Computer Science, West Pomeranian University of Technology,
Szczecin, Poland
{kmalecki,ppietruszka}@wi.zut.edu.pl
[2] Maritime University of Szczecin, Szczecin, Poland
s.iwan@am.szczecin.pl

Abstract. Optimal settings of traffic lights and traffic light cycles are important tasks of modeling a modern ordered traffic in smart cities. This article analyzes the comparative effectiveness of selected optimization algorithms for the identified area. In particular, it involves the comparison of the concepts of genetic algorithm using particle swarm optimization, the differential evolution and the Monte Carlo method with two new approaches: evolution strategy involving the adaptation of the covariance matrix and topology archipelago consisting of four islands - different algorithms to optimize the length of the phase in fixed time traffic signals. Developed simulation solutions allowed to achieve a quantitative improvement in the selection of the optimal durations of the phases of traffic lights for the tested roads with junctions.

Keywords: Intelligent transportation systems · Simulation platform · Traffic efficiency · Traffic simulators

1 Introduction

One important aspect of smart cities is the intelligent traffic management. In this area, the fundamental role is played by intelligent traffic lights. Modern trend in the management and optimization of intelligent traffic tends toward merging vehicles in streams and their control [1]. One practical way is to synchronize the traffic lights located in the analyzed section of the road. Assuming that the phase times of traffic signals are fixed with not big distances between them, it is possible to aggregate traffic through the synchronization of traffic lights by changing the phase shift of the green signal for the next junctions [2]. Assuming that there is a preferred direction with priority in terms of travel time in the particular streets, it is possible to create the effect of the so-called "green wave" [3, 4]. Phase shift of successive cycles must therefore equal to the average travel time between this and previous traffic signal [5].

Numerous studies indicate that in conditions of high traffic (or due to intersecting slip roads) the problem emerges when vehicles wait before traffic lights and form queues at the next junctions [2]. This means that the aggregate stream of vehicles will be stopped and partly merged with a stream of vehicles that are about to move. This interferes with the idea of a green wave in traffic. For this reason, the idea of setting a

© Springer International Publishing AG 2017
N.T. Nguyen et al. (Eds.): ACIIDS 2017, Part II, LNAI 10192, pp. 497–506, 2017.
DOI: 10.1007/978-3-319-54430-4_48

reverse cycle - cycle offset of the next traffic lights is set so as to remove the queues formed at further junctions, and only later allowing in a controlled way new aggregated stream of vehicles to enter the road [6, 7].

An important role in optimizing traffic lights is played by evolution algorithms [8] that use the mechanisms of evolution known in the natural sciences. A variation of evolution algorithms are genetic algorithms. The aim of the article is make the comparative analysis of the effectiveness of selected genetic algorithms. For this purpose, the authors analyzed literature, defined the area of simulation, as well as evaluated the effectiveness of these strategies and algorithms. The article is completed by conclusions from the research.

1.1   Related Work

The literature presents different methods and technologies that enable more efficient control of traffic lights. In [9] the authors propose using Simultaneous Perturbation Stochastic Algorithm (SPSA) to calculate in real time the best set of traffic lights in order to minimize travel time through the junction. Proposal for a new controller (EOM-ANN Controller) of lights based on neural networks is presented in [10]. Quite a different concept is presented in [11]. The authors use the video stream from the cameras at junctions to calculate the volume of traffic in real time. The results obtained are used by switching algorithm of lights depending on the density of vehicles on the road, which aims to reduce congestion, thereby reducing the number of accidents. This area of research also describes the concept of self-organizing traffic lights. For example, in [12] the authors present a distributed algorithm to control traffic lights. Hybrid algorithms were developed in [13] that combines Fuzzy Logic Controller (FLC) and Genetic Algorithms (GA) and examined its effectiveness. The theory of distributed scheduling strategy based on Lagrangian relaxation and subgradient method is dealt by the authors in [14]. One last suggestion is the use of warning lights to support the traffic lights in order to avoid congestion [15].

An important place in the effective control of traffic lights is taken by genetic algorithms. There have been many attempts to apply them. An example would be [16], where he described the use of genetic algorithm to determine the durations of the successive phases of traffic signals and offsets. Another derivative algorithm, which can be classified as evolution algorithm, is particle swarm optimization (PSO). This algorithm was also applied in research on the development of an optimal fixed time traffic light program [17].

However, in the conclusions of the research authors clearly indicate the narrow nature of the results obtained as well as the need for improvement and rating of optimization algorithms in case of effective control of the traffic light cycles. This is also the motivation for the research of the authors of this article.

2   Research Environment

One of the first methods used to coordinate traffic network for junctions is TRANSYT, a method developed by the Transport Research Laboratory (TRL) in the United Kingdom. This method allows setting the maximum length of the queue in front of each signaling device in order to prevent the formation of queues affecting blocking of the previous junctions. The procedure for optimization is to change the operating parameters of traffic lights: offsets (i.e., the times between the start of the whole system and the start of particular traffic lights) and the durations of the phases. Subsequent modifications of TRANSYT method allow, for example, to take into account in the process of optimization the lines of public transport and the use of additional parameters such as fuel consumption [18, 19].

2.1 The Traffic Simulator

The research involved the use of the traffic simulator SUMO (Simulation of Urban Mobility) [20–22]. An important reason was the popularity of SUMO in similar research, for example [17, 23–25]. For the purpose of this article Python language software was prepared that was integrated with SUMO simulator via TraCi (Traffic Control Interface) [26]. The interface allows retrieving data about the current status of objects and changes in their status, e.g. reading of data from motion detectors, reading of vehicle information, reading and modification of the state of the traffic lights.

SUMO can put on the road three types of motion detectors: inductive loops, area detectors, detectors with multiple inputs and outputs.

2.2 Simulation Scenarios and Effectiveness Measurement

For the purposes of research on the optimization of control algorithms of traffic lights simulation scenario consisting of a artificial map of the road network (Fig. 1), arrangement of detectors and definitions of generated vehicles were developed. Every road (between nodes) has a length of 200 m. Simulation applies to passenger vehicles.

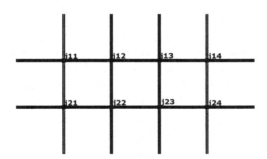

Fig. 1. Screenshot of SUMO simulator with open map of tested area marked with identifiers of junctions. Source: own research

Each of the traffic lights is based on a two-phase program: the first phase allows for the movement of vehicles in the direction north-south and south-north, while the second - the west-east and east-west.

The authors defined the movement of vehicles. Depending on the needs the following types traffic configuration were used: vehicles can go straight or turn right one time; vehicles can only go straight; vehicles can only go straight, the traffic on roads in the west-east direction is set to fixed high value for all such files, and the traffic on roads in the north-south direction is set for the specific file. According to [18] effectiveness of a specific algorithm of traffic light control is measured using the following criteria: throughput, probability of traffic relieve, etc.

In order to compare the tested algorithms, the number of vehicles that passed through the entire network of roads and vehicles stopping time (the time when the vehicle is stopped before the traffic lights or another vehicle) were used.

2.3 Methods of Testing the Efficiency of Algorithms for Fixed Time Traffic Lights

Evolution algorithms are a popular method for optimizing used to determine fixed time traffic light program. For this reason, this publication involved carrying out experiments designed to compare several selected types of evolution algorithms. These algorithms start from generating random parameters and strive with every step (generation) to find the parameters for which the rating function (also called adaptation function) returns closer value, the so-called optimum value.

For each of the algorithms the same number of units in a population was used: 25 and the same number of generations - 100. In the research configuration files were used, in which each of the vehicles can drive through the map straight or turn right one time. The traffic (the probability of generating a vehicle on each of the possible roads) is set to 0.282. The results of each unit are the arithmetic average of four ratings. The rating of a single simulation was calculated as a weighted average of the components such as the inverse of the sum of stopping times of vehicles and the number of vehicles that have reached their destination. Used components of the rating function have been normalized so that on the start with the default configuration generated by SUMO their values were close to 1. The aim of optimizing algorithms was to minimize the rating function. After the results were obtained, the algorithm with the best value of the rating function was selected and subject to a further test. This experiment consisted of simulation with the same settings of phase durations, but with files with the vehicle configuration for different traffic.

Figure 2 shows the structure of units that was used in the genetic algorithm. "F1" and "f2" are respectively the length of the first and second phase, "o" in the version

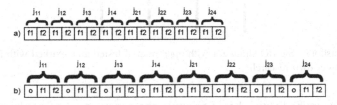

Fig. 2. Structure of a unit containing phase durations of traffic lights in the evolution algorithms, (a) version without the offset optimization, (b) version with the offset optimization. Source: own research

with the offset is the length of the offset. Vector of a unit value in a version without offset consisting of 16 elements – two phase durations for each of the eight junctions, while in the version with offset 24 elements - offset and durations of the two phases for each junction. For each element of unit vector from which it can take values the range was set to [1; 50].

Settings of genetic algorithm: crossover probability - 0.95, the probability of mutation - 0.02 elitism – yes, applied mutation - a Gaussian function with a standard deviation of 0.1, the selection algorithm - roulette.

Settings of particle swarm optimization: the weight of inertia - 0.7298, φ_1 - cognitive component - 2.05, φ_2 - social component - 2.05. The rest of this article involved the differential evolution, the Monte Carlo method and PSO algorithm [17] as methods to be compared with strategies proposed in this publication. The use of evolution strategies involving the adaptation of the covariance matrix and archipelago consisting of four islands in the optimization of the phase durations of traffic lights is a new solution.

Differential Evolution (DE). The algorithm of differential evolution [27] - similarly to the genetic algorithm - involves the steps consisting in crossover and mutation, but these operations are carried out for a randomly chosen unit elements, and do not depend (as in the case of crossover in a genetic algorithm) on mixing together the units.

This research involved the following parameters of differential evolution algorithm: the amplification factor - 0.8, the crossover probability - 0.9.

Covariance Matrix Adaptation Evolution Strategy (CMA-ES). Evolution strategy involving the adaptation of the covariance matrix is a special kind of evolution strategies, in which instead of the vector variance [28] for the individual variables covariance matrix representing the relationship between the pairs of values is used, and instead of the one-dimensional normal distribution a multi-dimensional distribution is applied. Its big advantage highlighted by the creator of this method is no need for parameter selection by the user - all parameters used in the method are automatically selected by itself [29]. In this publication it was the first time when evolution strategy involving the adaptation of the covariance matrix was used to optimize the length of the phases.

Monte Carlo Algorithm. The algorithm based on the draw of phase durations and checking the best result for the random sets of times. In the experiment, 2500 iterations involving drawing new times were conducted.

Archipelago. This is a method of connecting a group of algorithms. In such a model representation of a single algorithm is called an island. When creating the archipelago one should define the list of islands belonging to it and the topology of connections between them. For each connection you can optionally assign a weight indicating the likelihood of migration. Sample topologies are presented in [30]. The idea of the algorithm is so that each island performs optimization according to predetermined algorithm. Then, the migration process is performed, which means the transfer of solutions between interconnected islands. After migration, re-optimization is started, however, the first step of migrating random units is omitted (units used are those, which were on the island as a result of migration) [30] (Fig. 3).

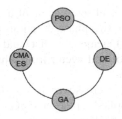

Fig. 3. The topology of the archipelago used to optimize the traffic light operation. Source: own research

Another new aspect of the experiment conducted in this publication is the creation of archipelago consisting of four islands in the following order: the particle swarm optimization (PSO), differential evolution (DE), genetic algorithm (GA), evolution strategy (CMA ES). It involved a use of a ring topology with non-directed edges (possible migration at each edge in both directions). Each algorithm works on a population of 25 units and creates 10 generations in each step of migration in the number of 20.

3   The Results of Experiment

The authors present the results of algorithms for selecting optimal durations of the phases of traffic lights on the test road network, without selecting the offset values (all traffic lights began to work at the same time from the same phase - the green light for the north-south direction and red for the west-east) - Table 1, and with offset (Table 2).

Both tables show the minimum value of the evaluation function reached, and the durations of the light phases at every junction (marked "j" with the number of junction) that were generated as optimal by the algorithm. In Table 2 first offset, and then phase durations were given for each junction.

Symbols algorithms used in the table involve: CMA-ES - evolution strategy involving the adaptation of the covariance matrix, DE - differential evolution, MC - Monte Carlo method, GA - genetic algorithm, PSO - particle swarm optimization, Archipelago - a combination of algorithms in the model based on the islands.

Table 1. Results of the fixed time traffic lights optimization with zero offset

Algorithm	Mark	Time							
		j11	j12	j13	j14	j21	j22	j23	j24
CMA-ES	0.7921	19/25	12/21	19/32	18/25	17/26	14/21	11/18	27/29
DE	0.9302	24/36	16/24	20/27	21/36	6/11	14/21	22/35	23/28
MC	1.1403	21/29	22/26	40/46	21/40	12/38	23/24	16/20	27/47
GA	0.9210	25/25	15/17	16/33	13/24	19/30	25/38	10/21	15/41
PSO	0.9468	19/38	17/23	16/25	24/35	17/28	24/37	18/29	34/35
Archipelago	0.9378	15/37	22/43	29/40	18/24	5/19	15/30	16/21	20/23

Table 2. Results of the fixed time traffic lights optimization with offset subject to optimization

Algorithm	Mark	Time							
		j11	j12	j13	j14	j21	j22	j23	j24
CMA-ES	0.8108	21	11	43	20	11	35	42	13
		19/27	18/33	14/23	18/27	14/24	12/24	16/26	23/30
DE	0.8813	7	49	17	4	18	41	40	29
		22/30	16/22	17/28	17/29	17/32	16/32	12/16	14/18
MC	1.1429	1	33	46	4	49	20	22	18
		38/34	8/15	28/43	42/49	10/26	10/26	38/48	10/16
GA	0.9195	24	22	42	12	11	32	42	15
		26/26	9/28	24/33	20/46	23/46	10/31	23/34	20/24
PSO	0.9145	1	49	26	46	1	49	23	22
		15/20	16/24	13/25	7/18	14/18	27/46	12/20	23/37
Archipelago	0.8658	39	35	10	19	40	38	23	39
		13/19	19/25	15/26	15/36	12/23	12/17	15/25	9/17

Figure 4 presents a graph of the evaluation function for the test algorithms with and without offset optimization. Black dashed line shows selected the value of the evaluation function calculated on the basis of the simulation with time settings taken from the map editor available in SUMO package.

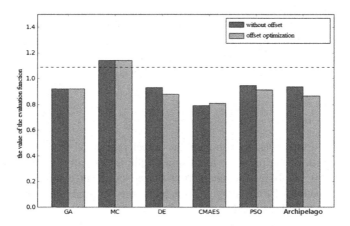

Fig. 4. A graph showing the value of the evaluation function for the best result of each algorithm Source: own research

Analysis of the results shows that all algorithms except Monte Carlo, managed to generate a solution rated better than the initial settings from the map editor. Evolution algorithms achieved better results, and among them was the best algorithm was evolution strategy with covariance matrix adaptation (CMA-ES).

In the presented results it can be seen that in most solutions the duration of the first phase is shorter than in the second phase. The reason why evolution of solutions has led to selection of such units is the fact that the map used in the simulation has two roads towards the west-east directions and four of them crossing the road in a north-south direction. The car passing the entire map straight in a single case has to drive through four junctions, and in the other only two. There is a risk that the car designed to drive through more junctions will have to stop completely more times (which would increase the main component of the evaluation - total stopping times). Therefore, probably algorithms developed in the evolution a solution to minimize this risk - longer phase with the green light for these cars.

One of the test procedures included the comparison of the effectiveness of algorithms tested only for the times and for the offsets and times. For most algorithms, the result obtained for the optimization of offset proved to be better than the result for the solution without offset. However, for CMA-ES algorithm solution without offset achieved better (lower) value of the evaluation function. The reason is the increase in the dimensionality (the number of elements of the vector solution) from 16 (two phases for each of the eight junctions) to 24 (two phases and offset for each junction) without changing the algorithm parameters - the number of generations, and the number of units.

A little disappointment is the result of the archipelago, which is combination of many algorithms. Despite the use of the four algorithms in it, including the best CMA-ES, the value of the evaluation function was lower than the value of a single CMA-ES algorithm. In this case, the poor result can be the reason for the use of the low number of generations between migration. For example, migration to the island representing PSO algorithm from the island representing any other algorithm must involve the assigning of new velocity vectors to migrating unit, since they are only part of the PSO algorithm. The migration of units between selected algorithms every 10 generations could interfere with the operation of different algorithms, which resulted in a weaker result of the entire archipelago of CMA-ES algorithm.

4 Summary

The aim of the article was to compare the effectiveness of selected algorithms to optimize the length of the phase in fixed time traffic lights with two new approaches: evolution strategy involving the adaptation of the covariance matrix and topology archipelago consisting of four islands. To achieve the set goal, software extension of SUMO simulator was developed.

To conduct optimization studies of the durations of the individual phases and offsets of fixed time traffic lights a group of selected evolution algorithms was applied: genetic algorithm, particle swarm optimization (PSO), the differential evolution and the Monte Carlo method. Evolution strategy involving the adaptation of the covariance matrix was applied, which had not been done for such tasks before. In addition to single algorithm, the research involved a model combining them in order to work together on a solution called a model based on the islands (each algorithm is represented by an island that is part of the archipelago). For a test case of certain traffic all the evolution algorithms reached better solutions than the original program of fixed time traffic lights generated by the editor supplied with SUMO simulator. The best result was obtained using evolution strategy involving the adaptation of the covariance matrix.

Further research on the algorithms to optimize traffic lights will include the creation of new algorithms, modification of existing ones, and the selection of appropriate parameters to allow the best efficiency. It is possible to carry out experiments using methods presented in this article, including the use of other test scenarios.

References

1. Cao, C., Cui, F., Guo, G.: Two-direction green wave control of traffic signal based on particle swarm optimization. Appl. Mech. Mater. **26–28**, 507–511 (2010)
2. Michalopoulos, P.G., Stephanopoulos, G.: Oversaturated signal systems with queue length constraints—II. Transp. Res. **11**(6), 423–428 (1977)
3. Tobita, K., Nagatani, T.: Green-wave control of an unbalanced two-route traffic system with signals. Physica A: Stat. Mech. Appl. **392**(21), 5422–5430 (2013)
4. Ye, B.-L., Wu, W., Zhou, X., Mao, W., Huang, Y.-S.: A green wave band based method for urban arterial signal control. In: Proceedings of the 11th IEEE International Conference on Networking, Sensing and Control, ICNSC (2014)
5. Zhang, Y., Huang, G.S.: Based on road green wave effect of collaborative strategy of signal timing fuzzy control. Appl. Mech. Mater. **321–324**, 1836–1841 (2013)
6. Nagatani, T.: Vehicular traffic through a sequence of green-wave lights. Physica A: Stat. Mech. Appl. **380**(1–2), 503–511 (2007)
7. Kong, X., Shen, G., Xia, F., Lin, C.: Urban arterial traffic two-direction green wave intelligent coordination control technique and its application. Int. J. Control Autom. Syst. **9** (1), 60–68 (2011)
8. Back, T.: Evolutionary Algorithms in Theory and Practice: Evolution Strategies: Evolutionary Programming, Genetic Algorithms. Oxford University Press, Oxford (1996)
9. Kwatirayo, S., Almhana, J., Liu, Z., Siblini, J.: Optimizing road intersection traffic flow using stochastic and heuristic algorithms. In: IEEE International Conference on Communications, ICC 2014 6883382, pp. 586–591 (2014)
10. De Oliveira, M.B.W., De Almeida Neto, A.: Optimization of traffic lights timing based on artificial neural networks. In: 17th IEEE International Conference on Intelligent Transportation Systems, ITSC 2014 6957986, pp. 1921–1922 (2014)
11. Kanungo, A., Sharma, A., Singla, C.: Smart traffic lights switching and traffic density calculation using video processing. In: Recent Advances in Engineering and Computational Sciences, RAECS 6799542 (2014)
12. Goel, S., Bush, S.F., Ravindranathan, K.: Self-organization of traffic lights for minimizing vehicle delay. In: 2014 – Proceedings International Conference on Connected Vehicles and Expo, ICCVE 7297692, pp. 931–936 (2015)
13. Odeh, S.M.: Hybrid algorithm: fuzzy logic-genetic algorithm on traffic light intelligent system. In: IEEE International Conference on Fuzzy Systems, 7338117 (2015)
14. Zhang, Y., Su, R., Gao, K.: Urban road traffic light real-time scheduling. In: Proceedings of the IEEE Conference on Decision and Control, 7402642, pp. 2810–2815 (2016)
15. Qi, L., Zhou, M., Luan, W.: An emergency traffic light strategy to prevent traffic congestion. In: ICNSC 2016 – 13th IEEE International Conference on Networking, Sensing and Control 7479013 (2016)

16. Kalganova, T., Russell, G., Cumming, A.: Multiple traffic signal control using a genetic algorithm. In: Dobnikar, A., Steele, N.C., Pearson, D.W., Albrecht, R.F. (eds.) Artificial Neural Nets and Genetic Algorithms, pp. 220–228. Springer, Vienna (1999)
17. García-Nieto, J., Alba, E., Olivera, A.C.: Swarm intelligence for traffic light scheduling application to real urban areas. Eng. Appl. Artif. Intell. **25**(2), 274–283 (2012)
18. Gaca, S., Suchorzewski, W., Tracz, M.: Traffic engineering. Theory and practice, Wydawnictwa Komunikacji i Łączności (2008) [in Polish]
19. Robertson, D., Bretherton, R.: Optimizing networks of traffic signals in real time - the SCOOT method. IEEE Trans. Veh. Technol. **40**(1), 11–15 (1991)
20. Krajzewicz, D., Hertkorn, G., Rössel, C., Wagner, P.: SUMO (Simulation of Urban MObility) - an open-source traffic simulation. In: Proceedings of MESM 2002 (2002)
21. Krajzewicz, D., Brockfeld, E., Mikat, J., Ringel, J.: Simulation of modern traffic lights control systems using the open source traffic simulation SUMO. In: Industrial Simulation (2005)
22. Behrisch, M., Bieker, L., Erdmann, J., Krajzewicz, D.: SUMO – simulation of urban mobility, an overview. In: The Third International Conference on Advances in System Simulation – SIMUL (2011)
23. Krajzewicz, D., Erdmann, J., Behrisch, M., Bieker, L.: Recent development and applications of SUMO - Simulation of Urban MObility. Int. J. Adv. Syst. Meas. **5**(3&4), 128–138 (2012)
24. Gu, W., Ito, T.: Optimization of road distribution for traffic system based on vehicle's priority. In: Booth, R., Zhang, M.-L. (eds.) PRICAI 2016. LNCS (LNAI), vol. 9810, pp. 729–737. Springer, Heidelberg (2016). doi:10.1007/978-3-319-42911-3_61
25. Cárdenas-Benítez, N., Aquino-Santos, R., Magaña-Espinoza, P., Edwards-Block, A., Cass, A.M.: Traffic congestion detection system through connected vehicles and big data. Sensors (Switzerland) **16**(5), 599 (2016). Open Access
26. SUMO Simulation of Urban MObility. http://sumo.dlr.de/wiki/Main_Page
27. Storn, R., Price, K.: Differential evolution a simple and efficient heuristic for global optimization over continuous spaces. J. Glob. Optim. **11**(4), 341–359 (1997)
28. Beyer, H.-G., Schwefel, H.-P.: Evolution strategies: a comprehensive introduction. Nat. Comput. **1**(1), 3–52 (2002)
29. Hansen, N.: The CMA evolution strategy: a tutorial. CoRR, abs/1604.00772 (2016)
30. Rucinski, M., Izzo, D., Biscani F.: On the impact of the migration topology on the island model. CoRR, abs/1004.4541 (2010)

Knowledge Representation Framework for Agent–Based Economic Systems in Smart City Context

Martina Husáková and Petr Tučník[✉]

University of Hradec Králové, Hradec Králové, Czech Republic
{martina.husakova.2, petr.tucnik}@uhk.cz

Abstract. The agent-based economic systems essentially need precise config-uration data and access to knowledge from various information sources in order to function properly. Main sources of such information are national statistical eco-nomic data, regional statistics, company performance indicators, etc. Generally, information sources of various formats and levels of detail are to be used. The main aim of the paper is to present a general framework for knowledge man-agement used in the smart city context, allowing efficient employment and dis-tribution of such data. The knowledge layer serves as an ontological intermediary between information resources and agents themselves, and is used mainly for improvement of the model efficiency especially in the following areas: (1) inter-agent communication, (2) system parameters configuration, (3) meta-data for improved search processes, and (4) unification of data exchange.

Keywords: Multi-agent system · Agent-based computational economy · Smart city · Knowledge representation · OWL ontology

1 Introduction

Depending on the intended purpose of the models, the design of the agent-based eco-nomic models can be approached by two different ways: (1) machine-oriented approach (MOA) and (2) simulation-oriented approach (SOA). While the former case is focused primarily on autonomous control, adaptation, and self-organizing aspects of multi-agent system's functioning while configurational aspect is slightly sidelined (it is usually expected that such system will adapt its parameters over time to the given situational settings), the SOA on the other hand requires very precise configuration of initial starting conditions, model parameters, and general settings, in order to provide reliable and usable results. In both cases, the use of various information sources is expected and desirable in order to create a model with suitable level of semantic precision with correct parameters settings, reflecting real-world in the required level of detail.

The emphasis on proper knowledge/information management is than even stronger for systems of larger scale, which consist of high number of (heterogeneous) agents working together in a shared environment. This is also the case when application of such agent-based economic system is considered in the context of a smart city. There is no generally accepted and complete definition of "smart city". Majority of definitions is

© Springer International Publishing AG 2017
N.T. Nguyen et al. (Eds.): ACIIDS 2017, Part II, LNAI 10192, pp. 507–516, 2017.
DOI: 10.1007/978-3-319-54430-4_49

used interchangeably. This fact is caused by the complexity of the smart city concept or various perspectives which can be used for investigation and deployment of this concept in praxis. Smart city is explained e.g. in [10] as "*a high-tech intensive and advanced city that connects people, information and city elements using new technologies in order to create a sustainable, greener city, competitive and innovative commerce, and an increased life quality.*" There is a common agreement that concept of utilization of smart technologies generally means rational, efficient use of resources, general system reliability and overall sustainability which should also consequently lead to "green", environment-friendly technologies. Communication and IT technologies are essential cornerstone of the smart approaches allowing design of the infrastructure for more effective resource management. It may be said that "smartness" is more likely an emergent phenomena, appearing when there is a sufficient number of supportive technologies in the given environment, rather than some specific ability to do something or status that may be precisely measured. For this reason, recognition of the "smart environment" is therefore a matter of subjective perception of what is freely available for use or automatically handled.

With smart technologies in general, there is always a strong emphasis on infrastructural aspect of communication. The basis which such systems are built upon are Internet of Things (IoT), Internet of Services (IoS), and embedded systems with ever-present network communication handling. The underlying idea is to connect various devices or software applications to work together more efficiently and positively contribute to services available to inhabitant of the smart environment. Because establishment of communication in such heterogeneous system is complicated and difficult due to compatibility issues, ontological representation is used to capture semantics required for efficient transition of data between various systems.

The agent-based economic system framework discussed in this paper is intended to be used only as a component of the whole "smart" municipal environment, but its principles are similar to other applications of larger scale since it faces the same challenges. The knowledge management framework, which is introduced in this paper, is used to provide efficient access to various information sources both for setting up initial model parameters and during its runtime. To do so, the ontological sub-system is designed to ease the information exchange. This knowledge layer serves as a type of filter between various data sources (most often in *.xls, *.xlsx formats, and system performance data/logs) and model logic itself.

2 Ontologies for Multi-agent Systems

At present, ontologies are frequently perceived as knowledge-based structures applied for conceptualisation of particular (often complex) application domain used in knowledge sharing, problem solving or decision making. The knowledge and data exchange are important components of every smart city framework. Smart city data are extensive, heterogeneous and potentially relevant and useful for all aspects of city management. Important role have citizens and their (pro-)active involvement is in many ways essential for efficient city functioning.

From this perspective, various contributions with certain similarities can be mentioned to highlight knowledge-related aspects of smart city frameworks' development. Aguilera [11] considers smart city citizens' to be vital for knowledge exchange and procurement and introduces concept of Internet-Enabled Services City. Citizens can actively contribute towards city knowledge through various mobile devices (smartphones) and effort is focused towards creation of platforms working with crowdsourced data. IES City concept allows construction of collaborative platforms allowing utilization of such data in various application areas. Similar stance is also adopted by Joshi [12]. In this case, the framework is called SMELTS, named by pillars considered to be essential for framework development: Social, Management, Economic, Legal, Technology and Sustainability. Joshi emphasises importance of (mobile) data interchange and processing, with Big Data and Internet of Things being key technological concepts in this effort. In this context we may also mention two case studies of smart cities of Seoul and San Francisco [13]. Authors provides description of the process of building smart cities from the practical perspective and their framework consists of 6 conceptual dimensions and 17 sub-dimensions of smart city practices.

There exist a lot of various applications of ontologies for the needs of multi-agent systems (MAS). As was shown by Malucelli and Oliveira [5], ontologies enable interoperation among intelligent agents existing in the electronic market. Ontologies provide semantic relationships existing between various vocabularies which are used by the agents in communication and negotiation for the formation and monitoring of virtual organisation. E-commerce OWL ontology is used during negotiations among agents which can uniformly interpret exchanged messages. Ontology mapping is applied for finding similarities among concepts used in such interactions.

Framework for knowledge management system (KMS) using MAS and ontologies is presented in [6]. Framework is practically applied in design of the KMS for e-commerce where XML-based ontologies are used for knowledge representation in knowledge bases and in content of messages exchanged among agents of the MAS.

Authors of the paper [7] propose MASRAM – multi-agent-based system for resource allocation and monitoring where FIPA-SL-based ontology is used for communication among the agents. The application of XML-based ontologies is presented in [8], where ontologies represent various products that are part of business. Buyer and seller should be able to communicate with the usage of the same vocabulary and the ontologies are suitable for this purpose. Multi-agent system helps the user to buy products directly or through the negotiation. Agents communicate on the basis of the messages written in the FIPA ACL. Case-based reasoning is used to reach conclusions during the negotiation process. The XML-based MAS called MAST (MAS for Traders) is proposed in [9]. The system supports B2C activities. MAS helps customer with determination of the most important deeds, selection of the most suitable product, selection of the best merchant and with operations related to the payment. Ontology model is used as a common language for all XML-based agents representing customers and merchants. The ontology is written in the XML-Schema and represents categories of products which are part of business among agents and customers.

3   Development of Agent-Based Economic Society

Development of the agent-based computation economy (ACE) used for simulation of particular economic processes requires systematic approach because of the complexity of the application domain. This is basically an initial pre-requisite for inclusion of ontologies in the system design since non-complex domains do not require sophisticated knowledge representation. The proposed conceptual framework has knowledge component which leads the analysts, designers and programmers during the process of ACE development. The framework consists of the five phases which are based on the synergy of ontologies and MAS. Content section of the framework (right side of the Fig. 1) specifies which approach or language is going to be used in our ACE project. Tool section (left side of the Fig. 1) introduces software products that are going to be used. It is obvious that different ACE-based projects can use various approaches and tools, depending on the purpose and nature of the project. The general framework consists of the following five phases:

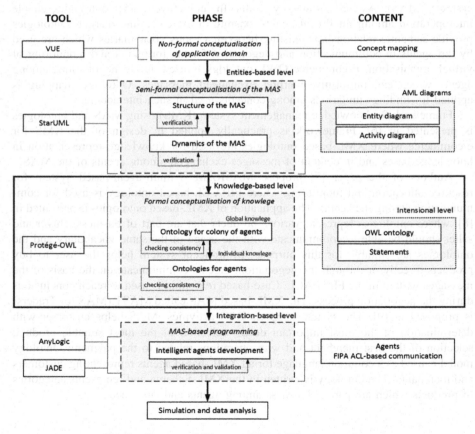

Fig. 1. Conceptual framework for ACE using ontology

1. *Non-formal conceptualisation of application domain:* The main aim of this introductory phase is to visualise crucial facts (pieces of knowledge) which will be covered by the MAS, especially which knowledge will be important for agents and their decision making. Conceptualisation helps with organisation of ideas about the structure of the developed MAS. Application of concept maps for conceptualisation in the ACE project is explained in the Sect. 4 in more details.
2. *Semi-formal conceptualisation of MAS*: Static and dynamic aspects of the MAS should be analysed and designed for facilitation of programming of the MAS. This should be ordinary practice in development of large-scale MAS containing hundreds or thousands of agents. The ACE is capable of utilization of such large-scaled systems. The AML (Agent Modelling Language) [2] is going to be used on the basis of positive experience with its usage [3, 4].
3. *Formal conceptualisation of knowledge*: If it is supposed that the agents need to dispose knowledge bases for problem solving or decision making, pieces of knowledge have to be encoded in a formal language for their machine-processing. The knowledge is not necessary for all situations solved by the agents, e.g. reactivity-based behaviour does not suppose the usage of knowledge structures. It depends on the purpose of the intended MAS whether the knowledge is necessary or not. The formal coding of knowledge is necessary for common understanding of concepts that are used by agents. Application of ontologies for formal conceptualisation in the ACE project is presented in the Sect. 5 in more details.
4. *MAS-based programming*: The fundamental building blocks are prepared during previous three phases and used for programming of the MAS. We suppose that intensive communication between agents of our "artificial economic society" will be realised. Agent-based communication is going to be based on the theory of speech acts, i.e. the FIPA-ACL language is going to be applied for formalisation of communication channels between agents.
5. *Simulation and data analysis*: Since this is generally the purpose of the whole model/system design, this part has fundamental importance. Experimental data are analysed and conclusions are formulated based upon obtained results.

These above mentioned phases are domain-independent with an exception of the third phase. Ontologies are not the only possibility when it comes to the knowledge representation. Conceptual graphs, frame-based systems, or semantic nets can also be used for knowledge representation, but the OWL ontologies are used for our purposes due to their ability to represent machine-processable semantics of concepts in sufficient levels of detail.

4   Non-formal Conceptualisation of Application Domain

The concept mapping is a conceptual approach for organisation and visualisation of information and knowledge [1]. Concept maps are informal structures that are valuable for understanding of complex phenomena, especially if the map is developed in cooperation with other person(s). Concept maps support active problem solving during brainstorming, decision making and capturing tacit knowledge of experts for expert systems development. This approach is the first phase of the conceptual framework. Concept maps visualise fundamental concepts and relations existing in the proposed "economic society" which is modelled by the MAS. The Fig. 2 depicts five concept maps. These maps represent semantic context of individual agents separately. Each one visualises particular agent of "economic society" with its relations to the other "economic agents" and related concepts. Only relevant and useful relations to other agents are visualised in the concept maps. The concept maps show a crucial similarity existing among "economic agents". All agents are part of the supply-demand relationships that is the most important relation existing among them. This fact leads to the simplification of the ontology that is based on concept mapping. Each agent has at least one relation which characterises it and is different from each other agents, it is indicated by dashed oval at the Fig. 2. This group of concept maps help with the representation of the MAS-based general structure and responsibilities of agents.

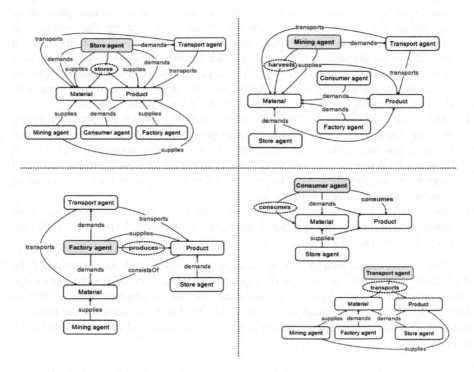

Fig. 2. Concept maps for building ontologies used by ACE

5   Global Ontology for Agent-Based Economic Society

Concept maps are not only used for visualisation of domain facts, but they are also used as a basis for formal modelling of the ontological structures used by "economic agents" for problem solving, decision making and communication. Two types of ontologies are distinguished in the ACE project. Colony is an "economic society" consisting of set of particular "economic agents": mining agent (MA), consumer agent (CA), transport agent (TA), factory agent (FA) and store agent (SA), see the [14] for more details. These agents should be aware of own responsibilities, responsibilities of other agents (because of making business contracts) and relations with other agents in the colony for clustering interested parties which solve problems in the ACE. Global ontology (the OntoForColony) covers knowledge related to the responsibilities and social roles of all "economic agents" existing in the colony. This ontology is shared by all agents in the colony. The OntoForColony ontology is based on the concept maps and is simplified because of the existence of similarities between agents identified during the phase of concept mapping. The OntoForColony ontology is the knowledge-layer covering "rules" for communication between agents. The agent will know who it can communicate with and make business contracts. Global ontology uses input data for fulfilment the knowledge base of the agent, see the Fig. 3. Input data contains particular identification data about agents with their responsibilities.

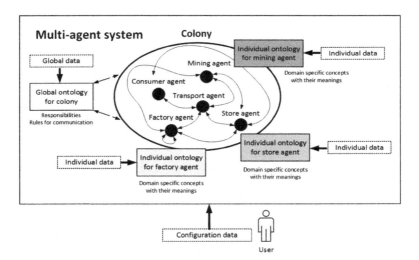

Fig. 3. Types of data and ontological structures for ACE

The OntoForColony is represented in the OWL 2 formal language in the Protégé-OWL editor (ver. 4.3.0). Semantics is mainly represented for particular agents in the ontology. The Fig. 4 depicts semantics of the StoreAgent class and SubjectOfContract class. The store agent can demand a subject of contract (industrial or consumer products). The store agent can find out possible suppliers of required products in the ontology and directly contact them. The OntoForColony ontology is normalised and contains 23 OWL classes, 10 object properties, 10 inverse object properties. Part-whole design pattern is used for modelling whole-part relations for contracts and colonies. Ontology contains two design patterns which model part-whole relations (contract and its parts, colony and its parts). The Fig. 5 visualises fragment of the OWL ontology

514   M. Husáková and P. Tučník

Store agent   **Subject of contract**

Fig. 4. Semantics of OWL classes in OntoForColony ontology

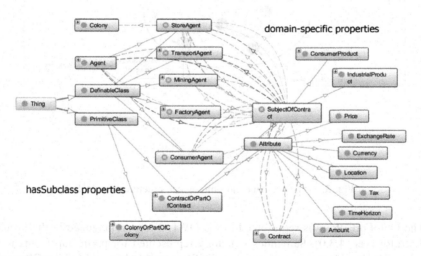

Fig. 5. Fragment of OntoForColony ontology visualised with OntoGraph plugin

containing domain-specific and hierarchical relations existing among particular OWL classes of the ontology. The OntoGraph plugin of the Protégé-OWL editor is used for this purpose.

6   Discussion and Future Work

The OWL ontology models the most fundamental knowledge which is relevant for initiation communication between "economic agents". The agents know semantics (responsibilities) of other agents existing in the colony and can decide with whom to make sense to sign a contract. Knowledge modelling has not been completed yet according to the conceptual framework, depicted at the Fig. 1. The one global ontology is not sufficient for provision of effective communication between "economic agents". The "economic agents" also use the ontology for finding out which rules of communication to follow in the colony. The global ontology does not integrate deep "knowledge complexity", because it does not contain particular domain knowledge that relates to concrete area of industrial production/services production, e.g. construction, chemical, automotive, electronics, food, fish, hospitality, software, cultural or entertainment. These areas of interest are going to be a part of individual ontologies that will be specific to "economic agents" in the colony, see the Fig. 3. Individual ontologies of agents have to be represented because of accurate modelling of "economic reality". It has to be obvious what particular kinds of services are part of the demand-supply relationship. The global ontology is not developed for this purpose.

Non-formal conceptualisation and ontology building is followed by the programming of the MAS. Other issues might arise, e.g. how to resolve conflicts in case of agents overloading (a risk of an inability to communicate), how to integrate and practically use the OWL ontologies by agents for communication in AnyLogic or how to ensure efficient ACL-based communication containing ontologies. Answers to these questions are not presented in the paper and will be addressed in future work. The paper presents initial steps that are necessary to be solved in modelling of agents-based computational economic systems requiring knowledge modelling.

7   Conclusion

The incorporation of knowledge-based component into economic system is especially useful for handling communication, configuration, compatibility, performance measurement, and information search tasks. For agent-based models of larger scale, it is often necessary to handle highly heterogeneous variety of agents and data formats. Efficient knowledge management framework improves system performance and helps handle various issues that may occur during the model runtime.

Acknowledgements. The financial support of GAČR scientific project GA15-11724S DEPIES - Decision Processes in Intelligent Environments is gratefully acknowledged.

References

1. Novak, J.: Concept maps and Vee diagrams: two metacognitive tools to facilitate meaningful learning. Instr. Sci. **19**, 29–52 (1990)

2. Červenka, R., Trenčanský, I.: The agent modeling language – AML a comprehensive approach to modeling multi-agent systems, vol. 366. Springer Science & Business Media, New York (2007)

3. Husáková, M.: The usage of the agent modeling language for modeling complexity of the immune system. In: Barbucha, D., Nguyen, N.T., Batubara, J. (eds.). SCI, vol. 598, pp. 323–332. Springer, Heidelberg (2015). doi:10.1007/978-3-319-16211-9_33

4. Husáková, M.: Combating infectious diseases with computational immunology. In: Núñez, M., Nguyen, N.T., Camacho, D., Trawiński, B. (eds.) ICCCI 2015. LNCS (LNAI), vol. 9330, pp. 398–407. Springer, Heidelberg (2015). doi:10.1007/978-3-319-24306-1_39

5. Malucelli, A., Costa Oliveira, E.: Ontology-services to facilitate agents' interoperability. In: Lee, J., Barley, M. (eds.) PRIMA 2003. LNCS (LNAI), vol. 2891, pp. 170–181. Springer, Heidelberg (2003). doi:10.1007/978-3-540-39896-7_15

6. Fu, R., et al.: An architecture of knowledge management system based on agent and ontology. J. Chin. Univ. Posts Telecommun. 15(4), 126–130 (2008)

7. Arora, M., Devi, M.S.: Ontology based agent communication in resource allocation and monitoring. IJCSI Int. J. Comput. Sci. 7(6), 28–32 (2010)

8. Jain, P., Dahiya, D.: An intelligent multi agent framework for e-commerce using case based reasoning and argumentation for negotiation. In: Dua, S., Gangopadhyay, A., Thulasiraman, P., Straccia, U., Shepherd, M., Stein, B. (eds.) ICISTM 2012. CCIS, vol. 285, pp. 164–175. Springer, Heidelberg (2012). doi:10.1007/978-3-642-29166-1_15

9. Rosaci, D., Sarne, G.M.L.: Multi-agent technology and ontologies to support personalization in B2C E-commerce. J. Electron. Commer. Res. Appl. 13(1), 13–23 (2014)

10. Bakici, T., Almirall, E., Wareham, J.: A smart city initiative: the case of Barcelona. J. Knowl. Econ. 2(1), 1–14 (2012)

11. Aguilera, U., Peña, O., Belmonte, O., López-de-Ipiña, D.: Citizen-centric data services for smarter cities. Future Generation Computer Systems. http://dx.doi.org/10.1016/j.future.2016.10.031, ISSN 0167-739X. Accessed 31 Oct 2016

12. Joshi, S., Saxena, S., Godbole, T.: Developing smart cities: an integrated framework. Procedia Comput. Sci. 93, 902–909 (2016)

13. Lee, J.H., Hancock, M.G., Hu, M.-C.: Towards an effective framework for building smart cities: lessons from Seoul and San Francisco. Technol. Forecast. Soc. Change 89, 80–99 (2014)

14. Bures, V., Tucnik, P.: Complex agent-based models: application of a constructivism in the economic research. E M Econ. Manag. 17, 152–168 (2014)

The Principles of Model Building Concepts Which Are Applied to the Design Patterns for Smart Cities

Katarzyna Ossowska[1(✉)], Liliana Szewc[1], and Cezary Orłowski[2]

[1] Faculty of Management and Economics,
Department of Applied Business Informatics, Gdansk University of Technology,
ul. Narutowicza 11/12, 80-233 Gdansk, Poland
`kossowska@zie.pg.gda.pl`
[2] WSB University in Gdansk, aleja Grunwaldzka 238A, 80-266 Gdansk, Poland

Abstract. The involvement of citizens into decision-making processes is one of the main features of smart cities. Such commitment is reflected in the form of requirements towards the city, and the benefits which are expected from the city. Requirements and benefits are thus the primary language of communication between decision-makers and urban residents. To develop such a language, it becomes necessary to develop design patterns for Smart Cities, that could integrate the requirements and benefits into ontological concepts referring to the rules describing design patterns.

The article proposes the construction of a conversion model of requirements and benefits, which are saved with the use of the natural language into ontological concepts of the principles referring to the patterns of Smarty Cities. The study verifies the model developed in the environment of an experiment. It applies ontologies for both languages: of benefits and of requirements. Then, it rates the mapping of both ontologies in relation to the sample requirements and benefits presented for Smart Cities. After that, the similarity of both ontologies is assessed and the concepts for the standard pattern rules are defined. This approach provides conditions for the development of Smart Cities patterns and for their use in decision-making processes which are so important for the development of Smart Cities.

Keywords: Smart Cities · Ontologies · Semantics · Requirements language · Design patterns · Controlled natural language · Benefits language

1 Introduction

Modern cities are committed to become smart cities. The [5] defines three levels of Smart Cities. Cities at the first level are characterised by the use of information technology to support decision-making. Cities Smart 2.0 are cities in which the decision-making process involves their residents. In turn, the concept of Smart 3.0 assumes inviting NGOs to participate in the decision-making process. At all the levels of the Smart Cities, it becomes necessary to exchange data and information between citizens and decision-makers of the city [5]. Policy makers try to communicate with the

© Springer International Publishing AG 2017
N.T. Nguyen et al. (Eds.): ACIIDS 2017, Part II, LNAI 10192, pp. 517–530, 2017.
DOI: 10.1007/978-3-319-54430-4_50

inhabitants using the requirement language. Residents see communication with decision-makers from the perspective of their expected benefits [12]. Therefore, the issue referring to the use of requirement and benefit language is so important to increase the levels of communication and thus to improve the Smart Cities.

It is also important from the perspective of the construction and use of design patterns of Smart Cities [1]. Design patterns are widely used in architecture (where they originate from), computer science, pedagogy, and they provide development opportunities for the design of large entities such as Smart Cities [10]. We define patterns as a method of documenting a solution to a repetitive (which is very important) design problem for a specific group of processes considered from the point of view of the Smart Cities development [4].

These processes are described with the use of the language of benefits and requirements [8]. Requirements and the requirement language are immediately associated with computer science, software engineering processes and requirements engineering which is a part of it [7]. These requirements specify the manner of software operation, considering its functionality (functional requirements are determined for that purpose) as well as its technical aspects (where non-functional requirements are determined) [3]. The IEEE 830 standard from 1998 defines the criteria for proper specification of requirements. They should be accurate, clear, complete, consistent, organized by importance / stability, verifiable and customizable [6].

In turn, the language of the benefits is associated mainly with the sale of products and services [2]. The benefit language analysed in the article will be associated with the products and services that residents receive during the implementation of the Smart Cities system. In other words, at the present stage of the development of Smart Cities, the benefit language is the primary language of communication between the inhabitants and the city.

So, there appears a question whether the use of ontology for the formalization of the benefit and requirement language will facilitate the communication process in the cities and in the same way will be translated into the development of Smart Cities. Therefore, the authors not only apply the ontology language to formalize both languages, but first of all, they present a method of comparing these two ontologies to indicate the necessity to use ontology and, first of all, the possibilities to map the ontologies with the use of the methods applied during the research work.

In this article the authors present a method they have developed to create and to compare the language of requirements and the benefits with the use of ontologies. It is an extension of syntactic analysis, focused on the pragmatic analysis of language. It may contribute to the development of the language of the Smart City patterns. It is carried out at four levels:

- morphological - including the structure of requirements and benefits which use regularities of both languages at the level of words.
- syntactic - including construction of the language of requirements and benefits
- semantic - presenting the meaning and context relations of both languages
- pragmatic - presenting how to use the wording: the benefits and requirements

These levels come as the components of the layers included into the model of the process referring to the acquisition of concepts for the development of the rules of design patterns. The authors use a design pattern *Man In The Loop-Resource conflict solver,* which is applied in a situation of limited resources (typical of the development of each city) [13].

2   The Model of the Process Referring to the Acquisition of Concepts for the Development of the Design Pattern Rules

The starting point for the preparation of the concept referring to the development of the model is the analysis of knowledge on the implementation of projects connected with the information systems for cities [11]. It has been found that during the process of obtaining requirements, requirements and benefits are frequently intertwined. They create a language which is difficult to accept by the designers, for whom it comes as the basis for the design of new solutions. Therefore, the authors propose a preliminary separation of requirements from benefits and the use of ontologies to generate concepts applied for the development of design pattern rules. These concepts are defined as classes and instances:

The structure of the c-class is defined as:

$$c = (Id^c, A^c, V^c) \tag{1}$$

where:

Id^c - unique class identifier (also referred to as a label)

A^c is a finite set of attributes assigned to a class, and V is a set of attribute fields from the set of N^c, which is defined as

$$V^c = \cup_{a \in A^c} V_a.$$

Each ontology, built for the classes defined in accordance with the formula 1, meets the following criteria: $\forall_{c \in C} A^c \subseteq A$

1. $\forall_{c \in C} V^c \subseteq V$

And it is referred to as ontology based on the world (A, V).

In turn, the instance of the C class is referred to the elements of the I set which take the form of triples

$$i = (id, A_i, v_i) \tag{2}$$

where id is a unique identifier of a given instance, A_i is a set of instance attributes, while v_i is a function of the following signature:

$v_i : A_i \rightarrow \cup_{a \in A_i} V_a$, which assigns the values from the sets of their fields to the instance attributes.

Assuming the existence of a certain class $c = (Id^c, A^c, V^c)$ we assume that

$$i = (id, v_i, A_i) \tag{3}$$

is an instance, and this class if the conditions are met:

1. $A^c \subseteq A_i$
2. $\forall_{a \in A_i} \bigcap A^c v_i(a) \in V^c$

The classes and their instances constitute the basis for defining the rules of design patterns

$$\text{If [body] then [head]} \tag{4}$$

Below there is an example of a rule generated for the class and the instance:

Smart - City - Pattern - Language **is a** pattern - language **and** operates - in Context - Of - Smart - City - Pattern - Language **and** has - name **equal−to'** *Smart City Pattern Language'*

$$\tag{5}$$

Context − Of − Smart − City − Pattern − Language **is a** context **and** has − description **equal− to'** *A single pattern is insufficient to deal with all Smart City problems at hand'*

$$\tag{6}$$

While defining rules for pattern (and introducing its concepts to the rules) we need to acquire knowledge for the requirements of these concepts. The paper proposes a model of acquiring these concepts which includes the processes of natural language processing, ontology designing and generating rules. Figure 1 shows a model of the process of acquiring concepts from the language of requirements and benefits which is used by residents and decision-makers.

At the beginning (Fig. 1), the assessment of independently acquired requirements and benefits in a layer of natural language is carried out. It becomes necessary to provide an analysis of both types of data at four levels (morphological, syntactic, semantic and pragmatic). This analysis comes as the basis for defining classes and their instances of ontology components, which are also independent from the requirements and benefits. At this level, the need of mapping both ontologies appears to determine

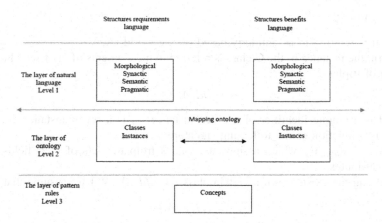

Fig. 1. Model referring to the process of concept acquisition for generating rules of design pattern

their consistency. Also at this level the concepts are defined (after discarding redundant ones) for those which will form the basis for defining the rules of design pattern. In turn, in the layer of the pattern rules some concepts are selected which are important during the development of decision rules for the specific patterns of Smart Cities.

3   The Use of a Model of the Process of Acquiring Concepts Required for Generating Rules of Design Pattern

In order to obtain the requirements and benefits expected by the urban population, the authors have conducted a questionnaire survey. The respondents are city dwellers. They have been asked to answer the question about their expectations referring to life in smart cities and what benefits they expect. As a result of this study 89 requirements and 79 expected benefits have been identified.

In order to provide an analysis of the collected requirements and benefits (according to [Fig. 1]), the authors have designed two ontologies (Fig. 1 Level 2) which allow them to systematize and to compare both sets. During the project, the Fluent Editor tool has been used to work on the ontologies. Currently available tools offer no support for the Polish language, which makes it difficult to check errors and to support strings of characters. The architecture of ontology organises the collected requirements and benefits by developing classes so that they are readable and possible to be compared. The conducted comparative concept makes it possible to find a part which is common for both ontologies, and thus to determine the significance of requirements implementation which will be translated into benefits expected by inhabitants.

The next stage (Fig. 1 Level 2) involves determination of the dependency hierarchy [code1] between various entities within the decision-making process. The presentation of the dependency structure allows us to indicate their place of occurrence in the whole decision-making process. The role of inhabitants, the role of authorities, the position of the smart city, requirements and benefits have been defined here.

```
Every requirement must realize expectancy.
Every requirement has a description-of-requirement.
Every requirement should be a benefit.

Every city has a inhabitant.
Every city has a management.
Every smart-city is a city.
Every smart- management is a management.
Every smart-city has a smart- management.

Every inhabitant define benefit.
Every smart- management define requirement.
Every benefit is not a requirement.
Every benefit has a description-of-benefit.
Every smart- management should realize benefit.
```

Code 1 The hierarchy of the relations of roles (Source: the authors' own study)

◢ ◎ ◆ description-of-benefit
   ▷ ◎ ◆ clean-city
   ▷ ◎ ◆ dissemination-of-information
   ▷ ◎ ◆ improve-the-standard-of-living
   ▷ ◎ ◆ new-technologies
   ▷ ◎ ◆ occupational-benefit
   ▷ ◎ ◆ pleasant-journey
   ▷ ◎ ◆ public-transport
   ▷ ◎ ◆ saving
   ▷ ◎ ◆ service-benefit

Fig. 2. Fragment presenting a taxonomic tree of benefits (Source: the authors' own study)

The further part of the design process of both ontologies is identical. The collected database is divided into sections by the nature of the problem concerning requirements or benefits. The requirements and benefits are grouped into those referring to urban logistics, services, education, infrastructure, labour, energy and those which generally improve the quality of inhabitants' life – Fig. 2.

The controlled language has been applied to determine the relations between particular requirements in the ontology of requirements and, respectively, between particular benefits in the ontology of benefits. The controlled language is the language applied to write ontologies, and it is used in the Fluent Editor tool (it has a form of a text file). The individual statements (axioms) are separated by a dot (.) [14]. While analysing the relations between various requirements in the ontology of requirements and particular benefits in the ontology of benefits, the function "Every B requirement is an A requirement" has been mainly used. An example of a text file applied to design the ontology is shown in Fig. 2. Such an approach can reduce the requirements and benefits to the dependencies hierarchy. Only a whole branch of the requirement hierarchy or the benefit hierarchy comes as a full description of the requirements or benefits.

The tools used to design both ontologies provide a graphical representation of the language applied to record the dependencies and descriptions. The presentation of ontologies "in the form of a graph – makes them easier for people to understand than their presentation in a natural language" [14]. This characteristics provides not only an accessible way to show the effect of designing ontology, but it also provides a possibility to control the correctness of further work. An example of a graphical presentation of the requirement ontology is presented in Fig. 3. Due to the readability of the graphic record language, the authors are able to analyze the results of further comparative analysis in the context of the actual state of the designed ontologies.

The layer of the pattern rules includes the comparative concepts of (Fig. 1 Level 3) both designed ontologies: the requirement ontology and benefit ontology. The comparison has been performed with the use of the edit distance methodology, also referred to as the Levenshtein algorithm [9]. It is a formal method "based on the similarity of character strings which uses the mechanisms of similarity calculation for two strings" [15]. The proposed approach is the most intuitive method applied to determine the relationship between the strings in different ontologies.

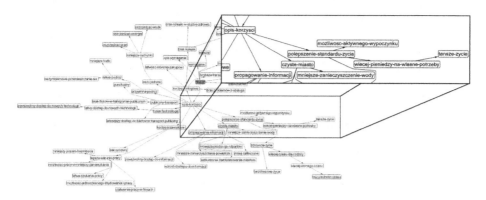

Fig. 3. Graphical presentation of the requirement ontology (Source: the authors' own study)

Used by the authors, the edit distance method consists in the search for the minimum number of a single string edition to obtain the final result: that is having the compared word identical with the word comparator. The operations which have been identified in the Levenshtein algorithm are: insertion, deletion and modification. It is assumed that each performed operation has its unit cost. The retaining of the symbol which is identical to the original version of the sequence cannot be treated as an operation. The sum of the minimum number of operations carried out in the sequence is the edit distance. The correct operation of edit distance, which has been applied in the study, is exemplified by the sequences of "lack-of-crowd-in-public-transport" and "crowd-control-in-public-transport", where the resulting edit distance is 11. On the basis of the obtained result, we have calculated a percentage change that is necessary to convert the string of characters in the A string into the B string. Based on this result, it is possible to determine whether the analysed string of characters can be classified as consistent or acceptably consistent with the compared ontology. It has been assumed that the A string may be edited up to 50% to be considered compatible with the B string.

During the comparative work on both ontologies, the authors have proposed not only the analysis of the characters in the individual requirements but also the analysis of the prevalence level of the particular requirement or benefit in the designed ontology. Considering the maximum number of levels in both ontologies, which translates into the maximum difference between the levels, the scale of individual differences scoring (Table 1) has been developed.

The compatibility of both ontologies is calculated basing on [formula 7], where:

$$\sum = \frac{\frac{(1-OL)}{n} * r}{N} \tag{7}$$

Σ -   *compliance of the benefit ontology with the ontology of requirements [%]*,

OL -   *edit distance between the analysed sequence of requirements, and the sequence of benefits,*

n -   *number of the sequence characters which describe benefits, r* -   *assessment of the level difference between the sequence of requirements in the ontology of requirements, and the sequence of benefits in the ontology of benefits,*

N -   *number of all the benefits included in the benefits ontology.*

Table 1. Scoring of level distance (Source: the authors' own study).

Distance levels	Rating
0	1
1	0,875
2	0,75
3	0,625
4	0,5
5	0,375
6	0,25
7	0,125
8*	0

*the maximum possible difference is 8

During the next stage, the authors have conducted a comparative analysis of compliance as regards the accumulated benefits and the collected requirements. In the discussed sets 26 string of characters with an edit distance equal to "0" and at the same level have been identified in both designed ontologies. Additionally, using the edit distance method, 8 strings of characters have been qualified as compliant or partially compliant. In total, all the categorised strings of characters in the ontology of benefits have been rated at 31.25 points, which means 50.41%.

At the level of the layer of the pattern rules, based on the comparative analysis of two ontologies, the benefits which are most compliant with the requirements have been identified. Then, the benefits have been presented in the way the requirements are presented: with the use of the Unified Modelling Language. They have been placed on the Use Case diagram (Fig. 4). It can be observed (Fig. 4), that the benefits are divided into four groups, hereinafter referred to as the levels, due to the fact that while satisfying one benefit we can, to some extent, satisfy another one, which is associated with first benefit.

In the above diagram we can see that the benefits of the higher levels are included in the benefits of the lower levels. For example, the benefits of the third level are included in the benefits of the second level, and the second level - in the benefits of the first level.

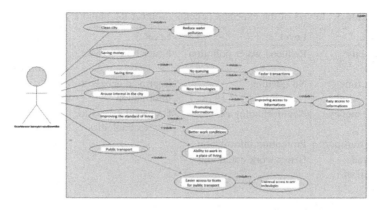

Fig. 4. The use case diagram for benefits expected by users (citizens or City decision makers) (Source: the authors' own study)

The next stage has involved the ranking of the benefits of Level 1 in terms of their importance, according to the following scale: the benefit which is the most important, very important, important, less important, unimportant; provided that only one benefit may be defined as the most important. This scale has been proposed on the basis of expertise resulting from the observation and the analysis of the behaviour of cities. The cities, due to their limited budget, must focus on meeting their residents' most expected benefits. A five-degree scale is designed to facilitate decision-making, so that it is possible to extract the expected benefits which must be satisfied in the first place. At this stage, it is also possible to reject the benefits which are unimportant (less important) or even insignificant (void) for the city and its inhabitants (Table 2).

Table 2. The priorities assigned to benefits (Source: the authors' own study)

Benefit	Importance
Clean city	**The most important**
Saving time	Very important
Saving money	Very important
Public transport	Important
Improved standard of living	Important
Increased interest in the city	Less important

Then, according to the levels shown in the Use Case diagram (Fig. 4), all the expected benefits have been transferred into Table 3 to the respective levels.

In the research it has been assumed that each expected benefit has been assigned with a number corresponding to the column and the row in which it is located, according to the diagram: the column number (level) the line number. In the table presenting the related benefits, each main benefit (level 1) is assigned with the included benefits of the levels 2 and 3 (Table 4).

Table 3. Levels of the expected benefits (Source: the authors' own study)

No	Level 1	Level 2	Level 3	Level 4
1	**Clean city**	Reduced water pollution	Faster transactions	Easier access to information
2	Saving time	No queuing	Improved access to information	
3	Saving money	New technologies	Universal access to new technologies	
4	Public transport	Promoting information		
5	Improved standard of living	Better work conditions		
6	Increased interest in the city	Ability to work in a place of residence		
7		Easier access to tickets for public transport		

Table 4. Related benefits (Source: the authors' own study)

Level 1	Level 2	Level 3	Level 4
1.1	2.1		
1.2	2.2	3.1	
1.3			
1.4	2.7		
1.5	2.5, 2.6		
1.6	2.3, 2.4	3.1, 3.2	4.1

The requirements towards the living in the smart city which have been defined during the questionnaire survey have enabled the authors of the study to verify which of them can meet the assumed benefits expected by the city residents. For this purpose, a supplementary table of the requirements and expected benefits can be provided (Table 5).

We have to pay attention to the order in which the above-presented table is filled in. It should be started with the highest level (in this case, level 3), since the benefit from the next level also comes as a benefit from the former level. In this example, the requirement "Symbols of the main attractions" can be defined as the fulfillment of the benefits numbered as 2.4, "The promotion of information", which is included in the benefit numbered as 1.6, namely: "Increased interest in the city." In this way, at the final stage of the completion of the level 1, it is possible to see which of the requirements will bring the benefits expected by the users and how many benefits will meet a specific requirement.

Table 5. Table of requirements and benefits (Source: the authors' own study)

The direction of filling the table →

Requirements	1.1	1.2	1.3	1.4	1.5	1.6	2.1	2.2	2.3	2.4	2.5	2.6	2.7	3.1	3.2	3.3	4.1	
Marking the most important attractions						x				x								1
Improving the flow of information on job					x						x							1
The ability to control water and energy consumption	x							x										1
Modern technologies in public transport operated by the passengers				x														1
Everywhere possible card payment				x								x						1
The increase in the number of intelligent buildings					x					x								1
Construction of the restaurant						x												1
A simple description of the city, important objects						x			x									1
Increased number of cameras																		0
Greater anonymity																		0
...																		0
...																		0
...																		0
...																		0
...																		0
...																		0
...																		0
	0	1	0	2	2	3												

The number of the benefits of Level 1, which satisfies the requirement

the sum of the number of meeting the requirements of the benefits of the level 1

The number of requirements which should be implemented is related to the importance of the benefits. Subsequently, in this way, the number of requirements which are necessary to provide the expected benefits is proposed as:

- the most important: equal to 100%,
- very important: min. 80%,
- important: min. 60%,
- less important: min. 50%,
- unimportant min: 30%.

It is also a proposal which is based on expert knowledge. The number of requirements for each level is related to the fact that there is a possibility of mutual exclusion of the requirements and it is necessary to choose which of them should be implemented by the city.

Using the proposed model Man-In-The-Loop-Resource-Resolver [11] some rules have been established [code2] to demonstrate the significance of decisions made by people in the decision-making process. The discussed rules are presented below; they have been generally developed for both designed ontology [code2].

The requirement that has-description equal-to
`What requirements expect the residents of the city?' is definiowac by
smart-managament.

The benefit that has-description equal-to `What benefits expect the
residents of the city?' is define by inhabitant.
Requirement-of-Smart-City is a requirement and has-description-of-
requirement equal-to *Requirement of Smart City* corresponds to the concept
of development of the city of improved living standards of its
inhabitants'.

Benefit-of-Smart-City is a benefit and has-description-of-benefit equal-to
``Benefit of Smart City responds to the expectations associated with the
development of the city by improving the living standards of its
inhabitants'.

Man-In-The-Loop-Resource-Resolver-Pattern is a context and has-description
equal-to `Smart-City Residents need to implement the requirements of
bringing the most benefits'.

Man-In-The-Loop-Resource-Resolver-Pattern is a pattern and has-name equal-
to *Man In The Loop Resource Resolver* and solves the problem that has-
description equal-to `What is the structure of the compliance requirements
and benefits?'.

Context-Of-Hierarchy-Of-Requirements-&-Benefits-Pattern is a context and
has-description   equal-to   'Without   a   hierarchy   of   determine   the
significance of requirements / benefits it is hard to manage a city'.

Hierarchy-Of-Requirements-&-Benefits-Pattern is a pattern and operates-in
Context-Of-Hierarchy-Of- Requirements-&-Benefits-Pattern and has-name
equal-to 'Hierarchy Of Requirements & Benefits and solves the problem that
has-description equal-to `How to place a hierarchy of implementation
requirements?' and is-part-of Smart-City-Pattern-Dictionary.

The problem that is solved by Hierarchy-Of- Requirements-&-Benefits-
Pattern is-part-of the problem that is solved by Smart-City-Pattern-
Language.

The force that has-description equal-to `often each of requirements
realized significand amount of expected benefits' is resolved by
Hierarchy-Of- Requirements-&-Benefits-Pattern and is prioritized by
Context-Of-Hierarchy-Of-Requirements-&-Benefits-Pattern.

Code 2 Proposed rules for the design pattern *Man-In-The-Loop-Resource-Resolver*

4   Conclusions

The paper presents a method applied to develop the patterns for Smart Cities. The method involves the integration of the benefits language and requirements language by using ontologies and checking the edit distance for the specific benefits and requirements. This paper example used a sample pattern of decision-making *Man In The Loop-Resource conflict solver*, which formed the rules for limited resources.

This approach requires considerable knowledge on the use of ontology and the respective research on edit distance. Therefore, the proposed solution may come as an example for the course of action in which the development of ontologies as well as the measurement of the edit distance must undergo the process of automation. The presented approach has become possible in the era of natural language processing, and the development of more and more advanced speech converters.

Solutions suggested by authors of this paper indicates the needs of permanent replacing the requirements language by the language. The results of research presented in the paper initiative the process of the language converter development for the Smart Cities pattern needs.

References

1. Alexander, C., Ishikawa, S., Silverstein, M.: A Pattern Language: Towns, Buildings, Construction (Center for Environmental Structure Series). Oxford University Press, Oxford (1977)
2. Bralczyk, J.: Language for sale (2008)
3. Chemuturi, M.: Requirements Engineering and Management for Software Development Projects. Springer, New York (2013)
4. Lasater, C.G.: Design Patterns. Wordware Publishing (Jones and Bartlett Publishers), Plano (2007)
5. Gotlib, D., Olszewski, R.: Spatial Information in Management of Smart City. PWN SA, Warsaw (2016)
6. http://www.cse.msu.edu/~chengb/RE-491/Papers/IEEE-SRS-practice.pdf
7. https://www.fastcoexist.com
8. Kotonya, G., Sommerville, I.: Requirements Engineering: Processes and Techniques. Wiley, Hoboken (1998). ISBN 0-471-97208-8
9. Meszaros, G., Doble, J.: A pattern language for pattern writing. In: Pattern Languages of Program Design 3, pp. 529–574. Addison-Wesley Longman Publishing Co. (1997)
10. Niewiadomski, A., Stanuszek, M.: The mechanism analysis of the similarity of short pieces of text, based on the Levenshtein distance. Cracow University of Technology (2013)
11. Norvig, P.: Design Patterns in Dynamic Languages (1998). http://www.norvig.com/design-patterns/design-patterns.pdf
12. Orłowski, C., Sitek, T., Ziółkowski, A., Kapłański, P., Orłowski, A., Pokrzywnicki, W.: Ontology of the design pattern language for smart cities systems. Trans. Comput. Collect. Intell. **25**, 76–100 (2016)
13. Orłowski, C., Ziółkowski, A., Orłowski, A., Kapłański, P., Sitek, T., Pokrzywnicki, W.: Implementation of business processes in smart cities technology. Trans. Comput. Collect. Intell. **25**, 15–28 (2016)

14. Orłowski, C.: Ontology of the Design Pattern Language for Smart Cities, Gdańsk University of Technology
15. Pastuszak, J., Orłowski, C.: Model of rules for IT organization evolution. In: Nguyen, N.T. (ed.) Transactions on Computational Collective Intelligence IX. LNCS, vol. 7770, pp. 55–78. Springer, Heidelberg (2013)
16. Pietranik, M.: Ontology mapping methods including semantics and rating of attributes. Wrocław University of Science and Technology, Department of Information Technology and Management, Information Technology Institute (2013)

Assessment and Optimization of Air Monitoring Network for Smart Cities with Multicriteria Decision Analysis

Aleksander Orłowski[1(✉)], Mariusz Marć[2], Jacek Namieśnik[2], and Marek Tobiszewski[2]

[1] Department of Management, Faculty of Management and Economics, Gdańsk University of Technology, 11/12 G. Narutowicza St., 80-233 Gdańsk, Poland
aorlowsk@zie.pg.gda.pl
[2] Department of Analytical Chemistry, Chemical Faculty, Gdańsk University of Technology, 11/12 G. Narutowicza St., 80-233 Gdańsk, Poland

Abstract. Environmental monitoring networks need to be designed in efficient way, to minimize costs and maximize the information granted by their operation. Gathering data from monitoring stations is also the essence of Smart Cities. Agency of Regional Air Quality Monitoring in the Gdańsk Metropolitan Area (pol. ARMAAG) was assessed in terms of its efficiency to obtain variety of information. The results on one-month average concentrations of seven parameters (data for three years) for eight monitoring stations were the input data to multicriteria decision analysis assessment and the least effective was detected.

Keywords: Smart Cities · Environmental monitoring · Monitoring network · PROMETHEE · Multicriteria decision analysis · Multicriteria · Air monitoring

1 Introduction

Currently, 54% of the people who live in the world inhabit city areas. According to the United Nations this is 3.5 billion people and is supposed to grow to 7 billion in 2045 [1]. The process might be most visible especially in North America (84% of the population living in urban areas) and Europe (73%).

The data shows that managing city areas is, and is going to be, more important with the growing number of inhabitants and limited area, in which the cities might and should (e.g. because of economic reasons) grow. What seems to be more important, from the perspective of current fast growing and developing cities, is the necessity to gain data to help properly manage the cities. Currently, the perspective of so called Smart Cities is gaining popularity, they are defined as cities that can generate, accumulate, use and share data [2]. Due to that fact, there is a need to analyze several factors (which can be grouped in categories like: environment, mobility, economy, governance, people, living [3]) of the cities environment.

Selection of sampling points/areas should be made in an optimal way, taking into account such aspects as: economic, logistic, geographical and local community problems. To take all that different aspects into consideration Multi criteria decision analysis

© Springer International Publishing AG 2017
N.T. Nguyen et al. (Eds.): ACIIDS 2017, Part II, LNAI 10192, pp. 531–538, 2017.
DOI: 10.1007/978-3-319-54430-4_51

(MCDA) is used. It is popular in various areas of business life, in which decision making processes are required, like heating systems [4] to stock exchange problems [5], waste management [6] to find the most sustainable solutions [7]. It has been successfully applied to select suitable remediation technique for contaminated mine restoration [8] or choosing priority areas for bottom sediments monitoring [9]. All these examples involve sustainability and technological criteria that are important during the selection of appropriate solution. PROMETHEE method was chosen from the range of all multicriteria decision analysis tools because it is easy to apply compared to others (such as AHP). It also ranks alternatives as well as identifies the best alternative, whereas the utility theory methods only identify the best alternative [8]. Ranking of alternatives seems to be the most important because it is not only important to know which is the best among all but to make it possible to compare all the alternatives and decide which might not be important. There are several rankings of MCDA methods published [10] and PROMETHEE is concerned to be one of the top in alternatives ranking, easiness of use and potential of future development.

The aim of the study is to prove the applicability of MCDA in assessment of air monitoring network performance. Such assessment methodology is important to identify the strong and weak points in the network to reduce the number of stations without decreasing the amount of information obtained from the network. It results in the significant economic savings and design of more efficient environmental monitoring networks.

2   Materials and Methods

2.1   ARMAAG Monitoring Network

Agency of Regional Air Quality Monitoring in the Gdańsk Metropolitan Area (pol. ARMAAG) was created in year 1993. Its appearance was the answer for the demand of society and local authorities for reliable information on the air quality in the Tri-city agglomeration (Gdańsk, Gdynia and Sopot). From year 1996 to 2001 there were located 8 monitoring stations (Fig. 1). The location of these stations was based on long term meteorological data for this area, especially wind directions, population density and the activities of point and linear emission sources.

2.2   The Input Data

In order to perform comparison of data obtained by respective monitoring stations the month average concentrations of SO_2, NO_x and PM_{10} were utilized. This choice was made on the basis that every of these parameters is measured on all ARMAAG stations. The data on BTEX concentrations were taken from measurements with passive sampling technique for collection of analytes from atmospheric air. To unify the interpreted results (on-line and passive sampling results) to allow their comparison the BTEX database was presented in the form of one month time weighted average (TWA) concentrations. All the input data are mean month average concentrations for the time period between January 2012 and December 2014 (three full years) (Table 1).

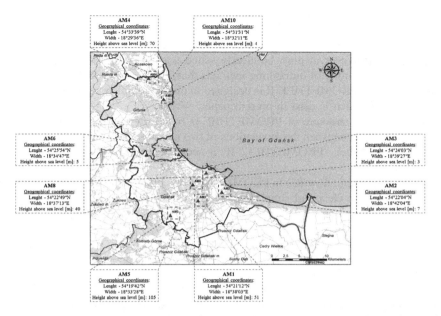

Fig. 1. Localization of ARMAAG air quality monitoring stations in the Tri-City agglomeration area (Gdańsk, Gdynia and Sopot).

Table 1. Analytical procedures used to determine respective air quality parameters.

Parameter	Applied method or technique (reference or alternative solution)
NO_X	The reference method for the measurement of nitrogen dioxide and oxides of nitrogen is that described in EN 14211:2013-02 - the measurement of the concentration of NO_x by chemiluminescence
SO_2	The reference method for the measurement of sulphur dioxide is that described in EN 14212:2005 - the measurement of the concentration of SO_2 by ultraviolet fluorescence
PM_{10}	The reference method for the sampling and measurement of PM10 is that described in EN 12341:1999 - gravimetric measurement method for the determination of the mass fraction of suspended particulate matter
Benzene	Non-reference technique – passive sampling technique [11–13]
Toluene	Sampling devices – radiello diffusive passive samplers (Fondazione
Ethylbenzene	SalvatoreMaugeri, Padova, Italy)
Sum of Xylenes	Analytes liberation technique – thermal desorption Final determination technique – GC-FID or GC-MS

Based in raw data in form of month average concentrations the calculation of determination coefficients (R^2) for respective parameters measured in the station and the nearest geographical neighbour and the second nearest geographical neighbour. Determination coefficients of respective measured parameters for closest and second closest neighbour were the input data to PROMETHEE analysis.

2.3   PROMETHEE Technique

The decision of PROMETHEE selection as a ranking tool was described in the previous chapter. However, it must be clearly stated that PROMETHEE is not a single method but it is a family of outranking methods that consist of PROMETHEE versions from I to VI [14]. PROMETHEE II is designed to obtain complete ranking of available alternatives and which is used in present paper.

The preference structure of PROMETHEE is based on pair wise comparisons, in which the deviation between the evaluations of two alternatives on a particular criterion is considered. When PROMETHEE II is considered, all the alternatives are comparable.

The net outranking flow (Ø) in PROMETHEE II is built on weights and preferences functions which are clear and simple preference information.

$$\emptyset(a) = \emptyset^+(a) - \emptyset^-(a)$$

The detailed mathematical description of PROMETHEE was presented by its developers Brans and Vincke [15].

For the current research VisualPromethee Academic software was used. This is a tool, dedicated for scientific purposes, that supports the usage of PROMETHEE and which was well tested in previous studies [16, 17].

The PROMETHEE using procedure was divided into three steps: Data preparation, assigning weights to respective criteria and application of PROMETHEE algorithm.

2.4   Creation of Scenarios and Weights Assigning

There were two scenarios defined based on different attempts to the location of nearest stations. In the first scenario, the first and the second closest stations had the same weight of importance of every station used. In the second scenario, it was decided to present the closest station as twice more important than in the second closest station. These assumptions were the basis for the calculation of weights applied in each scenario, during PROMETHEE assessment.

The usage of two scenarios makes it possible to present different attempts to the same problem: location of the stations. It might be considered that the overall result of the city/agglomeration is important which means that usage (and influence) of every single location of the station is same important so weights of the stations are equal (scenario no. 1). In other perspective it might be seen that locations of the stations should represent the maximum local area of the city/agglomeration so the further one station from another one is the smaller the influence one on another is and due to that different weights should be used to the stations (and that is represented in scenario no. 2).

"The set $\{w_j, j = 1, 2, \ldots, k\}$ represents weights of relative importance of the different criteria. These weighs are non-negative numbers, independent from the measurement units of the criteria. The higher the weight, the more important the criterion. There is no objection to consider normalized weights, so that:

$$\sum\nolimits_{j=1}^{k} w_j = 1$$

For the first run of the PROMETHEE algorithm two times higher weights were given for correlation parameters between variables measured for the closest stations than for correlation of variables measured for the second closest neighbouring station. For the second run of PROMETHEE algorithm different weights for measured variables were included to reflect their different importance. The relative importance of the measured parameters was assessed in the following way: benzene - 0.25; toluene - 0.20; ethylobenzene - 0.05; xylenes - 0.05; NOx - 0.20; SO2 - 0.10; PM10 - 0.15. Of course the application of weights is arbitrary but can be justified in the following way. The highest weight is given to benzene as this compound is carcinogenic and its determination is required by the directive. Lower but still high weight is given to toluene as it has relatively long atmospheric half-life and is emitted from fuel combustion in car engines. The necessity of its monitoring is also stated in the directive. Ethylobenzene and xylenes are given lower weights as their atmospheric half-lives are short and they originate from diffused sources. The weight for NOx is high as they are emitted from traffic related sources and are tropospheric ozone precursors. The weight given to SO2 is lower as this compound's concentration is near urban background level. The need for PM10 monitoring is also stated in the directive therefore weight for this parameter is relatively high. In this way weights were assigned for second PROMETHEE run, by subjective assessment of measured parameters importance.

3 Results and Discussion

The results of ranking of stations with PROMETHEE technique show how different are the results between neighbouring stations – the closest and the second closest neighbours. The monitoring stations with highest ranks measure the pollution originating from different sources than their neighbours. Oppositely, monitoring stations ranked as the last are characterized by high correlation of parameters with their neighbours. It can be concluded that they measure the emissions from the same processes as neighbouring stations or background concentration.

The results for the first set of weights (described in the previous section) are presented in the Table 2. The highest rank was gained by AM2 station, which is located in the vicinity of industrial area. The second emission source is activities during construction of the road underway passage in years 2014–2015. The construction resulted mostly in the emissions on PM10 generated by moving tracks and the measured concentrations of this parameter were ten times higher than normally. The second rank was given to AM5 station that is influenced by two significant sources – emissions from landfill site and neighbourhood of highway with heavy traffic density. The third rank was given to AM10 station, which is located next to the marine port and in the vicinity of railway stations. Stations AM1, AM4, AM3 and AM8 are considered to measure the processes typical for urban areas, mainly traffic intensity. The last rank was gained by AM6 station, what means that it has high correlation of results with its two closest neighbours. It happens despite the fact that AM6 station's closest neighbours

Table 2. The assessment results for the first run of PROMETHEE. Two times higher weights were given for correlation of variables between two closest stations.

Rank	Alternative	Phi
1	AM2	0.6369
2	AM5	0.3036
3	AM10	0.1369
4	AM1	0.0298
5	AM4	−0.0893
6	AM3	−0.1488
7	AM8	−0.3512
8	AM6	−0.5179

are geographically distant. This station, apart from the fact that it measures concentrations of pollutants from traffic, the concentrations are lower than in other stations. This may be also related to location of the station in local scale, as it is surrounded by the green areas.

The results of ranking, where different weights were assigned to criteria to differentiate their importance, are presented in the Table 3. The ranking itself is the same as in previous case but the Phi values are different. In case of Phi values in case of the first rank – AM2 station are 0.6369 and 0.581 in case of second scenario for application of weights. Because in second case higher weights were given to benzene, toluene and NO_x, the parameters related to traffic emissions, it means that AM2 station is affected by other sources than automobile emissions. In contrary, for station AM5 Phi value is higher for the second assessment, than for the first one. This supports the conclusion, that AM5 station is affected by pollution originating from highway. Again very low Phi value for AM6 indicate that the measured concentrations are urban background.

Table 3. The assessment results for the second run of PROMETHEE. The weights were given to discriminate the importance of parameters.

Rank	Alternative	Phi
1	AM2	0.581
2	AM5	0.3905
3	AM10	0.1952
4	AM1	0.1381
5	AM4	−0.0524
6	AM3	−0.2571
7	AM8	−0.3238
8	AM6	−0.6714

In this study we have included the assessment criteria that are referring to the effectiveness of monitoring network. However, PROMETHEE ranking tool allows to include in the assessment completely different criteria, that can be relevant to selection of location of monitoring stations. These can be population density, traffic intensity, land development or land value data. These kind of data can be relevant to monitoring network optimization and can be utilized during decision making process along with network efficiency data.

4 Conclusions

In this study we presented the method of assessment and optimization of environmental monitoring network that allows to simultaneously consider many criteria to assess rank each monitoring station in terms of efficiency. Being Smart in the criteria of Smart Cities means not only measuring but doing that most effectively which means generating necessary data with the most reasonable cost. Long term monitoring results, as input data to MCDA tool, allowed to select the monitoring station that is the least efficient. In this case study, it was the monitoring station that measures urban atmospheric background pollution. It is very easy to implement other criteria than those strictly referring to performance assessment, such as social and economic indicators (and that is going to be presented in next papers). These days, with current technology it is not a point how much data can be generated (from technological perspective) but how much data is necessary for the decision making processes and due to that the processes of optimization are necessary.

References

1. World Urbanization Prospects the 2014 Revision, Department of Economic and Social Affairs, United Nations (2014)
2. Smart Cities Council. http://smartcitiescouncil.com. Accessed 31 May 2016
3. European Smart Cities Report 2015, Vienna University of Technology. http://www.smart-cities.eu/index.php?cid=2&ver=4. Accessed 4 June 2016
4. Ghafghazi, S., Sowlati, T., Sokhansanj, S., Melin, S.: A multicriteria approach to evaluate district heating system options. Appl. Energy 87(4), 1134–1140 (2009)
5. Albadvi, A., Chaharsooghi, S., Esfahanipour, A.: Decision making in stock trading: an application of PROMETHEE. Eur. J. Oper. Res. 177(2), 673–683 (2007)
6. Vego, G., Kucar-Dragicevic, S., Koprivanac, N.: Application of multi-criteria decision-making on strategic municipal solid waste management in Dalmatia, Croatia. Waste Manag. 28(11), 2192–2201 (2007)
7. Tobiszewski, M., Orłowski, A.: Multicriteria decision analysis in ranking of analytical procedures foraldrin determination in water. J. Chromatogr. A 1387, 116–122 (2015)
8. Betrie, G.D., Sadiq, R., Morin, K.A., Tesfamariam, S.: Selection of remedial alternatives for mine sites: a multicriteria decision analysis approach. J. Environ. Manag. 119, 36–46 (2013)
9. Alvarez-Guerra, M., Viguri, J.R., Voulvoulis, N.: A multicriteria-based methodology for site prioritisation in sediment management. Environ. Int. 35, 920–930 (2009)
10. Cinelli, M., Coles, S.R., Kirwan, K.: Analysis of the potentials of multi criteria decision analysis methods to conduct sustainability assessment. Ecol. Ind. 46, 138–148 (2014)
11. Maré, M., Namieśnik, J., Zabiegała, B.: BTEX concentration levels in urban air in the area of the Tri-City agglomeration (Gdansk, Gdynia, Sopot), Poland. Air Qual. Atmos. Health 7, 489–504 (2014)

12. Marć, M., Bielawska, M., Wardencki, W., Namieśnik, J., Zabiegała, B.: The influence of meteorological conditions and anthropogenic activities on the seasonal fluctuations of BTEX in the urban air of the Hanseatic city of Gdansk, Poland. Environ. Sci. Pollut. Res. **22**, 11940–11954 (2015)

13. Marć, M., Bielawska, M., Simeonov, V., Namieśnik, J., Zabiegała, B.: The effect of anthropogenic activity on BTEX, NO2, SO2, and CO concentrations in urban air of the spa city of Sopot and medium-industrialized city of Tczew located in North Poland. Environ. Res. **147**, 513–524 (2016)

14. Brans, J.-P., De Smet, Y.: Promethee methods. In: Multiple Criteria Decision Analysis, pp. 187–219 (2016)

15. Brans, J.-P., Vincke, P.: A preference ranking organization method: the PROMETHEE method for multiple criteria decision-making. Manag. Sci. **31**, 647–656 (1985)

16. Tobiszewski, M., Pena-Pereira, F., Orłowski, A., Namieśnik, J.: A standard analytical method as the common good and pollution abatement measure. TrAC Trends Anal. Chem. **80**, 321–327 (2016)

17. Jędrkiewicz, R., Tobiszewski, M., Orłowski, A., Namieśnik, J.: Green analytical chemistry introduction to chloropropanols determination at no economic and analytical performance costs? Talanta **147**, 282–288 (2016)

Urban Air Quality Forecasting: A Regression and a Classification Approach

Kostas Karatzas[1(✉)], Nikos Katsifarakis[1], Cezary Orlowski[2], and Arkadiusz Sarzyński[3]

[1] Department of Mechanical Engineering,
Informatics Systems and Applications – Environmental
Informatics Research Group, Aristotle University, Thessaloniki, Greece
{kkara, nikolakk}@auth.gr
[2] Institute of Management and Finance,
WSB University in Gdańsk, Gdańsk, Poland
corlowski@wsb.gda.pl
[3] Faculty of Management and Economics,
Department of Applied Business Informatics,
Gdańsk University of Technology, Gdańsk, Poland
arek3108@gmail.com

Abstract. We employ Computational Intelligence (CI) methods to model air pollution for the Greater Gdańsk Area in Poland. The forecasting problem is addressed with both classification and regression algorithms. In addition, we present an ensemble method that allows for the use of a single Artificial Neural Network-based model for the whole area of interest. Results indicate good model performance with a correlation coefficient between forecasts and measurements for the hourly PM10 concentration 24 h in advance reaching 0.81 and an agreement index (Cohen's kappa) up to 54%. Moreover, the ensemble model demonstrates a decrease in Mean Square Error in comparison to the best simple model. Overall results suggest that the specific modelling approach can support the provision of air quality forecasts at an operational basis.

Keywords: Computational intelligence · Air pollution · Classification · Regression · Ensemble

1 Introduction

Air quality (AQ) forecasting plays an important role in urban environmental management as well as in contemporary smart city development [1, 2]). For this reason, a number of AQ forecasting models have been developed in the last decades following an analytic, deterministic approach [3], or a data-driven approach (as reported in [4] and in references therein). The objective of this paper is to use the available atmospheric quality data from the Greater Gdańsk Area (GGA) in Poland, in order to develop AQ forecasting models for operational environmental information provision to citizens and city authorities alike. Air pollutant concentrations are addressed on the basis of their numerical values, as well as on the basis of defined AQ classes. Concerning the AQ forecasting, we follow a twofold approach:

© Springer International Publishing AG 2017
N.T. Nguyen et al. (Eds.): ACIIDS 2017, Part II, LNAI 10192, pp. 539–548, 2017.
DOI: 10.1007/978-3-319-54430-4_52

(a) Each AQ monitoring station is treated individually, i.e. AQ models are developed and tested per station. Thus, the forecasting of the parameter of interest is performed initially as a regression problem and then as a classification problem.

(b) A generalized AQ regression model is being created based on ensemble modelling principles, which is then evaluated via its ability to forecast AQ levels at each monitoring station.

Particulate Matter of mean aerodynamic roughness of 10 μm (PM10) is chosen as the AQ parameter of interest, as this is the only parameter that was made available via the AQ monitoring network, but also as this is one of the most important pollutants in Poland [5]. The hourly concentration level of this pollutant 24 h in advance was the target of the forecasting models under development. This was done in order to be able to inform the citizens about the diurnal variation of the specific pollutant, thus allowing for timely information provision. The related research question posed in the current paper focuses on the investigation of data-driven AQ models types and their characteristics for operational PM10 forecasting in the GGA. In the rest of the paper we firstly present the materials of our study (Sect. 2), followed by the computational methods (Sect. 3). Then we proceed with the presentation and the discussion of the results in Sect. 4, and we draw our conclusion in Sect. 5.

2 Materials: Area of Study and Data Made Available

2.1 The Area of Interest

The city of Gdańsk is located on the Baltic coast in the south-west of the Bay of Gdańsk, in the northern part of Poland. The city merges with Gdynia and Sopot adding 750.000 residents in the GGA. The majority of air pollutant emissions originate from the industrial sector, the port activities and the city traffic [6]. The economy in Gdańsk is dominated by shipbuilding, petrochemicals and chemical industries, which are all concentrated close to the city center. The production of fuels and lubricants implicates emissions of pollutants due to the required heat production for the refining. The shipping activities in the port of Gdańsk are remarkably high [7]. In the case of Gdańsk and Gdynia the ports and city centers are also close together, a fact that may influence air quality in an even more direct way.

2.2 The Atmospheric Quality Data

The atmospheric quality data made available consist of hourly measurements for a period spanning from the 1st of January 2012 up to the first days of 2014 and for a total of nine monitoring stations in the area (Fig. 1). The following parameters were included in the dataset per station: PM10 ($\mu g/m^3$), air temperature (°C), dew point (°C), humidity (%), air pressure (mbar), wind direction (in 8 classes equally covering the 360 deg direction circle) and wind speed (m/s).

It is therefore obvious that the only pollutant for which concentration levels were available was PM10. It should be noted here that particulate matter is a pollutant further

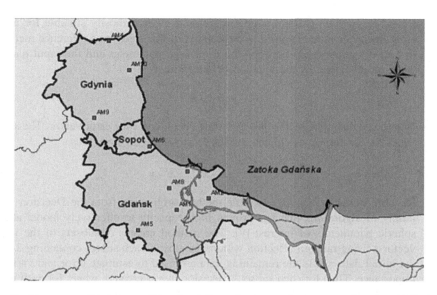

Fig. 1. The greater Gdańsk area. The nine AQ monitoring stations are mentioned as AM1-6 and AM8-10. Urban areas of the three cities of Gdańsk, Gdynia and Sopot are also shown.

characterized by the mean aerodynamic diameter of the particles. PM10 is thus particulate matter of mean diameter up to 10 μm (approx. 1/5th of the diameter of the human hair). Combustion processes, traffic and natural sources directly emit PM10, while in some regions the mechanical degradation of the road surface and of winter tires also contributes to its production. The air quality criterion for PM10 is its mean 24 h averaged concentration, and the limit value used equals 50 μg/m^3, not to be exceeded more than 35 times per calendar year.

3   Computational Methods

3.1   AQ Forecasting as a Regression Problem

The forecasting of the numerical value of PM10 concentration levels for 24 h ahead of the current hour of the day was the goal set for the development of relevant forecasting models. For this reason, we made use of the available datasets for each AQ monitoring station to develop individual (per station) AQ forecasting models.

Algorithms for Single Station Model Creation
The algorithms applied were selected based on computational experiments employing various CI methods, which were conducted with the aid of the WEKA computational environment [8]. On this basis, we chose the following three algorithms as the basis for AQ model development:

(i)   Linear Regression (LR), a simple regression method mostly suitable for modeling linear phenomena, used in the context of this study as a reference method. In LR, the relationship between the forecasted parameter and the input parameters are described by an equation of the form:

$$y = x \cdot \beta + \varepsilon \tag{1}$$

Where x is the input vector, β is the slope vector and ε the error vector. The slope vector is commonly calculated via the least square method, thus

$$\widehat{\beta} = (x' \cdot x)^{-1} \cdot x' \cdot y \tag{2}$$

(ii) Random Forests (RF), an ensemble method originating from the Decision Tree family of algorithms that has proven high capacity to effectively model atmospheric parameters of interest [9]. The method creates N subsets of the input vector x using random selection with replacement, each subset containing 2/3 of the initial data, while the remaining data are used to estimate error and variable importance. Then for each subset, a decision tree is created, where for each node the splitting is based on a (randomly) selected subset of m attributes that optimize a target function. In our case $m = int[log_2(Number\ of attributes) + 1]$ Each of the aforementioned random trees had an unlimited number of levels and nodes. The prediction created by each tree is averaged and thus the ensemble-based overall prediction of the RF (here the PM10 + 24 h concentration) is generated.

(iii) Artificial Neural Networks (ANNs), a powerful computation method capable of representing and estimating the behavior of nonlinear phenomena, like the ones involved in air quality management problems. In ANNs the input vector x for each neuron k, is weighted with the aid of a weighting vector w_k, and the result is summed (taking into account any bias) and then fed into a transfer function f to produce the overall output vector y_k:

$$y_k = f(w_k^T \cdot x) \tag{3}$$

The training of the ANN aims at reducing the error e_k between the model output y_k and the actual (real) value observed d_k, which here is the PM10 + 24 h concentration for each station.

$$e_k = \| y_k - d_k \| \tag{4}$$

This error reduction is based on a number of methods all of which aim at recalculating the initial weights so that the overall network error is minimized. In the case of the gradient descent method (which is the simples of all but nevertheless representative of the way that the weights are recalculated), the relationship between the updated and the initial weighting vector for all neurons k of the ANN, is given by

$$w(t+1) = w(t) - a(t)g(t) \tag{5}$$

Here t and $t+1$ denote the initial and the updated weights, while the error term is described by

$$g(t) = J^T(t) \cdot e(t) \tag{6}$$

where J^T is the (transposed) Jacobian and $e(t)$ is the overall error vector.

In this specific case a MultiLayer Perceptron Network was used, with an input layer consisting 11 nodes (i.e. all the input parameters per station), an output layer consisting of only one node (the PM10 concentration after 24 h) and a hidden layer with six nodes (i.e. the men value of nodes in the input and output layers). The sigmoid function is employed as the transfer function while the gradient descent algorithm is used for minimizing the error function.

Model Validation
In order to validate the results of the PM10 predictions, it is important to make use of as many of the available data as possible for the training as well as for the testing phase. For this reason we followed a 10-fold cross validation procedure [10]: we randomly divided the initial dataset into 10 equal subsets. Then 9 out of these datasets were used for training the model, while the 10th one was used for testing, This process was repeated 10 times, each time leaving a different subset out of the training phase and using it for the test phase. The overall model results are the mean values of the statistical indices of the 10 models developed.

Model results were evaluated based on the following statistical indices:

(a) Pearson's correlation coefficient r that describes the degree of linear relationship between forecasted and real PM10 concentration values.
(b) Mean Absolute Error which is a measure of the mean absolute distance between forecasted and real values and
(c) Root Mean Squared Error which is the square of the Mean Square Error and expresses the standard deviation of the differences between forecasted and actual values.

Creation of a Generalized Model
In addition to the above approach, we investigated the possibility to develop one generalized (ensemble-based) model to be common for all monitoring stations. For this reason, a feed-forward ANN was employed, with one input layer including 11 nodes (thus as many as the max number of input parameters available per station), a hidden layer with 20 nodes and an output layer with one node (for the PM10 + 24 h concentration value). The ANN made use of the sigmoid activation function and the Levenberg - Marquardt algorithm for calculating the weights, where

$$w(t+1) = w(t) + [H + \lambda I]g(t) \tag{7}$$

with H being the Hessian matrix of the error function, I the unit matrix and λ a regularization parameter. After the training phase, the synaptic weights w of each model were used to construct an ensemble of weights, based on

(a) the mean value of all weights so that $w_{overall} = \sum_{i=1}^{9} w_i$, thus the model was named Ensemble Mean (E.M.) and

(b) a weighted mean of the weights on the basis of the relevant importance (i.e. accuracy) of each model at each station, so that $w_{overall} = F_{weighted_{i=1}}^{9}(w_i)$, thus the model was named Weighted Ensemble (W.E.).

The resulting matrix of the weights in each one of the two aforementioned cases was then used for the development of a common ANN model, which was tested for each one of the 9 stations individually. The Matlab Neural Network Toolbox was used for the calculations.

3.2 AQ Forecasting as a Classification Problem

The approach used in the case of classification was the same as in the case of the regression problem with individual models per station, with the major difference being that the goal was to forecast a nominal (categorical) value for PM10 + 24 h, on the basis of the following transformation formulae:

$$
C_{PM10} = \begin{cases}
\text{"low" if} & C_{PM10_{Numeric}} \leq 50 \frac{\mu g}{m^3} \\
\text{"medium" if} & 50 \frac{\mu g}{m^3} < C_{PM10_{Numeric}} \leq 100 \frac{\mu g}{m^3} \\
\text{"high" if} & 100 \frac{\mu g}{m^3} < C_{PM10_{Numeric}} \leq 200 \frac{\mu g}{m^3} \\
\text{"very high" if} & C_{PM10_{Numeric}} > 200 \frac{\mu g}{m^3}
\end{cases}
$$

Here $C_{PM10_{Numeric}}$ is the numerical hourly concentration of PM10 for each one of the 9 stations.

In terms of algorithms, we employed again Random Forests and Artificial Neural Networks as classifiers in addition we included in the analysis the Random Tree (RT) and Random Committee (RC) algorithms. Random Trees are classifiers that consider a given number of random features at each tree node, while Random Committee is a meta-learning algorithm that builds an ensemble of basic classifiers and averages their predictions [11].

Model results were evaluated via (i) the Root Mean Squared Error (introduced in Sect. 3.1) and (ii) the Cohen's kappa: The Cohen's kappa index [12], is a measure of agreement between the classifications and the true classes, and it makes use of the observed accuracy A_o (what the classification model achieved) and the expected accuracy A_e. The index is then via Eq. 8:

$$
k = \frac{A_o - A_e}{1 - A_e} \tag{8}
$$

Cohen's kappa receives values between zero and one, and is considered to be very good if it is greater than 0.8, good if it is greater than 0.6, moderate if it is greater than 0.4 and fair to poor if lower, according to Altman [13].

4 Results and Discussion

4.1 Results of the Correlation Approach

On the basis of the model calculations performed as described in Sect. 3.1, the correlation coefficient r accompanied by the Mean Absolute Error (MAE) and the Root Mean Squared Error (RMSE) were calculated for the three models developed and for each one of the nine AQ monitoring stations for which data were available. Table 1 presents these results.

From Table 1 it is evident that the algorithm leading to the best (highest) correlation coefficient is Random Forest, with an *r* value ranging from 0.6937 for station AM3 up to 0.8162 for station AM4. The ANN algorithm can be ranked as 2nd, followed by Linear Regression. When it comes to MAE and RMSE criteria, Random Forest also presents with the best (lowest values) results, but here followed by Linear Regression and with ANNs ranked as 3rd. This demonstrates that Random Forest has a satisfactory performance, and may thus be used for operational purposes in order to provide numerical forecasts of the hourly PM10 concentration levels 24 h in advance for the GGA and based on the data made available in this study.

Table 1. Correlation Coefficient (*r*), Mean Absolute Error (MAE) and Root Mean Square Error (RMSE) for three models per monitoring station concerning the forecast of the hourly PM10 concentration 24 h in advance.

Datasets	Random forest			ANN (multilayer perceptron)			Linear regression (multivariate)		
	r	MAE	RMSE	r	MAE	RMSE	r	MAE	RMSE
AM1	0.8076	5.9285	9.5297	0.4107	9.7967	14.616	0.4526	9.21	14.029
AM2	0.7145	8.3093	13.407	0.3137	12.121	18.255	0.3331	11.863	17.740
AM3	0.6937	7.9227	12.644	0.26	10.898	17.382	0.2846	10.722	16.431
AM4	0.8162	5.8306	8.4723	0.3424	10.294	14.361	0.4527	8.8948	12.747
AM5	0.7566	5.886	9.0297	0.3852	9.357	13.354	0.4568	8.345	12.153
AM6	0.7131	6.3645	9.9836	0.3369	9.4424	13.836	0.3907	8.5981	12.969
AM8	0.7638	6.8537	11.442	0.3986	10.921	16.558	0.4596	9.8978	15.529
AM9	0.7736	6.2321	9.2116	0.3796	9.749	13.589	0.4553	8.7719	12.665
AM10	0.7499	9.7198	15.271	0.3109	15.505	22.539	0.3973	14.073	20.852

4.2 Results of the Ensemble Modelling in the Correlation Approach

Focusing on the ANNs and on the ensemble method used, the results presented in Table 2 for the Mean Square Error suggest that in the majority of cases, the Ensemble-based models improved the MSE of the forecast in comparison to the best individual models per station. Nevertheless, when the ensemble models were worse that the initial models, the degradation of the forecast was even higher, reaching $\sim$ 24% in the case of the AM8 monitoring station and for the Ensemble Mean model. The use of the Weighted Ensemble did not lead to such extreme values, thus "smoothing" the performance of the Ensemble

Table 2. The MSE for the Initial Model per station (I.M.) and for the Ensemble Mean (E.M.) and the Weighted Ensemble (W.E.) concerning PM10 forecasting 24 h in advance.

ID	MSE (I.M.)	MSE (E.M.)	MSE (W.E.)	% (E.M.)	% (W.E.)
1	219.27	211.44	198.41	3.57	9.51
2	257.36	281.22	270.46	−9.27	−5.09
3	279.27	251.38	252.08	9.99	9.74
4	143.19	140.94	134.82	1.57	5.84
5	125.89	117.54	127.63	6.63	−1.38
6	128.71	124.90	136.43	2.96	-6.00
8	208.91	258.45	200.03	−23.72	4.25
9	155.65	141.36	128.43	9.18	17.49
10	407.51	456.63	438.39	−12.05	−7.58
SUM	1925.75	1983.86	1886.68	−3.02	2.03
SUM (no extr.)				10.71	−7.88

Model. If the extremes (i.e. the best and the worst MSE statistic) are excluded from the comparison, then the overall improvement of the Weighted Ensemble becomes more pronounced.

4.3   Results of the Classification Approach

Concerning the classification problem, Table 3 summarizes the statistical indices describing the performance of the models used. It is interesting to note that Random Committee outperforms all other algorithms in terms of the kappa index for all monitoring station, reaching kappa values up to 0.5438 (for station #1). The ANN models are continuously the worse among those tested in terms of Cohen's kappa, with the exception of the RMSE, where they outperform Random Tree models. The confusion matrix corresponding to the best Random Committee model is the following (Table 4)

Table 3. Cohen's Kappa and the Root Mean Square Error (RMSE) for four models per monitoring station concerning the forecast of the hourly PM10 concentration 24 h in advance.

Data sets	Random forest		Random tree		Random committee		ANN (multilayer perceptron)	
	kappa	RMSE	kappa	RMSE	Kappa	RMSE	kappa	RMSE
AM1	0.5023	0.1128	0.4503	0.1573	**0.5438**	0.1162	0.2614	0.1388
AM2	0.3613	0.1536	0.3572	0.2081	**0.4076**	0.1618	0.1631	0.1764
AM3	0.3319	0.1295	0.3556	0.1769	**0.4037**	0.1349	0.1134	0.1522
AM4	0.4319	0.1242	0.4193	0.1671	**0.4838**	0.1308	0.1976	0.1486
AM5	0.4377	0.1105	0.4059	0.1535	**0.4792**	0.115	0.2068	0.1317
AM6	0.517	0.0935	0.4267	0.1297	**0.5174**	0.099	0.3165	0.1108
AM8	0.5398	0.1209	0.4938	0.1648	**0.5718**	0.1236	0.2732	0.1528
AM9	0.4521	0.1146	0.3972	0.1575	**0.4862**	0.1178	0.2251	0.1396
AM10	0.5009	0.1847	0.4398	0.2536	**0.5307**	0.1915	0.1609	0.2243

The Confusion Matrix suggests that the model completely fails to correctly predict the two "Very High" incidents of PM10 concentration values, suggesting "Medium" and "High" respectively. In addition, out of the 78 cases of actual "High" values, the model correctly classifies 25 of them, a percentage of 32%. Concerning medium values this percentage increases to 43%, while in the case of "Low" values (which are the most common ones), the correctly classified percentage reaches 81%. Overall, the models seems to be unable to predict the rarest of the events.

Table 4. The confusion matrix [11] of best random committee model.

		Forecasted values			
		Low	Medium	High	Very high
Real values	Low	**16539**	111	3	0
	Medium	384	**313**	23	1
	High	25	27	**25**	1
	Very high	0	1	1	**0**

5   Conclusions

In the current paper, we address the problem of air quality forecasting for the GGA via a regression and a classification approach, by developing area-specific models for each one of the monitoring stations under investigation. The best performance achieved in the case of regression comes from the Random Forest algorithm, reaching a correlation coefficient of r = 0.8. In addition, we attempted to develop a common model for all the monitoring stations, making use of the ensemble principle, and employing the recalculation of weights in an Artificial Neural Network. Results indicate that it is possible to improve the prediction ability of the best model developed per station, via the prediction ability of the common model based on an ensemble approach. This suggest that city authorities are able to expect good AQ model performance from a generalized AQ model, which has the advantage of easy tailoring to the decision making process. When coming to the classification approach, model performance was deemed as satisfactory for some of the cases, nevertheless bounded by the fact that events of high air pollution are rare in comparison to the overall air quality of the area, and thus difficult to depict via an AQ modeling classification approach. Overall, the paper contributes in the area of data-driven, smart city oriented AQ forecasting for operational purposes producing results that can be directly used by relevant city authorities.

References

1. Riffat, S., Powell, R., Aydin, D.: Future cities and environmental sustainability. Future Cities Environ. **2**, 1 (2016). doi:10.1186/s40984-016-0014-2
2. Webel, S.: Forecasting Software that's a Breath of Fresh Air. Pictures of the Future Siemens Magazine (2016). http://www.siemens.com/innovation/en/home/pictures-of-the-future/infrastructure-and-finance/smart-cities-air-pollution-forecasting-models.html

3. Kukkonen, J., Olsson, T., Schultz, D.M., Baklanov, A., Klein, T., Miranda, A.I., Monteiro, A., Hirtl, M., Tarvainen, V., Boy, M., Peuch, V.-H., Poupkou, A., Kioutsioukis, I., Finardi, S., Sofiev, M., Sokhi, R., Lehtinen, K.E.J., Karatzas, K., San José, R., Astitha, M., Kallos, G., Schaap, M., Reimer, E., Jakobs, H., Eben, K.: A review of operational, regional-scale, chemical weather forecasting models in Europe. Atmos. Chem. Phys. **12**, 1–87 (2012)

4. Karatzas, K., Kaltsatos, S.: Air pollution modelling with the aid of computational intelligence methods in Thessaloniki, Greece. Simul. Modell. Pract. Theory **15**(10), 1310–1319 (2007)

5. Juda-Rezler, K., Trapp, W., Reizer, M.: Modelling the impact of climate changes on particulate matter levels over Poland. In: Steyn, D.G., Rao, S.T. (eds.) Air Pollution Modeling and its Application XX, pp. 509–514. Springer, Netherlands (2010)

6. Szczepaniak, K., Astel, A., Bode, P., Sârbu, C., Biziuk, M., Raińska, E., Gos, K.: Assessment of atmospheric inorganic pollution in the urban region of Gdańsk. J. Radioanal. Nucl. Chem. **270**(1), 35–42 (2006)

7. Port of Gdańsk: Cargo Logistics (2016). https://www.portGdansk.pl/pg-eksploatacja-en

8. Hall, M., Frank, E., Holmes, G., Pfahringer, B., Reutemann, P., Witten, I.: The WEKA data mining software: an update. SIGKDD Explor. **11**(1), 10–18 (2009)

9. Breiman, L.: Random forests. Mach. Learn. **45**(1), 5–32 (2001)

10. Kohavi, R.: A study of cross-validation and bootstrap for accuracy estimation and model selection. In: Proceedings of the Fourteenth International Joint Conference on Artificial Intelligence, vol. 2, no. 12, pp. 1137–1143 (1995)

11. Witten, I.H., Frank, E., Hall, M.A.: Data Mining – Practical Machine Learning Tools and Techniques. Morgan Kaufmann, Burlington (2011)

12. Cohen, J.: A coefficient of agreement for nominal scales. Educ. Psychol. Measur. **20**(1), 37–46 (1960)

13. Altman, D.G.: Practical Statistics for Medical Research. Chapman and Hall, London (1991)

Analysis of Image, Video and Motion Data in Life Sciences

Ethnicity Distinctiveness Through Iris Texture Features Using Gabor Filters

Gugulethu Mabuza-Hocquet[1,2(✉)], Fulufhelo Nelwamondo[1,2],
and Tshilidzi Marwala[2]

[1] Modelling and Digital Science, Council for Scientific and Industrial Research,
Pretoria, South Africa
{gmabuza,fnelwamondo}@csir.co.za
[2] Engineering and the Built Environment, University of Johannesburg,
Johannesburg, South Africa
tmarwala@uj.ac.za

Abstract. Research in iris biometrics has been focused on utilizing iris features as a means of identity verification and authentication. However, not enough research work has been done to explore iris textures to determine soft biometrics such as gender and ethnicity. Researchers have reported that iris texture features contain information that is inclined to human genetics and is highly discriminative between different eyes of different ethnicities. This work applies image processing and machine learning techniques by designing a bank of Gabor filters to develop a model that extracts iris textures to distinctively differentiate individuals according to ethnicity. From a database of 30 subjects with 120 images, results show that the mean amplitude computed from Gabor magnitude and phase provides a correct ethnic distinction of 93.33% between African Black and Caucasian subjects. The compactness of the produced feature vector promises a suitable integration with an existing iris recognition system.

Keywords: Iris segmentation · Soft biometrics · Gabor filters · Iris texture extraction · Ethnic distinction

1 Introduction

The collective body of research in iris biometrics has been focused mainly on extracting minute iris features in order to generate an iris code that is used to uniquely recognize or identify an individual from a large database of enrolled individuals [1–3]. The iris recognition system (IRS) technology has evolved over the years such that its application and deployment has been witnessed in Aadhaar; India, to grant and distribute governmental benefits, since 2011. This technology is also used in the United Arab Emirates for security border control since 2003 to date. A classic IRS requires a person to be enrolled before any form of identification and verification can take place. Although the recognition system is automated, the capturing of other attributes such as ethnicity and gender is

© Springer International Publishing AG 2017
N.T. Nguyen et al. (Eds.): ACIIDS 2017, Part II, LNAI 10192, pp. 551–560, 2017.
DOI: 10.1007/978-3-319-54430-4_53

still done manually. Automated ethnic and gender detection and classification through iris images, are still the most crucial problems of computer vision in the research space [4]. So far, very little work has been done in using iris features to determine other soft biometrics such as ethnicity and gender.

The work presented here focuses on the extraction and analysis of texture features from iris images in order to distinctively differentiate the ethnicity of two racial groups. Ethnicity is a form of attribute referred to as a soft biometric. The objective of this work is not person recognition but rather; the integration of personal attributes that further give a specified description of an individual. The motivation is that; out of all available biometric modalities, the human iris comprises of informational features that span out to more than 200° of freedom amongst which can be effectively used to determine ethnicity. It is still considered the most stable, reliable and most accurate biometric throughout a persons lifetime [1,2]. Also, the use of iris features for individual identification is already well researched and deployed [1–3]. However the existing IRS methods have not been designed to determine soft biometrics such as ethnicity, for instance. The development of the proposed model will enhance and contribute to the existing IRS by (i) providing an automated method for capturing soft biometric data from the iris. (ii) Equipping the system with the competence and processing speed to perform a narrow and specified iterative search for an individual from a large database, thereby reducing the average search time. (iii) Offering the system the capability to automatically collect such data even before an individual is officially enrolled.

Currently, an IRS simply returns a non-match as a response for unenrolled persons, the proposed model however, would at least be able to automatically provide the basic demographic information about an individual merely through eye image scanning. This means while a person is enrolled, the soft biometric data will serve as an accuracy measure and confirmation of the information presented by the enrolling person. Applications of the proposed model will also be fitting in spotting imposters disguised under different genders or in the collection of statistical data without the need for person identification.

The remainder of this work is arranged as follows: Sect. 2 discusses related work on ethnic prediction using iris images. Section 3 presents the proposed method. Section 4 discusses experimental results and analysis. Section 5 is the conclusion.

2   Related Work

In 2006, Qui et al. [5] used the global texture information of iris images to develop a novel ethnic classification method. Their work argues that the texture of the iris is related to race. The experiment uses the publicly available iris databases from the Chinese Academy of Sciences Institute of Automation (CASIA v2), University of Palackecho and Olomopuc (UPOL) and University of Beira Interior (UBIRIS). The CASIA database consists of 2400 Asian eye images, the UPOL and UBIRIS respectively consists of 384 and 1198 European eye images. The utilised ethnic classification algorithm is based on three modules, namely; image pre-processing, global feature extraction and training. Image pre-processing involves the localization of the iris within the image, normalization and the elimination of eyelashes and eyelids. Only $\frac{3}{4}$ of the inner lower normalized iris is selected as the region of interest (ROI), from which features are extracted. The ROI is further equally divided into two regions referred to as region A and B respectively. The second module, extracts global features by convolving the ROI with multichannel 2D Gabor filters. The outputs of the even and odd symmetric Gabor filters are combined into a single quantity known as the Gabor Energy (GE). They report that region A has a rich texture for Asians than region B. For non-Asians however, both regions are reported to closely have the same rich texture. The training set with 1200 images has 600 random images from Asians and non-Asians respectively. The remaining 2782 images are used for testing. Experimental results report that using the GE features only, achieves a correct ethnic classification rate (CCR) of 79.44%. Also that using features from the Gabor energy ratio (GER) achieves 84.95% CCR. However, the combination of both the GE and GER features achieves an increased overall CCR of 85.95%. They blame most of the classification errors on the noisiness of the images, the ROI being affected by eyelid and eyelash occlusions as well as the illumination differences between the images from the different databases. From the results, Qui et al. still maintain that the global features of the iris seem to be genetically dependent. Also that on a larger scale, global features are similar for a specific race and therefore competent for ethnic classification.

Later, Qui et al. [6] continued the research in this space with a different approach from the previous work. Within the continued research, a novel and automated method to classify ethnicity based on analysing or learning the primitive appearance of iris images is proposed. With the previous notion, the continuation of their research is based on investigating the discriminative types of visual primitives from iris images that make Asians to be different from non-Asians. For the continued research work, the compact visual primitives of iris images are represented by micro structures also called Iris Textons. The same three module procedure as the previous work is conducted followed by learning the vocubulary of iris Textons. This is done by convolving the ROI with a 2D Gabor filter, where only 40 of the even symmetric Gabor filters; constituting of 8 orientations and five scales, are used. Employing K-means clustering algorithm, response vectors from the filtered images are further clustered into small sets of prototypes within which all the K-centers of each cluster are considered as iris Textons. The remaining filter response vectors are considered as appearance vectors. During the learning phase, 400 images respectively consisting of 200 images randomly selected from Asians and non-Asians are used. They obtain a total of 64 iris Textons. The mean of the vectors in a cluster represents each iris texton. Histograms of the iris texton are used to represent the visual appearance of the iris images. 2400 images from 60 subjects are used to evaluate their algorithm with 1200 images from CASIA database and the other half obtained in France from European subjects. The images are further divided into two sets; a training set of 1200 random images, with 600 from Asians and non-Asians respectively. The remaining 1200 images are used for training. From the results, an increased ethnic classification overall accuracy of 91.02% between Asian and non-Asians is achieved using the newly proposed method as compared to their previous work which achieved 85.95%.

Another work that proposes the analysis and possible use of iris texture features to predict ethnicity is the work of Lagree and Bowyer [7]. Their motivation is based on the assumption that a particular ethnic group might have the same iris texture features which differ from one ethnic group to another. The dataset of their study consists of 1200 images from the University of Notre Dame. From the 1200 images, 600 images respectively represent Caucasian and Asian subjects from 60 different subjects, with 5 images from each left and right eye of an individual. Their work proposes to eliminate the challenge of biased result by randomly dividing the whole dataset into 10 folds of 120 images each. For iris segmentation, the IrisBEE software from Notre Dame is used. Texture features are computed by applying six different filters at every non-masked pixel location of the normalized image. In addition to the six filters, two more filters referred to as S5S5 and R5R5 are created using Laws' texture measures. The added filters are convolved with the normalized image to obtain several types of textures responses. A feature vector that describes the texture of an image is computed by dividing the normalized image into ten 4-pixel horizontal bands and four 60-pixel vertical bands of neighbouring pixels. From each image, 630 features are obtained together with 5 calculated statistics from all 14 regions. The experimental results show that the Sequential Minimal Optimization (SMO) algorithm used for building a support vector machine classifier achieves the higher accuracy of 90.58% compared to other classifiers.

A year later [4] advanced their research in pursuit of using iris textures to consider the prediction of both ethnicity and gender. The recent work presents additional results on ethnicity prediction from their previous work. The difference or improvement to their previous work is the application of three filters that represent Laws' texture measures to the normalized image instead of two. The computation of texture features spanning from the boundary of the pupil and iris and out to the sclera is performed for eight 5-pixel horizontal bands and ten 24-pixel vertical bands of the normalized image. For each of the eight horizontal regions of the image, they compute six summary statistics for each of the nine basic texture filters. The calculated statistics give 432 features. They further calculate the first five summary statistics for the ten 24-pixel vertical bands to obtain 450 features, resulting to a total of 882 features. The experiment uses the same number of subjects and images per person as their previous work. Person disjoint and 10 fold cross validation is used to randomly divide the dataset into 10 folds, where each fold contains images for 6 persons of Asians and Caucasians. The person disjoint is done so that the train and test data do not contain the same images from the same person. From using the same SMO-SVM algorithm from the Weka package with default parameters, a 90.58% correct ethnic prediction is achieved. This is exactly the same result obtained in the previous work. However, running the experiment without any person disjoint enforced, they achieve an increased accuracy of 96.17%, which is a 5% increased ethnic prediction from their previous work. Now, the 5% increase achieved in the recent work might be due to the possible utilization of the same images from the same person and ethnicity within the training and testing data.

3 Proposed Method

Observation from the related work is that, so far, research focused on using iris textures to determine ethnicity, has only been conducted using Asian and Caucasian subjects from Asia and Europe respectively. This work focuses particularity on, Black males and Caucasian females within the African continent. The aim is to investigate the measure of distinction in iris textures by employing a bank of Gabor filters. The method proposed in this work is summarized by the block diagram shown by Fig. 1.

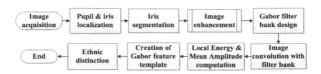

Fig. 1. Proposed method flow diagram

3.1 Image Acquisition, Segmentation and Enhancement

Due to the unavailability of an iris database for African subjects, all images used in this work have been self acquired under an uncontrolled environment from consenting participants consisting of 15 Black males and 15 White (Caucasian) females. Four images; constituting of two left and two right eyes were acquired from each participant using the Vista EY2 dual iris and face camera. The camera uses near-infrared illumination and produces 8-bitmap images of dimensions 640×480 pixels. The images are first converted to gray scale before any processing occurs. Bresenham's circle algorithm [8]; an algorithm based on the midpoint circle algorithm, is used to compute pupil and iris respective centers and to localize the pupil-iris boundaries. Iris segmentation is performed using the Chan-Vese algorithm [9]. This method traverses the spatial domain of an image to find a curve that appropriately separates an object of interest from the rest of the image. The method of Chan and Vese is "the minimization of energy based segmentation"; where the addition of regularizing terms minimizes the fitting term. The algorithm first defines an evolving curve C in Ω, as the boundary of an open subset ω of Ω. The region inside and outside the curve is respectively denoted by ω and $\Omega/\overline{\omega}$. The assumption is that an image u_0 is formed by two regions of distinct values u_0^i and u_0^o with piecewise-constant intensities. $u_0 \approx u_0^i$

and $u_0 \approx u_0^o$ respectively fall inside and outside the object boundary C_0. The algorithm adds the length of the curve C and the area of the region inside C as regularizing that are introduced to the energy functional to become:

$$F(c_1, c_2, C) = \mu \cdot Length(C) + \nu \cdot Area(inside(C))$$
$$+ \lambda_1 \int_{inside(C)} |u_0(x, y - c_1)|^2 dxdy + \lambda_2 \int_{outside(C)} |u_0(x, y - c_2)|^2 dxdy \quad (1)$$

Where $\mu \geq 0$, $\nu \geq 0$, λ_1, $\lambda_2 > 0$ are fixed parameters, constants c_1 and c_2 are averages of the image inside and outside C. As additional regularizing terms, this work uses the diameter, radii and centers of the pupil and iris respectively; parameters computed by Bresenham's algorithm. Not only do the parameters determine the boundaries C_0 of the initial contours but also influence the discontinuing of the evolving curve C, during the segmentation of the iris.

Image enhancement involves processing an input digital image by improving its quality so that the final output is more appropriate than the original image for use in specific applications [10]. The segmented iris is enhanced in two folds; by sharpening, in order to increase the contrast using the standard deviation of the Gaussian low pas filter and using contrast limited adaptive histogram equalization (CLAHE). CLACHE enhancement focuses on smaller and visually attractive regions of the iris as opposed to the whole image. The enhanced images are finally normalized to an equal size of 123×127.

3.2   Gabor Filter Design for Iris Texture Extraction

Before any texture features can be extracted from the enhanced iris, a bank of Gabor filters is designed at different wavelengths and orientations. A bank of Gabor filters employs frequency and orientation representations which make them effective for texture representation and discrimination [11]. In the spatial domain, a 2D Gabor filter is a Gaussian kernel modulated by a sinusoidal plane wave [12] and is represented as follows:

$$h(x, y) = \frac{1}{2\pi\sigma_x\sigma_y} \exp\left\{-\frac{1}{2}\left[\frac{x^2}{\sigma_x^2} + \frac{y^2}{\sigma_y^2}\right]\right\} \exp\left\{j2\pi Fx\right\}. \quad (2)$$

Where σ_x/σ_y is the Gaussian aspect ratio, the complex exponential has F and θ as the spatial frequency and orientation respectively. The response of the filter captures both the magnitude and phase in the entire frequency spectrum. Detailed reading on Gabor filters may be found in [11–13]. Depending on the specific implementation of Gabor filters, the tuning frequency or tuning period or wavelength establishes the kind of sinus wave that the filter will respond best. For this work, we design a Gabor filter array using three wavelengths (λ) of 3, 5, and 7 pixels per cycle as well as five orientations (θ) of 0, 30, 60, 90, 120 degrees. The choice of parameters is motivated by running a few trials of how dense we want the sinus wave (modulated with the input signal) to be. Since by varying the orientation, we can look for texture oriented in a particular direction.

This means smaller wavelengths give a denser sinus wave and larger wavelengths result in larger waves. The designed filter bank consists of 3 wavelengths × 5 orientations to produce a Gabor array of size 1 × 15. The resulting Gabor array is then convolved with the enhanced iris image at each wavelength and orientation. This process produces the magnitude and phase of the Gabor filter bank for the input image. From the magnitude and phase components we compute the mean amplitude (MA) and Local Energy (LE) as shown by Eqs. (14) and (15) in [13], to use as texture features.

4   Experimental Results and Analysis

The results are presented in the order of the obtained outputs from the segmentation stage to ethnicity prediction on the flow diagram.

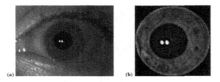

Fig. 2. Input image with segmented iris

Figure 2 shows iris segmentation from the input image using the Chan-Vese algorithm. Compared to the traditional integro differential equation used for iris segmentation by Daugman, the Chan-Vese algorithm performs faster, produces accurate and reliable results and is computationally inexpensive. Furthermore, enhancing the segmented iris highlights the smaller texture details of the iris edges needed for extraction.

Figure 3 shows the filter outputs of the designed Gabor envelope of size 1 × 15. Each filter output is convolved with the iris image shown at each specified wavelength and orientation in Fig. 4. The total magnitude and phase response of the designed filter bank for each iris image is shown in Fig. 5. This complex response is used to compute the Local Energy (LE) and Mean Amplitude (MA) to be used as iris texture features. The computed LE and MA of size 1 × 15 each are further horizontally concatenated to finally produce a feature vector of size 1 × 30 per individual. The resulting feature vector is further normalized and scaled through the computation of the mean and standard deviation. This means; we compute and scale the mean and standard deviation for all subjects per each of the 30 features belonging to one ethnic group. This results in a reduced feature dimension for both ethnicities. We started with a smaller number of 9 subjects consisting of 6 Black males and 5 White females. From the achieved results, two observations are made; (i) Blacks always fall on the negative of the z-plane while Caucasians fall on the positive. (ii) The distinction between both ethnicities is most clearly witnessed at the lower wavelengths of only the mean amplitude features. After

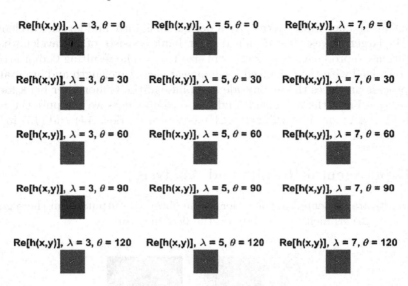

Re[h(x,y)], $\lambda = 3$, $\theta = 0$ Re[h(x,y)], $\lambda = 5$, $\theta = 0$ Re[h(x,y)], $\lambda = 7$, $\theta = 0$

Re[h(x,y)], $\lambda = 3$, $\theta = 30$ Re[h(x,y)], $\lambda = 5$, $\theta = 30$ Re[h(x,y)], $\lambda = 7$, $\theta = 30$

Re[h(x,y)], $\lambda = 3$, $\theta = 60$ Re[h(x,y)], $\lambda = 5$, $\theta = 60$ Re[h(x,y)], $\lambda = 7$, $\theta = 60$

Re[h(x,y)], $\lambda = 3$, $\theta = 90$ Re[h(x,y)], $\lambda = 5$, $\theta = 90$ Re[h(x,y)], $\lambda = 7$, $\theta = 90$

Re[h(x,y)], $\lambda = 3$, $\theta = 120$ Re[h(x,y)], $\lambda = 5$, $\theta = 120$ Re[h(x,y)], $\lambda = 7$, $\theta = 120$

Fig. 3. Gabor envelope at different wavelenghts and orientations

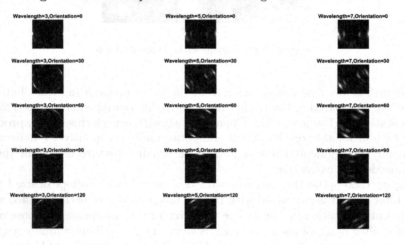

Fig. 4. Iris image convolved with Gabor array

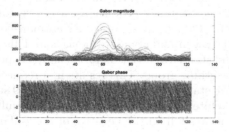

Fig. 5. Gabor magnitude and phase for one iris image

adding more subjects to have a total of 15 Blacks and Whites respectively, the stated observations still hold, with an error rate (ER) of 6% (Fig. 6).

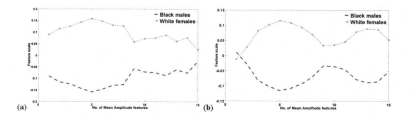

Fig. 6. Ethnicity distinction using Gabor mean amplitude features

Table 1. Method comparison

Authors	Database	Ethnicity	Technique	CCR
Qui et al.	CAS, UP, UB	Asian, non-Asian	Gabor filters	89.95%
Qui et al.	CAS, UP, UB	Asian, non-Asian	Iris Textons	89.95%
Lag, Bow	Notre Dame	Asian, Caucasian	9 Filters	90.58%
Proposed method	Independent	Black, Caucasian	Gabor filters	93.33%

5   Conclusion and Recommendations

This work presented an automated method for extracting and analysing useful iris texture features to determine the distinction between Black and Caucasian ethnic groups. The main objective of this work is not to address ethnic classification but rather to analyse Gabor features that efficiently provide differentiation between two racial groups. The design of a bank of Gabor filters with 3 wavelengths and 5 orientations convolved with enhanced iris images has been used to capture the global texture features from iris images. Through the computation of LE and MA as texture features from the Gabor magnitude and phase components; two observations are made (i) using the mean amplitude achieves a clear distinction between the Black and Caucasian subjects. (ii) The distinctive information between the two races lies on the lower wavelengths of the Gabor bank. Table 1 compares the CCR performance of the proposed method with those in Sect. 2. The proposed method shows an improvement of 2.75% from the work of Lagree ad Bowyer, however, discoveries articulated by the researchers confirm that iris texture is indeed related to race. Furthermore, ethnic distinction lies on the coarse scale texture features rather than on minute local features used for iris code generation in iris recognition. We therefore recommend the use of smaller wavelength values stretched between 3 and 5 pixels per cycle to achieve maximum ethnic distinction.

References

1. Daugman, J.: How iris recognition works. IEEE Trans. Circuits Syst. Video Technol. **14**(1), 21–30 (2004)
2. Wildes, R.P.: Iris recognition: an emerging biometric technology. Proc. IEEE **85**(9), 1348–1363 (1997)
3. Ma, L., Tan, T., Wang, Y., Zhang, D.: Efficient iris recognition by characterizing key local variations. IEEE Trans. Image Process. **13**(6), 739–750 (2004)
4. Lagree, S., Bowyer, K.W.: Predicting ethnicity and gender from iris texture. In: Technologies for Homeland Security (HST), pp. 440–445. IEEE (2011)
5. Qiu, X., Sun, Z., Tan, T.: Global texture analysis of iris images for ethnic classification. In: Zhang, D., Jain, A.K. (eds.) ICB 2006. LNCS, vol. 3832, pp. 411–418. Springer, Heidelberg (2005). doi:10.1007/11608288_55
6. Qiu, X., Sun, Z., Tan, T.: Learning appearance primitives of iris images for ethnic classification. In: 2007 IEEE, ICIP, vol. 2, p. II-405 (2007)
7. Lagree, S., Bowyer, K.W.: Ethnicity prediction based on iris texture features. In: MAICS, pp. 225–230 (2011)
8. Bresenham, J.: A linear algorithm for incremental display of circular arcs. Commun. ACM **20**(2), 100–106 (1977)
9. Chan, T., Vese, L.: Active contour models without edges. IEEE Trans. Image Process. **10**(2), 266–277 (2001)
10. Maini, R., Aggarwal, H.: A comprehensive review of image enhancement techniques. J. Comput. **2**(3), 8–12 (2010)
11. Haghighat, M., Zonouz, S., Abdel-Mottaleb, M.: Identification using encrypted biometrics. In: Wilson, R., Hancock, E., Bors, A., Smith, W. (eds.) CAIP 2013. LNCS, vol. 8048, pp. 440–448. Springer, Heidelberg (2013). doi:10.1007/978-3-642-40246-3_55
12. Clausi, D.A., Jernigan, M.E.: Designing Gabor filters for optimal texture separability. Pattern Recogn. **33**(11), 1835–1849 (2000)
13. Zheng, D., Zhao, Y., Wang, J.: Features extraction using a Gabor filter family. In: Proceedings of the Sixth Lasted, ICIP (2004)

Diminishing Variant Illumination Factor in Object Recognition

Ardian Yunanto[1(⊠)] and Iman Herwidiana Kartowisastro[2(⊠)]

[1] School of Computer Science, Bina Nusantara University,
Jl KH Syahdan 9, Jakarta 11480, Indonesia
ayunanto@binus.edu
[2] Doctoral Study Program in Computer Science, Bina Nusantara University,
Jl KH Syahdan 9, Jakarta 11480, Indonesia
imanhk@binus.edu

Abstract. Undetected object(s) from a camera due to a poor condition of light intensity, or shadow that appears are problems that can occurs in object detection. This can lead to a loss, especially when applied in the industrial world. The purpose of this research is to fix an illumination factor, particularly the shadow factor on an image that will be detected by combining two methods, namely adaptive single scale retinex and shadow removal. Smoothing from retinex and shadow removal process are performed after an image is captured. Accuracy of object detection obtained is 95.45%, using experimental image detection program and random sampling method from 22 images of two datasets used in this study. Namely "Shadow Removal Online Dataset and Benchmark for Variable Scene Categories" and "Klik BCA" which obtained from the simulation process. This method can be applied to real time conditions, where the speed of the process is stable and fast enough such that it can be applied into industrial companies to help their quality control.

Keywords: Image processing · Illumination · Shadow detection

1 Introduction

To make a computer able to detect an object, face or classify it as a common object recognizable by humans are the goals of object detection. Recently, there are several researches for optimizing the detection result. One of the research is about optimization of space object identification problem. They proposed a Multiview and Multiscale method to overcome the space object identification problem at very far distance [2]. In our best knowledge, it's quite difficult to get accurate object detection results when it is applied in accordance to the real-time situation. For example, there is a ball object in a room with a very high or low luminance factor and the surface of the ball can reflect light. The robot cannot easily detect the ball since the camera will almost likely take only the bright part or assumes there is no object passing through the track whenever the object covered by shadows. It can also be risky when the systems are applied where it can cause financial loss to an industrial company.

© Springer International Publishing AG 2017
N.T. Nguyen et al. (Eds.): ACIIDS 2017, Part II, LNAI 10192, pp. 561–571, 2017.
DOI: 10.1007/978-3-319-54430-4_54

To overcome the problem, one of solutions are by improving the quality of the camera or the accuracy of the algorithm [1]. Through this work, a new method is proposed by modifying & combining existing research methods, namely hybrid adaptive single scale retinex and shadow removal method. These methods are chosen to diminish the factor of the shadow on the object to be detected. The scope of research is around image filtering method in object recognition.

This research is conducted to answer two problems: How to diminish existing illumination and shadow factors on captured image until computer can properly detect it, and is the proposed method will be more accurate or faster compared to the previous research. Contribution of this study is to minimize the shadow factors. Hence, the benefits of this research are to speed up the algorithm process of the camera. Industrial company that uses robots can avoid the risk of loss due to the problems of the object detection in any lighting condition.

2   Literature Review

This section discusses previous research around object recognition, image filtering, image filtering: Adaptive Single Scale Retinex, image filtering: shadow removal.

2.1   Object Recognition

Object recognition nowadays are applied to a wide variety of technologies. For instance, CCTV [17], object monitoring tool [2, 5, 14, 16], the operation of industrial equipment [13], medical devices [7], even for artificial robot [17]. Many researchers have further optimized object recognition methods in terms of reduction of illumination factor [2–9, 22, 23], removal of shadow [10], removal of noise in the picture [11, 19], pattern recognition [14, 15, 20, 21, 24, 26], maximization of dataset comparator [18, 25, 27] and optimization of segmentation algorithm. The goals of researchers in developing object recognition are to speed up the process from a computer to recognize objects by size, color, shape, and so forth [1].

2.2   Image Filtering

The concept of this method is to fix contrast value from an image, so the quality of the image is enhanced [1, 3, 4, 8, 9]. The basic method to do that is modification of the histogram by stretches the gray values from each pixel in an image [12].

Image Filtering: Adaptive Single Scale Retinex
This method is created by combining two discontinuity measure methods for weight functions, namely gradient and local spatial inhomogeneity. Only these two-discontinuity measure from overall steps are highlighted because those two methods will be modified. To get the weight value of the spatial gradient, the equation used is as follows:

$$\nabla I(x,y) = \left[G_x, G_y\right] = \left[\frac{\partial I(x,y)}{\partial x}, \frac{\partial I(x,y)}{\partial y}\right] \tag{1}$$

Where I is the image with a size of x and y. G_x and G_y are the derivative from I and described in equation below:

$$G_x = I(x+1,y) - I(x-1,y) \tag{2}$$

$$G_y = I(x,y+1) - I(x,y-1) \tag{3}$$

After calculating the two coefficients of the gradient, the magnitude of its gradient vector is determined using this equation:

$$|\nabla I(x,y)| = \sqrt{G_x^2 + G_y^2} \tag{4}$$

The second method is local inhomogeneity, where the weaknesses of previous research as stated in [9] is located at here. The weakness of the local inhomogeneity is the calculation time process that takes too long. The concept is by calculating summarize of all average value of each pixels. Each pixel value is calculated from average value between the pixel with the pixel around it.

Image Filtering: Shadow Removal
Shadow is formed on the obstruction of light sources (direct light and environment) that leads to the object. There are four different additional methods in [10] that can be chosen to diminish the shadow region in an image. First, the additive method, where after the shadow region is detected, the program will compute and add the smooth mask that contains the difference value between shadow area and non-shadow area. In basic light model, luminance ratio in shadow and non-shadow area is computed. Then, each pixel's value is added by smooth mask value that contained by the luminance ratio. In the advanced light model, the luminance of direct and global light is computed. Afterwards, smooth mask is added in the image. Last is YCbCr, where an image is converted first from Red-Green-Blue (RGB) into YCbCr format. Additive method is used and the result is converted back to the RGB format [10].

3 Method

Research framework, methodology, and evaluation analyses are described in this section.

3.1 Research Framework

The overall of detection algorithm are explained. First is image snapshot simulation. Computer is used and attached to the camera via USB cable. Next, the object and the lamp is placed near each other. The lamp is turned on and adjusted based on the desired parameter. Industrial robot technology's parameter constraints such as camera range is also used as conditional variables for this research. To get the desired light parameter, light meter is used. Next is application of Object Recognition Algorithm.

```
Algorithm 1: Object Recognition Algorithm
Output: Detected Image by showing the polygon

      1.      Choose the camera that will be used.
      2.      Get dataset image of the object that will be detected
      3.      Convert the dataset into grayscale.
      4.      Get the feature point from the dataset using Speeded Up Robust
              Features (SURF)
      5.      While true
      6.      Snapshot the image taken from the camera and convert it to gray-
              scale
      7.      Filter the image using adaptive single scale retinex and shadow
              removal (Algorithm 2)
      8.      Get the filtered image feature point using SURF
      9.      Match the feature points between dataset and processed image.
      10.     If there is a matrix consist of the matched feature, then
      11.             Create a polygon to be shown at the image
      12.             Draw the line of the polygon.
      13.     Else
      14.             Continue to next snapshot
      15.     End While
```

Infinite loop is used to simulate the real process of the system, where a camera will continue to capture the image, then filter it using the proposed method.

Image Filtering: Hybrid Adaptive Single Scale Retinex and Shadow Removal Algorithm

Proposed method is applied by combining adaptive single scale retinex and shadow removal algorithm. It uses the algorithm core of adaptive single scale retinex whilst shadow removal algorithm is attached to the core.

```
Algorithm 2: Hybrid Adaptive Single Scale Retinex and Shadow Removal
Algorithm
Output: filtered image

      1.      Initialization operations
      2.      Compute Spatial Gradient in x and y directions
      3.      Start iterative convolution
      4.      Produce illumination invariant representation of input image
      5.      Call the Shadow Removal Function
```

3.2   Methodology

Some adjustments in the adaptive retinex were added. The first adjustment is applied in spatial gradient. New parameter k is applied to those equations:

$$G_x = k(I(x+1,y) - I(x-1,y)) \tag{5}$$

$$G_y = k(I(x,y+1) - I(x,y-1)) \tag{6}$$

This adjustment is conducted to see if the parameter has impact in spatial gradient. There are four types of approaches in a gradient filter [17]. First is the Basic Derivative Filter [9]. Second is Prewitt Gradient Filters, where the parameter k is set to 1/3. Third is Sobel Gradient Filters, where the parameter k is set to 1/4. Lastly is the Alternative Gradient Filters. Where the value of k freely determined. All the parameter k has the requisite range, between 0 and 1. Based on empirical observation, the optimal value for this case is around 0.5. Next, is modification of local inhomogeneity. Due to the nested looping of the algorithm, it will be contradicted with the purpose of this research. Therefore, to reduce the timing process, the value will be adjusted manually between 0 and 1:

$$\tau(x,y) = X \sim UNI(0,1), X \in R \tag{7}$$

Last adjustment is in shadow removal algorithm. Based on our empirical observation, the optimal method that will be used is advanced light models.

3.3 Evaluation Analyses

To measure the effectiveness of proposed method, it was compared with original adaptive single scale retinex [9] and shadow removal [10] using the same tool, data set and the object that will be detected. Effectiveness is measured in term of accuracy and computation time. Mean time process and standard deviation are also calculated and compared with the single scale retinex and shadow removal. Last analysis is the accuracy of the proposed method.

4 Results and Discussion

Two dataset were used in this work. First, simulation process to create dataset which conditioned to the goal of this research. The object used as an experiment is 'Klik BCA'. The dataset is divided to seven type of luminance factor units (Lux), namely 50, 100, 200, 400, 600, 800, and 1000 lx. Each luminance factor unit dataset contains set of image with the same object but different angles $(45^o, 90^o, 135^o, 180^o, 225^o, 270^o, and\ 315^o)$ (Fig. 1).

Next, using online benchmark dataset in accordance with similar studies conducted by previous researcher. Benchmark dataset used in this research is taken from Shadow Removal Online Dataset and Benchmark for Variable Scene Categories [28]. The dataset is collection of 214 images with various conditions (Fig. 2).

Ten images are used as experimental samples. Size of the shadow of the pictures is taken as consideration in selecting the candidate of the sample. The aim is to test how effective the shadow removal algorithms on diminishing the shadow region.

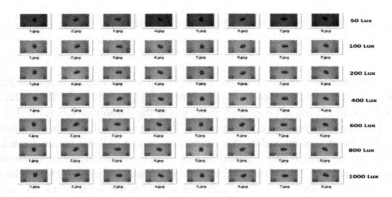

Fig. 1. Klik BCA dataset from simulation tools

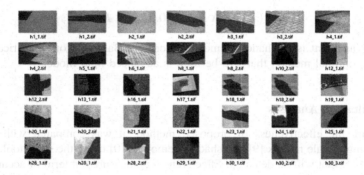

Fig. 2. Shadow removal online dataset and benchmark for variable scene categories [28]

Dataset 'Klik BCA' is used to test whether the program can detect objects as well using the object recognition algorithm. We label an image as detected or not detected per the feature match process. Polygon will be drawn if the label is detected (Fig. 3).

Fig. 3. The detection program test

The results obtained from the experiment are analyzed. First is an analysis of the proposed method's computing time process using 'Klik BCA' dataset.

56 images are taken from the dataset "Klik BCA" and its time span is recorded for each one image on every lux. If we look at the table, 50 lx give faster computation speed, especially at image ID 6, 7, 8. Other than that, different values between lux are just random calculation matters.

From Tables 1 and 2, results of mean and standard deviation are similar with the proposed method, while the difference is located at the standard deviation. The comparison method has higher standard deviation, meaning that the computational process is more unstable than the proposed method.

Table 1. Time process evaluation of proposed method using 'Klik BCA' (in milliseconds)

Time process (millisecond)							
Image ID	50 Lux	100 Lux	200 Lux	400 Lux	600 Lux	800 Lux	1000 Lux
1	790.633	809.09	892.582	952.788	881.131	886.95	820.802
2	713.05	777.622	809.235	800.805	758.055	812.307	834.945
3	776.153	741.21	835.978	755.038	776.84	776.63	827.902
4	656.778	751.512	770.745	815.648	875.335	851.613	788.544
5	645.581	751.73	764.039	793.909	783.042	957.997	796.005
6	538.382	742.433	822.235	778.874	738.992	815.847	807.806
7	539.166	736.821	720.098	790.695	762.776	799.807	792.409
8	505.396	736.18	765.096	828.887	795.483	793.359	803.541
Mean	645.642	755.825	797.501	814.581	796.457	836.8138	808.994
Std dev	110.346	25.3198	53.6486	60.1367	53.2763	60.16421	17.1694

Table 2. Time process evaluation of shadow removal [10] method using 'Klik BCA'

Time process (millisecond)							
Image ID	50 Lux	100 Lux	200 Lux	400 Lux	600 Lux	800 Lux	1000 Lux
1	1300.862	852.243	893.705	871.561	849.786	862.679	856.409
2	865.117	783.701	770.341	855.647	739.506	771.22	758.48
3	847.501	804.239	774.71	765.319	735.187	762.359	697.245
4	864.649	806.737	772.598	743.096	730.325	747.223	730.556
5	855.926	785.54	788.663	795.388	733.819	767.302	728.331
6	757.774	806.981	750.879	791.7	753.049	786.302	735.612
7	806.469	790.381	755.666	776.168	818.113	793.17	739.45
8	825.16	744.196	811.67	795.766	819.167	766.447	781.411
Mean	890.4323	796.7523	789.7790	799.3306	772.3690	782.0878	753.4368
Std dev	169.7479	30.3277	46.0715	43.6395	48.3521	35.4998	48.1604

In Table 3, the time processes are slower than proposed method and the shadow removal method. The local inhomogeneity is the main reason the computational time of the method is too high and cannot be implemented in industrial company cases (Table 4).

The proposed method's result in the average value of mean is 18.34 points better than shadow removal method [10], and 46454.77 points better than the retinex method [9]. While the average stability of the proposed method is 5.97 points better than shadow removal [10], and 2442.76 points better than the retinex method [9].

Table 3. Time process evaluation of compared retinex method [9] using 'Klik BCA'

Time process (millisecond)							
Image ID	50 Lux	100 Lux	200 Lux	400 Lux	600 Lux	800 Lux	1000 Lux
1	47,607.09	58,852.62	53,704.94	62,026.89	67,750.63	66,264.67	61,061.68
2	47,774.47	56,505.08	52,632.5	58,770.85	55,531.75	57,293.5	58,790.62
3	46,441.54	51,799.81	53,214.59	53,750.28	56,142.12	57,400.66	58,141.54
4	47,515.49	52,100.85	52,959.79	52,856.88	60,868.18	58,637.32	57,801.42
5	46,064.83	52,216.4	52,843.97	53,755.43	54,585.55	57,340.54	57,928.45
6	45,628.78	55,609.68	55,590.2	57,197.12	56,058.72	56,548.88	57,235.09
7	48,437.55	55,920.19	52,336.09	55,045.87	64,428.8	57,209.97	58,240.3
8	48,403.61	52,518.32	52,771.36	52,242.55	58,489.41	62,670.36	59,267.04
Mean	47,234.17	54,440.37	53,256.68	55,705.74	59,231.90	59,170.74	58,558.27
Std dev	1,062.26	2,628.81	1,026.72	3,375.30	4,746.44	3,455.46	1,184.33

Table 4. Mean and standard deviation of time process comparison between proposed method, shadow removal [10], and retinex [9]

	Proposed method	Shadow removal [10]	Retinex [9]
Average of all mean from Tables 1, 2 and 3	**779.4**	797.74	47,234.17
Average of all standard deviation from Tables 1, 2 and 3	**54.29**	60.26	2,497.05

Next Analysis is accuracy. 22 images containing seven desired images (supposed to detect object with poor luminance factor) & 15 random images with various condition are taken in this experiment. These 15 random images are consisting of nine images from 'Shadow Removal Online Dataset and Benchmark for Variable Categories Scene' dataset and six data from "Klik BCA" dataset (Fig. 4).

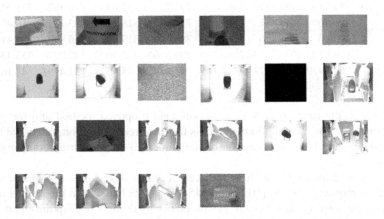

Fig. 4. Accuracy testing of proposed method using combined dataset

Table 5. Accuracy comparison between three methods

Image ID	Containing desired object	Proposed method	Shadow removal [10]	Retinex method [9]
1	No	No	No	No
2	No	No	No	No
3	No	No	No	No
4	No	No	No	No
5	No	No	No	No
6	No	No	No	No
7	Yes	Yes	Yes	Yes
8	Yes	Yes	Yes	Yes
9	No	No	No	No
10	Yes	Yes	Yes	Yes
11	Yes	**No**	**No**	**Yes**
12	Yes	Yes	Yes	Yes
13	No	No	No	No
14	No	No	No	No
15	No	No	No	No
16	No	No	No	No
17	Yes	Yes	Yes	Yes
18	Yes	Yes	Yes	Yes
19	No	No	No	No
20	No	No	No	No
21	No	No	No	No
22	No	No	No	No
Accuracy		**95.45%**	**95.45%**	**100%**

In the picture above, six out of the seven images from "Click BCA" could be detected correctly. Table 5 describes the result test for all methods:

5   Conclusions

There are various approaches to solve problems in object detection. Either by increasing the number of datasets for comparison, improving the quality of the image captured by the camera, as well as eliminating characteristics of the captured image. In this research, the approach is by reducing the level of the shadow of the object. Thus, the risk of undetected object is reduced because of shadow interference.

Unfortunately, the weaknesses found in this study is still a lack of optimization of the results of the diminished shadows, especially on images that has brightness value almost reached zero or has a luminance factor of less than 50 Lux. In contrary, there are trade-offs between accuracy and processing speed. Where after combining the two methods, the accuracy of the proposed method decreased compared with the adaptive single scale retinex method, while the speed of the process increases dramatically and faster than the comparison method.

From the results, we can conclude that the research framework and the experimental design can be used to answer the problem statement "How to diminish existing illumination and shadow factors on captured image until computer can properly detect it?", and "Is the proposed method that will be developed is more accurate or faster compared to the previous research?" is answered with the computation time test above. The algorithm proposed in this study can help increase the brightness level from an image. The shadow that covers the object is also reduced or even disappears.

As further development. Development in the object detection in this study can also be implemented in other cases. For instance, implementation to a moving robot which can recognize objects by giving probability of candidate objects. The proposed method in this research is expected to raise the probability percentage of guessed objects by the robot.

References

1. Andreopoulos, A., Tsotsos, J.K.: 50 years of object recognition: directions forward. Comput. Vis. Image Underst. **117**, 827–891 (2013)
2. Ding, H., Li, X., Zhao, H.: An approach for autonomous space object identification based on normalized AMI and illumination invariant MSA. Acta Astronautica **84**, 173–181 (2013)
3. Banerjee, P.K., Datta, A.K.: Class specific subspace dependent nonlinear correlation filtering for illumination tolerant face recognition. Pattern Recogn. Lett. **36**, 177–185 (2014)
4. Baradarani, A., Wu, Q.J., Ahmadi, M.: An efficient illumination invariant face recognition framework via illumination enhancement and DD-DT C WT filtering. Pattern Recogn. **46**, 57–72 (2013)
5. Bhaskar, H., Dwivedi, K., Dogra, D.P.: Autonomous detection and tracking under illumination changes, occlusions and moving camera. Sig. Process. **117**, 1–12 (2015)
6. Cao, X., Shen, W., Yu, L.G., Wang, Y.L., Yang, J.Y., Zhang, Z.W.: Illumination invariant extraction for face recognition using neighboring wavelet coefficients. Pattern Recogn. **45**, 1299–1305 (2012)
7. Fan, C.N., Zhang, F.Y.: Homomorphic filtering based illumination normalization method for face recognition. Pattern Recogn. Lett. **32**, 1468–1479 (2011)
8. Lin, Z., Wang, J., Ma, K.K.: Using eigencolor normalization for illumination-invariant color object recognition. Pattern Recogn. **35**, 2629–2642 (2002)
9. Park, Y.K., Park, S.L., Kim, J.K.: Retinex method based on adaptive smoothing for illumination invariant face recognition. Sig. Process. **88**, 1929–1945 (2008)
10. Blajovici, C., Kiss, P.J., Bonus, Z., Varga, L.: Shadow Detection and Removal from a Single Image (2011)
11. Constantin, J., Bigand, A., Constantin, I., Hamad, D.: Image noise detection in global illumination methods based on FRVM. Neurocomputing **164**, 82–95 (2015)
12. Jain, R., Kasturi, R., Schunck, B.G.: Machine Vision, vol. 5. McGraw-Hill, New York (1995)
13. FANUC America Corporation: Two Ultra-Fast Robots Pick & Place Batteries to Form Group Patterns - FANUC America, 22 December 2012. https://www.youtube.com/watch?v=tywZsEGm1xc

14. Mhamdi, M.A.A., Ziou, D.: A local approach for 3D object recognition through a set of size functions. Image Vis. Comput. **32**, 1030–1044 (2014)
15. Muselet, D., Macaire, L.: Combining color and spatial information for object recognition across illumination changes. Pattern Recogn. Lett. **28**, 1176–1185 (2007)
16. Nanni, L., Lumini, A.: Heterogeneous bag-of-features for object/scene recognition. Appl. Soft Comput. J. **13**, 2171–2178 (2013)
17. Zhang, S., Sui, Y., Yu, X., Zhao, S., Zhang, L.: Hybrid support vector machines for robust object tracking. Pattern Recogn. **48**, 2474–2488 (2015)
18. Kooij, J.F., Englebienne, G., Gavrila, D.M.: Identifying multiple objects from their appearance in inaccurate detections. Comput. Vis. Image Underst. **136**, 103–116 (2015)
19. Deng, Y., Duan, H.: Hybrid C2 features and spectral residual approach to object recognition. Optik Int. J. Light Electron Optics **124**, 3590–3595 (2013)
20. Matsukawa, T., Kurita, T.: Image representation for generic object recognition using higher-order local autocorrelation features on posterior probability images. Pattern Recogn. **45**, 707–719 (2012)
21. Li, Y., Wang, S., Tian, Q., Ding, X.: Feature representation for statistical-learning-based object detection: a review. Pattern Recogn. **48**, 3542–3559 (2015)
22. Guo, Y., Sohel, F., Bennamoun, M., Wan, J., Lu, M.: A novel local surface feature for 3D object recognition under clutter and occlusion. Inf. Sci. **293**, 196–213 (2015)
23. Bai, J., Wu, Y., Zhang, J., Chen, F.: Subset based deep learning for RGB-D object recognition. Neurocomputing **165**, 280–292 (2015)
24. Drew, M.S., Li, Z.N., Tauber, Z.: Illumination color covariant locale-based visual object retrieval. Pattern Recogn. **35**, 1687–1704 (2002)
25. Li, W., Dong, P., Xiao, B., Zhou, L.: Author' s accepted manuscript interest and optical bag of words model object recognition based on the region of interest and optical bag of words model. Neurocomputing **172**, 271–280 (2015)
26. Lian, Z., Er, M.J., Liang, Y.: A novel efficient local illumination compensation method based on DCT in logarithm domain. Pattern Recogn. Lett. **33**, 1725–1733 (2012)
27. Liu, Y.H., Lee, A.J., Chang, F.: Object recognition using discriminative parts. Comput. Vis. Image Underst. **116**, 854–867 (2012)
28. Cosker, H.G.D.: Shadow Removal Dataset and Online Benchmark for Variable Scene Categories, 28 June 2016. http://cs.bath.ac.uk/~hg299/shadow_eval/eval.php

Fast Moving UAV Collision Avoidance Using Optical Flow and Stereovision

Damian Pęszor[1,2(✉)], Marzena Wojciechowska[1,2], Konrad Wojciechowski[1,2], and Marcin Szender[1,2]

[1] Research and Development Centre, Aleja Legionów 2, 41-902 Bytom, Poland
damian.peszor@pja.edu.pl
[2] Polish-Japanese Academy of Information Technology,
Koszykowa 86, 02-008 Warsaw, Poland
http://bytom.pja.edu.pl

Abstract. Unmanned aerial vehicles are becoming popular, but their autonomous operation is constrained by their collision avoidance ability in high-velocity movement. We propose a simple collision avoidance scheme for fast, business-grade fixed-wing aircraft which is based on optical flow and stereovision. We calculate optical flow on parts of the image that are essential for collision avoidance and enlarge the analysed area only as long, as the framerate allows, thus avoiding the need to stretch calculations over several frames.

Keywords: Collision avoidance · Optical flow · Stereovision · Lazy flow · Scene flow · UAV

1 Introduction

Unmanned aerial vehicles (UAVs) are becoming more and more advanced and popular. Apart from their long good standing in terms of military applications, UAVs are currently used in areas such as photography, photogrammetry, meteorology, inspections, mapping, disasters management, damage assessment, search and rescue operations and more. Not only there are more and more of such vehicles in the air, but they also become more autonomous, which allows for reduction of costs. This means, that UAVs need to be able to avoid collisions in mid-air. For many of rotary-wing aircraft, the problem of collision avoidance can be solved using sophisticated algorithms calculating scene flow. However, when fast moving fixed-wing aircraft are considered, the velocity becomes a problem. Unlike typical aircraft, UAVs have limited computational performance and those two factors combined create a field for improvements.

Consider a vehicle moving at 100 km/h. If the vehicle is equipped with a camera (or cameras) working at 25 Hz, the distance travelled by vehicle between two consecutive frames is 1.1(1) m. Given the manoeuvrability of such vehicles, searching for obstacles and finding an obstacle-free course if one will be found cannot be spanned over many frames. And even if it is, the issue of correspondences between consecutive frames arise.

© Springer International Publishing AG 2017
N.T. Nguyen et al. (Eds.): ACIIDS 2017, Part II, LNAI 10192, pp. 572–581, 2017.
DOI: 10.1007/978-3-319-54430-4_55

Given above, we propose a simple collision avoidance solution for business-grade unmanned aerial vehicles based on not uniform analysis of optical flow calculated using stereo images.

2 Materials and Methods

2.1 UAV Shape Definition

Existing optical flow implementations tend to treat the analysed images as uniformly important data. In a case of fast moving unmanned aerial vehicles, it is not always the case. Since the processing time is limited, it is most important to make sure that there are no obstacles in a flight path of the UAV and to do that, the autonomous obstacle avoidance system has to take into account the specificity of vehicle's shape. Information about shape and course allows looking for collisions in part of the image where the course is blocked, which gives more time to plot a new course in the case of finding an obstacle.

Instead of using an actual projection of the vehicle, which would introduce additional computational overhead, we use simplified shape to define part of space in which UAV will find itself if the course will not change. There are two important factors; the wingspan of the vehicle d_{ws} and the height of the vehicle d_h. This information can be easily presented as either rectangle (wherein d_{ws} is the width of the rectangle and d_h is the height of the rectangle) or the ellipse (wherein d_{ws} is length of the major axis and d_h is length of the minor axis). While using ellipse requires additional computation, it reduces a number of analysed pixels, so we use the latter.

Vehicle dimensions define the ratio of ellipse's axes, but the initial scale factor s_i with relation to the image remains to be selected. If the initially analysed part of the image will be too small, there will be unnecessary computational overhead related with enlarging of the ellipse. If it will be too big, the image might not be analysed in time, before the next frame will be available. Instead, however, of changing s_i to cover for more and more pixels, one can use the precalculated index values to order the pixels in a way that will prioritise the more important pixels - those that are right in front of the body of the vehicle. Since the shape of the vehicle is defined as an ellipse, for the centre of each pixel it is possible to calculate at which scale the ellipse will contain it. From the smallest scale - the pixels that might present an obstacle that is most likely to hit the vehicle - to the largest, the pixels are ordered and can be analysed taking into account this ordering. In Fig. 1 one can see an example of scale calculated for typical 1080p frames. When axes of cameras are oriented to align with yaw and pitch axes of the aircraft and the centre of the frame corresponds to the course of the aircraft, as in most common scenario, this scale $s_{x,y}$ can be simply calculated as in Eq. 1, given the camera resolution of w width and h height.

$$s_{x,y} = \sqrt{\frac{(\frac{1-w}{2} + x)^2 d_h^2}{d_{ws}^2} + (\frac{1-h}{2} + y)^2} \qquad (1)$$

Fig. 1. Greyscale visualisation of pixel priority for 1920×1080 resolution for the wingspan of 5.5 and height of 1.3 resulting in the ratio of 4.23076923. The darker the pixel, the higher the priority and the sooner it is analysed.

Notice that we operate in discrete space here (thus the half pixel translation to the centre of the pixel). Once calculated, the pixels are sorted from the lowest scale thus obtaining ordering that prioritises the most important pixels, as seen in the example in Fig. 2.

2.2   Stereo-Enhanced Optical Flow

To detect a presence of an obstacle in analysed frames, a very fast algorithm has to be employed. The selection of such algorithm has to take few things into account:

– Algorithm has to be extremely fast due to the fact that the vehicle is moving with high velocity
– As consequence of above, algorithm has to be parallelisable in order to take advantage of 256 CUDA-cores (Compute Unified Device Architecture) available in NVidia Tegra X1
– In order to estimate depth, two cameras working in a stereo-rig are available onboard the UAV. The algorithm can benefit from using stereo data
– The vehicle contains onboard accelerometer and gyroscope. The algorithm can, again, benefit from this data

Having those in mind, an analysis of The KITTI Vision Benchmark Suite's [1, 2] evaluation of optical flow and scene flow algorithms indicates that Prediction-Correction Optical Flow (PCOF) combined with Adaptive Coarse-to-fine-stereo (ACTF, [3]) approach as presented in [4].

252	249	245	240	237	232	230	224	227	231	233	238	241	246	248	253
220	216	213	211	207	201	196	195	192	199	202	205	210	212	219	223
189	185	180	179	173	170	167	160	161	165	169	172	178	183	186	191
156	152	150	147	142	139	132	128	130	133	137	143	144	151	153	157
124	121	116	114	110	104	100	96	97	101	105	109	115	117	122	126
92	88	87	82	76	73	69	65	66	70	72	78	80	86	90	95
60	58	55	47	40	35	30	24	25	28	33	41	44	54	57	62
49	36	22	17	15	10	4	1	2	5	8	12	16	23	39	51
48	37	21	19	13	9	6	3	0	7	11	14	18	20	38	50
63	59	52	45	42	32	31	26	27	29	34	43	46	53	56	61
93	91	85	83	77	74	71	64	67	68	75	79	81	84	89	94
125	123	119	112	108	107	102	98	99	103	106	111	113	118	120	127
159	155	149	146	141	136	135	131	129	134	138	140	145	148	154	158
190	184	181	176	175	171	166	163	162	164	168	174	177	182	187	188
222	217	214	208	206	200	197	193	194	198	203	204	209	215	218	221
254	251	247	243	239	235	229	226	225	228	234	236	242	244	250	255

Fig. 2. An example of pixel ordering in 16×16 resolution for the wingspan of 5.5 and height of 1.3 resulting in the ratio of 4.23076923. The lower the index, the higher the priority of given pixel and the sooner it is analysed.

The first step of an algorithm presented by Derome et al. is to estimate ego-motion using visual odometer eVO proposed in [7]. It is also the only part of the algorithm proposed by Derome et al. which is done on CPU rather than GPU. In our case, however, the vehicle contains accelerometer and gyroscope, which are used by the UAV. Instead of estimating camera motion from images, we can, therefore, use already available information, thus reducing the Prediction-Correction Optical Flow algorithm to the part that can be computed on GPU.

The second important step of the solution presented in [4] is Adaptive coarse-to-fine matching. Instead of performing the entirety of algorithm for every pixel of currently analysed pair of frames, we use a lazy approach; local disparity optimisation is calculated only when needed, as not the entirety of the frame is analysed in our approach. When we refer to obstacle test for given pixel, we perform necessary local disparity optimisation for given pixel.

The last part of PCOF is computing the optical flow. The obvious performance advantage of eFOLKI ([8,9] and most importantly [10]) over other algorithms, as presented in [4] convinces us, that it is appropriate for fast-moving vehicles as in our case. Similarly to ACTF, here too we use a lazy approach. The structure of algorithm requires to use bigger neighbourhoods than just the pixels we decide to test for obstacles, but once obtained information can be used for next pixels.

Let us consider the usage of priority information on an example of part of PCOF when it is most applicable, namely the optical flow. In order to generalize the description to entirety of Lucas-Kanade-based algorithms using inverse additive (see [5] for classification), we will describe it in terms of PyramLK (as in [6]) rather than FOLKI, as the presentation of PyramLK is much clearer and it still allows for inclusion of changes made by FOLKI-GPU and eFOLKI. Let us consider the structure of optical flow algorithm, and how to utilise pixel priority ordering. In order to present the influence of pixel ordering, we look at the algorithm from the end. To calculate displacement d^0, that is displacement d^L on the finest level $L = 0$ of Gaussian pyramid, series of K iterations has to be performed that come down to solving a local system of equations that might be presented as in Eq. 2.

$$v^k(x,y) = v^{k-1}(x,y) + \mathbf{G}(\mathbf{x},\mathbf{y})^{-1}\boldsymbol{b}_k(x,y) \tag{2}$$

Wherein $\mathbf{G}(\mathbf{x},\mathbf{y})$ is a pseudo-Hessian valued function (as called in terms of eFOLKI) or spatial gradient matrix (in terms of PyramLK). The vector $v^k(x,y)$ is the displacement as calculated in k-th iteration and $\boldsymbol{b}_k x,y$ is image mismatch vector (or right-hand side vector c in terms of eFOLKI). Given values of $\mathbf{G}(\mathbf{x},\mathbf{y})$ calculated for the entirety of iterative process for given level of Gaussian pyramid, and $v^{k-1}(x,y)$ being past step, this does not require anything else to be known and can be calculated in order of importance. To go further, one would prefer to calculate $\boldsymbol{b}_k(x,y)$ only for currently analysed pixels. Let us consider the calculation of this vector in Eq. 3.

$$\boldsymbol{b}_k(x,y) = \sum_{m=x-r}^{x+r}\left\{\sum_{n=y-r}^{y+r}\begin{bmatrix}\delta I_k(m,n)I_x(m,n)\\ \delta I_k(m,n)I_y(m,n)\end{bmatrix}\right\} \tag{3}$$

Wherein $I_x(m,n)$ and $I_y(m,n)$ are derivatives calculated only once per level and $\delta I_k(m,n)$ is the image difference (or ϵ) between current frame and the current estimation of next frame calculated as in Eq. 4.

$$\delta I_k(m,n) = I^L(m,n) - J^L(m + g_m^L + v_m^{k-1}, n + g_n^L + v_n^{k-1}) \tag{4}$$

Calculation of $\boldsymbol{b}_k(x,y)$ is therefore dependent on interrogation window of $(2r + 1)^2$ pixels, where r denotes the radius of the window (radius excluding column x and row y). One can, therefore, notice that given specified number of pixels with high priority, only the limited number of surrounding gradient values is important. In terms of devices such as Tegra TX1, where the calculations can be highly parallelized, but are still limited in a number of parallel operations, this information can be useful.

Let us assume that a device can handle o parallel threads. In order to prioritise the most important pixels, the optical flow calculations are done in batches of o pixels. At every level of Gaussian pyramid, a supporting region around those o pixels is needed. Since pixels are ordered in order of priority, one can notice, that neighbourhood of pixels is next in order after the o pixels. If for x and y values of o-th pixel current scale is $s_o = s(x, y)$, then using the interrogation window radius of r one can notice that the scale of neighbourhood is described by Eq. 5 similar to Eq. 1.

$$s_{x,y}(r) = \sqrt{(\frac{d_h}{d_{ws}})^2 max(|\frac{1-w}{2} + x| - r, 0)^2 + max(|\frac{1-h}{2} + y| - r, 0)^2} \quad (5)$$

In both Eqs. 1 and 5 the square root does not change the order, so it can be safely dropped, however, this does not really matter, as the order of pixels can be calculated once per resolution and size of the UAV and thus is not in any way related to the performance of the algorithm.

One could assume, that such approach could only work on the last level of Gaussian pyramid, and the entirety of previous levels have to be calculated. This is not the case, as the only values used from previously calculated level are displacements. Those displacements might be big enough to affect calculation of image difference ($\delta I_k(m, n)$ or ϵ in FOLKI terms) value, but the actual value used in calculations is just an intensity (or rank, if eFOLKI is used), so that it is not needed to calculate displacement for anything other than 1-pixel (due to the upsampling of displacement values) neighbourhood of the previous level in Gaussian pyramid. Partial calculation of surroundings of the previous level is needed only as far, as the radius goes. See Figs. 3, 4 and 5 to see example of calculations needed for different levels.

Since the supporting region is bigger than o, it can be extended (usually to twice the number of o), so that additional threads could be used simultaneously using SIMD (Single Instruction, Multiple Data) instructions. At each level of the pyramid, the calculated data is stored. Once the o pixels' flow values are calculated, the next batch in order of priority is processed. Since some of the gradient data are already calculated, there is no need to repeat those calculations, instead, next gradients in priority can be calculated. In this way, the most important data can be analysed before calculations for entire image are finished.

2.3   Collision Detection

At first, a region defined by scaled ellipse is tested for obstacles. At the scale of s_i, this region corresponds to the position of UAV in somehow distant future. Before that, however, the vehicle will pass through space that is projected onto a greater part of the image plane. Assuming continuous collision detection, there are two reasons that an obstacle might be present in the previously analysed part of the image. First is the fact that vehicle might change its course. After each

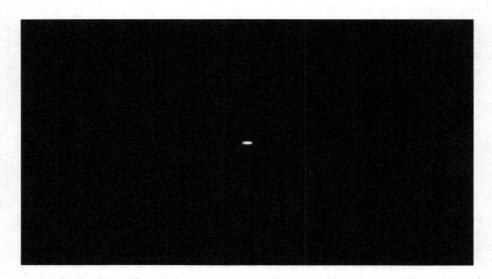

Fig. 3. The first level of the pyramid, 1920 by 1080 pixels, see Fig. 5.

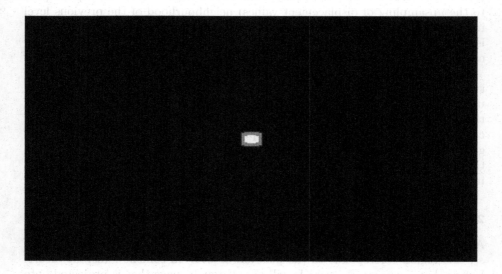

Fig. 4. The third level of the pyramid, 480 by 270 pixels, see Fig. 5.

Fig. 5. White represents the first batch of 256 pixels, light grey is supporting area needed by image mismatch vector using radius 3, dark grey is supporting region for gradient calculation (1 pixel). The fifth level of the pyramid, 120 by 68 pixels with the wingspan of 5.5 and height of 1.3 resulting in the ratio of 4.23076923.

course correction, the entire frame is analysed, despite calculations being longer than the time between frames. This is in order to make sure that the course is correct and does not rely on assumption that the static part of the image does not represent an obstacle. The second source of obstacles is other objects that move fast in relation to the vehicle, due to them being a dynamic part of scene themselves, or due to the fact that they are close to the UAV.

In order to detect dynamic obstacles, following scheme is assumed. After testing for obstacles in ellipse defined by d_{ws}, d_h and s_i, the scale factor is increased, so that the ellipse projection would contain a number of additional, surrounding pixels. These pixels are then analysed for obstacles, and if none is found, the scale is increased again. This goes on until one of two happens:

- Next frame is available. We prioritise analysis of newest frame over complete analysis of a lower number of frames. This is because some of the obstacles that were in course of UAV (i.e. other vehicles) might not longer be present in its trajectory, so the correction, of course, is no longer necessary. Second, more important reason is that with a fast moving vehicle, new obstacles might be present in the current frame. Complete analysis with a lower number of frames per second would mean, that UAV is able to travel further without detecting an obstacle, which could result in a collision.
- One of the pixels is detected as an obstacle. At this point, the usual scheme is paused, new frames are ignored until a part of space that does not result in a collision is found by collision avoidance algorithm, after which the avoidance manoeuvre takes place. After the change of course, collision detection is resumed as normally.

2.4   Collision Avoidance

Once the collision is detected, an avoidance manoeuvre has to be employed. Depending on specific kind of unmanned aerial vehicle, different manoeuvres might be available. In general (as in our case of fixed-wing aircraft rather than rotary-wing aircraft), there are six elementary manoeuvres available:

1. *Up manoeuvre.* In a case of fixed-wing aircraft, it corresponds to positive pitch that raises vehicle's nose and lowers the tail.
2. *Down manoeuvre.* In a case of fixed-wing aircraft, it corresponds to negative pitch that lowers vehicle's nose and raises the tail.
3. *Right manoeuvre.* In a case of fixed-wing aircraft, it corresponds to positive yaw that turns the nose of the vehicle to the right.
4. *Left manoeuvre.* In a case of fixed-wing aircraft, it corresponds to negative yaw that turns the nose of the vehicle to the left.
5. *Clockwise manoeuvre.* In a case of fixed-wing aircraft, it corresponds to positive roll that turns the vehicle clockwise.
6. *Counter-clockwise manoeuvre.* In a case of fixed-wing aircraft, it corresponds to negative roll that turns the vehicle counter-clockwise.

Depending on current speed and orientation (especially in the case of aircraft being currently in roll manoeuvre), wind direction and speed, and different environmental and structural factors, the vehicle might have different rates of performing different elementary manoeuvres. All of those rates, or costs, are relative to each other, so specific values are irrelevant. We assume following variables:

1. c_u – Cost of *up manoeuvre*. Maximum angle rate of positive pitch.
2. c_d – Cost of *down manoeuvre*. Maximum angle rate of negative pitch.
3. c_r – Cost of *right manoeuvre*. Maximum angle rate of positive yaw.
4. c_l – Cost of *left manoeuvre*. Maximum angle rate of negative yaw.
5. $c_c w$ – Cost of *clockwise manoeuvre*. Maximum angle rate of positive roll.
6. $c_c cw$ – Cost of *counter-clockwise manoeuvre*. Maximum angle rate of negative roll.

Once an obstacle is detected at a specific scale, the lowest cost of avoiding collision is calculated, and then the target ellipse is transformed according to the found avoidance manoeuvre. All the pixels in transformed ellipse are tested for obstacles. If no obstacles are found, the manoeuvre is performed by the vehicle. If, however, another obstacle is found, the initial avoidance manoeuvre is ignored and a new one with the lowest cost is computed, which takes both obstacles into account. This is done until a free path is found and avoidance manoeuvre is performed.

3   Discussion

We presented a simple collision avoidance solution designed for business-grade fast moving unmanned aerial vehicles, especially for fixed-wing aircraft. While the proposed scheme is simple, it is fast enough for such vehicles and as such solves the issue of optical flow algorithms proving to be too slow for such an application. Also, it serves as a proof of usefulness and robustness of new, fast, optical flow algorithm presented by Derome et al. in [4].

The presented work is the initial part of greater research regarding UAV collision avoidance. Although it responds to the critical issue of performance in real-time video analysis, further work is planned in order to ensure that the obstacle detection based on optical flow is robust enough to be used where human intervention is not possible.

Acknowledgements. This work has been supported by the National Centre for Research and Development, Poland in the frame of project POIR.01.02.00-00-0009/2015 "System of autonomous landing of an UAV in unknown terrain conditions on the basis of visual data".

References

1. Menze, M., Geiger, A.: Object scene flow for autonomous vehicles. In: Conference on Computer Vision and Pattern Recognition (CVPR) (2015)
2. Menze, M., Heipke, C., Geiger, A.: Joint 3D estimation of vehicles and scene flow. In: ISPRS Workshop on Image Sequence Analysis (ISA) (2015)
3. Sizintsev, M., Kuthirummal, S., Samarasekera, S., Kumar, R., Sawhney, H.S., Chaudhry, A.: GPU accelerated realtime stereo for augmented reality. In: 3DPVT (2010)
4. Derome, M., Plyer, A., Sanfourche, M., Besnerais, G.: A prediction-correction approach for real-time optical flow computation using stereo. In: Rosenhahn, B., Andres, B. (eds.) GCPR 2016. LNCS, vol. 9796, pp. 365–376. Springer, Heidelberg (2016). doi:10.1007/978-3-319-45886-1_30
5. Baker, S., Matthews, I.: Lucas-Kanade 20 years on: a unifying framework. IJCV **56**(3), 221–255 (2004)
6. Bouguet, J.-Y.: Pyramidal implementation of the Lucas Kanade feature tracker. Intel Corporation, Microprocessor Research Labs (2000)
7. Sanfourche, M., Vittori, V., Le Besnerais, G.: eVO: a realtime embedded stereo odometry for MAV applications. In: IROS (2013)
8. Le Besnerais, G., Champagnat, F.: Dense optical flow by iterative local window registration. In: ICIP (2005)
9. Plyer, A., Le Besnerais, G., Champagnat, F.: Folki-GPU: a powerful and versatile cuda code for real-time optical flow computation. In: GPU Technology Conference (2009)
10. Plyer, A., Le Besnerais, G., Champagnat, F.: Massively parallel Lucas Kanade optical flow for real-time video processing applications. J. Real-Time Image Proc. **11**(4), 713–730 (2016)

Towards the Notion of Average Trajectory of the Repeating Motion of Human Limbs

Sven Nõmm[1(✉)], Aaro Toomela[2], and Ilia Gaichenja[1]

[1] Department of Computer Science, Tallinn University of Technology,
Akadeemia Tee 15a, 12618 Tallinn, Estonia
sven.nomm@ttu.ee
[2] School of Natural Sciences and Health, Tallinn University,
Narva mnt. 25, 10120 Tallinn, Estonia
aaro.toomela@tlu.ee

Abstract. Average trajectory of the repeating motion of the limb joint defined in this paper. Majority of the existing results communicate analysis of the motions by means of different numeric parameters which does not necessary provide desirable feedback for those without deep knowledge of the subject. The notion of the average trajectory defined as the pair, consisting of trajectory, describing the shape of the joint motion, and pipe-shaped neighbourhood, describing variability of the observed motions. Proposed notion allows visualisation in three-dimensional space, which is easily interpretable and in turn may be used as feedback communicating results of the training or therapy session. Numeric parameters are associated with the average trajectory to validate proposed definition.

Keywords: Average trajectory · Limb motion · Motor performance

1 Introduction

Medicine, psychology and sport are the main areas where analysis of human motor functions used to determine condition of the individual (trainee or patient). Trainees or patients (undergoing rehabilitation of their motor functions) may benefit from the visualization of a training or therapy session. Another direction demonstrating growing interest towards evaluation of motor performance is the learning of specific movements, for example in laparoscope surgery [3,5] or military training [4]. It is assumed that in these areas condition (either level of training or stage of the disease) of the individual related to the motor performance. Motion capture system records the motions of the joints of interests, observed during one session. Such sessions usually consist of number repeated exercises, which constitute object of the research reported in this paper.

S. Nõmm—The work of Sven Nõmm and Ilia Gaichenja was supported by Tallinn University of Technology through the direct base funding project B37.
A. Toomela—The work of Aaro Toomela was partially supported by the Estonian Research Council through the Grant IUT3-3.

© Springer International Publishing AG 2017
N.T. Nguyen et al. (Eds.): ACIIDS 2017, Part II, LNAI 10192, pp. 582–591, 2017.
DOI: 10.1007/978-3-319-54430-4_56

The notion of *average trajectory of the motion* introduced in the present paper. Its main purpose is to describe overall shape and deviation of the repeated motion captured during training or therapy session in an easily interpretable way. Nowadays majority of techniques describing motions of the human limbs relay on different numeric parameters. Main drawback of this approach is that it may be difficult to interpret and explain to those without scientific background of the field. Majority of the existing results, describing and comparing motor performance, are based on feature extraction techniques [2], where some parameters are assigned to the certain time instances. Another approach [6] proposed a set of so called motion mass parameters to describe amount and smoothness of the motion observed over a certain time interval and not at a particular time instance. Projections of the trajectories onto two dimensional planes and even coordinate wise time parameterizations are analyzed [7]. The main drawback of such approach is that it is difficult to interpret which in turn motivates the idea of more understandable geometric representation of the results. While the idea of the geometric approach is not new, majority of the available results achieved in the domain of video processing and not suited for evaluation of motor performances. In [1] along with parametric approach, trajectories of a certain joints presented in the form of 3D diagram. In [10,11] the idea of motion region is introduced. The latest, by its nature, is the closest to the results reported in the present paper.

The paper organized as follows. Formal problem statement, background behind this research, conditions and limitations are explained in Sect. 2. Average trajectory formally defined in Sect. 3. Illustrative examples are presented in Sect. 4. Section 5 discusses different aspects of proposed notion. Concluding remarks presented in the last section.

2   Background and Problem Statement

Formally, the problem may be stated in the following way.

– Let T is the set of n trajectories corresponding to the n captures of the motion of interest observed during one session. Define a trajectory $\bar{T}$, such that its each point would be the closest to the corresponding points of n captured trajectories, with respect to the certain parameter.
– For each point of the trajectory $\bar{T}$ define a convex shape describing the variance of the corresponding points of the captured trajectories.
– Associate the set of numeric parameters enabling to evaluate proposed technique.

Last condition refers the necessity to validate achieved results. And will be discussed in Sect. 3.

One of the motivations behind present research was the desire to visualise changes in motor functions caused by training or therapy. Within the frameworks of the present research, learning of a new motor activity will be used to illustrate proposed notions and evaluate their suitability. The choice of the learning process instead of rehabilitation or progressing disease imposes fewer restrictions and allows concentrating attention on the proposed idea. The action of the ball throwing into the basket will be considered. The shape of the basket, weight of the ball and distance to the basket chosen to make learning (training) necessary. On the one hand it is easy to distinguish successful and unsuccessful trials and easy to capture. On the other hand, the action requires learning (or it is always possible to adjust conditions in a way that learning (training) is required). The learning process is organised into sessions separated by the equal time intervals. Each session consists of ten trials - attempts to throw the ball into the basket. Trainees are not allowed to practice between the sessions. Motion capture and necessary post processing of the captured data for each session will result ten numeric arrays describing motion trajectories of the joints of interest. Since trajectory (ordered set of points) is used to describe the movement of the joint, it is obvious that trajectory may be used to describe generalization of the joint movements observed during given session. In this case, one may compare observed trajectories to the trajectory representing generalized movement of the joint and determine variance at each particular point.

When talking about averaging sequential or (time)-series data Dynamic Time Warping (DTW) and its variations usually mentioned first. While DTW suits well as the measure of similarity for two trajectories, its application for the averaging trajectories does not satisfy the goals of the present paper. Also unlike the case of similarity computation there is no common agreement about DTW averaging, number of different approaches exists [8].

2.1 Motion Mass Parameters

Motion Mass parameters is relatively new notion. In addition, here it is applied in slightly extended form. In order to make this paper self-sufficient let us remind the notion as it introduced in [6]. Let J be the set of human body joints, such that each point j_i represent one joint. $J = \{j_1, \ldots, j_n\}$, n is the number of the joints in consideration. With each joint j_i three following parameters are associated. The length of trajectory T_{j_i} observed during the movement. *Acceleration mass A_{j_i}* the sum of the absolute values of the accelerations observed at each observation point. Euclidean distance E_{j_i} computed between the locations of the joint in the beginning end ending of the movement.

$$T_J = \sum_{i=1}^n T_{j_i}; \quad A_J = \sum_{i=1}^n A_{j_i}; \quad E_J = \sum_{i=1}^n E_{j_i} \tag{1}$$

Motion mass of the set J is defined as tuple $M_j = \{T_J, A_J, E_J, t\}$. Unlike the original paper [6] present research also considers velocities mass (defied by analogy to the acceleration mass) and ratios of the trajectory length to the Euclidean distance and acceleration mass to the Euclidean distance. Within the frameworks of present contributions motion mass parameters used to demonstrate the difference between the states of the motor function in the beginning and in the ending of the training.

3   Average Trajectory

Let us suppose that there is n captures of the same motion. Without loss of generality, assume that motion takes place the most along one of the coordinates. This does not imply that there is no movement along the other coordinates. This coordinate will be referring as *leading coordinate*. Imagine normal plane sliding along the leading coordinate from the point where motion begins to the point marking ending of the motion. At each point of this interval there will be n points where trajectories T_i intersect with the normal plane. Averaging the coordinates of those points would provide corresponding point of the average trajectory as shown in Figs. 1 and 2.

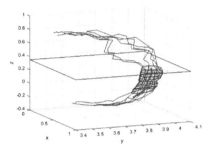

Fig. 1. ith iteration of building average trajectory. (Color figure online)

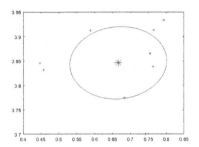

Fig. 2. Intersection points and their average.

In the case of Fig. 1, z is chosen as the leading coordinate, thin blue lines represent captured trajectories, and yellow plane is the normal to z, finally bold red line represent average trajectory.

More formally:

- **Step 1.** For each captured motion determine the interval on the leading axis where trajectory is present. Find intersection of these intervals, (in other words find the interval where all the trajectories are present).
- **Step 2.** begin cycle: for each consequent point of the interval defined in Step 1
    - Define the normal plane to the leading coordinate.
    - For each plane find points where trajectories intersect with this plane. Compute an "average point" (coordinate wise).
    - Ellipse is used to represent variability of the points with respect to their average. Compute the covariance matrix (of the intersection points) and perform its eigendecomposition. Eigenvalues of the covariance matrix represent squared radiuses of the ellipses and the columns of the eigenmatrix describes orientation of the radiuses Fig. 2.

- **Step 3.** Average points (computed for each plane) define the average trajectory. The curve depicts generalized shape of the motion. Pipe-shaped neighbourhood is defined by the boundaries of the ellipses and depicts variability between the motions.

In Fig. 2 '+' represent the intersection points of the individual trajectories with the normal plane. '*' represents the point of intersection between the average trajectory and the normal plane. Ellipse represent variability of the intersection points. One may argue about the choice of the ellipse to describe variability of the intersection points in Step 2. On the viewpoint of the authors, this is most natural and intuitive choice. The arguments to support this choice, its drawbacks and other possible choices are discussed in Sect. 5. Here and after the pair consisting of the trajectory derived by the algorithm and its enclosing pipe-shaped neighborhood would be referred as the *average trajectory* of the motion.

Proposed algorithm leads to the purely geometric representation that totally excludes time from the consideration. On the first view, such approach may seem strange, since timing is a natural way of motion parametrization. Also time frequently used as the parameter of the motion both in sportive training and in medical exercises. The latest is the main reason to consider time-less representation of the motion and demonstrate other properties, which vary significantly during learning of new motor activity or medical therapy.

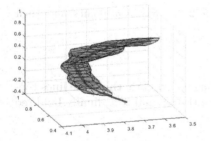

Fig. 3. General structure of the *average trajectory*

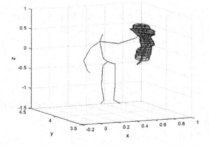

Fig. 4. Average trajectory of the human wrist during the ball throwing exercise.

Average trajectory derived on the basis of the trajectories representing successful ball throwing in the beginning of the training (learning) process is demonstrated in Fig. 4. Figure 3 depicts position of the average trajectory with respect to the standing human. There are two numeric parameters, which may be naturally associated with the average trajectory. The first one is the length of the average trajectory and the second one volume of the pipe-like neighborhood. While the length of the trajectory will not necessarily change much during the training, the volume should decrease. This is because goal-directed movements to visual targets consist of an initial impulse towards the target and a later corrective adjustment near the target to compensate for initial trajectory errors [9]. We hypothesized that during learning the primary movement needs less and less adjustment and thus the overall volume of the movements decreases. One of the possible ways to validate the proposed definition is to demonstrate that volume decrease during the training.

4   Illustrative Examples

Let us now consider the learning process of the ball throwing explained in Sect. 2. The process divided into ten sessions whereas each session consists of ten attempts (ball throwing). In the beginning of the training motion mass parameters computed for the trajectories of successful trials are indistinguishable from those of failed trials, (corresponding parameters of the motion mass are available from the authors upon request). In the end of training process, three of the motion mass parameters of the successful trials were distinguishable from those of the failed. Testing results are presented in Table 1.

Table 1. Test results for the mean values of MM parameters comparing successful and failed trials in the end of the training

Parameter	Reject H_0	p-value	t-statistic
V_J	1	0.0338	-2.1842
T_J	1	0.0355	-2.1629
A_J	0	0.6922	-0.3981
E_J	0	0.6791	-0.4162
T_J/E_J	0	0.0732	1.8315
A_J/E_J	0	0.5779	0.5601
t	1	0.0356	-2.1622

In addition, statistical hypothesis testing has demonstrated that four motion mass parameters computed for the successful trials in the end of the training were clearly distinguishable from those computed in the beginning. See Table 2 for details. This demonstrate that learning of the new motor activity caused changes in the way motions are performed.

Let us now turn attention to the average trajectories computed for the successful and failed attempts in the beginning and in the end of the training. In the beginning, there were no visible difference between the average trajectories computed for the successful and failed attempts Figs. 5 and 6.

The difference between the average trajectories corresponding to the successful trials in the beginning and the ending of the training depicted in Fig. 7, where narrow (red) pipe corresponds to the end of the training and wider (green) pipe to the beginning.

Table 2. Test results for the mean values of MM parameters comparing beginning and ending of the training.

Parameter	Reject H_0	p-value	t-statistic
V_J	1	0.0125	-3.2068
T_J	1	0.0134	-3.1581
A_J	0	0.2492	-1.2423
E_J	0	0.7886	0.2772
T_J/E_J	1	0.0212	2.8570
A_J/E_J	0	0.2069	1.3733
t	1	2.2007e$-$04	-6.3522

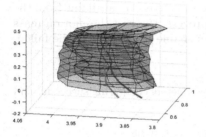

Fig. 5. Average trajectories corresponding to the successful and failed trials in the beginning of the training process.

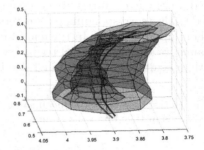

Fig. 6. Average trajectories corresponding to the successful and failed trials on the end of the training process.

In the beginning of the learning, trainee was not able to make more than two successful attempts, which is not enough to generate pipe-shaped neighbourhood of the average trajectory. In order to overcame this problem the session results considered pair wise. Evolution of the volumes of average trajectories during the learning process is summarized in Table 3. One may see that in course of the training, volume of the average trajectory computed for the successful attempts decreases as training progresses. At the same time the volume of the average trajectories computed for the failed attempts does not demonstrate any trend in its behavior. The other participants of the pilot research group demonstrated similar results.

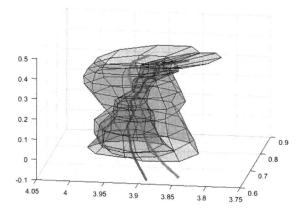

Fig. 7. Average trajectories of the successful trials in the beginning and ending of the training. (Color figure online)

Table 3. Evolution of the volumes and success rates.

	Success rate	Volume of the average trajectory of:	
		Success attempts	Failed attempts
Trials 1–2	3	0.0042	0.0049
Trials 3–4	5	0.0085	0.0315
Trials 5–6	5	0.0051	0.0054
Trials 7–8	6	0.0011	0.0064
Trials 9–10	3	0.0027	0.0064

5 Discussion

In Sect. 3 variability of the intersection points between the normal plane and captured trajectories represented by means of the ellipse. On the one hand, this is common way to describe deviation from the average in two-dimensional space. On the other hand, this method may lead to undesirable results. If the motion was repeated just a few times. For example, some patients could not perform the exercise more than three times. Specific distribution of the trajectories also may cause that ellipses will become degenerate and would not produce any pipe-shaped neighbourhood. In those cases ellipses may be replace by a circles. Whereas the radius may be chosen as the maximum of ellipsoid radiuses or as the average or maximal distance between the average point and intersection points. Most probably, the way to describe variability is not universal and depends on the particular application.

In order to illustrate the notion of the average trajectory relatively short, in time, learning process was observed. On the one hand, small number of observation points does not allow validating proposed notion in a stronger way. For example by demonstrating significant correlation between the success rate and volume of the average trajectory. On the other hand, it allows avoiding effects caused by fatigue and other complications of lengthy processes. Another tuple of the parameters, which may be used for comparison of different trajectories, is the sequence constitute by the areas of the ellipses on the secant planes. In order to perform comparison one should build the trajectories within the same limits. This will result in the same number of secants for both cases. Which leads the possibility to compare pairwise areas of the corresponding ellipses. While this option seems attractive on the first view its results are not strait forward interpretable and therefore left of future studies. One may suggest using principal component analysis (PCA) to find the direction of maximum variance. Application of PCA means coordinate rotation, which would tangle interpretation of the results. Therefore, the choice of the leading coordinate is left to the coach or practitioner.

6 Conclusions

The notion of the *average trajectory* of the repeating motion is introduced in the present paper. The pair of elements describe average trajectory. The first one represents the shape of the motion and the second one represents variability. Being strictly geometric, the notion allows visualizing the results of the training or therapy session, in an easily understandable form. Volume of the average trajectory is the most natural numeric value to be associated with average trajectory. It was demonstrated that during the learning process volume of the average trajectory (computed for the successful trials (attempts)) is tend to decrease. Based on results of this pilot research one may conclude that type of activity is a defining factor to choose the way to describe variability of the trajectories. This will be first main problem for further studies.

References

1. Ahmadi, A., Destelle, F., Monaghan, D., O'Connor, N., Richter, C., Moran, K.: A framework for comprehensive analysis of a swing in sports using low-cost inertial sensors. In: 2014 IEEE SENSORS, pp. 2211–2214, November 2014
2. Alexiadis, D., Daras, P.: Quaternionic signal processing techniques for automatic evaluation of dance performances from MoCap data. IEEE Trans. Multimed. **16**(5), 1391–1406 (2014)
3. Estrada, S., O'Malley, M., Duran, C., Schulz, D., Bismuth, J.: On the development of objective metrics for surgical skills evaluation based on tool motion. In: 2014 IEEE International Conference on Systems, Man and Cybernetics (SMC), pp. 3144–3149, October 2014
4. Kwak, Y.S., Jung, S.K.: Recognition of visual signals and firing positions for virtual military training systems. In: 2013 The 6th International Conference on Human System Interaction (HSI), pp. 656–658, June 2013
5. Lin, Z., Uemura, M., Zecca, M., Sessa, S., Ishii, H., Tomikawa, M., Hashizume, M., Takanishi, A.: Objective skill evaluation for laparoscopic training based on motion analysis. IEEE Trans. Biomed. Eng. **60**(4), 977–985 (2013)

6. Nõmm, S., Toomela, A.: An alternative approach to measure quantity and smoothness of the human limb motions. Est. J. Eng. **19**(4), 298–308 (2013)

7. Payeur, P., Nascimento, G., Beacon, J., Comeau, G., Cretu, A.M., D'Aoust, V., Charpentier, M.A.: Human gesture quantification: an evaluation tool for somatic training and piano performance. In: 2014 IEEE International Symposium on Haptic, Audio and Visual Environments and Games (HAVE), pp. 100–105, October 2014

8. Petitjean, F., GanSarski, P.: Summarizing a set of time series by averaging: from Steiner sequence to compact multiple alignment. Theoret. Comput. Sci. **414**(1), 76–91 (2012)

9. Suzuki, M., Kirimoto, H., Sugawara, K., Kasahara, Y., Kawaguchi, T., Ishizaka, I., Yamada, S., Matsunaga, A., Fukuda, M., Onishi, H.: Time course of change in movement structure during learning of goal-directed movement. J. Med. Biol. Eng. **35**(1), 113–124 (2015)

10. Takai, M.: Extracting method of characteristic posture from human behavior for surveillance camera. In: ICCAS-SICE 2009, The International Joint Conference on Instrumentation, Control and Information Technology, Fukuoka, Japan, pp. 159–164 (2009)

11. Takai, M.: Production of body model for education of dance by measurement active quantity. In: The 1st IEEE Global Conference on Consumer Electronics 2012, pp. 212–216, October 2012

Interfered Seals and Handwritten Characters Removal for Prescription Images

Wen-Hong Zhang[1], Teng-Hui Tseng[2], and Chun-Ming Tsai[1(✉)]

[1] Department of Computer Science, University of Taipei,
No. 1, Ai-Kuo W. Road, Taipei 100, Taiwan, ROC
z227389@gmail.com, cmtsai2009@gmail.com
[2] Department of Communication Engineering, Oriental Institute of Technology,
New Taipei City 220, Taiwan, ROC
alex@mail.oit.edu.tw

Abstract. Text detection in prescription images is a challenging problem because the prescription images are captured by heterogeneous cameras and under different illuminations. Further, the text is always interfered with by affixed seals and by handwritten characters. In this paper, a binarization method is proposed based on Niblack's method to remove the interfering affixed seal and handwritten characters in prescription images. Experimental results show that the proposed method can threshold the text and remove interference effectively with an accuracy recognition rate of 93.38%. This experimental result is compared with four others methods.

Keywords: Interfered seals · Interfered handwritten characters · Prescription image · Niblack's method

1 Introduction

In March 1995, Taiwan implemented the "National Health Insurance" [1]. After two years, it implemented a "separate medical system" whereby the patients could have their prescriptions issued by doctors after medical treatment, then take their prescriptions to a health insurance pharmacy to get their medicines. As demographic aging progresses and life expectancy increases, long-term care and uses of drugs are increasingly needed to stabilize the condition and provide better care for an increasing number of people with chronic diseases. However, pharmacies are not like 7–11s that are found everywhere. So it is not very convenient for chronic disease patients to have prescriptions filled.

Today, if patients with chronic diseases need to receive a continuous flow of prescription drugs for a chronic condition (2nd take medicine) from a pharmacy, they must take the original prescription to the pharmacy one week in advance. When the pharmacy receives of the prescription, they enter it into their file, and order the medications from the pharmaceutical company. When they receive it, they adjust the prescription, prepare a listed drug bag, and then notify the patient to come to take medicine. The patient therefore must wait until all this process is completed and they receive notification from the pharmacy, and then return to the pharmacy and collect their drugs.

N.T. Nguyen et al. (Eds.): ACIIDS 2017, Part II, LNAI 10192, pp. 592–601, 2017.
DOI: 10.1007/978-3-319-54430-4_57

Further complicating the process, a large number of chronic disease patients need medications. The pharmacy will be overcrowded, and the patients have to wait a significant time to collect medicines. In order to let chronic diseases patients have a more convenient way to fill their prescription, without having to go to the pharmacy to deliver the prescriptions physically and then travel again to the pharmacy to collect their medications each time they need a renewal, reducing the round-trip time and improve customer satisfaction, pharmacies need an automatic identification system to identify the prescription contents. Such a system could dispense medicine automatically and deliver the drugs to the chronic disease patients automatically.

The prescription script, that is the doctor's instruction of the medicine the patient needs, contains the medical information and medical orders that are essential for the patient to receive and use the medication properly. Therefore, in any prescription identification system, the first stage is to capture the prescription image. Thanks to the development of intelligent phones, this is now possible. The chronic diseases patients could use their phones to take a photograph of the prescription image and upload it to the cloud system for the pharmacy to process.

However, prescriptions are not easy for a text detection system to read. There are many seals in the prescription form, and these seals often are overlapped with the texts of the medical information and medical order. These seals will impact the text setting forth the medical information and medical order that must be detected and recognized correctly. Furthermore, there are many handwritten characters in the prescription. These handwritten characters will also influence accurate detection and recognition of text in prescriptions.

Figure 1 shows two examples of interferences. From Fig. 1(a), a blue seal has been stamped over part of the some texts of the medical information. This blue seal will affect the text to be detected and recognized. From Fig. 1(b), part of the text of the medical order is written in handwritten digits – which are themselves partly covered by a red seal. In particular, the number of the clinic – "4001020028" – is interfered with by the blue handwritten digits.

(a)   (b)

Fig. 1. Two examples of interferences in prescription. (a) Interfered by an affixed blue seal. (b) Interfered by handwritten digits and an affixed red seal. (Color figure online)

Our survey of the literature shows that there are very few papers that address these problems. Most of the related papers addressed the removal of interfering strokes and lines. Govindaraju and Srihari [2] proposed a method to remove interfering strokes. Liang et al. [3] proposed a method to extract the words from printed documents that have interference and other strokes (enclosing curves, underlines etc.). Yu and Jain [4] proposed a method to separate filled-in characters or symbols that are either touching or crossing from form frames. Huang et al. [5] proposed a stroke tracing approach for recovering the strokes from the handwriting images. Han et al. [6] proposed a method to analysis the shape types of overwritten characters and proposed a method to restore characters broken by line removal. Tseng and Lee [7] proposed a two-step method for removing interfering-lines from form documents.

A major drawback of these methods is that most character recognition engines are trained by sample characters without interference, so those engines will fail to recognize characters with interference. In this paper, a seal and hand-writing character removing method is proposed to remove the interfering seals and handwritten characters in prescriptions.

2   Related Binarization Methods

Binarization methods can be divided into four classes: global, regional, local, and hybrid [8]. Global thresholding methods determine a threshold from the information of the entire image. Otsu thresholding method provides good results among the global thresholding techniques. However, it neglects the spatial relationships among pixels and it needs much time for multilevel threshold selection [9]. Region-based thresholding divides the image into fixed $r \times r$ or intelligent regions and decides how to threshold each region. Chou et al. [10] proposed learning rules to threshold document images. Their method divided a document image into 3×3 regions. Each region is thresholded by decision rules. Tsai [8] proposed an intelligent region-based thresholding method for binarization the color document images with highlighted regions. His method intelligently divides a document image into several foreground regions and decides the background range for each foreground region. Local thresholding methods determine an individual threshold for pixels according to the intensities of each pixel and its neighbor. Niblack's thresholding method [11] produced better and faster OCR results among the local thresholding techniques. Hybrid thresholding methods combine both global and local methods. Tseng and Lee [12] proposed a two-layer block extraction method for document image binarization.

Applying the above-mentioned binarization methods to threshold prescriptions with interfered seals and interfered handwritten characters cannot produce satisfactory results. In order to remove interfered seals and interfered handwritten digits, a method based on the Niblack thresholding method [8] is proposed. The Niblack-based thresholding methods are discussed as follows.

2.1   Niblack Thresholding Method

In Niblack's approach, the pixel-wise threshold is computed by moving a kernel window of different sizes over the grayscale image. Different threshold value is obtained from the local mean $m(x, y)$ and the standard deviation $\sigma(x, y)$ at a window. The threshold value T in a kernel window is obtained from:

$$T(x,y) = m(x,y) + k \cdot \sigma(x,y), \tag{1}$$

where k is a constant for determining the number of edge pixels considered as object pixels and is fixed to -0.2 by the author. The window size should be small to reflect the local illumination and large to include both objects and background.

2.2 Sauvola's Thresholding Method

Sauvola's approach [13] is improved from Niblack's approach. This method computed the threshold value by adapting the contribution of the standard deviation. The threshold value T in a small window is obtained from:

$$T(x,y) = m(x,y) \cdot \left[1 + k \cdot \left(\frac{\sigma(x,y)}{R} - 1 \right) \right], \tag{2}$$

where R is defined as the dynamic range of the standard deviation based on experimental results and k is a positive value and allows the method to capture background in a window which helps to achieve a clean background far from foreground objects. Authors suggested R is 128 and k is 0.5.

2.3 Wolf's Thresholding Method

Wolf's approach [14] is improved from Sauvola's approach. This method calculates image contrast, gray level mean and standard deviation in a window and over whole image. The threshold value T in a window is obtained from:

$$T(x,y) = (1-k) \cdot m(x,y) + k \cdot M + k \cdot \frac{\sigma(x,y)}{R} (m(x,y) - M), \tag{3}$$

where k is fixed to 0.5, M is the minimum gray level of the whole image and R is the maximum gray level standard deviation and obtained from all over the local windows.

2.4 Feng's Thresholding Method

Feng's approach [15] is improved in turn from Wolf's approach. This method uses two local windows: a primary and a secondary window with one contained within other. The values of standard deviation, σ, the minimum gray level, M, and local mean, m, are calculated in the primary local window and the dynamic range of standard deviation R_σ is calculated in the secondary larger window. The threshold value T in a local window is obtained from:

$$T(x,y) = (1 - \alpha_1) \cdot m(x,y) + \alpha_2 \cdot \left(\frac{\sigma(x,y)}{R_\sigma} \right) \cdot (m(x,y) - M) + \alpha_3 \cdot M, \tag{4}$$

where $\alpha_2 = k_1(\sigma(x,\ y)/R_\sigma)^\gamma$, $\alpha_3 = k_2(\sigma(x,\ y)/R_\sigma)^\gamma$, and α_1, γ, k_1, and k_2 are positive constants. γ is 2 and α_1, k_1, and k_2 are in the ranges of 0.1–0.2, 0.15–0.25, and 0.01–0.05, respectively.

2.5   NICK Thresholding Method

NICK's approach [16] improves the binarization results of lighted and low images. The threshold value T in a kernel window is obtained from:

$$T(x,y) = m(x,y) + k\sqrt{\frac{\left(\sum p_i^2 - m(x,y)^2\right)}{NP}},\tag{5}$$

where k varies between -0.1 and -0.2 according to application need, m is the average gray level, p_i is the gray level of pixel i, and NP is the total number of pixels. In their tests, the authors used a window of size 19×19.

3   Proposed Method

Figure 2 shows the diagram of the proposed method. The proposed method includes input color prescription images, transforms RGB color space into HSV color space, removes white background color, computes the noise factor, applies the Niblack's thresholding method, and uses the morphological operation to obtain binary images.

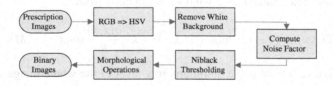

Fig. 2. The flow diagram of the proposed system

In order to remove the white background color, the *value* (v) in HSV color space is greater than 75%, the background color is white [16]. When applying the Niblack's thresholding method to threshold image, the constant k needs to be set up first. If this constant is not set properly, the thresholding result will not be satisfactory. In this paper, the constant k is called the noise factor (*NF*), and is defined as follows:

$$NF = \left|\left(\mu_{<75\%} - \left(\frac{\mu}{\sigma_{<75\%}}\right)\right)\right|/255,\tag{6}$$

where μ is the total mean of the prescription image, $\mu_{<75\%}$ and $\sigma_{<75\%}$ are, respectively, the mean and the standard deviation of the prescription image which the background white color has been removed. After obtaining the noise factor, the Niblack's thresholding method is applied. The thresholding method is shown as follows:

$$T(x, y) = m(x, y) + NF \cdot \sigma(x, y), \tag{7}$$

Noise factor (NF) will control the noises in the thresholded images. If the threshold value $T(x, y)$ is large, the thresholded results will have many noises. If the threshold value $T(x, y)$ is small, the thresholded results will have less noise.

After applied Eqs. (6) and (7), the binary result still has some noises; the morphological operations are applied to remove these retained noises.

4   Experimental Results

The proposed method was performed using Matlab R2016a on a 64-bit Windows 8.1 platform of Intel® Core(TM) i5-3230 M CPU @ 2.60 GHz ASUS laptop. To demonstrate the performance of the proposed method, 10 prescription images with interference by an affixed seal and 5 interfered by handwritten characters and by an affixed seal were used to experiments.

The binarization results of the proposed method were compared with Sauvola's, Wolf's, Feng's, and NICK's methods. Figure 3 shows the binarization results for two sample prescriptions, Fig. 1(a) and (b), applying Sauvola's, Wolf's, Feng's, NICK's, and the proposed methods. The binarization results for Fig. 1(a) and (b) with Sauvola's method are shown in Fig. 3(a) and (b). The binarization results for Fig. 1(a) and (b) with Wolf's method are shown in Fig. 3(c) and (d). The binarization results for Fig. 1(a) and (b) with Feng's method are shown in Fig. 3(e) and (f). The binarization results for Fig. 1(a) and (b) with NICK's method are shown in Fig. 3(g) and (h). Finally, the binarization results for Fig. 1(a) and (b) with proposed method are shown in Fig. 3(i) and (j).

These binarization results show that the blue interfered seal in Fig. 1(a) has been removed by the proposed method. The interferences in Fig. 1(b) by the handwritten digits and by the red affixed seal have also been removed by the proposed method. For comparison, in Sauvola's and Wolf's binarization results, while the blue interfered seal in Fig. 1(a) is removed, the stroke of the text is broken and many noises are produced. None of the comparison methods can remove the interference in Fig. 1(b).

Figure 4(a)–(f) show the original image, the binarization results of Sauvola's, Wolf's, Feng's, NICK's, and the proposed method, respectively. Sauvola's result (Fig. 4(b)) has many broken strokes. The binarization result of the Feng's method (Fig. 4(d)) is all white. The NICK's result (Fig. 4(e)) shows both the text and the red seal. Wolf's (Fig. 4(c)) and the proposed methods both threshold the prescription image without the interference by the red affixed seal. However, the stroke width in the Wolf's result is less than in the proposed method.

Figure 5 shows the original image and the results for the methods of Sauvola, Wolf, Feng, NICK, and the proposed method. The original image is interfered by a red affixed seal and by red and blue handwritten digits and a handwritten character. The binary results of Sauvola's, Wolf's, and Feng's method, Fig. 5(b)–(d), respectively, are not

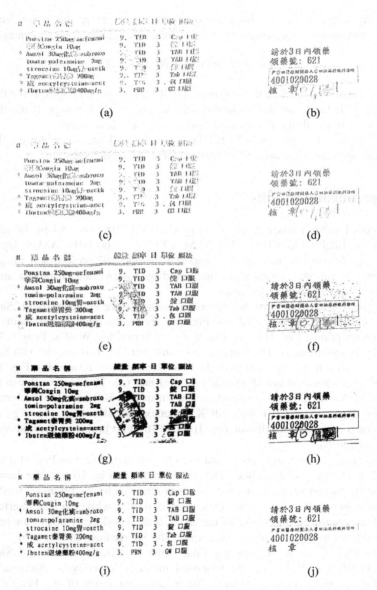

Fig. 3. Binarization results of Fig. 1(a) and (b) by using five methods. (a) Sauvola's binary result of Fig. 1(a). (b) Sauvola's binary result of Fig. 1(b). (c) Wolf's binary result of Fig. 1(a). (d) Wolf's binary result of Fig. 1(b). (e) Feng's binary result of Fig. 1(a). (f) Feng's binary result of Fig. 1(b). (g) NICK's binary result of Fig. 1(a). (h) NICK's binary result of Fig. 1(b). (i) The proposed method's binary result of Fig. 1(a). (a) The proposed method's binary result of Fig. 1(b).

sufficiently accurate to be useful because the text cannot be thresholded. The binary results of NICK's (Fig. 5(e)) and the proposed method (Fig. 5(f)) each threshold the text. However, the interference by the red affixed seal and by the red and blue

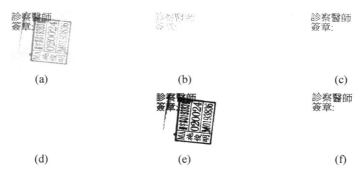

(a)   (b)   (c)

(d)   (e)   (f)

Fig. 4. Binarization results of the portion prescription which is interfered by a red affixed seal. (a) Original image. (b) Sauvola's result. (c) Wolf's result. (d) Feng's result. (e) NICK's result. (f) The proposed method's result. (Color figure online)

handwritten digits and character is also thresholded by NICK's method. As a result, the text "第" and "3日內" – parts of the medical information and medical order in the prescription – are interfered by the handwritten character "S", so the interfered text cannot be easily recognized by the OCR. By contrast, the binarization result from using the proposed method removes the interference by both the seal and the handwritten digits and character.

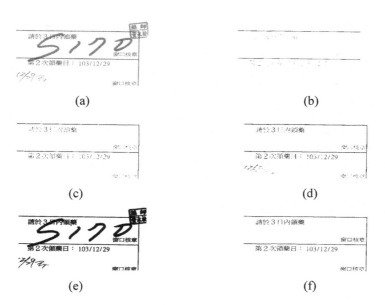

(a)   (b)

(c)   (d)

(e)   (f)

Fig. 5. Binarization results of the portion of a prescription which is interfered by a red affixed seal and also by red and blue handwritten digits and a character. (a) Original image. (b) Sauvola's result. (c) Wolf's result. (d) Feng's result. (e) NICK's result. (f) The proposed method's result. (Color figure online)

Table 1. Recognition rate achieved by the Snap2PDF.

Image	Total characters	Number of characters correctly recognized				
		Sauvola	Wolf	Feng	NICK	Proposed method
Figure 1(a)	225	58	55	48	58	205
Figure 1(b)	43	23	18	14	23	43
Figure 4(a)	6	0	6	0	0	6
Figure 5(a)	28	0	4	5	14	28
Total	302	81	83	67	95	282
RR		26.82	27.48	22.19	31.46	**93.38**

To quantify the efficiency of the proposed binarization method, an OCR experiment was also performed on the binarization results of the comparison methods. The character recognition used the well-known APP Snap2PDF [17]. The OCR result is quantified by character recognition rate and is defined as follows:

$$recognition\ \ rate\,(RR) = \frac{Number\ \ of\ \ correctly\ \ \det ected\ \ characters}{Total\ \ characters\ \ in\ \ original\ \ prescription} \quad (8)$$

The recognition rate for Figs. 1(a) and (b), 4(a) and 5(a) images are given in Table 1. From this table, it is shown that the proposed method achieved a higher OCR recognition rate than any of the other methods.

5   Conclusions

In this research, we proposed a binarization method to threshold prescription images with interfered characters. The proposed method modified the Niblack's method to remove interfering seals and handwritten characters. The binarization results from experiment applying the proposed method to collected images show the proposed method has a higher accuracy recognition rate than other methods. Further, the experimental results demonstrate the proposed method is feasible.

In the future, more test prescription images will be collected and will be used to test to obtain a still higher recognition accuracy rate, and the proposed method will be set to complete the prescription identification system.

Acknowledgements. The authors would like to express his gratitude to Walter Slocombe and Dr. Jeffrey Lee, who assisted editing the English language for this article. This work was supported by the Ministry of Science and Technology, R.O.C., under Grants MOST 104-2221-E-845-003.

References

1. Ministry of Health and Welfare. http://www.nhi.gov.tw/english/index.aspx?menu=8&menu_id=30&webdata_id=0&WD_ID=30

2. Govindaraju, V., Srihari, S.N.: Separating Handwritten Text from Interfering Strokes. From Pixels to Features III: Frontiers in Handwriting Recognition. Elsevier Science Publisher, Amsterdam (1992). pp. 17–28

3. Liang, S., Ahmadi, M., Shridhar, M.: Segmentation of interference marks using morphological approach. In: Proceedings of the Third ICDAR, pp. 1042–1046 (1995)

4. Yu, B., Jain, A.K.: A generic system for form dropout. IEEE Trans. PAMI **18**(11), 1127–1134 (1996)

5. Huang, W., Rong, G., Bian, Z.: Strokes removing from static handwriting. In: Proceedings of the Third ICDAR, pp. 861–864 (1995)

6. Han, S.Y., Yoo, J.Y., Kim, M.K., Kwon, Y.B.: Line removal and restoration of handwritten characters on the form documents. In: Proceedings of the 4th ICDAR, pp. 128–131 (1997)

7. Tseng, Y.H., Lee, H.J.: Interfered-character recognition by removing interfering-lines and adjusting feature weights. In: Proceedings of the 14th ICPR, pp. 1–3 (1988)

8. Tsai, C.M.: Intelligent region-based thresholding for color document images with highlighted regions. PR **45**, 1341–1362 (2012)

9. Otsu, N.: A thresholding selection method from gray-scale histogram. IEEE Trans. SMCs **9**, 62–66 (1979)

10. Chou, C.H., Lin, W.H., Chang, F.: A Binarization method with learning-built rules for document images produced by cameras. PR **43**(4), 1518–1530 (2010)

11. Niblack, W.: An Introduction to Digital Image Processing, pp. 115–116. Prentice Hall, Englewood Cliffs (1986)

12. Tseng, Y.H., Lee, H.J.: Document image Binarization by two-stage block extraction and background intensity determination. PAAs **11**, 33–44 (2008)

13. Sauvola, J., Pietikainen, M.: Adaptive document image binarization. PR **33**(2), 225–236 (2000)

14. Wolf, C., Jolion, J.M.: Extraction and recognition of artificial text in multimedia documents. PAAs **6**(4), 309–326 (2003)

15. Feng, M.L., Tan, Y.P.: Contrast adaptive Binarization of low quality document images. IEICE Electron. Express **1**(16), 501–506 (2004)

16. Androutsos, D., Plataniotis, K.N., Venetsanopoulos, A.N.: A novel vector-based approach to color image retrieval using a vector angular-based distance measure. CVIU **75**(1/2), 46–58 (1999)

17. Penpower Inc., Snap2PDF. https://play.google.com/store/apps/details?id=com.penpower.snap2pdf.play&hl=zh_TW

Neuromuscular Fatigue Analysis of Soldiers Using DWT Based EMG and EEG Data Fusion During Load Carriage

D.N. Filzah P. Damit[1,2(✉)], S.M.N. Arosha Senanayake[1],
Owais A. Malik[1], and Nor Jaidi Tuah[1]

[1] Faculty of Science, Universiti Brunei Darussalam, Gadong, Brunei Darussalam
nurhayatul.damit@mindef.gov.bn, 13h0355@ubd.edu.bn
[2] Performance Optimisation Centre, Ministry of Defence,
Bandar Seri Begawan, Brunei Darussalam

Abstract. This research reports peripheral and central fatigue of soldiers during load carriage on a treadmill. Electromyography (EMG) was used to investigate peripheral fatigue of lower extremity muscles and electroencephalography (EEG) was used for central fatigue detection on frontal lobe of the brain. EMG data were processed using Db5 and Rbio3.1 discrete wavelet transforms with a six levels of decomposition and EEG data were iteratively transformed into multi-resolution subsets of coefficients using Db8 wavelet function to perform the power spectrum analysis of alpha, beta and theta waves. Peak alpha frequency (PAF) was also calculated for EEG signals. A majority of significant results ($p < 0.05$) from EMG signals were observed in the lower extremity muscles using Db5 wavelet function at all conditions. While, significant changes were only observed during unloaded conditions at the frontal cortex. Significant changes ($p < 0.05$) in the PAF was also detected at certain conditions in the pre-frontal and frontal cortex. A significant increase in heart rate and rating of perceived exertion values were seen at all conditions. Hence, peripheral fatigue was the cause of the exhaustion sustained by the soldiers during load carriage which sends signals to the brain for decision making as to stop the exercise.

Keywords: Load carriage · EMG · EEG · Fatigue · Military

1 Introduction

Fatigue induced by prolonged foot marches for soldiers with load can lead to injuries affecting soldiers' mobility [1]. Fatigue can be defined as the sensation of tiredness affecting muscular performance [2]. Neuromuscular fatigue refers to reduction in the power production of a muscle despite increases in perception of effort and occurs on both central and peripheral mechanisms [2]. For central mechanisms, fatigue developed causes cognitive alterations in the brain which influences the excitement and recruitment of skeletal muscle and these changes in the brain activity can be assessed using electroencephalography (EEG) [2–5]. While the peripheral failure theory is where fatigue occurs directly at the muscles [2] which can be investigated using electromyography (EMG) that analyzes myoelectric signals in neuromuscular activation of

© Springer International Publishing AG 2017
N.T. Nguyen et al. (Eds.): ACIIDS 2017, Part II, LNAI 10192, pp. 602–612, 2017.
DOI: 10.1007/978-3-319-54430-4_58

muscles. Therefore, the relationship between central and peripheral fatigue is a safety mechanism, whereby motor unit firing rate is reduced by the central nervous system in order to avoid excessive damage of the muscle fibers or central fatigue is a response to afferent input from peripheral organs in order to prevent injury or death as reported in [2]. Discrete wavelet transform (DWT) of EMG has been reported in [6] to evaluate muscle fatigue during dynamic contractions, as it can simultaneously elucidate local spectral and temporal information from a signal and accurately indicate muscle fatigue in EMG signal [6]. The wavelet functions chosen are Daubechies 5 (Db5) and Reverse Biorthogonal 3.1 (Rbio3.1) which are proven to accurately predict induced fatigue spectral changes caused by repetitive exertions in muscles [6]. DWT has also been used in EEG for feature extraction due to its non-stationary and join time-frequency resolution nature and the wavelet function chosen is Daubechies 8 (Db8) due to its near optimal time-frequency localization properties and similarities to waveforms of EEG signals [7]. For this study, 8 muscles were selected due to their influence in walking and have shown signs of fatigue during load carriage. Previous studies on brain waves activity during high intensity exercises showed significant changes in theta, alpha and beta frequencies and fatigue index level at frontal cortex [3, 8–12]. Peak alpha frequency (PAF) corresponds to the discrete frequency showing the highest power within the alpha oscillation range and a putative marker of an individual's state of arousal and attention which can indicate fatigue [13, 14].

Hence, the aim of this research is to determine the effects of neuromuscular fatigue on soldiers' performance during submaximal load carriage exercise by addressing peripheral fatigue from EMG activity of lower extremity muscles and central fatigue from EEG activity in the frontal cortex of the brain using discrete wavelet transform.

2   Methodology

The overall data flow of the system is illustrated in Fig. 1. The details of each of these components are described below.

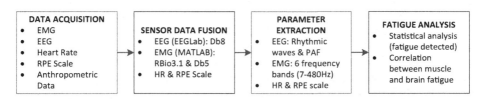

Fig. 1. Overall system overview of fatigue analysis during load carriage

2.1   Participants

Ten healthy male Officer Cadets with age 25.2 ± 1.6 years, height 170.7 ± 5.5 cm, weight 68.2 ± 5.5 kg (mean $\pm$ SD), participated in this study. All soldiers have no history of significant musculoskeletal injuries and cardiopulmonary pathology and has trained in carrying military load carriage system. All participants gave informed consent to participate in the study and investigators adhered to the Universiti Brunei Darussalam's ethical policies for their participant.

2.2 Experimental Protocol

Participants were asked to warm-up prior to experiment without load. For load trials, participants undergo a familiarization protocol of walking with the prescribed load for 5 min on a self-selected pace. There was a rest period of 5 to 10 min before the experiment starts to prevent carryover effects and was conducted in a climate-controlled laboratory. Before the trial, subjects' EEG data were recorded for 3 min while sitting down with eyes closed. The test is then followed by the trial of marching or walking on the treadmill with a speed of 6.4 km/h at 1% elevation. At every 5 min interval, the elevation was increased by 2% to induce fatigue. EMG data were recorded throughout the trial at 5 min intervals. Subjects continue this trial until they reach volitional exhaustion and the trial is stopped. Upon finishing, EEG data was recorded again for 3 min with subject sitting down and eyes closed. Heart rate (HR) was monitored at every 1 min interval until the subject stopped and rating of perceived exertion (RPE) was recorded at start, before each inclination and at the end of the trial.

2.3 Experimental Setup and Data Acquisition

The EMG parameters were recorded using two 8 channel wireless BioRadio units (Cleveland Medical Devices Inc, USA), surface EMG Ag/AgCl electrodes and sampling rate of 960 Hz. Foam snap electrodes were placed on eight muscles bilaterally: erector spinae (ES), multifiduus (MF), biceps femoris (BF), semitendinosus (SM), vastus lateralis (VL), gastrocnemius lateralis (GL), peroneus longus (PL) and tibialis anterior (TA). These signals were viewed and stored using BioCapture software package. EEG parameters were recorded using 14 channel wireless Emotiv EPOC Neuroheadset (Model 1.0, Emotiv, USA). Moist foam electrodes were placed on 14 equally distributed scalp locations (based on International 10–20 locations): AF3, AF4, F3, F4, F7, F8, FC5, FC6, P7, P8, O1, O2, T7, T8 and two reference electrodes: CMS and DRL. The sampling rate was set to 128 Hz, signals were viewed and stored using EmotivXavier TestBench software. The data from both sensing units were simultaneously recorded and transmitted to the same computer and the stored signals (raw EMG and EEG data) were exported to the files for further processing by MATLAB (R2012B, The MathWorks, Inc., Natick, USA). For load carriage trial, subjects were instructed to carry a 15 kg loaded military backpack (standard issue Royal Brunei Armed Forces, RBAF, vest and backpack). HR monitor (Polar ST4 watch, Polar, Kempele, Finland) was strapped to their chest to monitor changes in HR throughout the trial. Exertion levels were assessed using Borg's 6–20 Rating of Perceived Exertion (RPE) scale [15]. All participants wore the RBAF issued jungle boots and individual compressive sports pants and shirt.

2.4   EMG and EEG Data Processing

DWT analysis was carried out for the EMG and EEG signals recorded using wavelet toolbox from MATLAB (R2012b). The time windows corresponding to these exercise were was 30 s before the first inclination and 30 s before the subject stopped the exercise for EMG and 30 s before the start and after the exercise for EEG. At each DWT decomposition, the time-domain EMG and EEG signal ($E_{(t)}$) is passed through various low-pass and high-pass filters to obtain high and low frequencies [6]. The samples are passed through a high-pass filter of 480 Hz for EMG and 64 Hz for EEG and a full wave rectification was applied before the signal goes through a low-pass filter of 3 Hz for EMG and 1 Hz for EEG to obtain the approximation coefficient subsets and detail coefficients subsets [6]. The procedure was repeated for subsequent decompositions of the EMG and EEG signals into six levels. Frequency bandwidth (B) at different decomposition levels is obtained using the equation reported in [16]. The decomposed signal was reconstructed through synthesis filters [6] and the constructed signal could be expressed using the following vector:

$$E_{(t)} = (cA_6.cD_6.cD_5.cD_4.cD_3.cD_2.cD_1)$$

Where A_6 represents approximation coefficients at the 6th construction level and cD_1–cD_6 represents detail coefficients at the 1st to 6th reconstruction levels, respectively. The aptitude of DWT to extract features from the signal is reliant on the suitable option of mother wavelet function [6]. The wavelet functions used in this study were Db5 and Rbio3.1 for EMG and Db8 for EEG. These wavelet functions have been used previously in studies to extract muscle and brain activity features as well as denoise signals [6]. Prior to DWT analysis of EEG signals, movement artifacts (unusual data) were first removed by eye from continuous data using EEGLAB module and toolbox. Further artifacts, such as blinking, eye movement and motion artifacts were removed using independent component analysis. The decomposed EEG signal is further processed using Fast Fourier Transform in order to calculate the PAF, which is the frequency band where the peak alpha power value is found [8].

2.5   Parameter Extraction

The power distribution level from EMG and EEG signals were computed for loaded and unloaded conditions by performing DWT analysis. For EMG data, the power distribution was computed for the 8 muscles at 6 frequency bands (7–15 Hz, 15–30 Hz, 30–60 Hz, 60–120 Hz, 120–240 Hz and 240–480 Hz) at start and end of the trial using Db5 and Rbio3.1 wavelet functions. For EEG data, power distribution levels and fatigue index were computed using Db8 wavelet function for unloaded and loaded conditions before and after the trial. Eight electrode positions (AF3, AF4, F3, F4, F7, F8, FC5, and FC6) were analyzed focusing on prefrontal, mid-frontal and frontal cortex to provide theta, alpha and beta waves power levels including the calculation of fatigue index (ratio of alpha/beta, which indicates the state of arousal) where it was measured in dB. The PAF was also extracted from these conditions and measured in Hz. Mean

(SD) values for all the parameters were computed. Additionally, the HR, RPE and time taken for subjects to reach exhaustion for each trial session were extracted.

2.6   Statistical Analysis of Extracted Parameters

Paired two-tailed Wilcoxon test (α = 0.05) was performed using Statistical Package for the Social Sciences V. 22 (SPSS, Inc., Chicago) to calculate differences in EMG and EEG parameters. Paired two-tailed t-test for HR and RPE scale parameters. Spearman's correlation was performed on results that displayed significant changes in the low frequency values (7–30 Hz) for EMG and alpha and beta values for EEG.

3   Results

3.1   Time Differences

Unloaded condition, the average time taken for subjects to reach exhaustion was 34 min 15 s while the average time taken for subjects to reach exhaustion during loaded conditions was 19 min 37 s.

3.2   Heart Rate and RPE Values

At unloaded conditions, mean (SD) values of HR at start is 104 (11.12) and at end is 174 (8.35), while loaded it is 146 (18.32) and 180 (8.53) respectively. RPE mean (SD) values varied as well with 7.8 (1.55) and 16.9 (3.03) during no load and 10.1 (2.77) and 16.8 (2.10) during load respectively. All physiological data shows an increase of value between the start and end of the trial which were significant ($p < 0.001$).

3.3   EMG

3.3.1   No Load and Load (Start vs. End)

Db5. At unloaded conditions, all eight muscles showed significant changes ($p < 0.05$) between the start and the end of the trial (as shown in Table 1), mostly in the low (7 to 30 Hz) and mid (30 to 60 Hz) frequency range. Whereas during loaded conditions, seven muscles (ES, MF, SM, BF, VL, GL and PL) showed significant changes ($p < 0.05$) between the start and the end of the trial, mostly in the mid frequency range.

Rbio3.1. At unloaded conditions, six muscles (MF, SM, BF, VL, GL and PL) showed significant results ($p < 0.05$), mostly in the low and high frequency range (120 to 480 Hz). At load trial, four muscles (ES, BF, VL and GL) showed significant changes ($p < 0.05$) between the start and the end of the trial, majority in the high frequency.

Table 1. EMG p-values for Db5 during unloaded conditions (start vs. end) for all eight muscles (*statistical significance, p-value < 0.05)

Freq Band	ES	MF	SM	BF	VL	GL	TA	PL
7–15	0.046*	0.091	0.018*	0.028*	0.091	0.116	0.116	0.028*
15–30	0.046*	0.080	0.018*	0.028*	0.028*	0.028*	0.398	0.028*
30–60	0.075	0.091	0.046*	0.028*	0.043*	0.028*	0.398	0.018*
60–120	0.600	0.046*	0.028*	0.018*	0.028*	0.028*	0.237	0.128
120–240	0.398	0.091	0.630	0.018*	0.091	0.018*	0.018*	0.249
240–480	0.028*	0.028*	0.116	0.028*	0.046*	0.018*	0.173	0.091

3.3.2 No Load vs. Load (Start and End)

Db5. At the start of the trial, six muscles (ES, MF, SM, VL, GL and TA) showed significant changes when compared between unloaded and loaded conditions (Table 2), mostly in the low frequency range. At the point of exhaustion all eight muscles showed significant changes ($p < 0.05$) and majority in the low frequency range.

Table 2. EMG p-values for Db5 between both conditions at start time for all muscles (*p-value < 0.05)

Freq Band	ES	MF	SM	BF	VL	GL	TA	PL
7–15	0.043*	0.043*	0.237	0.063	0.237	0.028*	0.028*	0.128
15–30	0.128	0.128	0.237	0.128	0.046*	0.176	0.249	0.091
30–60	0.046*	0.917	0.398	0.345	0.735	0.237	0.398	0.345
60–120	0.028*	0.345	0.043*	0.500	0.063	0.063	0.612	0.249
120–240	0.128	0.138	0.043*	0.345	0.138	0.116	0.176	0.063
240–480	0.018*	0.028*	0.075	0.080	0.080	0.028*	0.499	0.116

Rbio3.1. At start (no load vs. load), 6 muscles (ES, MF, SM, BF, GL and TA) showed significant changes ($p < 0.05$) mostly in the low frequency range. At the time of exhaustion, six muscles (ES, MF, SM, BF, VL and GL) have also shown significant results ($p < 0.05$), mostly in the low frequency range.

3.4 EEG

No Load. AF4 showed significant increase ($p < 0.05$) in theta power level (Fig. 2) at end of trial. AF4, F4 and F7 have also showed significant increase in alpha power. Significant changes were also seen in beta power for AF4 and fatigue index for FC5 with a decrease in the fatigue index as subjects reach exhaustion.

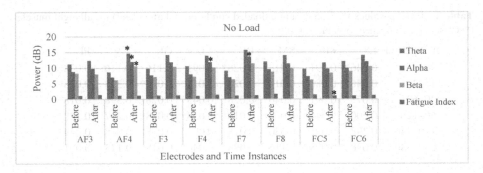

Fig. 2. Changes in EEG rhythmic waves power level (mean value) for 8 electrodes during unloaded conditions (*statistical significance, *p*-value < 0.05)

Load. There were no significant changes seen in any of the electrodes for all waves, despite changes were seen in the power level between before and after the trial.

No Load vs. Load. Before the trial, there were no significant differences seen between unloaded and loaded conditions for all electrode positions at all waves. However after the trial, theta and alpha power level at FC6 electrode were significantly lower (p < 0.05) at loaded conditions as compared to unloaded (as shown in Fig. 3). F7 has also shown a significantly higher fatigue index (p < 0.05) during unloaded conditions.

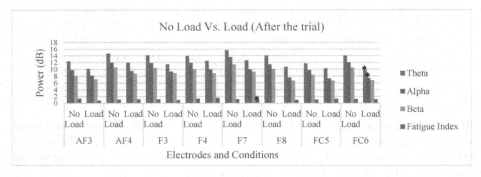

Fig. 3. Changes in EEG rhythmic waves power level (mean value) for 8 electrodes between unloaded and loaded conditions after the trials (**p*-value < 0.05).

Peak Alpha Frequency. During unloaded conditions, PAF significantly (p < 0.05) shift to lower frequencies at AF3 and F7 electrodes between before and after the trial (Table 3). AF4 has also shown significantly higher PAF during loaded conditions after trial.

Table 3. EEG results in p-values for peak alpha frequency (*p-value < 0.05)

Conditions (time instances)	AF3	AF4	F3	F4	F7	F8	FC5	FC6
No load (before vs. after)	0.038*	0.074	0.052	0.139	0.015*	0.241	0.092	0.444
load (before vs. after)	0.477	0.723	0.058	0.373	0.477	0.477	0.859	0.674
No load vs. load (before)	0.160	0.285	0.360	0.233	0.085	0.477	0.374	0.859
No load vs. load (after)	0.109	0.011*	0.091	0.482	0.635	0.341	0.440	0.260

3.5   EMG vs. EEG (Spearman's Correlation)

At unloaded condition, Spearman's correlation was performed on muscles that have shown significant changes (ES, SM, BF and PL) at 7 to 15 Hz against alpha values from F7 electrode. A significant negative correlation ($r_s = -0.786$, p $= 0.036$) was seen between ES muscle and alpha waves at F7 at point of exhaustion. Spearman's correlation was also performed on muscles that have shown significant changes (ES, SM, BF, VL, GL and PL) at 15 to 30 Hz against beta values at AF4 electrode at no load. A significant negative correlation ($r_s = -0.929$, p $= 0.003$) was reported between VS muscle and beta waves at AF4 at the start of trial. There were no significant correlation between EMG results (loaded trial) at point of exhaustion between muscles: SM, BF, VL and PL at 7 to 15 Hz against alpha values from FC6.

4   Discussion

The important results of this study was the detection of fatigue in both the muscle and the brain during either unloaded or loaded carriage trial. The overall outcomes of this research work is consistent with conclusions reached in previous literatures based on the comparative analysis done using Table 4. DWT analysis proved to be a useful tool for detecting fatigue induced based on the spectral changes in both EMG and EEG data especially using Db5 and Rbio3.1 wavelet functions. However, the spectral changes detected were found to be dependent on the type of wavelet function used. For EMG, Db5 showed a more consistent and statistically significant trend for all conditions. Most of the significant trends were also detected in the lower and middle frequency bands and a majority of the muscles showed significant changes in Db5 as compared with Rbio3.1. Statistically significant changes observed in the power distribution at the lower and middle frequency, with the development of fatigue shows that muscle fatigue results in a shift to lower frequency range.

EEG data obtained from DWT analysis using Db8 wavelet function, between before and after the trial, showed most of the significant changes present in the brain waves during unloaded condition. In F6, the significant decrease in theta spectral power during loaded conditions with the development of fatigue, can be due to decrease in executive attention, thus causing problems in monitoring and modulating sensory inputs where subjects felt a sensation of fatigue earlier and stops early as compared to unloaded conditions. However, significant increase in theta power of AF4 during unloaded conditions might affect vigilance leading to decrease of cortical arousal

Table 4. Parameters that have shown consistent results with those of previous studies [3–6, 8–11, 13, 17, 18].

	Parameters experimented in this research	Hussain et al. (2012) [17]	Chowdhury et al.(2013)[6]	Nielsen et al. (2001) [9]	Bailey et al. (2008) [10]	Kacem et al. (2014) [11]	Baumeister et al (2012) [3]	Mechau et al. (1998) [5]	Schneider et al (2009) [4]	Brummer et al. (2011) [18]	Ng & Raveendran (2007) [8]	Gutmann et al. (2015) [13]
EMG	Shift to low frequencies with muscle fatigue	X	X									
EMG	Increase in power at low frequencies		X									
EMG	Inconsistent trends at high frequencies		X									
EEG (unloaded)	Increase in theta at AF4			X	X	X						
EEG (unloaded)	Increase in beta AF4 unloaded			X	X				X			
EEG (unloaded)	Increase in AF4, F4, F7 alpha			X	X			X	X	X		
EEG (loaded)	Decrease in FC6 alpha and theta with load						X					
EEG (loaded)	PAF at AF4 higher in load vs. unloaded										X	X

during long monotonous task requiring sustained attention and thus, decreasing performance based on previous research reported. The significant increase in AF4 alpha power during unloaded condition was due to decreases cortical activation [5]. Significant increase monitored in AF4 beta power during unloaded condition was due to greater attentional demand, the resistance to movements and negative feelings of emotional intensity as also reported previously. PAF measured in AF3 and F7 during unloaded conditions led to significant decrease was an implication of reduced speed mental processes and physical fatigue set in. Subjects has also felt more fatigue during loaded conditions as compared to unloaded, evident in the increase in PAF at AF4. The significant negative correlation monitored between muscles and brain waves was due to fatigue detected in the muscles correlated with changes observed in the frontal cortex. These correlation results imply previously triggered fatigue stored in memory leading to tolerance levels and hence, decision making occurs in order to stop the exercise [4].

Therefore, the results of this study proves that submaximal unloaded and loaded carriage trial has led to neuromuscular fatigue; in particular lower back and lower limb muscles, prefrontal, mid-frontal and frontal areas of the cortex. Minimal changes in the brain activation level implies fatigue inducing at the peripheral level than central.

5   Conclusions

The findings report in this research provide new insights into peripheral and central fatigue in the neuromuscular system during high intensity exercise of soldiers during unloaded and loaded carriage trial. Moreover, the results prove peripheral failure theory played a critical role in neuromuscular fatigue compared to the central failure theory. Soldiers also stopped the trial earlier during loaded trials compared to unloaded due to lack of fatigue observed without allowing enough time to induce fatigue consistently. Thus, the overall outcome of this research demonstrates that fatigue induces in the muscles first leading to the brain and body to stop the exercise.

Acknowledgements. Authors would like to thank all participants for their time and effort. At the same time, authors appreciate Officer Cadet School, Royal Brunei Armed Forces and the Royal Brunei Armed Forces, Brunei for their contribution and support.

References

1. Knapik, J., Reynolds, K.L., Harman, E.H.: Soldier load carriage: historical, physiological biomechanical and medical aspects. Mil. Med. **169**, 1–45 (2004)
2. Abiss, C., Laursen, P.: Models to explain fatigue during prolonged endurance cycling. Sports Med. **35**(10), 865–898 (2005)
3. Baumeister, J., Reinecke, K., Schubert, M., Schade, J., Weiss, M.: Effects of induced fatigue on brain activity during sensorimotor control. Eur. J. Appl. Physiol. **112**, 2475–2482 (2012)
4. Schneider, S., Askew, C., Diehl, J., et al.: EEG activity and mood in health orientated runners after different exercise intensities. Physiol. Behav. **96**, 709–716 (2009)
5. Mechau, D., Mucke, S., Weib, M., Liesen, H.: Effect of increasing running velocity on electroencephalogram in a field test. Eur. J. Appl. Physiol. **78**, 340–345 (1998)
6. Chowdhury, S.K., Nimbarte, A.D., Jaridi, M., Creese, R.: Discrete wavelet transform analysis of surface electromyography for the fatigue assessment of neck and shoulder muscles. J. Electromyogr. Kinesiol. **23**, 995–1003 (2013)
7. Murugappan, M., Ramachandran, N., Sazali, Y.: Classification of human emotion from EEG using discrete wavelet transform. J. Biomed. Sci. Eng. **3**, 390–396 (2010)
8. Ng, S.C., Raveendran, P.: EEG peak alpha frequency as an indicator for physical fatigue. In: Jarm, T., Kramar, P., Zupanic, A. (eds.) Medicon, IFMBE Proceedings, vol. 16, pp. 517–520. Springer, Heidelberg (2007)
9. Nielsen, B., Hyldig, T., Bidstrup, F.: Brain activity and fatigue during prolonged exercise in the heat. Eur. J. Physiol. **442**, 41–48 (2001)
10. Bailey, S., Hall, E., Folger, S., Miller, P.: Changes in EEG during graded exercise on a recumbent cycle ergometer. J. Sports Sci. Med. **7**, 505–511 (2008)

11. Kacem, A., Ftaiti, F., Chamari, K., Dogui, M., Grelot, L., Tabka, Z.: EEG-related changes to fatigue during intense exercise in the heat in sedentary women. Health **6**, 1277–1285 (2014)
12. Robertson, C.V., Marino, F.E.: Prefrontal and motor cortex EEG responses and their relationship to ventilator thresholds during exhaustive incremental exercise. Eur. J. Appl. Physiol. **115**(9), 1939–1948 (2015)
13. Gutmann, B., Mierau, A., Hulsdunker, T., et al.: Effects of physical exercise on individual resting state EEG Alpha peak frequency. Neural Plast. **2015** (2015)
14. Klimesch, W.: EEG alpha and theta oscillations reflect cognitive and memory performance: a review and analysis. Brain Res. **29**(2–3), 169–195 (1999)
15. Borg, G.: Psychophysical bases of perceived exertion. Med. Sci. Sports Exerc. **14**(5), 377–381 (1982)
16. Cong, F., Huang, Y., Kalyakin, I., Li, H., Huttunen-Scott, T., Lyytinen, H., et al.: Frequency-response-based wavelet decomposition for extracting children's mismatch negativity elicited by uninterrupted sound. J. Med. Biol. Eng. **32**, 205–213 (2012)
17. Hussain, M.S., Mamun, Md.: Effectiveness of the wavelet transform on the surface EMG to understand the muscle fatigue during walk. Meas. Sci. Rev. **12**(1), 28–33 (2012)
18. Brummer, V., Schneider, S., Abel, T., Vogt, T., Struder, H.K.: Brain cortical activity is influenced by exercise mode and intensity. Med. Sci. Sports Exerc. **43**, 1863–1872 (2011)

Manifold Methods for Action Recognition

Agnieszka Michalczuk[1,2], Kamil Wereszczyński[1,2], Jakub Segen[1],
Henryk Josiński[1,2], Konrad Wojciechowski[1], Artur Bąk[1],
Sławomir Wojciechowski[1], Aldona Drabik[1], and Marek Kulbacki[1(✉)]

[1] Polish-Japanese Academy of Information Technology,
Koszykowa 86, 02-008 Warszawa, Poland
mk@pja.edu.pl
[2] Institute of Informatics, Silesian University of Technology,
Akademicka 16, 44-100 Gliwice, Poland

Abstract. Among a broad spectrum of published methods of recognition of human actions in video sequences, one approach stands out, different from the rest by not relying on detection of interest points or events, extraction of features, region segmentation or finding trajectories, which are all prone to errors. It is based on representation of a time segment of a video sequence as a point on a manifold, and uses a geodesic distance defined on manifold for comparing and classifying video segments. A manifold based representation of a video sequence is obtained starting with a 3d array of consecutive image frames or a 3rd order tensor, which is decomposed into three $3 \times k$ arrays that are mapped to a point of a manifold. This article presents a review of manifold based methods for human activity recognition and sparse coding of images that also rely on a manifold representation. Results of a human activity classification experiment that uses an implemented action recognition method based on a manifold representation illustrate the presentation.

Keywords: Manifold methods · Action recognition

1 Introduction

Recognition of human actions [1] and tracking moving objects [2] in video sequences are active research areas in computer vision. Most approaches to action recognition are based on identifying localized entities in an image or video, such as objects, points of interest, space-time events, or trajectories, forming a description of a video sequence using the identified entities and classifying actions based on such descriptions [3,4]. An exception to this trend are methods of action recognition that represent consecutive segments of a video sequence as points on a manifold and compare them using manifold based distance measures [9]. This approach doesn't require image segmentation, feature detection or trajectory formation.

Section 2 gives a brief description of background notions of tensors, manifolds, the Higher Order Singular Value Decomposition (HOSVD), and geodesic

N.T. Nguyen et al. (Eds.): ACIIDS 2017, Part II, LNAI 10192, pp. 613–622, 2017.
DOI: 10.1007/978-3-319-54430-4_59

distance defined on a manifold. Section 3 describes two methods of action classification, an object tracking method, and sparse coding and learning of a dictionary, based on manifolds. Section 4 discusses the results of a manifold based classification experiment.

2 Background

2.1 Manifold

A manifold is a topological space where each point has a neighborhood that is homeomorphic to the Euclidean space [5,7]. A Grassmann manifold is defined [6] as:

Definition 1. *Grassmann manifold. For natural numbers m and n with $m > n$, we define Grassmann manifold $\mathcal{G}(m,n)$ to be the space of all n-dimensional linear subspaces of the real vector space $\mathbb{R}^m$ of dimension m.*

Since any n dimensional linear subspace of $\mathbb{R}^m$ is a point on a Grassmann manifold, any $n \times n$ full rank matrix can be mapped to a point of a Grassmann manifold, as the linear subspace spanned by the row vectors of the matrix.

2.2 Principal Angles

Points on the manifold can be compared using the geodesic distance, which is defined on a $\mathcal{G}(n,p)$ using *principal angles* [12]. Principal (or canonical) angles were defined in [11] based on Hotelling [10] work on relations between two sets of variables.

Definition 2. *Let $\mathcal{X} \subset \mathbb{R}^n$ and $\mathcal{Y} \subset \mathbb{R}^n$ be subspaces; $dim(\mathcal{X}) = p$, and $dim(\mathcal{Y}) = q$. The principal angles between $\mathcal{X}$ and $\mathcal{Y}$: $\Theta(\mathcal{X}, \mathcal{Y}) = [\theta_1, ..., \theta_m]$, where $\theta_k \in [0, \pi/2], k = 1, ..., m$ are defined recursively as:*

$$cos(\theta_k) = max_{x \in \mathcal{X}} max_{y \in \mathcal{Y}} |x^T y| = |x_k^T y_k|$$

where x and y are unit vectors.

2.3 Tensor Space

A video sequence can be represented as 3rd order tensor where the first mode is frame height, the second one is width and the third one is time (or frame number).

Definition 3. *Tensor. Let $I_1, I_2, ..., I_N \in \mathbb{N}$ denote index upper bounds. A tensor $\mathcal{A} \in \mathbb{R}^{I_1 \times I_2 \times ... \times I_n}$ of order N is an N-dimensional array where elements $y_{i_1, i_2, ..., i_n}$ are indexed by $i_n \in 1, 2, ..., I_n$ for $1 \leq n \leq N$.*

A tensor fiber is a one-dimensional fragment of a tensor, obtained by fixing all indices except for one [8]. Tensor is often represented by matrices by applying the unfolding operation.

Definition 4. *Unfolding. The n-mode unfolding of tensor $\mathcal{A} \in \mathbb{R}^{I_1 \times I_2 \times ... \times I_n}$ is denoted by $A^{(n)}$ and arranges the n-mode fibers into columns of a matrix.*

2.4 Higher Order Singular Value Decomposition (HOSVD)

HOSVD is a general decomposition method for multidimensional arrays where each unfolded matrix can be factorized using the Singular Value Decomposition (SVD). Let $\mathcal{A}$ be a 3rd order tensor representing a video sequence. It is unfolded along all three modes of tensor to $A^{(1)}, A^{(2)}$ and $A^{(3)}$ matrices. Each of these matrices can be factorized using SVD as:

$$A^{(k)} = U^{(k)} \Sigma^{(k)} V^{(k)^T}$$

where Σ is a diagonal matrix, $U^{(k)}$ and $V^{(k)}$ are orthogonal matrices spanning the column and row spaces of A respectively. Each video sequence represented by 3rd order tensor can be expressed as:

$$\mathcal{A} = \mathcal{S} \times_1 V^{(1)}_{horizontal\_motion} \times_2 V^{(2)}_{vertical\_motion} \times_3 V^{(3)}_{appearance}$$

where: $\mathcal{S}$ is a core tensor, $V^{(k)}$ are orthogonal matrices from HOSVD decomposition [12] and $\times_k$ denotes k-mode multiplication. Since the matrix $V^{(k)}$ is orthogonal, its rows or columns can be used as a basis of a linear vector subspace, and mapped onto a point on a Grassmann manifold.

3 Methods Description

3.1 Action Classification on Product Manifolds

Manifold based method of action recognition proposed by Lui and Beveridge in [9] and shown in Fig. 1 uses Principal Angles as a measure of distance between sequences. Knyazev in [11] proved in Theorem 2.1 that increasingly ordered vector of cosines of principal angles is equal to vector of singular values of $X^T Y$, where X and Y are matrices of column vectors from orthonormal bases for $\mathcal{X}$ and $\mathcal{Y}$, respectively. Let $X^T Y = U \Sigma V^T$. Knyazev proved that principal vectors corresponding with $\mathcal{X}$ and $\mathcal{Y}$ are given by first m column of XU and YV. Using knowledge introduced above, principal angles computation method could be defined based on unfolded tensor created from video sequence. As it could be seen on schema in Fig. 1, principal angles are computed using $V^{(k)}$ matrices from unfolded tensors. After this operation three angles are obtained (one for each unfolding slice): $\theta^{(k)}, k = 1, 2, 3$. Because the space of Grassmann manifold is curved, Lui in [9] defined chordal distance between two video sequences:

$$dist(\mathcal{A}, \mathcal{B}) = ||sin(\Theta)||_2$$

where: $\mathcal{A}$ and $\mathcal{B}$ are N order tensor and elements of $\Theta = (\theta_1, \theta_2, ..., \theta_N)$ are principal angles for each of tensor mode. Nearest neighbor classifier with chordal distance was used for action recognition in video sequences.

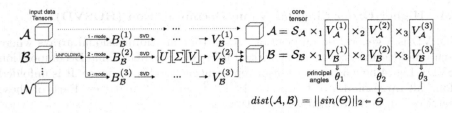

Fig. 1. Action classification on product manifold schema.

3.2   Linear Least Squares Regression Model in Grassmann Manifold

On the basis of structures described above Lui in [12] proposed another app-
roach for video sequences classification using Linear Least Squares Regression
(LLSQ) concept for Grassmann manifolds with an extension to non-linear case.
In Euclidean spaces LLSQ could be written as:

$$R(\beta) = ||y - A\beta||^2$$

where $y = A\beta \in \mathbb{R}^n$ is regression value, $A(a_1|a_2|...|a_k|) \in \mathbb{R}^{n \times k}$ is training
examples matrix and $\beta \in \mathbb{R}^k$ is fitting parameter which could be computed by
minimizing the residual sum-of-square. Therefore, after minimization, regression
equation for the training set has a form: $\widehat{y} = A\widehat{\beta}$ where:

$$\widehat{\beta} = (A^T A)^{-1} A^T y$$

One method for transformation of linear structures to non-linear ones is to use
a kernel function. Lui in [12] proposed the RBF kernel, which has a form:

$$x \star y = e^{-\Sigma_k \theta_k / 2}$$

where x, y are the points on manifold, θ_k is the principal angle computed for x
and y, and $\star$ is kernel operator. In that case regression could be re-written in
manifold form as:

$$\Psi(y) = A^{(k)}(A^{(k)} \star A^{(k)})^{-1}(A^{(k)} \star y^{(k)})$$

where $A^{(k)}$ is a set of orthogonal matrices factorized from HOSVD made
for one mode of unfolding, k is tensor mode. Each element of $A^{(k)}$ represents one video sequence from training set for one mode. Similarly to Euclidean case fitting parameter, here called weighted vector, could be defined as:
$w = (A^{(k)} \star A^{(k)})^{-1}(A^{(k)} \star y^{(k)})$.

This vector represents a set for which regression has been made. For each
group of video sequences containing the same action (e.g. walking) such representation is computed. For sequence from behind of training set the closest w is
founded and this video is considered as containing action connected with w.

The problem is, that $w \in \mathcal{V}$ is a vector but each element of A is on manifold
(factorized from HOSVD), therefore regression equation needs to be modified to

have the same domain. Lui [12] proposed to use operator $\bullet$ which maps point from vector space to factor manifold. This operator is composition of function $\mathcal{G} \circ \mathcal{H}$ where $\mathcal{H} : \mathcal{M} \to \mathcal{V}$ and $\mathcal{G} : \mathcal{V} \to \mathcal{M}$; $\mathcal{V}$ - vector space, $\mathcal{M}$ - manifold.

Now, regression formula for manifold case could be written as follows:

$$\Psi(y) = A^{(k)} \bullet (A^{(k)} \star A^{(k)})^{-1}(A^{(k)} \star y^{(k)}) = A^{(k)} \bullet w$$

Generally, there are some training samples on manifold and a weighted vector w for each training sample to find such $y \in \mathcal{M}$ that is as close as possible to all training samples. Lui in [12] used Karcher mean [13,14] introducing modification employing tangent space considering intrinsic geometry of manifold.

To implement this algorithm two mappings should be defined: standard logarithmic map and exponential map for crossing from manifold to tangent space and backward.

Dreisigmeyer in [14] announced algorithm for computing logarithmic map for Grassmann manifolds, for given points $p, q \in \mathcal{G}(n, k)$. It contains 3 steps:

1. Find CS decomposition $p^T q = VCZ^T, p_\perp^T q = WSZ^T$ where V, W, Z are orthogonal matrices and C, S are diagonal matrices fulfilling the following condition: $C^T C + S^T S = I$
2. Delete or add zero rows from S so that it is a square matrix, delete corresponding columns from W or add zero columns to W to make sizes of W and S compatible with each other;
3. Let $U = p_\perp W$ and $\Theta = \arctan(SC^{-1})$.

The interesting output of this algorithm are U, V and Θ, from which logarithm of $p, q \in G(n, k)$ could be formulated as:

$$Log_p(q) = U\Theta V^T$$

Dreisigmeyer in [14] gives a formulation and proof for this algorithm.

Mapping from tangent space to manifold is also given by the same author:

$$\exp_p(\omega) = pV \cos(\Theta) + U \sin(\Theta)$$

where a point $\omega \in T_p \mathcal{G}(n, k)$ in the tangent space to manifold $\mathcal{G}(n, k)$ at point $p \in \mathcal{G}(n, k)$ has decomposition SVD $\omega = U\Theta V^T$. Now, following [14] Karcher Mean algorithm for Grassmann manifolds could be formulated as follows:

Initialize tolerance $\delta > 0$ and set $q = p_1$
while $\omega <= \delta$ **do**
$\quad$ $\omega = 1/m \sum_{i=1}^{m} Log_q(p_i)$;
$\quad$ **if** $\| \omega \| > \delta$ **then**
$\quad\quad$ $\omega = U\Sigma V^T$;
$\quad\quad$ $q = qV \cos(\Sigma) + U \sin(\Sigma)$;
$\quad$ **end**
end

Each action is represented by three sub-regression models $(\Psi_j^{(1)}, \Psi_j^{(2)}, \Psi_j^{(3)})$ for the class j and each tensor mode. In the test phase the 1NN classifier was used where distance between the regression output and evaluated sequence was determined.

3.3 Sparse Coding and Dictionary Learning

Sparse coding of signals means a method of signal representation using a few values (atoms) from a whole signal domain. If atoms are tied with objects which have been labeled, the direct method could be used. If each atom has a specific label, prediction could be introduced as finding the closest atom to the given object and labeling this object using found atom's label. In opposite case, if atoms just represent domain without having specific labels, for each training and predicted samples sparse codes could be computed and used for fitting well-known classifiers, e.g. SVM, kNN.

Riemannian Manifold Representation of Positive Defined Matrices. Harandi et al. in [15] proposed using set of Riemannian manifold's points as a space for creating dictionary and sparse codes. They show exemplary implementation to: (1) face recognition, (2) texture classification and (3) person re-identification. Authors of the current paper are working on implementing this method for action recognition. Lui in [16] mentioned that for action classification feature-based model could be used as alternative for SVD-based video representation. This approach is based on computing covariance descriptor [17] which is positive-defined matrix actually. Therefore, it could be identified as point on Riemannian manifold.

The video sequence could be expressed by positive defined matrices, therefore Harandi et al. in [15] shows connection between such matrices divergence and metric in Riemannian manifold. They pointed to a similarity between the Stein divergence and AIRM (Affine Invariant Riemannian Metric) introduced by Pennec in [18]) on symmetric positive defined matrices of dimension d, which is a Riemannian manifold Sym_+^d of negative curvature. Harandi's [15] reasoning relies on *Bregmann matrix divergence* $D_\varsigma(X,Y)$ concept and its *Jensen-Shannon* symmetrisation.

Additionally, for two points $X, Y \in Sym_+^d$ the AIRM is defined as: $d_g^2 = ||log_X(Y)||_X^2 = Tr\{\log^2(X^{-\frac{1}{2}}YX^{-\frac{1}{2}})\}$ where $log_X(Y) = X^{\frac{1}{2}}\log(X^{-\frac{1}{2}}YX^{-\frac{1}{2}})$ $X^{\frac{1}{2}}$. Harandi et al. in [15] pointed out the relation establishes a bound between the geodesic distance in Riemannian manifold and Stein divergence. The second relation proves that Stein divergence along geodesic curves behaves similar to true Riemannian geometry. All together it allows implementation of Riemannian manifold concept to sparse coding problem for object that could be represented by matrices especially for action recognition in video sequences. The following is presented by Sra [19], after Harandi et al. [15].

Kernel Sparse Coding. Let $\mathbb{D} = \{D_1, D_2, ..., D_N\}; D_i \in Sym_+^d$ be a Riemann dictionary and so called embedding function: $\phi : Sym_+^d \rightarrow \mathbb{H}$, where $\mathbb{H}$ is a vector space. For $X \in Sym_+^d$ sparse vector $v \in \mathbb{R}^N$ could be created using $\phi(X)$. Vector v is linear combination over $\{\phi(D_1), \phi(D_2), ..., \phi(D_N)\}$. So this is the minimization problem.

Learning Approaches. If atoms in dictionary $\mathbb{D}$ are associated with class objects the sparse codes could be used for classification. This approach is based on searching the closest atom in dictionary and associating its label to object being classified. Authors proposed so called residual error using for finding the closest atom:

$$\epsilon_i(X) = ||\phi(X) - \sum_{j=1}^{N} v_j \phi(D_j)\delta(l(j) - i)||^2$$

where (l_j) is label of class D_j and $\delta(x)$ is discrete Dirac function.

If atoms in dictionary $\mathbb{D}$ are not labeled, sparse codes became feature vectors and are used for fitting Euclidean based classifiers like SVM.

Creating Dictionaries. Given set $\Omega = \{X_1, X_2, ..., X_m\}; X_i \in Sym_+^d$ learning of dictionary $\mathbb{D}$ could be formulated as the problem of minimizing the energy function:

$$J = (||\phi(X) - \sum_{i=1}^{N} v_i \phi(D_i)||^2 + \lambda ||v||_1)$$

Harandi et al. in [15] used derivative $\frac{\partial J}{\partial D_r}$ of J and fact that $\bigtriangledown_X S(X, Y) = (X+Y)^{-1} - \frac{1}{2}X^{-1}$. They proposed iterative method of computing atoms. Finally atoms are normalized by their second norm in each step.

Implementation for Video Sequences. For implementation of the method described above, a positive defined matrix is needed. Such an object was presented by Tuzel et al. in [17] and called a covariance descriptor. Let F be $W \times H \times d$ where d is dimension of feature image got from image I: $F(x, y) = \phi(I, x, y)$, where ϕ is any mapping made for pixel (x, y).

Definition 5. *Covariance descriptor. Let $R \subseteq F$ be a rectangular region on feature image F, $\{z_k\}_{k=1,...,n}$ be d-dimensional feature point belonging to R. Covariance matrix for region R is:*

$$C_R = \frac{1}{n-1} \sum_{k=1}^{n} (z_k - \mu)(z_k - \mu)^T$$

where μ is the mean of the points.

This definition is made for images and has to be extended for video sequence case. Normalized video sequence could be imagined as $W \times H \times T$ cube. For each point in this cube (x, y, t) descriptor of length d could be computed. It could be intensity, color, gradients, filter responses etc. After concatenation in time axis we got $W \times H \times T \cdot d$ feature image extracted for cube. In that case z_k represents $T \cdot d$-dimensional feature point being concatenation of descriptors; C_R has also greater dimensions: $T \cdot d \times T \cdot d$. Such a matrix could be used as input for sparse coding.

4   Results

Authors are in the process of creating their own implementation of the above-described methods. Method presented in subsection Action Classification on Product Manifolds was tested on the KTH database. It contains six types of human action. Actions represent "gait" (walking, jogging, running) or "no gait" (boxing, handwaving and handclapping). Each action is represented in database by 25 subjects performing four different scenarios: outdoors ($s1$), outdoors with scale variation ($s2$), outdoors with different clothes ($s3$) and indoors ($s4$). In our test for faster calculations we used 5 subjects in three scenarios: s1, s3 and s4. Because gait sequences contain the subject's gait repeated, each sequence was divided into single gait and time series were normalized to 32 frames. Sequences without gait were also divided into 32 frames. Each frame has been converted to grayscale and scaled to 32×32 matrix using Bicubic interpolation. In result we obtained 360 sequences represented by $32 \times 32 \times 20$ tensors. As a measure of the similarity between two sequences chordal distance was used and we also treated vectors of principal angles as feature vectors (FV). Sequence similarity was tested for each tensor and each n-mode of tensor to check which mode of tensor contains the highest individual characteristics (appearance or time series). Table 1 includes results of action recognition where each next experiment has a smaller training set. The best result of classification (94%) was obtained using the nearest neighbor in the first experiment where the largest training set was used and measure of similarity was a feature vector. A little bit worse result (93% of correctly classified sequences) was obtained for feature vector being 3-rd order tensor which represents spatial-temporal dimension.

Table 1. The results of action recognition using the KTH database.

Id	Validation method	Chordal distance	Angle 1-mode	Angle 2-mode	Angle 3-mode	FV 1-mode	FV 2-mode	FV 3-mode	FV all mode
1	Leave-one-out	0.73	0.53	0.70	0.71	0.87	0.89	0.93	**0.94**
2	50% training	0.67	0.43	0.66	0.64	0.74	0.77	0.89	0.87
3	25% training	0.62	0.42	0.61	0.63	0.79	0.74	0.9	0.90
4	12,5% training	0.51	0.34	0.5	0.51	0.68	0.64	0.77	0.80

Table 2 represents a confusion matrix in which false detections are visible. All of false detections of action refer to running vs jogging in all performed experiments. In experiments where 12.5% of sequences were in training set, the most of mis-classified cases are related to jogging vs running actions, which are so similar that distinguishing between them is problematic also for human. It could be observed in confusion matrix (Table 3). Described actions connected with "gait" and "no gait" are pretty good distinguished from each other. For all

Table 2. Confusion matrix - leave-one-out

	Boxing	Handclapping	Handwaving	Jogging	Running	Walking
Boxing	56	0	0	0	0	0
Handclapping	0	60	0	0	0	0
Handwaving	0	0	60	0	0	0
Jogging	0	0	0	56	4	0
Running	0	0	0	16	44	0
Walking	0	0	0	0	0	64

315 sequences only one gait sequence (walking) was incorrectly classified as "no gait" sequence (hand clapping). Actions' distinguishing in the set of "no gait" (boxing, handclapping, handwaving) is greater: 91% than for "gait" actions: 70%.

Table 3. Confusion matrix - 12,5% training set

	Boxing	Handclapping	Handwaving	Jogging	Running	Walking
Boxing	46	0	1	0	0	0
Handclapping	1	53	11	0	0	1
Handwaving	0	1	39	0	0	0
Jogging	0	0	0	41	23	11
Running	0	0	0	12	27	1
Walking	0	0	0	1	1	45

5   Conclusion

The authors conclude that methods described above are useful for future work in the project Intelligent video analysis system for behavior and event recognition in surveillance networks project for two purposes:

1. As pre-selector distinguishing general classes of action like "gait" and "no gait".
2. As one of weak classifiers in ensemble learning processing pipeline. Such classifier has to exclude some classes with high confidence level. Because all described methods concern measurement of differences between video sequences, such confidence could be defined.

Acknowledgement. This work has been supported by the National Centre for Research and Development (project UOD-DEM-1-183/001 "Intelligent video analysis system for behavior and event recognition in surveillance networks").

References

1. Cheng, G., Wan, Y., Saudagar, A.N., Namuduri, K., Buckles, B.P.: Advances in Human Action Recognition: A Survey. arXiv preprint arXiv:1501.05964v1 (2015)
2. Li, M., Cai, Z., Wei, C., Yuan, Y.: A survey of video object tracking. Int. J. Control Autom. **8**(9), 303–312 (2015)
3. Wang, H., Ullah, M.M., Klaser, A., Laptev, I., Schmid, C.: Evaluation of local spatio-temporal features for action recognition. In: BMVC (2009)
4. Uijlings, J., Duta, I.C., Sangineto, E., Sebe, N.: Video classification with densely extracted HOG/HOF/MBH features: an evaluation of the accuracy/computational efficiency trade-off. Int. J. Multimed. Info. Retr. **4**, 33–44 (2014)
5. Wells, R.O.: Differential Analysis on Complex Manifolds. Springer, New York (2008)
6. Sat, H.: Algebraic Topology: An Intuitive Approach. American Mathematical Society, Providence (1999)
7. Guillemin, V., Pollack, A.: Differential Topology. Prentice-Hall, Upper Saddle River (1974). pp. 2–5
8. Cichocki, A., Zdunek, R., Phan, A.H., Amari, S.: Nonnegative Matrix and Tensor Factorizations: Applications to Exploratory Multiway Data Analysis and Blind Source Separation. Wiley, Hoboken (2009)
9. Lui, Y.M., Beveridge, J., Kirby, M.: Action classification on product manifolds. In: Proceedings of the IEEE Conference on CVPR, pp. 833–839 (2010)
10. Hotelling, H.: Relations between two sets of variates. Biometrika **28**, 321–377 (1936)
11. Knyazev, A.V., Zhu, P.: Principal Angles Between Subspaces and Their Tangents. Technical report TR2012-058, Mitsubishi Electric Research Laboratories (2012)
12. Lui, Y.M.: Human gesture recognition on product manifolds. J. Mach. Learn. Res. **13**(1), 3297–3321 (2012)
13. Karcher, H.: Riemannian center of mass and mollifier smoothing. Commun. Pure Appl. Math. **30**(5), 509–541 (1977)
14. Dreisigmeyer, D.W.: Direct Search Algorithms Over Riemannian Manifolds (2007). http://ddma.lanl.gov/Documents/publications/dreisigm-2007-direct.pdf
15. Harandi, M.T., Sanderson, C., Hartley, R., Lovell, B.C.: Sparse coding and dictionary learning for symmetric positive definite matrices: a kernel approach. In: Fitzgibbon, A., Lazebnik, S., Perona, P., Sato, Y., Schmid, C. (eds.) ECCV 2012. LNCS, vol. 7573, pp. 216–229. Springer, Heidelberg (2012)
16. Lui, Y.M.: Advances in matrix manifolds for computer vision. Image Vis. Comput. **30**(6), 380–388 (2012)
17. Tuzel, O., Porikli, F., Meer, P.: Region covariance: a fast descriptor for detection and classification. In: Leonardis, A., Bischof, H., Pinz, A. (eds.) ECCV 2006. LNCS, vol. 3952, pp. 589–600. Springer, Heidelberg (2006). doi:10.1007/11744047_45
18. Pennec, X.: Intrinsic statistics on Riemannian manifolds: basic tools for geometric measurements. J. Math. Imaging Vis. **25**(1), 127–154 (2006)
19. Sra, S.: Positive definite matrices and the symmetric Stein divergence. arXiv:1110.1773 (2012)

Optical Flow Based Face Anonymization in Video Sequences

Kamil Wereszczyński[1,2], Agnieszka Michalczuk[2], Jakub Segen[1],
Magdalena Pawlyta[1], Artur Bąk[1], Jerzy Paweł Nowacki[1],
and Marek Kulbacki[1(✉)]

[1] Polish-Japanese Academy of Information Technology,
Koszykowa 86, 02-008 Warszawa, Poland
mk@pja.edu.pl
[2] Institute of Informatics, Silesian University of Technology,
Akademicka 16, 44-100 Gliwice, Poland

Abstract. In this paper we present a method of anonymization of people's faces in video. Results are analyzed on the basis of optical flow methods. Anonymization bases on face detection. Because of mistakes made by such detectors in video sequences, gaps and false detections appear. They are recognized using the results of face detections and optical flow analysis. In this paper we describe: face detectors and the results of method of analysis of optical-flow based detector. We present novel method of filing gaps and false detection recognizing with use of optical flow. Then we present visual results.

Keywords: Face detection · Anonymization · Optical flow

1 Introduction

In the Research and Development Center of the Polish-Japanese Academy of Information Technology there are many video recordings collected by measurements systems from various laboratories: motion capture, surveillance cameras network, video cameras etc. There is a need of anonymization of video data due to community standards requirements. In other words, faces should not be recognizable which can be done by recognizing and masking of faces. This can be done by semi-automatic application for video editing like After Effect [7] or [8]. Although such solution gives reasonable results, they need big amount of work made by human. Therefore we decided to create solution that will work more automatically.

Using face detectors alone is not sufficient since some faces are left undetected and in other places the are false detections. In each sequence there are frames for which there was no detection at all or there was a false detection (the situation when detector does not recognize face on frame will be called "gap"). The problem comes down to gap and false detection recognizing which was resolved using optical flow methods. If face was recognized in the current frame, then computing the average optical flow for the face area allows to estimate a face location in the next frame. Repeating this procedure will create a path will make a path from one frame to another and allows to fill the gaps and recognize false detection.

© Springer International Publishing AG 2017
N.T. Nguyen et al. (Eds.): ACIIDS 2017, Part II, LNAI 10192, pp. 623–631, 2017.
DOI: 10.1007/978-3-319-54430-4_60

There are two processing steps: (1) Face detection using known face-detectors and (2) Post-processing step which has two targets: (a) filling the gaps and (b) recognizing the false detections.

For face detection the cascade classifier announced by Viola and Johns [1] and improved by Lienhart et al. in [2] was used. For computing optical flow we used the Farnebäck [3] method.

Presentation of the method is completed with an example.

2   Background

2.1   Cascade Face Detector

Cascade face detector announced by Viola and Johns in [1] is based on trained ensemble classifier AdaBoost [4,5] working on input Haar-like feature extracted from image.

Features. Method consists in dividing the image in small rectangles (e.g. 24×24 size). For each rectangle the feature value is computed. Feature is called Haar-like because it is generalization of Haar wavelet [6] into 2 dimensions. There are three types of such features:

1. Rectangle is divided in half. The feature value is equal to difference between sum of these halfs.
2. Rectangle is divided in three identical tiles. Feature is the sum of two outside rectangles subtracted from the middle one.
3. Rectangle is divided in four identical tiles as a chessboard. The feature response is difference between diagonal parts of rectangle.

Lienhart et al. in [2] extended the list of feature types to 15. It is done by rotating 1-st and 2-nd type of original feature by angle $+/-45°$. Second extension is introducing rectangles where one tile surrounds another one. All are shown on Fig. 1.

Learning is based on AdaBoost algorithm in three variants: Discrete [4], Real and Gentle [5].

Training examples consist of responses of Haar-like feature x_i and label y_i indicating if pattern described by feature contains the object or not: $TE = \{(x_i, y_i) : x_i \in \mathbb{R}^k \wedge y_i \in \{-1, 1\}\}$. Certainly on realistic environment there should not be expected that all faces has the same size. Therefore multiscalling technique was implemented consisting of changing size of Haar-like feature prototypes.

The basic concept of cascade classifier tells that at each stage a classifier should eliminate certain fraction of object that is not of class of interest with minimal false eliminations. In that case false elimination means rejecting object that is of interest class actually. In fact, classifiers being part of cascade (called weak) have to be a little bit better than simple random choice only.

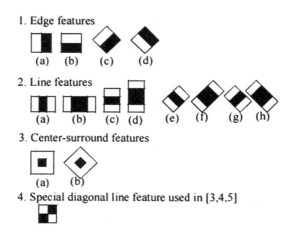

1. Edge features

(a) (b) (c) (d)

2. Line features

(a) (b) (c) (d) (e) (f) (g) (h)

3. Center-surround features

(a) (b)

4. Special diagonal line feature used in [3,4,5]

Fig. 1. All types of feature prototypes used in extended version of Haar-like features for object recognition.

Lienhart et al. [2] used very simple weak-classifiers named feature-based, because they take one feature and make a decision using binary threshold operation on weighted sample set. At each stage weights are updated and normalized. AdaBoost types differs by types of weak classifiers. While Discrete version fits two-class classifier, Gentle type fits the regression function.

2.2   Farneback Optical Flow

The main idea of this method (announced by Farnebäck in [3]) is based on the so called Parameters Displacement Fields with A Priori Knowledge incorporated. The motion, called by author "displacement" is estimated from two frames only, with compensation for the background motion. Firstly let us consider a signal approximated by the following polynomial:

$$f(\mathbf{x}) \approx \mathbf{x}^T \mathbf{A} \mathbf{x} + \mathbf{b}^T \mathbf{x} + c$$

where $\mathbf{A}$ is a symmetric matrix, $\mathbf{b}$ is a vector and c is a scalar. Coefficients $\mathbf{A}$, $\mathbf{b}$ and c are parameters of polynomial and have similar meaning like in linear quadratic function. These coefficients are estimated from a weighted least squares fit to the values of pixels located in a considered neighborhood. Displacement is computed on the basis of the following observation. Let $f_1(\mathbf{x}) = \mathbf{x}^T \mathbf{A_1} \mathbf{x} + \mathbf{b_1}^T \mathbf{x} + c_1$ be an approximation of one signal. Approximation of another signal f_2 could be constructed by global displacement (which is estimated optical flow vector as well) by $\mathbf{d}$: $f_2(\mathbf{x}) = f_1(\mathbf{x} + \mathbf{d})$. Hence the following formula could be constructed:

$$\mathbf{x}^T \mathbf{A_1} \mathbf{x} + (\mathbf{b_1} - 2\mathbf{A_1}\mathbf{d})^T \mathbf{x} + \mathbf{d}^T \mathbf{A_1} \mathbf{d} - \mathbf{b_1}^T \mathbf{d} + c_1 = \mathbf{x}^T \mathbf{A_2} \mathbf{x} + \mathbf{b_2}^T \mathbf{x} + c_2$$

Using this equality with assumption that $\mathbf{A_1}$ is non-singular, $\mathbf{d}$ could be solved from:

$$2\mathbf{A_1}\mathbf{d} = -(\mathbf{b_2} - \mathbf{b_1})$$

$$\mathbf{d} = -\frac{1}{2}\mathbf{A_1}^{-1}(\mathbf{b_2} - \mathbf{b_1})$$

In the real world assumption that a difference between two approximated signals is a simple displacement $\mathbf{d}$ not dependent on $\mathbf{x}$, is not correct. That's why spatially variation should be introduced to displacement which is modified to a form $\mathbf{d}(x)$. It is estimated using equation:

$$\mathbf{A}(\mathbf{x})\mathbf{d}(\mathbf{x}) = \Delta\mathbf{b}(\mathbf{x})$$

where $\mathbf{A}(\mathbf{x}) = \frac{\mathbf{A_1}(\mathbf{x})+\mathbf{A_2}(\mathbf{x})}{2}$ and $\Delta\mathbf{b}(\mathbf{x}) = -\frac{1}{2}(\mathbf{b_2}(\mathbf{x}) - \mathbf{b_1}(\mathbf{x}))$ - this is the equation of global displacement with a spatially varying field. Right now we can turn to approximation of neighborhood I of pixel x. The assumption that displacement field varies slowly is reasonable because greater variation means smaller similarity of pixels. Therefore optical flow computing problem could be expressed as minimization problem of function:

$$\sum_{\Delta\mathbf{x}\in I} w(\Delta\mathbf{x})\|\mathbf{A}(\mathbf{x} + \Delta\mathbf{x})\mathbf{d}(\mathbf{x}) - \Delta\mathbf{b}(\mathbf{x} + \Delta\mathbf{x})\|^2$$

where $w(\Delta\mathbf{x})$ is a weight for points in neighborhood. Consequently, displacement could be formulated as:

$$\mathbf{d}(\mathbf{x}) = \left(\sum w\mathbf{A}^T\mathbf{A}\right)^{-1}\sum w\mathbf{A}^T\Delta\mathbf{b}$$

and the minimum value is given by:

$$e(\mathbf{x}) = \left(\sum w\Delta\mathbf{b}^T\Delta\mathbf{b}\right) - \mathbf{d}(\mathbf{x})^T\sum w\mathbf{A}^T\Delta\mathbf{b}$$

Authors improved robustness of this method introducing affine motion model - the eight parameters model:

$$d_x(x, y) = a_1 + a_2x + a_3y + a_7x^2 + a_8xy$$

$$d_y(x, y) = a_4 + a_5x + a_6y + a_7xy + a_8y^2$$

which could be rewritten as follows:

$$\mathbf{d} = \mathbf{S}\mathbf{p}$$

$$\mathbf{S} = \begin{pmatrix} 1 & x & y & 0 & 0 & 0 & x^2 & xy \\ 0 & 0 & 0 & 1 & x & y & xy & y^2 \end{pmatrix},$$

$$\mathbf{p} = (a_1 a_2 a_3 a_4 a_5 a_6 a_7 a_8)^T$$

Authors give the final solution for $\mathbf{p}$, where i is used for indexing of neighborhood pixels:

$$\mathbf{p} = \left(\sum_i w_i\mathbf{S}_i^T\mathbf{A}_i^T\mathbf{A}_i\mathbf{S}_i\right)^{-1}\sum_i w_i\mathbf{S}_i^T\mathbf{A}_i^T\Delta\mathbf{b}_i$$

3   Optical Flow Analysis

Let $f_1, f_2, ..., f_s$ be consecutive frames of video sequence. For each f_i detector computes areas containing face, which are defined by $t_l(i, j)$ and $b_r(i, j)$ which are top-left and bottom-right vertices of detected areas. As was mentioned before algorithm has two purposes: (1) Cleaning false detections and (2) Estimating missing detections - filling gaps.

3.1   Face Motion Estimation

Let's assume that one of all face detections j in frame f_i is given by an rectangle $R_{i,j}(t_l(i, j), r_b(i, j))$. Face motion estimation leads to assign $R_{i+1,j}(t_l(i + 1, j), b_r(i + 1, k))$. Below we introduce implemented method of face motion computing (Fig. 2).

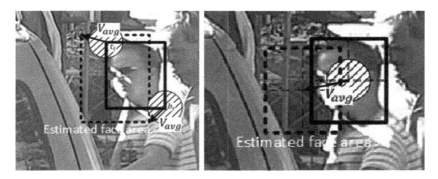

Fig. 2. All types of feature prototypes used in extended version of Haar-like features for object recognition.

Face Area Estimation is limited to obtaining new position of $t_l(i + 1, j)$ and $b_r(i + 1, j)$. Let $\mathcal{B}(x, \epsilon)$ be a circle with center in point x and radius equal to ϵ, a $v(x) \in \mathbb{R}^2$ is computed optical flow for point x. In that case

$$V_{avg}(x, \epsilon) = \frac{\sum_{x \in \mathcal{B}(x,\epsilon)} v(x)}{\pi \epsilon^2}$$

Therefore estimated face area could be computed as:

$$R_{i+1,j} = (t_l(i, j) + V_{avg}(t_l(i, j), \epsilon), \epsilon) + V_a vg(b_r(i, j), \epsilon))$$

This method has a limitation, because vertex is rather placed outside of face oval. That's why another method was introduced.

Using $V_{avg}(x, \epsilon)$ estimation of center of face $(R_{i,j})$ area could be made. Let $C(R_{i,j}) = C(i, j) = t_l(i, j) + \frac{b_r(i,j) - t_l(i,j)}{2}$ be center of detected face candidate. New face center is estimated by: $C(R_{i+1,j}) = C(R_{i,j}) + V_{avg}(C(R_{i,j}), \epsilon)$. While in the first approach both place and size of area could be changed, the second one concerns place solely. Problem of changing size will be resolved in other way described further.

Detection Path is a sequence of points got from consecutive frames that are equal to face center displacement computed using optical flow. The simple fact that if close frames are considered, the difference between face places would rather be smaller, should be respected by creating detection path. Let k and m be the starting and ending frame of such sequence segment $S_{m,k}$ that exists $R_{k,j}$ and $R_{m,j}$; $k < i < m$ will be index of frame being considered for which exists (detected or estimated before) $R_{i-1,j}$; $w_k = \frac{C(R_{k,j}) - C(R_{i,j})}{m-k-1}$; $w_m = \frac{C(R_{m,j}) - C(R_{i,j})}{m-k-1}$. So right now three vectors are considered: V_{avg}, w_k and w_m. In the middle of set G optical flow should have more influence but close to begin and end f_k or f_m are more important. To achieve this situation three weights are introduced: Let $l = m - k - 1$ in that case $\omega_k = 1 - \frac{i-k}{l}$, $\omega_m = 1 - \frac{m-i}{l}$ and $\omega_{avg} = 1 - \frac{|i - \lceil \frac{1}{2} l \rceil|}{\lceil \frac{1}{2} l \rceil}$. These weights should be normalized so that $\widehat{\omega}_k + \widehat{\omega}_m + \widehat{\omega}_{avg} = 1$. In that case detection path $DP(k,m)$ from frame k to m is a sequence of points p_i on each of consecutive frames defined iteratively below:

$$p_i = (\frac{1}{3}\widehat{\omega}_k w_k + \widehat{\omega}_m w_m + \widehat{\omega}_{avg} V_{avg}) p_{i-1}, p_0 = C(R_{k,j})$$

is called detection path of face center $C(R_{k,j})$.

Gap Filling and False Detection Recognition could be based on detection path computing (see Fig. 3). Let $G_j = \{R(k+1,j), R(k+2,j), ..., R(m-1,j)\}; k < i < m$; $G_j.first = R(k+1,j) \in G_j$, $G_j.last = R(m-1,j)$, which means first and last element in G_j and $|R(k+i,j)| = i$.

Algorithm of filling gaps:
for $i = 1$ **to** $video.length$ **do**
| $F = detect\_faces(f_i)$;
end
create: $\mathcal{G} = \{G_1, ..., G_N\}$ using F;
for $i, j = 1$ **to** N **do**
| **if** $i \neq j$ AND $|G_i.last| \leq |G_j.first|$ **then**
| | $\widehat{\mathcal{G}} \leftarrow (G_i, G_j)$;
| **end**
end
foreach $(G_i, G_j) \in \widehat{\mathcal{G}}$ **do**
| compute: $DP(|G_i.last|, |G_j.first|)$;
| **if** $||p_{|G_i.last|} - G_j.first||_2 < \tau_1$ **then**
| | $G_{new} =$ fill gaps adding all estimated detection from
| | $DP(|G_i.last|, |G_j.first|)$
| | delete from $\widehat{\mathcal{G}}$ all (G_i, G_k) and (G_j, G_k); replace in $\widehat{\mathcal{G}}$ all (G_k, G_j)
| | with (G_k, G_{new}).
| **end**
end

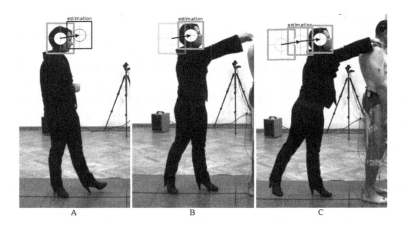

Fig. 3. Detection path creation.

After detecting faces in each frame of a video sequence the algorithm creates set $\mathcal{G}$ which contains sequences of detections of the same face and will be called *path*. It is simply done by finding the closest detections in the sibling frames. If they are close enough to each other, they are considered as concerning the same face and they belong to one $G_i \in \mathcal{G}$. A few elements of $\mathcal{G}$ could concern the same face if there are gaps in detections of one face. Then cartesian product is created $\mathcal{G} \times \mathcal{G}$. For each element (G_i, G_j) of this product detection path $DP(|G_i.last|, |G_j.first|)$ is computed. If the end of this path resides close enough to the begin G_j of new path of G_i, $DP(|G_i.last|, |G_j.first|)$ and G_j. All pairs containing G_i or G_j on the first place are removed from $\mathcal{G}$ and in all pairs containing G_j on second place G_j is replaced with new path.

Algorithm of false detection recognition is created basing on assumption that faces are already detected and set $\mathcal{G}$. Original detections are considered solely:

foreach $G_i \in \mathcal{G}$ **do**
    compute: $DP(|G_i.first|, |G_i.last|)$;
    foreach $\widehat{R}(k,i) \in DP$ **do**
        if $||C(\widehat{R}(k,i)) - C(R(k,i))||_2 > \tau_2$ **then**
            $\{G_j, G_k\}$ =divide G_i throwing out $R(k,i)$ from G_i;
            $G_i := G_j$;
            push back G_k to $\mathcal{G}$
        end
    end
end

This algorithm goes through all computed paths from $\mathcal{G}$ and compares detection from path with that got from $DP(|G_i.first|, |G_i.last|)$. If the divergence between computed and detected face position is big enough, algorithm divides the given path and the second part of new path adds to the end of $\mathcal{G}$.

4   Results

The created method was used in two laboratories in Research and Development Center of the Polish-Japanese Academy of Information Technology: Computer Seeing Lab for anonymization sequences from PTZ surveillance network and in Human Motion Lab (HML) for anonymization of video sequences recorded simultaneously with motion capture recordings, whose exemplary output frames are shown in the Fig. 4. Although the proposed method makes anonymization automatically some actions have to be done manually:

1. Training of face detector. For each environment (lab) another training set is used. Our examination with common training set indicated significantly worst detection results. For HML there were created two separate data sets for frontal and profile faces. For surveillance network one training set respects different views perspective of faces.
2. Some sequences need to be manually tuned up. This problem will be the objective of future work on this method.

Fig. 4. Exemplary anonymized frames. Black rectangles are made manually over the automatically generated blur-mask for better visibility of results.

5   Conclusion

The presented method of improving the face detection in video sequences basing on optical flow analysis excludes majority of false detection got from face detector. The method could be developed by using more face detectors. After implementing in goal system, this method could be used for automatic face detection on video sequences used in our research on video surveillance.

Acknowledgement. This work has been supported by the National Centre for Research and Development (project UOD-DEM-1-183/001 "Intelligent video analysis system for behavior and event recognition in surveillance networks").

References

1. Viola, P., Johns, M.: Rapid object detection using a boosted cascade of simple features. In: Computer Vision and Pattern Recognition Conference (2001)
2. Lienhart, R., Kuranov, A., Pisarevsky, V.: Empirical analysis of detection cascades of boosted classifiers for rapid object. MRL Technical report, May 2002, revised December 2002
3. Farnebäck, G.: Two-frame motion estimation based on polynomial expansion. In: Bigun, J., Gustavsson, T. (eds.) SCIA 2003. LNCS, vol. 2749, pp. 363–370. Springer, Heidelberg (2003). doi:10.1007/3-540-45103-X_50
4. Friedman, J.H., Hastie, T., Tibshirani, R.: Additive logistic regression: a statistical view of boosting. Technical report, Department of Statistics, Stanford University (1998)
5. Shapire, R.E., Singer, Y.: Improved boosting algorithms using confidence-rated predictions. Mach. Learn. **37**(3), 297–336 (1999)
6. Haar, A.: Zur Theorie der orthogonalen Funktionensysteme. Mathematische Annalen **69**(3), 331–371 (1910)
7. http://www.adobe.com/pl/products/aftereffects.html
8. http://sensarea.software.informer.com/

An Analysis of the Centre of Mass Behavior During Treadmill Walking

Henryk Josiński[1]($\boxtimes$), Adam Świtoński[1], Agnieszka Michalczuk[2],
Konrad Wojciechowski[1], and Jerzy Paweł Nowacki[1]

[1] Polish-Japanese Academy of Information Technology,
Koszykowa 86, 02-008 Warszawa, Poland
{hjosinski,aswitonski,kwojciechowski,nowacki}@pjwstk.edu.pl
[2] Institute of Informatics, Silesian University of Technology,
Akademicka 16, 44-100 Gliwice, Poland
agnieszka.michalczuk@polsl.pl

Abstract. The authors present the preliminary results of the analysis
of the centre of mass behavior during treadmill walking by means of
the sample entropy which quantifies a regularity of a time series. The
research is focused on the centre of mass trajectories in the mediolateral,
anteroposterior and longitudinal axes recorded using the motion capture
technique. From among several entropy measures the sample entropy was
chosen for the purpose of assessment of the influence of both walking
speed and ground inclination on a regularity in movements of the centre
of mass. The results were compared with the sample entropy values for
periodic, chaotic and stochastic signals.

Keywords: Nonlinear time series analysis · Centre of mass · Sample
entropy · Human motion analysis

1 Introduction

A human body is perceived as a highly nonlinear dynamical system [1]. Its
behaviors that evolve over time, can be successfully investigated using nonlinear analysis methods based on a dynamical systems approach (e.g. the largest
Lyapunov exponent). A sequence generated by an observed dynamical system
at consecutive equidistant time instants is called a time series. Entropy – commonly known as a measure of disorder or a measure of degree of uncertainty –
is considered here as a loss of information in a time series [2]. Different types
of entropy (e.g. approximate entropy ($ApEn$) [3] and sample entropy ($SampEn$)
[4]) are measures that can quantify the predictability or regularity of a time
series – lower values reveal a more regular or even periodic behavior, whereas
the bigger ones correspond to greater irregularity.

Human gait is a rhythmical oscillation [5], so the entropy measures were
successfully applied in several gait studies. The degree of predictability in trunk
acceleration time series was determined by means of the $SampEn$ to quantify

© Springer International Publishing AG 2017
N.T. Nguyen et al. (Eds.): ACIIDS 2017, Part II, LNAI 10192, pp. 632–640, 2017.
DOI: 10.1007/978-3-319-54430-4_61

the effect of impaired cognition and dual tasking on gait variability and stability in geriatric patients [6]. The *ApEn* technique revealed that during walking the anterior cruciate ligament (ACL) deficient knee exhibits more regular patterns than the contralateral intact knee [5].

A centre of mass (CoM) reflects a motion of a whole human body. Consequently, a trajectory of a CoM is an important factor in an analysis of human gait – an alteration in a CoM trajectory may help maintain a gait stability as well as indicate an underlying pathology [7]. The study of Arif et al. [1] can serve as an example of how the CoM data (acceleration) can be analyzed in the context of a gait stability (of young vs elderly subjects) by means of the *ApEn*.

Benefiting from a motorized treadmill application and introducing the aspect of a treadmill inclination the present study is focused on an analysis of the CoM trajectories using the *SampEn* measure on motion capture data from treadmill walking recordings. The context of walking speed is also taken into consideration.

2 The Sample Entropy

Entropy measures quantify the unpredictability of fluctuations in a time series and reflect the probability that similar patterns of observations will not be followed by additional similar observations [8]. The *ApEn* concept has some shortcomings [2], so the authors focused on the *SampEn* technique, which counteracts them. Let's assume that a vector $p_m(i)$ is a pattern consisting of m successive points from an N-point time series $x_k = x(k\tau_s)$, $k = 1, \ldots, N$, where τ_s is the sampling time:

$$p_m(i) = [x_i, x_{i+1}, \ldots, x_{i+m-1}]. \tag{1}$$

Two different patterns: $p_m(i)$, $p_m(j)$ are *similar*, if the maximum absolute difference d between any pair of corresponding points is less than the value of the similarity criterion r:

$$d(p_m(i), p_m(j)) \leq r, i \neq j. \tag{2}$$

Assuming that the set P_m of all patterns of length m:

$$P_m = \{p_m(1), p_m(2), \ldots, p_m(N - m + 1)\} \tag{3}$$

includes $n_{im}(r)$ patterns that are similar to the pattern $p_m(i)$, $i = 1, \ldots, N - m + 1$ (excluding itself), the fraction $C_{im}(r)$ of such patterns is described by the following formula [8]:

$$C_{im}(r) = \frac{n_{im}(r)}{N - m + 1}. \tag{4}$$

Let $C_m(r)$ denote the mean of all $C_{im}(r)$ values calculated for each pattern from P_m as follows:

$$C_m(r) = \frac{1}{N - m + 1} \sum_{i=1}^{N-m+1} C_{im}(r). \tag{5}$$

$C_m(r)$ can be interpreted as the probability that any two patterns of length m in the N-point time series are similar within the tolerance r. The *SampEn* is defined as the negative of the natural logarithm of the conditional probability that similar patterns of length m will also remain similar when one point is added to each pattern and self-matches are excluded from calculating the probability [4]. Taking into account the aforementioned interpretation of both quantities $C_m(r)$ and $C_{im}(r)$, the *SampEn* can be computed according to the following formula [9]:

$$SampEn(m, r, N) = \ln \left(\frac{\sum_{i=1}^{N-m} n_{im}(r)}{\sum_{i=1}^{N-m} n_{im+1}(r)} \right). \tag{6}$$

However, it is necessary to mention, that given an N-point time series, only the first $(N - m)$ patterns of length m are considered, ensuring thereby that the pattern $p_{m+1}(i)$, $i = 1, \ldots, N - m$ of length $m + 1$ is also defined.

Entropy algorithms are very sensitive to their input parameters [2]. This remark concerns in particular m (length of a pattern being compared) and r (similarity criterion). The recommendations from the literature [4], [5] suggest that r should be chosen from the range of $[0.1, 0.25]$ times the standard deviation of the entire time series. As far as m is concerned, in all human motion studies value of 2 was used. Although the *SampEn* measure has been assessed as independent of N (length of a time series) [4], N has been advised to be greater than 200 [2].

The *SampEn* values for exemplary periodic, chaotic and stochastic time series (see: Fig. 1), that were computed using MATLAB, are presented in Table 1. The following values were assigned to the *SampEn* parameters: $m = 2$, $r = 0.25 \cdot SD$ (SD denotes standard deviation), $N = 5000$ points.

Table 1. The *SampEn* values for different signals

System	SampEn
$y = sin(x)$	0.0178
Lorenz chaotic system	
$\dot{x} = \sigma(y - x)$	x: 0.1101
$\dot{y} = -xz - \rho x - y$	y: 0.1258
$\dot{z} = xy - \beta z$	z: 0.2255
$\sigma = 10$ $\rho = 28$ $\beta = 8/3$	
$x_0 = 1$ $y_0 = z_0 = 0$	
$1/f$ (pink) noise	1.4845
white Gaussian noise	1.9579

The aforementioned relationship between regularity and sample entropy was completely confirmed: from values close to 0 that are assigned to regular patterns in form of periodic signals, through values in the approximate range of

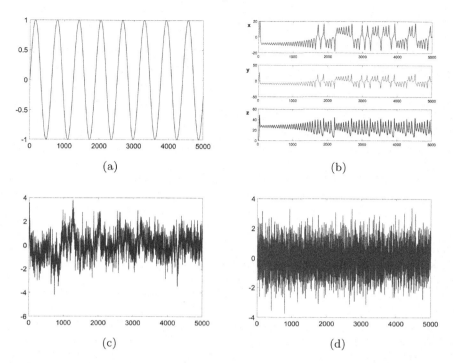

Fig. 1. Exemplary periodic, chaotic and stochastic time series: (a) sine wave, (b) Lorenz chaotic system, (c) $1/f$ (pink) noise, (d) white Gaussian noise. (Color figure online)

[0.1, 0.25] that are related to deterministic chaos, to significantly higher values (here: nearing 2) that are specific for various types of noisy random data.

3   Experimental Research

Gait kinematic data were acquired in a continuous way over sufficiently big number of consecutive strides (several dozen seconds of walking) using a treadmill in different variants of walking speed and treadmill inclination (see: Table 2). The recordings took place in the Human Motion Laboratory (HML) of the Polish-Japanese Academy of Information Technology (http://bytom.pja.edu.pl/) which is equipped with the Vicon motion capture system.

Table 3 includes mean values of age, weight, height and preferred walking speed for 17 healthy subjects (6 women and 11 men), that participated in the experiments.

At the start of each variant a subject practiced until she/he was able to walk comfortably. Next, 3 gait sequences were recorded for every person with short intermissions in between. Taking into account 5 modes of treadmill inclination and walking speed, the total number of sequences recorded for every subject was equal to 15.

Table 2. Variants of walking speed and treadmill inclination

Variant	*Normal*	*Faster*	*Slower*	*Normal Up*	*Normal Down*
Walking speed	Preferred (*PWS*)	1.2 · *PWS*	0.8 · *PWS*	*PWS*	*PWS*
Treadmill inclination	0°	0°	0°	+7°	−3°

Table 3. Mean values of subjects' age, weight, height and preferred walking speed

	Age [years]	Weight [kg]	Height [cm]	PWS [m/s]
Mean ± SD	27.18 ± 8.72	71.14 ± 10.06	174.47 ± 6.46	2.38 ± 0.63

Initially, motion data were filtered and repaired (e.g. in the case of occluded markers) using methods that are built in to the Vicon software, e.g. a quintic spline (Woltring filtering routine). A preliminary analysis showed that a stride interval varied across subjects and tested variants of ground inclination and walking speed.

The research is focused on an analysis of the CoM trajectories in the mediolateral (x), anteroposterior (y) and longitudinal (z) axes (movements of the elements of the kinetic chain of the lower extremity in the sagittal plane, i.e. dorsiflexion/plantarflexion angle of the ankle, knee flexion/extension angle and hip flexion/extension angle were studied in [10]). It is worthwhile to mention, that the current CoM position was always computed by the Plug-In Gait software as the centre of mass of all the modeled segments on the basis of markers' positions. Three exemplary CoM time series (*Normal* mode) are presented in Fig. 2a–c. A single stride was extracted from the time series and shown in Fig. 2d–f. The unit on each vertical axis is 'mm' and the negative values of CoM_x and CoM_y result from the treadmill location in regard to the origin of the laboratory coordinate system.

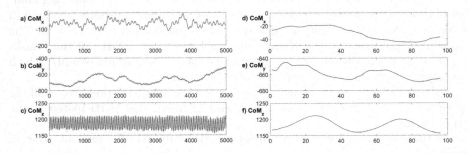

Fig. 2. Recorded time series representing movements of the CoM (a–c) and a single stride (d–f) in: (a, d) mediolateral axis, (b, e) anteroposterior axis, (c, f) longitudinal axis.

The CoM displacements are affected by the application of a treadmill in particular along the anteroposterior axis (Fig. 2e). Lateral displacements from side to side are caused by the change of the weight-bearing leg (Fig. 2d). Along the longitudinal axis the CoM is displaced twice during a stride creating a smooth trajectory with 2 cycles for every stride (Fig. 2f).

Several descriptive statistics (mean, standard deviation, median, minimum and maximum) of the *SampEn* were computed for each of 5 variants of walking speed and treadmill inclination separately, using the following set of parameters: $m = 2$, $r = 0.25 \cdot SD$, $N = 5000$ points. The results are included in Tables 4, 5 and 6 for x, y, and z axes, respectively. Additionally, all mean values are presented in Fig. 3.

Table 4. Basic descriptive statistics of the sample entropy for x-axis

	Normal	*Faster*	*Slower*	*Normal Up*	*Normal Down*
Mean	0.120	0.104	0.129	0.124	0.115
SD	0.029	0.029	0.032	0.028	0.029
Median	0.128	0.104	0.143	0.130	0.118
Min.	0.050	0.051	0.049	0.045	0.049
Max.	0.162	0.159	0.162	0.162	0.167

Table 5. Basic descriptive statistics of the sample entropy for y-axis

	Normal	*Faster*	*Slower*	*Normal Up*	*Normal Down*
Mean	0.049	0.055	0.048	0.049	0.055
SD	0.021	0.026	0.019	0.019	0.023
Median	0.045	0.049	0.047	0.050	0.051
Min.	0.010	0.019	0.010	0.014	0.014
Max.	0.104	0.107	0.099	0.107	0.113

The results from Table 6 illustrate a big leeway of the CoM vertical movements that are typical for natural walking. They could be juxtaposed with outcomes for a race walking, where an amplitude of vertical fluctuations should be limited resulting in increase of a race-walker speed.

Bearing in mind, that the smaller the entropy, the more regular and predictable the system, the *SampEn* values suggest the presence of repetitive patterns in the CoM time series of young healthy adults recorded during treadmill walking and, by extension, a high level of predictability. However, the mean values for particular axes are clearly different: in the case of y-axis the mean value

Table 6. Basic descriptive statistics of the sample entropy for z-axis

	Normal	*Faster*	*Slower*	*Normal Up*	*Normal Down*
Mean	0.334	0.325	0.336	0.339	0.330
SD	0.029	0.029	0.028	0.023	0.030
Median	0.334	0.323	0.334	0.342	0.332
Min.	0.273	0.271	0.275	0.281	0.262
Max.	0.393	0.388	0.398	0.379	0.381

correspond to entropy of a periodic time series, the mean value for x-axis suggests a slightly lesser regularity, whereas the corresponding value for z-axis seems to indicate that chaotic fluctuations are present in the gait patterns. However, it is worthwhile to mention that in the case of a similar experiment for $m = 3$ the *SampEn* values for z-axis were significantly lower. Thus, the results should be treated with caution, as preliminary outcomes. Nevertheless, as far as human gait analysis is concerned, $m = 2$ is explicitly recommended in the literature.

Comparing the mean *SampEn* values for *Slower* and *Faster* modes, the biggest difference between them (in favor of the slow walking) appears for x-axis. Following the interpretation of entropy, the lateral trajectories in the *Faster* variant are more regular, what could be related to the observation in [7], according to which the CoM displacements in the lateral direction are decreased during fast walking. Considering the variants of treadmill inclination (*Normal Down* vs *Normal Up*), the mean *SampEn* value for both x and z axes in walking downwards is less than upwards. Hence, the lateral and vertical trajectories are more regular in the *Normal Down* mode, perhaps due to a slightly rigid way of walking downwards.

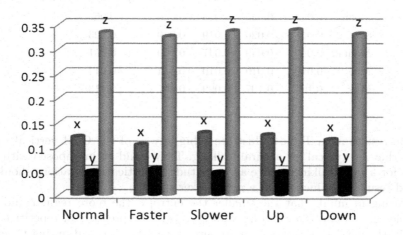

Fig. 3. Mean values of the sample entropy for each direction.

4   Conclusion

The authors present the preliminary results of the analysis of the centre of mass behavior during treadmill walking by means of the sample entropy, taking into account the CoM trajectories in the mediolateral, anteroposterior and longitudinal axes. The results were compared with entropy values for exemplary periodic, chaotic and stochastic time series, thereby facilitating the introductory interpretation of the CoM behavior in different variants of walking speed and treadmill inclination.

It's worth mentioning that entropy quantifies also the degree in which complexity is present in a movement – a time series of the knee joint flexion/extension of the ACL deficient knee, containing many repetitive patterns, thus characterized by lower entropy, has been interpreted in [5] not only as more predictable but as less complex as well. According to [8,11], complexity is visible in a time series in chaotic fluctuations that may be necessary to adapt to changing conditions (i.e. everyday stresses) in a flexible way. Bearing in mind that the largest Lyapunov exponent (LLE) quantifies the local dynamic stability of walking by indicating the presence of chaotic behavior in gait kinematic data, the juxtaposition of outcomes of both approaches – *SampEn* and LLE – supported by an appropriate statistical analysis will constitute the next stage of the research.

Acknowledgements. The work is supported by the following projects: the "Virtual Physiotherapist" (TANGO1/269419/NCBR/2015) of The Polish National Centre for Research and Development and BK/Rau2/2016.

References

1. Arif, M., Ohtaki, Y., Nagatomi, R., Inooka, H.: Estimation of the effect of cadence on gait stability in young and elderly people using approximate entropy technique. Meas. Sci. Rev. **4**(2), 29–40 (2004)
2. Yentes, J.M., Hunt, N., Schmid, K.K., Kaipust, J.P., McGrath, D., Stergiou, N.: The appropriate use of approximate entropy and sample entropy with short data sets. Journal Articles, Paper 44 (2013). http://digitalcommons.unomaha.edu/biomechanicsarticles/44
3. Pincus, S.M., Goldberger, A.L.: Physiological time-series analysis: what does regularity quantify? Am. J. Physiol. - Heart Circ. Physiol. **266**(4), 1643–1656 (1994)
4. Richman, J.S., Moorman, J.R.: Physiological time series analysis using approximate entropy and sample entropy. Am. J. Physiol. - Heart Circ. Physiol. **278**(6), 2039–2049 (2000)
5. Georgoulis, A.D., Moraiti, C., Ristanis, S., Stergiou, N.: A novel approach to measure variability in the anterior cruciate ligament deficient knee during walking: the use of the approximate entropy in orthopaedics. J. Clin. Monit. Comput. **20**, 11–18 (2006)
6. Lamoth, C.J., van Deudekom, F.J., van Campen, J.P., Appels, B.A., de Vries, O.J., Pijnappels, M.: Gait stability and variability measures show effects of impaired cognition and dual tasking in frail people. J. NeuroEng. Rehabil. **8**(2) (2011)

7. Jurčevič Lulič, T., Sušič, A., Kodvanj, J.: Biomechanical analysis of walking: effects of gait velocity and arm swing amplitude. Periodicum Biologorum **112**(1), 13–17 (2010)
8. Decker, L.M., Cignetti, F., Stergiou, N.: Complexity and human gait. Journal Articles, Paper 91 (2010). http://digitalcommons.unomaha.edu/biomechanicsarticles/91
9. Costa, M., Goldberger, A.L., Peng, C.-K.: Multiscale entropy analysis of biological signals. Phys. Rev. E **71**, 021906 (2005)
10. Piórek, M., Josiński, H., Michalczuk, A., Świtoński, A., Szczęsna, A.: Quaternions and joint angles in an analysis of local stability of gait for different variants of walking speed and treadmill slope. Inf. Sci. (2016). http://dx.doi.org/10.1016/j.ins.2016.08.069
11. Lipsitz, L.A.: Dynamics of stability: the physiologic basis of functional health and frailty. J. Gerontol. **57**, 115–125 (2002)

A Bayesian Framework for Chemical Shift Assignment

Adam Gonczarek[✉], Piotr Klukowski, Maciej Drwal, and Paweł Świątek

Department of Computer Science, Wrocław University of Science and Technology,
Wybrzeże Wyspiańskiego 27, 50-370 Wrocław, Poland
{adam.gonczarek,piotr.klukowski,maciej.drwal,pawel.swiatek}@pwr.edu.pl

Abstract. Nuclear magnetic resonance (NMR) spectroscopy is one of
the techniques used in structural biology and drug discovery. A critical
step in analysis of NMR images lies in automation of assigning NMR
signals to nuclei in studied macromolecules. This procedure is known
as sequence-specific resonance assignment and is carried out manually.
Manual analysis of NMR data results in high costs, lengthy analysis and
proneness to user-specific errors. To address this problem, we propose
a new Bayesian approach, where resonance assignment is formulated as
maximum a posteriori inference over continuous variables.

Keywords: NMR spectroscopy · Resonance assignment · Chemical
shifts · Optimization

1 Introduction

Identification of NMR signals, later called as resonances or chemical shifts,
is a prerequisite for studying protein structures, interactions and dynamics.
The problem of automation of resonance assignment has been studied for over
20 years [7–9], and it yielded in over 40 published methods that are based on e.g.
exhaustive search [1,2,5,11], best-first [10,14], Monte Carlo [13], metaheuris-
tics [15,18] and probabilistic inference [3]. Nevertheless, existing methods are
still inaccurate enough for practical application, unless careful preprocessing,
known as peak picking, is carried out. However, high quality peak picking is
usually intractable and therefore chemical shift assignment is performed manually, which is extremely time consuming. According to the records of Protein
Data Bank very few automated procedures were used in practice [8].

In this paper we present a new Bayesian framework that allows representing
the uncertainty in the chemical shifts in the form of posterior probability distributions. Thereby, unlike other methods, in our approach perfect peak picking is
not required. We formulate automated chemical shift assignment as a continuous
optimization problem, where discrepancy between expected and observed peak
positions is minimized. The presented framework is inspired by the probabilistic formulation of the pictorial structure model [6], widely used in the field of
computer vision.

© Springer International Publishing AG 2017
N.T. Nguyen et al. (Eds.): ACIIDS 2017, Part II, LNAI 10192, pp. 641–649, 2017.
DOI: 10.1007/978-3-319-54430-4_62

2   Chemical Shift Assignment Problem

An underlying assumption of NMR experiment is that a studied macromolecule can be described by a vector of chemical shifts $\mathbf{z} \in \mathbb{R}^M$, where each element z_i encodes the resonating frequency of a single nucleus in a protein sequence. The main goal of the chemical shift assignment is to determine $\mathbf{z}$, given a set of spectral images $\mathcal{I} = \{I^{(1)}, \ldots, I^{(K)}\}$. Using probabilistic modeling, we can formulate this issue as a problem of finding MAP estimator of $\mathbf{z}$:

$$\mathbf{z}_{\text{MAP}} = \arg\max_{\mathbf{z}} p(\mathbf{z}|\mathcal{I}). \tag{1}$$

Further, we can apply the Bayes' theorem to decompose the model as follows:

$$p(\mathbf{z}|\mathcal{I}) \propto p(\mathcal{I}|\mathbf{z})p(\mathbf{z}). \tag{2}$$

Hence, the modeling task is reduced to the problem of proposing a prior distribution over chemical shifts $p(\mathbf{z})$ and a spectral image likelihood model $p(\mathcal{I}|\mathbf{z})$. While proposing a proper prior is relatively easy, there are a few reasons why modeling the likelihood as well as finding MAP estimator is challenging.

First, each spectral image $I^{(j)} \in \mathcal{I}$ is a result of a different type of NMR experiment (e.g. HNCA, HNCACB, TOCSY, etc.) and contains only partial information about $\mathbf{z}$. Depending on the type of an experiment, a spectrum is usually two- or three-dimensional image composed of signals associated with pairs or triplets of chemical shifts, respectively. For example, in HNCA experiment we observe two signals (peaks) for every amino acid i in the protein sequence. The peaks are associated with triplets (H_i, N_i, C_i^α) and $(H_i, N_i, C_{i-1}^\alpha)$[1]. Each of these triples corresponds to some three elements in $\mathbf{z}$. Notice that no information about chemical shifts of other atoms (e.g. C^β) is included in the HNCA spectrum. Therefore, different types of experiments are required to complete the information about $\mathbf{z}$.

Second, a chemical shift of a single atom is usually observed in more than one type of spectral images. However, due to measurement noise the peak positions may slightly vary on different spectra. Thus, our likelihood model must be robust to these small perturbations.

Third, spectral images $\mathcal{I}$ are noisy and contain redundant signals caused by the presence of additional compounds (e.g. solvent) as well as artifacts originating from imperfections of measurement setup. Usually, these signals visually differ from the signals that encode protein atoms. This property may be utilized in the pre-processing step, called *peak picking*, where selection of the signals of interest is performed.

Finally, a vector $\mathbf{z}$ is high-dimensional (usually from a few hundred up to a few thousand dimensions, depending on the size of a protein), and determining its values in MAP estimation procedure usually leads to solving a non-trivial and highly non-convex optimization problem.

---

[1] C^α denotes carbon alpha in the amino acid.

3   Prior Distribution

Prior distribution on vector $\mathbf{z}$ models constraints and dependencies between chemical shifts, which are independent of the type of spectral image. More precisely, these constraints and dependencies are the consequence of the physico-chemical nature of amino acids. In particular, the simplest distribution could have a form:

$$p(\mathbf{z}) = \prod_{i=1}^{M} \mathcal{U}(z_i | a_i, b_i), \tag{3}$$

which is a product of independent uniform distributions, i.e., every shift can take a value from interval $[a_i, b_i]$. Parameters a_i, b_i could be assumed to correspond to the range of possible chemical shift values for different atoms within amino acids.

However, chemical shifts of atoms are Gauss- or Laplace-distributed rather than uniformly. Therefore, we suggest to use the following form of the prior:

$$p(\mathbf{z}) = \prod_{i=1}^{M} \mathrm{Lap}(z_i | \mu_i, b_i), \tag{4}$$

where μ_i is a location parameter and $b_i > 0$ is a scale parameter. Obviously, the model parameters can be estimated based on manually assigned chemical shifts.

However, typically for side chains in the protein sequence, many chemical shifts may be correlated. To overcome this problem, we can utilize experimental data to learn more sophisticated models that can capture these dependencies. We suggest to use e.g. products of Mixture Models or Restricted Boltzmann Machines.

4   Spectral Image Likelihood

In order to model a spectral image likelihood certain assumptions have to be made. First, we can define a likelihood as a product of independent likelihoods for each spectral image:

$$p(\mathcal{I}|\mathbf{z}) = \prod_{j=1}^{K} p(I^{(j)}|\mathbf{z}). \tag{5}$$

This assumption can be made without the loss of generality, since spectral images are only dependent through $\mathbf{z}$.

Second, we must take into consideration the fact that chemical shift for a given atom may slightly vary across different spectra. Therefore, we construct a likelihood model from two blocks, i.e., a *peak generative model* and an image likelihood for a given list of peaks:

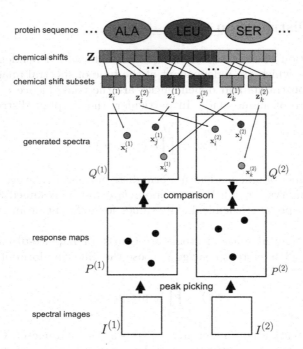

Fig. 1. Overview of proposed chemical shift assignment procedure. Input images $I^{(j)}$ undergo peak picking, which allows to generate response maps $P^{(j)}$. At the same time, corresponding spectra $Q^{(j)}$ are generated based on the given values of chemical shifts. Intuitively, vector $\mathbf{z}$ is estimated by minimizing discrepancy between all $P^{(j)}$ and $Q^{(j)}$.

$$
\begin{aligned}
p(I^{(j)}|\mathbf{z}) &= \int p(I^{(j)}, \mathbf{X}^{(j)}|\mathbf{z})\mathrm{d}\mathbf{X}^{(j)} \\
&= \int p(I^{(j)}|\mathbf{X}^{(j)}, \mathbf{z})p(\mathbf{X}^{(j)}|\mathbf{z})\mathrm{d}\mathbf{X}^{(j)} \\
&= \int p(I^{(j)}|\mathbf{X}^{(j)})p(\mathbf{X}^{(j)}|\mathbf{z})\mathrm{d}\mathbf{X}^{(j)} \\
&= \mathbb{E}_{\mathbf{X}^{(j)}|\mathbf{z}}\left[p(I^{(j)}|\mathbf{X}^{(j)})\right],
\end{aligned}
\tag{6}
$$

where $\mathbf{X}^{(j)}$ is a list of peak positions observed in the spectral image $I^{(j)}$. Since $\mathbf{X}^{(j)}$ is given, the distribution over images $I^{(j)}$ is independent of $\mathbf{z}$. Thus, we can express the spectral likelihood model as an expectation of $\mathbf{X}^{(j)}$ from the image likelihood for a given list of peaks. For brevity, we can combine (5) and (6) into the following expression:

$$p(\mathcal{I}|\mathbf{z}) = \prod_{j=1}^{K} \int p(I^{(j)}|\mathbf{X}^{(j)})p(\mathbf{X}^{(j)}|\mathbf{z})\mathrm{d}\mathbf{X}^{(j)}$$

$$= \int p(\mathcal{I}|\mathcal{X})p(\mathcal{X}|\mathbf{z})\mathrm{d}\mathcal{X}$$

$$= \mathbb{E}_{\mathcal{X}|\mathbf{z}}\left[p(\mathcal{I}|\mathcal{X})\right], \tag{7}$$

where $\mathcal{X} = \{\mathbf{X}^{(1)}, \ldots, \mathbf{X}^{(K)}\}$ is a set of all peak lists.

Finally, to complete the spectral likelihood model we must specify the form of peak generative models $p(\mathbf{X}^{(j)}|\mathbf{z})$ and image likelihoods $p(I^{(j)}|\mathbf{X}^{(j)})$.

4.1   Peak Generative Model

Let $\mathbf{x}_i^{(j)}$ (an element of $\mathbf{X}^{(j)}$) denote a position of the peak i in the spectral image j. The position depends only on a small number of elements of vector $\mathbf{z}$ (typically 2 or 3). For each peak expected in a spectrum there is a known subset of indices in the vector $\mathbf{z}$ that correspond to the shifts determining its position. Consequently, we denote by $\mathbf{z}_i^{(j)}$ a vector containing a group of values from $\mathbf{z}$ that correspond to the true peak position $\mathbf{x}_i^{(j)}$ (see Fig. 1). Finally, we can model the dependency as:

$$\mathbf{x}_i^{(j)} = \mathbf{z}_i^{(j)} + \boldsymbol{\varepsilon}_i^{(j)}, \tag{8}$$

where $\boldsymbol{\varepsilon}_i^{(j)} \sim \mathcal{N}(\boldsymbol{\varepsilon}_i^{(j)}|0, \ \mathrm{diag}(\boldsymbol{\sigma}_j^2))$ is independent Gaussian noise. Notice that by adding random noise, we can model the fact that peaks in different spectra, obtained from the same group of chemical shifts, may slightly vary in their position. Variances $\boldsymbol{\sigma}_j^2$ depend mostly on the image resolution, and for every image $I^{(j)}$ need to be individually estimated.

Thus, the generative model for the ith peak in the jth spectral image has the following form:

$$p(\mathbf{x}_i^{(j)}|\mathbf{z}_i^{(j)}) = \mathcal{N}(\mathbf{x}_i^{(j)}|\mathbf{z}_i^{(j)}, \mathrm{diag}(\sigma_j^2)), \tag{9}$$

which leads to the formulation of a generative model for all true peaks in a single spectrum:

$$p(\mathbf{X}^{(j)}|\mathbf{z}) = \prod_{i=1}^{N_j} p(\mathbf{x}_i^{(j)}|\mathbf{z})$$

$$= \prod_{i=1}^{N_j} p(\mathbf{x}_i^{(j)}|\mathbf{z}_i^{(j)}), \tag{10}$$

where N_j denotes the number of visible peaks in the spectrum j. The second equality in (10) is the consequence of the fact that peak positions are determined only by small subsets of chemical shifts within the vector $\mathbf{z}$.

Finally, it is worth to emphasize that the list of peaks $\mathbf{X}^{(j)}$ depends on the type of the jth spectrum. Fortunately, it is known a priori which peaks (or in other words, which pairs or triplets of chemical shifts) are likely to be visible in a spectrum of a given type.

4.2   Peak Picking

We define a *response map* $P^{(j)}$ as a tensor of the same order and dimensions as the spectral image $I^{(j)}$, which contains probabilities $p(y^{(j)} = 1|\mathbf{x})$, where $y^{(j)}$ is a binary variable denoting an occurrence of true peak at location $\mathbf{x}$. The response map is constructed from the spectral image through the procedure called *peak picking* (Fig. 1). Its role is to pre-process the spectral image in order to distinguish signals originating from the protein atoms (true peaks) from noise and artifacts. The response map is generated as follows:

1. Local optima $\mathbf{x}_1, \ldots, \mathbf{x}_n$ are selected in the spectral image $I^{(j)}$.
2. For each point $\mathbf{x}_i$, tensor element $P^{(j)}(\mathbf{x}_i)$ is set to $p(y^{(j)} = 1|\mathbf{x}_i)$ using peak classification model (e.g. Convolutional Neural Network [12,16,17]).
3. Tensor $P^{(j)}$ is convolved with smoothing kernel (e.g. Gaussian), making further inference more robust to small perturbations.

4.3   Image Likelihood for a Given Peak List

In order to define the image likelihood model for a given peak list $p(I^{(j)}|\mathbf{X}^{(j)})$ we will use the response map $P^{(j)}$ generated from the image $I^{(j)}$. By $Q^{(j)}$ we denote a tensor with the same shape as $P^{(j)}$, generated from the peak list $\mathbf{X}^{(j)}$ (see Fig. 1). The procedure of constructing $Q^{(j)}$ is as follows:

1. Initialize tensor $Q^{(j)}$ to all zeros.
2. Project each point $\mathbf{x}_i^{(j)} \in \mathbf{X}^{(j)}$ to the nearest voxel in $Q^{(j)}$, setting its value to 1.
3. Apply smoothing filter to the resulting tensor.
4. Normalize the tensor to unity.

We define the log-likelihood model to be proportional to the distance between tensors $P^{(j)}$ and $Q^{(j)}$:

$$\ln p(I^{(j)}|\mathbf{X}^{(j)}) = -\gamma d(P^{(j)}, Q^{(j)}) + \text{const}, \tag{11}$$

where d is a distance measure, and γ is a parameter determining the smoothness of the likelihood function. Function d must satisfy certain properties, such as $d(P,Q) \geq 0$ and $d(P,P) = 0$, for all response maps P and Q. One possible choice of d is the ℓ_1-norm, $d(P,Q) = ||P - Q||_1$, which leads to:

$$p(I^{(j)}|\mathbf{X}^{(j)}) = \frac{1}{Z} \exp\left\{-\gamma||P^{(j)} - Q^{(j)}||_1\right\}. \tag{12}$$

The ℓ_1-norm is more robust in terms of handling outlier points than the Euclidean distance. However, the use of arbitrary distance function in further inference brings the need for computing the normalization constant Z. In general, its value depends on the choice of peaks $\mathbf{X}^{(j)}$.

For the choice of likelihood with ℓ_1-norm, the normalization constant is given as follows:

$$Z = \int_{[0,1]^N} \exp\left\{-\gamma||P - Q^{(j)}||_1\right\} dP, \tag{13}$$

where N is the total number of voxels in the spectral image. Denoting by $q_n^{(j)}$ the nth voxel[2] in the generated spectrum $Q^{(j)}$, we can determine the value of the constant as follows:

$$
\begin{aligned}
Z &= \int_{[0,1]^N} \exp\{-\gamma \sum_{n=1}^{N} |p_n - q_n^{(j)}|\} dP \\
&= \int_{[0,1]^N} \prod_{n=1}^{N} e^{-\gamma|p_n-q_n^{(j)}|} dP \\
&= \prod_{n=1}^{N} \int_{[0,1]} e^{-\gamma|p_n-q_n^{(j)}|} dp_n \\
&= \prod_{n=1}^{N} \left(\int_0^{q_n^{(j)}} e^{-\gamma(-p_n+q_n^{(j)})} dp_n + \int_{q_n^{(j)}}^1 e^{-\gamma(p_n-q_n^{(j)})} dp_n \right) \\
&= \prod_{n=1}^{N} \left(e^{-\gamma q_n^{(j)}} \int_0^{q_n^{(j)}} e^{\gamma p_n} dp_n + e^{\gamma q_n^{(j)}} \int_{q_n^{(j)}}^1 e^{-\gamma p_n} dp_n \right) \\
&= \prod_{n=1}^{N} \left(\frac{1}{\gamma}(1 - e^{-\gamma q_n^{(j)}}) - \frac{1}{\gamma}(e^{\gamma(q_n^{(j)}-1)} - 1) \right) \\
&= \prod_{n=1}^{N} \left(\frac{2}{\gamma} - \frac{1}{\gamma}(e^{-\gamma q_n^{(j)}} + e^{\gamma(q_n^{(j)}-1)}) \right). \quad (14)
\end{aligned}
$$

It is recommended to use log-likelihood, and consequently the logarithm of the normalization constant, in order to prevent the numerical rounding errors in the above product.

5   Optimization Problem

Putting together the prior (4) and the spectral image likelihood (7), we obtain the following posterior distribution by applying Bayes' theorem (2):

$$
\begin{aligned}
p(\mathbf{z}|\mathcal{I}) &\propto p(\mathbf{z})p(\mathcal{I}|\mathbf{z}) \\
&= p(\mathbf{z})\mathbb{E}_{\mathcal{X}|\mathbf{z}}[p(\mathcal{I}|\mathcal{X})] \\
&= p(\mathbf{z}) \prod_{j=1}^{K} \mathbb{E}_{\mathbf{X}^{(j)}|\mathbf{z}}\left[p(I^{(j)}|\mathbf{X}^{(j)})\right]. \quad (15)
\end{aligned}
$$

Our goal is to find the MAP estimator of $\mathbf{z}$. By taking the logarithm of RHS of (15), the MAP estimation problem can be stated as follows:

$$
\max_{\mathbf{z}} \; \ln p(\mathbf{z}) + \sum_{j=1}^{K} \ln \mathbb{E}_{\mathbf{X}^{(j)}|\mathbf{z}}\left[p(I^{(j)}|\mathbf{X}^{(j)})\right]. \quad (16)
$$

---

[2] For brevity, we use a single value to index the voxel number.

The main computational difficulty in (16) is the expected value, which in general cannot be calculated analytically. One way of dealing with this problem is to use Monte Carlo approximation, which involves sampling a matrix $\mathbf{X}^{(j)}$ from a probability distribution $p(\mathbf{X}^{(j)}|\mathbf{z})$. Subsequently, given a sequence of sampled peak lists, we evaluate the expected value according to formulas (12)–(13).

The second difficulty lies in calculation of the normalization constant directly from the Eq. (13). It may lead to numerical instabilities, since the product over voxels can rapidly decrease to zero. In that case however, we may use Jensen's inequality to get a lower bound on the objective function and then apply Monte Carlo sampling:

$$\sum_{j=1}^{K} \ln \mathbb{E}_{\mathbf{X}^{(j)}|\mathbf{z}} \left[p(I^{(j)}|\mathbf{X}^{(j)}) \right]$$

$$\geq \sum_{j=1}^{K} \mathbb{E}_{\mathbf{X}^{(j)}|\mathbf{z}} \left[\ln p(I^{(j)}|\mathbf{X}^{(j)}) \right]$$

$$\approx \frac{1}{L} \sum_{j=1}^{K} \sum_{l=1}^{L} \left[-\gamma \| P^{(j)} - \tilde{Q}_l^{(j)}) \|_1 - \ln Z \right], \tag{17}$$

where $\tilde{Q}_l^{(j)}$ is a spectrum generated from a sample from the peak generative model. The number of samples L should be selected appropriately, as it depends on the variance of the model (the greater the variance the greater the required L). Finally, we can replace the likelihood term in (16) by its lower bound (17), and then optimize the resulting lower bound of the log-posterior.

6   Conclusions

In this paper we have presented a novel formulation of the chemical shift assignment problem, which leads to continuous optimization rather than combinatorial search. The proposed framework is based on Bayesian statistics, where determining the chemical shifts for a protein is stated as finding MAP estimator. This requires solving the optimization problem (16), which is in general difficult and highly non-convex. Therefore, our future work would focus on finding optimization algorithms that make it possible to solve the problem efficiently.

Acknowledgements. The research conducted by the authors has been partially co-financed by the Ministry of Science and Higher Education, Republic of Poland, namely, Adam Gonczarek: grant No. 0402/0075/16.

References

1. Andrec, M., Levy, R.M.: Protein sequential resonance assignments by combinatorial enumeration using $^{13}C\alpha$ chemical shifts and their $(i, i - 1)$ sequential connectivities. J. Biomol. NMR **23**(4), 263–270 (2002)

2. Atreya, H.S., Sahu, S.C., Chary, K.V., Govil, G.: A tracked approach for automated NMR assignments in proteins (TATAPRO). J. Biomol. NMR **17**(2), 125–136 (2000)
3. Bahrami, A., Assadi, A.H., Markley, J.L., Eghbalnia, H.R.: Probabilistic interaction network of evidence algorithm and its application to complete labeling of peak lists from protein NMR spectroscopy. PLoS Comput. Biol. **5**(3), e1000307 (2009)
4. Bartels, C., Billeter, M., Güntert, P., Wüthrich, K.: Automated sequence-specific NMR assignment of homologous proteins using the program GARANT. J. Biomol. NMR **7**(3), 207–213 (1996)
5. Coggins, B.E., Zhou, P.: PACES: protein sequential assignment by computer-assisted exhaustive search. J. Biomol. NMR **26**(2), 93–111 (2003)
6. Felzenszwalb, P.F., Huttenlocher, D.P.: Pictorial structures for object recognition. Int. J. Comput. Vis. **61**(1), 55–79 (2005)
7. Gronwald, W., Kalbitzer, H.R.: Automated structure determination of proteins by NMR spectroscopy. Prog. Nucl. Magn. Reson. Spectrosc. **44**(1), 33–96 (2004)
8. Guerry, P., Herrmann, T.: Advances in automated NMR protein structure determination. Q. Rev. Biophys. **44**(3), 257–309 (2011)
9. Gntert, P.: Automated structure determination from NMR spectra. Eur. Biophys. J. **38**(2), 129 (2008)
10. Hyberts, S.G., Wagner, G.: IBIS - a tool for automated sequential assignment of protein spectra from triple resonance experiments. J. Biomol. NMR **26**(4), 335–344 (2003)
11. Jung, Y.S., Zweckstetter, M.: Mars - robust automatic backbone assignment of proteins. J. Biomol. NMR **30**(1), 11–23 (2004)
12. Krizhevsky, A., Sutskever, I., Hinton, G.E.: ImageNet classification with deep convolutional neural networks. In: Advances in Neural Information Processing Systems, 1097–1105 (2012)
13. Lemak, A., Steren, C.A., Arrowsmith, C.H., Llinąs, M.: Sequence specific resonance assignment via Multicanonical Monte Carlo search using an ABACUS approach. J. Biomol. NMR **41**(1), 29–41 (2008)
14. Moseley, H.N., Monleon, D., Montelione, G.T.: Automatic determination of protein backbone resonance assignments from triple resonance nuclear magnetic resonance data. Methods Enzymol. **339**, 91–108 (2001)
15. Schmidt, E., Güntert, P.: A new algorithm for reliable and general NMR resonance assignment. J. Am. Chem. Soc. **134**(30), 12817–12829 (2012)
16. Simonyan, K., Zisserman, A.: Very deep convolutional networks for large-scale image recognition. arXiv preprint arXiv:1409.1556 (2014)
17. Szegedy, C., et al.: Going deeper with convolutions. In: Proceedings of the IEEE Conference on Computer Vision and Pattern Recognition (2015)
18. Volk, J., Herrmann, T., Wüthrich, K.: Automated sequence-specific protein NMR assignment using the memetic algorithm MATCH. J. Biomol. NMR **41**(3), 127–138 (2008)

Modern Applications of Machine Learning for Actionable Knowledge Extraction

Traditional vs. Machine Learning Techniques: Customer Propensity

Mamta A. Rajnayak[✉], Snigdha Moitra, and Charu Nahata

Accenture Services Private Limited, Building No. 8, Tower C, DLF Cyber City,
DLF Phase 2, Gurgaon 122002, Haryana, India
mamtarajnayak@gmail.com, snigdhamoi@gmail.com,
charu.nahata@accenture.com

Abstract. In today's world there is a need of speedy tools and techniques to convert big data into information. Traditional techniques are robust but might carry inherent bias of data scientist's modeling skills. Machine Learning (ML) models makes the machine learn from the data and might not suffer from human bias. Some researchers argue that ML techniques can turn around model faster for big data, while others debate traditional techniques would still outperform when data is not so big. This paper tests arguments raised above in Oil & Gas (ONG) industry for actionable knowledge extraction on not so big sample customer data to predict customer's campaign response. Our experiment reveals that on our data ML doesn't outclass traditional modeling but is either slightly better or at par in terms of model accuracy i.e. percentage of instances classified correctly. It also establishes that though ML results don't improve much on accuracy but some of them can be developed much faster.

Keywords: Traditional statistical modeling · Random Forests · Logistic regression · Generalized linear model · Neural Network · Gradient Boosting

1 Introduction

While conducting any statistical modeling, there are a number of dimensions to get a sense of what data exploration and predictive techniques can be applied viz. - what is the size of the data sample? What is the dimensionality of the feature space? Is the problem expected to be linearly separable? Are features independent? Are features expected to be linearly dependent with the target variable? Is overfitting expected to be a problem? What are the system's requirement in terms of speed/performance/memory usage? And few more. Choosing the best model for the data is one of the first questions any analyst deals with. The good news though is, just as for many problems in life, you can use the least complicated algorithm that can achieve your objective and only go for something more complicated if strictly necessary.

Logistic regression (LR) models have low complexity especially when no or few interaction terms and variable transformations are used. Performing variable selection is a way to reduce a model's complexity and consequently decrease the risk of overfitting. The contribution of parameters in logistic regression (coefficients and intercept) can be interpreted (Dreiseitl and Ohno-Machado 2002). A final advantage of LR is that the output can be interpreted as a probability (Kohler and Kreuter 2005).

© Springer International Publishing AG 2017
N.T. Nguyen et al. (Eds.): ACIIDS 2017, Part II, LNAI 10192, pp. 653–663, 2017.
DOI: 10.1007/978-3-319-54430-4_63

On the other hand, ML is the science of getting computers to act without being explicitly programmed (Arthur Samuel 1959). The iterative aspect of ML is important because as models are exposed to new data, they are able to independently adapt. ML is used to reproduce known patterns and knowledge, apply those to new data, and further automatically apply the results to decision making and actions. Ensemble methods are learning algorithms that construct a set of classifiers and then classify new data points by taking a (weighted/voted) prediction (Dietterich 2000). Results are less dependent on features of a single model and training set (reduced variance) and combination of multiple classifiers may produce more reliable classification than single classifier (reduced bias). Concepts like bagging (being relatively impervious to overfitting (Maclin and Opitz 1997)) helps in reducing overall variance of predictions using voting/averaging and completely grows the tree finding the best split for the given sample. However redundant attributes may add noise and reduce overall model accuracy.

Random Forest (RF) is an ensemble classifier using many decision tree models based on bootstrapped sample of data and random selection of variables. It randomly selects variables at each node, leading to choice from all available features and selection of best feature that splits the data. RF cross-validate themselves and naturally deal with nonlinearity without needing to invoke polynomial transformations of the data. It provides the class of dependent variable based on many trees, combining "bagging" and random selection of features. Bootstrapping is done on the training data (63% of total). Each tree is grown on training data and remaining 37% of the data is used for out of bag validation. At each node, number of variables are randomly selected to create the tree. A large number of trees are developed using bootstrapping & sampling. The trees are then combined via voting/averaging for prediction. There are typically two parameters in RF to tune - number of trees and no. of features to be selected at each node.

Gradient Boosting Model (GBM) is a method of producing regression models used as an ensemble where learners are learned sequentially. It builds trees one at a time, where each new tree helps to correct errors made by previously trained tree. With each tree added, the model becomes even more expressive. There are majorly three parameters that require tuning - number of trees, depth of trees and learning rate, and each tree built is generally shallow.

An Artificial Neural Network (ANN) is an information processing paradigm that is inspired by the way biological nervous systems, such as the brain, process information. The key element of this paradigm is the novel structure of the information processing system. It is composed of a large number of highly interconnected processing. The network is composed of a large number of highly interconnected processing elements (neurones) working in parallel to solve a specific problem. Neural networks learn by example. They cannot be programmed to perform a specific task (Shalma 2015). If the data arrives in a stream, incremental updates with stochastic gradient descent (unlike decision trees, which use inherently batch-learning algorithms) can be performed. When the features are heterogeneous, weights and updates will all be on different scales which needs input data standardization.

In this paper, we have tried to apply as well as compare results of all above mentioned modeling techniques to predict coupon redemption rates for our ONG. Below sections outline the problem, data, results and our recommendations.

2 Literature Review

Over the last few years, there has been a lot of research on the application of ML techniques in the analytical world across industries. ML techniques like RF, GB, NN, Support Vector Machine (SVM) etc. have been explored, researched and grilled since its existence in late 90s. An early example is deep diving on bagging concept of RF Modeling (Breiman 1996), where to grow each tree a random selection (without replacement) is made from the examples in the training set which forms a basis of RF ML technique. Another example is random split selection (Dietterich 1998) where author talks about the split at each node by selecting at random from among the K best splits. Amit and Geman (1997) define a large number of geometric features and search over a random selection of these for the best split at each node. Ho (1998) has written a number of papers on "the random subspace" method which does a random selection of a subset of features to use to grow each tree. Breiman (1999) in his paper generates new training sets by randomizing the outputs within the original training set. Another approach is to select the training set from a random set of weights on the examples in the training set. Breiman (2001) in his paper formally calls this technique of creating a combination of tree predictors such that each tree depends on the values of a random vector sampled independently and with the same distribution for all trees in the forest as Random Forests. Livingston (2005), implemented Breiman's RF algorithm in Java and variable importance into Weka. Further to this Biau (2012) performed an in-depth analysis of a RFs Model suggested by Breiman and concluded that RF are fast and easy to implement, produce highly accurate predictions and can handle a very large number of input variables without overfitting. In fact, they are considered to be one of the most accurate general-purpose learning techniques available.

Another ML technique viz. GBM was interrogated by Schapire (2002) on AdaBoost algorithm and interpreted it as a genuine boosting algorithm and a procedure based on functional gradient descent. Dietterich et al. (2008), described a new algorithm for training CRFs via gradient tree boosting. Yiefi (2013) developed a nonparametric model for survival analysis and called it GBMCI (gradient boosting machine for concordance index).

ML has been theoretically compared with traditional approach in several different forums. Studies have been comparing traditional modeling techniques like linear/logistic regression with the more modern ML algorithms such as RF, NN, SVM and GB. Several studies covering implementation of NN have been discussed by academicians. For instance, Lisboa et al. (2006) assessed the benefit of artificial NNs as decision making tools in the field of cancer and further Lisboa et al. (2008) compared the linear model with NNs approach showing the benefit of using the NN framework especially for patients at high risk. Lichen et al. (2009) described a biomedical application of ML approaches for modeling heterogeneous data using several medical data sets and demonstrated the advantages and limitations of these new approaches, relative to standard inductive SVM classifiers. Wasserman (2012) could be credited with setting the agenda for a formal debate about comparing traditional statistical modeling with ML in his blog. Wasserman concluded a striking overlap between the two by citing an example of Reproducing Kernel Hilbert Space (RKHS) methods which are hot in ML but they began in Statistics. However, he distinctively talked about the differences in terminology used by the two branches of study. Here are some examples (Table 1):

Table 1. Statistics vs. ML terminology

Statistics	ML
Estimation	Learning
Classifier	Hypothesis
Data point	Example/instance
Regression	Supervised learning
Classification	Supervised learning
Covariate	Feature
Response	Label

Traditionally logistic regression is used extensively in various industries to predict customer's likelihood of an action like buying a product, repeat purchase, attrition etc. However, there is very limited work in using ML techniques to model propensity behavior. Lee et al. (2010) examined the performance of various CART-based propensity score models using simulated data and asserted that using these techniques to estimate propensity can greatly reduce bias across a range of sample sizes, scenarios, and propensity score application methods. They concluded that these techniques offer a number of advantages over LR. Westreich et al. (2010) further enhanced this work and compared most of the ML techniques with LR theoretically to state the advantages and disadvantages.

From the above studies, we realize that no empirical study has been done so far to compare LR with other ML techniques like GBM and NN. Though Westreich et al. did compare all these techniques with LR but their entire research was theoretically. Thus through this paper, our attempt is to bridge these gaps and provide recommendation based on our data. Using this data we also intend to validate Lee's conclusions for CART based algorithms.

3 Study and Results

The objective of our experiment was to predict the redemption rate for a point reward coupon distributed during a marketing campaign, i.e., to predict which customers (who have been issued the coupon) will redeem. Our entire experiment is conducted in an open source project R.

Our data consisted of all the customers who were issued this coupon during 3 months' campaign period. All those accounts who's redeemed the coupon during this period were tagged as 1 (dependent variable), while the others were flagged 0. A host of more than 120 Independent variables were created ranging across various coupon types redeemed and issued in the past few months, points gathered in a visit, frequency of visit and the number of sites visited in past few months, fuel and non-fuel volume purchased, share of premium fuel purchased, fuel and non-fuel transactions, maturity and tenure of account, days since last visit etc. These were created using last 1-year data.

A total of almost 10 K accounts formed our model data which was split into 4 equal datasets to perform robust validation. Each time one dataset was used to develop the model and the rest 3 were used for validation. Here is the snapshot of all 4 datasets on label (Table 2).

Table 2. Data snapshot on label

Dataset	Non redeemers	Redeemers	% redeemers
Sample 1	1,261	1,197	49%
Sample 2	1,293	1,165	47%
Sample 3	1,249	1,209	49%
Sample 4	1,294	1,167	47%
Total	5,097	4,738	48%

Once the model data and variables were created, the next step was to assess what modeling techniques to apply to get the best predictions. Standard missing value treatment was applied on the variables (in our case replaced with zero). Before fitting any models, we tried to select the best variables by applying a variable reduction techniques.

For the purpose of our study, we tried the standard LR model, as well as some of the more advanced ML techniques – RF, NN and GB. All 4 samples (explained above) were leveraged across all the models to ensure consistent and direct comparison. Following model selection stats are used to gauge model accuracy and robustness.

- Accuracy: The ability of a test to correctly identify those with and without the disease (True rate) i.e. (TP + TN)/(TP + FN + TN + FP)

  Where, TP = True Positive, TN = True Negative, FN = False Negative and FP = False Positive

- AUC of ROC: Area under Receiver Operating Characteristic curve. It is a measure to determine models predictive power for classification problem. It is mainly used to select best models out of shortlisted models.

For the LR model (GLM), we adopted the standard Varclus and VIF functions for variable reduction, making sure only uncorrelated variables were kept, and then selected the best variables based on their significance levels as well as business importance. The best model contained 16 out of the 120 + initial variables (after multiple iterations). The same set of these 16 variables were also used for building our NN models. This would help us understand which technique performs best in a scenario where same data and variables are selected. Also, before applying NN models, variable scaling/standardization was performed so as to get best results. Since the above variable reduction techniques are not mandatory for implementing RF, we used all variables for this techniques and proceeded on variable selection through Mean Decreasing Gini & Mean Decreasing Accuracy and without doing any correlation treatment. In Parallel, we also tried another RF model using the above selected 16 uncorrelated variables from GLM as inputs. GBM also don't need variable reduction hence we proceeded with 16 variables shortlisted in GLM to perform our analysis. Thus, four different types of models were tried on our data to gauge which one yielded best predictions for propensity to redeem a coupon.

We started with the traditional logistic regression, followed standard variable reduction and transformation techniques (Varclus and VIF), tried numerous iterations to achieve best model (high accuracy, high AUC and lowest AIC, making sure the selected set of 16 variables in the chosen model were statistically significant and made business sense). The final list of variables ranged across maturity of account, fuel volume per visit, total number of visits, monthly spend, kind of coupons issued/ redeemed in the past etc. Although the course of variable and model selection is iterative and manual, it gave us a strong model in terms of Accuracy and AUC both. Moreover, we had control over the kind of variables that we could select and the functioning of the algorithm. Training and test data model KPIs obtained from the GLM Model have been shown below (Tables 3 and 4).

Table 3. GLM results

GLM/logistic	Sample 1	Sample 2	Sample 3	Sample 4
Accuracy	0.775	0.795	0.783	0.781
AUC	0.851	0.861	0.853	0.860

Table 4. Validation accuracy

Model data	Sample 1 (validation)	Sample 2 (validation)	Sample 3 (validation)	Sample 4 (validation)
Sample 1		0.781	0.783	0.775
Sample 2	0.779		0.785	0.785
Sample 3	0.773	0.783		0.779
Sample 4	0.777	0.782	0.780	

It can be seen clearly from the above table that the GLM model is quite consistent on all the 4 samples in terms of Accuracy and AUC. Next, we decided to move to the popular RF technique and compare the predictions to that obtained from the logistic model. RF are much easier to train. Generally, they have two tuning parameters mtry and ntrees. Mtry is number of variables chosen randomly from the set of input variables and ntrees is number of trees to grow. Now default value of mtry in R (sort of n) generally gives good results. As for ntree, you can increase the value of variable and it will lead to better fit (More number of trees means less variance). Now a days with so much computing power, training a RF with even 5000 trees is quite fast. So essentially there is only one parameter to tune mtry which with its default value gives good results.

Variable reduction was done by selecting on the basis of importance (Mean Decreasing Accuracy and Mean Decreasing Gini), and standard correlation check or variable transformation was not applied. Optimal number of trees was selected as 800 (basis least OOB error and best accuracy). Using the set of 32 selected variables, as well as the selected number of ntree and mtry, we obtained a model with accuracy and AUC marginally better than that achieved from the logistic model. Training and test data model KPIs obtained from the RF Model have been shown below (Tables 5 and 6).

Table 5. RFM1 results (32 variables)

RFM1	Sample 1	Sample 2	Sample 3	Sample 4
Accuracy	0.794	0.798	0.792	0.795
AUC	0.865	0.879	0.871	0.872

Table 6. RFM1 validation accuracy

Model data	Sample 1 (validation)	Sample 2 (validation)	Sample 3 (validation)	Sample 4 (validation)
Sample 1		0.793	0.797	0.798
Sample 2	0.791		0.796	0.798
Sample 3	0.792	0.790		0.797
Sample 4	0.797	0.798	0.799	

In order to compare RF v/s logistic on the same set of independent variables, another RF Model was built using the earlier 16 variables that were in the final Logistic model (achieved on the basis of VIF and Varclus). Again, the results were only marginally better than the GLM model results in terms of accuracy, and very comparable to the random forest model built where 32 variables were used basis variable importance (Tables 7 and 8).

Table 7. RFM2 results (16 variables)

RFM2	Sample 1	Sample 2	Sample 3	Sample 4
Accuracy	0.784	0.798	0.787	0.780
AUC	0.857	0.872	0.861	0.863

Table 8. RFM2 validation accuracy

Model data	Sample 1 (validation)	Sample 2 (validation)	Sample 3 (validation)	Sample 4 (validation)
Sample 1		0.794	0.798	0.781
Sample 2	0.795		0.806	0.788
Sample 3	0.791	0.802		0.799
Sample 4	0.786	0.793	0.796	

It can be seen from the above tables that the model results from GLM, RFM1 and RFM2 and quite similar with very slight (almost negligible) increase in the Accuracy in RFM models. However, while building the models RF models takes far less time as compared to GLM due to inherit robustness automatically coming by ensembling approach. Also, variable reduction is not required for RF models which also saves analysts time. Moving on, the third kind of model that we tried for our predictions was building a NN basis the 16 variables shortlisted basis our variable reductions applied. Scaling/Standardization of these variables was performed before using them in the model. While trying various iterations, the value of hidden layers and threshold were altered so as to arrive at the best model, giving high prediction accuracy on the validation data. We achieved a test data accuracy of around 0.79 which was at par with the GLM and RF models. However, the AUC was slightly lowers for NN models. Also, the accuracy parameter for NN validation samples was not as consistent with the test as it was in previous models. Which clearly indicates that NN models are more prone to overfitting (Tables 9 and 10).

Table 9. NN results

NN	Sample 1	Sample 2	Sample 3	Sample 4
Accuracy	0.797	0.794	0.799	0.795
AUC	0.856	0.853	0.859	0.853

Table 10. NN validation accuracy

Model Data	Sample 1 (validation)	Sample 2 (validation)	Sample 3 (validation)	Sample 4 (validation)
Sample 1		0.779	0.775	0.779
Sample 2	0.798		0.776	0.784
Sample 3	0.789	0.778		0.776
Sample 4	0.802	0.781	0.781	

Finally, we tried to see if a GBM would be able to outperform the earlier ones. GBM generally have 3 parameters to train shrinkage parameter, depth of tree, number of trees. Now each of these parameters should be tuned to get a good fit. And you cannot just take maximum value of ntree in this case as GBM can over fit higher number of trees. But if you are able to use correct tuning parameters, they might give somewhat better results than RF. Model was tuned on the basis of number of trees (500 trees selected by cv fold method for our model), shrinkage (0.005) value, distribution, tree depth (5) and number of minimum observations in a node (10). A test data accuracy similar to all previous models was obtained through GBM as well (Tables 11 and 12).

Table 11. GBM results (16 variables)

NN	Sample 1	Sample 2	Sample 3	Sample 4
Accuracy	0.776	0.782	0.781	0.776
AUC	0.856	0.863	0.860	0.858

Table 12. GBM validation accuracy

Model data	Sample 1 (validation)	Sample 2 (validation)	Sample 3 (validation)	Sample 4 (validation)
Sample 1		0.777	0.775	0.765
Sample 2	0.765		0.783	0.767
Sample 3	0.768	0.779		0.773
Sample 4	0.765	0.770	0.773	

In totality, our experiment showed that the results (prediction accuracy) of all the models tried was more or less same, and the more sophisticated ML algorithms added only marginal, if at all, value to the model fit.

Below table compares the model accuracy on test data achieved from the different models we tried on our small set of data (GLM/RF/NN/GBM) (Table 13).

Table 13. Test data traditional vs ML models' accuracy

	Sample 1	Sample 2	Sample 3	Sample 4
GLM	0.775	0.795	0.783	0.781
RFM1 (32 vars)	0.794	0.798	0.792	0.795
RFM2 (16 vars)	0.784	0.798	0.787	0.780
NN	0.797	0.794	0.799	0.795
GBM (16 vars)	0.776	0.782	0.781	0.776

It is quite evident from the above table that traditional vs. ML models more or less delivered similar results. ML techniques like RF which doesn't need variable reduction perform better when more variables are considered.

4   Conclusions and Recommendations

The study conducted on our small data set in the ONG industry data is a crude example of the fact that applying sophisticated ML techniques may not add much value in terms of predictive accuracy to all kinds of data and problems. A lot depends on the size of the data and how the model has been tuned. Traditional regression techniques, although involving more manual intervention, can sometimes give results at par or even better than some of the ML techniques.

Since in our case, RF provided best results in least time without overfitting (given that it didn't involve any correlation checks/variable standardization/pruning based on multiple parameters), for our study we recommended coupon distribution for the campaign based on the results from this model. RFM1 included 32 variables which were selected based on Gini and Accuracy, and optimal number of trees selected based on accuracy and OOB. RFMs overcome the problems of bias, variance, over-fitting and are easy to tune. The key to accuracy in RF is low correlation and bias. Each tree is grown to maximum size without pruning. Individual trees might over-fit but combination of trees takes care of overfitting. Runs efficiently on large data and can handle thousands of input variables without variable deletion. However, RFs do come with their own limitations. There is a need of more studies conducted on data from various industries to establish clear advantages.

However, as expected, and as suggested by most studies conducted by researchers across industries, more sophisticated applications of GBM and NN didn't fetch the best model for our data and variables. Also, they take a longer time to tune and implement, and are also more difficult to interpret. These techniques come handier when we deal with very large data sets, ensuring that minimal time and manual intervention is involved. GBM training generally takes longer as trees are built sequentially after learning from previous run. They are prone to overfitting, however there are methods to overcome same and build more generalized trees using a combination of parameters like learning rate (shrinkage) and depth of tree. Generally, the two parameters are kept on the lower side to allow for slow learning and better generalization.

As a next step to this study, we would like to try other ML techniques like SVM, KNN etc. as well as across other clients in different industries, so as to assess if the results may vary significantly when there is more/different data available and also when the nature of input variables changes across different industries.

References

Breiman, L.: Bagging predictors. ML **26**(2), 123–140 (1996)

Amit, Y., Geman, D.: Shape quantization and recognition with randomized trees. Neural Comput. **9**, 1545–1588 (1997)

Ho, T.K.: The random subspace method for constructing decision forests. IEEE Trans. Pattern Anal. Mach. Intell. **20**(8), 832–844 (1998)

Dietterich, T.: An experimental comparison of three methods for constructing ensembles of decision trees: bagging, boosting and randomization. ML **40**(2), 1–22 (1998)

Breiman, L.: Using adaptive bagging to debias regressions. Technical report 547, Statistics Department UCB (1999)

Breiman, L.: Random Forests (2001). www.stat.berkeley.edu/~breiman/randomforest2001.pdf

Schapire, R.E.: The Boosting Approach to ML An Overview. In: MSRI Workshop on Nonlinear Estimation and Classification (2002)

Livingston, F.: Implementation of Breiman's random forest ML algorithm. ECE591Q ML J. Pap. (2005)

Lisboa, P.J., et al.: The use of artificial neural networks in decision support in cancer: a systematic review. Neural Netw. **19**(4), 408–415 (2006)

Lisboa, P.J., et al.: Time-to-event analysis with artificial neural networks: an integrated analytical and rule-based study for breast cancer. Neural Netw. **21**(2–3), 414–426 (2008)

Dietterich, T.G., et al.: Gradient tree boosting for training conditional random fields. J. ML Res. **9**, 2113–2139 (2008)

Lichen, L., et al.: Predictive learning with structured (grouped) data. Neural Netw. **22**(5–6), 766–773 (2009)

Biau, G.: Analysis of a random forests model. J. ML Res. **13**, 1063–1095 (2012)

Wasserman, L.: Statistics vs. Machine Learning (2012). https://normaldeviate.wordpress.com/2012/06/12/statistics-versus-machine-learning-5-2/

Yiefi, C., et al.: A gradient boosting algorithm for survival analysis via direct optimization of concordance index. Comput. Math. Methods Med. **2013**, 8 p. (2013). Article ID 873595

Dreiseitl, S., Ohno-Machado, L.: Logistic regression and artificial neural network classification models: a methodology review. J. Biomed. Inform. **35**(5–6), 352–359 (2002)

Kohler, U., Kreuter, F.: Data Analysis Using Stata (2005). ISBN 1–59718–007–6

Dietterich, T.G.: Ensemble methods in machine learning. In: Kittler, J., Roli, F. (eds.) MCS 2000. LNCS, vol. 1857, pp. 1–15. Springer, Heidelberg (2000). doi:10.1007/3-540-45014-9_1

Maclin, R., Opitz, D.: An empirical evaluation of bagging and boosting. In: The Fourteenth National Conference on Artificial Intelligence, Providence, Rhode Island. AAAI Press (1997)

Shalma, B.A., et al.: A study on neural networks. Int. J. Innov. Res. Comput. Commun. Eng. **3**(12), (2015)

Westreich, D., et al.: Propensity score estimation: machine learning and classification methods as alternatives to logistic regression. J Clin. Epidemiol. **63**(8), 826–833 (2010)

Lee, B.K., et al.: Improving propensity score weighting using machine learning. Stat. Med. **29**, 337–346 (2010)

Analytics on the Impact of Leadership Styles and Leadership Outcome

Waseem Ahmad[(⊠)] and Muhammad Akhtaruzamman

Waiariki Institute of Technology, Rotorua, New Zealand
{waseem.ahmad,akhtar.zaman}@waiariki.ac.nz

Abstract. Data mining and machine learning approaches are helping organisations to streamline and optimise their processes and assisting management to devise more effective and efficient business strategies. In organisations, management or leadership style play an important role in employees motivation, work satisfaction, work commitment and productivity. Moreover, leadership style shapes up the organisational culture. The aim of this paper is to explore the impact of various leadership styles on employees using data mining techniques. A significant research has taken place in this area using simple statistical methods such as correlation and regression analysis. However, no research has been conducted in this area which used data mining algorithms to extract useful information from the data to examine how leadership styles influence/affect employees. In this research, rule and decision tree based algorithms are used to extract actionable information from the collected data.

Keywords: Data mining · Leadership styles · Rule based algorithms · Decision tree algorithm · Management

1 Introduction

In recent years, technological advancements has enabled organizations to store large operational and non-operational data. The exponential growth in data required efficient exploratory tools which can help in extracting useful information from data. Moreover, this extracted information can assist various segments of businesses in decision making process. This need for efficient data analysis led to the emergence of data mining disciple. Data mining is concerned with extracting useful information from the raw data. Data mining approaches are widely used in various business domains such as stock market, retail industry, customer relationship management and telecommunication sectors [1–4].

Data mining is a subfield of machine learning that deals with finding useful patterns from the raw data which are hidden from the naked eye. These underpinning novel relationships among various features of data allow businesses to achieve competitive advantage as well as used to make future forecasts. In today's market companies are competing against each other to acquire new customers, retain their existing customers and finding ways to decrease their operating costs. One such example is last mile logistics [5], where retailers compete to provide fast delivery of products and services from their warehouse to customers. Companies are using various optimization

© Springer International Publishing AG 2017
N.T. Nguyen et al. (Eds.): ACIIDS 2017, Part II, LNAI 10192, pp. 664–675, 2017.
DOI: 10.1007/978-3-319-54430-4_64

algorithms to find robust and efficient routes to deliver products to customers faster. Other examples include market segmentation, e-marketing, e-commerce, market basket analysis, fraud detection, customer loyalty programs and customer buying behavior, where data mining approaches are helping businesses make strategic decisions.

There are number of data mining techniques which have been used to extract implicit, useful and actionable information from the data such as classification (decision trees, Naïve Bayes, K-nearest neighbors and neural networks) [6], clustering (k-means, hierarchical and density based clustering algorithms) [7] and association (one dimensional, multidimensional and multilevel association) [8]. Classification is an example of supervised learning, where data consists of input attributes and discrete class labels (output attribute). The task of any classification algorithm is to find hidden relationship between input attributes and output class labels so that the discovered relationship can be used to predict class association of unknown instances. Fraud detection, credit approval and customer relationship management are few examples of application of classification algorithms. Clustering algorithms are used to perform market segmentation, where customers are grouped together based on their buying behavior to identify their common characteristics. The aim of clustering approaches is to find grouping in data which share common properties and each group must have minimum intragroup dissimilarity and maximum intergroup dissimilarity. There approaches are successfully used to find customers' buying behavior to devise appropriate sale campaigns [9].

Data mining algorithms are successfully applied to improve the organizations relationship with their external stakeholders, however there is less research on the application of data mining techniques on improving the relationship between different facets of internal stakeholders. One such example is the relationship of supervisors with their employees. Extensive research has taken place in the field of social sciences which concludes that relationship between employees and management play an important role in organizational success [10, 11]. Researchers have used simple statistical techniques such as correlation and regression analysis to highlight this relationship [12, 13]. However, no efforts are made to use data mining algorithms to extract useful and actionable information. The focus of this proposed research is to explore the relationship of supervisors' leadership styles on employees using data mining techniques.

Leadership has become an important component in organizations. Leadership is a process by which one person influences the thoughts, attitudes, and behaviors of others [14]. Each organization has a unique personality and organizations take up the personality of their leaders. In the literature various leadership styles are proposed. However, these various leadership styles can be grouped into five distinct leadership styles. These five main leadership styles as stated by Eid *et al.* [15] are Autocratic, Laissez faire, Participative, Transactional, and Transformational leadership styles. There is certain degree of overlapping in definitions of these leadership styles (meaning these styles are not mutually exclusive). In Autocratic leadership style the leader is the sole decision maker and there is no opinion seeking. The leader imposes his will on others, no one challenges the decision of the leader. In Laissez faire, the leader does not supervise their employees and no feedback is given to them. This type of style works well when the employees are highly skilled and needs no supervision. Participative leadership is also called democratic leadership style, contrary to autocratic style the leader welcomes ideas from team members but the final decision making right rests with the leader. Transactional leadership style provides tasks for the employees and reward or punishment is determined on the basis of the performance of the subordinates. Finally in Transformational leadership style, the leader encourages employees to perform well and increase the efficacy by communication and superior visibility, the leader also delegates the work to subordinates. This style requires a considerable amount of involvement from management to meet organizational goals [15].

The outcome of leadership styles is measured using extra effort, effectiveness and satisfaction of employees [13]. Extra effort is the measure of willingness of employees to exert extra effort. Effectiveness determines the degree to leaders' usefulness in the perception of employees and satisfaction is the level of employees' satisfaction on leader's capabilities. The leadership outcome is measured as employees intend to exert extra effort, employee satisfaction and effectiveness. In literature number of researchers have studied the impact of various leadership styles on the employee outcome [13, 16–18]. The emphasis of earlier approaches was only to investigate the relationship of various leadership styles on leadership outcomes. In order to achieve this, researchers have used statistical analysis tools such as correlation and regression analysis. Correlation analysis can explain the mutual relationship between two variables and regression analysis examines the relationship between one dependent variable and one or more independent variables. However, these approaches cannot characterize dependencies at an abstract level or why these dependencies exits. Moreover, these techniques cannot provide a justification of underlying relationships in the form of a higher level logic description. The aim of this paper is to fill this gap in the literature by investigating impact of leadership styles on employees (leadership outcome) using state of the art data mining techniques. Moreover, this study will explore the usage of data mining algorithms to extract useful information, which can be easy to understand and action plans can be devised based on the information provided by these techniques.

2 Literature Review

Aiden *et al.* [19] find that statistical modelling is a popular data analysis method in social science, where 78% studies conducted used correlation and 54% studies used regression analysis. These techniques are popular due to their sound theoretical basis. However, such models are based on the assumptions of normality, independency, linear additive and constant variance [20]. The authors argued that data in social science do not adhered to these four assumptions, therefore, other empirical modelling approaches must be used to analyses data. An extensive review of various data mining algorithms used to model educational data in the tertiary education sector was conducted in [19]. The contribution of various data mining algorithms such as support vector machines, decision tree, Bayesian networks and artificial neural networks were highlighted. However, the focus of most of these researchers were to highlighting model accuracies and no efforts were made to extract knowledge from the models.

The success of every business is governed by its employees urge to succeed through their efforts, commitment, engagement and practice. Hence the leader's prime responsibility is to motivate their subordinates with their leadership styles and keep their aspirations high and achievable [21]. Motivation, whether intrinsic or extrinsic, the basic instrument of regulating the work performance of the staff and is also the mainspring in chasing individuals own needs. Motivated employees are outcomes of effective leadership. Various leadership styles has impact on employees/subordinates, which has direct or indirect significance on the behaviors and attitudes of the employees [22]. The influence and effects of leadership styles (transformational and transactional) on different factors such as job performance, employee satisfaction, organizational commitment were studied in depth by Norwawi [23] and revealed that leadership styles were vital to organizational development. The author also concluded that by using appropriate leadership style, job satisfaction, stress, employee turnover, job satisfaction and productivity could be controlled effectively. Wang et al. [24] found in their studies that transformational leadership has a positive effect on organizational justice and job characteristics. The authors concluded that transformational leadership has an affirmative implication on motivation.

A study conducted by Mah'd Alloubani et al. [13] focused on the leadership styles (Laissez faire, transactional and transformational) and their effects on employees' motivation to exert extra effort, effectiveness and satisfaction. The study was conducted in the healthcare sector in Jordan. The authors found that transformational and Laissez faire leadership styles has statistical significant effects on employees' willingness to exert extra effort, effectiveness and satisfaction. The authors found that transformational leadership style had positive correlation with all three dependent variables (willingness to exert extra effort, effectiveness and satisfaction) and laissez faire leadership style had negative correlation with all three dependent variables.

It has been observed that the leaders with transformational leadership style who set the goal, motivate the employees with their clear communication and inspires their subordinates to achieve more than that is expected [25]. Employees are challenged when the goals are difficult, which induces them to put extra effort to achieve those goals [26]. This results in enhancing the intrinsic motivation of the employee which drives employees to exert extra effort [25]. On the other hand, transactional leaders has clear expectations which are communicated to their employees regarding goal attainment, the rewards upon fulfilment of the same clarifies their expectations as contractual responsibilities [25]. Transactional leaders use of contingent rewards inspires the employees directly or indirectly to exert extra effort for goal attainment. At certain times a combination of transactional and transformational leadership is used by the leaders to increase motivation and productivity, this hybrid leadership style is becoming very popular in organizations [25].

Leadership also has its impact on organizational commitment and job performance [27]. Authors empirically found that there is a strong correlation between leadership style and organizational commitment. Moreover, it was observed in the study that appropriate leadership style considerably enhanced the ability of problem solving and passion for innovation [27]. It is important to match leadership style with the features of the followers [21]. Therefore, situational leadership theory was proposed, according to which each leader must exhibit multiple leadership styles based on the situation and subordinates it is working with. Situational leadership theory has a direct relevance as contemporary organizations are encouraging diversity in workforce. Therefore, multiple leadership style must be applied to get optimal performance from employees. This is one of the reasons why data mining approaches are most suitable to the area of social science. Data mining approaches can be used to extract complex relationships between independent and dependent variables.

3 Methodology

This study employed a five-point Likert scale questionnaire to collect quantitative data. In the questionnaire various statements concerning leadership styles and leadership outcome were stated and participants had to respond by selecting 1 to 5 Likert scale point (where 1 meant strongly disagree and 5 represented strongly agree). The data was collected from the employees of small to medium enterprises (SMEs) in Bay of plenty region of New Zealand. Total 202 useable data samples were gathered from 24 various organizations. The participants answered questionnaires that portray the style of leadership of their immediate level manager. The goal of this research was to find and compare leadership styles of managers and their implications from the employee's point of view.

Eighteen questions were designed to collect feedback on three leadership styles (Laissez faire, transactional and transformation leadership styles) with each style has 6 questions. The response of all 6 questions were added to obtain a score of each leadership style. Three questions each for extra effort, effectiveness and satisfaction were designed and responses were calculated by taking the mean of each category questions. The leadership outcome was derived by taking the mean of extra effort, effectiveness and satisfaction. In this study, dependent variable is leadership outcome which is calculated based on the scores of extra effort, satisfaction and effectiveness. The independent variables are three leadership styles: Laissez faire, transactional and transformation.

There are number of data mining algorithms available in literature. However, in this research we will be using decision tree and rule based algorithms. The rationale of using these algorithms is that these algorithms provide output in the form of either rules or decision trees, which are easy to interpret. Moreover, basic statistical data analysis techniques such as correlation and regression analysis will also be performed of the collected data. The comparison of results obtained from regression analysis will be compared with the results of data mining approaches. The algorithms used in this study are OneR, Modlem and J48.

OneR stands for One Rule and this algorithm only select one variable to classify all data [28]. The selected variable is one that provides maximum classification accuracy. In order to create rules for each predictor (variable) it constructs a frequency table. This algorithm is selected due to its simplicity and reasonably good classification accuracy. Modlem algorithm [29] generates tree based on rough set theory. It is a sequential covering algorithm where attribute's space is being searched to find the best rule condition during rule induction. More detail on this model can be found at [29]. J48 is a decision tree based classifier [30]. According to this algorithm a subset of features (attributes) are selected based on their importance to construct decision tree. The entropy or information gain formula is used to select important features in data. In each cycle, feature with most information gain is selected to construct decision tree.

The dependent variables (extra effort, satisfaction, effectiveness and leadership outcome) are transformed into categorical variables to run rule based and decision tree algorithms. The values of independent variables varies between 1 (lowest) and 5 (highest). These values are transformed into Low, Medium and High where Low ranges from 1 to 3, Medium from greater than 3 to 4 and High from greater than 4 to 5. Table 1 highlights the frequency of each class label.

Table 1. Frequency distribution of extra effort, effectiveness, satisfaction and (leadership) outcome after performing data transformation

Labels/measure	Extra effort	Effectiveness	Satisfaction	Outcome
Low	57	33	42	33
Medium	102	102	94	94
High	43	67	66	75
Total	202	202	202	202

4   Experimental Results

Multiple regression analysis was performed on the data, where dependent variable was leadership outcome and independent variables were three leadership styles (transformational, laissez faire and transactional leadership styles). The results are stated in Table 2. The multiple regression analysis between the leadership outcome and the leadership styles demonstrates that transformational leadership has a statistical significant association with leadership outcome. The slope coefficient values of regression model is represented in the Table as coefficient. The regression model has adjusted R2 value of 0.54, which means approximately 54% of the variance in the data can be explained using this regression model. In other words, holding other things constant, approximately 54% of the leadership outcome can be explained by the leadership styles. More specifically, between the alternative choices of leadership styles, the coefficient estimates of multiple regression results suggest, ceteris paribus, leadership outcome is predicted to increase by 14% if transformation leadership is employed. It is also evident in the data that correlation between leadership outcome and transformational leadership is the highest/largest (0.74) amongst the all.

Table 2. Multiple regression of leadership outcome

	Laissez faire	Transactional	Transformational
P	0.325	0.566	5.655E-30
Coefficient	−0.010	0.007	0.142
Correlation	0.042	0.353	0.738

The regression analysis is helpful in establishing a relationship between independent and dependent variables, however, it does not explain how instances of these independent and dependent variables were associated with each other. To extract this deeper relationship, advanced data mining techniques were used.

The first data mining algorithm used was OneR. The effectiveness of this algorithm (as stated above) is that it partitioned the data based on single variable. The data consists of all three leadership styles (Laissez faire, Transactional and Transformational) as independent variables and extra effort, effectiveness, satisfaction and outcome (one at a time) as dependent variables (class labels). The rules extracted from this algorithms are stated in Fig. 1. The output of OneR model states that transformational leadership style is most important in classifying the data. This finding strongly supports the results found in multiple regression analysis provided in Table 2.

Extra Effort (66% Accuracy,0.70 Precision and 0.66 Recall)
Transformational: < 17.5→ LOW (26/30)
Transformational: < 25.5→ MEDIUM (91/149)
Transformational: >= 25.5→ HIGH (16/23)
Effectiveness (63% Accuracy, 0.69 Precision and 0.63 Recall)
Transformational: < 16.5→ LOW (14/23)
Transformational: < 25.5→ MEDIUM (93/156)
Transformational: >= 25.5→ HIGH (20/23)
Satisfaction (60% Accuracy, 0.63 Precision and 0.60 Recall)
Transformational: < 16.5→ LOW (17/23)
Transformational: < 21.5→ MEDIUM (36/75)
Transformational: < 22.5→ HIGH (10/20)
Transformational: < 25.5→ MEDIUM (40/61)
Transformational: >= 25.5→ HIGH (19/23)
Leadership Outcome (64% Accuracy, 0.76 Precision and 0.64 Recall)
Transformational: < 16.5→ LOW (17/23)
Transformational: < 25.5→ MEDIUM (90/156)
Transformational: >= 25.5→ HIGH (23/23)

Fig. 1. Rules extracted using OneR algorithm (these rules are given in order and that, for instance, a second rule is only triggered if the first rule is not triggered)

These rules are easy to interpret. For example, rules in extra effort states that if the value of transformational leadership style is less than 17.5, then employees tend to exert low amount of extra effort, if transformational leadership style value is in between 17.5 and 25.5 the level of extra effort will be medium and if transformational leadership style value is greater of equal to 25.5 then employees are likely to exert extra effort. This set of rules has an accuracy of 66% meaning out of 202 instances in the data, these rules can correctly classify 133 instances. The individual accuracy of each rule is also provided in the Fig. 1. The accuracy can be read as: (correctly classified instances/total number of instances).

Modelm is second rule based algorithm used in this research. These rules are extracted using leadership outcome data vs leadership styles. Total 72 rules are extracted from the data. However, few selected rules are shown in Fig. 2. Managers can be evaluated on these rules and strategies for moving from one class to another can be

Rules for Low
Rule 59. (Transformational < 16.5) & (Laissez Faire < 14.5) → (OUTCOME = LOW)   (7/7, 24.14%)
Rule 60. (Transformational < 14.5) & (Laissez Faire < 18.5) → (OUTCOME = LOW)   (8/8, 27.59%)
Rule 61. (Transformational < 16.5) & (Transactional in [14.5, 17.5]) → (OUTCOME = LOW)   (5/5, 17.24%)
Rule 62. (Transactional < 13.5) & (Laissez Faire >= 15.5) → (OUTCOME = LOW)   (4/4, 13.79%)
Rule 63. (Transformational < 19.5) & (Laissez Faire in [12.5, 13.5]) → (OUTCOME = LOW)   (5/5, 17.24%)
Rule 69. (Transformational < 18.5) & (Transactional < 16.5) & (Laissez Faire >= 16.5) → (OUTCOME = LOW)   (5/5, 17.24%)
Rules for Medium
Rule 1. (Laissez Faire in [21.5, 22.5]) → (OUTCOME = MEDIUM)   (5/5, 5.88%)
Rule 2. (Laissez Faire >= 18.5) & (Transformational < 20.5) & (Transactional in [13.5, 16.5]) → (OUTCOME = MEDIUM)   (5/5, 5.88%)
Rule 5. (Transformational in [18.5, 20.5]) & (Transactional in [15.5, 16.5]) → (OUTCOME = MEDIUM)   (7/7, 8.24%)
Rule 7. (Transformational in [19.5, 21.5]) & (Laissez Faire < 16.5) & (Transactional < 16.5) → (OUTCOME = MEDIUM)   (6/6, 7.06%)
Rule 9. (Transformational in [20.5, 25.5]) & (Transactional in [19.5, 20.5]) & (Laissez Faire in [14.5, 19.5]) → (OUTCOME = MEDIUM)   (8/8, 9.41%)
Rule 10. (Laissez Faire >= 18.5) & (Transactional in [20.5, 21.5]) & (Transformational >= 21.5) → (OUTCOME = MEDIUM)   (6/6, 7.06%)
Rule 11. (Transformational in [21.5, 22.5]) & (Laissez Faire in [16.5, 23.5]) → (OUTCOME = MEDIUM)   (6/7, 7.06%)
Rule 14. (Transactional in [17.5, 19.5]) & (Transformational in [21.5, 22.5]) → (OUTCOME = MEDIUM)   (6/6, 7.06%)
Rules for High
Rule 34. (Transformational >= 25.5) → (OUTCOME = HIGH)   (23/23, 35.38%)
Rule 35. (Transactional >= 25.5) → (OUTCOME = HIGH)   (4/4, 6.15%)
Rule 36. (Laissez Faire >= 24.5) → (OUTCOME = HIGH)   (5/5, 7.69%)
Rule 37. (Transactional in [21.5, 22.5]) & (Laissez Faire >= 12.5) → (OUTCOME = HIGH)   (8/8, 12.31%)
Rule 38. (Transactional >= 24.5) & (Laissez Faire >= 16.5) → (OUTCOME = HIGH)   (6/6, 9.23%)
Rule 41. (Transformational >= 24.5) & (Transactional < 18.5) → (OUTCOME = HIGH)   (8/9, 12.31%)
Rule 47. (Laissez Faire < 12.5) & (Transformational >= 22.5) → (OUTCOME = HIGH)   (6/6, 9.23%)

Fig. 2. Selected rules of Modelm Algorithm (91% Accuracy, 0.91 Precision and 0.91 Recall)

devised based on some proximity measure. For example, a supervisor falls in low class and belongs to rule 60, where his transformation leadership style value is 14 and Laissez fair is 18. The top management can suggest him to move from low to medium by choosing either rule 1 or rule 11, depending on his/her personality traits. By following rule 1, the supervisor has to enhance his laissez fair leadership skills by 4 points. Alternatively, supervisor can follow rule 11, according to which he only has to increase his transformational leadership skills and not changing his laissez faire skills. The supervisor has to increase his transformational leadership skills by 8 points. An appropriate strategy can be adopted by top management with consultation of supervisor. Same can be implemented to move from medium to high level. This model has

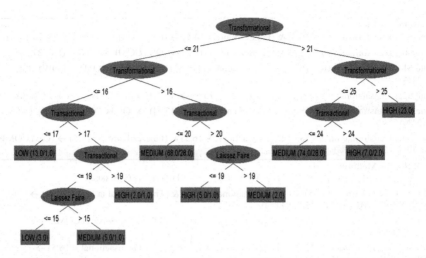

Fig. 3. The decision tree obtained using J48 algorithm (69% Accuracy, 0.77 Precision and 0.69 Recall)

produced 91% accurate rules on surveyed data (184 instances are correctly classified out of 202 total instances).

The final data mining algorithm used in this research is J48. The output of this algorithm is in the form of a tree, which are easy to interpret and if required can be converted into rules. This decision tree achieved an overall classification accuracy of 69% (140/202 correct instances). The original decision tree can be seen in Fig. 3. Few examples of rules extracted from decision tree can be seen below:

Rule 1: Transformational ≤ 17 (OUTCOME = LOW) (13/1) [out of total/incorrect]

Rule 2: Transformational between [16, 21] & Transactional ≤ 20 (OUTCOME = MEDIUM) (68/28)

Rule 3: Transformational > 25 (OUTCOME = HIGH) (23/0)

Rule 4: Transformational ≤ 25 & Transactional ≤ 24 (OUTCOME = MEDIUM) (74/28)

Rule 5: Transformational ≤ 25 & Transactional > 24 (OUTCOME = HIGH) (7/2)

A step-wised greedy feature selection algorithm was used to demonstrate the importance of each leadership style when leadership outcome is considered as class labels. Table 3 demonstrates the individual and collective variable importance on J48 and Modelm algorithms. When considering single variable importance Table 10 clearly suggests that transformational leadership style is better than transactional or laissez faire as it has more accuracy both algorithms (J48 and Modlem). When two variable model was generated using first variable as transformational and second variable as either transactional or laissez faire. It was found that transactional + transformational variables produced better classification. The model accuracy when all three variables

Table 3. Classification accuracy (in percentage) of J48 and Modlem on selected leadership style variables against leadership style outcome. (Tr, Tf and Lf abbreviated for Transactional, Transformational and Laissez Fair Leadership styles, respectively)

	Tf	Tr	Lf	Tf + Tr	Tf + Lf	Tf + Tr + Lf
J48	**64.36%**	58.42%	53.47%	**66.34%**	64.36%	69.31%
Modlem	**14.85%**	2.48%	7.92%	**71.78%**	70.30%	91.09%

are used can be seen in Table 10. The step-wise experiment conducted in Table 3 suggests that it is rationale to use all three leadership styles to model the leadership outcome as the accuracies of three variables are higher than individual or two variable based models. In other words, each of these variables contribute in determining the leadership outcome.

In this experiment, we have used three algorithms, namely, OneR, J48 and Modlem to analyse our surveyed data. Our experiments demonstrated that data mining algorithms can be useful in understanding the data effectively. The rules extracted or decision trees generated using data mining algorithms can help devise effective management strategies. This information once implemented can leads to knowledge discovery.

Our data mining approaches find support to existing literature that transformational leadership is most effective because this specific style motives and inspire employees to achieve their best, increases employees' engagement and gives an opportunity to businesses to produce managerial talent pool of future business leaders [31]. The second most important leadership style is transactional leadership style and Laissez faire is the least important in this research. The results presented in Table 3 also support Fiedler's contingency leadership model [32] as each leadership style is adding some information to the models (J48 and Modlem). According to contingency model leadership effectiveness is based on the situation in hand.

5   Conclusion

In this paper, impact of leadership styles on leadership outcome are explored using basic statistical methods and data mining techniques. The statistical methods used were correlation and regression analysis and datamining techniques such as J48, Modlem and OneR algorithms were used. The focus of the experimental results presented in this paper was not to compare and contrast rule based algorithms with tree based algorithm (J48), but to explore multiple ways to extract actionable information from the data. This paper has successfully demonstrated that data mining techniques can be implemented in organisation to improve employee relationships with management. The feedback gathered from employees can be analysed effectively to extract actionable information using state of the art data mining techniques.

Business information systems consist of four main components, namely, input, process, output and feedback. In the context of our surveyed data, independent and dependent variables are input, correlation coefficient and p values obtained are output and modelling methods (correlation & regression analysis) can be considered as processes. In the context of a complete information system, output of any information system must provide a feedback to either input or process so that output can be corrected or improved. The feedback provided by regression analysis is limited as it only established an underlying relationship between independent and depend variables. This technique does not explain how instances of independent variables are associated with dependent variable. Moreover, in strategic business management, various alternatives to top management must be provided so that they can establish better performance based relationship with their subordinates. In this paper data mining techniques were used to extract useful information from the data that could be helpful for top management to make informed decisions.

References

1. Fiol-Roig, G., Miró-Julià, M., Isern-Deyà, A.P.: Applying data mining techniques to stock market analysis. In: Demazeau, Y., et al. (eds.) Trends in Practical Applications of Agents and Multiagent Systems. AISC, vol. 71, pp. 519–527. Springer, Heidelberg (2010). doi:10. 1007/978-3-642-12433-4
2. Ahmed, S.R.: Applications of data mining in retail business. In: International Conference on Information Technology: Coding and Computing (ITCC) (2004)
3. Apte, C., et al.: Business applications of data mining. Commun. ACM 45(8), 49–53 (2002)
4. Ngai, E.W.T., Xiu, L., Chau, D.C.K.: Application of data mining techniques in customer relationship management: a literature review and classification. Expert Syst. Appl. 36(2), 2592–2602 (2009)
5. Lee, H.L., Whang, S.: Winning the last mile of e-commerce. MIT Sloan Manag. Rev. 42(4), 54 (2001)
6. Kotsiantis, S.B., Zaharakis, I., Pintelas, P.: Supervised machine learning: a review of classification techniques (2007)
7. Jain, A.K., Murty, M.N., Flynn, P.J.: Data clustering: a review. ACM Comput. Surv. 31, 265–323 (1999)
8. Zhang, C., Zhang, S.: Association Rule Mining: Models and Algorithms. Springer, Heidelberg (2002)
9. Sumathi, S., Sivanandam, S.N.: Introduction to Data Mining and Its Applications, vol. 29. Springer, Heidelberg (2006)
10. Elena, Å.e.A., How to improve employee motivation and group performance through leadership conceptual model. The Annals of the University of Oradea, p. 1086
11. WCDHRP Need: Human Resource Management: Gaining a Competitive Advantage (2006)
12. Jung, D.I., Chow, C., Wu, A.: The role of transformational leadership in enhancing organizational innovation: hypotheses and some preliminary findings. Leadersh. Q. 14(4), 525–544 (2003)
13. Mah'd Alloubani, A., et al.: Impact of leadership styles on leadership outcome (effectiveness, satisfaction and extra effort) in the private healthcare sector in Jordan. Eur. Sci. J.
14. Daft, R.L., Pirola-Merlo, A.: The Leadership Experience. Cengage Learning, Australia Pty Limited, Mason (2009)
15. Eid, J., et al.: Situation awareness and transformational leadership in senior military leaders: an exploratory study. Mil. Psychol. 16(3), 203 (2004)
16. Limsila, K., Ogunlana, S.O.: Performance and leadership outcome correlates of leadership styles and subordinate commitment. Eng. Constr. Archit. Manag.t 15(2), 164–184 (2008)

17. Lam, C.S., O'Higgins, E.R.E.: Enhancing employee outcomes: the interrelated influences of managers' emotional intelligence and leadership style. Leadersh. Org. Dev. J. **33**(2), 149–174 (2012)
18. Hamidifar, F.: A study of the relationship between leadership styles and employee job satisfaction at Islamic Azad University branches in Tehran, Iran (2015)
19. Aiden, C., et al.: A review of psychometric data analysis and applications in modelling of academic achievement in tertiary education. J. Learn. Anal. **1**, 57–106 (2014)
20. Nisbet, R., Elder, J., Miner, G.: Statistical Analysis and Data Mining Applications. Academic Press, Waltham (2009)
21. Naile, I., Selesho, J.: The role of leadership in employee motivation. Mediterr. J. Soc. Sci. **5**(3), 175 (2014)
22. Asrar-ul-Haq, M., Kuchinke, P.: Impact of leadership styles on employees' attitude towards their leader and performance: empirical evidence from Pakistani banks. Future Bus. J. **2**(1), 54–66 (2016)
23. Norwawi, S.: Leadership Styles: A Comparative Analysis of PTD Officers and DG Officers in the Ministry of Education Malaysia (2010)
24. Wang, X., Ma, L., Zhang, M.: Transformational leadership and agency workers' organizational commitment: the mediating effect of organizational justice and job characteristics. Soc. Behav. Personal.: Int. J. **42**(1), 25–36 (2014)
25. Quintana, T., Park, S., Cabrera, Y.: Assessing the effects of leadership styles on employees' outcomes in international luxury hotels. J. Bus. Ethics **129**, 469–489 (2015)
26. Kayemuddin, M.D.: Leadership in small business in Bangladesh. Int. J. Entrep. **16**, 25–35 (2012)
27. Chu, L., Lai, C.: A research on the influence of leadership style and job characteristics on job performance among accountants of county and city government in Taiwan. Publ. Pers. Manag. **40**(2), 101–118 (2011)
28. Holte, C.R.: Very simple classification rules perform well on most commonly used datasets. Mach. Learn. **11**, 63–91 (1993)
29. Stefanowski, J.: On rough set based approaches to induction of decision rules. Rough Sets Knowl. Discov. **1**(1), 500–529 (2008)
30. Quinlin, J.R.: C4.5: Programs for Machine Learning. Morgan Kaufmann, Burlington (1993)
31. Hoon-Song, J., et al.: Role of transformational leadership in effective organizational knowledge creation practices: mediating effects of employees' work engagement. Hum. Resour. Dev. Q. **23**(1), 65–101 (2012)
32. Fiedler, F.E.: The contingency model and the dynamics of the leadership process. Adv. Exp. Soc. Psychol. **11**, 59–112 (1978)

An Investigation into the Relationship of Strategic Planning Practices and Organizational Performance Using Advanced Data Mining Techniques

Philip Bright[(✉)], Waseem Ahmad, and Uswa Zahra

Toi Ohomai Institute of Technology, Tauranga, New Zealand
{Philip.Bright, Waseem.Ahmad,
Uswa.Zahra}@toiohomai.ac.nz

Abstract. This paper presents an investigation of the strategic planning practices of small businesses in Rotorua. The Rotorua district has a population of around 70,000 residents, located in the heart of New Zealand's North Island. Tourism is Rotorua's largest employer contributing around $593 million per year, or 10% of the district's economy. The strong link between planning and performance is understood for larger organisations but less so for small and medium sized firms. This study explores the degree of strategic planning undertaken by Rotorua-based businesses. A combination of quantitative and qualitative data collection via a structured questionnaire and a semi-structured interview, comprised the research methodology. Collected data was analyzed using advanced machine learning (classification) techniques. Findings suggest, a firm's industry sector influences the degree of formal planning. Other factors that influence the degree of planning formality include the age of the business, annual revenue, and the number of employees. More significantly, this study observes that formal business training and an age of a business are two features responsible for separating more formal planners from less formal planners.

Keywords: Strategic planning · Organizational performance · Data mining · Rule based learning · Decision tree algorithms

1 Introduction

The link between strategy and performance in large firms has been extensively studied [1–3]. Organizations using complex strategic management processes tend to be more successful than organizations that do not [4]. Schwenk and Shrader [5] examined 14 research studies investigating planning and performance relationships in small firms and found "a significant link" between formal planning activities and performance. Strategic planning leads to increased firm performance [6]. While performance-variation for small business has been a subject of some research, Verreynne [4] also asserts that strategic processes in small firms is a field of study that has been under-explored. This research aims to address two research questions. Firstly, is formal strategic planning a significant factor in the performance of small and medium size

© Springer International Publishing AG 2017
N.T. Nguyen et al. (Eds.): ACIIDS 2017, Part II, LNAI 10192, pp. 676–687, 2017.
DOI: 10.1007/978-3-319-54430-4_65

firms based in Rotorua? Secondly, what aspects of strategy are important success factors? For example, how important is it for an organisation to establish a vision; or to engage in external monitoring of its environment; or to understand its internal capabilities? Bracker and Pearson [7] identified the following six components to strategic planning:

- Having a sense of direction – where does an organisation see itself in the near future? – often articulated via a Vision Statement
- Formalized planning – shorter-term, specific goals that support the vision, written down and shared with relevant stakeholders. Has a Mission Statement, describing the purpose of the firm, been produced?
- Environmental analysis - Is environmental scanning and analysis undertaken? Does the firm understand the influences that external factors such as the economy, technological innovations and the political scene and other factors have on its operations?
- Internal Capability analysis - Does the firm understand the internal capabilities of the business? Are the firm's strengths and weaknesses identified and managed? Does the firm understand how it is adding-value throughout its operations?
- Financial forecasting - Does the firm forecast by projecting its financial performance and does the firm analyze past financial performance?
- Quality assurance - Does the firm monitor quality and have appropriate control systems in place to correct performance when necessary?

Small and Medium size Enterprise (SME) research undertaken in New Zealand has been confined to either: a particular sector, such as managed and unmanaged retail firms [8] or only looked at a specific aspect of strategy, such as strategy-making process. [4] Other New Zealand studies were simply parts of larger studies involving overseas firms, and these were conducted some time ago [9]. If these strategic aspects are important contributors to the success of Rotorua-based small businesses, are some aspects more important than others? Most small firms, especially very small firms, do not engage in strategic planning or only do so in a very limited way [10]. Planning is conspicuously absent in small firms. In a review of over 50 planning related small business studies it was concluded that for most small businesses planning was unstructured, irregular, and uncomprehensive.

To better understand the degree of planning and the relationship with the success of small businesses in Rotorua, a study was conducted by interviewing the owners of an appropriate variety, of Rotorua based small businesses. Two approaches were initially considered, a case study approach whereby a very limited number of small Rotorua-based firms might be studied in-depth, or a more descriptive approach using a combination of quantitative and qualitative methods. It is this latter approach that was determined as more appropriate for this study as a larger sample was likely to better represent the SME profile of Rotorua-based businesses. Most of the earlier work conducted in this field was analyzed either using simple statistical analysis such as mean, standard deviation and percentage analysis or through correlation & regression analysis. In this paper an attempt is made to analyze collected data using advanced data mining techniques. Rule based algorithms and decision tree algorithms [11–15] are used to analyze data. The rationale of these algorithms is that the final model created by these techniques are easy to understand and can be used to devise appropriate strategic decision making. These techniques are contrary to Black Box techniques (i.e. Artificial Neural Networks) where it is nearly impossible to extract actionable information from the final constructed model.

The data mining algorithms used in the paper are OneR, Part, Modlem, J48 and LMT. OneR stands for One Rule and this algorithm only generate one rule based on only one selected variable from the data [11]. The selected variable is one that provides maximum classification accuracy. In order to create rules for each predictor (variable) it constructs a frequency table. According to Part, rules are generated one rule at a time by generating partial decision tree from the data. At each stage, decision tree on the data is generated and a leaf with largest coverage is used to generate rule. The tree is discarded once rule is generated. Rest of the data is used again to construct decision tree. This process is repeated until all instances are allocated to various decision rules. More details of this algorithm can be found at [12]. Modlem algorithm generates set of rules based on rough set theory. It is a sequential covering algorithm where attribute's space is being searched to find the best rule condition during rule induction. More detail on this model can be found at [13]. J48 is a decision tree based classification algorithm [14]. In this algorithm a subset of variables are selected based on some cost function. The entropy or information gain criteria is used as a cost function to select most important variables. This is an iterative approach, in each cycle, feature with highest information gain is selected to construct decision tree. The algorithm finished when all leaves cannot be further divided. In logistic model trees (LMT) a simple decision tree is generated and on each leave regression function is performed to classify data [15].

2   Methodology

A combination of quantitative and qualitative data was collected via a semi-structured interview and a written questionnaire. This allowed the informality necessary to establish rapport and thus gain the trust and candour of participating business owners (Newton 2010). Participants in the study comprised owners of small businesses in Rotorua city. This included retail and service firms in the CBD and service, manufacturing and other firms operating in the city surroundings. Any small business operating within 5 km of Rotorua's CBD was considered eligible for inclusion in the study.

The Australian and New Zealand Standard Industrial Classification (ANZSIC) system was used to ensure participating firms were selected from a cross-section of industry types. The ANZSIC classification system was developed by Statistics New Zealand and the Australian Bureau of Statistics in the 1990s to reflect the structure of Australian and New Zealand industries. The owners of small firms were approached via telephone, email or face to face and requested to participate in the study. A personal, face to face contact was the preferred method of initial communication. The first few minutes or so entailed the formal, quantitative aspect of the interview obtaining information regarding several contingency variables. These variables were determined by examining past research projects regarding SME's; it was hoped to find strong evidence of common contingency factors among the 22 research projects reviewed. A wide variety of factors were identified by the researchers but some factors appeared common to several projects. Wiklund and Sheperd [16] suggest business size, and different organisational and environmental characteristics, as well as the firm's *main line of business*, were control variables. They also considered the firm's age, and number of employees as contributing factors to a firm's degree of formal planning. Haber and Reichel [17] explored, the number of employees, a firm's revenue, and revenue growth, other variables considered include *the human capital* of the firm's owner, such as the educational level, and a business skills index; constructed on the basis of six questions. Finally the location of the business was also a factor considered. Another study [18], considered the formalization of routines, other variables included: gender and age of the entrepreneur (business owner), plus their educational background, such as whether they had a tertiary level education or only a primary-secondary level education. The industry sector, and the number of employees were also important variables in this study.

Common factors include the size of the business, measured by both the number of employees, and the firm's revenue. Other factors include, the gender of the owner(s), the environmental conditions, including location, the firm's principal industry sector, the age of the business, and the educational background of the owner. Therefore, the following nine contingencies were considered as possible influencers of strategic planning formality:

- Industry type (as per ANZSIC classification)
- Business structure – sole trader, partnership, company, etc.
- Business option – independent, franchise, etc.
- Years in business
- Number of employees
- Annual revenue (or revenue band if owner is reluctant to reveal concise data)
- Business location
- Gender(s) of principal leader(s)
- The degree of formal business training, educational qualification, or corporate background

1. **Industry Type:** The 90 participating businesses encompass 11 industry sectors as identified by the Australian and New Zealand Standard Industrial Classification code (ANZSIC). These industries were (1) Agriculture, forestry and fishing (2) manufacturing (3) construction (4) wholesale trade (5) retail trade (6) accommodation and food service (7) rental, hiring and real estate services (8) professional, scientific and technical services (9) healthcare and social assistance (10) public administration and safely and (11) other services.

2. **Location:** A deliberate attempt was made to collect data from a number of different locations. These locations consists of (1) CBD area (2) industrial area (3) commercial area (4) outlying shopping district (5) central mall and (6) other areas.

3. **Age:** Information about the age of a business was collected. Categories were (1) under 1 year (2) between 1–2 years (3) between 3–4 years (4) 5–6 years (5) 7–8 years and (6) 10+ years.

4. **Number of employees:** Categories in collected data were (1) zero employees (2) 1–3 employees (3) 4–7 employees (4) 8–10 employees (5) 11–15 employees (6) 16–20 employees and (7) 20+ employees.
5. **Annual Revenue:** categories for annual revenue were (1) less than 99K [99,000] (2) 100–249K (3) 250–499K (4) 500–749K (5) 750–999K (6) 1–2 million (7) 2–5 million and (8) 5+ millions.
6. **Business nature:** Collected data had five categories (1) independent (2) cooperative (3) franchise (4) buying group and (5) dealership.
7. **Business Structure:** categories were (1) sole traders (2) company (3) partnership (4) incorporation society and (5) charter trust.
8. **Gender:** Female (F), Male (M) and MF, where business was owned by Male and Female partners.
9. **Formal Business Training:** Yes or No on whether formal business training was provided or not.

Strategic planning component has 6 elements and each has a score 1. So, this component can has a total score of 6. This variable is considered as class label for our classification based model development. As per Bracker and Pearson's [7] study, strategic planning was broken into six components:

1. **Vision** – does the business have a <u>written</u> long-term objective, usually evidenced by way of a vision statement via the organisation's website and/or other marketing material? Often explained as a 'sense of direction' – where the firm sees itself in a few years' time.
2. **Mission and Written Plans** – does the firm have a mission statement *and* <u>written</u> plans that detail how the mission is to be achieved? The mission is described as 'the organisation's purpose. This needs to be supported with written shorter-term objectives that support the vision and outline how the mission will be achieved.
3. **External Analysis** – does the firm engage in formal analysis of its environment? This is much more than irregular discussions with staff or customers. For a *yes* response, the firm needs to have predetermined meetings with at least one stakeholder such as other owners or staff with the expressed purpose of evaluating the environment. The findings need to be written and at least one strategic analysis tool is likely to be used such as a SWOT or PEST analysis.
4. **Internal Capability and Resource Analysis** – does the business set aside time to discuss its strengths and weaknesses, its internal capabilities and determine what's working well, as well as where there might be opportunities for the organisation to add value to its offerings? Again, this would need to be in a formal setting and findings recorded, other strategic tools other than a SWOT might be used such as a VRIO analysis.
5. **Financial Analysis and Forecasting** – does the firm analyze past financial performance, does the firm do any financial forecasting, such as predicting future sales?
6. **Quality Assurance** – for a *yes* response an organisation needs to have written policy and procedures and operational control measures in place. This would be evidenced by a performance manual, either hard copy or in electronic form.

A questionnaire was used for the quantitative aspect of the study. The remainder of the interview was less formal, qualitative data was collected during this time. This included information about the nature of strategic planning in the small business, in particular: Whether or not planning was undertaken within the business; and the degree of planning formality, if planning was undertaken. It was important to determine the degree of strategy formalization for each participating firm. A strategic formalization score was determined for each business. Perry [19] utilized an extent of planning questionnaire to help determine the degree of planning formality using the six strategic decision-making dimensions suggested by Bracker. Both Bracker and Perry gave equal weighting to each element of strategic planning; this study also accorded equal weighting for each aspect of strategic planning to determine the strategic formalization score for small business in Rotorua.

Based on the responses to the interview questions, participating organisations received a score of 1 for each of the activities they engaged with, and a score of 0 for each activity that they did not engage with; this totaled to a score out of 6. A total score of 0 indicated that the small business did not engage in any strategic planning and thus was determined as a *non-planner*. A strategic formalization score between 1 and 5 indicated the firm used some strategic planning processes so was deemed an *informal planner* while an organisation scoring across all of the six strategic components, received a strategic formalization score of 6 and was considered a *formal strategic planner*. This scale is very similar to the formality score [1], derived from work conducted by Bracker et al. [20] and Robinson and Pearce [10]. Correlation analysis was used to determine the strength and direction of the relationship between each variable and a firm's degree of formal planning, as measured by the SFS (Strategic Formalization Score).

3 Experimental Results

Nine independent and one dependent variables were considered for this study. Industry type, location, business age, number of employees, annual revenue, business nature, ownership structure, owner's gender, and previous exposure to strategic planning concepts and practices via tertiary level business education or corporate experience were independent variables and strategic formalization score (SFS) was dependent variable (class label). For these experiments Weka version 3.7.13 was used.

In this paper, data is analyzed on three hierarchical level. At first level, 2 classes are created using SFS variable. Data instances with 0–5 SFS values are considered as informal planners and SFS score of 6 is regarded as formal planners. At the second level, informal planner class (SFS score 0–5) is further divided into two classes. SFS score of zero represents non-planner class and SFS values 1–5 represents informal planners. Finally at third level, informal planners are divided into more informal planners (SFS 1–2) and less informal planners (SFS values 3–5). At first hierarchical level we will have 2 classes (formal and informal planers), at second level, we will have three classes (non-planners, informal planners and formal planners) and finally at level 3 we will have four classes (non-planners, more informal planners, less informal planners and formal planners). This scheme can be seen in Fig. 1.

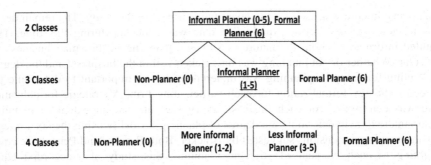

Fig. 1. A divisive hierarchical method used in this paper to split class labels to analyze data

In the first experiment, SFS data was categorized into two classes (1) informal planner and (2) formal planner. SFS score from 0–5 was labelled as informal planner and score of 6 was labelled as formal planner. Informal and formal planning had 69 and 21 instances respectively. The data was run on five algorithms, namely, OneR, Part, Modlem, J48 and LMT. The 10 fold cross validation accuracy results obtained on all five algorithms are stated in Table 1. Rules obtained by OneR algorithm were listed in Table 2.

Table 1. Model accuracy results using 2 class data

	OneR	Part	Modlem	J48	LMT
Accuracy	74.4%	76.7%	**78.9%**	-	77.8%

Table 2. Rule generated by OneR algorithms on 2 class data.

Annual Revenue:	
< $99,000	→ Informal Planning
250–499k	→ Informal Planning
500–749k	→ Informal Planning
2–5 mill	→ Informal Planning
1–2 mill	→ Informal Planning
100–249k	→ Informal Planning
750–999k	→ Informal Planning
5+ mill	→ Formal Planning

According to these results, annual revenue is most important feature to differentiate between formal and informal planners. Business earning more than 5 million dollars in revenue are implementing formal planning methods and all other groups of revenue are coming under informal planners. The model generate by OneR has an accuracy of 74.4% (using 10 fold cross validation approach). The same data was run on Part rule based algorithm and the generated rules are mentioned below:

AnnualRevenue = 500–749k → Informal Planning (11.0/1.0)
employees = 1–3 emp → Informal Planning (18.0)
AnnualRevenue = 2–5 mill → Informal Planning (20.0/5.0)
FormalBusinessTraining = No → Informal Planning (28.0/6.0)
Gender = M → Formal Planning (9.0/2.0)
BusinessNature = independent → Informal Planning (2.0)
All others → Formal Planning (2.0)

The model generated using Part algorithm has an accuracy of 76.7% (using 10 fold cross validation approach). According to first rule if annual revenue is between 500–749K then the organizations will be following informal planning. According to rule 4, if organisations has no formal business training it implies that those organisations will be following informal planning. Male owners are more likely to follow formal planning approach than female owners (rule 5). *Gender may be a contentious issue. The authors acknowledge that while gender may be a contributing factor in this project, they accept that gender is highly unlikely to be a determinant of planning formality. In this project, based in a single small city, the sample included fewer female business owners, and many of these were running small, hobby-type enterprises requiring little formal planning. The authors accept that a larger sample is likely to show no difference in the relationship between gender and the degree of strategic planning.*

In the second set of experiments, 3 class labels were obtained from the data. Those labels were non-planner, informal planner and formal planner labels. A SFS score of zeros indicated no-planner, score of 1–5 indicated informal planner and score of 6 represented formal planners. In this data, we had 9, 60 and 21 instances representing non-planner, informal planner and formal planner respectively. The 10 fold cross validation results are listed in Table 3. In Table 3, rows 3–5 represents models generated on paired classes. The pair-wised class modeling is helpful in establishing relationship between sub-class samples. This provides deeper insight to the data and how variables are associated with one another. For this data, it will be interesting to see what variables are responsible for differentiating non-planners from informal planners. The decision tree generated by LMT algorithm when data related to non-planners and informal planners was used is shown in Table 5. According to Table 5, number of employees and business structure are two variables responsible for finding point of difference between non-planners and informal planners. OneR algorithm was used on the same data. The rule generated indicate that if business structure is sole traders, the business will be non-planner and for all other business structures (company, partnership and chartered trust based businesses) it will be informal planners.

Table 3. Model accuracy results using 3 class data (0 non-planner, 1 informal planner and 2 formal planner)

	OneR	Part	Modlem	J48	LMT
All 3 classes	67.8%	65.6%	70%	65.6%	**72%**
0, 1 classes	85.5%	85%	84%	85.5%	**89.9%**
0, 2 classes	**83.3%**	70%	80%	70%	83.3%
1, 2 classes	72.8%	66.7%	**75.4%**	74%	70.4%

Table 4. Rules generated by Modelm algorithm (0 non-planner, 1 informal planner and 2 formal planner)

No.	Rule	Correctness
1	(BusinessStructure in {Sole Trader}) ⇒ (class 3 = 0)	3/3
2	(AnnualRevenue in {1–2 mill, <$99,000, 500–749k, 100–249k}) & (Gender in {M, MF}) ⇒ (class 3 = 1)	20/22
3	(employees in {0 emp, 8–10 emp, 16–20 emp}) & (AnnualRevenue in { < $99,000, 2–5 mill, 100–249k}) & (BusinessStructure in {Company, Partnership}) ⇒ (class 3 = 1)	11/11
4	(IndustryClassification in {8, 1}) & (BusinessStructure in {Company}) ⇒ (class 3 = 1)	6/6
5	(Location in {CBD, other}) & (IndustryClassification in {6}) ⇒ (class 3 = 1)	5/5
6	(Location in {CBD}) & (Age in {3–4 yrs, 7–8 yrs}) ⇒ (class 3 = 1	6/6
7	(Location in {CBD}) & (employees in {4–7 emp, 0 emp, 20+ emp}) & (Age in {10+}) & (BusinessStructure in {Company, Char Trust}) ⇒ (class 3 = 1	8/8
8	(AnnualRevenue in {2–5 mill, 100–249k}) & (Age in {10+}) & (employees in {11–15 emp, 9+ emp}) & (FormalBusinessTraining in {Yes}) ⇒ (class 3 = 1)	6/6
9	(IndustryClassification in {7, 4, 9}) ⇒ (class 3 = 2)	4/4
10	(BusinessNature in {Dealership}) ⇒ (class 3 = 2)	3/4
11	(Location in {Com}) & (IndustryClassification in {5}) => (class 3 = 2)	3/3
12	(AnnualRevenue in {5 + mill}) & (FormalBusinessTraining in {Yes}) => (class 3 = 2)	5/5

Table 5. Decision tree generated by LMT using non-planner and informal planner data

Class non-planner:
−1.28 + [employees = 1–3 emp] * 0.67 + [BusinessStructure = Sole Trader] * 1.82

Class informal planner:
1.28 + [employees = 1–3 emp] * −0.67 + [BusinessStructure = Sole Trader] * −1.82

Modlem algorithm was used to model data (all three classes) and following rules were extracted using 10-fold cross validation accuracy of 70% are mentioned in Table 4. This algorithm produced 31 rules however, in Table 4 only few rules are listed.

According to Table 4 (rule 2), if annual revenue is 1–2 million or < 99,000 or 500–749K or 100–249K and owner is male or co-owned by male and female then it will fall under informal planner. A total 22 instances fall under this rule, out of which 20 belongs to informal and 2 instances belongs to other than informal class. The extracted rules provide insight of the gathered data and these rules can be used to formulate inferences. Seven rules in Table 4 (rules 2–8) represents 71% of the data. Another algorithm used to extract information from the data was LMT (logistic model

Table 6. Model accuracy results using 4 class data (0 is non-planner, 1, more informal planner, 2 less informal planner and 3 formal planner)

	OneR	Part	Modlem	J48	LMT
All 4 classes	46.7%	42.2%	38.9%	37.8%	47.8%
0, 1 classes	67.6	67.6	55.9	67.6	70.6
0, 2 classes	84.1	86.4	75	84.1	79.6
1, 2 classes	56.7	65	66.7	71.7	71.7

trees). This algorithm performed logistic regression at the leaves. The tree generated by this algorithm has an accuracy of 72.2% and rules are as below:

Class No Planning:

$$-0.79 + [\text{Business Structure} = \text{Sole Trader}] * 2.79$$

Class Informal Planning:

$$1.2 + [\text{Annual Revenue} = 5 \; + \; \text{mill}] * -1.66$$

Class Formal Planning:

$$-0.51 + [\text{Annual Revenue} = 5 \; + \; \text{mill}] * 1.69$$

According to this decision tree, business structure (sole trader subtype) is useful in classifying no-planner data and informal and formal planning classes can be recognized using annual revenue variable.

In the third set of experiments, 4 class labels were generated from the data. Those labels were non-planner (SFS score 0), more informal planner (SFS score 1–2), less informal planner (SFS score 3–5) and formal planners (SFS score 0). In this data, we had 9, 25, 35 and 21 instances representing non-planner, more informal planner, less informal planner and formal planners respectively. The 10 fold cross validation results are listed in Table 6. The models generated by all 5 algorithms had an accuracy less than 50%. In Fig. 2, J48 model of more informal and less informal planners' data is

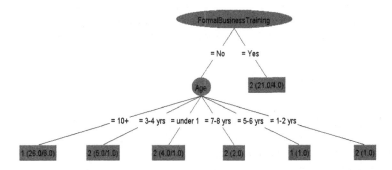

Fig. 2. J48 on more informal planners (class label '1') and less formal planners (class label '2')

presented. According to this model, formal business training and age are most important variable that differentiate these two classes.

4 Conclusion and Discussion

It was hoped that this project would answer questions around factors determining the degree of formal planning in Rotorua based small businesses. Nine variables were considered for this study, industry type, location, business age, number of employees, annual revenue, business nature, ownership structure, owner's gender, and previous exposure to strategic planning concepts and practices via tertiary level business education or corporate experience. Most small businesses in Rotorua are not formal planners. Just 21 of the 90 participating firms; 23.3% attained a Strategic Formalisation Score (SFS) of 6, indicating all six components of strategic planning were evident. Some of the main findings of data analysis performed in this paper are as below:

- Location, firm size by revenue, and number of employees, and the strategic planning experience of a firm's owner are factors that seem to influence the degree of strategic planning in small firms.
- Annual revenue of organisations are decisive factor in determining formal and informal planning organisation (Table 2).
- Organisations owned by male tend to be more likely to be a formal planners.
- Number of employees and business structure are factors responsible for separating non-planners from informal planners (Table 5).
- Formal Business training and age are two variables responsible for separating more informal planners from less formal planners (Fig. 2).

However these variables need further investigation, for example, location to some extent will be limited to some industry types, either because of local council zoning rules or marketing and supply-chain considerations. Franchisees and buying-group members may engage in more formal planning activities, not due to the business owner's determination, but because of contractual obligations. This work is useful in that it prompts further thinking about the strategy-performance link for small Rotorua based businesses. This opens up the possibility of a business case-study project, and possibly a more longitudinal research looking at the outcome of organisational performance as the firms engage in more formal planning. This should help inform the planning/performance link. Past organisational performance can be compared with new organisational performance post the introduction of formal planning processes.

References

1. Baird, L.S., Lyles, M.A., Orris, J.B.: Formalised planning in small business: increasing strategic choices. J. Small Bus. Manag. **32**(1), 48–59 (1994)
2. Schoeffler, S., Buzzell, R.D., Heany, D.: Impact of strategic planning on profit performance. Harv. Bus. Rev. 137–145 (1974)

3. O'Regan, N., Ghobadian, A.: Strategic planning – a comparison of high and low technology manufacturing small firms. J.: Technovation **25**(10), 1107–1117 (2005)
4. Verreynne, M.: Strategy-making process and firm performance in small firms. enterprise and innovation. Auckland University of Technology, Faculty of Business (2005). ISSN 1176-1997, Paper 20-2004
5. Schwenk, C.R., Shrader, C.B.: Effects of formal strategic planning on financial performance in small firms: a meta-analysis. Entrep. Theory Pract. **17**(3), 53–64 (1993)
6. Wang, C., Walker, E., Redmond, J.: Explaining the lack of strategic planning in SMEs: the importance of owner motivation. Int. J. Organ. Behav. **12**(1), 1–16 (2007)
7. Bracker, J.S., Pearson, J.N.: Planning and financial performance of small, mature firms. Strateg. Manag. J. **7**(6), 503–522 (1986)
8. Cox, C., Vos, E.: Small business failure rates and the New Zealand retail sector. Small Enterp. Res. **13**(2), 46–59 (2005)
9. Clark, D.N.: Strategic management tool usage: a comparative study. Strateg. Change **6**(7), 417–427 (1997)
10. Robinson, R., Pearce, J.: Research thrusts in small firm strategic planning. Acad. Manag. Rev. **9**(1), 128–137 (1984)
11. Holte, R.C.: Very simple classification rules perform well on most commonly used datasets. Mach. Learn. **11**, 63–91 (1993)
12. Frank, E., Witten, I.H.: Generating accurate rule sets without global optimization. In: 15th International Conference on Machine Learning, pp. 144–151 (1998)
13. Stefanowski, J.: The rough set based rule induction technique for classification problems. In: 6th European Congress on Intelligent Techniques and Soft Computing, pp. 109–113 (1998)
14. Quinlan, R.: C4.5: Programs for Machine Learning. Morgan Kaufmann Publishers, San Mateo (1993)
15. Landwehr, N., Hall, M., Frank, E.: Logistic model trees. Mach. Learn. **95**(1–2), 161–205 (2005)
16. Wiklund, J., Shepherd, D.: Entrepreneurial orientation and small business performance: a configurational approach. J. Bus. Ventur. **20**, 71–91 (2005)
17. Haber, S., Reichel, A.: The cumulative nature of the entrepreneurial process: the contribution of human capital, planning an environmental resources to small venture performance. J. Bus. Ventur. **22**, 119–145 (2007)
18. Baltar, F.: A suitable "GPS" for SME's: the strategic planning and organisational learning nexus. J. Knowl. Manag. Econ. Inf. Technol. **3**(2) (2013)
19. Perry, S.: The relationship between written business plans and the failure of small businesses in the U.S. J. Small Bus. Manag. **39**(3), 201–208 (2001)
20. Bracker, J., Keats, B., Pearson, J.: Planning and financial performance among small firms in a growth industry. Strateg. Manag. J. **9**(6), 591–603 (1988)

Mathematics of Decision Sciences and Information Science

A Ranking Procedure with the Shapley Value

Aleksei Kondratev and Vladimir Mazalov[✉]

Russian Academy of Sciences, Karelian Research Center, Institute of Applied
Mathematical Research, Pushkinskaya. 11, 185910 Petrozavodsk, Karelia, Russia
{kondratev,vmazalov}@krc.karelia.ru
http://mathem.krc.karelia.ru/

Abstract. This paper considers the problem of electing candidates for
a certain position based on ballots filled by voters. We suggest a voting
procedure using cooperative game theory methods. For this, it is necessary to construct a characteristic function via the preference profile
of voters. The Shapley value serves as the ranking method. The winner
is the candidate having the maximum Shapley value. And finally, we
explore the properties of the designed procedures.

Keywords: Tournament matrix · Shapley value · Preference aggregation
rule · Voting procedure · Condorcet criterion · Characteristic function

1 Introduction

The present paper focuses on the ranking problem of candidates for a certain
position based on ballots filled by their supporters. Such problem arises during
elections of a president, a company's CEO, a professor in a chair and many other
positions. By an assumption, elections are free, honest and open. In the course of
voting, electors fill ballots and specify their preferences for existing candidates.
Generally, the number of ballots appreciably exceeds the number of candidates.
The winner is defined on the basis of all filled ballots. Here a major role belongs
to the winner selection method. This preference aggregation rule must possess a
series of positive properties. We will believe that elections choose an appropriate
candidate from at least two ones. Voters have to fill a ballot and specify their
relative preferences for given candidates. As a matter of fact, there exist different
ways of ballot filling.

Ranked elections dictate placing all candidates in a ballot in the descending
order of voter's preferences. Several techniques to count the votes are applicable here, and they have different sensitivity to possible variations in voters'
preferences.

In the case of plurality voting, electors have to specify a most preferable candidate. The winner is the candidate receiving the largest number of votes. This
voting rule appears widespread, has easy numerical implementation and requires
reasonable computations. However, it neglects voting aspects in situations when
several candidates are equally preferable for electors.

© Springer International Publishing AG 2017
N.T. Nguyen et al. (Eds.): ACIIDS 2017, Part II, LNAI 10192, pp. 691–700, 2017.
DOI: 10.1007/978-3-319-54430-4_66

Plurality with runoff voting is used, for example, to elect the president of France and Russia. The two candidates that are ranked first by most voters face off in a majority runoff. Instant-runoff voting (single-transferable vote) is used, for example, to elect members of the Australian House of Representatives, the president of India. Candidates with the fewest votes are sequentially eliminated. Ballots assigned to the eliminated candidate are recounted and added to the totals of the remaining candidates based on who is ranked next on each ballot. The Borda procedure can be combined with an instant-runoff procedure to create hybrid election methods that are called Nanson method and Baldwin method. Unfortunately, last four procedures are not monotonic.

Under approval voting, an elector specifies only trusted candidates [3, 4]. Therefore, each voter assigns 1 to trusted candidates and 0 to others. The candidate with the maximum number of labels 1 wins the elections. The idea with three labels $0, 1, 2$ was pioneered in the paper [8]. An axiomatization of such score voting procedures was suggested in [7]. According to the majority judgement voting, each candidate is associated with some group of preferences, e.g., A, B, C, D, [2]. Such method depends on the number of groups and entails difficulties in the course of ballot handling, but seems computationally easy. Ranked elections and approval elections are combined in hybrid voting method, preference-approval voting [3].

In this work, we examine ranking procedures based on the tournament matrix of pairwise comparisons. Consider given sets of candidates and voters. Each voter fills a ballot by ranking all candidates in the descending order of its preferences. It is required to define the winner based on the filled ballots. There exist a series of ranking procedures and single-winner election methods. Here we mention methods of Borda, Copeland [6], Simpson (minimax Condorcet), Schulze (Schwartz sequential dropping, beatpath), see [9], Kemeny-Young (votefair popularity ranking), see [11], Tideman (ranked pairs), see [10].

The present paper suggests involving some methods of cooperative game theory as ranking procedures. The idea is to construct a special-form characteristic function of such game using filled ballots, with subsequent weighting of each player (e.g., by the Shapley value). The stated procedure enjoys remarkable properties, as it takes into account not just the correlation of two candidates, but the correlation of any candidates depending on their belonging to certain coalitions. The procedure yields weights for all candidates and is therefore applicable to models, where one has to rank all candidates (instead of a single winner).

2   The Criteria for Ranking Procedures

Consider the voting problem, where $n \geq 2$ electors have to choose a winner among $m \geq 2$ candidates. The preferences of each voter are defined by a linear order on the set of candidates. Denote by $A = \{a, b, c, \dots\}$ the set of candidates. Based on joint preferences, a voting procedure (comprising a ballot and a counting rule of votes) leads to a result of voting. The procedure aims at assigning ranks from 1 to m to all candidates; note that such ranking can be nonstrict. In other words, we construct so-called social choice function or preference aggregation rule [1].

Below we give an example of a **preference profile**.

Example 1. There are $n = 45$ voters and $m = 5$ candidates. The existing preferences of voters are defined by the following Table 1.

Table 1. The preference profile and the tournament matrix

5	5	8	3	7	2	7	8
a	a	b	c	c	c	d	e
c	d	e	a	a	b	c	b
b	e	d	b	e	a	e	a
e	c	a	e	b	d	b	d
d	b	c	d	d	e	a	c

	a	b	c	d	e
a		20	26	30	22
b	25		16	33	18
c	19	29		17	24
d	15	12	28		14
e	23	27	21	31	

Designate by $h(i, j)$ the number of ballots, where candidate i is preferable to candidate j. Let us compile a matrix from the values of the function $h(i, j)$. Such matrix is called **a tournament matrix**. For instance, the tournament matrix in Example 1 takes the form above.

Candidate x is termed the **Condorcet winner** if x beats any other candidate in the case of pairwise comparison. This means that for any $y \in A \setminus \{x\}$ over one-half of the voters rank x higher than y, i.e., $h(x, y) > n/2$.

Denote by $w(x)$ and $l(x)$ the sets of candidates losing to and winning against candidate x, respectively, in the case of pairwise comparison. Interestingly, candidate x represents the Condorcet winner iff $w(x)=A\setminus\{x\}$ and $l(x)=\emptyset$. We say that candidate x **covers** candidate y if $w(x) \supseteq w(y)$, $l(x) \subseteq l(y)$ and $h(x, y) > n/2$. Uncovered set is one of generalizations for the Condorcet winner concept [5].

Generally, tournament matrices serve for the final choice of winners. In the sequel, we will construct characteristic functions using tournament matrices. It is desired that the existing preference aggregation rules and single-winner election methods meet a series of properties (criteria, principles).

Anonymity. Voters are equal. The permutation for names of voters does not change a collective ranking.

Neutrality. Candidates are equal. The permutation for names of candidates does not change a collective ranking.

Homogeneity. Proportional changing of preference profile (or tournament matrix) does not change a collective ranking.

Unanimity (Pareto efficiency). If each voter ranks cand. x higher than cand. y, then x appears not lower than y in the collective preference.

Monotonicity. Suppose that in its individual preference a voter moves cand. x by one position up (down) under a fixed ranking of all other candidates; then in the collective preference x does not decrease (increase, respectively) its rank.

The Condorcet criterion. If a candidate is the Condorcet winner, then it ranks first in the collective preference.

The cover criterion. If x covers y, then in the collective preference candidate x ranks not lower than candidate y. Moreover, if x covers all the remaining candidates, then x ranks first alone.

The majority criterion. If one candidate is preferred by a majority (more than one half) of voters, then that candidate must win.

The next lemma is well-known in the literature.

Lemma 1. *(1) A ranking procedure meeting the cover criterion also satisfies the Condorcet criterion. (2) A ranking procedure meeting the Condorcet criterion also satisfies the majority criterion.*

In the forthcoming sections, we will check these properties for a new ranking procedure. This procedure involves a certain cooperative game associated with the preference profile.

Let $K \subseteq A$ indicate a coalition of candidates. Suppose that each coalition K is assigned with a nonnegative monotonic function $v(K)$ (a characteristic function in the terminology of cooperative game theory). Find the characteristic function $v(K)$ for a given voter's preference profile. Then candidate ranking can be performed on the basis of cooperative game theory criteria adopted in voting problems analysis. In the sequel, the role of such criterion belongs to the candidate's power in the form of the Shapley value. In this case, candidate ranking runs according to the Shapley values for a given characteristic function. We will define characteristic functions using tournament matrices.

3   Characteristic Function as the Value of a Constant-Sum Game

Define the characteristic function as follows. Consider a coalition K and its complement $A \backslash K$. Assume that the coalition K proposes for elections a common candidate $i \in K$, whereas the coalition $A \backslash K$ nominates its representative $j \in A \backslash K$. The candidate receiving over half of the votes becomes the winner; otherwise, the elections are drawn. The payoff in this game makes up

$$H(i,j) = I\left(h(i,j) - \frac{n}{2}\right),$$

where the Heaviside step function $I(z) = 1$ for $z > 0$, $I(z) = 1/2$ for $z = 0$ and 0 for the rest values of z.

This mixed strategy game has an equilibrium according to the von Neumann theorem; moreover, its value gives the value of the characteristic function $v(K)$ in the cooperative game. Therefore, the payoff $v(K)$ of the coalition K makes the equilibrium payoff in the constant-sum game of the coalition K against the countercoalition $A \setminus K$. A mixed strategy of the coalition K is a vector $p = (p_i)_{i \in K}$. The coalition K proposes its common candidate $i \in K$ with the probability $p_i \geq 0$, where $\sum_{i \in K} p_i = 1$. A strategy of the coalition $A \setminus K$ is a vector $q = (q_j)_{j \in A \setminus K}$ such that $q_j \geq 0$ for all $j \in A \setminus K$, where $\sum_{j \in A \setminus K} q_j = 1$. Then the characteristic function v acquires the form

$$v(K) = \max_p \min_q \sum_{i \in K} \sum_{j \in A \setminus K} H(i,j) p_i q_j = \max_p \min_{j \in A \setminus K} \sum_{i \in K} H(i,j) p_i, \quad v(A) = 1,$$

and, the second characteristic function in pure strategies equals

$$v_0(K) = \max_{i \in K} \min_{j \in A \setminus K} H(i,j), \quad v_0(A) = 1.$$

Note that $v(K) + v(A \setminus K) = 1$, $v_0(K) + v_0(A \setminus K) \leq 1$.

For instance, calculate the payoff v for the coalitions ac and bde in Example 1. The payoff matrix of the coalition ac against the coalition bde is defined by

$$\begin{array}{c} \\ a \\ c \end{array} \begin{array}{c} b \quad d \quad e \\ \begin{pmatrix} 0 & 1 & 0 \\ 1 & 0 & 1 \end{pmatrix} \end{array}.$$

In mixed strategies, the coalition ac has the payoff 0.5; hence, $v(ac) = v(bde) = 0.5$. In pure strategies, the guaranteed payoff equals zero, $v_0(ac) = v_0(bde) = 0$. A coalition wins at least by 23 affirmative votes. The coalitions ce, abe and acd are minimal winning coalitions. Candidates c and e possess the highest power under such voting procedure.

The characteristic function can be calculated for all 2^m coalitions of candidates. After construction of the characteristic function, it is possible to evaluate the candidate's power using the Shapley value $\varphi(v)$.

The Examples section provides the corresponding table with all values for the characteristic functions v, v_0 in Example 1.

Further exposition employs two auxiliary lemmas. The next lemma is well-known in the literature.

Lemma 2. *Let (h_{ij}) be a payoff matrix of dimensions $n \times m$ in the constant-sum game with a value v^*. Then the game with a matrix $\widehat{h}$ such that $\widehat{h}_{ij} \geq h_{ij}$ for all i, j has a value not smaller than v^*.*

Lemma 3. *Suppose that for some pair of candidates x, y and any coalition $S \subseteq A \setminus \{x, y\}$ the characteristic functions v and $\widehat{v}$ meet the following conditions:*

$$v(S) = \widehat{v}(S), \quad v(S \cup y \cup x) = \widehat{v}(S \cup y \cup x),$$

$$v(S \cup y) \geq \widehat{v}(S \cup y), \quad v(S \cup x) \leq \widehat{v}(S \cup x).$$

Then transition from v to $\widehat{v}$ does not decrease the Shapley value φ_x for candidate x and does not increase the Shapley value φ_y for candidate y. For any other candidate $z \in A \setminus \{x, y\}$, the Shapley value increment $\varphi_z(\widehat{v}) - \varphi_z(v)$ is not greater than the increment $\varphi_x(\widehat{v}) - \varphi_x(v)$ for x and not smaller than the increment $\varphi_y(\widehat{v}) - \varphi_y(v)$ for y. Consequently, candidate x does not decrease its rank and candidate y does not increase its rank in the collective preference as the result of replacing the characteristic function v by $\widehat{v}$.

Proof. The premises of the lemma directly imply that the Shapley value does not decrease (increase) for candidate x (for candidate y, respectively). Now, demonstrate that for any other candidate $z \in A \setminus \{x, y\}$ the Shapley value increment does not exceed that for x.

The premises of the lemma also directly imply that $\widehat{v}(K \cup x) \geq v(K \cup x)$, $\widehat{v}(K \cup z) \leq v(K \cup z)$ for any coalition $K \subset A \setminus \{x, z\}$. Hence, it appears that

$$\varphi_x(\widehat{v}) - \varphi_z(\widehat{v}) = \sum_{K \subset A \setminus \{x,z\}} \frac{k!(m-k-2)!}{(m-1)!} \left(\widehat{v}(K \cup x) - \widehat{v}(K \cup z)\right)$$

$$\geq \sum_{K \subset A \setminus \{x,z\}} \frac{k!(m-k-2)!}{(m-1)!} \left(v(K \cup x) - v(K \cup z)\right) = \varphi_x(v) - \varphi_z(v),$$

and we have that $\varphi_x(\widehat{v}) - \varphi_x(v) \geq \varphi_z(\widehat{v}) - \varphi_z(v)$.

Similarly, $\widehat{v}(K \cup z) \geq v(K \cup z)$, $\widehat{v}(K \cup y) \leq v(K \cup y)$ for any coalition $K \subset A \setminus \{y, z\}$. Thus, the inequalities

$$\varphi_z(\widehat{v}) - \varphi_y(\widehat{v}) = \sum_{K \subset A \setminus \{y,z\}} \frac{k!(m-k-2)!}{(m-1)!} \left(\widehat{v}(K \cup z) - \widehat{v}(K \cup y)\right)$$

$$\geq \sum_{K \subset A \setminus \{y,z\}} \frac{k!(m-k-2)!}{(m-1)!} \left(v(K \cup z) - v(K \cup y)\right) = \varphi_z(v) - \varphi_y(v),$$

also yield $\varphi_z(\widehat{v}) - \varphi_z(v) \geq \varphi_y(\widehat{v}) - \varphi_y(v)$.

Summarizing the outcomes, we have argued that Lemma 3 is proved.

Theorem 1. *The characteristic function v (v_0) is nonnegative and monotonic. Ranking procedure $\varphi(v)$ $(\varphi(v_0))$ based on the Shapley value for the function v (v_0) satisfies anonymity, neutrality, homogeneity, unanimity, monotonicity, majority, Condorcet and cover criteria.*

Proof. The nonnegativity and monotonicity of the function v is obvious from the definition. The preference aggregation rule $\varphi(v)$ is based on the tournament matrix and the Shapley value, hence, directly from the definition it satisfies anonymity, neutrality and homogeneity criteria.

First, we show the property of unanimity. Suppose that candidate x is preferable to candidate y for all voters. It suffices to verify the inequality $v(y \cup S) \leq v(x \cup S)$ for any coalition $S \subseteq A \setminus \{x, y\}$. Denote $K = A \setminus \{x, y\} \setminus S$. Interestingly, $h(i, x) \leq h(i, y)$ and $h(y, j) \leq h(x, j)$ for any i, j. This means that $H(i, x) \leq H(i, y)$ and $H(y, j) \leq H(x, j)$ for any i, j. Compare the payoff matrix of the coalition $y \cup S$ against $x \cup K$

$$\begin{array}{c} \\ y \\ s_1 \\ \dots \\ s_l \end{array} \begin{pmatrix} x & k_1 & \dots & k_r \\ 0 & H(y, k_1) & \dots & H(y, k_r) \\ H(s_1, x) & H(s_1, k_1) & \dots & H(s_1, k_r) \\ \dots & \dots & \dots & \dots \\ H(s_l, x) & H(s_l, k_1) & \dots & H(s_l, k_r) \end{pmatrix} \qquad (1)$$

with the payoff matrix of the coalition $x \cup S$ against $y \cup K$

$$
\begin{array}{ccccc}
 & y & k_1 & \ldots & k_r \\
x & \begin{pmatrix} 1 & H(x, k_1) & \ldots & H(x, k_r) \\ H(s_1, y) & H(s_1, k_1) & \ldots & H(s_1, k_r) \\ \ldots & \ldots & \ldots & \ldots \\ H(s_l, y) & H(s_l, k_1) & \ldots & H(s_l, k_r) \end{pmatrix}
\end{array}
\tag{2}
$$

Clearly, elements in the lower matrix are not smaller than their counterparts in the upper matrix. By virtue of Lemma 2, we have $v(y \cup S) \leq v(x \cup S)$. And so, the Shapley value for candidate x is not less than that for candidate y.

Next, let us prove the monotonicity of the ranking procedure. Assume that in a ballot candidate x moves by one position up, whereas candidate y goes by one position down. Designate by $\widehat{v}$ the characteristic function resulting from such transformation. Obviously, for any coalition $S \subseteq A \setminus \{x, y\}$ the conditions of Lemma 3 take place. According to Lemma 3, candidate x (candidate y) does not decrease (increase, respectively) its rank in the collective preference.

Third, check the Condorcet criterion. Imagine that candidate x represents the Condorcet winner and compare it with any other candidate y. For any set $S \subseteq A \setminus \{x, y\}$, the coalition $x \cup S$ proposes the common candidate x and wins. Consequently, we have $1 = v(x \cup S) > v(y \cup S) = 0$, whence it follows that the Shapley value is higher for candidate x than for candidate y.

And finally, establish the cover criterion. Let x covers y, i. e., $w(x) \supseteq w(y)$, $l(x) \subseteq l(y)$ and $h(x, y) > n/2$. Then $I(h(i, x) - \frac{n}{2}) \leq I(h(i, y) - \frac{n}{2})$, $I(h(y, j) - \frac{n}{2}) \leq I(h(x, j) - \frac{n}{2})$ for any i, j. Verify the inequality $v(y \cup S) \leq v(x \cup S)$ for any set $S \subseteq A \setminus \{x, y\}$. Designate $K = A \setminus \{x, y\} \setminus S$. The payoff matrices of the coalition $y \cup S$ against $x \cup K$ coincide with the matrices (1) and (2), see the proof of unanimity. Then Lemma 2 brings to the condition $v(y \cup S) \leq v(x \cup S)$. Consequently, for candidate x the Shapley value is not less than for candidate y.

The function v satisfies the cover criterion, *ergo* the majority criterion (see Lemma 1). This concludes the proof of Theorem 1.

The characteristic function v takes into account merely the win of one candidate over another under pairwise comparison. Here the advantage of one vote and unanimity are equivalent. Such ranking method reflects candidate's capability for creating coalitions that propose the Condorcet winner. If the Condorcet winner is among all candidates, the Shapley value vector makes up $(1, 0, \ldots, 0)$.

Remark. To calculate the functions v and v_0, we have utilized the payoff matrix $H(i, j)$ composed of zeros and unities. In the same manner we introduce new characteristic functions u and u_0 as the guaranteed payoff in the game with payoff matrix $h(i, j)$ among mixed and pure strategies, respectively. Therefore, the characteristic functions u and u_0 are defined by

$$
u(K) = \max_p \min_q \sum_{i \in K} \sum_{j \in A \setminus K} h(i, j) p_i q_j = \max_p \min_j \sum_{i \in K} h(i, j) p_i, \quad u(A) = n,
$$

$$
u_0(K) = \max_{i \in K} \min_{j \in A \setminus K} h(i, j), \quad u_0(A) = n.
$$

For any coalition K, we have the equality $u(K) + u(A \setminus K) = n$.

Theorem 2 below studies the properties of the Shapley value-based ranking procedure for the characteristic functions u, u_0 (Table 2).

Theorem 2. *The characteristic functions u, u_0 are nonnegative and monotonic. The Shapley value-based ranking procedure for the functions u, u_0 possesses the criteria presented in the table below (compared with Borda, Simpson and Copeland procedures).*

Table 2. Satisfaction of the properties

Criterion	$\varphi(v_0),$ $\varphi(v)$	$\varphi(u_0),$ $\varphi(u)$	Bo	Co	Si	Criterion	$\varphi(v_0),$ $\varphi(v)$	$\varphi(u_0),$ $\varphi(u)$	Bo	Co	Si
anonymity	yes	yes	yes	yes	yes	monotonicity	yes	yes	yes	yes	yes
neutrality	yes	yes	yes	yes	yes	majority	yes	yes	no	yes	yes
homogeneity	yes	yes	yes	yes	yes	Condorcet	yes	yes	no	yes	yes
unanimity	yes	yes	yes	yes	yes	cover	yes	no	no	yes	no

Proof. The statements on the nonnegativity and monotonicity of the functions u, u_0 follow directly from their definition. Criteria of anonymity, neutrality, homogeneity are performed, because the tournament matrix is anonymous, neutral, homogeneous for any preference profile. The statements for unanimity and the monotonicity criteria can be proved as in Theorem 1.

Next, check the Condorcet criterion. Imagine that candidate x represents the Condorcet winner and compare it with any other candidate y. For any coalition $S \subseteq A \setminus \{x, y\}$ it is clear that $u(x \cup S) > \frac{n}{2} > u(y \cup S)$, $u_0(x \cup S) > \frac{n}{2} > u_0(y \cup S)$. Therefore, the Shapley value is higher for candidate x than for candidate y. Hence, by Lemma 1 the majority criterion is done for $\varphi(u), \varphi(u_0)$.

The cover criterion fails for the functions $\varphi(u)$, $\varphi(u_0)$ in the case of the preference profile with $m = 5$ candidates and $n = 7$ voters. Four of them choose the ranking $a > b > c > d > e$, and the rest three voters prefer $e > a > b > c > d$. Candidate e loses to all opponents under pairwise comparison, but $\varphi(u = u_0) = (4, 0, 0, 0, 3)$. The proof of Theorem 2 is completed.

4   Numerical Ranking

The table with all values of the characteristic functions v, v_0, u, u_0 in Example 1 can be found below (Table 3).

Compare the results derived in Example 1 with other ranking procedures and single-winner election methods involving tournament matrices.

A well-known ranking technique of m candidates is the Borda rule: a candidate receives $m-1$ points for rank 1, $\ldots$, 0 points for rank m in a ballot. The winner is the candidate having the maximum total points. For candidate i, the

Table 3. The characteristic functions v, v_0, u, u_0 in Example 1

K	v	u	u_0	v_0
$\emptyset$	0	0	0	0
a	0	20	20	0
b	0	16	16	0
c	0	17	17	0
d	0	12	12	0
e	0	21	21	0
ab	0	22	22	0
ac	0.5	346/15	20	0
ad	0	20	20	0
ae	0.5	23.5	21	0
bc	0.5	21.5	18	0
bd	0	17.5	16	0
be	0	21	21	0
cd	0	19	19	0
ce	1	23	23	1
de	0.5	329/15	21	0
abc	0.5	346/15	22	0
abd	0	22	22	0
abe	1	26	26	1
acd	1	24	24	1
ace	1	27.5	27	1
ade	0.5	23.5	21	0
bcd	0.5	21.5	19	0
bce	1	25	25	1
bde	0.5	329/15	21	0
cde	1	23	23	1
abcd	1	24	24	1
abce	1	33	33	1
abde	1	28	28	1
acde	1	29	29	1
bcde	1	25	25	1
abcde	1	45	45	1

total scores have the form $\sum_{j \in A \setminus \{i\}} h(i,j)$. In Example 1, the Borda rule yields the ranking $e > a > b > c > d$ with the total points of $102, 98, 92, 89$ and 69, respectively.

The Copeland method [6] proceeds from scores calculated by $\sum_{j \in A \setminus \{i\}} H(i,j)$. In our case, the method brings to the ranking $e > a = b = c > d$ with the total points of $3, 2, 2, 2$ and 1, respectively.

According to the Simpson rule (the minimax Condorcet), the total scores of candidate i are defined by $\min_{j \in A \setminus \{i\}} h(i,j)$. Clearly, we obtain the ranking $e > a > c > b > d$ with the total points of $21, 20, 17, 16$ and 12, respectively (Table 4).

Table 4. The ranking in Example 1

	a	b	c	d	e	Ranking
$\varphi(v) = \varphi(v_0)$	0.167	0.083	0.333	0.083	0.333	$e = c > a > b = d$
$\varphi(u)$	10.994	7.9	9.161	5.306	11.639	$e > a > c > b > d$
$\varphi(u_0)$	10.95	7.867	9.033	5.367	11.783	$e > a > c > b > d$
Borda	98	92	89	69	102	$e > a > b > c > d$
Copeland	2	2	2	1	3	$e > a = b = c > d$
Simpson	20	16	17	12	21	$e > a > c > b > d$

5   Conclusion and Remarks

This paper has employed cooperative game theory methods to solve the ranking (social choice) problem of candidates for a certain position. For this, it is necessary to construct a characteristic function using the filled ballots of voters; such function defines the payoff of each coalition. As input this procedure requires every voter's preference in the form of ranked list of candidates, ties are allowed. The tournament matrix of pairwise comparisons is constructed.

The next step is to find the Shapley value which serves as the ranking method. Note that the stated ranking procedure takes into account the weight of each candidate in all possible coalitions. Calculating the Shapley values involves solving games in which every non-empty subset of the candidates faces its complement. The aim was to show that new method works for single-winner elections and for more general ranking problem. For these purposes we verify basic criteria: anonymity, neutrality, homogeneity, unanimity, monotonicity, majority, Condorcet. And finally, the paper has compared this method with other well-known tournament matrix-based ranking procedures.

Acknowledgments. This work is supported by the Russian Humanitarian Science Foundation (grant 15-02-00352_a) and the Russian Fund for Basic Research (project 16-51-55006 China_a).

References

1. Arrow, K.J.: Social Choice and Individual Values, vol. 12. Yale University Press, New Haven (2012)
2. Balinski, M., Laraki, R.: A theory of measuring, electing, and ranking. Proc. Natl. Acad. Sci. **104**(21), 8720–8725 (2007)
3. Brams, S.J.: Mathematics and Democracy: Designing Better Voting and Fair-Division Procedures. Prinston University Press, Princeton (2007)
4. Brams, S.J., Fishburn, P.C.: Going from theory to practice: the mixed success of approval voting. Soc. Choice Welf. **25**(2–3), 457–474 (2005)
5. Brandt, F., Brill, M., Harrenstein, P.: Tournament solutions. Handbook of Computational Social Choice (2009)
6. Copeland, A.H.: A reasonable social welfare function (mimeo). University of Michigan, Ann Arbor (Seminar on Application of Mathematics to the Social Sciences) (1951)
7. Gaertner, W., Xu, Y.: A general scoring rule. Math. Soc. Sci. **63**(3), 193–196 (2012)
8. Hillinger, C.: The case for utilitarian voting. Homo Oeconomicus **22**(3), 295–321 (2005)
9. Schulze, M.A.: New monotonic, clone-independent, reversal symmetric, and condorcet-consistent single-winner election method. Soc. Choice Welf. **36**(2), 267–303 (2011)
10. Tideman, T.N.: Independence of clones as a criterion for voting rules. Soc. Choice Welf. **4**(3), 185–206 (1987)
11. Young, H.P.: Group choice and individual judgements. In: Mueller, D. (ed.) Perspectives on Public Choice: A Handbook, pp. 181–200. Cambridge University Press, Cambridge (1997)

Optimal Design of Robust Combinatorial Mechanisms for Substitutable Goods

Maciej Drwal[(✉)]

Department of Computer Science, Wrocław University of Science and Technology,
Wybrzeże Wyspiańskiego 27, 50-370 Wrocław, Poland
`maciej.drwal@pwr.edu.pl`

Abstract. In this paper we consider multidimensional mechanism design problem for selling discrete substitutable items to a group of buyers. Previous work on this problem mostly focus on stochastic description of valuations used by the seller. However, in certain applications, no prior information regarding buyers' preferences is known. To address this issue, we consider uncertain valuations and formulate the problem in a robust optimization framework: the objective is to minimize the maximum regret. For a special case of revenue-maximizing pricing problem we present a solution method based on mixed-integer linear programming formulation.

Keywords: Algorithmic game theory · Min-max regret · Pricing · Mathematical economics

1 Introduction

We consider the following setup with a monopolist seller who wants to sell a set of substitutable items to a group of buyers. Each buyer has their own preferences over the offered items, and has a fixed demand. Each item is unique, and thus can be sold to one buyer (this is without the loss of generality, since the seller may be offering for sale several copies of the same item). The seller wants to determine the prices of items in order to maximize his revenue. However, his knowledge of buyers' preferences is limited. Indeed, the item valuations are private information of each buyer, who may want to strategically misreport them in order to receive items ahead of the others, or simply pay less than certain item is worth (according to the buyers' knowledge). In problems of this kind it is usually assumed that the seller has some probabilistic model of buyers' preferences, and wants to maximize his expected revenue, subject to the information he can extract from the available data. However, in many practical situations it is not possible to have any reliable statistical data to build such a model. Instead, a risk-averse seller may want to assume as little as possible about buyers' preferences, but enough to obtain a profit from the sale. One reasonable approach, motivated by a firm axiomatic basis [1,2], is to use robust optimization approach and assign prices to the items, so that the maximum regret of the revenue is minimized. We examine this approach in this paper for the considered combinatorial mechanism design problem.

© Springer International Publishing AG 2017
N.T. Nguyen et al. (Eds.): ACIIDS 2017, Part II, LNAI 10192, pp. 701–710, 2017.
DOI: 10.1007/978-3-319-54430-4_67

1.1   Related Work

The contribution of the paper would be summarized by contrasting it with the following related well-established research areas.

Combinatorial Auctions. In combinatorial auctions buyers are described by valuation functions $v(S)$, where S is a subset of items that seller has in the offer [3]. Typically, the goal of the auction is to allocate subsets of items to the buyers so that the sum of resulting utilities (the total welfare) is maximized. The computed allocation is a function of bids given by all buyers. A good design incentivizes truthful bidding. Combinatorial auction model is similar to the one presented in this paper, however, we consider not only the allocation problem, but also the problem of determining prices that maximize seller's revenue. Moreover, while combinatorial auction model allows for very general model of utility based on subsets of items, in this paper we consider a specific class of utility functions based on multidimensional valuation vectors.

Multidimensional Mechanism Design. The problem considered in this paper is motivated by the works [4–7] and others, that build on the ideas initiated in [8]. In [9] a similar setup to ours is also presented, but with buyers' types being single-dimensional, and utility functions defined on subsets of items.

However, the Bayesian approach proposed in these works is not adequate for the setup of interest in this paper. In particular, the Bayesian approach requires prior beliefs on buyers' valuations and/or historical data for parameter estimation. These assumptions do not always hold in practice, especially when a set of new products is introduced to the market, or the seller offers goods that are unique for the given buyers. Consequently, the *robust mechanism design* has already been proposed in recent publications [10–13]. However, these prior works focus on single item pricing, and to the best of our knowledge, min-max regret approach has not yet been applied to the multidimensional mechanism design for discrete goods. This paper investigates an approach that is complimentary to the Bayesian analysis, when robustness of solutions is needed.

2   Problem Formulation

In this section we give definitions of optimal combinatorial mechanism design and pricing problems for discrete substitutable items.

2.1   Preliminaries

We consider a seller which has M substitutable items for sale, and N concurrent buyers, each with demand for $D_i \leq M$ items. Each buyer is described by a vector of type $\mathbf{x}_i \in \Omega \subset \mathbb{R}^M$, where x_{ij} is ith buyer's valuation of jth item.

A (*direct-revelation*) *mechanism* is a pair of functions (ϕ, ψ), where ψ maps a profile of buyers' valuations $\mathbf{X} = (\mathbf{x}_1, \dots, \mathbf{x}_N)$ into a matrix of allocations of items $\mathbf{Q} = (\mathbf{q}_1, \dots, \mathbf{q}_N)$, and ϕ maps allocations of items $\mathbf{Q}$ into the vector of payments $\mathbf{p}$ that each ith buyer makes for receiving items according to the allocation $\mathbf{Q}$. Function ϕ is called the *payment* function, and ψ is called the *allocation* function.

In this paper we only consider mechanisms of a specific kind. The allowed allocation matrices $\mathbf{Q}$ are only such that each buyer receives up to D_i items, and each item is assigned to up to a single buyer. Formally, $\psi : \Omega^N \to \mathcal{Q}$, where:

$$\mathcal{Q} = \{\mathbf{Q} \in \{0,1\}^{MN} : \sum_{i=1}^{N} q_{ij} \le 1, \ \sum_{j=1}^{M} q_{ij} \le D_i\}.$$

Moreover, prices of items are independent of the buyer's index, thus the payment function $\phi : \mathcal{Q} \to \mathbb{R}_+^N$ can be defined by a vector $\mathbf{p} = (p_1, \ldots, p_M)$, so that $\phi(\mathbf{Q}) = (\phi_1(\mathbf{Q}), \ldots, \phi_N(\mathbf{Q}))$, where:

$$\phi_i(\mathbf{Q}) = \sum_{j=1}^{N} p_j q_{ij}.$$

The seller's revenue is defined as the sum of payments all the buyers make, given the type profile $\mathbf{X}$:

$$r(\mathbf{p}, \psi; \mathbf{X}) = \sum_{i=1}^{N} \sum_{j=1}^{M} p_j \psi_{ij}(\mathbf{X}).$$

Buyer's utility is defined by a quasi-linear function: $u_i(\mathbf{p}, \mathbf{q}_i, \mathbf{x}_i) = \mathbf{q}_i \cdot (\mathbf{x}_i - \mathbf{p})$. A mechanism is called *incentive-compatible* (IC), if the allocations corresponding to the true valuations maximize utilities of each buyer. Then a dominant strategy of each buyer is to report his true (private) valuations to the seller. Formally, mechanism $(\mathbf{p}, \psi)$ is IC if:

$$\forall_i \ \forall_{\mathbf{x}' \in \Omega} \ \ u_i(\mathbf{p}, \psi_i(\mathbf{X}), \mathbf{x}_i) \ge u_i(\mathbf{p}, \psi_i(\mathbf{x}', \mathbf{X}_{-i}), \mathbf{x}_i),$$

which in the considered case is equivalent to:

$$\forall_i \ \forall_{\mathbf{x}' \in \Omega} \ \ (\psi_i(\mathbf{X}) - \psi_i(\mathbf{x}', \mathbf{X}_{-i})) \cdot (\mathbf{x}_i - \mathbf{p}) \ge 0. \tag{1}$$

Notation $(\mathbf{x}', \mathbf{X}_{-i})$ denotes the matrix $\mathbf{X}$ with i-th column $\mathbf{x}_i$ replaced with $\mathbf{x}'$.

It is further assumed that buyers are *individually rational* (IR), which means that they participate in the mechanism only when the utility gained is at least equal to a certain threshold $u_{i,0}$:

$$\forall_i \ u_i(\mathbf{p}, \psi_i(\mathbf{X}), \mathbf{x}_i) \ge u_{i,0}. \tag{2}$$

Henceforth, we normalize the threshold for all buyers to zero, i.e., $u_{i,0} = 0$ for $i = 1, \ldots, N$. We denote the set of all *feasible* mechanisms by:

$$\mathcal{M} = \left\{ (\mathbf{p}, \psi) : \ \mathbf{p} \in \mathbb{R}_+^M, \ \forall_{\mathbf{X} \in \Omega^N} \ \psi(\mathbf{X}) \in \mathcal{Q} \text{ and satisfy (1) and (2)} \right\}. \tag{3}$$

2.2 Robust Mechanism Design

We assume that vectors of buyers' types are not known to the seller. Instead, there is a known set of possible type scenarios $\mathcal{S} \subset \Omega^N$. Two most widely considered special cases of $\mathcal{S}$ are discrete scenario sets and interval uncertainty sets. The *discrete scenario* set is defined as $\mathcal{S} = \{\mathbf{X}^{(1)}, \ldots, \mathbf{X}^{(S)}\}$. The *interval uncertainty* set is defined as $\mathcal{S} = \{[x_{ij}^-, x_{ij}^+] \subset \mathbb{R}_+ : i = 1, \ldots, N, \ j = 1, \ldots, M\}$.

For a fixed scenario $\mathbf{X} \in \mathcal{S}$ the regret of a mechanism $(\mathbf{p}, \psi)$ is defined as:

$$R(\mathbf{p}, \psi; \mathbf{X}) = \left(\max_{(\mathbf{p}', \psi') \in \mathcal{M}} r(\mathbf{p}', \psi'; \mathbf{X}) \right) - r(\mathbf{p}, \psi; \mathbf{X}),$$

i.e., the difference between the revenue generated by an optimal mechanism for known valuations $\mathbf{X}$, and the revenue generated by the mechanism $(\mathbf{p}, \psi)$. The objective of the seller is to determine a mechanism $(\mathbf{p}^*, \psi^*)$ that minimizes the maximum regret over the set of scenarios:

$$(\mathbf{p}^*, \psi^*) \in \arg \min_{(\mathbf{p}, \psi) \in \mathcal{M}} \max_{\mathbf{X} \in \mathcal{S}} R(\mathbf{p}, \psi; \mathbf{X}). \tag{4}$$

The problem of finding an optimal robust mechanism (4) can be seen as three-level optimization problem, in which we minimize (over the space of all feasible mechanisms) the objective function involving unconstrained optimization over the set of all type profiles $\mathbf{X} \in \mathcal{S}$, with objective function that contains a term involving constrained optimization of revenue for a fixed type profile (again over the space of all feasible mechanisms). The innermost sub-problem of finding revenue-maximizing prices and allocations is usually referred to as the *deterministic* problem, as it assumes fixed $\mathbf{X}$ given as an input. The outermost problem will be referred to as the *robust* problem.

Similarly to the Bayesian variant of the multidimensional mechanism design, the robust version of the problem is difficult to solve. Since we model buyers' valuations using subsets of multidimensional real space, the set (3) of feasible mechanisms $(\mathbf{p}, \psi)$ contains allocations ψ expressed as functions on possibly continuous domain. This makes the problem (4) a functional optimization problem (this is also the case of general optimal mechanism design problem [4]). Note that, given the total function solution ψ, the seller would be able to optimally assign items for all buyers' types $\mathbf{X} \in \Omega^N$, and that would cover also non-truth telling buyers (which do not comply to the incentive-compatibility assumption).

2.3 Robust Pricing

Instead of computing an optimal robust mechanism, in some applications it may be enough for the seller to determine only an optimal robust set of prices. In this slightly simplified solution concept, buyers' reported types will not map directly into an allocation rule. Instead, reported types may be the basis for constructing the uncertainty set $\mathcal{S}$, given as an input to the seller. If buyers are aware of the fact that the seller will price the items so that each buyer can receive their utility-maximizing item, truth telling will remain to be an equilibrium strategy. Note that in this approach the solution is also more resistant to the buyers' types misreporting caused by their inherent uncertainty of item valuations (i.e., *non-strategic* misreporting).

It is enough to consider unit demands $D_i = 1$ for all buyers $i = 1, \ldots, N$, since if in the original problem a buyer have arbitrary demand $D_i > 1$, we substitute them with D_i identical buyers of unit demands. Consequently, we can assume that matrix $\mathbf{X}$ is square. Thus if there are less items than buyers we add zero-value dummy items, and interpret matching a dummy item to buyer as not allocating any items to that buyer. If there are less buyers than items we add all-zero valuations buyers, and interpret assigning to such a buyer as not selling an item. Let $K = \max\{M, N\}$.

Consequently, the optimal (deterministic) pricing problem can be defined for any $\mathbf{X} \in \Omega^K$ as the follows:

$$\max_{\mathbf{p} \geq 0, \mathbf{Q} \in \mathcal{Q}} r(\mathbf{p}, \mathbf{Q}), \tag{5}$$

subject to:

$$\forall_i \forall_j \ \sum_{k=1}^{K} q_{jk}(x_{jk} - p_k) \geq \sum_{k=1}^{K} q_{ik}(x_{jk} - p_k), \tag{6}$$

$$\forall_j \ \sum_{k=1}^{K} q_{jk}(x_{jk} - p_k) \geq 0, \tag{7}$$

where $r(\mathbf{p}, \mathbf{Q}) = \sum_{i=1}^{N} \mathbf{p} \cdot \mathbf{q}_i$ is the revenue.

Constraint (6) forces the assignment of items to be utility-maximizing for each buyer. The problem (5)–(7) is a mixed-integer nonlinear program. Introducing new variables u_i, and substituting them for each ith buyer's utility $u_i = \sum_{k=1}^{K} q_{ik}(x_{ik} - p_k)$, we can transform the nonlinear formulation into a mixed-integer linear one as follows:

$$\max_{\mathbf{u} \geq 0, \mathbf{Q} \in \mathcal{F}(\mathbf{u}, \mathbf{X})} \sum_{i=1}^{K} \left(\sum_{j=1}^{K} q_{ij} x_{ij} - u_i \right), \tag{8}$$

where:

$$\mathcal{F}(\mathbf{u}, \mathbf{X}) = \{\mathbf{Q} \in \mathcal{Q} : \forall_i, \forall_j \ u_j - u_i \geq \sum_{k=1}^{K} q_{ik}(x_{jk} - x_{ik})\}. \tag{9}$$

The equivalence of the constraint sets given by (6)–(7) and $\mathcal{F}$ can be seen by observing that the lefthand side of (6) is equal to u_j, thus by subtracting u_i from both sides of this constraint, and expanding the right-hand side, we obtain the inequality defining (9). We note that this new constraint is a special case of more general condition that incentive-compatible allocations must correspond to utility functions that are convex continuous (see, e.g., [5]). These inequalities are a discretized equivalent of the convexity constraint imposed on utility functions. Constraint (7) becomes a standard non-negativity constraint $\mathbf{u} \geq 0$. Given optimal solution $(\mathbf{u}^*, \mathbf{Q}^*)$, the vector of optimal prices $\mathbf{p}^*$ can be computed from the definition of utility.

Before we give the robust formulation of this problem, we need to observe that the feasibility of a particular solution $(\mathbf{u}, \mathbf{Q})$ depends on the actual scenario $\mathbf{X}$. Most notions of robustness require that solution should be given before the true scenario is realized [14]. Consequently, we will require from the robust solution that it is unconditionally feasible, that is, regardless of the scenario. Formally, solution must be *robust feasible*, i.e., from the set:

$$\mathcal{R} = \bigcap_{X \in \mathcal{S}} \{(\mathbf{u}, \mathbf{Q}) : \ \mathbf{u} \geq 0, \ \mathbf{Q} \in \mathcal{F}(\mathbf{u}, \mathbf{X})\}.$$

Note that in many cases robust feasible solutions require that not every item is sold, i.e., matrix $\mathbf{Q}$ does not have a full rank. The robust formulation of the pricing problem is the following:

$$\min_{(\mathbf{u},\mathbf{Q}) \in \mathcal{R}} \max_{\mathbf{X} \in \mathcal{S}} \left(\max_{(\mathbf{u}',\mathbf{Q}') \in \mathcal{F}(\mathbf{u}',\mathbf{X})} \sum_{i=1}^{K} \left(\sum_{j=1}^{K} q'_{ij} x_{ij} - u'_i \right) - \sum_{i=1}^{K} \left(\sum_{j=1}^{K} q_{ij} x_{ij} + u_i \right) \right).$$
(10)

3   Solution Method for Interval Uncertainty

In this section we present an algorithm for solving (10) for the interval uncertainty case of valuations set $\mathcal{S} = \{[x_{ij}^-, x_{ij}^+] : \ i, j \in \{1, \ldots, K\}\}$. The algorithm is based on Benders cut generation method. It makes use of the mixed-integer linear program for deterministic pricing problem (8) and (9). A similar solution method was used for the robust assignment problem in [15], however problem (10) introduces additional constraints on feasible assignments.

The algorithm proceeds as follows:

1. Let $\mathcal{A} \leftarrow \emptyset$, $LB \leftarrow -\infty$, $UB \leftarrow +\infty$.
2. Solve:

$$\min_{\mathbf{Q} \in \mathcal{Q}, \mathbf{u} \geq 0, \theta \geq 0} \left(\theta - \sum_{i=1}^{K} \left(\sum_{j=1}^{K} q_{ij} x_{ij}^- - u_i \right) \right),$$

subject to:

$$\forall_{i,j} \ u_j - u_i \geq \sum_{k=1}^{K} q_{ik} (x_{jk}^+ - x_{ik}^-)$$

and:

$$\forall_{(\mathbf{u}',\mathbf{Q}') \in \mathcal{A}} \ \theta \geq \sum_{i=1}^{K} \left(\sum_{j=1}^{K} q'_{ij} \left(x_{ij}^+ + (x_{ij}^- - x_{ij}^+) q_{ij} \right) - u'_i \right).$$

Let $(\hat{\mathbf{u}}, \hat{\mathbf{Q}}, \hat{\theta})$ be an optimal solution, and $\hat{v}$ be the value of this solution.

3. If $\hat{v} > LB$ then $LB \leftarrow \hat{v}$.
4. Solve the deterministic problem (8) and (9) for scenario:

$$x_{ij} = \begin{cases} x_{ij}^-, & \hat{q}_{ij} = 1, \\ x_{ij}^+, & \hat{q}_{ij} = 0. \end{cases}$$

Let $(\bar{u}, \bar{\mathbf{Q}})$ be an optimal solution.
5. Compute value $\tilde{v}$ of solution $(\hat{\mathbf{u}}, \hat{\mathbf{Q}}, \bar{\mathbf{u}}, \bar{\mathbf{Q}})$ for (10). If $\tilde{v} < UB$ then $UB \leftarrow \tilde{v}$.
6. If $UB \leq LB$ then STOP.
7. Add $(\bar{\mathbf{u}}, \bar{\mathbf{Q}})$ to the set $\mathcal{A}$ and go to Step 2.

The procedure starts from relaxing constraints that define set $\mathcal{F}(\mathbf{u}, \mathbf{Q})$, restricting the solution to be robust feasible, $(\mathbf{u}, \mathbf{Q}) \in \mathcal{R}$. An initial solution is found by optimizing the relaxed problem. Such a solution, however, is usually not feasible for the problem (10), but only provides a lower bound (variable LB). However, given this solution, we determine the worst-case scenario $\mathbf{X} \in \mathcal{S}$, which is an extreme scenario, consisting of only lower or upper interval bounds for each valuation. A deterministic pricing problem (8) and (9) is then solved for that worst-case scenario, which corresponds to the inner maximization sub-problem in (10). However, since it is computed only for one particular choice of optimization variables $(\mathbf{u}, \mathbf{Q})$, the solution of this sub-problem gives only a lower bound on the first term in the objective function of (10), which is represented by a new optimization variable θ. Given a solution of the sub-problem, we obtain a feasible solution of (10), which provides an upper bound on an optimal one (variable UB). We create a Benders cut from the solution of the sub-problem, and add its indexing variable to the set $\mathcal{A}$. Note that in order to completely describe the feasible set of (10) we would require all robust feasible solutions to be contained in $\mathcal{A}$. But then $\mathcal{A}$ would have exponential size. However, it is very often that most of these constraints are superfluous, and an optimal solution of (10) can be found by taking only very small fraction of these constraints into consideration.

3.1   Experimental Results

For an experimental evaluation we have prepared interval uncertainty sets $\mathcal{S}$ by randomly generating lower bounds of valuations uniformly from range $[x_{\min}^-, x_{\max}^-]$, and then for the corresponding upper bound by adding number uniformly from range $[0, \Delta]$, independently for each buyer and item pair. Table 1 shows the summary of results. Problem instances were solved using cut-generating algorithm described in the previous section. For each problem size K the values given are averaged over 10 repetitions of experiment. Columns contain, respectively: the problem size (number of buyers and items), value of an optimal regret, revenue generated by robust optimal solution, resulting welfare (sum of buyers' utilities), number of items sold in the robust optimal solution, and approximate computation time in seconds. Solutions for $K \leq 30$ are optimal with absolute error allowed 0.05. Solutions for $K \geq 40$ were computed with time limit of 1 h, and in most cases no optimal solution was found until then, thus results for best feasible solutions are reported. In a vast majority of cases it was enough to generate only 2–4 cuts in order to find a robust optimal solution.

Notice that for a large number of items and competing buyers the revenue generated by robust solutions starts to decrease, after reaching its highest level. We may conclude that, depending on the range of uncertainty sets, there is a certain number of items that the seller should try to sell concurrently, in order to safely profit from the competition between buyers. However, above that threshold the buyers' utilities quickly decrease and so the guaranteed revenue drops down (indicated by a robust solution), thus it is best to reduce the size of the product line.

Table 1. Experimental results. Highest values of average revenue generated by robust solutions are marked in bold.

K	Optimal regret	Robust revenue	Robust welfare	Sold items	Time
$x_{min}^- = 10$, $x_{max}^- = 500$, $\Delta = 30$					
5	111.27	1452.45	133.72	4.18	11.94
10	623.27	3518.00	96.00	7.91	17.79
20	3116.27	5687.00	72.00	12.00	287.04
30	7464.82	**6393.45**	57.64	13.27	7565.10
40	13324.82	5505.18	19.09	11.27	10946.47
50	18785.75	5282.00	11.00	10.75	10262.33
$x_{min}^- = 10$, $x_{max}^- = 500$, $\Delta = 50$					
5	200.09	1390.00	84.18	3.73	11.73
10	1148.73	3037.36	116.18	6.91	16.60
20	5024.36	**4012.45**	50.18	8.45	203.89
30	10534.55	3660.55	15.55	7.55	8451.96
40	16941.36	2349.09	9.54	4.81	8452.51
50	22709.09	1876.00	0.45	3.82	10433.85
$x_{min}^- = 100$, $x_{max}^- = 500$, $\Delta = 50$					
5	284.27	1453.27	90.55	3.73	13.18
10	1611.36	2736.64	80.36	6.09	14.11
20	6126.09	**3161.00**	28.00	6.55	145.86
30	12098.45	2365.91	2.91	4.82	6733.55
40	17864.73	1695.55	2.91	3.45	9937.67
50	23881.71	987.43	0.00	2.00	10466.73
$x_{min}^- = 100$, $x_{max}^- = 1000$, $\Delta = 50$					
5	175.82	2937.45	269.91	4.27	10.94
10	742.64	7422.36	329.09	8.55	15.97
20	5437.09	12043.00	218.09	12.82	361.68
30	12198.00	**15526.60**	159.10	16.10	9316.22
40	24506.00	12933.67	54.33	13.33	9345.68
50	31608.00	6151.00	86.00	16.50	10804.96

3.2   Algorithm for Deterministic Sub-Problem

Finally, we present a fast heuristic for solving sub-problem (8) and (9) for a fixed type profile. It can be used as a sub-routine in the outer problem of minimizing the maximum regret, when the problem size is prohibitive for applying exact algorithm from the previous subsection. Note that this problem can be also solved directly using mixed-integer linear programming. The presented procedure is based on solving assignment problem, using e.g., Hungarian algorithm [16].

Observe that the revenue-maximizing item allocation is upper-bounded by the value of maximum weight matching of items to buyers. Indeed, if $\mathbf{Q} \in \mathcal{Q}$ is one-to-one matching, and p_j is the price of jth item for some (unique) buyer i, such that $Q_{ij} = 1$, then the revenue $\mathbf{Q} \cdot \mathbf{p}$ is also the value of the matching. However, the vector of prices $\mathbf{p}$ corresponding to the maximum weight matching may violate constraints defining (9). The idea of solution algorithm is to adjust the prices so that to make sure that these constraints are satisfied, while at the same time keeping the assumed allocation $\mathbf{Q}$ corresponding to the maximum weight matching. Note however, that such allocation may not be optimal for any choice of prices $\mathbf{p}$.

The algorithm is the following:

1. Find a maximum weight assignment for the matrix $\mathbf{X}$ of buyers' valuation profile (each column $\mathbf{x}_i$ corresponds to buyer's i valuations, and each row $\mathbf{X}_j^T$ corresponds to an item j). Denote the matching by permutation matrix $\mathbf{A}$.
2. Let $\mathbf{p} \leftarrow (p_1, \ldots, p_M)$, where $p_j = x_{ij}$, such that $A_{ij} = 1$ for some $i = 1, \ldots, N$.
3. For each buyer $n = 1, \ldots, N$:
   (a) Let U_i be the utility of matched item j_0, i.e., $U = x_{ij_0} - p_{j_0}$, where $A_{ij_0} = 1$.
   (b) Let $\hat{U}_i$ be the maximum utility of buyer i given the current prices, i.e., $\hat{U}_i = x_{ij'} - p_{j'}$, where j' maximizes $\{x_{ij} - p_j : j = 1, \ldots, M\}$. Here j' is the item that client prefers under the current prices.
   (c) If $\hat{U}_i > U_i$ then $p_{j_0} \leftarrow p_{j_0} - \hat{U}_i - U_i$.
   (d) If $p_{j_0} < 0$ or $U_{j_0} < 0$ set $A_{ij_0} = 0$ (i.e., item j_0 is not sold).
4. Check if $\hat{U}_i = U_i$ for all buyers $i = 1, \ldots, N$. If not, then go to Step 3.

The procedure can be further improved. For example, assignment matrix A can be recomputed when initial item match turns out to be infeasible for a given buyer in Step 3d (in a new valuation matrix we would set the buyer's valuation of that item to zero).

4   Conclusions and Further Work

The problem of optimal mechanism design, while relatively easy for single-dimensional types, becomes more difficult to solve in the case of multidimensional type space and specific constraints regarding allowed allocations of goods. Different formulations using Bayesian models of valuation uncertainty are subject to active ongoing research. In this paper we presented formulation in which no prior beliefs regarding multidimensional valuations are assumed, and the objective is to design a robust mechanism in a min-max regret sense. We provided a cut-generation based algorithm for solving a special case of pricing mechanism for interval uncertainty.

There are many directions of interesting further research. One is to assume budget-constrained buyers with arbitrary demands. Then we would obtain a formulation of the deterministic problem that contains a generalized assignment problem as a special case, thus the problem becomes even more complex. Another idea is to introduce a regulated monopoly: in such case it is required to design a mechanism that maximizes a weighted sum of seller's revenue and buyers' welfare. Finally, more efficient solution algorithms would allow to extend the range of possible practical applications for the class of mechanisms under consideration.

References

1. Milnor, J.: Games against nature. Technical report, DTIC Document (1951)
2. Rosenhead, J., Elton, M., Gupta, S.K.: Robustness and optimality as criteria for strategic decisions. J. Oper. Res. Soc. **23**(4), 413–431 (1972)
3. Nisan, N., Ronen, A.: Algorithmic mechanism design. In: Proceedings of the 31st Annual ACM Symposium on Theory of Computing, pp. 129–140 (1999)
4. Rochet, J.C., Chone, P.: Ironing, sweeping, and multidimensional screening. Econometrica **66**(4), 783–826 (1998)
5. Manelli, A.M., Vincent, D.R.: Multidimensional mechanism design: revenue maximization and the multiple-good monopoly. J. Econ. Theor. **137**(1), 153–185 (2007)
6. Balcan, M.-F., Blum, A., Mansour, Y.: Item pricing for revenue maximization. In: Proceedings of the 9th ACM Conference on Electronic Commerce, pp. 50–59 (2008)
7. Cai, Y., Daskalakis, C., Weinberg, S.M.: On optimal multidimensional mechanism design. ACM SIGecom Exch. **10**(2), 29–33 (2011)
8. Myerson, R.B.: Optimal auction design. Math. Oper. Res. **6**(1), 58–73 (1981)
9. Ülkü, L.: Optimal combinatorial mechanism design. Econ. Theor. **53**(2), 473–498 (2013)
10. Bergemann, D., Schlag, K.H.: Pricing without priors. J. Eur. Econ. Assoc. **6**(2–3), 560–569 (2008)
11. Handel, B.R., Misra, K.: Robust new product pricing. Market. Sci. **34**(6), 864–881 (2015)
12. Caldentey, R., Liu, Y., Lobel, I.: Intertemporal pricing under minimax regret, Available at SSRN 2357083
13. Drwal, M.: Multidimensional monopolist pricing problem with uncertain valuations (to appear)
14. Ben-Tal, A., El Ghaoui, L., Nemirovski, A.: Robust Optimization. Princeton University Press, Princeton (2009)
15. Pereira, J., Averbakh, I.: Exact and heuristic algorithms for the interval data robust assignment problem. Comput. Oper. Res. **38**(8), 1153–1163 (2011)
16. Papadimitriou, C.H., Steiglitz, K.: Combinatorial Optimization: Algorithms and Complexity. Courier Corporation, North Chelmsford (1982)

Communication and KP-Model

Takashi Matsuhisa[1,2]($\boxtimes$)

[1] Ibaraki Christian University, Ohmika 6-11-1, Hitachi-shi, Ibaraki 319-1295, Japan
`takashimatsuhisa.mri.bsbh@gmail.com`
[2] Institute of Applied Mathematical Research, Karelia Research Centre, RAS,
Pushkinskaya Ulitsa 11, Petrozavodsk, Karelia 185910, Russia
`takashi.matsuhisa@krc.karelia.ru`

Abstract. This paper treats a Bayesian routing problem from the epistemic point of view. We discuss on the role of communication among all users about the users' individual conjectures on the others' selections of channels in the network game. In this paper we focus on the expectations of social costs and its individual conjectures, and we show that, in a revision process of all users' conjectures on the expectations of social costs by communication through the message on the conjecture among the all users, the process yields a Nash equilibrium for social cost in the based KP-model.

Keywords: Bayesian KP-model · Communication · Conjecture · Expected delay equilibrium · Information partition · Knowledge revision · Rational expectations equilibrium · Social cost

1 Introduction

This paper is a continuation of the paper Matsuhisa [6], in which a Bayesian routing problem will be investigated from the epistemic point of view. We focus on the role of communication among all users about the users' individual conjectures on the others' selections of channels in the network game introduced by Koutsoupias and Papadimitriou [5].

The equilibrium notion of 'rational expectations equilibrium' for a social cost is introduced in Matsuhisa [6], and it is shown that 'common-knowledge' assumption on the equilibrium plays essential role in yielding a Nash equilibrium for social cost in KP-model. However the assumption is actually very strong, because common-knowledge is introduced by the infinite regress of interactions among individual knowledges. So we would like to remove it out in our framework.

There seems to be several ways to improving this point, here the communication process introduced by Parikh and Krasucki [9] is adopted instead common-knowledge. We would like to sketch it as follows.

The model will be given as follows. Each user has the conjectures about the other users' choices of the channels given his/her information, and so he/she uses the common social cost function. Then the users can obtain rational expectations equilibrium for the social cost according the conjectures. In this set up, the users

© Springer International Publishing AG 2017
N.T. Nguyen et al. (Eds.): ACIIDS 2017, Part II, LNAI 10192, pp. 711–720, 2017.
DOI: 10.1007/978-3-319-54430-4_68

communicates privately his/her conjecture about the other users' choices through messages among all users according to the communication network, where the message is information about his/her individual conjecture about the others' choices. The recipient of the message updates her/his belief. Precisely, at every stage in the communication, each user communicates privately not only his/her conjectures about the others' choices but also his/her rationality as messages according to the communication, i.e., he/she send the message about his/her information on rational expectations equilibrium for the social cost). Then the recipient updates their private information, and she/he revises her/his conjecture and sends the revised information to the other user. Such revision process of conjectures will have to proceed infinitely often.

In the circumstance we can show

Main Theorem. *The profile of the limiting conjectures on rational expectations for social cost function after long run communication leads to a Nash equilibrium for the social cost.*

Garing et al. [3] first investigated Bayesian Nash equilibriums in KP-model, in which Bayesian extension of routing game is specified information structure in their model by the type-space model of Harsanyi [4]. They obtained several results, especially the upper bounds of the price of anarchy for specific types of social function associated with Bayesian Nash equilibria are given.

In this paper we shall modify their model by adopting arbitrary partition structure by Aumann [1] instead of the type-space model. The merit of adopting this structure lies in getting the close connection to computational logic,[1] and also in increasing the range of its applications in various fields.

The paper is organized as follows. Following Matsuhisa [6], we shall review KP-model in Sect. 2.1, and we present a Bayesian extension of the KP-models in Sect. 2.2. On highlighting users' conjectures on the others' selections, we introduces the two equilibrium notions (expected delay equilibrium and rational expectations equilibrium). In Sect. 3, we will proceed to introduce the new model in this paper; the communication model modifying that by Parikh and Krasucki [9], and limiting conjectures. We will state precisely the main theorem (Theorem 1) with its sketchy proof. Finally, we conclude some remarks on agenda for further research.

2   Basic Model[2]

In this paper a simple type of the *KP-model* as below is only considered[3]. Let $m, n \in \mathbb{N}$ with $m, n \geq 2$.

---

[1] See Fagin et al. [2].

[2] All materials in this section are taken from Matsuhisa [6].

[3] Koutsoupias and Papadimitriou [5], Mazalov [7], Chap. 9, pp. 314–351.

2.1   KP-Model

A *KP-model*, denoted by KP, is the structure $KP = \langle S, N, L, W, C, (\lambda_i^{l_i})_{i \in N} \rangle$, in which S is one *storage*, N is a set of n *users* (clients) $\{1, 2, \cdots, i, \cdots, n\}$, L a set of m *channels* (providers) $\{l_i | k = 1, 2, \cdots, m\}$, W a set of *volumes* $\{w_i | i \in N\}$, C a set of *capacities* $\{c_l | l \in L\}$ and $\lambda_i^{l_i}$ is the *delay* for a user i selecting a channel l_i. Specifically each user i has to use one of m *channels* (providers) to connect the storage. Each channel (or provider) $l = 1, 2, \cdots, m$ has a given capacity $c_l \geqq 0$. User i intends to send/receive information with volume w_i to/from the storage S through provider l_i. User i's actions are choices of the channels, so i's action set $L = \{l_i | l_i = 1, 2, \cdots, m\}$. Let L^n denote $L \times L \times \cdots \times L$ (n times). Each member $l = (l_i)_{i \in N}$ of L^n is a *pure strategies profile*, and then the *delay* for a user i selecting a channel l_i is defined by

$$\lambda_i^{l_i} = \frac{1}{c_{l_i}} \{ w_i + \sum_{k \in N \setminus i; l_k = l_i} w_k \}.$$

By a *social cost* $S(w, l)$ we mean a real valued function of pure strategies $l = (l_i)_{i \in N}$.

Example 1. The below is typical of social costs:

Linear cost: $LSC(w, l) = \sum_{i=1}^{n} \frac{1}{c_{l_i}} \left(\sum_{k; l_k = l_i} w_k \right)$;

Quadratic cost: $QSC(w, l) = \sum_{i=1}^{n} \frac{1}{c_{l_i}} \left(\sum_{k; l_k = l_i} w_k \right)^2$;

Maximal cost: $MSC(w, l) = \text{Max}_{i=1}^{n} \frac{1}{c_{l_i}} \left(\sum_{k; l_k = l_i} w_k \right)$.   □

Throughout we assume that any volume bundle $w = (w_1, w_2, \cdots, w_n)$ is given a priori and each w_i is *indivisible*.

Let $\Delta(L)$ be the set of all probability distributions on L, and $\Delta(L^n) = \Delta(L)^n$. Each member $\sigma = (\sigma_i)_{i \in N}$ of $\Delta(L^n)$ is called a *mixed strategy*. The user i's *expected delay* of channel l_i for a mixed strategy $\sigma = (\sigma_i)_{i \in N}$ is

$$\lambda_i^{l_i}(\sigma) = \frac{1}{c_{l_i}} \{ w_i + \sum_{k \in N \setminus i} w_k \sigma_k(l_i) \}$$

By a *Nash equilibrium* for the model, KP we mean a mixed strategy $\sigma = (\sigma_i)_{i \in N}$ with the properties: For any user $i \in N$ and for any channel $l_i \in L$ adopted by i,

$$\lambda_i^{l_i}(\sigma) = \text{Min}_{j \in L} \lambda_i^j(\sigma) \text{ when } \sigma_i(l_i) \geqq 0 \text{ and}$$
$$\lambda_i^l l_i(\sigma) \geqq \text{Min}_{j \in L} \lambda_i^j(\sigma) \text{ when } \sigma_i(l_i) = 0$$

Let NE(KP) denote the set of all Nash equilibria for KP.

The *expected social cost* according to a mixed strategy $\sigma = (\sigma_i)_{i \in N}$ is

$$SC(w, \sigma) = \sum_{l = (l_k)_{k \in N} \in L^n} SC(w, l) \prod_{k \in N} \sigma_k(l_k).$$

A *Nash equilibrium* for $SC(w, l)$ is meant a mixed strategy $\sigma = (\sigma_i)_{i \in N}$ such that $SC(w, \sigma) \leqq SC(w, (l_i, \sigma_{-i}))$ for any user $i \in N$ and for any channel $l_i \in L$ with $l_i \in \mathrm{Supp}(\sigma_i)$.[4]

Let $\mathrm{NE}(SC)$ be the set of all Nash equilibria for $SC(w, l)$. According to Nash [8] it can be shown that *there exists a Nash equilibrium for each KP-model KP and a Nash equilibrium for any social cost $SC(w, l)$; i.e., $\mathrm{NE}(KP) \neq \emptyset$, $\mathrm{NE}(SC) \neq \emptyset$.*[5]

2.2   Bayesian Extension Model

We will present *Bayesian KP-models* as KP-models under uncertainty, and we will only treat the network topology provided that *every user connects directly to the unique storage.*

Partition Information Structure. By this we mean a structure $\langle \Omega, \mu, (\Pi_i)_{i \in N} \rangle$ in which (1) Ω is a non-empty *finite* set, called *state-space*, whose members are called *states*, whose subsets are *events*, (2) μ is a probability measure on Ω with full support[6], and (3) $(\Pi_i)_{i \in N}$ is a class of user i's *information partition* functions $\Pi_i : \Omega \to 2^\Omega$.

The intended interpretation will be given as follows: $\Pi_i(\omega)$ is the set of all the states of nature that i knows to be possible at ω, or as the set of the states that i cannot distinguish from ω, and we call $\Pi_i(\omega)$ i's *information set* at ω. The last postulate **BP** means that 'user i knows absolutely his/her selection of channel l'.

Bayesian KP-model. By this we mean a structure

$$BKP = \langle KP, \Omega, \mu, (\Pi_i)_{i \in N}, (\mathbf{l}_i)_{i \in N} \rangle$$

which is a Bayesian extension of a KP-model KP equipped with a partition information structure $\langle \Omega, \mu, (\Pi_i)_{i \in N} \rangle$ together with random variables (r.v.s) $\mathbf{l}_i : \Omega \to L$, satisfying the below properties:

P1 $\{\Pi_i(\omega) | \omega \in \Omega\}$ is a partition of Ω;
P2 $\omega \in \Pi_i(\omega)$,
BP $\Pi_i(\omega) \subseteq [l]_i = [\mathbf{l}_i = l]$ for any $l \in L$ and for any $\omega \in [l]_i$. [7]

Denote by $\mathbf{B}(KP)$ the set of all the Bayesian KP-models.

---

[4] $\mathrm{Supp}(\sigma_i)$ is defined by $\mathrm{Supp}(\sigma_i) = \{l \in L \mid \sigma_i(l) \neq 0\}$.
[5] Proposition 1 in Matsuhisa [6].
[6] I.e., $\mathrm{Supp}(\mu) = \Omega$.
[7] Where $[l]_i$ is defined by $[l]_i = [\mathbf{l}_i = l] = \{\omega \in \Omega | \mathbf{l}_i(\omega) = l\}$.

Individual Conjecture. From Bayesian point of view we assume that each user i knows his/her choice of channel l, but he/she never know the other user's choices. By the above postulate **BP** we have already formulated the former assumption. For the latter we will introduce i's *conjecture on the other user k's selection of channel*. By this we mean a probability distribution q_i on the others' section set $(L^n)_{-i}$ of channels $(l_k)_{k \in N \setminus i}$, i.e., $q_i \in \Delta((L^n)_{-i})$. Denoting $l \in L$ by l_k if user k selects l, the marginal probability $q_i(l_k)$ on the other user k's selection set L is the i's conjecture on k's selections. By the random variable (r.v.), we obtain

$$\mathbf{q}_i([l]_k; \omega) = \mu([l]_k | \Pi_i(\omega))$$

we define the events concerned of the conjecture q_i as follows:

$$[q_i(l_k)] = \{\omega \in \Omega | \mathbf{q}_i([l]_k; \omega) = q_i(l_k)\} \quad \text{and}$$
$$[q_i] = \cap_{(l_k)_{k \in N \setminus i} \in L^{n-1}} [q_i((l_k)_{k \in N \setminus i})]$$

Individual Expected Delay. A user i's *expectation of delay* λ_i^l for channel $l \in L$ selected by i according to his/her conjecture q_i is defined by

$$\mathbf{E}_i[\lambda_i^l; q_i] = \frac{1}{c_l} \sum_{k \in N} w_k q_i(l_k). \tag{1}$$

In viewing of **BP** it follows that $\mathbf{q}_i([l]_i; \omega) = 1$, and so

$$\mathbf{E}_i[\lambda_i^l; q_i] = \frac{1}{c_l} \{w_i + \sum_{k \in N \setminus i} w_k \mathbf{q}_i([l]_k; \omega)\} = \frac{1}{c_l} \{w_i + \sum_{k \in N \setminus i} w_k q_i(l_k)\}.$$

Now we can present the notion of an equilibrium for expected delay associated with individual conjecture:

Definition 1. By user i's *rational expected delay* for BKP at state $\omega \in [q_i]$ we mean an i's conjectures $q_i \in \Delta((L^n)_{-i})(i \in N)$ such that

$$\mathbf{E}_i[\lambda_i^{l_i}; (q_i)_{i \in N}] = \text{Min}_{j \in L} \mathbf{E}_i[\lambda_i^j; (q_i)_{i \in N}] \tag{2}$$

for any channel $l_i \in L$ selected by $i \in N$.

We denote by $\text{ED}^{\text{BKP}}(q_i)$ the set of all the states in which q_i is i's rational expected delay for $BKP \in \mathbf{B}(KP)$. Moreover,

Definition 2. A profile of conjectures $q = (q_i)_{i \in N} \in (\Delta((L^n)_{-i}))^n$ is called an *expected delay equilibrium* (e.d.e.) for BKP at state $\omega \in [q_i]$ if $\omega \in \cap_{i \in N} \text{ED}^{\text{BKP}}(q_i)$.

We denote by $\text{EDE}^{\text{BKP}}(q)$ the set of all the states in which $q = (q_i)_{i \in N}$ is an e.d.e. for $BKP \in \mathbf{B}(KP)$.

Rational Expectation. User i's *expectation of social cost SC* according to his/her conjecture q_i is

$$\mathbf{E}_i[SC(w, (l_i, 1_{-i})); q_i] = \sum_{\xi \in \Omega} SC(w, (l_i, 1_{-i}(\xi))) q_i(1_{-i}(\xi)) \tag{3}$$

It follows that for $\omega \in [q_i]$,

$$\mathbf{E}_i[SC(w, (l_i, 1_{-i})); q_i] = \sum_{l_{-i} \in (L^n)_{-i}} SC(w, (l_i, l_{-i})) \mathbf{q}_i([l_{-i}]; \omega).$$

We can introduce another notion of equilibriums for expected social cost associated with individual conjecture.

By user i's *rational expectation* for a social function $SC(w, l)$ at $\omega \in \Omega$ we mean i's individual conjecture q_i provided that $\omega \in [q_i]$ and that

$$\mathbf{E}_i[SC(w, (\mathbf{l}_i(\omega), 1_{-i})); q_i] \leqq \mathbf{E}_i[SC(w, (l_i, 1_{-i})); q_i] \tag{4}$$

for any channel $l_i \in L$ adopted by i.

We denote by $\mathrm{RE}^{\mathrm{BKP}}(SC(w, l) : q_i)$ be the set of all the states in which $q_i (i \in N)$ is a rational expectation for a social cost $SC(w, l)$ for $BKP \in \mathbf{B}(KP)$.

Definition 3. A profile of all users' conjectures $q = (q_i)_{i \in N}$ is called an *rational expectations equilibrium* (r.e.e.) for a social function $SC(w, l)$ at $\omega \in \Omega$ if

$$\omega \in \cap_{i \in N} \mathrm{RE}^{\mathrm{BKP}}(SC(w, l) : q_i)$$

We denote by $\mathrm{REE}^{\mathrm{BKP}}(SC(w, l) : q)$ the set of all the states in which $q = (q_i)_{i \in N}$ is a r.e.e. for a social cost $SC(w, l)$ for $BKP \in \mathbf{B}(KP)$.

In these circumstances we have already know that, *for any* $\mathrm{BKP} \in \mathbf{B}(KP)$, *there exist an e.d.e.* $q' = (q_i')_{i \in N}$ *and a r.e.e.* $q = (q_i)_{i \in N}$ *for any social function* $SC(w, l)$ *simultaneously at everywhere; i.e.,* $\mathrm{REE}^{\mathrm{BKP}}(SC(w, l) : q) = \mathrm{EDE}^{\mathrm{BKP}}(q') = \Omega$.[8]

Actually we have interested in the converse as follows.

Problem 1. Under what conditions can a r.e.e. for $SC(w, l)$ actually lead a Nash equilibrium for $SC(w, l)$? Or/and how about an e.d.e.?

We gave an affirmative answer to the problem under common-knowledge assumption by Theorem 2 in Matsuhisa [6]. In the next section we will give another answer without this assumption.

3   Communication Model

In this section we will present the communication process modified the model of Parikh and Krasucki [9].

---

[8] Theorem 1 in Matsuhisa [6].

3.1   Protocol

We assume that the users communicate by sending *messages*. Let T be the time horizontal line $T = \{0, 1, 2, \cdots t, \cdots\}$.

A *protocol* is a mapping $\Pr : T \to N \times N, t \mapsto (s(t), r(t))$ such that $s(t) \neq r(t)$. Here t stands for *time* and $s(t)$ and $r(t)$ are, respectively, the *sender* and the *recipient* of the communication which takes place at time t. We consider the protocol as the directed graph whose vertices are the set of all users N and such that there is an edge (or a path) from i to j if there are infinitely many t such that $s(t) = i$ and $r(t) = j$. The communication is assumed to proceed in *rounds*.[9] A protocol is said to be *fair* if the graph is strongly connected; in words, every user in this protocol communicates directly or indirectly with every other user Infinitely often. It is said to be *acyclic* if the graph contains no cyclic path.

3.2   Communication Process

The *belief revision process* (b.r.p.) π for r.e.e. (or e.d.e. respectively) is a triple $\langle \Pr, (\Pi_i^t)_{(i,t)}, (\mathbf{q}_i^t)_{(i,t)} | (i,t) \in N \times T \rangle$, in which

1. $\Pr(t) = (s(t), r(t))$ is a fair protocol such that for every t, $r(t) = s(t+1)$, and
2. $\Pi_i^t : \Omega \to 2^\Omega$ is defined inductively in the following way:
   - $\Pi_i^0(\omega) := \Pi_i(\omega)$.
   - Suppose Π_i^t is defined. Let $\mathbf{q}_i^t(l_{-i}; \omega) \in \Delta((L^n)_{-i}) \times \Omega$ be the conditional probability

$$\mathbf{q}_i^t(l_{-i}; \omega) := \mu([l_{-i}] | \Pi_i^t(\omega)),$$

   - $M_i^t(\omega)$ is the message of user i when i is a sender at t, which means his/her rationality:

$$M_i^t(\omega) = \begin{cases} \mathrm{ED}^{\mathrm{BKP}}(\mathbf{q}_i^t(\omega)), \text{ if } \pi \text{ is a b.r.p. for e.d.e.;} \\ \mathrm{RE}^{\mathrm{BKP}}(SC(w, l) : \mathbf{q}_i^t(\omega)) \text{ if } \pi \text{ is a b.r.p. for r.e.e.} \end{cases}$$

Let $\mathcal{M}_i^t$ be the partition of Ω generated by $\{M_i^t(\omega) | \omega \in \Omega\}$.
3. Then Π_i^{t+1} is defined as the refinement $\Pi^t \vee \mathcal{M}_i^t$ if i is a recipient of a message at time $t + 1$, otherwise it is still the same as Π^t; i.e.,

$$\Pi_i^{t+1} = \begin{cases} \Pi_i^t \vee \mathcal{M}_{s(t)}^t, \text{ if } i \text{ is a recipient of a message at time } t + 1; \\ \Pi_i^t, \text{ if } i \text{ is not a recipient of a message at time } t + 1. \end{cases}$$

Specifically the sender j sends to i the message of his/her conjecture $\mathbf{q}_i^t([l_j]_{j \in N \setminus i}; \omega)$ about the other users' selections of channels together with his/her rationality, which is formally that

$$\Pi_i^{t+1}(\omega) = \begin{cases} \Pi_i^t(\omega) \cap M_{s(t)}^t(\omega), \text{ if } i \text{ is a recipient of a message at time } t + 1; \\ \Pi_i^t(\omega), \text{ if } i \text{ is not a recipient of a message at time } t + 1. \end{cases}$$

---

[9] There exists a time m such that for all t, $\Pr(t) = \Pr(t + m)$. The *period* of the protocol is the minimal number of all m such that for every t, $\Pr(t + m) = \Pr(t)$.

Let $(q_i^t \,|i \in N, t \in T)$ be a stochastic process with $q_i^t \in \Delta(L_{-i}^n)$. We denote by $[q_i^t]$ the event

$$[q_i^t] = \cap_{l_{-i} \in (L^n)_{-i \in N \setminus i}} \{\xi \in \Omega \mid \mathbf{q}_i^t([l_{-i}]; \xi) = q_i^t(l_{-i})\}.$$

Definition 4. A stochastic process $(q_i^t \,|i \in N, t \in T)$ is called a *communication process* (c.p.) according to a b.r.p. π if there is a sequence $\{\omega^t\}_{t=0,1,2,\cdots}$ such that $\Pi_i^t(\omega^t) \neq \emptyset$ and $\omega^t \in \cap_{i \in N}[q_i^t]$ for every $t = 0, 1, 2, \cdots$.

Because Ω is finite, the descending chain $\{\Pi_i^t(\omega) \,|\, t = 0, 1, 2, \dots \}$ is finite, and so it must be stationary, and the limit $\Pi_i^\infty(*)$ exists in any states.

Definition 5. By the *limiting conjecture* of user i at ω we mean $\mathbf{q}_i^\infty(l_{-i}; \omega) = \mu([l_{-i}] | \Pi_i^\infty(\omega))$. Furthermore, if $(q_i^t \,|i \in N, t \in T)$ is a c.p. then each q_i^∞ is defined as $q_i^\infty = \mathbf{q}_i^\infty(l_{-i}; \omega)$ in which $\omega \in \cap_{i \in N}[q_i^t]$ for sufficiently large $t \in T$.

4 Result

4.1 Communication to Equilibrium

Under the above circumstances we can now prove

Theorem 1. *In the communication process, the below statements are true:*

(i) *The profile of the limiting conjectures after long-run communication for r.e.e. yields a Nash equilibrium for $SC(w, l)$ in the based KP-model, and*

(ii) *The profile of the limiting conjectures after long-run communication for e.d.e. yields a Nash equilibrium for the based KP-model.*

The illustrating examples will be given in the conference talk, because of the limitation of the pages of the proceedings papers.

4.2 Key Lemma

Before proceeding in the proof of Theorem 1, we need the below lemma, and the proof will be given in the final version of this paper.:

Lemma 1. *Notation and assumptions are the same in Theorem 1 as above. For any members $i, j \in N$, their conjectures $\mathbf{q}_i^\infty$ and $\mathbf{q}_j^\infty$ on $L^n \times \Omega$ must coincide; that is, $\mathbf{q}_i^\infty(l; \omega) = \mathbf{q}_j^\infty(l; \omega)$ for every $l \in L^n$ and $\omega \in \Omega$.* $\square$

4.3 Proof of Theorem

Let $\omega \in \cap_{i \in N}[q_i^\infty]$. In viewing of Lemma 1, it follows by summing over l_{-i} that $q_j^\infty(l_i) = q_k^\infty(l_i)$ for any $j, k, \neq i$, and hence $q_j^\infty(l_i)$ is independent of $j \neq i$; which means that $q_j^\infty(l_i)$ depending only on i makes a probability distribution on $\Delta(L_{(i)})$. So we let $(\sigma_i)_{i \in N} \in (\Delta(L_{(i)}))^n$ be $\sigma_i(l_i) = q_i^\infty(l_i)$ for any $j \neq i$.

We denote by $\Gamma(i)$ the set of all the players who directly receive the message from $i \in S$; i.e., $\Gamma(i) = \{ j \in N \mid (i,j) = \mathrm{Pr}(t) \text{ for some } t \in T\}$.

Let W_i denote $[q_i^\infty] := \bigcap_{l_{-i} \in (L^n)_{-i}} [\mathbf{q}_i^\infty(l_{-i}; *) = q_i^\infty(l_{-i})]$, and note $\omega \in W_i \cap W_j \neq \emptyset$ for each $i \in N$ and for any $j \in \Gamma(i)$.

For completing the proof it suffices to show that for every $l \in \prod_{i \in N} \mathrm{Supp}(\sigma_i)$,

$$q_i^\infty(l_{-i}) = \prod_{j \in N \setminus i} \sigma_j(l_j). \tag{5}$$

In fact. For (i): Viewing Eq. (3) together with the above Eq. (5) we can observe that $\mathbf{E}_i[SC(w, (l_i, \mathbf{1}_{-i})); q_i] = SC(w, l_i, \sigma_i)$, and thus (i) follows immediately from the inequality (4) for any $i \in N$.

For (ii): Viewing Eqs. (1) and (5) we obatin that $\mathbf{E}_i[\lambda_i^l; q_i] = \lambda_i^l(\sigma)$, and so (ii) follows immediately from Eq. (2) for any $i \in N$.

Now it remains to prove Eq. (5). Unfortunately, because of the limitation of pages, we have to postpone giving the proof in the final version of the paper again. $\qquad\square$

5   Concluding Remarks

We have been keeping many important agenda left, so it will end this paper well by remarking some of them.

5.1   Nash Equilibrium and Topology of Network

We start a technical remark. In Theorem 1, the Nash equilibriums to which the communication process leads depend on the protocol Pr (communication network). So it seems to be depend on the topology of the network graph. Is it actually true? If so, can we determine all the protocol that yields each Nash equilibrium in the communication model?

5.2   Expected Price of Anarchy

Much more important agenda is that we have to tackle to investigate *social costs with its price of anarchy* from epistemic point of view, especially to give the upper bound of the price of anarchy for special types of social functions.

Let us extend them into our framework. User i's *individual expected social cost* $\mathbf{E}_i[SC; q_i^t]$ according to q_i^t at time $t = 0, 1, 2, \cdots$ is

$$\mathbf{E}_i[SC; q_i^t] = \sum_{l_i \in L_{(i)}} \mathbf{E}_i[SC(w, (l_i, \mathbf{1}_{-i})); q_i^t]$$

and the *expected social cost* at time t is $\mathbf{E}[SC; q^t] = \sum_{i \in N} \mathbf{E}_i[SC; q_i^t]$. The *expected price* of anarchy at time t would be given by

$$\mathbf{E}[PA]^{BKP}(SC)^t = \frac{\sup_{q^t \in \mathrm{REE}^{\mathrm{BKP}}(SC(w,l))} \mathbf{E}[SC; q^t]}{\inf_{q^t \in (\Delta((L^n)_{-i}))^n} \mathbf{E}[SC; q^t]}$$

$$\mathbf{E}[PA]^{BKP}(SC)^\infty = \frac{\sup_{q^\infty \in \mathrm{REE}^{\mathrm{BKP}}(SC(w,l))} \mathbf{E}[SC; q^t]}{\inf_{q^\infty \in (\Delta((L^n)_{-i}))^n} \mathbf{E}[SC; q^\infty]}$$

In this set up it can be easily seen that

$$\mathbf{E}[PA]^{BKP}(SC)^{\infty} = \lim_{t \to \infty} \mathbf{E}[PA]^{BKP}(SC)^t$$

The interesting problems are to study the behavior of $\mathbf{E}[PA]^{BKP}(SC)^t$ as t tends to ∞, and to determine an upper bound of the expected price of anarchy at each $t = 0, 1, 2, \cdots, \infty$. Further, how about it for the cases $SC(w, l)$ would be one of $LSC(W, l), QSC(w, l)$ and $MSC(w, l)$?

5.3   General Information Partition

The information partition structure coincides with the model for the multi-modal logic **S5n**. In this regards it seems important to extend all the results in this paper into the non-partition information structure, which is a model for the multi-modal logic **S4n**.

References

1. Aumann, R.J.: Agreeing to disagree. Ann. Stat. **4**, 1236–1239 (1976)
2. Fagin, R., Halpern, J.Y., Moses, Y., Vardi, M.Y.: Reasoning About Knowledge. MIT Press, Cambridge (1995)
3. Garing, M., Monien, B., Tiemann, K.: Selfish routing with incomplete information. Theory Compt. Syst. **42**, 91–130 (2008)
4. Harsanyi, J.C.: Games with incomplete information played by Bayesian players, I, II, III. Manag. Sci. **14**, 159–182, 320–332, 468–502 (1967)
5. Koutsoupias, E., Papadimitriou, C.: Worst-case equilibria. In: Meinel, C., Tison, S. (eds.) STACS 1999. LNCS, vol. 1563, pp. 404–413. Springer, Heidelberg (1999). doi:10.1007/3-540-49116-3_38
6. Matsuhisa, T.: Common-knowledge and KP-model. In: Nguyen, N.T., Trawiński, B., Fujita, H., Hong, T.-P. (eds.) ACIIDS 2016. LNCS (LNAI), vol. 9621, pp. 490–499. Springer, Heidelberg (2016). doi:10.1007/978-3-662-49381-6_47
7. Mazalov, V.: Mathematical Game Theory and Applications. Wiley, Hoboken (2014)
8. Nash, J.F.: Equilibrium points in n-person games. Proc. Natl. Acad. Sci. Unit. States Am. **36**, 48–49 (1950)
9. Parikh, R., Krasucki, P.: Communication, consensus, and knowledge. J. Econ. Theory **52**, 78–89 (1990)

Scalable Data Analysis in Bioinformatics and Biomedical Informatics

Orchestrating Task Execution in Cloud4PSi for Scalable Processing of Macromolecular Data of 3D Protein Structures

Dariusz Mrozek$^{(\boxtimes)}$, Artur Kłapciński, and Bożena Małysiak-Mrozek

Institute of Informatics, Silesian University of Technology,
ul. Akademicka 16, 44-100 Gliwice, Poland
dariusz.mrozek@polsl.pl

Abstract. The growing amount of biological data, including macromolecular data describing 3D protein structures, encourages the scientific community to reach for computing resources of the Cloud in order to process and analyze the data on a large scale. This applies, among many different analytical processes performed in bioinformatics, to protein structure alignment and similarity searching. In this paper, we show a parameter sweep-based approach for scheduling computations related to massive 3D protein structure alignments performed with Cloud4PSi system working on Microsoft Azure public cloud.

Keywords: Bioinformatics · Proteins · 3D protein structure · Alignment · Cloud computing · Microsoft azure · Scheduling · Parameter sweep

1 Introduction

3D protein structure alignment is one of the basic processes performed in structural bioinformatics. The process aims at finding common structural regions of protein structures, which may reflect functional similarity of compared protein molecules, allow to find evolutionary relationships between organisms, validate predicted protein models, with the general purpose of understanding molecular machinery of organisms, as complex biological systems. Massive protein structure alignment allows to find similarities between many molecules at the same time. This can be done by parallel pair-wise comparisons of 3D protein structures. These comparisons are usually time-consuming, due to complex nature of protein structures and complexity of algorithms that are used to align protein molecules. Moreover, regarding the continuous increase in the volume of the macromolecular data in public databases, like the Protein Data Bank [4], large scale and massive 3D protein structure alignments require dedicated, scalable architectures to accommodate the growth of data and increasing demand for computing power needed for its analysis.

Cloud computing can be one of the solutions for the problem, as it delivers a shared pool of configurable computing resources that can be quickly provisioned

© Springer International Publishing AG 2017
N.T. Nguyen et al. (Eds.): ACIIDS 2017, Part II, LNAI 10192, pp. 723–732, 2017.
DOI: 10.1007/978-3-319-54430-4_69

and released on demand [11]. With the use of computing resources of the Cloud we are able to design and develop scientific workflows for computing-intensive and data-intensive problems not only in bioinformatics, but also in many other scientific fields that deal with Big Data.

Cloud4PSi [17] is the system that uses Cloud resources to store macromolecular data of 3D protein structures and perform scalable structural alignments of protein molecules. Cloud4PSi was developed to work on Microsoft Azure public cloud, where it can be dynamically scaled out to handle growing amount of workload. Good scalability of the system was proved in a series of performance tests and results from proper orchestration of task execution coordinated by middle-layer Manager role in the temporally decoupled system.

Scheduling computations through proper orchestration of task execution is an important part of every system that processes large data sets on a distributed computer infrastructure. The parameter-sweep application model has emerged as a killer application model for developing high-throughput computing (HTC) applications for processing on highly distributed computing environments [1], like Grids or Clouds. This model essentially assumes processing different data sets by n independent tasks on k distributed compute units, where n is typically much larger than k. This high-throughput parametric computing model is simple, yet powerful enough to enable efficient distributed execution of computations performed in various areas related to bioinformatics and Life sciences, such as: gene expression and SNPs analysis [6,9], genomic and protein sequence alignment [2,7], protein structure similarity searching [13,16], protein folding [10,15], molecular modeling for drug design [3], radiation equipment calibration analysis [1], tomography [21], brain activity analysis [19], and others.

In this paper, we show how we schedule computations in the Cloud4PSi system for protein structure similarity searching and massive structural alignments with the use of parameter-sweep computing model. We present the approach that is based on the division of the main job into a set of tasks created for data packages containing a fixed number of protein structures (*fixed number of proteins package*-based, FNPP-based approach). We also show results of experiments that allowed us to determine the number of proteins that should be assigned to the package in order to reduce the whole computation time.

2   Related Works

The parameter-sweep computing model is sufficient for pleasingly parallel problems, especially, if there are no additional constraints, like application execution makespan, computation budget limitations, or hardware limitations (e.g., GPU memory [13]). Scheduling computations in the model is also relatively simple in implementation. For example, CloudPSR [16] uses two schemes, $S1$ and $S2$, for scheduling computations related to protein structure similarity searching on Microsoft Azure cloud. In both schemes, a number of tasks is created for generated data packages, which are parts of repository. The number of tasks is equal to the number of data packages. Scheduling scheme $S1$ assumes that the whole repository of protein structures (the whole data set) is divided into equal parts. The number of parts of the whole data set and the number of tasks is equal to the number of workers (compute units). Thus, each Worker is assigned only one task with one large data package to be processed. Scheduling scheme $S2$ assumes that the whole repository of protein structures is divided into small packages, each one containing only one protein structure, and each package has a corresponding task. Then, each Worker is assigned multiple such tasks, and performs multiple executions of the same process for each such a small package. The number of task executions on each of the compute unit is usually not the same, since protein structures vary in size and processing times may vary significantly.

Interesting observations can be also made for Hadoop-based implementations for massive protein structure alignment and similarity searching, e.g., HDInsight4PSi [14], H4P [18] and Hung and Lin's system [8]. MapReduce applications proposed by Hung and Lin [8] and developed in H4P system [18] working in one-to-one comparison scenario compare protein structures in pairs (given protein against candidate protein), generating one Map task for each candidate protein from the repository. Therefore, the number of Map tasks is equal to the number of proteins in the repository. These Map tasks are then scheduled for execution on nodes of the Hadoop cluster by JobTracker/ApplicationMaster. This approach is similar to scheduling scheme $S2$ used in CloudPSR system. HDInsight4PSi and H4P working in one-to-many comparison scenario use *sequential files* that group proteins in large fixed-sized packages of 64 MB, in order to fit the Hadoop Distributed File System (HDFS) block size and speed up computations. The number of Map tasks then depends on the number of packages (Hadoop splits) resulting from packing structures into 64 MB sequential files. This approach works on the basis of fixed-sized packages (FSP-based approach).

3   System Architecture

Cloud4PSi is a cloud, Software as a Service (SaaS) system that allows to perform parallel structural alignments of 3D protein structures. Cloud4Psi has been implemented with the use of abstract objects of Microsoft Azure programming model, like roles and queues, for developing applications as Cloud services for the Azure cloud (Fig. 1). In Cloud4PSi, users launch parallel structural alignments through the Web role, where among various parameters, they specify the input, query protein structure and alignment algorithm. After launching the alignment on the Cloud, the Web role places the user's request (so called *search request* or *job*) in the Input queue, where the request waits for being consumed by Manager role. The Manager role consumes search requests one by one and schedules parallel alignments by generating so called *task descriptors*. Task descriptors are created on the basis of the information included in the search request, and they store information on how the task should be executed by Searcher roles, e.g., query protein structure, alignment algorithm, the number of structures that should be processed. Idle Searcher roles consume task descriptors, decode them, and search for protein similarities by performing structural alignments against the repository of candidate proteins located in Azure BLOB Storage, according to parameters extracted from the decoded task descriptor. At this stage of the development of the system, we have parallelized two alignment algorithms in Cloud4PSi - jCE and jFATCAT [20]. Results of the similarity searches are stored in Azure Table Storage on the Cloud.

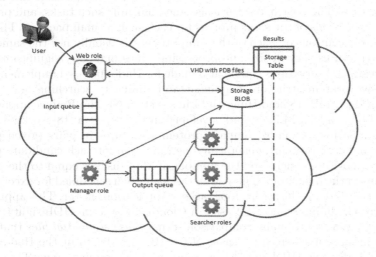

Fig. 1. Architecture of Cloud4Psi – a Microsoft Azure cloud-based, SaaS system for protein structure similarity searching. Reproduced from [12]

Working in the Azure cloud, the systems can theoretically be infinitely scaled out on the Searcher roles with a growing amount of protein macromolecular data and growing demand for computing power. In practice, scaling of the Cloud4PSi is limited only by a possessed subscription for the Azure cloud and by capabilities of the Cloud.

4   Scheduling Computations in Cloud4PSi

The Manager role has a specific functionality as one of the Worker roles in the Cloud4PSi. It schedules tasks for execution on idle Searcher roles based on requests received from the Web role, passes parameters, arranges the scope of the similarity searching, and manages associated computational load between instances of the Searcher role. Manager is also responsible for the preparation of the read-only, virtual hard drive located on Azure BLOB Storage, which stores all candidate protein structures, and eventually, uploaded given query protein structure, all used by instances of the Searcher role when performing parallel structural alignments.

As a standard activity, the Manager role implements the pseudocode of Algorithm 1. The role listens if there are any search requests in the Input queue (line 2). Incoming requests are immediately captured by the Manager role, which divides the whole range of candidate proteins into packages (lines 3–5). Packages contain a small number of protein structures from the main repository R of PDB files that should be compared with user's query protein by a single Searcher role.

Algorithm 1. Manager role: Search request processing and task creation

---
1: **while** true **do**
2:     Check messages in the Input queue
3:     **if** exists a message **then**
4:         Retrieve the message (search request) and extract parameters
5:         Divide repository R into data packages according to specified configuration
6:         **for each** package $P_i \subset P$ **do**
7:             Create task descriptor as output message
8:             Encode package metadata and other parameters in the output message
9:             Enqueue the output message in the Output queue
10:        **end for**
11:    **end if**
12: **end while**

Let $R = R_1, R_2, ..., R_m$ be the repository of m candidate protein structures to be compared and aligned by k Searcher roles ($m \gg k$), and $P = P_1, P_2, ..., P_n$ be a finite set of n packages ($n < m, n \gg k$) that satisfies the following three conditions:

$$\forall_{P_i \in P} \ P_i \subseteq R, \tag{1}$$

$$\exists_{f_1} \ R \xrightarrow{f_1} P, \tag{2}$$

$$|P_i| = const. \tag{3}$$

The last condition tells that the number of proteins in the package P_i is fixed, and we can state that:

$$R = \bigcup_{i=1}^{n} P_i \quad \text{and} \quad \sum_{i=1}^{n} |P_i| = m, \tag{4}$$

where P_i is the i-th package of protein structures, and n is the number of packages that should be processed by all k instances of the Searcher role.

Packages satisfy the following relationship:

$$\forall_{1 \le i,j \le n} \ i \ne j \implies P_i \cap P_j = \emptyset. \tag{5}$$

The number of packages n determines the number of all search tasks $T = T_1, T_2, ..., T_n$ that will be executed on Searcher roles $S = S_1, S_2, ..., S_k$. The number of packages and tasks (n) depends on the number of protein structures in the repository R and the number of protein structures in package P_i:

$$n = \left\lceil \frac{m}{|P_i|} \right\rceil. \tag{6}$$

The number of proteins in the package P_i should be chosen experimentally in order to minimize the whole computation time, when processing search requests. Descriptors of successive tasks with sweep parameters that determine package content are sent by Manager role as messages to the Output queue, where they wait for being processed (lines 6–9). In such a way, Manager role creates search tasks T that are scheduled for execution. Given the number of protein structures in each package is fixed, we call these packages as *fixed number of proteins packages*, and the scheduling scheme will be called *fixed number of proteins package-based* (FNPP-based) scheduling scheme. Since protein structures have various sizes (various numbers of chains, amino acids in each chain, and atoms), sizes of packages may vary significantly, and processing time of each package may differ.

Format of the message containing task descriptor that is sent to the Output queue is shown in Fig. 2. Task descriptor starts with *search request ID* (also called *job identifier*), which is a unique identifier of the search request the task belongs to. The identifier is followed by *package size* expressed as the number of proteins contained in the package ($|P_i|$). The next field *repository size* consists of the information on the number of protein structures in the whole repository (m). The field *query protein* consists of the PDB ID code of the query protein, if it is also present in the repository, or name of the uploaded file with macromolecular data describing input protein structure. The following *upload flag* field tells if the query protein was uploaded as a file ($True$) or is one of the proteins in the repository ($False$). The next field, *start point*, defines the sweep starting parameter for each instance of the Searcher role that consumes the task - Searchers process $|P_i|$ successive protein structures from the repository (by aligning query protein structure to candidates from the repository) starting from the *start point*. In Fig. 3 each square can be interpreted as a part of repository (a package) that will be retrieved by a single instance of the Searcher role while executing task, which description is retrieved from the Output queue. For each task descriptor that is generated by the Manager role the *start point* is incremented by the value of the *package size*. The *message time* field stores the information on the time the task descriptor was created, which is needed to determine the task processing time in the system. Additionally, task descriptor consists of the URL address of the shared virtual hard drive image that contains repository of protein structures (*snapshot URI* field), encoded name of the alignment algorithm that should be used when comparing proteins (*algorithm* field), and *byChain* flag that informs Searcher whether to align single chains of protein structures or whole molecules.

Task descriptors are encoded as messages, where successive fields are separated by the | character. Sample message from the Output queue is presented below:

```
4aa284b0-d1aa-42bc-9f1d-4cfcf98b2641|10|100000|1bsn.A|True|20|
3/10/2016 8:53:55 AM|
http://prot.blob.core.windows.net/drives/pdb.vhd?snapshot=...|1|True
```

search request ID	package size $|P_i|$	repository size m	query protein	upload flag	start point	message time	snapshot URI	algorithm	byChain flag

Fig. 2. Format of the task descriptor sent to the Output queue.

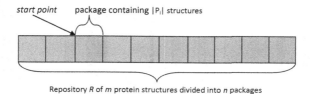

start point   package containing $|P_i|$ structures

Repository R of m protein structures divided into n packages

Fig. 3. Division of the repository R of m proteins into n packages of the size $|P_i|$ and the *start point* for each instance of the Searcher role.

5   Experimental Results

The FNPP-based scheduling scheme that we have implemented in the Cloud4PSi assumes division of the whole repository into packages. These packages have fixed number of proteins. Package size determines how many protein structures will be compared by every single instance of the Searcher role in the execution of a single task. Since instances of the Searcher role may handle various search requests at any time, they retrieve the structure of the query protein for each task T_i and each package P_i of protein structures that must be processed. However, as could be expected and was confirmed by our experiments, the number of protein structures in the package affects the average time of processing the package. Therefore, while testing Cloud4PSi we checked how the assumed number of protein structures in the package influences the total time of the similarity searching. Tests were performed in Microsoft Azure public cloud with the use of macromolecular data repository containing a sample of 1,000 protein structures from the Protein Data Bank. Cloud4PSi was configured to use 4 Medium-sized (2 core CPUs, 3.5 GB RAM, 489 GB HDD) instances of the Searcher role (multi-core Searchers consumed multiple tasks). The input, query protein structure (PDB ID code: 1A0T, chain P) used in tests was 413 amino acids long and is a medium-sized protein in the Protein Data Bank, containing three chains [5].

For both algorithms (jCE and jFATCAT) we measured total execution time for the following numbers of protein structures in the package $|P_i|$: 1, 5, 10, 20, 50, 100, and 200 protein structures. Smaller packages with fewer structures processed in one task seems to be more flexible solution. In theory, this should allow for more effective management of free processing power in the system. Moreover, for small packages (and more tasks) the distribution of tasks and packages among Searchers should be more balanced. Large packages decrease the number of tasks scheduled for processing within a single search request, and cause that one processing unit is occupied relatively longer, while other units may be idle because of the absence of tasks to be executed. On the other hand, more tasks (smaller packages) cause that each instance of the Searcher role

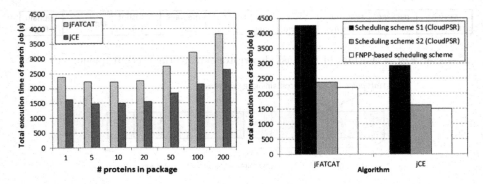

Fig. 4. (left) Total execution time as a function of the number of proteins in package. (right) Comparison of execution times when using various scheduling schemes in Cloud4PSi – the FNPP-based presented in the paper, and scheduling schemes $S1$ and $S2$ used in CloudPSR. Results for jCE and jFATCAT algorithms parallelized on the cloud, obtained for 4 medium-sized instances of the Searcher role.

must often repeat some fixed actions in its operating cycle, such as retrieving the query protein structure from repository, creation of internal structures for storing results, passing results to be saved, and others. This may also negatively affect the processing time.

Our experiments allowed to verify how the system behaves for various values of $|P_i|$. In Fig. 4 (left) we present total execution time for various number of protein structures in a package. Results confirmed that for small packages the total execution time of the entire search job was shorter than for large packages. For packages containing 20 protein structures or more, the total execution time gradually increased due to lower flexibility in allocating tasks and packages to individual Searchers. We found that the most effective searches were achieved for packages of 5 structures (2,217 s for jFATCAT, and 1,476 s for jCE) or 10 structures (2,200 s for jFATCAT, and 1,496 s for jCE). For 200 structures per package the whole execution time of the search job increased to 3,826 s for jFATCAT and 2,617 s for jCE. On the basis of our experiments, we decided to use ten protein structures per package in the FNPP-based scheduling scheme and such a value was implemented as a default one in the Cloud4PSi system.

We also compared the FNPP-based scheduling scheme implemented in Cloud4Psi with two other scheduling schemes ($S1$ and $S2$) implemented in CloudPSR competitive system for protein structure similarity searching. Actually, $S1$ and $S2$ scheduling schemes are two special cases of FNPP-based approach - in $S1$ scheme $|P_i^{S1}| \gg |P_i|$ and depends on the number of compute units in use, and in $S2$ scheme $|P_i^{S2}| = 1$. All computations were performed with the use of Cloud4PSi and three scheduling schemes – the FNPP-based presented in the paper for the package size set to 10 protein structures ($|P_i| = 10$), and scheduling schemes $S1$ and $S2$ from CloudPSR. Results of the comparison are presented in Fig. 4 (right). As can be observed the FNPP-based scheduling scheme implemented in Cloud4PSi was much more efficient than scheduling scheme $S1$, and slightly better than scheduling scheme $S2$ that were used in CloudPSR system.

6   Discussion and Concluding Remarks

Proper orchestration of task execution while performing computations related to massive alignments of 3D protein structures on the public cloud has a significant impact on the costs of the computations. By minimizing the total execution time of the process, we can reduce the costs of the resource provisioning, which is important for long running processes, such as protein structure similarity searching performed for the whole repository of protein molecules. The FNPP-based scheduling scheme presented in the paper and implemented in Cloud4PSi allows to choose proper amount of protein molecules and reduce the execution time. The value of ten protein structures per package, which we have selected as a default package size in our system, was chosen on the basis of a consensus by taking into account the total execution time for two algorithms for protein structure alignment, jFATCAT and jCE, parallelized in Cloud4PSi. This value provides the possibility to flexibly distribute tasks and data among Searcher role instances regardless of the number of the instances currently in use, which was a weak point of scheduling scheme $S1$ in CloudPSR. This value also prevents from repeating some common tasks required before starting computations related to protein structure alignments for each package (like loading input structure, preparation of internal storage structures, etc.), which was a weakness of scheduling scheme $S2$ in the same system. Moreover, the value allows for proper load balancing for 3D protein structure similarity searches performed on the Cloud without the necessity to pre-process the macromolecular data of proteins, which is required, e.g., in HDInsight4PSi that uses Hadoop and fixed-sized sequential files stored on Hadoop Distributed File System or in GPU-CASSERT [13] that extracts only selected features of protein structures and prepares data packages adjusted to the amount of global memory of the GPU device.

Acknowledgments. This work was supported by The National Centre for Research and Development grant No. PBS3/B3/32/2015 and by Microsoft Research within Microsoft Azure for Research grant.

References

1. Abramson, D., Giddy, J., Kotler, L.: High performance parametric modeling with Nimrod/G: killer application for the global grid? In: Proceedings of the International Parallel and Distributed Processing Symposium (IPDPS 2000), pp. 1–5. IEEE Computer Society Press, Los Alamitos (2000)
2. Al-Absi, A., Kang, D.: Long read alignment with parallel MapReduce cloud platform. Biomed. Res. Int. 1–13 (2015). Article ID 807407
3. Beberg, A., Ensign, D., Jayachandran, G., Khaliq, S., Pande, V.: Folding@home: lessons from eight years of volunteer distributed computing. In: 2009 IEEE International Symposium on Parallel and Distributed Processing, pp. 1–8 (2009)
4. Berman, H., et al.: The Protein Data Bank. Nucleic Acids Res. **28**, 235–242 (2000)
5. Forst, D., Welte, W., Wacker, T., Diederichs, K.: Structure of the sucrose-specific porin ScrY from salmonella typhimurium and its complex with sucrose. Nat. Struct. Biol. **5**(1), 37–46 (1998)

6. Hung, C.L., Chen, W.P., Hua, G.J., Zheng, H., Tsai, S., Lin, Y.L.: Cloud computing-based TagSNP selection algorithm for human genome data. Int. J. Mol. Sci. **16**(1), 1096–1110 (2015)

7. Hung, C.L., Hua, G.J.: Local alignment tool based on Hadoop framework and GPU architecture. Biomed. Res. Int. 1–7 (2014). Article Id 541490

8. Hung, C.L., Lin, Y.L.: Implementation of a parallel protein structure alignment service on cloud. Int. J. Genomics 1–8 (2013). Article Id 439681

9. Inda, M.A., Belloum, A.S.Z., Roos, M., Vasunin, D., de Laat, C., Hertzberger, L.O., Breit, T.M.: Interactive workflows in a virtual laboratory for e-Bioscience: the SigWin-detector tool for gene expression analysis. In: Proceedings of the Second IEEE International Conference on e-Science and Grid Computing, E-SCIENCE 2006, pp. 19–26. IEEE Computer Society, Washington, DC (2006)

10. Leaver-Fay, A., Tyka, M., Lewis, S., Lange, O., Thompson, J., Jacak, R., et al.: ROSETTA3: an object-oriented software suite for the simulation and design of macromolecules. Methods Enzymol. **487**, 545–574 (2011)

11. Mell, P., Grance, T.: The NIST Definition of Cloud Computing. Special Publication 800-145 (2011). http://nvlpubs.nist.gov/nistpubs/Legacy/SP/nistspecialpublication800-145.pdf. Accessed 24 Sept 2016

12. Mrozek, D.: High-Performance Computational Solutions in Protein Bioinformatics. SpringerBriefs in Computer Science. Springer, Heidelberg (2014)

13. Mrozek, D., Brożek, M., Małysiak-Mrozek, B.: Parallel implementation of 3D protein structure similarity searches using a GPU and the CUDA. J. Mol. Model **20**, 2067 (2014)

14. Mrozek, D., Daniłowicz, P., Małysiak-Mrozek, B.: HDInsight4PSi: boosting performance of 3D protein structure similarity searching with HDInsight clusters in Microsoft Azure cloud. Inf. Sci. **349–350**, 77–101 (2016)

15. Mrozek, D., Gosk, P., Małysiak-Mrozek, B.: Scaling Ab initio predictions of 3D protein structures in Microsoft Azure cloud. J. Grid Comput. **13**, 561–585 (2015)

16. Mrozek, D., Kutyła, T., Małysiak-Mrozek, B.: Accelerating 3D protein structure similarity searching on Microsoft Azure cloud with local replicas of macromolecular data. In: Wyrzykowski, R., Deelman, E., Dongarra, J., Karczewski, K., Kitowski, J., Wiatr, K. (eds.) PPAM 2015. LNCS, vol. 9574, pp. 254–265. Springer, Heidelberg (2016). doi:10.1007/978-3-319-32152-3_24

17. Mrozek, D., Małysiak-Mrozek, B., Kłapciński, A.: Cloud4Psi: cloud computing for 3D protein structure similarity searching. Bioinformatics **30**(19), 2822–2825 (2014)

18. Mrozek, D., Suwała, P., Małysiak-Mrozek, B.: High-throughput and scalable protein function identification with Hadoop and Map-only pattern of the MapReduce processing model. J. Knowl. Inf. Syst. (submitted for publication)

19. Olabarriaga, S.D., Nederveen, A.J., O' Nuallain, B.: Parameter sweeps for functional MRI research in the "Virtual Laboratory for e-Science" project. In: Proceedings of the 7th IEEE International Symposium on Cluster Computing and the Grid, CCGRID 2007, pp. 685–690. IEEE Computer Society, Washington, DC (2007)

20. Prlić, A., Yates, A., Bliven, S., et al.: BioJava: an open-source framework for bioinformatics in 2012. Bioinformatics **28**, 2693–2695 (2012)

21. Smallen, S., Casanova, H., Berman, F.: Applying scheduling and tuning to online parallel tomography. In: Proceedings of the 2001 ACM/IEEE Conference on Supercomputing, SC 2001, p. 12. ACM, New York (2001)

Probabilistic Neural Network Inferences on Oligonucleotide Classification Based on Oligo: Target Interaction

Abdul Rahiman Anusha[1](✉) and S.S. Vinodchandra[2]

[1] Department of Computational Biology and Bioinformatics,
University of Kerala, Thiruvananthapuram 695581, Kerala, India
anushapraveenkhan@gmail.com
[2] Computer Centre, University of Kerala, Thiruvananthapuram 695034, India
vinodchandrass@gmail.com

Abstract. Oligonucleotides are small non-coding regulatory RNA or DNA sequences that bind to specific mRNA locations to impart gene regulation. Identification of oligonucleotides from other small non-coding RNA sequences such as miRNAs, piRNAs etc. is still challenging as oligos exhibit a notable overlap in sequence length and properties with these RNA categories. This work focuses on a probabilistic oligonucleotide classification method based on its distinct underlying feature vectors to identify oligos from other regulatory classes. We propose a computational approach developed using a probabilistic neural network (PNN) based on oligo: target binding characteristics. The performance measure showed promising results when compared with other existing computational methods. Role and contribution of extracted features was estimated using the receiver operating curves. Our study suggests the potentiality of probabilistic approaches over non-probabilistic techniques in oligonucleotide classification problems.

Keywords: Oligonucleotides · Oligo: mRNA duplex · Probabilistic neural networks

1 Introduction

Regulatory transcripts or riboregulators are small non-coding RNAs that directly involve in the gene regulatory mechanisms inside the cell. These RNA class constitutes mainly microRNAs (miRNAs), oligonucleotides, small nuclear RNAs (snRNAs), small nucleolar RNAs (snRNAs) and piwi interacting RNAs (piRNAs). Like all regulatory RNAs, oligos also proven their efficiency as biological tools in gene knockdown procedures [1]. These are 7–30 nts long short fragments that can hybridize target mRNAs by Watson-Crick complementary base pairing principle [2]. Oligos are drawing significant attention since they can be synthetically designed and modified to induce mRNA: target stability, gene specific protein synthesis suppression, transcriptional and post transcriptional silencing [3]. However, oligos share its sequence length and certain properties with other small regulatory RNA classes. Hence incorrect mRNA

© Springer International Publishing AG 2017
N.T. Nguyen et al. (Eds.): ACIIDS 2017, Part II, LNAI 10192, pp. 733–740, 2017.
DOI: 10.1007/978-3-319-54430-4_70

targeting is often witnessed and these large set of false positives cause hindrance in oligo probe designing and therapeutic applications [4, 5]. Computational identification of oligonucleotides is more efficient than time consuming and expensive experimental methods. Computational methods are usually employed through machine learning techniques as they perform optimal classifications with less algorithmic complexity and give accurate results.

Literatures reported many computational approaches in oligo classification problems. Most machine learning works focused around categorizing the oligos into high or low profile sequences by accounting its efficiency in mRNA binding. Giddings and his team developed possible combinations of 4-mer motifs and classified a pool of oligos using an Artificial Neural Network (ANN) model [6]. Chalk and Erik combined different parameter set collected from all previous reports and built an ANN classifier [7]. A Support Vector Machine (SVM) based soft computing approach was attempted by Gustavo by optimally selecting energy related and motif related features using correlation analysis, mutual information and recursive feature elimination methods [8]. Enhanced accuracy was observed in another ANN based method that considered sequence motif related, structural and thermodynamical features [9].

Statistical methods were also implemented in oligo researches. The first attempt of this type was made by Matveeva group by identifying the frequency of 3-mer or 4-mer motifs in certain positions of oligo sequences that contribute towards oligonucleotide efficacy [10]. Ding and Lawrence conducted their work by analyzing the secondary structures of mRNA target [11]. However, none of these works reported a computational model featuring a stable oligo: mRNA interaction.

Highly efficient computational models survive from a balanced accuracy and sensitivity [12–14]. In any of the previous oligo prediction works mentioned above, sensitivity measures were never in agreement with the accuracy results. In this study a Probabilistic Neural Network (PNN) is proposed to build the mol combining positional, thermodynamic and structural properties of oligo: mRNA duplex. PNN is so chosen that it allows and handles huge dataset without iterative training producing accurate results.

2   PNN Paradigm

The probabilistic neural networks (PNN) are feed-forward multilayer perceptron classifiers. For large dataset, PNN converges to a probabilistic Bayesian classifier thereby reducing the risk of over training and incorrect classification. Thus probabilistic neural nets can be trained to learn faster with high signal to noise ratio with less computational and time complexity. It maps any number of input variables to specific class based on Bayes rule and a Kernel Fisher discriminate analysis algorithm [15, 16].

The standard PNN architecture is depicted in Fig. 1. It is a four layered neural network which has input layer, pattern layer, summation layer and output layer [17].

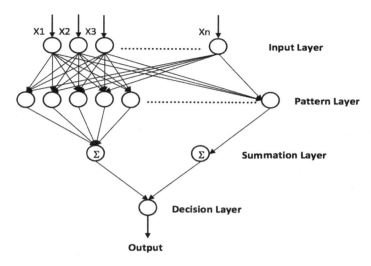

Fig. 1. Standard PNN architecture

Input Layer: The first layer with many nodes with each node representing one feature. All input nodes are fully interconnected with a weight of 1 to the next layer nodes to pass on normalized vectors.

Pattern Layer: The second layer is the pattern layer where the number of nodes is determined by the number of instances in the training set. For a new event, pattern layer calculates the Euclidian distance between input vector and training vectors. Each node finds the dot product of the input vector from previous layer and the corresponding weight vector. This dot product is transformed using an activation function called Radial Basis Function (RBF) and a smoothening parameter (σ). The smoothening parameter determines the change in the function for any kind of an increase or decrease in distance from a point. An RBF activation output is shown in Fig. 2.

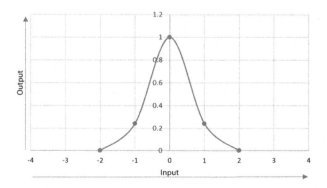

Fig. 2. RBF activation output

The output of each pattern node is given as:

$$\phi^I_J(x) = \frac{1}{(2\pi)^{d/2}\sigma^2} \exp\left(-\frac{1}{2\sigma^2}(x - x^i_j)^t(x - x^i_j)\right)$$

where x^i_j is the jth training set vector, σ is the smoothing factor, x is the new input and d is the number of input feature vectors. These nodes are partially connected to the third summation layer to which the resulting value is carried depending on the category of input pattern.

Summation Layer: The third layer of PNN architecture is the summation layer or class layer that accepts hidden layer output and sums it up to produce a probability density function (pdf) output.

The output of each summation node in terms of pdf is given as:

$$P(C_i|x) = \frac{1}{(2\pi)^{d/2}\sigma^d} \frac{1}{N_i} \sum_{j=1}^{N_i} \exp\left(-\frac{1}{2\sigma^2}(x - x^i_j)^t(x - x^i_j)\right)$$

where N_i is the total training set belonging to class C_i.

Decision Layer: This output layer has one node that choose the maximum among the probability densities corresponding to each class it receives from the pattern layer node. The binary neuron takes the advantage of Bayes decision rule to finalize the decision of the classification process. The decision layer node output of PNN architecture with m number of input classes can be written as:

$$\hat{C}(x) = arg_{j=1,...,m} \max\left\{\frac{1}{(2\pi)^{d/2}\sigma^d} \frac{1}{N_i} \sum_{j=1}^{N_i} \exp\left(-\frac{1}{2\sigma^2}(x - x^i_j)^t(x - x^i_j)\right)\right\}$$

3   Materials and Methods

3.1   Datasets and Feature Set Selection

The work was conducted as an extensive version of previously reported non-probabilistic artificial neural network method to investigate the inferences of probabilistic response in oligo classification [9]. Hence the same dataset along with the extracted structural and thermodynamic oligo: mRNA characteristics were adopted to build the present probabilistic model. A total of 493 datasets were accumulated to train and test the model. The positive sets were collected from experimentally published literatures [6, 8, 18, 19]. Negative sets were constructed manually by extending nucleotide rearrangements by insertion or deletion of nucleotides in target mRNA sequences. These negative set were further checked for its validity by aligning them against target mRNAs to confirm non oligo: mRNA association. Moreover, all duplicated and matched negative sequences were totally avoided from the list to assure a streamlined negative dataset. The list of oligo: target binding features selected in oligonucleotide prediction are tabulated in Table 1. Structural features were derived based on the secondary structure analysis of the duplex and thermodynamic features were derived based on the depth of oligo: target stability. The oligo stacking interactions determine the enthalpy and entropy values whereas the Gibbs free energy determines the minimum folding energy required for duplex stability. All these estimations were performed computationally with nearest neighbor (NN) parameters for Watson- Crick base pairs.

Table 1. List of features selected for oligo prediction

Sl. no.	Features	Category
1	CCAC	Sequence motif
2	TCCC	Sequence motif
3	ACTC	Sequence motif
4	GCCA	Sequence motif
5	CTCT	Sequence motif
6	Length	Structural
7	GC content	Structural
8	AU content	Structural
9	AU pairs	Structural
10	GC pairs	Structural
11	Seed score	Structural
12	Enthalpy (ΔH)	Thermodynamic
13	Entropy (ΔS)	Thermodynamic
14	Free energy (ΔG)	Thermodynamic
15	Melting temperatures	Thermodynamic
16	Molecular weight	Thermodynamic

Probabilistic Neural Network (PNN) Oligonucleotide Classifier

A multilayer feed forward four layered probabilistic neural network (PNN) was trained to develop the classifier model. The architecture of proposed PNN is depicted in Fig. 1. The input layer has 16 nodes to represent selected oligo features that are fully connected to second layer nodes. The pattern layer performs a non-linear operation on the input with radial basis activation function. A sigma parameter of 0.9 was fixed as the radial basis function showed significant peak at this values selected for experimentation. The output layer has one node to measure the desired output as a score of either 0 or 1. Thus for a given non-coding RNA sequence, the classifier distinguishes oligonucleotide sequence with a score of 1 from a non-oligo sequence with a score 0 at the output.

4   Results and Discussions

Model validation was performed using the test dataset elements exempted from training set. The test results can fall in four categories namely true positive (TP), true negative (TN), false positive (FP) and false negative (FN). The proposed PNN response was measured in general terms of accuracy, sensitivity and specificity as mentioned below.

Accuracy = (TP + TN)/(TP + FP + TN + FN)
Sensitivity = TP/(TP + FN)
Specificity = TN/(TN + FP)

The accuracy was found to be 96.99% with 98.72% sensitivity and 94.55% specificity. The proposed probabilistic neural network results were compared with the previously reported non-probabilistic oligonucleotide prediction methods. Table 2 lists this comparison between the proposed PNN and the existing architectures. From the analysis, it is learned that PNN method outperformed the previous attempts though the parameter contribution towards oligo identification remained the same suggesting the supremacy of probabilistic methods in oligonucleotide prediction strategies.

Table 2. Comparison of PNN with previous architectures in oligonucleotide prediction

Work	Method	Accuracy
Giddings et al.	ANN	53%
Chalk et al.	ANN	92%
Gustavo et al.	SVM	83.3%
Anusha et al.	ANN	92.48%
Proposed	PNN	96.99%

With the same test dataset, an additional validation of the proposed model was performed to scrutinize the real contributions of selected training parameters. This was accomplished with a permutation test conducted with various combinations of input

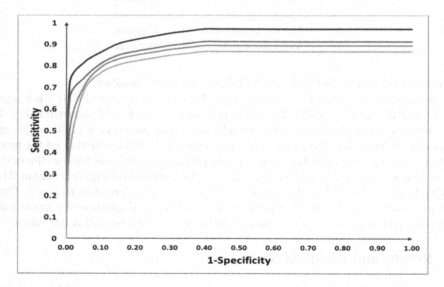

Fig. 3. ROC: oligonucleotide prediction. Blue line represents the ROC generated for all parameter set. Red line represents the ROC for thermodynamic parameters. Grey line represents the ROC for positional parameters and yellow line represents the ROC for structural parameters. (Color figure online)

feature vectors. The accuracy fluctuations showed decayed results in the respective ROCs compared with the original PNN accuracy.

The accuracy decay shift was prominent in case of structural parameters by 11%. It got further reduced by 8% when only sequence based parameters are included. Meanwhile the shift observed was only 6% with thermodynamic parameters alone which further strengthen the fact that thermodynamic features rule over structural parameters in oligo detection problems. The compared accuracy shifts for various parameter sets are depicted in Fig. 3.

5 Conclusions

The primary objective of the study concentrated on developing an optimal computational model that distinguish oligo nucleotides from other small non-coding RNA sequences. A probabilistic neural network based oligonucleotide identification methodology was developed with sixteen relevant parameters extracted based on oligo: target duplex interactions. On comparison with the previous non-probabilistic approaches using SVM and ANN, the new probabilistic method uniquely classifies oligos from non-oligos with promising accuracy of 96.99%. The high efficiency of the model can be accounted to the multilevel characteristics of probabilistic networks used instead of ANN and SVM since PNN conceives training samples heavily and can bend to Bayesian nature for any kind of over training. Also, the influence of relevant heterogeneous feature set selected may be considered as the second reason for this supreme accuracy. This study could also reveal the influence and significance of structural and thermodynamic features associated with oligo predictions.

Acknowledgements. This study is conducted at Department of Computational Biology and Bioinformatics, University of Kerala and financially supported by Kerala State Council for Science, Technology and Environment (KSCSTE).

References

1. Stein, C.A., Krieg, M. (eds.): Applied Antisense Oligonucleotide Technologies. Wiley-Liss, Hoboken (1998)
2. Crooke, S.T.: Progress in antisense technology. Annu. Rev. Med. **55**, 61–95 (2004)
3. Vickers, T.A., Wyatt, J.R., Freier, S.M.: Effects of RNA secondary structure on cellular antisense activity. Nucleic Acids Res. **28**, 1340–1347 (2000)
4. Evertsz, E.M., Au-Young, J., Ruvolo, M.V., Lim, A.C., Reynolds, M.A.: Hybridization cross-reactivity within homologous gene families on glass cDNA microarrays. Biotechniques **31**, 1182, 1184, 1186 (2001). Passim
5. Xu, W., Bak, S., Decker, A., Paquette, S.M., Feyereisen, R., Galbraith, D.W.: Microarray-based analysis of gene expression in very large gene families: the cytochrome P450 gene superfamily of Arabidopsis thaliana. Gene **272**, 61–74 (2001)

6. Giddings, M.C., Shah, A.A., Freier, S., Atkins, J.F., Gesteland, R.F., Matveeva, O.V.: Artificial neural network prediction of antisense oligodeoxynucleotide activity. Nucleic Acids Res. **30**, 4295–4304 (2002)
7. Chalk, A.M., Sonnhammer, E.L.: Computational anti-sense oligo prediction with a neural network model. Bioinformatics **18**, 1567–1575 (2002)
8. Gustavo, C., Chalk, A.M., Serrano-López, A.J., Martín-Guerrero, J.D., Sonnhammer, E.L.: Profiled support vector machines for antisense oligonucleotide efficacy prediction. BMC Bioinformatics **5**, 135 (2004)
9. Anusha, A.R., Vinodchandra, S.S.: Prediction of antisense oligonucleotides using structural and thermodynamic motifs. Bioinformation. **8**, 1162–1166 (2012)
10. Matveeva, O., Felden, B., Tsodikov, A., Johnston, J., Monia, B.P., Atkins, J.F., Gesteland, R.F., Freier, S.M.: Prediction of antisense oligonucleotide efficacy by in vitro methods. Nat. Biotechnol. **16**, 1374–1375 (1998)
11. Ding, Y., Lawrence, C.E.: Statistical prediction of single-stranded regions in RNA secondary structure and application to predicting effective antisense target sites and beyond. Nucleic Acids Res. **29**, 1034–1046 (2001)
12. Vinodchandra, S.S., Reshmi, G., Achuthsankar, S.N., Sreenathan, S.M., Radhakrishnapillai, M.: MTar: a computational microRNA target prediction architecture for human transcriptome. BMC Bioinform. **10**, 1–9 (2010)
13. Rahiman, A.A., Ajitha, J., Vinodchandra, S.S.: An integrated computational schema for analysis, prediction and visualization of piRNA sequences. In: Huang, D.-S., Bevilacqua, V., Prashan, P. (eds.) ICIC 2015. LNCS, vol. 9225, pp. 744–750. Springer, Heidelberg (2015). doi:10.1007/978-3-319-22180-9_75
14. Vinodchandra, S.S., Rejimoan, R., Shalini, R.: An ANN model for the identification of deleterious nsSNPs in tumor suppressor genes. Bioinformation **6**, 41–44 (2011)
15. Vinodchandra, S.S., Anand, H.S.: Artificial Intelligence and Machine Learning, 368 p. PHI Publishers, New Delhi (2014)
16. Mika, S., Rätsch, G., Weston, J., Schölkopf, B., Müller, K.R.: Fisher discriminant analysis with kernels. Neural Netw. Sig. Process. **9**, 41–48 (1999)
17. Mao, K.Z., Tan, K.C., Ser, W.: Probabilistic neural network structure determination for pattern classification. IEEE Trans. Neural Netw. **11**, 1009–1016 (2000)
18. Matveeva, O.V., Tsodikov, A.D., Giddings, M., Freier, S.M., Wyatt, J.R., Spiridonov, A.N., Shabalina, S.A., Gesteland, R.F., Atkins, J.F.: Identification of sequence motifs in oligonucleotides whose presence is correlated with antisense activity. Nucleic Acids Res. **28**, 2862–2865 (2000)
19. Xiaochen, B., Shaoke, L., Daochun, S., Wenjie, S., Jing, Y., Shengqi, W.: Selection of antisense oligonucleotides based on multiple predicted target mRNA structures. BMC Bioinform. **7**, 122 (2006)

Scalability of a Genomic Data Analysis in the BioTest Platform

Krzysztof Psiuk-Maksymowicz[1,2]($\boxtimes$), Dariusz Mrozek[3], Roman Jaksik[1,2],
Damian Borys[1,2], Krzysztof Fujarewicz[1,2], and Andrzej Swierniak[1,2]

[1] Institute of Automatic Control, Silesian University of Technology,
ul. Akademicka 16, 44-100 Gliwice, Poland
`krzysztof.psiuk-maksymowicz@polsl.pl`
[2] Biotechnology Centre, Silesian University of Technology,
ul. Krzywoustego 8, 44-100 Gliwice, Poland
[3] Institute of Informatics, Silesian University of Technology,
ul. Akademicka 16, 44-100 Gliwice, Poland

Abstract. BioTest platform is dedicated for the processing of biomedical data that originate from various measurement techniques. This includes next-generation sequencing (NGS), that focuses the attention of researchers all of the world due to its broad possibilities in determining the structure of the DNA and RNA. However, the analysis of data provided by NGS requires large disk space, and is time-consuming, becoming a challenge for the data processing systems. In this paper, we have analyzed the possibility of scaling the BioTest platform in terms of genomic data analysis and platform architecture. Scalability tests were carried out using next-generation sequencing data and relied on methods for detection of somatic mutations and polymorphisms in the human DNA. Our results show that the platform is scalable, allowing to significantly reduce the execution time of performed calculations. However, the scalability capabilities depend on the experiment methodology and homogeneity of resources required by each task, which in NGS studies can be highly variable.

Keywords: Scalability · Bioinformatics · NGS data analysis · Computational platform · HPC cluster · Workload manager

1 Introduction

In recent years huge amounts of biomedical data are generated by biological laboratories as a result of medical examinations and screening processes performed in hospitals. Patients suspected of having a certain disease, *e.g.* cancer, undergo a preliminary screening usually based on blood examination or medical imaging techniques like endoscopy or mammography. Patients diagnosed with a certain disease undergo a set of clinical-analytical processes in order to provide more insights on the type of the disease and its location, helping to select an appropriate treatment scenario. A typical workflow consists of several steps that

© Springer International Publishing AG 2017
N.T. Nguyen et al. (Eds.): ACIIDS 2017, Part II, LNAI 10192, pp. 741–752, 2017.
DOI: 10.1007/978-3-319-54430-4_71

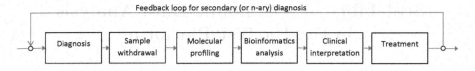

Fig. 1. Typical workflow for oncological diagnostics of patients.

are presented in Fig. 1. One of the important steps of the workflow is molecular profiling, which may comprise various single molecule-oriented methods or in many cases high-throughput technologies based on genome, transcriptome or proteome profiling, including microarray experiments, next-generation sequencing (NGS), mass spectrometry, and others. Results of the molecular profiling must be processed and analyzed in dedicated pipelines by means of specialized algorithms provided by bioinformatics community. These pipelines are dedicated to perform a certain type of data processing usually comprising of the methods used to control the quality of data, standardize the measurements and select features that show significant variations in the signal intensity. This provides a way to identify unique biosignatures and biomarkers that can be used in the diagnosis, prognosis, and therapeutic prediction of a disease. Patients' treatment may be then adapted to a specific case, and may include targeted therapy or combination of therapies by following a specific therapeutic pattern.

Undoubtedly, effective and efficient analysis of data coming from molecular profiling is crucial for the detailed diagnosis, prognosis, and therapeutic prediction of a disease. Meaningful conclusions can be drawn only by the use of sophisticated methods for biomedical and molecular data analysis. However, for many of the biological laboratories that are hosted in hospitals the analysis of such huge volumes of biomedical data can be a challenge. Frequently, these laboratories do not possess appropriate hardware infrastructure to process and analyze terabytes of biomedical data in a reasonable time or they do not hire qualified personnel that is able to perform such analyzes. For those reasons, various systems were created, including those presented in [1,6,10,15–17], that support the scientists in their biomedical analyses. Such systems are capable of storing large amounts of various types of biomedical data and adapting to the current processing requirements by scaling computations on high-performance computing (HPC) clusters, located in private data centers, on virtualized platforms, or public Clouds. The BioTest platform [18] is one of these systems, integrating unique capabilities of keeping raw data in a dedicated repository, building complex data processing workflows with Galaxy Server, storing multidimensional structured data for analysis and reporting in biomedical data warehouse, providing hard and long-running computations on its HPC cluster in virtualized environment, and delivering sophisticated reports through graph analysis and Apache Spark computations. The platform helps scientists in organizing data from many biomedical experiments and from many resources, allowing for integrated genomic, transcriptomic and proteomic analysis. The results of multi-step analysis (built in the system as data workflows) are kept in data warehouse, which helps in higher order studies. It enables verification of various scientific hypotheses by checking whether the use of the same bioinformatics tools but with different parameters or different version of reference data will give the same results. It might be particularly useful in the case of launching earlier analyzes with up-to-date settings. An important feature of the platform is that it can be integrated with hospital information system (HIS) in order to provide anonymized clinical data. Another important feature is that the BioTest platform allows for addition of custom analysis tools and workflows.

One of the key features of the BioTest platform is its capability of scaling computations for various types of biomedical data. In the field of the DNA sequencing the topic of scalability is very actual, especially when it comes to the whole genome sequencing that provides the largest amounts of data. Many recent, scalable computational tools for genomic analysis rely on various techniques of parallelization of computations, including the use of MapReduce programming paradigm [4] and/or Cloud computing architectures. Genome Analysis Toolkit (GATK) [11] that is used for secondary analyses of NGS data, supports an automatic shared-memory parallelization on a single multi-CPU/multi-core machine and distributed parallelization scheme on large computational clusters. In our works reported in this paper, we make use of both parallelization techniques while scaling our computations on the BioTest platform. Distributed parallelization scheme was also utilized in Halvade [3] that implements GATK-based DNA sequencing analysis pipeline for variant calling on 15-node Hadoop cluster with 360 CPUs in total, bringing a significant reduction of the analysis time. Scalability issue for NGS data analyses was also rised in Refs. [10,20]. These solutions are devoted to *ad hoc* secondary analyses of NGS data with MapReduce procedures on Apache Spark computational framework [20] and tertiary analyses of NGS data with Apache Hadoop [10]. In both cases, authors reported a significant reduction of processing time while scaling their computations.

In this paper, we show scalability of computations performed on the BioTest platform during NGS data analysis, involving detection of somatic point mutations and identification of inherited polymorphisms. This paper is organized as follows. Section 2 describes available computational resources and applied methodology. Section 3 contains the results of performance tests, and Sect. 4 a short discussion of obtained results.

2   Computational Resources and Methods

2.1   Platform Architecture

BioTest platform is comprised of a few essential components: a software building the whole context of the experiment, Galaxy Server, computational cluster, graph database, and data warehouse. The general architecture of the system is presented in Fig. 2.

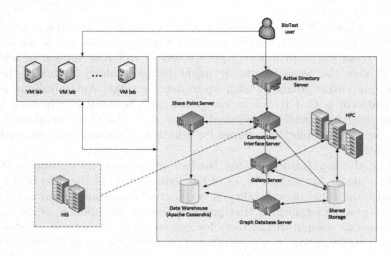

Fig. 2. Architecture of BioTest platform.

A user can start working with the system by building the context of the study, connecting experimental data into a single coherent project. This is done in the Context User Interface view. Next step involves choosing an appropriate workflow, for the data provided by experiments conducted in the study, using the Galaxy Server interface. If needed, the version of reference databases used by individual tools can be specified at this stage. The most computationally expensive calculations are done by placing them in the queue system on the HPC computer cluster, which is connected to the Galaxy Server. The essential part of the partial or final results of calculations is stored in Apache Cassandra data warehouse to perform further analysis or aggregations of the results across many dimensions. Results can be visualized and distributed by Windows Share Point server, for example in Excel Power Pivot add-on or through the graph database located near the Context User Interface. The project and context data are stored for Galaxy Server and HPC cluster using a shared file system. The user management is simplified using domain services (MS Active Directory) to store users information across all internal platforms of the system. The user can access the system directly through the Context User Interface or by using virtual laboratories. Those laboratories are virtual machines specially prepared to facilitate usage of the system and where all tools needed to fully exploit the system are installed. The number of those virtual accesses will change depending on the number of system users. All machines mentioned above, which are the basis of presented architecture, are virtualized. This solution allows us to scale resources needed to keep the system responsive as the number of users grows up. When the performance of the system decreases, physical resources can be added to run additional virtual machines. The system ensures also scalability on the level of computational cluster resources. This scalability is implemented by increasing the quantity of computational nodes that will vary with the load of the system.

2.2   Data Processing Algorithms

In the paper we present two types of NGS analysis. The first one concerns finding somatic point mutations in the entire genome of cancer cells (whole genome sequencing - WGS). This is done by comparing the newly sequenced genome of both normal and cancer cells, from a single individual, to the reference genome used in all studies. This allows to identify mutations that are specific to the cancer cells, acquired by the patient during his life.

Second analysis, focuses on detecting single nucleotide polymorphisms and indels (nucleotide insertions and deletions) based on whole exome sequencing data (WES), which allows to identify inherited variations. Unlike the previous analysis this is a complete study which also includes quality control and data pre-processing. However the input data are much smaller in size, mainly due to the fact that only a part of the genome formed by exons was sequenced. The comparison between WES with the most recent PCR-free WGS sequencing method is presented in [13].

Detection of Somatic Point Mutations. Somatic point mutation (SPM) calling from matched tumor-normal patient samples is a critical step for cancer genome characterization and clinical genotyping. Over the last few years, NGS has become a popular strategy for genotyping, enabling more precise mutation detection compared to the traditional methods due to its high resolution and high throughput [14]. Nevertheless, detection of somatic mutations is still challenging, especially for low-allelic-fraction variants caused by tumor heterogeneity, copy number alteration, and sample degradation. In our analysis, we used MuTect algorithm [2] that accounts for all those factors, showing high precision and sensitivity of somatic variant detection. There are many methods for finding SPMs, including: GATK UnifiedGenotyper [5], MuTect [2], Strelka [19], SomaticSniper [8] and VarScan2 [7] and others, based on distinct scoring functions and variant filtering approaches. The interested reader is referred to the work by Xu *et al.* [21] where a comparison of mentioned methods is performed.

MuTect is a method developed for detecting somatic mutations with very low allele fractions in NGS data using a Bayesian classifier approach. The method includes pre-processing aligned reads separately in tumor and normal samples and post-processing resulting variants by applying an additional set of filters. MuTect takes as input NGS data from tumor and normal samples and, after removing low-quality reads, determines whether there is evidence for a variant beyond the expected random sequencing errors. Variant detection in the tumor sample is based on a Bayesian classifier. For each site the reference allele is denoted by $r \in \{A, C, G, T\}$ and the base of read i ($i = 1...d$) that covers the site and the probability of error of that base call are denoted by b_i and e_i. To call a variant in the tumor the data are expresed by two models: (i) model M_0 in which there is no variant at the site and all nonreference bases are explained by sequencing noise, and (ii) model M_f^m in which a variant allele m truly exists at the site with an allele fraction f and, as in M_0, reads are also subject to sequencing noise. M_0 is equivalent to M_f^m with $f = 0$. The likelihood of the model M_f^m is given by

$$L(M_f^m) = P(\{b_i\}|\{e_i\}, r, m, f) = \prod_{i=1}^{d} P(\{b_i\}|e_i, r, m, f), \qquad (1)$$

assuming the sequencing errors are independent across reads. If all substitution errors are equally likely, that is, occur with probability $e_i/3$, we obtain

$$P(\{b_i\}|e_i, r, m, f) = \begin{cases} f\frac{e_i}{3} + (1-f)(1-e_i) & \text{if } b_i = r \\ f(1-e_i) + (1-f)\frac{e_i}{3} & \text{if } b_i = m \\ \frac{e_i}{3} & \text{otherwise} \end{cases}. \qquad (2)$$

Variant detection is performed by comparing the likelihood of both models and their ratio, that is, the LOD score, with a decision threshold ($\log_{10} \delta_T$)

$$LOD_T(m, f) = \log_{10}\left(\frac{L(M_f^m)P(m, f)}{L(M_0)(1 - P(m, f))} \right) \geq \log_{10} \delta_T. \qquad (3)$$

If the LOD score exceeds the decision threshold, m is declared as a candidate variant at the site. Setting δ_T to 2 ensures that we are at least twice as confident that the site is variant as compared to noise. The rewritten LOD score is as follows

$$LOD_T(m, f) = \log_{10}\left(\frac{L(M_f^m)}{L(M_0)} \right) \geq \log_{10} \delta - \log_{10}\left(\frac{P(m, f)}{(1 - P(m, f))} \right) = \theta_T. \qquad (4)$$

Specific determination of necessary parameters is present in [2].

Candidate variant sites are then passed through six filters (Proximal gap, Poor mapping, Triallelic site, Strand bias, Clustered position, Observed in Control) to remove artifacts. Next, a panel of normal samples filter is used to screen out remaining false positives caused by rare error modes only detectable in additional samples. Finally, the somatic or germ-line status of passing variants is determined using the matched normal sample. The entire process is performed for sequencing data in BAM format, that includes the information about short sequence fragments (reads), 30–100 nucleotides long, and their position in the reference genome. For this analysis we used 77 sample pairs (tumor, normal tissue) that originate from both WES and WGS studies, with sequencing read numbers ranging from 10^7 to over 10^9. Since MuTect-based data processing considers all possible positions in the human genome ($\sim$3 billion) the analysis is very time consuming, additionally depending on the number of sequencing reads gathered for each sample.

NGS Data Preprocessing and Variant Discovery. Analysis of inherited DNA sequence variants has become one of the basic tools that allows to identify relationship between the genotype and phenotype. By comparing the genomes of individuals with different disease history it is possible to determine genetic factors involved in their development and effectiveness of the therapy. This further allows to determine the susceptibility of specific persons to certain malignancies, and classify individuals into risk groups, covered by a selected prevention program. The development of individualized therapy is, among other things, also based on variants discovery, allowing to design drugs that target specific cellular mechanism, and at the same time minimizing side effects, which are often patient-specific. Data processing methodology used for variant discovery in this study is based on Genome Analysis Toolkit (GATK) guidelines [11], which are currently widely used in the largest genomic projects. The most important processing steps of the workflow are as follows:

1. Align sequencing reads obtained for individual patients to the reference genome, using BWA mem algorithm [9] - this is one of two most time consuming steps, which involves determining the position of short sequencing reads in the reference genome. This later allows to identify differences between the DNA of analyzed individual and selected reference genome.
2. Remove duplicated reads based on the MarkDuplicates algorithm from the Picard tool set - this step allows to mark reads that are likely an artifact of the DNA amplification process and omit them in the subsequent processing steps.
3. Recalibrate read quality scored using BaseRecalibrator, which is a part of the GATK - each base read in the sequencing process has a quality score assigned that determines the probability of making a mistake, which is assigned by the sequencing machine. However such scores are subject to various sources of systematic technical error, which is reduced by applying the BaseRecalibrator algorithm.
4. Identify SNVs (single nucleotide variants) and indels (nucleotide insertions and deletions) using GATKs HaplotypeCaller - this is the second most time consuming step. It utilizes an algorithm that works on a similar principle to MuTect. However, it is not based on case-control comparison (e.g. tumor vs normal cells) comparing the DNA obtained for a single patient to the reference genome during the variant detection.
5. Annotate all variants using Variant Effect Predictor [12] - this step allows to determine the association of all variants with genes, and the effect each variant has on the protein encoded by this gene.

All of the steps are conducted independently for each patient allowing them to be executed simultaneously. Further steps are conducted based on combined sample knowledge. However, due to their relatively short execution time and limited parallelization possibilities, they are omitted in this work. The data used in this study were obtained from 37 individual patients whose exome was sequenced using Illumina HiSeq 4000 sequencer providing 133 GB of compressed data.

2.3   Parallelization Techniques

Testing performance and scalability of the calculations related to NGS data analysis was carried out on the BioTest platform integrated with the Ziemowit HPC computing cluster (www.ziemowit.hpc.polsl.pl). The cluster makes use of both, TORQUE and SLURM workload managers that use two independent sets of nodes. We performed two types of tests, in the first one each task represents analysis conducted on an individual sample using the MuTect algorithm. In the second test each sample undergoes a complete processing according to the GATK recommendations using a set of various tools. Each task is assigned to an individual node that has a total of 12 cores. MuTect and all of the tools used in the GATK workflow that make up more than 95% of the processing time are parallelized, and use all of the cores available in a single node. However, since the number of cores is highly limited the scalability tests are based on parallelizing the execution of tasks across individual nodes, independently for each sample. The tasks were added to the queue in a random order, however we also consider a scenario in which they are ordered according to the LPT rule (Longest Processing Time), determined based on the number of reads in the input files (highest number of reads first).

3   Results

The first series of performance tests was carried out for the set of 77 sample pairs analyzed with MuTect. In Fig. 3 (left panel) we can observe changes of the execution time depending on the number of nodes of the computing cluster when scaling computations on the BioTest platform. We can see a significant decrease of the execution time while scaling the cluster from one to sixteen nodes. Above the sixteen nodes the increase in the computation speed is negligible, which is a result of differentiated execution times for the respective tasks. The reduction in total computation time was achieved by analyzing individual samples in parallel on separate cluster nodes. The algorithm itself was also parallelized, however due to its limitations only by multi-threaded execution on multi-core CPUs of the cluster, using individual cores of each node. The ability to parallelize the GATK workflow are even more limited, since it comprises of a series of processing steps that need to be conducted in a specific order, and each step needs to be finished before the next one can be executed. In many cases the tasks processed at the end do not use all of the available resources, since the number of remaining tasks at some point is lower than the number of nodes.

The second series of performance tests was carried out with the use of 37 samples and TORQUE workload manager. Results of the performance test, when scaling out computations on the HPC cluster in BioTest platform, are shown in Fig. 3 (right panel). Significant decrease of the execution time was achieved while scaling the cluster from one to twenty two nodes. Above the twenty two nodes the increase in the computation speed is no longer significant, which again is a result of differentiated execution times for the respective tasks.

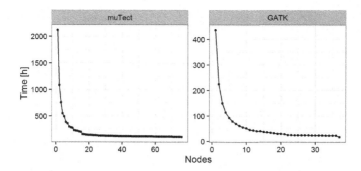

Fig. 3. The execution time of the MuTect algorithm (left panel) and the GATK work-flow (right panel) depending on the number of HPC cluster worker nodes.

The specificity of the execution times for both types of analysis are shown in Fig. 4. It can be clearly seen that execution times for the MuTect algorithm are much more unbalanced comparing to the execution times of the GATK algorithm. Most of the tasks in MuTect have short execution times with some showing a 10-fold increase, whereas task execution times for GATK algorithm are more uniformly distributed. The differences in execution times of MuTect result from the fact that some samples originate from whole genome (WGS) and others from whole exome sequencing (WES), with execution times for WGS being much longer.

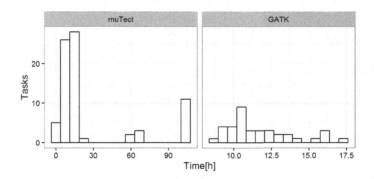

Fig. 4. Task execution times for MuTect and GATK algorithms.

Distribution of task execution times for particular algorithms, shown in Fig. 4, have a direct impact on the speedup curves (Fig. 5). This impact is even more noticeable in the case when the tasks are launched not randomly, but in an ordered way, when scheduled according to LPT rule. The order is consistent with a length of execution time of the task. Ordering of the tasks is only possible when we can *a priori* assume the execution time for successive tasks, which in this

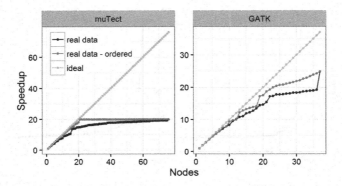

Fig. 5. Scalability of MuTect and GATK algorithms depending on different number of worker nodes. Each plot contains ideal, linear speedup; speedup curve for real data; and speedup curve for ordered data (LPT).

case is possible, since it is proportional to the number of reads in the input files. The decrease in a total processing time resulting from LPT scheduling is more significant for the first dataset analyzed with MuTect. This results from lower homogeneity of the execution times, which can lead to a situation, where one of the longest tasks is executed at the end extending the entire analysis. However, even with scheduling, the scalability is highly limited and the gain can be no longer observed above 21 nodes, since despite increasing the number of nodes the longest tasks prolong the analysis to the level at which all remaining tasks of shorter length can be executed without affecting the total execution time.

4   Discussion and Concluding Remarks

Next-generation sequencing is a technique for large-scale sequencing of genomic sequences that delivers information of genetic changes that can be crucial for the diagnosis and treatment of patients affected by serious diseases, like cancer. However, it produces large volumes of data that must be processed and analyzed in an efficient way. Parallelization of computations related to the processing and analysis of the NGS data, involving detection of somatic point mutations and identification of inherited genetic variations, is one of the primary approaches that may speed up the entire process of getting useful information that is hidden in long genomic sequences. Results of our performance experiments for MuTect and GATK analysis methods show that due to the nature of the NGS data and the nature of performed analysis processes, we can significantly accelerate computations on parallel platforms, like BioTest. Best speedups were achieved when scaling computations on HPC cluster up to sixteen nodes for MuTect, and up to twenty two nodes for GATK. Similarly to many other systems that process various types of biomedical data, like those presented in [3,15–17,20], scalability of the BioTest is not ideal. Speedups for both methods are sublinear due to the overhead introduced by the necessity of managing the growing number of processes in the HPC cluster. Uneven distribution of workload resulting from large differences in execution time of individual tasks is the most significant limiting factor of the scalability of BioTest platform. Despite that, significant time savings that can be achieved are invaluable in the process of medical diagnostics and treatment. BioTest is capable of running pre-designed workflows with a constant set of parameters and reference libraries for each sample provided, additionally monitoring any custom changes that can affect the final outcomes. This feature is very important for clinical applications as it allows to control the reproducibility of the studies. The system supports methods commonly used in basic sciences to identify regulatory interactions and genomic alterations of the malignant cells by utilizing high throughput measurement methods. However, the functionality offered by the system is also oriented on studying various effects on a single molecule scale, important in the clinical applications.

Acknowledgements. This work was partially supported by The National Centre for Research and Development grant No. PBS3/B3/32/2015 and Strategmed2/267398/4/ NCBR/2015. Presented system was developed and installed on the infrastructure of the Ziemowit computer cluster (www.ziemowit.hpc.polsl.pl) in the Laboratory of Bioinformatics and Computational Biology, The Biotechnology, Bioengineering and Bioinformatics Centre Silesian BIO-FARMA, created in the POIG.02.01.00-00-166/08 and expanded in the POIG.02.03.01-00-040/13 projects.

References

1. Bensz, W., et al.: Integrated system supporting research on environment related cancers. In: Król, D., Madeyski, L., Nguyen, N.T. (eds.) Recent Developments in Intelligent Information and Database Systems. SCI, vol. 642, pp. 399–409. Springer, Heidelberg (2016). doi:10.1007/978-3-319-31277-4_35
2. Cibulskis, C., Lawrence, M.S., Carter, S.L., Sivachenko, A., Jaffe, D., Sougnez, C., Gabriel, S., Meyerson, M., Lander, E.S., Getz, G.: Sensitive detection of somatic point mutations in impure and heterogeneous cancer samples. Nat. Biotechnol. **31**, 213–219 (2013)
3. Decap, D., Reumers, J., Herzeel, C., Costanza, P., Fostier, J.: Halvade: scalable sequence analysis with MapReduce. Bioinformatics **31**(15), 2482–2488 (2015)
4. Dean, J., Ghemawat, S.: MapReduce: simplified data processing on large clusters. Commun. ACM **51**, 107–113 (2008)
5. DePristo, M.A., Banks, E., Poplin, R., Garimella, K.V., Maguire, J.R., Hartl, C., Philippakis, A.A., del Angel, G., Rivas, M.A., Hanna, M., McKenna, A., Fennell, T.J., Kernytsky, A.M., Sivachenko, A.Y., Cibulskis, K., Gabriel, S.B., Altshuler, D., Daly, M.J.: A framework for variation discovery and genotyping using next-generation DNA sequencing data. Nat. Genet. **43**, 491–498 (2011)
6. Hung, C.L., Lin, Y.L.: Implementation of a parallel protein structure alignment service on cloud. Int. J. Genomics **439681**, 1–8 (2013)
7. Koboldt, D.C., Zhang, Q., Larson, D.E., Shen, D., McLellan, M.D., Lin, L., Miller, C.A., Mardis, E.R., Ding, L., Wilson, R.K.: VarScan 2: somatic mutation and copy number alteration discovery in cancer by exome sequencing. Genome Res. **22**, 568–576 (2012)

8. Larson, D.E., Harris, C.C., Chen, K., Koboldt, D.C., Abbott, T.E., Dooling, D.J., Ley, T.J., Mardis, E.R., Wilson, R.K., Ding, L.: SomaticSniper: identification of somatic point mutations in whole genome sequencing data. Bioinformatics **28**, 311–317 (2011)

9. Li, H.: Aligning sequence reads, clone sequences and assembly contigs with BWA-MEM. arXiv:1303.3997 (2013)

10. Masseroli, M., Canakoglu, A., Ceri, S.: Integration and querying of genomic and proteomic semantic annotations for biomedical knowledge extraction. IEEE/ACM Trans. Comput. Biol. Bioinf. **13**(2), 209–219 (2016)

11. McKenna, A., Hanna, M., Banks, E., Sivachenko, A., Cibulskis, K., Kernytsky, A., Garimella, K., Altshuler, D., Gabriel, S., Daly, M., DePristo, M.A.: The genome analysis toolkit: a mapreduce framework for analyzing next-generation DNA sequencing data. Genome Res. **20**, 1297–1303 (2010)

12. McLaren, W., Gil, L., Hunt, S.E., Riat, H.S., Ritchie, G.R.S., Thormann, A., Flicek, P., Cunningham, F.: The ensembl variant effect predictor. Genome Biol. **17**(1), 122 (2016)

13. Meienberg, J., Bruggman, R., Oexle, K., Matyas, G.: Clinical sequencing: is WGS the better WES? Hum. Genet. **135**, 359–362 (2016)

14. Metzker, M.L.: Sequencing technologies - the next generation. Nat. Rev. Genet. **11**(1), 31–46 (2010)

15. Mrozek, D., Małysiak-Mrozek, B., Kłapciński, A.: Cloud4Psi: cloud computing for 3D protein structure similarity searching. Bioinformatics **30**(19), 2822–2825 (2014)

16. Mrozek, D., Gosk, P., Małysiak-Mrozek, B.: Scaling Ab initio predictions of 3D protein structures in Microsoft Azure cloud. J. Grid Comput. **13**, 561–585 (2015)

17. Mrozek, D., Daniłowicz, P., Małysiak-Mrozek, B.: HDInsight4PSi: boosting performance of 3D protein structure similarity searching with HDInsight clusters in Microsoft Azure cloud. Inf. Sci. **349–350**, 77–101 (2016)

18. Psiuk-Maksymowicz, K., Placzek, A., Jaksik, R., Student, S., Borys, D., Mrozek, D., Fujarewicz, K., Swierniak, A.: A holistic approach to testing biomedical hypotheses and analysis of biomedical data. Commun. Comput. Inf. Sci. **616**, 449–462 (2016)

19. Saunders, C.T., Wong, W.S., Swamy, S., Becq, J., Murray, L.J., Cheetham, R.K.: Strelka: accurate somatic small-variant calling from sequenced tumor-normal sample pairs. Bioinformatics **28**, 1811–1817 (2012)

20. Wiewiorka, M.S., Messina, A., Pacholewska, A., Maffioletti, S., Gawrysiak, P., Okoniewski, M.J.: SparkSeq: fast, scalable, cloud-ready tool for the interactive genomic data analysis with nucleotide precision. Bioinformatics **30**(18), 2652–2653 (2014)

21. Xu, H., DiCarlo, J., Satya, R.V., Peng, Q., Wang, Y.: Comparison of somatic mutation calling methods in amplicon and whole exome sequence data. BMC Genom. **15**, 244 (2014)

Quantifying the Effect of Metapopulation Size on the Persistence of Infectious Diseases in a Metapopulation

Cam-Giang Tran-Thi[1]([✉]), Marc Choisy[2], and Jean Daniel Zucker[3]

[1] UPMC Paris, UMI 209 UMMISCO (IRD-UPMC), Bondy Cedex, France
`thi.tran11@etu.upmc.fr`
[2] UMR MIVEGEC, 34394 Montpellier, France
`marc.choisy@ird.fr`
[3] UMI 209 UMMISCO, IRD Bondy, Bondy Cedex, France
`Jean-Daniel.Zucker@ird.fr`

Abstract. We investigate the special role of the three-dimensional relationship between periodicity, persistence and synchronization on its ability of disease persistence in a meta-population. Persistence is dominated by synchronization effects, but synchronization is dominated by the coupling strength and the interaction between local population size and human movement. Here we focus on the quite important role of population size on the ability of disease persistence. We implement the simulations of stochastic dynamics in a susceptible-exposed-infectious-recovered (SEIR) metapopulation model in space. Applying the continuous-time Markov description of the model of deterministic equations, the direct method of Gillespie [10] in the class of Monte-Carlo simulation methods allows us to simulate exactly the transmission of diseases through the seasonally forced and spatially structured SEIR meta-population model. Our finding shows the ability of the disease persistence in the meta-population is formulated as an exponential survival model on data simulated by the stochastic model. Increasing the meta-population size leads to the clearly decrease of the extinction rates local as well as global. The curve of the coupling rate against the extinction rate which looks like a convex functions, gains the minimum value in the medium interval, and its curvature is directly proportional to the meta-population size.

1 Introduction

Infectious diseases are today one of the leading causes of death worldwide among children and adolescents, particularly in low income countries. Finding solutions for extincting infectious diseases has become major interest for scientists. The nature of the extinction is strongly dependent on the relationship between population size and individual density in population. Estimating well a population threshold for extinction or a critical community size (CCS) below which infections do not persist, leads to the proper implementations for controlling emerging and re-emerging pathogens, in particular for preventing stochastic fade-out

© Springer International Publishing AG 2017
N.T. Nguyen et al. (Eds.): ACIIDS 2017, Part II, LNAI 10192, pp. 753–764, 2017.
DOI: 10.1007/978-3-319-54430-4_72

during the early stages of an epidemic. Since the 1960s, there have been many studies about the level of CCS of infectious diseases. As for measles, the finding of Black in Island studies shows a CCS of above 500,000 [6], while the data of England and Wales pointed out a lower CCS of 250,000–300,000 [4]. We fought out there are contradictions with CCS for the same disease in different communities. Because the abundance of diseases depends not only linearly on host population size [13], but also affected by many different factors, such as demographic and social structures [9], birth and death rates which are main driver of disease persistence [7]. Therefore, we will use epidemic models to examine the relationships between extinction and population size for spatial heterogeneity in transmission probability, in terms of a meta-population of multi-subpopulation. We will analyze the level of CCS for a meta-population and at the same time, and explore the key ingredients in determining the level of the CCS of a meta-population: seasonality, spatial structure and the phase difference in the infectious and latent stages among sub-populations. Local population dynamics and spatial connection strength are two key factors determining the ability of disease persistence. Understanding persistence and its mechanistic determinants are two golden keys to predict either the eradication of disease or even further the conservation of endangered populations in a general ecological context.

The probability of disease persistence in a meta-population strongly depends on the spatial heterogeneity and control. This is proven in the "cities and villages" models of Anderson and May [2]. The nature of this models are of a meta-population models where there are coupled sub-populations of different host densities. Measles spatial dynamics suggests that the infection of measles really spread in space and from cities to surrounding areas. Then, the infection rate of the cities of high host densities is bigger than those of the lower host densities and the diseases have the tendency to diffuse from the cities of the high densities to the areas of the lower densities. However, the obtained results of the disease transmission from one region to the other are only analyzed by basing on the weekly measles notifications and associated demographic data from England and Wales [11,14]. There are not any exact meta-population models that has large enough spatial and long temporal scales to clearly model the migration among areas and particularly the disease diffusion. This is why we will build a big meta-population model that represents the complex regional patterns among areas of a nation. We will point out that the persistence probability not only depends on the the spatial and demographic heterogeneity, but also can be measured and formulated by these heterogeneity. In addition to the result of Grenfell [14], the different areas of the same population size has the same propensity for local extinction of infection, we will present the local and global extinction of infection is not only dominated by the population size, but also is changed by the seasonal and spatial heterogeneity.

Here, we use the seasonally forced stochastic SEIR meta-population model to quantify the effect of the distribution of population size on the extinction probabilities in a meta-population. We consider the effect of the phase difference of seasonality on the persistence probability. In order to strengthen the exactitude of simulations, we set the population size around the estimated level of CCS by previous authors [2,11,14]. We use the parameter values that are estimated for the measles epidemic. Because, the persistence and the extinction of the measles have been studied for more than 100 years [2,3,16,20,23], human is still faced with measles with more complex morbidity, particularly in low income countries [9,20]. The disease persistence depends on both the local population size and the the spatial connectivity. That is the reason why in this work, we do not focus on finding the level of CCS for a meta-population. We will concentrate on measuring the ability of the disease persistence against the meta-population size when the sub-populations are in either the synchrony or the asynchrony, as well as when the connection strength changes. We can formulate the relation between the extinction rates and the meta-population size in different models.

2 Methods

2.1 Deterministic Meta-population Model

We study a seasonally forced and spatially structured stochastic meta-population model of n sub-populations. At a sub-population i of the population size N_i, its disease fluctuation can be deterministically presented by the following set of differential equations [2] with environmental and movement-based transmission as follows:

$$\frac{dS_i}{dt} = \mu N_i - \lambda_i S_i - \mu S_i \tag{2.1}$$

$$\frac{dE_i}{dt} = \lambda_i S_i - \mu E_i - \sigma E_i \tag{2.2}$$

$$\frac{dI_i}{dt} = \sigma E_i - \mu I_i - \gamma I_i \tag{2.3}$$

$$\frac{dR_i}{dt} = \gamma I_i - \mu R_i \tag{2.4}$$

where S_i, E_i, I_i and R_i are the numbers of susceptible, exposed, infectious and recovered respectively. Birth and death rates are equal of μ. Susceptible individuals become infected with the force of infection λ_i, infectious after a latency period of an average duration of $1/\sigma$ and recover at the rate γ.

The force of infection λ in a meta-population is a quite important parameter that is strongly affected by many different factors: seasonality, periodicity and demographic factors. Here, we successfully formulated the force of infection of a sub-population i in a meta-population that closely includes the seasonally forced and spatially structured features:

$$\lambda_i(t) = \sum_j \rho_{ij} \kappa_j \log \left[1 - \sum_{k=1}^{M} \left(\frac{|I_k(t)|}{N_k(t)} \times c_{ik} \times \xi_{jk} \right) \right] \tag{2.5}$$

where $c_{i,k}$ is the probability that a susceptible individual native from i being in contact with another infected individual native from k gets infected. ξ_{jk} refers to the probability that an individual y meeting x in the sub-population C_j comes from the sub-population C_k. κ_j is the average number of contacts per unit of time a susceptible will have when visiting city j. $\rho_{i,j}$ is denoted as the probability that an individual from sub-population i visits sub-population j. Of course, $\sum_{j=1}^{M} \rho_{ij} = 1$ and $0 \leqslant c_{ij}, \xi_{ij}, \rho_{ij} \leqslant 1$.

Describing the strength of connection ρ in a meta-population (sub-populations connected by individual movement) is a quite difficult and complex work. Because, we need a large knowledge about geographic features. Here we simplify the conditions of the spatial structures. We investigate the null model (model 0) that looks like the original Levins's model [22]. Where all the sub-populations are at the same distance from each other and have the same population size N.

Additionally, to present the environmental forcing, we focus on describing the seasonal periodicity and the seasonal phase difference among sub-populations in a meta-population. Seasonality is an annually periodic function of time [15], and the average number of contacts per unit of time κ_i is seasonally forced [1]. As a result, for the sub-population i:

$$\kappa_i(t) = \kappa_{i0} \left[1 + \kappa_{i1} \cos \left(\frac{2\pi t}{T} + \varphi_i \right) \right] \qquad (2.6)$$

where t is the time, κ_{i0} and κ_{i1} are the mean value and amplitude of the average contact rate κ_i at which a susceptible will have when visiting city i per unit of time, T and φ_i are the period and the phase of the forcing. The cosine function is used to illustrate the seasonal periodicity and the phase of the forcing to present the seasonal phase differences of the different areas in a meta-population.

We use a new parameter $\varphi_{\max}$ ($0 \leq \varphi_{\max} \leq \pi$) to express the phase differences among sub-populations. It is called a synchrony parameter. A given synchrony value $\varphi_{\max}$ in radian, the interval $[0, \varphi_{max}]$ is divided into a set of $(n-1)$ equal samples for the meta-population of n the number of sub-populations. The value of the forcing phase of the ith sub-population is correspondent to i^{th} value in the set.

By using the sinusoidally forced SEIR meta-population model, we will focus on studying the effect of the population size on the disease persistence in a meta-population, more particularly the 3D relationship among the meta-population size N, the number of sub-population, and the coupling rate.

2.2   Stochastic Meta-population Model

We focus on studying demographic and environmental stochasticities in epidemic models. Demographic stochasticity expresses fluctuations in population processes that are functioned by the random nature of events at the level of the individual. Hence, we use stochastic approach that is built by depending upon the corresponding deterministic model of different equations. The first stochastic approach published by Gillespie in 1976 [10] is an exact stochastic simulation approach for chemical kinetics. This method of Gillespie has become the standard procedure of the discrete-event modelling by taking proper value of the available randomness in such a system. To implement the standard SEIR stochastic model, there are nine events that can occur, each causing the numbers in the relative groups to go up or down by one. Below, Table 1 lists all the events of the model, occurring in sub-population i:

Table 1. Events of the stochastic version of the model of Eqs. 2.1–2.4, occuring in subpopulation i.

Events	Rates	Transitions
Birth	μN_i	$S_i \leftarrow S_i + 1$ and $N_i \leftarrow N_i + 1$
Death of a susceptible	μS_i	$S_i \leftarrow S_i - 1$
Death of an exposed	μE_i	$E_i \leftarrow E_i - 1$
Death of an infected	μI_i	$I_i \leftarrow I_i - 1$
Death of an immune	μR_i	$R_i \leftarrow R_i - 1$
Infection	$\lambda_i S_i$	$S_i \leftarrow S_i - 1$ and $E_i \leftarrow E_i + 1$
Becoming infectious	σE_i	$E_i \leftarrow E_i - 1$ and $I_i \leftarrow I_i + 1$
Recovery	γI_i	$I_i \leftarrow I_i - 1$ and $R_i \leftarrow R_i + 1$

To initialize the variables S, E, I, R for sub-populations, we establish a stationary meta-population model. First for a given meta-population size N, we calculate the equilibrium values of these variables. We use the equilibrium values to simulate the dynamics of a single population with the size N. The simulation time is 100 years to obtain stationary regime of the disease dynamics. We harvest the stationary values of variables at the 100th point. After that, for a meta-population of n sub-populations, the initial state of the system is established by dividing the above stationary values by n. The initial values of variables of all sub-populations are identical. Hence, we have successfully established the stationary and stochastic SEIR meta-population model just at the initial step of simulation.

2.3   Measuring Local and Global Extinction Rates in Meta-population

The probability of disease persistence in a meta-population was functioned as an exponential survival model [7,21]. There are many methods to estimate this ability. In 2002, Keeling and Grenfell [19] proposed two methods. The first was for a meta-population without migration by calculating the expected extinction time or the extinction rate during a given period. This method was only a theoretical measure and can not be used in the real life. Because, no real dataset exists to compare with model results. The second was for a meta-population with migration by computing the number or the total duration of extinctions. After some years, in 2010, Conlan et al. [8] proposed also two different methods:

"mean annual fade-out" and "fade-outs post epidemic". These methods function by counting the proportion or on the frequency of zero reports in a given reporting interval. The same point of these four methods is easy to implement them in all models from simple to complex. However, the obtained results are not exact and unreliable. Because, the plot of the disease persistence are Kaplan-Meier survival curves that illustrate the lifetime of every sub-population. Thus, the probability of disease persistence is formulated like a survival function and really, it was defined as an exponential survival model. In this paper, we introduce a new method to quantifying the disease persistence. We measure the ability of persistence via Kaplan-Meier survival curves and survival functions. We believe that obtained results will be more exact and more reliable. However, some previous findings have pointed out that there are recurrent waves of extinction and recolonization in spatial dynamics within sub-populations [12]. The problem measuring exactly the ability of the disease persistence is still open and no exact formula can calculate this ability. Hence, instead of estimating global persistence, we focus on estimating local and global extinction rates as well as recolonization rate that represent the relationship between the spatial structure and the seasonality in a meta-population.

We start by simulating a meta-population of n sub-populations. We get n independent dynamics. After, we accumulate the value of each variables in all sub-populations at each sample time. We compute the average value across the entire time series for each meta-population. Finally, in the case we simulate m different meta-populations, we stock m dates t of global disease extinction. These dates allow to draw Kaplan-Meier survival curves from which we estimate the global extinction rates χ:

$$M(t) = \exp(-\chi t) \tag{2.7}$$

where $M(t)$ ($0 \leqslant M(t) \leqslant m$) is the number of meta-populations in which the disease is not extinct at time t. Then, we apply the parametric survival model for the exponential distribution by using available functions in the R package 'survival' [24]. In order to estimate extinction rates, we consider that extinctions follow a Poisson process and thus fitting an exponential distribution to times to extinctions. The same way is for recolonisation rates.

3   Result

3.1   Effect of Meta-population Size on Extinction Rates

We start with the synchrony model by setting $\varphi_{\max} = 0$, the sub-populations are together coupled and in-phase. It means that the ability of rescue effect among sub-populations are small, they simply get extinct. Inversely, $\varphi_{\max} \neq 0$ leads to the asynchrony of sub-populations. The synchrony is interrupted, the phase dynamics of the coupled sub-populations are out of phase. The decreased similarity among sub-populations leads an increase of the chance of recolonization. A sub-population has even reached local extinction, but the disease comes easily back due to the recolonization from other sub-populations. Figure 1 reveals that the rates of extinction local as well as global in the case $\varphi_{\max} \neq 0$ is lower that in the case $\varphi_{\max} = 0$, and reduce with increasing population size. The extinction rates of the synchrony is better than that of the asynchrony.

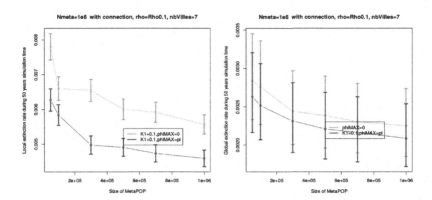

Fig. 1. Influence of the meta-population size on the extinction rate in the meta-population model. $n = 7$, $\rho = 0.1$, $\kappa_1 = 0.1$, the fixed meta-population size N from $5e + 04$ to $1e + 06$, and $\varphi_{\max} = 0$ or π.

3.2 Effect of the Meta-population Size on the Extinction Rates in the Coupled Model and the Uncoupled Model

In order to study the gap of extinction rates in the model coupled as well as uncoupled, we plot the extinction rates against the meta-population size from $5e + 04$ to $1e + 06$. We repose two models in the same state. We just change the value of the parameter ρ that represents the interaction strength among sub-populations. For the uncoupled model, we set $\rho = 0$, and inversely, $\rho = 0.1$ for the coupled model. The parameter $\varphi_{\max} = 0$, it means that the meta-population model is installed in synchrony.

The extinction rates of the isolated model $\rho = 0$ is higher than those of the coupled model $\rho = 0.1$ (Fig. 2). Because, in the isolated model, the sub-populations are independent, no exchange, and no immigration. Thus, there is no occurred recolonization in this model. Inversely, for the coupled model, due to the connection among sub-populations in the meta-population, there are migrations among them, then the disease is again saved in sub-populations that have gone locally extinct. Hence, increasing the meta-population size draws an increase of the sub-population size, and the extinction rates have the tendency to reduce when the meta-population size rises.

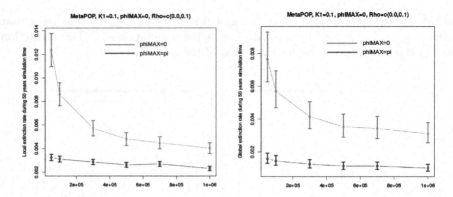

Fig. 2. Plotting extinction rates against population size in the coupled and uncoupled models. $n = 3$, $\kappa_1 = 0.1$, $\varphi_{\max} = 0$. The red line is for the seasonally forced model of uncoupled sub-populations with $\varphi_{\max} = 0$ and the blue line is for the seasonally forced model of coupled sub-populations with $\varphi_{\max} = 0$. (Color figure online)

3.3   3D Relation Among the Meta-population Size, the Coupling Rate, and the Estimated Extinction Rate

Figure 3 points out that the line plotting the coupling rate against the estimated local extinction rate is a convex curve and obtains minimum when the coupling rate is medium.

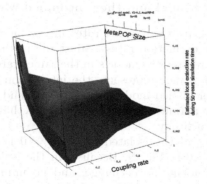

Fig. 3. Three-dimensional relation among the meta-population size, the estimated local extinction rate and the coupling strength. N from $1e + 05$ to $1e + 06$, ρ from 0 to 1, $n = 5$, $K_1 = 0.1$, and $\varphi_{\max} = 0$.

Here, the dispersal rate ρ represents migration strength. The disease transmission speed fast grows when coupling rate goes up in meta-populations. Similar to that, global disease persistence surges also. In our experiment, we permit coupling rate change from weak to strong in a meta-population of five sub-populations. The rate ρ is divided into three intervals: low, intermediate and high. In each interval, we chose some coupling rates that highlight the coupling strength among sub-populations. With each value ρ, we estimate local extinction rate. When ρ is small from 0.0 to 0.001, the extinction rate significantly decreases. However, this rate is minimum when ρ has medium values from 0.001 to 0.01. Lastly, the extinction rate augments back when the coupling rate is very strong from 0.01 to 1.0. As shown in Fig. 3, the local extinction rate in a meta-population is one humped function for the coupling rate. The medium coupling rate (from 0.01 to 0.1) minimizes the extinction rate of disease. Because in the case of the small and medium coupling rates, the coupling rate and the speed of migration among sub-populations are directly proportional. The dispersal speed increases. Thereby the local recolonization speed rises, the duration of persistence grows, the local extinction rate goes down. However, this trend of local extinction with decreasing coupling rate, is not right any more when the dispersal rate is strong. The meta-population has tendency to become one big population. In this case, the phase difference or the recolonization among sub-populations are no longer significant. Hence, the local extinction rate goes up. Besides, increasing the meta-population size draws an increase of the sub-population size, then the local extinction rate goes significantly down. Additionally, by looking over the 3D Fig. 3, the curvature of this curve depends on the meta-population size. Increasing the meta-population size drives the increase of the curvature.

3.4 3D Relation Among the Meta-population Size, the Number of SubPOP, and the Estimated Extinction Rate

An other result, we describe a three-dimensional relation among the meta-population size, the number of sub-populations and the local extinction rate estimated. Figure 4 shows that the local extinction rate decreases with increasing population size. Inversely, the extinction rate increases with increasing number of sub-populations.

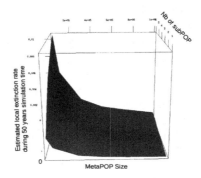

Fig. 4. Three-dimensional relation among the meta-population size, the number of sub-populations and the local extinction rate estimated. N from $1e + 05$ to $1e + 06$, n from 1 to 6, $\kappa_1 = 0.1$, $\rho = 0.1$, and $\varphi_{\max} = 0$.

When the sub-population number is fixed, we increase the meta-population size, we increase the sub-population size. Hence, the time of disease persistence augments, so the local extinction rate goes down. Inversely, when the meta-population size is fixed, we increase the number of sub-populations, it means that we decrease the sub-population size. Hence, the persistence time significantly declines, so the extinction rate goes up.

4 Discussion

In this paper, our main aim is to observe the effect of the population size on the disease persistence. But the aim of Conlan et al. [8] is to find a threshold population size above which disease will persist. Although these aims are different, our results and theirs are identical. The extinction rate is really against the population size in a meta-population of connected sub-population. In addition, we showed the effect of the population size on extinction rates in many distinct aspects and special cases. We proved that although the estimation methods of the local extinction rate and the global extinction rates are different, all extinction rates inversely scales with the population size when sub-populations are synchronous as well as asynchronous as in Fig. 1. More particularly, we gain the same finding when we exploited spatially isolated model and spatially coupled model. For the spatially isolated model, sub-populations are independent without no migration. So, no recolonization of disease occurs in this model as in Fig. 2. There are no other outside factors affecting the ability of disease persistence, but the level of disease extinction still goes down when the population size goes up. In the research of Bjørnstad et al. [5], they took into account the effect of the population size and the population density on measles transmission rates by using the measles data of England and Wales. Fast talk about these two terms that population size is the number of individuals in a population, and population density is the number of individuals per unit area, i.e., population size divided by total land area. Different from the extinction rates, the disease transmission rate is directly proportional to the population density and has no relation with population size. This is explained that when the number of individuals increases into a given area, then the contact rate will be greatly increased. In our model, we have not considered the effect of the population density on the extinction rate. We just performed the null model. Hence, the population density are the same at all areas in the meta-population. An interesting work for future is better than if we can quantify the effect of the population density on the extinction rates.

Grenfell and Bolker [14] in 1998 focused on the effect of the spatial heterogeneity in transmission probability, in term of urban-rural hierarchies in infection rate by using the pre-vaccination measles data of England and Wales. They showed that the infection rate in large cities is higher than in small towns. Particularly, local extinction of infection is unchanged when the population sizes of the urban and rural regions are the same. In addition to their finding, we examined the special relationship between the number of sub-populations and the extinction rates. The meta-population size is fixed, but the number of sub-populations changes. With the same meta-population, but the spatial structures are different, the extinction rates are different. The extinction rates in the meta-population of the higher number of sub-populations is bigger than those in the meta-population of the smaller number of sub-populations (Figs. 3 and 4).

In addition to examining the effects of the seasonal factors and the model structure (such as the number of sub-populations, the coupled model), we strongly explored one of important factors. This is the connection strength ρ. We implemented the simulations with the value of ρ from 0 to 1.0. Zero is for the isolated model. One is the maximum value of the interaction in which meta-population becomes a big population. The obtained result in Fig. 3 showed that there are three intervals of the parameter ρ: small from 0 to 0.001, medium from 0.001 to 0.01, and high from 0.01 to 1.0. The disease persistence in meta-population persists the most neither in the small interval nor in the high interval, but in the medium interval. This finding is similar to those of Huffaker [18], Holyoak and Lawler [17], when they exhaustively explored the disease persistence behavior of many different meta-population models. One more time affirmed that the interaction rate ρ in meta-population models are significant from 10^{-3} to 0.1 [20]. In addition, our results revealed that the curve of the coupling rate against the extinction rate looks like as a convex functions that gains the minimum value in the medium interval (in the Fig. 3). Then, by looking over the 3D Fig. 3, the curvature of this curve depends on the meta-population size. Increasing the meta-population size drives the increase of the curvature.

In short, we successfully simulated the stochastic seasonally forced SEIR meta-population model for measles, and well quantified the effect of the distribution of population size on the extinction probabilities in a meta-population of coupled sub-populations. We also consider the effect of the phase difference of seasonality on the relationship of the persistence probability and the meta-population size. Particularly, we pointed out not only the influence of the coupling strength among sub-populations on the extinction rates, but also revealed a new feature of the tri-partite relationship: the meta-population size, the extinction rate, and the interaction strength as presented above. Our obtained results coincide with previous authors [2,11,14]. Our findings have implications for detecting the basic features of the disease persistence in a meta-population of multisubpopulation, and for future theoretical work on spatial vaccination strategies we can apply these features to optimize vaccination policies. However, these results are not enough. We only analyzed the $SEIR$ meta-population model with a null spatial structure. In the future, we can analyze the meta-population model in more complex structure, exploit a meta-population of coupled sub-populations with different sub-population sizes. We can consider sub-populations with large size urban and sub-populations with small size rural areas.

References

1. Altizer, S., Dobson, A., Hosseini, P., Hudson, P., Pascual, M., Rohani, P.: Seasonality and the dynamics of infectious diseases. Ecol. Lett. **9**(4), 467–484 (2006)

2. Anderson, R.M., May, R.M.: Infectious Diseases of Humans: Dynamics and Control. Oxford University Press, Oxford (1992)
3. Bailey, N.T.J., et al.: The mathematical theory of epidemics (1957)
4. Bartlett, M.S.: The critical community size for measles in the united states. J. Roy. Stat. Soc. Ser. A (Gen.) 37–44 (1960)
5. Bjørnstad, O.N., Finkenstädt, B.F., Grenfell, B.T.: Dynamics of measles epidemics: estimating scaling of transmission rates using a time series sir model. Ecol. Monogr. **72**(2), 169–184 (2002)
6. Black, F.L.: Measles endemicity in insular populations: critical community size and its evolutionary implication. J. Theor. Biol. **11**(2), 207–211 (1966)
7. Conlan, A.J.K., Grenfell, B.T.: Seasonality and the persistence and invasion of measles. Proc. Biol. Sci. **274**(1614), 1133–1141 (2007)
8. Conlan, A.J.K., Rohani, P., Lloyd, A.L., Keeling, M., Grenfell, B.T.: Resolving the impact of waiting time distributions on the persistence of measles. J. R. Soc. Interface (2010)
9. Ferrari, M.J., Grais, R.F., Bharti, N., Conlan, A.J.K., Bjørnstad, O.N., Wolfson, L.J., Guerin, P.J., Djibo, A., Grenfell, B.T.: The dynamics of measles in sub-Saharan Africa. Nature **451**(7179), 679–684 (2008)
10. Gillespie, D.T.: Exact stochastic simulation of coupled chemical reactions. J. Phys. Chem. **81**(25), 2340–2361 (1977)
11. Grenfell, B., Harwood, J.: (Meta)population dynamics of infectious diseases. TREE **12**, 395–399 (1997)
12. Grenfell, B.T., Bjørnstad, O.N., Kappey, J.: Travelling waves and spatial hierarchies in measles epidemics. Nature **414**(6865), 716–723 (2001)
13. Grenfell, B.T., Bjørnstad, O.N., Finkenstädt, B.F.: Dynamics of measles epidemics: scaling noise, determinism, and predictability with the TSIR model. Ecol. Monogr. **72**(2), 185–202 (2002)
14. Grenfell, B.T.: Cities and villages: infection hierarchies in a measles metapopulation. Ecol. Lett. **1**, 68–70 (1998)
15. Grenfell, B.T., Bolker, B.M., Klegzkowski, A.: Seasonality and extinction in chaotic metapopulation. R. Soc. **259**, 97–103 (1995)
16. Hamer, W.H.: The Milroy Lectures on Epidemic Disease in England: The Evidence of Variability and of Persistency of Type. Bedford Press, London (1906)
17. Holyoak, M., Lawler, S.P.: Persistence of an extinction-prone predator-prey interaction through metapopulation dynamics. Ecology 1867–1879 (1996)
18. Huffaker, C.B.: Experimental studies on predation: dispersion factors and predator-prey oscillations. Hilgardia **27**, 343–383 (1958)
19. Keeling, M.J., Grenfell, B.T.: Understanding the persistence of measles: reconciling theory, simulation and observation. Proc. Biol. Sci. **269**(1489), 335–343 (2002)
20. Keeling, M.J., Rohani, P.: Modeling Infectious Diseases in Humans and Animals. Princeton University Press, Princeton (2008)
21. Kleinbaum, D.G.: Survival Analysis (2005)
22. Levins, R.: Some demographic and genetic consequences of environmental heterogeneity for biological control. Bull. Entomol. Soc. Am. **15**, 237–240 (1969)
23. Soper, H.E.: The interpretation of periodicity in disease prevalence. J. R. Stat. Soc. **92**(1), 34–73 (1929)
24. Therneau, T.M.: A Package for Survival Analysis in S, 2014. R Package Version 2.37-7

Large-Scale Data Classification System Based on Galaxy Server and Protected from Information Leak

Krzysztof Fujarewicz[1,2]($\boxtimes$), Sebastian Student[1,2], Tomasz Zielański[1,2], Michał Jakubczak[1,2], Justyna Pieter[1,2], Katarzyna Pojda[1,2], and Andrzej Świerniak[1,2]

[1] Institute of Automatic Control, Silesian University of Technology, ul. Akademicka 16, 44-100 Gliwice, Poland
`krzysztof.fujarewicz@polsl.pl`
[2] Biotechnology Centre, Silesian University of Technology, ul. Krzywoustego 8, 44-100 Gliwice, Poland

Abstract. In this work we present SPICY (SPecialized Classification sYstem) application for a supervised data analysis (feature selection, classification, model validation and model selection) with the structure preventing the data processing work-flow from so called information leak. The information leak may result in optimistically biased classification quality assessment, especially for large-scale, small-sample data sets. The application uses the Galaxy Server environment that originally allows the user to manual data processing and is not prevented from the information leak. The way how the classification model is built by the user and the specific structure of all implemented methods makes the information leak impossible. The lack of information leak in the presented supervised data analysis tool is demonstrated on numerical examples, where synthetic and real data sets are used.

Keywords: Machine learning · Information leak · Galaxy Server · Classification · Feature selection · Model validation · Model selection · Large-scale data · Small-sample data · Genomic data · Proteomic data

1 Introduction

Galaxy [1] is an open, web-based platform for data intensive bio-medical research. The idea of Galaxy is to provide a platform for experiment construction in a number of well described analysis steps described as workflows. Workflows concept is easier to configure than traditionally scripting. We can record the whole workflow and therefore make it possible to repeat the whole analysis pipelines. It is easy to change the part of the analysis sequence by changing the selected module, or selecting other method parameters. In each analysis we can use a framework for accessing and executing remote Web Services and run all computation on remote server or computer cluster. It is particularly important in

© Springer International Publishing AG 2017
N.T. Nguyen et al. (Eds.): ACIIDS 2017, Part II, LNAI 10192, pp. 765–773, 2017.
DOI: 10.1007/978-3-319-54430-4_73

large-scale classification analysis, where we are dealing with complex and big datasets. Galaxy is constructed as an easy to use flexible and complex workflow management system. In this system we can connect different compatible tools in one workflow. All the tools collected in the same toolset are designed to have compatible connections, particularly data input and output formats. In the standard data analysis the connected tools in a one pipeline are executed sequentially on computer server or distributed cluster. In the data classification tasks this scheme can lead to serious problem with improper classification quality assessment.

One of the biggest problems in classification using sequential tool execution is a risk of an "information leak" also called an "incomplete cross validation". In the standard way after data import the feature selection module is executed on whole dataset. For example we can create a big dataset with 1000 features, all generated from the Gaussian distribution $N(0, 1)$ and randomly assigned two class labels. After execution of a feature selection module for this whole totally non-informative dataset the classifier can generate optimistically biased classification accuracy, sometimes better than 90%. Since this data is non-informative and balanced, the accuracy should be around 50% no matter what method is used. This serious problem, especially visible for small-sample data sets (where number of features much greater than number of observations), has been discussed for example in [2,11,13,17]. Ntzani and Ioannidis [8] and Michiels et al. [7] report that, at least in studies up to 2003, most of the 84 considered studies lacked appropriate validation of derived significant feature sets.

2   Properties of the Application

The presented classification system called SPICY (SPecialized Classification sYstem) is created in Galaxy Server environment and can be obtained freely from the website www.spicy.polsl.pl. SPICY is also used as a part of a bigger computer platform for remote testing and analyzing of biomedical data BioTest [10], which is currently developed and will be finally available as a web-based tool www.biotest.polsl.pl.

The main window of the application is presented in Fig. 1. It uses the Galaxy Server environment, however implemented methods of feature selection, classification etc. are mainly written in R language. The user can choose between different feature selection/extraction methods, for example Partial Least Squares (PLS), Principal Component Analysis (PCA), t-test, ANOVA, different classification methods, for example Support Vector Machines (SVM), Linear Discriminant Analysis (LDA), Random Forests (RF) and different model validation methods: Bootstrap, k-fold cross–validation, Leave-One-Out, etc. The examples of the analysis obtained using the presented classification system can be found in our recent works [15,16]. The same idea of the workflow and organization of calculations implemented in different programming environments has been also used in our previous works [3–6,12].

Fig. 1. The main window of the application.

Galaxy offers a finite set of tools on many Galaxy servers for different scientific areas, but only limited tools for data classification. None of them can deal with described problem. Correct classification result can be obtained only when we use proper cross-validation scheme and a non sequential pipeline. In our approach we use the Galaxy in non standard way using cross validation loop right after data is loaded into the classification system [14].

3 Protection from Information Leak

A very common problem in classification task is information leak on different analysis stages. As "information leak" we mean situation, when we use data information obtained from learning data in testing task in any of the whole analysis pipeline. The most dangerous mistake is to use system without separate learning and testing data in each analysis step. This case is named here as scheme 1 and is presented in Fig. 2. Such an approach is also called resubstitution.

Sometimes researchers make this mistake only in some analysis stages, mainly in the feature selection step. This is the case of the validation scheme 2 and is shown in Fig. 3. The information leak comes from the fact that the feature selection step is done for the whole original dataset. The data partitioning and validation is applied only for the classification step.

In our system we use program code structure (described in the next section), that is protected against the information leak. In this case all analysis pipeline steps are executed with separate learning dataset in learning mode and using test data in testing mode. We demonstrate an example of the used structure, defined as scheme 3, in Fig. 4.

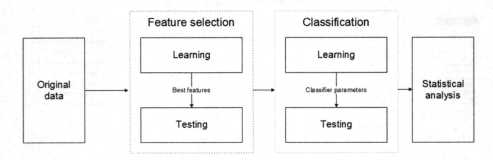

Fig. 2. Improper model validation scheme (resubstitution)—scheme 1.

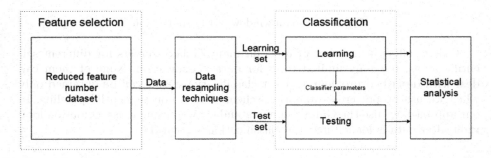

Fig. 3. Improper model validation scheme (incomplete cross-validation)—scheme 2.

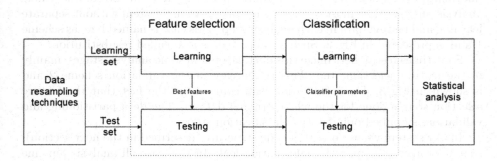

Fig. 4. Proper model validation scheme—scheme 3.

This exemplary scheme contains only two steps: feature selection and classification, but in general the number of processing steps may be arbitrary. For example the input data may be additionally re-scaled before the feature selection.

4   Structure of One Module/Method

It is very important to stress that one method used in the data processing pipeline performs different action depending on the "mode" of the method, which can be *learning* or *testing*. For example, a feature selection method based on Student's t-test in *learning* mode finds indexes of features with the highest absolute value of t-statistics. The same method working in *testing* mode does not calculate any statistics but only chooses features with indexes found in *learning* mode. Similarly, any classification method in *learning* mode finds classifier's parameters based on a learning set and in *testing* mode these parameters are used for classification of the test set.

All this means that implementing any processing method there is a need to create a computer code for both modes and usually in available programming tools these modes are implemented as two separate functions—see Fig. 5a. In our classification system we create only one common function for each implemented method, see Fig. 5b. It can run in both modes depending on the "Mode" switch. In *learning* mode the parameter "Parameters" is an output from the function and in *testing* mode the parameter "Parameters" is an input (argument) of the function.

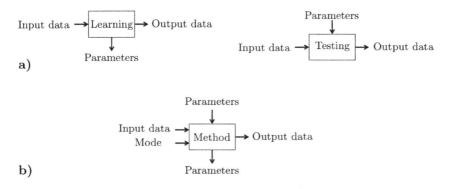

Fig. 5. Two possibilities of a processing method implementation: a—as two distinct functions, one for learning and one for testing; b—as one common function performing learning or testing depending on a "Mode" switch

5   Numerical Experiment

In this study we use 3 different data sets to show the differences between described different classifier validation system scenarios. First of all we generate the synthetic random 2-class data. We use 500 features generated from uniform distribution for 200 samples with randomly assigned class labels. As the second dataset we used the publicly available dataset with 2-class merged genomic and proteomic data of human breast tumors (BRCA). The dataset can be freely downloaded from The Cancer Genome Atlas. The results shown here are in whole based upon data generated by the TCGA Research Network: http://cancergenome.nih.gov/. The third multi-class dataset (4 classes) comes from microarray experiment for different high grade gliomas tissues with 4434 features and 50 samples [9].

In all our numerical experiments we have used the same analysis workflow: selection method based on t-test, classification method based on Diagonal Linear Discriminant Analysis (DLDA) and bootstrap model validation method with 50 bootstrap iterations.

Table 1. Table of bootstrap based accuracy rate for different analysis schemes

Dataset	Analysis scheme	Feature number	Mean accuracy rate	95% confidence interval
Randomly generated data	scheme 1	30	0.8	–
	scheme 2	30	0.768	(0.677, 0.83)
	scheme 3	30	0.559	(0.455, 0.676)
2-class BRCA data	scheme 1	30	0.980	–
	scheme 2	30	0.958	(0.815, 0.993)
	scheme 3	30	0.864	(0.653, 0.961)
4-class glioma data	scheme 1	30	0.843	–
	scheme 2	30	0.824	(0.747, 0.870)
	scheme 3	30	0.763	(0.632, 0.828)

As we can see in Table 1 for randomly generated synthetic data only the system scheme 3 is acceptable (see Fig. 4). Only in this case the 95% confidence interval include the expected 0.5 accuracy rate value which is correct result for such non-informative data. One can see that the results for system schemes 1 and 2 are optimistically biased and show false accuracy greater than 0.5 (Fig. 6).

In case of both real 2-class and multi-class datasets we also observe that the smallest accuracy rate is obtained for the proper workflow scheme 3 and for workflows 2 and 3 the obtained accuracy is optimistically biased (see Figs. 7 and 8).

For all datasets (synthetic and real) the most optimistic bias is observed for the workflow scheme 1 (resubstitution), where no data split is applied.

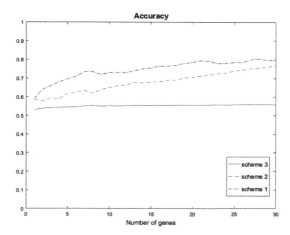

Fig. 6. Bootstrap based classification accuracy for randomly generated synthetic data and different validation schemes

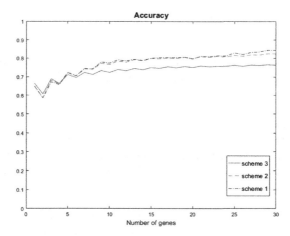

Fig. 7. Bootstrap based classification accuracy for BRCA dataset and different validation schemes

It is important to note that in all numerical experiments the same data analysis methods (selection and classification) are used. This means that the observed differences in accuracy rate are only due to the choice of the model validation scheme. The only non-biased results are obtained for the proper scheme 3 while schemes 1 and 2 give more or less optimistically biased accuracy assessment.

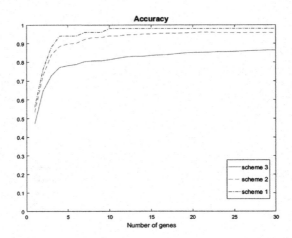

Fig. 8. Bootstrap based classification accuracy for multiclass glioma dataset and different validation schemes

6   Conclusions

In this work we present an application for a supervised data analysis (feature selection, classification, model validation and model selection) which structure makes impossible to manual data processing. All steps of the data processing must be included as separate stages of the specific Galaxy pipeline. Moreover, each method is implemented in a special two-mode (learning and testing) manner. This prevents the overall data processing from the information leak which can result in wrong (too optimistic) classification quality assessment. The provided numerical examples confirm the lack of information leak in the presented large-scale classification system.

Acknowledgment. This work was supported by the NCBiR under Grants DZP/ PBS3/2441/2014 (KF, SS, TZ, MJ, AS) and Strategmed2/267398/4/NCBR/2015 (JP, KP). Calculations were performed using the infrastructure supported by the computer cluster Ziemowit (www.ziemowit.hpc.polsl.pl) funded by the Silesian BIO-FARMA project No. POIG.02.01.00-00-166/08 and expanded in the POIG.02.03.01-00-040/13 in the Computational Biology and Bioinformatics Laboratory of the Biotechnology Centre at the Silesian University of Technology.

References

1. Afgan, E., Baker, D., van den Beek, M., Blankenberg, D., Bouvier, D., Cech, M., Chilton, J., Clements, D., Coraor, N., Eberhard, C., Grning, B., Guerler, A., Hillman-Jackson, J., Von Kuster, G., Rasche, E., Soranzo, N., Turaga, N., Taylor, J., Nekrutenko, A.: The galaxy platform for accessible, reproducible and collaborative biomedical analyses: 2016 update. Nucl. Accids Res. **44**, w3–w10 (2016)

2. Ambroise, C., McLachlan, G.J.: Selection bias in gene extraction on the basis of microarray gene-expression data. PNAS **99**(10), 6562–6566 (2002)
3. Eszlinger, M., Wiench, M., Jarzab, B., Krohn, K., Beck, M., Luter, J., Gubaa, E., Fujarewicz, K., Swierniak, A., Paschke, R.: Meta- and reanalysis of gene expression profiles of hot and cold thyroid nodules and papillary thyroid carcinoma for gene groups. J. Clin. Endocrinol. Metab. **91**, 1934–1942 (2006)
4. Fujarewicz, K., Kimmel, M., Rzeszowska-Wolny, J., Swierniak, A.: A note on classification of gene expression data using support vector machines. J. Biol. Syst. **11**(1), 43–56 (2003)
5. Fujarewicz, K., Jarzab, M., Eszlinger, M., Krohn, K., Paschke, R., Oczko-Wojciechowska, M., Wiench, M., Kukulska, A., Jarzab, B., Swierniak, A.: A multi-gene approach to differentiate papillary thyroid carcinoma from benign lesions: gene selection using support vector machines with bootstrapping. Endocr. Relat. Cancer **14**, 809–826 (2007)
6. Jarzab, B., Wiench, M., Fujarewicz, K., Simek, K., Jarzab, M., Oczko-Wojciechowska, M., Wloch, J., Czarniecka, A., Chmielik, E., Lange, D., Pawlaczek, A., Szpak, S., Gubala, E., Swierniak, A.: Gene expression profile of papillary thyroid cancer: sources of variability and diagnostic implications. Cancer Res. **65**, 1587–1597 (2005)
7. Michiels, S., Koscielny, S., Hill, C.: Prediction of cancer outcome with microarrays: a multiple random validation strategy. Lancet **365**(9458), 488–492 (2005)
8. Ntzani, E.E., Ioannidis, J.P.A.: Predictive ability of DNA microarrays for cancer outcomes and correlates: an empirical assessment. Lancet **362**(9394), 1439–1444 (2003)
9. Nutt, C.L., Mani, D.R., Betensky, R.A., Tamayo, P., Cairncross, J.G., Ladd, C., Pohl, U., Hartmann, C., McLaughlin, M.E., Batchelor, T.T., Black, P.M., von Deimling, A., Pomeroy, S.L., Golub, T.R., David Louis, D.N.: Gene expression-based classification of malignant gliomas correlates better with survival than histological classification. Cancer Res. **63**(7), 1602–1607 (2003)
10. Psiuk-Maksymowicz, K., Placzek, A., Jaksik, R., Student, S., Borys, D., Mrozek, D., Fujarewicz, K., Swierniak, A.: A holistic approach to testing biomedical hypotheses and analysis of biomedical data. Commun. Comput. Inf. Sci. **616**, 449–462 (2016)
11. Ruschhaupt, M., Huber, W., Poustka, A., Mansmann, U.: A compendium to ensure computational reproducibility in high-dimensional classification tasks. Stat. Appl. Genet. Mol. Biol. **3**(1), 1–26 (2004)
12. Simek, K., Fujarewicz, K., Swierniak, A., Kimmel, M., Jarzab, B., Wiench, M., Rzeszowska, J.: Using SVD and SVM methods for selection, classification, clustering and modeling of DNA microarray data. Eng. Appl. Artif. Intell. **17**, 417–427 (2004)
13. Simon, R., Radmacher, M.D., Dobbin, K., McShane, L.M.: Pitfalls in the use of dna microarray data for diagnostic and prognostic classification. J. Natl. Cancer Inst. **95**(1), 14–18 (2003)   .
14. Student, S., Fujarewicz, K.: Stable feature selection and classification algorithms for multiclass microarray data. Biol. Direct **7** (2012). Article ID. 33
15. Student, S., Pieter, J., Fujarewicz, K.: Multiclass classification problem of large-scale biomedical meta-data. Proc. Technol. **22**, 938–945 (2016)
16. Student, S.: Breast cancer prognostic 2-class classification of multidimensional molecular data. In: Prusty, R.M. (eds.) IRAJ, Hungary, pp. 59–62 (2016)
17. Wessels, L.F., Reinders, M.J., Hart, A.A., Veenman, C.J., Dai, H., He, Y.D., van't Veer, L.J.: A protocol for building and evaluating predictors of disease state based on microarray data. Bioinformatics **21**(19), 3755–3762 (2005)

Technological Perspective of Agile Transformation in IT organizations

Agents of RUP Processes Model for IT Organizations Readiness to Agile Transformation Assessment

Włodzimierz Wysocki[1], Cezary Orłowski[2], Artur Ziółkowski[2(✉)], and Grzegorz Bocewicz[1]

[1] Technical University of Koszalin, Koszalin, Poland
wlodzimierz.wysocki@tu.koszalin.pl,
bocewicz@ie.tu.koszalin.pl
[2] WSB University in Gdańsk, Gdańsk, Poland
{corlowski, Artur.Ziolkowski}@wsb.gda.pl

Abstract. A significant problem in modern software engineering is maturity assessment of organizations developing software.

We propose the use of process model of RUP development methodology as a pattern for comparing it with the tested project. Percent values of accordance coefficient determine the task accordance of the tested project with the pattern of activities flow. This RUP model concept is based on a multi-agent based simulation (MABS). It presents agents and their behaviours as well as objects placed in the agent system environment. The behaviour of agents is presented as a set of Finite State Automatons.

The usefulness of the method for assessment of organization's maturity was examined in a two-part experiment. The result of the first part of the experiment was used in the second part as the process pattern to determine the accordance of a sample project to the result of simulated model. The results confirmed the usefulness of the model in maturity assessment.

Keywords: IT organization maturity · Rational Unified Process · RUP · Maturity assessment · Software development processes · Software process simulation · Multi-agent based simulation · MABS · JADE · Finite State Automaton · FSA

1 Introduction

In recent years, there is an increased interest in the transformation [11] of traditional methodologies such as Waterfall [5] and RUP [6] to agile methodologies like Scrum [1] and XP [13]. This is the result of the failure of a large number of IT projects in organizations that produce software. The transformation to agile methodologies is seen as a solution to the situation, but the transformation processes are burdened with a high risk of failure. IT organizations are looking for ways to increase the effectiveness of the transformation process. They conducted research on transformation processes that focus on the analysis of transformation paths (technology, process, design and organizational culture), and the search for methods of determining an organization's

© Springer International Publishing AG 2017
N.T. Nguyen et al. (Eds.): ACIIDS 2017, Part II, LNAI 10192, pp. 777–786, 2017.
DOI: 10.1007/978-3-319-54430-4_74

readiness to start the process of agile transformation [3, 7–10]. The authors examine the relationship of manufacturing processes maturity to the state of readiness of organizations for agile transformation.

For testing the maturity of the development processes the use of an agent-object software development process model is proposed by the authors. As an example of the traditional process, Rational Unified Process (RUP) has been chosen because it precisely defines development and management processes. This is an excellent example of a formal description that forms the basis for the simulation. RUP divides the project into stages, performed in iterations, described by the flow of activities in each of the disciplines. The activities consist of tasks aimed at the manufacture of artefacts - documents, requirements, architecture, testing, and program code. RUP Project roles perform the task of determining the competence of agents representing the project team.

The initial plan was to use the system by project managers for scheduling of software development projects. In the course of work on IT organization's process transformation the use of the system for evaluating organization's state of readiness for transformation turned out to be more valuable than simple project management. Such usefulness of the agent-object system led the authors to propose the use of this system as a model for agile transformation process readiness measurement.

The following is the next article in the series of papers on the assessment of readiness for agile transformation utilizing the agent-object system [15, 16]. In this article we focus on the behaviour of agents dynamically created from Finite State Automata (FSA).

The article is subdivided into three main parts. In Sect. 2 we present an agent-object model of the RUP process simulated using FSA. In Sect. 3 we present the experiment confirming the usefulness of this method. In Sect. 4 we summarize the results of the experiment, draw conclusions and present future work.

2   Agent-Object RUP Process Model

The multi-agent system [14] is used to simulate the software process. Agents operate independently of each other, have the ability to communicate, observe the environment, can alter their environment and are capable of decision making. Agents are used to map Members of the project team performing the tasks assigned to them, in corresponding project roles, according to the project plan and activities flow.

Agents perform the tasks, processing the input artefacts from the repository into output artefacts and saving them to the repository. Figure 1 shows the cooperating elements of the agent-object model.

2.1   Model Components

In the model dynamic components of the software development process are represented by the agents immersed in an agent environment. Static components are represented in

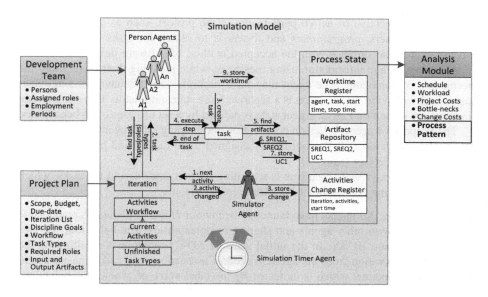

Fig. 1. The agent-object model of Rational Unified Process

the environment by objects on which agents perform operations. Model consists of the following classes of agents and objects:

Person Agents. Project team model is a set of person agents representing team members:

$$G = \{g_i | i = 1 \ldots p\} \tag{1}$$

where the agent g_i is characterized by the sequence:

$$g_i = (\{r_{i,k} | k = 1 \ldots s\}, \{b_{i,k} | k = 1 \ldots r\}) \tag{2}$$

which defines the role set of an agent, as well as its behaviour set.

The behaviour of the finite state automaton is defined by these ordered five:

$$b = (S, s_0, L, T, F) \tag{3}$$

where: S is a finite set called the states, $s_0 \in S$ is the start state, L is the finite set of labels, $T \subseteq (S \times L \times S)$ is the transition set, $F \subseteq S$ is set of finish states.

The project team includes people actively performing tasks and a passive person - to help others in the performance of their tasks. Passive participants are stakeholders of contracting organization ordering the system, which consists of the board, managers and ordinary users. They participate in various forms of activities aimed at meeting the requirements of an IT project.

The main objective of active participants in the project, such as system analysts, architects, designers, programmers and testers is to search for tasks to perform and execute them in accordance with the accepted method of software development.

Agents exchange information using messages, while pursuing their goals using the mechanism of behaviour. The agent may have multiple active behaviours simultaneously. Each behaviour of the agent is a separate finite state automaton. Behaviours state diagrams are shown on Fig. 2. Agent state is defined by a set of states of its behaviours.

The purpose of agents relies on cooperation and helping other agents and is carried out by *HelperBehaviour* behaviour that is assigned to all (active and passive) agents - members of the project team.

The aim of active team members is searching for and carrying out the tasks, which is performed by the behaviour of *TaskManagerBehaviour*.

HelperBehaviour. All agents have launched behaviour *Helper Behaviour*, responsible for cooperation with other agents and sharing responsibilities. The behaviour has two states, READY and BUSY. In the initial state READY behaviour is awaiting on the proposal of cooperation from the agent in need of assistance, let's call it master agent. If it starts to cooperate, which is indicated by the label **start_of_work**, this behaviour

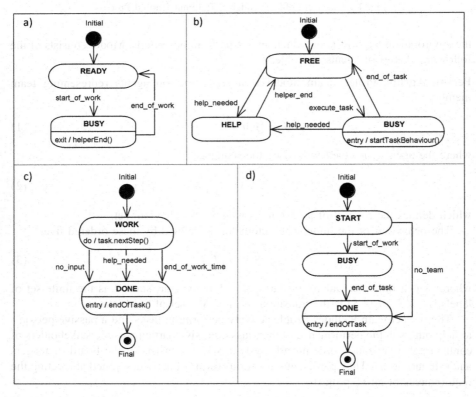

Fig. 2. Finite state automatons diagrams: (a) *HelperBehaviour*, (b) *TaskManagerBehaviour*, (c) *TaskBehaviour*, (d) *WorkshopBehaviour*.

becomes a BUSY state, in which it waits for information from the master agent for completion of the collectively performed task.

After the task is completed the Master agent stores information about the type and time of the task in the Work Time Registry. It then sends a message of cooperation completion **endOfHelp()** to the auxiliary agents. In *HelperBehaviour* behaviour of an auxiliary agent reception of the message **endOfHelp()** is indicated by the label **end_of_work**, causing the transition behaviour into READY state. Exit from the state BUSY sends a message **helperEnd()**, which switches the behaviour *TaskManagerBehaviour* from the HELP state to FREE.

TaskManagerBehaviour. Active participants belonging to the project team have launched two behaviours: *HelperBehaviour* and *TaskManagerBehaviour*. The latter is responsible for searching tasks that can be performed by the agent and start performing the desired task. *TaskManagerBehaviour* behaviour is a three-state automaton with states FREE, HELP and BUSY.

In the initial FREE state a check is performed whether another agent requires assistance in completing the task. If so, the state is switched to HELP and it remains in this behaviour until the behaviour *HelperBehaviour* cooperates with the master agent. Completion of cooperation is marked with the label **helper_end**. Then behaviour based on the roles performed by the project participant looks for tasks to be performed. Behaviour checks whether the task can run and if it can, it starts this task.

The task can start if the following conditions are met:

- there are input artefacts unprocessed by the task,
- the time remaining to the end of workday is sufficient to complete the task if the task has a minimum execution time (e.g. requirements workshop).

Once the type of task that can be done is found - the label **execute_task**, the current behaviour goes into BUSY state and method **startTaskBehaviour()** is started. The method creates a behaviour *TaskBehaviour* that is performing the task. The new behaviour is added to the behaviour of an agent and is running. If during the task another agent needs help, and the task can be interrupted - the label **help_needed**, the task is completed, and behaviour changes from BUSY state to HELP.

TaskBehaviour. It is a universal behaviour which performs tasks. The behaviour of the two-state finite state machine with states WORK and DONE. Initial state is a state WORK, wherein in each clock cycle of the simulation the next step of the task runs. Behaviour changes from WORK to DONE state when one of the conditions is met:

- **no_input** - the task has been completed, meaning that all the inputs artefacts had been processed,
- **help_needed** - a cooperation with another participant in the project started, and the current task is performed independently,
- **end_of_work_time** - the agents workday has finished.

The transition to the state DONE sends a completion message to the agent **endOfTask()**, which causes the switching of *TaskManagerBehaviour* behaviour from the state BUSY to the state FREE. If the job ended due to start of cooperation with another agent, this behaviour *TaskManagerBehaviour* changes from BUSY state to HELP. In the final state behaviour is removed from the set of active agent behaviour.

WorkshopBehaviour. Performing the tasks requiring the cooperation of agents have specialized behaviour. An example of such a behaviour is used to run the task ElicitStakeholderRequestsOnWorkshop.

Simulator Agent. Manages the implementation of the project plan. It is required to:

- initialize the agent environment and the create agents
- switch activities and iterations in the project plan
- store information about changes of current activities and iterations in Activity Change Register
- end the simulation, remove agents, and store the course and final state of the simulation

Simulation Clock Agent. It synchronizes operation of all agents and determines the simulation time in the system. One simulation clock cycle represents fifteen minutes of project time because this is the shortest work period intended for the task.

Project Plan Object. Project plan model is a set of iterations performed during the project:

$$PP = \{I_i | i = 1 \ldots n\} \tag{4}$$

where: I_i – i-th iteration of project plan PP.

Iteration I_i consist of activities $O_{i,k}$ defined as subsets of elementary tasks E, according to RUP terminology:

$$I_i = \{O_{i,k} | k = 1 \ldots m\} \tag{5}$$

where: $O_{i,k} \subset E$

Elementary task set E contains indivisible operations which can be implemented in a given development environment, i.e. Detail a Use Case. Task $e_l \in E$ is described by a sequence: $e_l = (r_l, ai_l, ao_l, d_l)$ defining the role of an agent (r_l), which can accomplish the task, a set of input artefacts necessary to start the task $(ai_l \subseteq A)$, a set of artefacts produced after finishing of the task $(ao_l \subseteq A)$, the duration of the task $(d_l$ - measured in units of agreed time, u.a.t.).

Artefact Repository Object. The repository is a part of the model state. It is responsible for storing instances of artefacts grouped by type. The ability to carry out the elementary tasks e_l depends on availability of the required artefacts ai_l. A pair defined in the repository determines the types and number of available artefacts in the simulation system:

$$R = (A, \varphi) \tag{6}$$

where: A – is a set of artefacts types in the model, $\varphi(a, t) = \varphi_{a,t}$ - function specifying the number of type a artefacts $a \in A$ at a discrete point in time t.

Activity Change Register Object. This is the model output and an element of model state. This register stores changes of activities in the flow of the project. The active iteration of the project, current activities, and time of the change are stored.

In accordance with above definitions, simulation system SS is represented as a sequence of the project plan PP which is defined on the set of elementary tasks E, artefact repository R and agent set G:

$$SS = (PP, E, R, G) \tag{7}$$

Behaviour of the simulation system is a set of tasks performed during the project simulation:

$$S = \{W_i^S | i = 1 \ldots n\} \tag{8}$$

where: W_i^S – i-th task of simulation S

The simulation is related to the schedule for completion of elementary tasks:

$$X^s = \{x_{i,j,k} | i = 1 \ldots n, j = 1 \ldots m, k = 1 \ldots z\} \tag{9}$$

where: $x_{i,j,k}$ – and start time of the i-th task, in regard to activity j-th and elementary task k-th.

Based on the schedule of elementary tasks, a schedule of activities can be defined, which is used as a process pattern to examine the accordance of the actual design process with the RUP model:

$$\overline{X^S} = \{x_{i,j} | i = 1 \ldots n, j = 1 \ldots m\} \tag{10}$$

where: $x_{i,j} = min\{x_{i,j,1} \ldots x_{i,j,z}\}$.

2.2 Implementation of the Agent-Object Model System in a Software Environment

JADE was chosen for the implementation of the agent-object system described by Eqs. (1)–(10). It is a multi-agent system written in Java as open source [17]. It had to be adapted for use as a runtime multi-agent simulation (MABS) [12]. The JADE system lacks the solution for simulation of time. Therefore we use a proprietary solution of the central timing simulation with a fixed step, which is an extension of the concept of clock simulation in the Jadex package [2]. Simulation clock has been extended to determine the order of sending time information to groups of agents, with the introduction of a multi-phase clock cycle. This solution has reduced entropy and produced a better result reproducibility of RUP process simulation.

3 Experimental Use of the Agent-Object System to an Actual RUP Project

To test the usefulness of the model we have used it to examine readiness of organization to agile transformation. The experiment consisted of two parts. The first stage was to measure the size of the project, prepare information about completed tasks and tune parameters of the internal model to the actual project.

The second stage was planned to use the well-calibrated simulation as a pattern to assess the maturity of the development process in the actual project.

Detailed data on the course of the development process of a large information system in the field of insurance was used as the input for the experiment.

3.1 Phase One – Calibrating the Model

The aim of the first stage is to obtain a process pattern matching the RUP process. To do this, we had to estimate the size of the project and adjust internal parameters of the simulation to reproduce effort of actual process. Results of model calibrating phase are shown in Table 1.

Table 1. Results of the first part of the experiment

Output variable	Actual value	Simulation result	Relative error
Number of artefacts	15 517	14 945	3,70%
Effort of project	319 130	320 456	0,42%

3.2 Phase Two - Using the Calibrated Agent-Object Model to Assess Software Process Maturity

The second stage of the experiment consisted of comparing results of the simulated process as a pattern with the course of the actual project.

The actual project is defined as a set of tasks performed during the project:

$$P = \{W_i^P | i = 1 \ldots n\} \tag{11}$$

where: W_i^P – i-th task of project P, defined as the simulation results (7).

The project is related to its schedule:

$$X^P = \{x_{i,k} | i = 1 \ldots n, k = 1 \ldots m\} \tag{12}$$

where: $x_{i,k}$ – start time of i-th task, k-th activity of project P.

The accordance of the actual task with the simulated task is marked $\equiv$ and defined as such:

$$W_i^P \equiv W_i^S \text{ if } \exists o_i \in O_i^S, \text{ for which } (x_{i,j}^P \geq \overline{x_{i,j}^S}) \wedge (x_{i,j}^P \leq \overline{x_{i,j}^S} + d_{i,j}^S) \tag{13}$$

where: O_i^S - activity set, $d_{i,j}^S$ - timespan of i-th task, with regards to j-th activity in the simulation result.

Table 2. The results of the comparison actual project with the process pattern

Variable	Description	Value
Lz	Number of tasks in accordance	38008
Lw	Total number of tasks	51701
Z	Activity accordance coefficient	74%

Activity accordance Z is defined as the percentage ratio of the number of tasks in accordance L_z the total number of tasks L_w.

$$Z = 100\% \frac{L_z}{L_w} \tag{14}$$

4 Conclusions and Future Work

The aim of the research was to analyse the RUP process in an IT organization using the agent-object system developed by the authors. It was assumed that the assessment of the maturity of an IT organization using the system will show the readiness of an organization for eventual transformation process. The article presents the design of the model focusing on the behaviours of the agents as composed of finite state automata. The latter part describes verification environment and the use of the system in the simulation of an actual IT project. The result of the first phase of the experiment was a well-calibrated model, which was used to check the test project and showed high readiness of the organization for agile transformation. Readiness was later confirmed by the successful transformation of the organization to the use of agile methodologies software development.

The model, which consists of mapping development methodologies and maintain the project team as a dynamic, cooperating set of finite automatons, reproduces IT projects very well. Use of the model allows automate and hence to reduce costs and shorten process maturity assessment.

RUP methodology contained in the model can be easily exchanged for agile methodology because its description is composed of abstract elements, which all methodologies are built upon. It is planned to extend the model, among others, by the Scrum methodology.

The behaviour of agents allows to dynamically build finite state machines, depending on the roles of design and tasks executed by the agent. Because of the configuration changes to automata and cooperation between them, it is difficult to analyse the ongoing relationship between them and the resulting action. It is planned to select a typical configuration of cooperating agents and verify the resulting finite state automaton. Automata will be described using the Promela language and verified using the SPIN method [4].

References

1. Attanasio, F.: Scrum Mastering Reloaded. Lulu Press Inc., Raleigh (2015)
2. Braubach, L., Pokahr, A.: The Jadex project: simulation. In: Ganzha, M., Jain, L.C. (eds.) Multiagent Systems and Applications, vol. 45, pp. 107–128. Springer, Heidelberg (2013)
3. Deręgowski, T., et al.: Building project and project team characteristic for creating hybrid management processes. In: Software Engineering: Improving Practice through Research. Polish Information Processing Society (PTI), Wrocław (2016)
4. Holzmann, G.: Spin Model Checker, the: Primer and Reference Manual. Addison-Wesley Professional, Boston (2003)
5. Khan, A.I., et al.: A Comprehensive Study of Commonly Practiced Heavy and Light Weight Software Methodologies (2011)
6. Kroll, P., Kruchten, P.: The Rational Unified Process Made Easy: A Practitioner's Guide to the RUP. Addison-Wesley, Boston (2003)
7. Kurzawski, M., et al.: Trigger-based model to assess the readiness of IT organizations to agile transformation. In: Software Engineering: Improving Practice through Research. Polish Information Processing Society (PTI), Wrocław (2016)
8. Orłowski, C., et al.: A model for shaping IT project management process to raise the readiness of an IT organization for agile transformation. Presented at the XIX Conference on Innovation in Production Management and Engineering, Zakopane (2016)
9. Orłowski, C., et al.: The use of complexity measures of the project to assess the state of evolution of the IT organization. Presented at the XIX Conference on Innovation in Production Management and Engineering, Zakopane (2016)
10. Ossowska, K., et al.: The aggregate evaluation measure of the organization's state of maturity in the process of its evolution. Presented at the XIX Conference on Innovation in Production Management and Engineering, Zakopane (2016)
11. Tripp, J.F., Armstrong, D.J.: Agile methodologies: organizational adoption motives, tailoring, and performance. J. Comput. Inf. Syst., 1–10 (2016)
12. Uhrmacher, A., Weyns, D. (eds.): Multi-agent Systems: Simulation and Applications. CRC Press/Taylor & Francis, Boca Raton (2009)
13. Unhelkar, B.: The Art of Agile Practice: A Composite Approach for Projects and Organizations. CRC Press, Boca Raton (2016)
14. Wooldridge, M.J.: An Introduction to Multiagent Systems. Wiley, Chichester (2009)
15. Wysocki, W., Orłowski, C., Ziółkowski, A., Bocewicz, G.: Efficiency and maturity assessment model of RUP process in IT organizations. In: Wilimowska, Z., Borzemski, L., Grzech, A., Świątek, J. (eds.). AISC, vol. 524, pp. 209–219. Springer, Heidelberg (2017). doi:10.1007/978-3-319-46592-0_18
16. Wysocki, W., Orłowski, C., Ziółkowski, A., Bocewicz, G.: Model of RUP processes maturity assessment in IT organizations. In: Madeyski, L., Śmiałek, M., Hnatkowska, B., Huzar, Z. (eds.). AISC, vol. 504, pp. 187–199. Springer, Heidelberg (2017). doi:10.1007/978-3-319-43606-7_14
17. Jade Site: Java Agent DEvelopment Framework. http://jade.tilab.com/

Evaluation of Readiness of IT Organizations to Agile Transformation Based on Case-Based Reasoning

Cezary Orłowski[1(✉)], Tomasz Deręgowski[2], Miłosz Kurzawski[3], and Artur Ziółkowski[1]

[1] Department of Information Technology Management,
WSB University, Gdansk, Poland
{corlowski,aziolkowski}@wsb.gda.pl
[2] Acxiom Corporation, Gdańsk, Poland
Tomasz.Deregowski@acxiom.com
[3] Faculty of Management and Economics,
Gdańsk University of Technology, Gdansk, Poland
mkurzawski@zie.pg.gda.pl

Abstract. Nowadays many of IT organization decides to change the way of delivering from classic, waterfall approach to agile. This transition is called "agile transformation" (AT). The problem of this process is that part of companies started AT without any analysis. This causes that many of transitions fails and organizations must return to old methods of delivering. Cost of return is significant and number of projects with violated project management triangle is bigger than before. In this paper authors described the results of conducted research and the model of evaluating of readiness to agile transformation based on case-based reasoning.

Keywords: Agile transformation · Case-based reasoning · Associative rules

1 Introduction

Today's IT world is changing very fast. IT organizations must look for solutions, which allow to build competitive advantage. Many companies decide to change the approach of delivering from classic, waterfall to agile. This transition is called "Agile Transformation" (AT) [1].

AT is very complex process therefore it was divided into four perspectives [1]: project perspective, processes perspective, organizational culture perspective and technology perspective. Perspectives are the areas of functioning of organization [2]. Between them are some dependencies, but in view of complexity, authors decided to consider each of them separately. This article focused only on perspective of processes, which scope includes changes in project management techniques.

Beside perspectives, during agile transformation process, there is another key element called trigger. This is the factor that lead to the decision to abandon current, classic way of delivering and to replace it by agile methodologies [1]. During previous

© Springer International Publishing AG 2017
N.T. Nguyen et al. (Eds.): ACIIDS 2017, Part II, LNAI 10192, pp. 787–797, 2017.
DOI: 10.1007/978-3-319-54430-4_75

research, based on observations authors identified four, main groups of triggers: effectiveness triggers, forcing triggers, project triggers and motivational triggers [1]. Decision to start agile transformation is often taken without any analyse in example when the decision is taken by company management. This causes that many of agile transformations fails and IT organization must return to the previous way of delivering [3]. In consequence more projects than before AT finishes with violation of project management triangle. In addition, returning to previous methods is expensive. These are the reasons why it is so important to evaluate readiness of organization to agile transformation before process of changes starts.

This article, which is the continuation of research conducted last years by authors, presents the model of evaluation of readiness to agile transformation of IT organizations, based on case-based reasoning method. This paper is divided into four chapters. The first one is the introduction to the topic. Second contains obtained results during quantitative research and the method of data reduction. In chapter three authors have described model based on case-based reasoning method and verification of the model. The last chapter is the summary of research.

2   Associative Data Analysis

During the studies on agile transformation authors decided to conduct quantitative research. As a result, data about IT organizations has been gained. In further steps, because of the number of data, authors decided to use associative data analysis to find dependencies among data. In next subsection, there are described obtained results.

2.1   Qualitative Research and Results

Qualitative research included in its scope preparation of questionnaire and handing it over to team leaders, managers, directors and board members of IT organizations. The survey was divide into three main sections [1]:

- **Organization** – questions about characteristics of organization i.e. size or number of development teams.
- **Processes** – the aim of this section was to verify the level of processes in organization. Authors have decided to use definition of processes from Capability Maturity Model Integration (CMMI) standard [4]. For each of processes definition four-point scale has been used:
  Process is not defined
  Process is defined but not implemented
  Process is defined and partially implemented
  Process is defined and fully implemented
- **Agile transformation** – questions about processes of transformation, key factors, problem and used solutions to resolve occurred problems.

As a result of research authors have obtained data from twelve IT organizations. Among these companies, nine were after agile transformation process. The scope of data was very wide, therefore the decision to reduce data has been taken. Process and method of reduction is described in next subsection.

2.2 Association Analysis and Data Reduction

Associative analysis is the method of extraction of item sets for discovery of association rules [9]. First of all, it is necessary to build associative rules. The single rule has following form (2.1) [7]:

$$X \rightarrow Y \qquad (2.1)$$

which means **IF X THEN** Y. This method assumed that are two sets of data: set of observations (2.2) and set of elements which occurred in observations (2.3).

$$O = \{O_1, O_2 \ldots O_n\} \qquad (2.2)$$

O_n - single observation

$$E = \{E_1, E_2 \ldots E_n\} \qquad (2.3)$$

E_n - single element

Each of single observations can contain any combination of elements from set E. This is single subset (2.4) [5].

$$\sigma(X) = |\{O_n | X \subseteq O_n, O_n \in O\}| \qquad (2.4)$$

Important parameters which characterizes the power of association between X and Y are support (2.5) and confidence (2.6). Support specifies how often associative rule is applicable in set O. Confidence specifies how frequently Y occurs in transactions from set E, which contains X [8].

$$s(X \rightarrow Y) = \frac{\sigma(X \cup Y)}{N} \qquad (2.5)$$

$$c(X \rightarrow Y) = \frac{\sigma(X \cup Y)}{\sigma(X)} \qquad (2.6)$$

Based on support and confidence it is possible to limit the set of elements. Low value of support and confidence may indicate that there is no association between elements.

During the analysis of association, it is necessary to determine lowest acceptable value of support (mins) and confidence (minc). In this article authors assumed that set O contains 12 elements (2.7). Every single element of set is the observation from quantitative research. Set E contains 22 elements (2.8). Every single element it is the process described in CMMI standard).

$$O = \{O_1, O_2, O_3, O_4, O_5, O_6, O_7, O_8, O_9, O_{10}, O_{11}, O_{12}\} \qquad (2.7)$$

where:

O_1 = answers of respondent 1

...

O_{12} = answers of respondent 12

$$E = \{E_1, E_2, E_3, E_4, E_5, E_6, E_7, E_8, E_9, E_{10}, E_{11}, E_{12}, E_{13}, E_{14}, E_{15}, E_{16}, E_{17}, E_{18}, E_{19}, E_{20}, E_{21}, E_{22}\} \qquad (2.8)$$

E = {Project Monitoring and Control (PMC), Requirements Development (RD), Project Planning (PP), Configuration Management (CM), Process and Product Quality Assurance (PPQA), Supplier Agreement Management (SAM), Measurement and Analysis (MA), Organizational Process Definition (OPD), Organizational Process Focus (OPF), Organizational Training (OT), Integrated Project Management (IPM), Risk Management (RSKM), Product Integration (PI), Requirements Management (REQM), Technical Solution (TS), Validation (VAL), Verification (VER), Decision Analyst and Resolution (DAR), Organizational Process Performance (OPP), Quantitative Project Management (QPM), Causal Analysis and Resolution (CAR), Organizational Process Performance (OPP)}.

Authors wanted to check if there is any association between level of each process and success of agile transformation. Based on this, 22 association rules (2.9) have been built.

$$X \rightarrow Y \qquad (2.9)$$

where:

$X =$ $\{E_1, E_2, E_3, E_4, E_5, E_6, E_7, E_8, E_9, E_{10}, E_{11}, E_{12}, E_{13}, E_{14}, E_{15}, E_{16}, E_{17}, E_{18}, E_{19}, E_{20}, E_{21}, E_{22}\}$

$Y =$ success of agile transformation

The rule can be understood as: **IF** process E_n is implemented in organization **THEN** agile transformation is success. For example, for process E_1, there is following rule: **IF** Project Monitoring and Control process is implemented **THEN** agile transformation is success.

Table 1 is the matrix of answers of respondents. Value 0 for process (E_n) means that answer to the question about this process was: "Process is not defined" or "Process is defined but not implemented". Value 1 for process means that answer to question about this process was: "Process is defined and partially implemented" or "Process is defined and fully implemented". Value 0 for Y means that organization is before agile transformation or transformation failed. Value 1 for Y means that organization is after agile transformation and has implemented agile partially (some principles were not implemented) or fully (methodology implemented in accordance with all principles).

Table 1. The matrix of answers of respondents

	E_1	E_2	E_3	E_4	E_5	E_6	E_7	E_8	E_9	E_{10}	E_{11}	E_{12}	E_{13}	E_{14}	E_{15}	E_{16}	E_{17}	E_{18}	E_{19}	E_{20}	E_{21}	E_{22}	Y
O_1	0	0	0	0	0	0	0	0	0	0	1	0	0	0	0	0	0	0	0	0	0	0	1
O_2	0	0	0	0	0	0	1	0	0	0	1	0	1	0	0	0	0	0	0	0	0	0	0
O_3	1	1	1	1	1	0	1	1	0	1	0	1	0	1	0	0	0	0	0	0	0	0	0
O_4	1	1	1	1	0	0	0	0	0	0	0	0	0	0	1	1	1	0	0	0	0	0	1
O_5	0	0	1	1	0	0	0	0	0	1	0	0	0	1	0	0	0	0	0	0	0	0	1
O_6	0	0	0	0	0	0	0	0	0	0	0	0	0	0	0	0	0	0	0	0	0	0	1
O_7	0	0	0	0	0	0	0	0	0	1	0	0	0	0	0	0	0	0	0	0	0	0	1
O_8	0	0	0	0	0	0	0	0	0	0	0	0	0	0	0	0	0	0	0	0	0	0	1
O_9	0	0	0	0	0	0	0	0	0	0	0	0	0	0	0	0	0	0	0	0	0	0	0
O_{10}		0	0	0	0	0	0	0	0	0	0	0	0	0	0	0	0	0	0	0	0	0	1
O_{11}	0	0	0	0	0	0	0	0	0	0	0	0	0	0	0	0	0	0	0	0	0	0	1
O_{12}	1	1	1	0	0	0	1	0	0	1	0	0	0	0	0	0	0	0	0	0	0	0	1

For each process (E_n) and observation (O_n) authors have calculated the value of support and confidence. Table 2 presents obtained results. As minimum value of mins assumed 0,15 and for minc 0,60.

Based on results from Table 2 it was possible to limit the set of processes. Only 5 from 22 elements comply with conditions mins = 0,15 and minc = 0,60. New set E (2.10) contains five processes.

$$E = \{E_1, E_2, E_3, E_4, E_{10}\} \tag{2.10}$$

Associative rules allowed to limit the set of processes which are significant during evaluating a readiness of organization to agile transformation. Next step of research was to create generic case, build database of cases (with limited set of elements) and use case-based reasoning (CBR) to predict the result of agile transformation and problems which probably occur during AT. Next chapter of this article contains the description of CBR method and built database.

Table 2. Values of support and confidence for elements

E_n	$\sigma(X)$	$\sigma(X \cup Y)$	$s(X \to Y)$	$c(X \to Y)$	E_n	$\sigma(X)$	$\sigma(X \cup Y)$	$s(X \to Y)$	$c(X \to Y)$
E_1	3	2	0,1667	0,6667	E_{12}	2	0	0,0000	0,0000
E_2	3	2	0,1667	0,6667	E_{13}	0	0	0,0000	N/A
E_3	4	3	0,2500	0,7500	E_{14}	2	1	0,0833	0,5000
E_4	3	2	0,1667	0,6667	E_{15}	1	1	0,0833	1,0000
E_5	1	0	0,0000	0,0000	E_{16}	1	1	0,0833	1,0000
E_6	0	0	0,0000	N/A	Ei_{17}	1	1	0,0833	1,0000
E_7	3	1	0,0833	0,3333	Ei_{18}	0	0	0,0000	N/A
E_8	1	0	0,0000	0,0000	Ei_{19}	0	0	0,0000	N/A
E_9	0	0	0,0000	N/A	Ei_{20}	0	0	0,0000	N/A
E_{10}	6	4	0,3333	0,6667	Ei_{21}	0	0	0,0000	N/A
E_{11}	0	0	0,0000	N/A	Ei_{22}	0	0	0,0000	N/A

3   Evaluation Based on Case-Based Reasoning

Case – based reasoning (CBR) is a method for finding similarity and solutions for problems based on cases from past. The biggest advantage of this method is that is based on knowledge and experiences from past, not on theory. CBR cycle consist of four stages: retrieve, reuse, revise and retain [6]. Authors decided to use Case – Base Reasoning to find similarity between IT organizations deciding to change the way of delivering. Construction of database and generic case are described in next subsection.

3.1   Generic Case and Database of Cases

At first stage of preparation of database authors decided to prepare generic case (Table 3) Prepared base contains 10 records (cases), each described by 28 variables. Table 3 presents variables, types and possible values. Figure 2 is the screenshot, which presents ready-made base of cases (Figure 1).

Each case is described by 28 variables grouped into 5 main categories:

- Organizational characteristics – variables such as size, number of team member etc.
- Processes – boolean variables which describes if each process is implemented in organization.
- Triggers – boolean variables which describes if each triggers occurs in organization.
- Agile transformation – previous methodology, chosen methodology and status of agile transformation.
- Problems – boolean variables which describes if some problems occur during agile transformation.

Model of evaluating consist of three layers: input, processing layer and output. In the input there are 19 variables (every variable from Table 3 excluding Agile Transformation Status and Problems). Processing layer compares inputted organization with cases in database. Output is percentage similarity of inputted organization with cases in the base. Figure 3 is a graphical presentation of the model.

To verify whether variables have been properly chosen authors decided to make two tests. Results of verification are described in next subsection.

3.2   Verification

Authors randomly chosen two cases to verify if cased-based model works properly. First organization is company after agile transformation, second where agile transformation is in progress. Variables for both companies inputted into model are presented in Figs. 3 and 4.

Table 4 presents results of comparison of evaluated organizations and cases from base. Hit indicates how assessed organization is similar to organizations stored in data base.

Table 3. Generic case characteristics

Variable	Type	Possible values
Organization Size (V1)	String	Micro, Small, Medium, Big
Geographical dispersion of team members (V2)	String	No, Yes but less than 6 h, Yes and more than 6 h
International Environment (V3)	Bool	True, False
Number of developers teams (V4)	String	1 team, 2–5 teams, >5 teams
Type of client (V5)	String	External, Internal
Organization Type (V6)	String	Product – Oriented, Service – Oriented
[Process] Project Monitoring and Control (P1)	Bool	True, False
[Process] Requirements Development (P2)	Bool	True, False
[Process] Project Planning (P3)	Bool	True, False
[Process] Configuration Management (P4)	Bool	True, False
[Process] Organizational Training (P5)	Bool	True, False
[Trigger] Need to reorganize organization (T1)	Bool	True, False
[Trigger] Violation of project management triangle (T2)	Bool	True, False
[Trigger] Improve efficiency of delivering (T3)	Bool	True, False
[Trigger] Decision-Makers decision (T4)	Bool	True, False
[Trigger] Employees initiative (T5)	Bool	True, False
[Trigger] Changing philosophy of organization (need to be Agile) (T6)	Bool	True, False
Used Methodology (V7)	String	None, RUP, MSF, PRINCE2, other options
Chosen Methodology (V8)	String	SCRUM, Agile, XP, other options
Agile Transformation status (V9)	String	No (failed), Yes but partially, Yes, fully)
[Problem] Reluctance of employees (PR1)	Bool	True, False
[Problem] Low level of knowledge about chosen methodology (PR2)	Bool	True, False
[Problem] Low management commitment (PR3)	Bool	True, False
[Problem] Lack of description of transformation processes (PR4)	Bool	True, False
[Problem] Problem with availability of team members (PR5)	Bool	True, False
[Problem] Lack of tools (PR6)	Bool	True, False
[Problem] Lack of common vision (PR7)	Bool	True, False
[Problem] Change the way of thinking of employees (working in short iterations) (PR8)	Bool	True, False

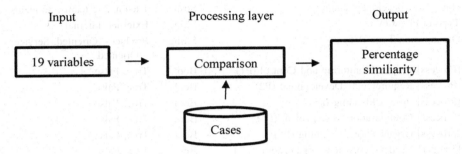

Fig. 1. Screenshot of cases database

Input   Processing layer   Output

| 19 variables | → | Comparison | → | Percentage similiarity |

↑

Cases

Fig. 2. Scheme of the model

Fig. 3. Variables and values for case 1 - organization after AT

For Case 1 three cases have been taken to further analysis, because percentage match was the same for three cases from database. For Case 2 two because percentage match was the same for two cases from database (Table 5).

As shown in Table 6 algorithm predicted that agile transformation will succeed partially and it is probable that three problems occur. In reality organization has implemented partially Scrum and two problems of predicted occurred. Percentage level of matching of predicted and real result is 88%. This value may indicate that model is working properly and accuracy of prediction is high.

As shown in Table 6 there is probability that organization should successfully finish agile transformation but it is also probable that four problems occurs. Organization should implement corrective actions to decrease likelihood of problems. In further steps it is planned to compare real result of AT in this organization with prediction.

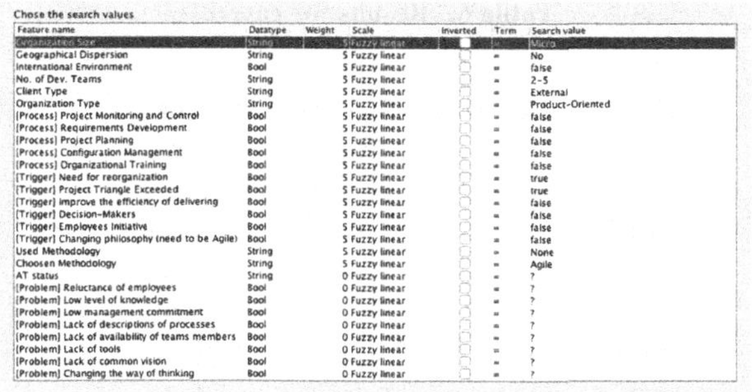

Chose the search values

Feature name	Datatype	Weight	Scale	Inverted	Term	Search value
Organization Size	String	5	Fuzzy linear		=	Micro
Geographical Dispersion	String	5	Fuzzy linear		=	No
International Environment	Bool	5	Fuzzy linear		=	false
No. of Dev. Teams	String	5	Fuzzy linear		=	2-5
Client Type	String	5	Fuzzy linear		=	External
Organization Type	String	5	Fuzzy linear		=	Product-Oriented
[Process] Project Monitoring and Control	Bool	5	Fuzzy linear		=	false
[Process] Requirements Development	Bool	5	Fuzzy linear		=	false
[Process] Project Planning	Bool	5	Fuzzy linear		=	false
[Process] Configuration Management	Bool	5	Fuzzy linear		=	false
[Process] Organizational Training	Bool	5	Fuzzy linear		=	false
[Trigger] Need for reorganization	Bool	5	Fuzzy linear		=	true
[Trigger] Project Triangle Exceeded	Bool	5	Fuzzy linear		=	true
[Trigger] improve the efficiency of delivering	Bool	5	Fuzzy linear		=	false
[Trigger] Decision-Makers	Bool	5	Fuzzy linear		=	false
[Trigger] Employees Initiative	Bool	5	Fuzzy linear		=	false
[Trigger] Changing philosophy (need to be Agile)	Bool	5	Fuzzy linear		=	false
Used Methodology	String	5	Fuzzy linear		=	None
Choosen Methodology	String	5	Fuzzy linear		=	Agile
AT status	String	0	Fuzzy linear		=	?
[Problem] Reluctance of employees	Bool	0	Fuzzy linear		=	?
[Problem] Low level of knowledge	Bool	0	Fuzzy linear		=	?
[Problem] Low management commitment	Bool	0	Fuzzy linear		=	?
[Problem] Lack of descriptions of processes	Bool	0	Fuzzy linear		=	?
[Problem] Lack of availability of teams members	Bool	0	Fuzzy linear		=	?
[Problem] Lack of tools	Bool	0	Fuzzy linear		=	?
[Problem] Lack of common vision	Bool	0	Fuzzy linear		=	?
[Problem] Changing the way of thinking	Bool	0	Fuzzy linear		=	?

Fig. 4. Variables and values for case 2 - organization after AT

Table 4. Results of comparison for cases

Case 1		Case 2	
Case number	Hit [%]	Case number	Hit [%]
1	48,7011	1	54,1169
2	48,7011	2	54,1169
3	48,7011	3	48,7011
4	39,3023	4	48,7011
5	39,3023	5	43,8049
6	35,1114	6	43,8049
7	31,1753	7	43,8049
8	31,1753	8	35,1114
9	31,1753	9	31,1753

Table 5. Results for case 1

Variable	Evaluating organization	Case 1	Case 2	Case 3	Average	Degree of matching [1, 0]
V9	**Yes, partially**	Yes, partially	Yes, partially	Yes, partially	**Yes, partially**	**1**
P1	**True**	True	True	True	**True**	**1**
P2	**True**	True	True	False	**True**	**1**
P3	**False**	True	True	True	**True**	**0**
P4	**False**	False	False	True	**False**	**1**
P5	**False**	False	False	False	**False**	**1**
P6	**False**	False	False	False	**False**	**1**
P7	**False**	True	False	False	**False**	**1**
P8	**False**	False	False	False	**False**	**1**
						88%

Table 6. Results for case 2

Variable	Case 1	Case 2	Average
V9	Yes, fully	Yes, partially	**Yes**
P1	True	True	**True**
P2	True	False	**True**
P3	False	True	**True**
P4	False	True	**True**
P5	False	False	**False**
P6	False	False	**False**
P7	False	False	**False**
P8	False	False	**False**

4   Summary

This paper presents the case-based model to evaluate the readiness of IT organization to agile transformation. Article summarizes quantitative research made by authors and reduction of data by using associative rules. Based on selected variables, generic case has been built. This allowed to create database, which contains ten cases. Built model has been verified by two test cases – the first one was the organization which finished agile transformation with success, and the second one was organization in which AT is currently in progress. For first test case, model predicted success and three potential problems. Comparing to real result of transformation in this organization it is accuracy of prediction on level 88,89%. For second case the model predicted success and four potential problems. In further steps, when organization will finish transformation processes, authors want to compare predicted result with obtained in reality. This article is part of research on agile transformation. In next stages authors want to verify model in another organizations and add new cases into database. Furthermore, next version of model will not only return potential agile transformation status and problems but also proposition of its solutions based on experience from other organizations.

References

1. Orłowski, C., Deręgowski, T., Kurzawski, M., Ziółkowski, A., Chrabski, B.: Trigger-based model to assess the readiness of IT organizations to agile transformation. In: Hnatkowska, B., Warszawa, M.S. (eds.) Software Engineering: Improving Practice Through Research, pp. 207–221. Polish Information Processing Society (2016)
2. Orłowski, C., Deręgowski, T., Kurzawski, M., Ziółkowski, A., Chrabski, B.: Building project and project team characteristics for creating hybrid management processes. In: Hnatkowska, B., Warszawa, M.S. (eds.) Software Engineering: Improving Practice Through Research, pp. 241–255. Polish Information Processing Society (2016)
3. Orłowski, C., Deręgowski, T., Kurzawski, M., Ziółkowski, A., Chrabski, B.: The reference model of tools adaptation in the perspective of technological agile transformation in IT organizations. In: Hnatkowska, B., Warszawa, M.S. (eds.) Software Engineering: Improving Practice Through Research, pp. 223–240. Polish Information Processing Society (2016)

4. CMMI Product Team, CMMI for Acquisition, Version 1.3. Carnegie Mellon University (2013)
5. Osowski, S.: Metody i narzędzia eksploracji danych. Wydawnictwo BTC, Legionowo (2013)
6. Aamodt, A., Plaza, E.: Case-based reasoning: foundational issues, methodological variations and system approaches. AI Commun. **7**, 39–59 (1994)
7. Agrawal, R., Srikant, R.: Fast Algorithms for Mining Association Rules. IBM Almaden Research Center
8. Tan, P.-N., Steinbach, M., Kumar, V.: Introduction to Data Mining. University of Minnesota (2010)
9. Leskovec, J., Rajaraman, A., Ullman, J.D.: Mining of Massive Datasets. Cambridge University Press, Cambridge (2010)

Building Dedicated Project Management Process Basing on Historical Experience

Cezary Orłowski[1(✉)], Tomasz Deręgowski[2], Miłosz Kurzawski[3],
and Artur Ziółkowski[1]

[1] Department of Information Technology Management,
WSB University, Gdańsk, Poland
{corlowski,aziolkowski}@wsb.gda.pl
[2] Acxiom Corporation, Gdańsk, Poland
Tomasz.Deregowski@acxiom.com
[3] Faculty of Management and Economics, Gdańsk University of Technology,
Gdańsk, Poland
mkurzawski@zie.pg.gda.pl

Abstract. Project Management Process used to manage IT project could be a key aspect of project success. Existing knowledge does not provide a method, which enables IT Organizations to choose Project Management methodology and processes, which would be adjusted to their unique needs. As a result, IT Organization use processes which are not tailored to their specific and do not meet their basic needs. This paper is an attempt to fill this gap. It describes a method for selecting management methodologies, processes and engineering practices most adequate to project characteristic. Choices are made on the basis of organization's historical experience. Bespoken Project Management process covers technical and non-technical aspects of software development and is adjusted to unique project challenges and needs. Created process is a hybrid based on CMMI for Development Model, it derives from different sources, uses elements of waterfall and Agile approaches, different engineering practices and process improvement methods.

Keywords: Project Management · Standards · Methodologies · Waterfall approach · Agile · Case Based Reasoning

1 Introduction

Contemporary IT Organizations need to implement and support wide range of projects. Very often IT organization is responsible for providing simple services such as maintenance of existing solutions, where project activities are limited to bug fixing and making minor changes. At the same time the same IT organization actively develops existing systems and products, integrates products and system provided by external companies or work on rewriting existing solutions using new, more efficient technologies. In addition to project types listed above, the same IT organization is often engaged in developing new, unique products for internal and external clients.

© Springer International Publishing AG 2017
N.T. Nguyen et al. (Eds.): ACIIDS 2017, Part II, LNAI 10192, pp. 798–810, 2017.
DOI: 10.1007/978-3-319-54430-4_76

Not only the type of the projects can differ as a part of the same IT organization. Even if projects are of the same type, they parameters such as budget, schedule, requirements, quality expectations, project risks, used technology and team structure may vary. These parameters are also crucial factors which shape unique project characteristic. Each IT organization and its members have also unique experience resulting from previous projects. All these different factors should be taken into consideration when choosing the way in which project will be managed.

There is no single methodology which allows effectively manage a wide range of unique IT projects. Authors of this paper see the need to expand Agile mindset and use Agile approach not only to develop projects but also to develop processes used to manage these projects. The way project management process is build could be in many ways similar to the process of building a software. Project Management process should constantly adapt to changing requirements and environment. The choice of Project Management Methodology and Processes should be based on project characteristic and historical experience. All these principles are derived from Agile approach.

The shape of Project Management Process used to manage IT project is very often a key to project success. Unfortunately, existing knowledge does not provide a method, which enables IT organizations to choose proper project management methodology, adjusted to unique project specific.

This paper is an attempt to fill this gap. It describes a method of selecting management methodologies, processes and engineering practices most adequate to project needs. Decisions are based on organization's historical experience. Knowledge of processes, methodologies and tools which worked for similar projects in the past, helps design optimal process for newly started projects. The goal is to build bespoke Project Management process, which covers all different aspects of software development (technical and non-technical), adjusted to unique project challenges and needs. Newly created process is from definition a hybrid, it derives from different sources, uses elements of waterfall and Agile methodologies, different engineering practices and process improvement methodologies.

Building dedicated, hybrid Project Management processes, adjusted to the needs of particular project, is an important part of Agile Transformation. Agile Transformation is understood in the context of this paper as applying Agile mindset to the process of creation and adaptation of project management processes and methodologies. Agile mindset is understood as constant adaptation to changing requirements, circumstances and environment. Organizations that haven't yet begun Agile Transformation Process use same approach to manage all different types of projects or don't use any methodology at all.

Organization which has begun Agile Transformation starts adapting project management processes to the needs of particular project. Only newly started projects are managed using dedicated, bespoke project management process, but organization is aware of the need to use dedicated solutions and starts collecting metrics on historical projects and projects in progress. It also starts improving ongoing projects and adapts their management processes to the specific needs of managed project.

Organization which has completed Agile Transformation is ready to implement various projects with different specific using dedicated, bespoken Project Management process. The course of Agile Transformation is summarized in the figure below (Fig. 1).

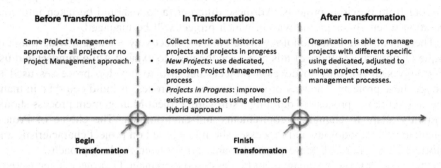

Fig. 1. Agile Transformation Process

A problem associated to lack of tools and methods, which allow choosing adequate, processes and methodologies, adjusted to unique needs of the project was described in one of earlier publications "Model for building Project Management processes as a way of increasing organization readiness for Agile transformation" (Deręgowski Tomasz 2016). Proposed in this article Model for Designing Hybrid Management Processes (DHMP) defines concept of building custom project management processes, which are adjusted to unique project needs. Processes are based on results of project, client and delivery organization analysis and they try to fill all the gaps in existing processes and address specific needs resulting from project characteristic.

This paper is a continuation of earlier research. It concentrates on the process of creating hybrid, bespoken project management process based on organization's historical experience. It also describes experience with an introduction of hybrid approach and how it impacted project performance and project result. Described in this paper method for defining project management process based on characteristic of the project and historical experience is an integral part of Model for Designing Hybrid Management Processes (DHMP) defined in previous publications.

2   Case Based Reasoning as a Tool for Constructing Project Management Process

Described in this paper method for selecting most appropriate, adjusted to unique project needs, Project Management methodology and management processes is based on Case Based Reasoning (CBR) method (Fig. 2).

CBR is a method for solving new problems based on the solutions of similar past cases. It has been chosen for two main reasons. Firstly, CBR method is based on team's practical knowledge and previous experience, not on theoretical models or general knowledge. CBR knowledge-based systems retrieve and reuse solutions that have

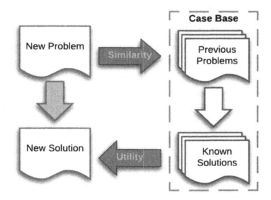

Fig. 2. Case Based Reasoning method

worked for similar cases in the past. A new problem is solved by finding a similar past cases and reusing them in the new problem situation.

Secondly, CBR enables incremental learning. New experience is retained each time a problem has been solved and is added to Case Base so it could be reused to solve future problems.

A widely accepted model of the CBR processing is the CBR cycle proposed by Aamondt and Plaza (1994) (see Fig. 3). There are four stages in CBR processing defined in this model: Retrieve, Reuse, Revise and Retain.

At Retrieve Stage CBR system looks for historical cases relevant to target problem which is to be solved. Similar cases are extracted from the system and transferred for further processing. Relevant cases are to some extend similar to target problem: most of their key aspects and parameters are similar to key aspects and parameters of target

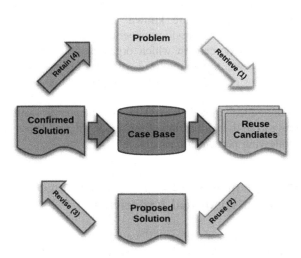

Fig. 3. Case Based Reasoning cycle

case. At Reuse Stage experience described in the corresponding, historical cases, which were retrieved in the previous step of CBR processing, is used to solve target problem. In most cases experience and knowledge contained in historical cases need to be adapted to the needs of target case. During Reuse Stage the solution for target problem is defined. In the next step of CBR processing called Revise Stage, applicability of defined in previous step solution is revised and evaluated in practice. At this stage it is checked how solution created in Reuse stage impacted target case and if it helped with solving it. Last, fourth step of CBR processing is Retain. During Retain Stage solution generated and revised in previous stages is archived in CBR system so it could be reused to solve future problems.

Description defines the problem which was solved (past Cases stored in Case Base) or is about to be solved. It defines not only the problem, but also state of the world when the problem occurred. Problem discussed in this paper concerns building dedicated, adjusted to unique project needs, Project Management processes which would maximize the probability of project success. Project is characterized by seven factors, each of them is defined by the level of its complexity. Parameters used to describe project characteristic are budget, requirements, schedule, risk, technology and project team.

The last part of case definition determines outcome, effect of applying described solution to given problem. What was the reaction of the system after the solution was applied, if the problem was totally or partially solved. In case of problem discussed in the paper outcome is understand as project result. Project result can be defined in many different ways. Traditionally project success is defined as delivering project on time, within the budget, with all features which were defined at the beginning of the project. Project could be also considered as success if particular level of client satisfaction was achieved. The question of how to define the success of a project is not relevant to this paper and will not be discussed.

In order to be available for future reuse, cases are organized in a Case Base. Case Base contains of collection of historical cases which were solved in the past and for which the solution and outcome are known. Case Base contains organization's knowledge which result from its previous experience. It is also often the source of knowledge about business domain in which organization operates. To draw useful conclusions, Case Base needs to be initialized with set of past cases. It also contains a mechanism which enables effective comparison of historical cases and drawing useful conclusions.

The usage of theoretical model defined in this chapter as well as findings and conclusions resulting from the application of this model in production environment, are described in the following part of this paper.

3 Design and Implementation of Project Management Process Based on Historical Experience

Pilot study was carried out in a software development company. The company has more than 4,000 employees, distributed among different localizations in United States, South America, Australia, Asia and Europe, including Poland. Solutions and products provided by company are used to support multichannel marketing (Arikan 2008), data mining (Witten 2011) and Big Data (Marz 2015).

The main goal of pilot study was to check if it is possible to improve team's performance and probability of project success by using dedicated, bespoke, project management processes. It was also important to prove, that such a process can be created on the basis of company's historical project experience.

Prior to the pilot study Case Base was initialized with information about eleven past projects which were carried out by described organization. Survey developed and described in "Building Project and Project Team Characteristic for creating Hybrid Management Process" (Deręgowski Tomasz 2016) paper was used to build characteristic of each of analyzed projects. Survey was used to gather information about methodologies and project management processes used to manage each of examined projects. Because all surveyed projects have been completed, it was possible to collect information about projects' results and their success or failure factors which impacted project's overall result.

In order to verify model for building project management processes based on historical experience one of the projects carried out by described organization has been chosen. Project which has been chosen for the model verification was carried out for one of the biggest and most important clients of European branch of the company. Cooperation with this client lasted for three years, delivery team consisted of 15 associates. Authors of pilot study decided to choose this particular project because of repeatability of project environment. Every four months project team delivered functional release which contained of similar number of Change Requests (CR). Each functional release was provided by the same project team, used the same technologies and had similar scope. Repeatability of experiment environment allowed to measure the impact of each change on project overall performance. Similarity between functional releases allowed to minimize the impact of external factors.

3.1   Retrieving from Case Base Projects with Similar Characteristic

In the first step of CBR processing, cases which are similar to the test case for which project management process is to be designed, are extracted from Case Base. There are various different approaches to determine the level of similarity between two projects. One of them are numeric measures. In context of described experiment, to describe the characteristic of the project, seven key parameters of the project were identified. Each parameter describes different aspect of the project. Projects aspects that are used to determine the degree of similarity between two different cases are budget, schedule, requirements, quality, risk, technology and project team.

Each parameter can be one of three values: Low Complexity (1), Medium Complexity (2) and High Complexity (3). The comparison result of two projects is the result of a comparison of individual parameters that make up the characteristics of the project. There are three possible results of parameters comparison:

- Parameters are equal; both of them have either high, medium or low complexity. When both parameters are equal, the result of the comparison is 1.

– Parameters are similar; it means that there are not equal, but at the same time they are also not significantly different. Two parameters are similar when one of them have low complexity and the second have medium complexity, or when one of the parameters have medium complexity when complexity of the second parameter is high. When two parameters are similar, the result of the comparison is 0.5.
– Parameters are different when their complexity is significantly different. It means that one parameter has low and second high complexity. The result of comparison of two parameters which are different is 0.

The overall result of comparison of individual parameters defines the level of similarity between two projects. Dedicated weights can be assigned to each of the parameters. The higher the weigh, the greater the impact of particular parameter on the summary result of comparison. The degree of similarity between two projects is brought to a value which is the average similarity of individual parameters.

For a numeric representation of similarity, where each aspect which makes up the total result has attribute-value representation, a similarity measure is the generalized Hamming measure. It combines the importance of each attribute of project characteristic with its local similarity value. Global similarity for two projects P_1 and P_2 is the sum of local similarities $sim_j(P_1, P_2)$ multiplied by relevance factor w_j.

$$sim(P_1, P_2) = \sum_{j=0}^{n} w_j \cdot sim_j(P_1, P_2)$$

sim_j - local similarity for aspect j
w_j - relevance (weight) of aspect j of the project
n - number of aspects describing project.

In the context of described study four projects with characteristic similar to the characteristic of Test Project P_T were retrieved from Case Base. Authors of this study assumed, that similar project is a project for which the level of similarity is greater than 66%. The degree of similarity differed from project to project. The characteristic of first project (P_A) was identical to the characteristic of Test Project P_T and the degree of similarity was 100%. The degree of similarity for two next projects $(P_B$ and $P_C)$ was 69%. The degree of similarity for last projects (P_D) was 75%.

Characteristic of each project and detailed results of comparison are described in the table below (Table 1).

The experience gained during the implementation of projects P_A, P_B, P_C and P_D have been used to design Project Management Process for project P_T which was about to begin.

3.2   Reuse of Proven Management Solutions

In the course of the survey, interviewed expert listed Process Areas which had the most significant impact on project result. List of processes was based on Processes Areas

Table 1. The result of comparison of analogical projects

Process Area	w_j	Test Proj. P_T	Project P_A		Project P_B		Project P_C		Project P_D	
			Comp.	sim_j	Comp.	sim_j	Comp.	sim_j	Comp.	sim_j
Budget	0.125	Med.	Med.	1,0	Low	0,5	Med.	1,0	Med.	1,0
Schedule	0.125	Med.	Med.	1,0	Med.	1,0	Med.	1,0	High	0,5
Requirements	0.125	Med.	Med.	1,0	Med.	1,0	High	0,5	High	0,5
Quality	0.125	Low	Low	1,0	Low	1,0	Low	1,0	Low	1,0
Risk	0.125	Med.	Med.	1,0	Med.	1,0	Med.	1,0	Med.	1,0
Technology	0.125	Low	Low	1,0	High	0,0	Low	1,0	Low	1,0
Team Comp.	0.25	Low	Low	1,0	Med.	0,5	High	0,0	Med.	0,5
sim (P_T, P_A)			**100%**		**69%**		**69%**		**75%**	

defined in CMMI for Development model (CMMI-DEV) (Chrissis et al. CMMI for Development: Guidelines for Process Integration and Product Improvement 2011).

CMMI is a process improvement method, which integrates all process and procedures existing in IT Organization into single coherent model and is often used to assess the quality and maturity of processes implemented in IT organisations. CMMI is also used to identify potential gaps and shortcomings of existing processes and procedures. CMMI provides holistic approach, processes described in CMMI model cover full Software Development Lifecycle, from gathering requirements through software development to maintenance and support.

Another important feature of CMMI is the fact that it is method and tool agnostic. It does not concentrate on specific methodologies or tolls. CMMI defines which process should be implemented in an organization and why, but it doesn't define how they should be implemented. Thanks to that it can be used with all types of methodologies, models and techniques, both Agile and traditional.

Within the CMMI-DEV model 22 process areas are defined. Each Process Area describes a key aspects related to particular area of software delivery. CMMI processes are grouped into 4 categories, each category relates to different level of organisation maturity. Seven first Process Areas are related to the Managed Maturity Level. On Managed Level projects are planned and managed in accordance with policy, however the policy might vary between different projects carried out by the same organisation. On Managed Level polices are limited to the most important aspects of Software Development Process such as Configuration Management, Project Planning and Monitoring, Quality Assurance and Requirements Management. For the purpose of this study only first seven Process Areas associated with Managed Maturity Level were used. Analyzing all twenty-two Process Areas would be too time-consuming and would require significant expertise from people participating in pilot study.

As mentioned in previous chapter, four projects had characteristic similar to the characteristic of Test Project P_T. Projects Areas and their impact on the success of each of the projects are listed in Table 2. Two Process Areas which had the biggest impact on project result in all four cases were Configuration Management and Process and Product Quality Assurance. Team responsible for delivering Test Project P_T had already decided to implement Process and Product Quality Assurance Requirements.

Table 2. Project Areas and their impact on project result

Process Area	Success factor				
	Project A	Project B	Project C	Project D	Avg.
Configuration Management	High	Med.	High	High	**High**
Measurement and Analysis	Low	Low	Low	Low	Low
Project Monitoring and Control	Med.	Med.	Med.	Med.	Med.
Project Planning	Med.	Med.	Med.	Med.	Med.
Process and Product Quality Assurance	High	Med.	High	Med.	**High**
Requirements Management	Med.	Med.	Med.	Med.	Med.
Supplier Agreement Management	Low	Med.	High	Low	Med.

Based on survey results it has been decided to also implement additional Process Area Configuration Management. The introduction of this process wasn't originally planned.

Other Process Areas were of minor importance and weren't taken into consideration when designing project management process for Test Project P_T.

Establishing Configuration Management Process requires several actions which are described in detail in CMMI-DEV model. The first action involves identifying and documenting the characteristics of objects which configuration will be managed. In described case Project Team decided to formally manage versions of application code and application environments. To do it effectively, it was important to prevent any unauthorised changes in application and its runtime environments. Every new release of code and change in environment configuration was done through Release Management Team which used a set of tools for release automation and configuration management. It helped to prevent any unauthorised changes made outside of the process, without being tracked, reviewed and formally approved. Before any change was migrated to new environment it was audited and the possibility of negative impact on the system behaviour was eliminated. Right after migration full set of automated regression and performance tests was run. Every issue introduced by new release was immediately discovered, faulty release was withdrawn and issue was immediately fixed. New processes and procedures significantly increased the quality of code and let the team discover potential issues right after they were introduced, when they were easier and cheaper to fix.

To implement requirements related to Configuration Management Process Area, team decided to use practices and tools related to DevOps movement (Kim, 2016) such as Infrastructure as a Code (Humble 2010), Virtualisation (Humble 2010), Continuous Integration (Duvall 2007), Continuous Delivery (Humble 2010) and Test Automation (Bisht 2013). Team also adapted some of Extreme Programming techniques such as Unit Testing (Beck 2004), Code Reviews (Beck 2004) and Static Code Analyze (Chess 2007).

The impact of implementation of new process was measured in the next step of CBR processing called Revise. Results are described in the next chapter.

3.3 The Impact of Changes in Project Management Processes on the Course of the Project

Project Management Process that has been used to manage Test Project P_T was based on CBR model and historical experience. One of the key recommendations concerned the formal implementation of Configuration Management Process Area. Configuration Management processes were adapted to the needs of P_T Project. Impact of changes in project performance which were caused by changes in Project Management process are described in this chapter.

Before establishing proper Configuration Management process four engineers of different specialisation were performing new software releases. Each engineer was responsible for releasing changes in particular technology, all releases were performed manually. Java releases were performed by Java developers, .NET releases by .NET developer and PLSQL releases by DBA. Release Process was coordinated by Release Engineer. After Transformation releases for all three technologies were automated and performed by Release Engineer. Human cost of software release was reduced by 75%.

Implementation of proper, fully automated Configuration Management process had also impact on time needed to perform release. In case of Production Releases, release time was reduced by 90%. In case of System Test and QA release time was reduced by 96%. Number of releases performed each week was increased by 1000%. Impact of implementing proper Configuration Management was summarised in Table 3.

Table 3. Impact of changes in Configuration Management on Release Management performance

	Before change	After change	Improvement rate
Number of engineers performing release	4	1	75%
Time to perform Production Release [days]	5	0,5	90%
Time to perform System Test/QA Release [days]	3	0,125	96%
Number of release per week	1	10	1000%

To test the quality of each release, project team introduced test automation. After every code commit full set of Unit Tests and Smoke Tests were run. In couple of minutes developer who committed the code got feedback from Continuous Integration system if his or her change didn't break the build and if all tests are passing. Even if one of the tests didn't pass, fixing the code and code build was main responsibility of developer.

Automated tests were also a key step in release process. Creation of new software package was preceded by the launch of full set of integration, regression and performance tests. To automate test process team used Robot Framework. This new approach guaranteed that only high quality code which passed all types of tests was migrated to test and production environments. Thanks to test automation, the time needed to perform full set of integration, regression and performance tests was reduced significantly.

Prior to test automation, the team needed five days to run full set of tests. After test automation was introduction application could be tested in four hours. This means that time and effort required to perform full set of tests was reduced by 90%.

To improve the quality of code, project team introduced Code Reviews. Every time developer finished working on new feature, he or she was obligated to organise Code Review. During Code Review meeting all new and updated code was reviewed by other developer and architect. Thanks to this new practice, many potential defects and issues were discovered and eliminated even before code was passed to testers. During Code Review developer also presented the result of Static Code Analyse. SonarQube was a tool used to perform static code analyse. Thanks to SonarQube, in couple of minutest tens of thousands of lines of code were checked against hundreds of rules. It let the team discover and fix many potential security and performance issues long before code was passed for testing.

All these new practices had significant impact on quality of the code and number of defects. When compared to analogical project from the past, significantly less defects were discovered during development, testing and in production. Number of defects found during Development and System Test was reduced by 42%. During User Acceptance Tests 75% less defects were found. The team was particularly proud of the fact, that no defects were found after the system was migrated into production. Changes in the number of defects and their distribution were presented in the Fig. 4.

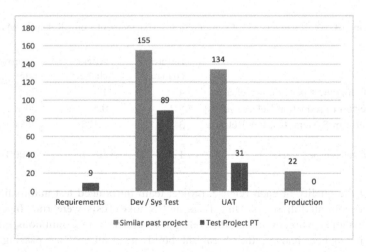

Fig. 4. Reduction in number of defects as the result of Configuration Management Process Area introduction

After the solution has been successfully adapted to the target problem, processes definition and information on its impact on project performance were stored as a new case in Case Base. Experience gained during pilot study has become available for future projects.

The impact of introducing Configuration Management Process Area was so significant, that it was decided to introduce it in all projects conducted by organization in which the pilot study was carried out.

4 Conclusions

Example of Test Project P_T shows that changes in Project Management processes could significantly impact project performance. Thanks to introduction of Configuration Management Process Area, the quality of software was much higher than the quality of software created in analogical projects which didn't manage configuration in the formal manner. It's also visible, that changes which caused such a significant improvement in project performance have been initiated by the usage of CBR model. Therefore, it is worth to continue research on adapting management processes to the specific of project characteristic.

The selection of analytical tools is a matter to be decided. Case Based Reasoning seems to be effective tool when it is used to analyse relatively small and simple set of data. Such solution seems to be sufficient, when IT organisations build recommendations sets based on their own, limited experience. Expanding knowledge base and using experience of other organisations will require more sophisticated and effective analytical tools. Another limitation of CBR method is inability to learn and make predictions on data.

The usage of selected elements of Machine Learning (Kelleher et al. 2015) might be the right direction for future research. Introduction of Pattern Recognition and use of algorithms that can learn from historical data and make data-driven predictions or decisions could bring discussed solution to the next level. Authors of this paper intend to address these issues in their further studies and develop described in this paper method.

References

Aamodt, A., Plaza, E.: Case-based reasoning: foundational issues, methodological variations and system approaches. AI Commun. **17**(1), 39–59 (1994)

Arikan, A.: Multichannel Marketing: Metrics and Methods for On and Offline Success. Sybex, Indianapolis (2008)

Beck, K.A.: Extreme Programming Explained: Embrace Change, 2nd edn. Addison-Wesley, Boston (2004)

Bisht, S.: Robot Framework Test Automation. Packt Publishing, Birmingham (2013)

Chess, B.: Secure Programming with Static Analysis. Addison-Wesley Professional, Boston (2007)

Chrissis, M.B., Konrad, M., Shrum, S.: CMMI for Development: Guidelines for Process Integration and Product Improvement. Addison-Wesley Professional, Boston (2011)

Deręgowski Tomasz, O.C.: Building Project and Project Team Characteristic for creating Hybrid Management Process. KKIO, Wrocław (2016)

Deręgowski Tomasz, O.C.: Model for building project management processes as a way of increasing organization readiness for Agile transformation. Konferencja Innowacje w Zarządzaniu i Inżynierii Produkcj, Tom II (2016)

Duvall, P.M.: Continuous Integration. Improving Software Quality and Reducing Risk. Pearson Education, Upper Saddle River (2007)

Humble, J.F.: Continuous Delivery. Reliable Software Releases through Build, Test, and Deployment Automation. Pearson Education, Upper Saddle River (2010)

Kelleher, J.D., Namee, B.M., D'Arcy, A.: Fundamentals of Machine Learning for Predictive Data Analytics: Algorithms, Worked Examples, and Case Studies. The MIT Press, Cambridge (2015)

Kim, G.: The DevOps Handbook: How to Create World-Class Agility, Reliability, and Security in Technology Organizations. IT Revolution Press, Singapore (2016)

Marz, N.: Big Data: Principles and Best Practices of Scalable Realtime Data Systems. Manning Publications, Greenwich (2015)

Witten, I.H.: Data Mining: Practical Machine Learning Tools and Techniques. Morgan Kaufmann, Burlington (2011)

Describing Criteria for Selecting a Scrum Tool Using the Technology Acceptance Model

Gerard Wagenaar[1]($\boxtimes$), Sietse Overbeek[2], and Remko Helms[3]

[1] Avans University of Applied Sciences,
Breda, The Netherlands
g.wagenaar@avans.nl
[2] Utrecht University, Utrecht, The Netherlands
s.j.overbeek@uu.nl
[3] Open University, Heerlen, The Netherlands
remko.helms@ou.nl

Abstract. Scrum teams extensively use tools to support their processes, but little attention has been given to criteria a Scrum team applies in its selection of such a tool. A greenfield approach was used to explore these criteria. To this extent twelve Scrum teams were asked to list criteria and assigned weights in their decision processes. After having chosen and used a tool for a number of Sprints, the teams also evaluated the selected tools. Using the Technology Acceptance Model to structure findings, two major categories were identified: Perceived usefulness, alias criteria directly related to Scrum, and perceived ease of use. Most teams listed more or less the same criteria. Within the categories several specific subcategories were distinguished, for instance burn-down chart support or multi-platform aspects. Teams evaluated more issues, positive or negative, within the Scrum-related criteria. The findings indicate that Scrum teams prefer perceived usefulness over perceived ease of use. In other words: Specific support of Scrum, especially its artefacts, are of greater value to a team than general tool considerations.

Keywords: Agile · Scrum · Supporting tool · Technology Acceptance Model (TAM) · Weighted criteria

1 Introduction

Agile software development (ASD) prefers *"Individuals and interactions over processes and tools"* with an annotation: *"... there is value in the items on the right ..."* [1]. Tools have proven to be valuable, as, for instance, global Scrum projects depend on a wide range of tool support [2]. This includes communication and collaborative tools for issue or bug tracking, backlog management, and burn-down chart visualization. Also in a collocated project a tool proved to effectively manage a Scrum project [3]. In general, communication and collaboration tools, whether used face-to-face or global, are crucial in modern agile teams [4].

Agile tools come in abundance [5]; a choice has to be made by an agile team. However, knowledge of criteria applied by a team in this choice is anecdotal and

© Springer International Publishing AG 2017
N.T. Nguyen et al. (Eds.): ACIIDS 2017, Part II, LNAI 10192, pp. 811–821, 2017.
DOI: 10.1007/978-3-319-54430-4_77

fragmented, whereas such knowledge gives insight in where agile teams need support. This in turn deepens our theoretical understanding of agile processes through the explicit articulation of important ASD elements by its practitioners, while at the same time practically supporting them in choosing appropriate support. Because of our approach, a greenfield one, our results would especially be useful for novel teams involved in an agile transformation.

The remainder of this paper is organised as follows. In Sect. 2 we discuss related work; in Sect. 3 we present our research method. Results are shown in Sect. 4, followed by a discussion and conclusions in Sect. 5, which is the final section.

2   Related Work

Research on criteria agile teams use for tool support has not been extensive so far. Fragmented research has been presented, which can be captured under two headings: (1) Individual experiences [6–8] and (2) Research explicitly addressing the need for a more general approach [9, 10]. However, all tend to assess tools on the basis of existing criteria rather than to tackle the problem of first defining criteria for the selection of tools thereafter. We now first discuss both categories and introduce a more general theoretical viewpoint afterwards.

2.1   Experience Reports

Experience reports describe a selection process or, somewhat more general, a transition process towards ASD. Uy and Rosendahl [6] discuss a case where a technology team, responsible for a corporate website as well as a number of other products and information systems, entered a migration trajectory from SharePoint to a tool better suited for Scrum. Three critical factors were used to distinguish between two short-listed alternatives (Rally, VersionOne): Usability, functionality, and configurability. Additional technical considerations concerned were in the areas of enterprise infrastructure, architecture and quality assurance.

Møller et al. [7] designed a Scrum tool dedicated to assist in a daily Scrum meeting with four overall requirements: Intuitive user interface, high accessibility, commitment to Scrum, and project history. The requirements were found on the basis of experiences of three Scrum teams.

Engum et al. [8] report on a small North European agile company, which transitioned to a tool for managing Product and Sprint Backlog and distilled lessons from its experiences. Impediments the tool should address included task description and tracking, specifically in order to help organize, specify and prioritize the product backlog and track development, and time usage.

Whether phrased as guidance [8], tips [3] or lessons learned [6] all experience reports acknowledge the fact the criteria used are situational, that is, driven by the specific circumstances of the organization, the project, or the team. This is no surprise, since it is already known that a software development process itself is already regarded as being dependent on the situational characteristics of individual software development settings [11]. If the choice of a process itself is already situational, the choice of a tool to support the process is situational too.

2.2   Criteria Models

To move beyond the experiences from a single case, more recent research explicitly addresses the need for this general approach, especially modelling criteria for tool evaluation. Azizyan et al. [9] investigated (dis)satisfactory aspects of agile tools through a survey of agile tool usage and needs. The criteria used to evaluate agile tools were: Ease of use, integration with other systems, customizability, availability of reports, and price.

Taheri and Sadjad [10] used a comparison chart with criteria in a selection process for a tool: Lifecycle coverage, Simplicity & ease of use, Collaboration, analytics, visibility & reporting, workspace & process, program management, deployment, and integrity & security. The criteria originated, beside from previous research, for a large part from surveys, sponsored by vendors of agile tools. This introduces a bias, since vendors are of course unlikely to use unfavourable criteria.

2.3   Theoretical Viewpoint

Both criteria models, and also the individual experiences, draw heavily upon existing surveys, where their resources are subject to bias. To have a more independent viewpoint the Technology Assessment Model, TAM [12, 13] can be used. Its central proposition is that user acceptance of technology depends on perceived usefulness and perceived ease of use. Perceived usefulness is *"the degree to which a person believes that using a particular system would enhance his or her job performance"*; perceived ease of use *"the degree to which a person believes that using a particular system would be free of effort"* [12, p. 320]. TAM theorizes that the effects of external variables (e.g., system characteristics, development process, and training) on intention to use are mediated by perceived usefulness and perceived ease of use [14]. Both categories are also included in the Unified Theory of Acceptance and Use of Technology (UTAUT) as performance and effort expectancy, as are 2 others: social influence and facilitating condition [15]. TAM thus provides a useful first distinction in modelling criteria for agile tool support. This is confirmed, although not explicitly by the use of TAM, in an evaluation of four agile project management tools, where both usability evaluation criteria and a task-oriented usability inspection were used [16]; these categories reflect very well the TAM categories. This supports the use of dichotomy in TAM in modelling criteria that agile teams use for tool support.

3   Research Method

In our research we have chosen for a greenfield approach, in analogy with the greenfield project: *"In software engineering jargon, a greenfield is a project which lacks any constraints imposed by prior work"* [17, p. 21]. Although this approach may be considered as situational as any other, it provides a relatively unbiased way to explore features a Scrum team thinks of as being important in its support. It at least assures that prior organizational or commercial influences are excluded. To implement this approach we used student groups, where we phrased our research question as follows:

Which criteria do Scrum teams adopt when choosing computer-based support for their Scrum process?

We characterize our research as a variant of case study research with the teams being the cases, but explicitly instructed through their assignments.

In the next two paragraphs we will describe, first, the educational context of the student groups and, second, the assignments the groups were provided with.

3.1 Context

The teams in this study are teams of second year students in a four year Information Technology study programme leading to a Bachelor degree. Students have experience with software development methods, some basic experience with waterfall-like development, followed by other methods, such as RUP (Rational Unified Process). In ten weeks students were to develop an application, applying Scrum as ASD method.

In the first three weeks of the course students acquainted themselves with Scrum. After a lecture on agile methods in general students were referred to the Scrum guide [18] for self-study and conducted a test consisting of twenty multiple choice questions afterwards. Furthermore, they were provided with training on all major Scrum elements, which are roles, artefacts and events.

From week four onwards the students made three Sprints of four days each, working fulltime on their application. During a Sprint teams held all Scrum meetings: Sprint Planning Meeting (Monday morning), Daily Scrum (Tuesday till Thursday morning), Sprint Review, followed by an informal Sprint Retrospective (Thursday afternoon). Teams were stimulated to collectively work at the same place, for instance a computer or a conference room. A few teams had their members working (more) individually, sometimes from home. From week seven onwards students continued to work on the application individually using team results as a basis. This had didactical reasons and although still working in a Scrum-like way, this part of the course is left out of consideration here, if only because communication between team members did not take place anymore.

3.2 Assignments

Students were divided into twelve Scrum teams, with each team between four and six members. Before starting their first Sprint, teams were asked to select a tool to support them in their Scrum process as a first assignment. In more detail teams were asked to specify at least five weighted criteria they thought relevant for their Scrum tool, three different alternatives (Scrum tools), and a score on each criterion for each alternative. In addition, teams were asked to motivate all of the alternatives, criteria and scores.

As a second assignment, teams were asked for an evaluation after their last Sprint. They were specifically asked to answer two questions: On which characteristics did the tool (not) meet your expectations? In other words, where did the tool (not) effectively and efficiently support the Scrum process?

3.3 Data Collection and Analysis

For data collection, we first collected reports with regard to the first assignment from all twelve groups at once. We analysed the reports according to Grounded Theory (GT). GT is a systematic research method known for the generation of theory derived from systematic and rigorous analysis of data [19, 20]. This does not mean that there is not a specific problem, but rather problems and their key concerns emerge during data analysis [20].

To analyse the data, we used three abstraction levels: category, concept, code [21]. For the highest level, category, we adopted TAM: Perceived usefulness and perceived ease of use. Applying open and axial coding [22] we found at the next level concepts like support of a product or sprint backlog (perceived usefulness) or multi-platform support (perceived ease of use). Through coding different wordings were mapped onto the same (sub-)criteria. For instance, the criterion "Ease of use" was phrased as: "*The users must be able to use it easily*", "*The cognitive load should be low*", and "*Often used functions should be clearly visible*". Coding was a fairly straightforward exercise, because the assignment by its nature already led to structured reporting.

Our analysis proceeded from one report to another; we coded a report for criteria and weights, adding to our abstraction levels where necessary. After each change we scrutinized previous reports with respect to the most recent supplements. This was for instance necessary when we met a criterion previously considered as sub-criterion, or vice versa. This overlap between criterion and sub-criterion happened more often and has been resolved by using our judgement in the analysis, supported by the classification already applied by the teams themselves and the number of teams that used the one or the other.

We also collected reports for the second assignment. We analysed them for positive and negative remarks, but only superficially coded them, because most teams did not explicitly refer to their (sub-)criteria, thus allowing coding at the category and concept level only.

4   Results

The results from our research are described next. We first describe the criteria which were used by the teams to evaluate the tools and show the weights teams attached to them. We then provide a summary of the tools considered by the teams and present a brief overview of the teams' evaluation after having used their tool.

4.1 Criteria

All twelve Scrum teams reported criteria, alternatives, scores as well as considerations leading to them. They reported their sources as (Internet) research, acquaintance with a tool (for instance through experience from a part time job), advice from experts, including teaching staff, or a combination. Since the teams worked independently, we applied coding to highlight similarities and discrepancies, especially between (sub-) criteria, which includes homonyms and synonyms.

Six teams used a flat list, that is, all criteria were on the same level of abstraction. The others had a hierarchical approach, in which a classification of criteria with sub-criteria was used, yielding a two level hierarchy. After analysis a hierarchical list of criteria and sub-criteria arose, structured at the top level according to TAM (Table 1);

Table 1. Overview of categories, criteria and sub-criteria

Category Criterion *Sub-criterion*	#[1]	Category Criterion *Sub-criterion*	#
Perceived usefulness	1	**Perceived ease of use**	
Contains visualization task board	5	Ease of use	9
split in to do, doing, testing & done	*7*	*can be personalized*	*3*
Supports backlog(s)	-	*have an easily accessible lay out*	*3*
contain product backlog (items)	*5*	*have short loading & response time*	*2*
contain sprint backlog (items)	*5*	*have user guide / tutorial*	*1*
assign priorities to items	*2*	Multiple platform support	7
split items in tasks	*3*	*have a website*	*1*
link items/tasks to team members	*4*	*be app/application/board*	*4*
allow comments on item/task	*5*	*be physical*	*1*
allow documents per item/task	*2*	*can be reached from other locations*	*1*
register time per task	*1*	Communication support (e-mail, chat)	2
sort backlog items on priority	*2*	Security mechanism support	-
show status backlog item	*1*	*require log in*	*2*
show activities per member	*1*	*administer member rights*	*3*
show deadlines for tasks	*3*	Integration with other tools	-
show effort (per member)	*1*	*Visual Studio*	*1*
show statistics	*4*	*GIT*	*2*
couple sprint backlog to releases	*1*	Acceptable price (or trial available)	10
Supports burn-down chart	8		
show team	*1*		
show individual	*1*		
Has additional features	1		
Has Scrum compatibility	2		
allow (user) stories	*1*		
have predictions	*1*		
support planning poker	*1*		

---

[1] # is the number of occurrences of the (sub-)criterion.

this list is the superset of all individual lists. All (sub-)criteria could be classified as a TAM category, perhaps with the exception of price, which we included under perceived ease of use, but might have had its own category.

The frequency for criteria, not being sub-criteria, represents the number of times the criterion itself is mentioned directly, i.e. it does not in any way accumulate frequencies of sub-criteria.

All criteria and sub-criteria were rated with relative importance by the teams, see Table 2.

Table 2. Overview of weights of criteria

Category	Criterion	Teams	Weight[2]	σ^3
Perceived usefulness	Contains visual task board	8	12,5	14
	Supports backlog(s)	10	27,8	19
	Supports burn-down chart	9	8,3	6
	Has additional features	1	1,3	4
	Has Scrum compatibility	3	3,4	6
Perceived ease of use	Ease of use	10	17,2	16
	Multiple platform support	11	11,8	9
	Communication support (e-mail, chat)	2	2,1	5
	Security mechanism support	3	2,3	5
	Integration with other tools	3	3,3	7
	Acceptable price (or trial available)	10	10,0	7

[2]All numbers are rounded to 1 decimal.
[3]σ: Standard deviation, rounded to an integer.

In Table 2 the number of teams refers to teams that have listed the criterion or one of its sub-criteria. The weight is calculated as the sum of the relative weights for teams who listed the criterion and/or one of its sub-criteria divided by the total number of teams (twelve). To this extent the scale of weights for each individual team has been normalized to a scale of 0–100. Furthermore, we have restricted ourselves to the level of criteria in Table 2. Incorporating sub-criteria in the table would have little value, because the vast majority of sub-criteria is mentioned three times or less, which would in general result in lots of weights smaller than one.

4.2   Choice of Tools and Evaluation

Teams were free in their choice of alternatives. A total of eighteen different Scrum tools was considered. All tools were fairly dedicated Scrum tools with the exception of Trello (a tool to organize anything), and Proofhub and Dapulse (project management tools). Five of them were finally chosen by the teams. Visual Studio Online (Team Foundation Server) was the most chosen one, five times, followed by Trello, four times. The other three teams used Scrumwise, ScrumDo and QuickScrum.

In the evaluation almost all teams reported to be satisfied with their choice of a tool. Their satisfaction was backed up by both positive and negative remarks of (the use of) their tools. We counted the number of remarks, related them to our two (TAM-) categories and subdivided them into positive or negative feedback (Table 3).

Table 3. Feedback of chosen tools

		Team												Total
		1	2	3	4	5	6	7	8	9	10	11	12	
Perceived usefulness	Positive feedback	7	2	4	1	1	1	4	3	2	3	1	5	34
	Negative feedback	3	3	2	3	–	–	1	1	2	1	1	–	17
Perceived ease of use	Positive feedback	2	1	–	4	5	5	3	–	1	1	2	–	24
	Negative feedback	2	–	–	1	–	–	–	1	–	1	1	2	8

We have related remarks to the category only, because the evaluation data most often was not related to only one of the (sub-)criteria. Some examples of remarks are:

- *"The basic features of the tool, such as the task board, the description of Sprint Backlog items and the planning of Sprints, were easy to use".*
- *"We would have liked planning poker points instead of hours to measure velocity".*

One team switched from tool after Sprint one. It had chosen for Trello, but started using Visual Studio Online for Sprint two and three.

5   Discussion and Conclusions

In this section we first compare our results with previous research. We then discuss the (sub-)criteria in detail and analyse the teams' preferences. Finally, we discuss limitations of our study.

5.1   Comparison with Previous Models

The teams collectively came up with 45 criteria, including sub-criteria, far beyond and more detailed than any of the previous models [9, 10]. However, our sheer amount still does not warrant a claim to completeness. And we do notice a match between criteria from previous research and our current research, at least at the level of criteria. Furthermore, there was consensus among the teams. Six criteria were mentioned by at least 75% of the teams (visual task board, backlog(s), burn-down chart, ease of use, multiple platform, acceptable price), with the remaining ones all below 25% (see Table 1).

We conclude that our Scrum teams, most likely unaware, followed theory (TAM) and practice (experience reports and models) in the constitution of their list of relevant (sub-)criteria, but added lots of detail concerning practical considerations, such as wishes to prioritize and sort backlog items or to integrate with GIT. This level of detail was not established in previous research and in this way our results are especially attractive for teams in agile transformation, when adopting agile tools.

5.2   Classification of Criteria

We found criteria could easily be classified according to the TAM, with no overlap between Perceived usefulness and Perceived ease of use. Notwithstanding the high standard deviations in the attribution of weights (Table 2), a separation between '(sub-) criterion mentioned often/high weight' and '(sub-)criterion mentioned seldom/low weight' is clear and cuts through the categories: Teams value support of Scrum arte-facts most, as is shown by the number of criteria devoted to the Product and Sprint Backlog (and the burn-down chart), and acknowledged by the weights they attached to those criteria. This observation is also confirmed by the evaluation where teams devoted more remarks, positive or negative, to Perceived usefulness than Perceived ease of use, approximately in a ratio 5:3 (Table 3). This suggests Scrum teams value perceived usefulness over perceived ease of use.

However, this is not confirmed in the scores they assign to their alternatives/criteria, and hence the choice of their tool. Trello is the second best chosen tool, and Trello may certainly be suited for the task, but it is a general tool and certainly not a Scrum specific one. However, the one team that switched from Trello to Visual Studio Online (VSO), when comparing the two, commented: "*VSO seems, as far as burndown chart and overall look & feel are concerned, far more professional than Trello*". "*Professional*" here also implies the existence of more and better Scrum features in VSO. Probably Trello's ease of use and flexibility lured some teams away from more specific Scrum tools.

We conclude that Scrum teams prefer perceived usefulness over perceived ease of use. Especially support of Scrum's artefacts is of greater value to a team than general tool considerations.

5.3   Limitations

We excluded commercial and organizational influences by involving student Scrum teams with basic and theoretical knowledge of Scrum. A drawback of this greenfield approach is that our findings apply to novice Scrum practitioners. Generalizability of results is therefore limited.

The teams tended to prefer freely available Scrum tools, or a trial version thereof. Although this may have restricted their choice of a tool, it did not influence the constitution of the list with criteria, which was our main research goal.

Acknowledgements. This research would not have been possible without the efforts of students taking the course. The first author also wants to express his gratitude to Avans University of Applied Sciences for facilitating and supporting this research.

References

1. Beck, K., Beedle, M., Van Bennekum, A., Cockburn, A., Cunningham, W., Fowler, M., et al.: Agile Manifesto (2001). agilemanifesto.org. Accessed 24 Sept 2012
2. Hossain, E., Babar, M.A.: Risk identification and mitigation processes for using Scrum in global software development: a conceptual framework. In: Proceedings of the Asia-Pacific Software Engineering Conference, Penang, Malaysia, pp. 457–464. IEEE (2009)
3. Schatz, B., Abdelshafi, I.: Primavera gets agile: a successful transition to agile development. IEEE Softw. 22(3), 36–42 (2005)
4. Crowder, J.A., Friess, S.: Productivity tools for the modern team. In: Crowder, J.A., Friess, S. (eds.) Agile Project Management: Managing for Success, pp. 43–48. Springer, Heidelberg (2015)
5. Alyahya, S., Alqahtani, M., Maddeh, M.: Evaluation and improvements for agile planning tools. In: Proceedings of the International Conference on Software Engineering Research, Management and Applications, Baltimore, USA, pp. 217–224. IEEE (2016)
6. Uy, E., Rosendahl, R.: Migrating from SharePoint to a better Scrum tool. In: Proceedings of the Agile Conference, Toronto, Canada, pp. 506–512. IEEE (2008)
7. Møller, L.S., Nyboe, F.B., Jørgensen, T.B., Broe, J.J.: A Scrum tool for improving project management. In: Wouters, I., Man, Fl. K., Tieben, R., Offermans, S., Nagtzaam, H. (eds.) Flirting with the Future: Prototypes Visions By The Next Generation - Proceedings of the Student Interaction Design Research Conference, Eindhoven, The Netherlands, pp. 30–32 (2009)
8. Engum, E.A., Racheva, Z., Daneva, M.: Sprint planning with a digital aid tool: lessons learnt. In: Proceedings of the Conference on Software Engineering and Advanced Applications, Patras, Greece, pp. 259–262. IEEE (2009)
9. Azizyan, G., Magarian, M.K., Kajko-Mattson, M.: Survey of agile tool usage and needs. In: Proceedings of the Agile Conference, Salt Lake City, USA, pp. 29–38. IEEE (2011)
10. Taheri, M., Sadjad, S.M.: A feature-based tool-selection classification for agile software development. In: Proceedings of the International Conference on Software Engineering and Knowledge Engineering, Pittsburgh, USA, pp. 700–704 (2015)
11. Clarke, P., O'Connor, R.V.: The situational factors that affect the software development process: towards a comprehensive reference framework. Inf. Softw. Tech. 54(5), 433–447 (2012)
12. Davis, F.D.: Perceived usefulness, perceived ease of use, and user acceptance of information technology. MIS Q. 13(2), 319–340 (1989)
13. Davis, F.D., Bagozzi, R.P., Warshaw, P.R.: User acceptance of computer technology: a comparison of two theoretical models. Manag. Sci. 35(8), 982–1003 (1989)
14. Venkatesh, V., Davis, F.D.: A theoretical extension of the technology acceptance model: four longitudinal field studies. Manag. Sci. 46(2), 186–204 (2000)
15. Venkatesh, V., Morris, M.G., Davis, G.B., Davis, F.D.: User acceptance of information technology: toward a unified view. MIS Q. 27(3), 425–478 (2003)
16. Alomar, N., Almobarak, N., Alkoblan, S., Alhozaimy, S., Alharbi, S.: Usability engineering of agile software project management tools. In: Marcus, A. (ed.) DUXU 2016. LNCS, vol. 9746, pp. 197–208. Springer, Heidelberg (2016). doi:10.1007/978-3-319-40409-7_20
17. Gupta, R.M.: Project Management. PHI Learning Pvt. Ltd., Delhi (2011)
18. Schwaber, K., Sutherland, J.: The Scrum guide – The Definitive Guide to Scrum: The Rules of the Game (2013). www.scrum.org. Accessed 17 Sept 2013
19. Glaser, B.G., Strauss, A.L.: The Discovery of Grounded Theory: Strategies for Qualitative Research. Aldine, Chicago (1967)

20. Glaser, B.G.: Basics of Grounded Theory Analysis: Emergence vs Forcing. Sociology Press, Mill Valley (1992)
21. Allan, G.: The use of grounded theory as a research method: warts and all. In: Brown, A. (ed.) Proceedings of the European Conference on Research Methodology for Business and Management Studies, pp. 9–19. Management Centre International Ltd. (2003)
22. Corbin, J., Strauss, A.: Basics of Qualitative Research - Techniques and Procedures for Developing Grounded Theory. Sage Publications Inc., Thousand Oaks (2008)

Author Index